MW00684523

INTRODUCTORY PLANT SCIENCE

HENRY T. NORTHEN

UNIVERSITY OF WYOMING

•

THIRD EDITION

John Wiley & Sons

New York • Santa Barbara • Chichester • Brisbane • Toronto

ISBN 0 471 06816-0

Library of Congress Catalog Card Number: 68–13473

PRINTED IN THE UNITED STATES OF AMERICA

10 9 8 7 6 5 4

Preface

My purpose in writing this textbook of botany is to present to the beginning student a view of the plant world that will leave a lasting and vivid impression. The presentation aims, by its direct approach, to give the student an understanding not only of the plants themselves but also of their relationship to people here and around the world. The contributions of botany to our past, its accelerated contributions to the present-day world, and the urgent need for basic research to fill the needs of the future are interwoven throughout the book with discussions of the established fundamentals. I have tried to alert the student to such worldwide problems as hunger, disease, pollution, and thirst, and the role of botany in the relief of human suffering. I hope to make the subject come alive to the general student and develop a sound and enthusiastic foundation for the potential botanist.

To make the book teachable and readable, the material is presented in a form commensurate with the beginning student's background. Knowledge gained in the fields of electron microscopy and molecular biology enables us more fully to understand cell structure, genes, gene action, function, plant development, and evolution. Such knowledge, obtained from the study of both higher and lower plants, is included in this revision, and I believe that the topics have been written and illustrated in an understandable manner, one that does not presuppose organic chemistry or biochemistry.

New information about space biology, biological clocks and calendars, phytochrome, flowering, respiration, photosynthesis, growth regulators, tissue culture, chemical taxonomy, the marine habitat, and world plant formations has been included. In line with the more sophisticated equipment now available for laboratory work, I have substituted quantitative methods for some qualitative ones in studying photosynthesis, respiration, and transpiration.

I have tried to maintain a balance between the molecular and the descriptive, and I have not favored the one over the other. The book is a balanced presentation of all aspects of botany. The plant as a whole and its relationship to the environment have been a central theme. The chapters on plant communities and conservation have been expanded.

The illustrations continue to be a vital teaching tool. Many new drawings have been added; electron micrographs enhance the visualization of structures, and the combining of illustrations increases visual comparisons.

Laboratory Studies in General Botany by William M. Carlton (The Ronald Press Company) is particularly adapted to this text in both its organization and its content. It is an excellent and accurate companion for laboratory instruction.

The work of revision has been greatly aided by the suggestions and criticisms of colleagues and students here and in the colleges and universities where the Second Edition has been in use. It has been gratifying to have their suggestions as to expansions and additions that would enhance the text.

I am grateful to Mr. William Eastman, Mr. I. V. Tobler, Mrs. John Reed, Mrs. Arthur Carroll, and Mr. Don Wiest for their art work and to the many generous botanists who supplied photographs for this and previous editions of *Introductory Plant Science.* My warm appreciation goes to my wife and to Dr. John Reed, whose help with the original edition was invaluable, to my colleagues for personal suggestions, and to the reviewers who read the manuscript critically.

HENRY T. NORTHEN

Laramie, Wyoming
January, 1968

Contents

INTRODUCTORY
PLANT SCIENCE

1

Botany in Human Affairs

Plants are all around us, no matter where we live. In the tropics they grow with an almost visible speed, hiding the earth beneath our feet and, in the jungle, often hiding the sky. During the brief summer of the Arctic zone, rock crevices and pockets where the ice melts give forth brilliant flowers. Deserts, which are often depicted as sandy wastes, become vivid with bloom when moisture strikes them. Prairies and plains are carpeted with grasses and flowers. High above the line where trees can grow on mountain tops, small plants find the means to survive and reproduce in harmony with their rigorous environment. In lakes and running rivers and to a depth of about 900 feet in the ocean, green plants thrive.

There are other plants too, such as the fungi (singular: fungus), which lack green coloring. The dustlike spores, or reproductive bodies, of these are in the air about us at all times. The plants themselves have various forms, sometimes of microscopic dimensions. They may appear as a bluish mold upon old shoes, as mildew on a lilac leaf, as rust on wheat or smut on corn, as a grayish web which helps to reduce dead plant remains to their primal substances, or they may be the invisible cause of a skin infection such as athlete's foot. They give us succulent mushrooms for our table, put the flavor in our cheese, make our bread rise, and ferment the grapes for our wine; they have become the source of one of the most effective weapons of medical practice—the antibiotics.

These are the plants of the world—the green ones of garden, field, and forest, desert, prairie, and sea, and the nongreen plants which surround us, often without our being aware of them. Animals, whether they are cattle, whales, moose, wolves, trout, or human beings, all depend ultimately or directly on plants to furnish their nutritional requirements: carbohydrates, fats, proteins, vitamins, and minerals. If the vegetation is deficient in one or another dietary requirement, the animals do not remain in good health. If the crops of fields fail, animal life suffers; if forests are decimated, man is affected in many ways, through loss of raw materials, power, water, and protection from erosion of his lands.

As you study this book, you will see how plants not only feed, clothe, shelter, and warm us, and gratify our senses, but also how they provide the primary sustenance for every living creature, from microscopic units of the ocean's plankton to man himself and his domestic animals. Plants furnish medicines and perfumes and flavorings. About 50 per cent of the 300 million prescriptions filled each year in the United States contain a drug of plant origin. The sales of drugs from botanical sources increased more than fivefold from 1950 to 1960, and the trend is sharply upward. Products from green-growing things and from nongreen plants go into much of today's manufacturing.

Thus plants are among man's greatest allies, for without them he could not survive. Some kinds, however, are his enemies—those that produce diseases in his garden plants, in his crops, and in himself; those that poison; and those that grow so much more persistently than wanted plants that they are classed as weeds.

To have a knowledge of plants—how they grow and reproduce, what uses they have, how they may be improved or controlled—is one of the most rewarding of all accomplishments derived from study, for it provides an understanding of the fundamentals of all life. The science which is concerned with the study of plant life is called **botany**, and those who work in this field are **botanists.** Botany is one of the two branches of **biology,** the science of life. The other is **zoology,** the science of animal life.

SIMILARITIES BETWEEN PLANTS AND ANIMALS

We are very conscious today of "new models," but did you ever stop to think that we ourselves are new models, as are other mammals and the seed plants? Just as with cars, though, we living things are not entirely different from the older models. The new models of cars have different lines and a few new innovations, but basically they retain the fundamental mechanisms of previous models. Old and new alike have wheels, tires, spark plugs, transmission, crankshaft, flywheel, and many other features; old and new alike require fuel and lubrication. So, too, the new models of life follow the fundamental structure and have the same working parts as the older ones. Hundreds of millions of years ago there were developed mechanisms that have not been much improved upon since. The basic features of the simple plants and animals of long ago proved so reliable that they were retained in the great variety of new models which stemmed from them.

What are these features found in all, or nearly all, of the millions of different kinds of plants and animals present on earth today? Even though there is great diversity in external form among the plants and animals of the land and of the waters, there is on the microscopic level a oneness of form, for all plants and animals are composed of similar units called **cells.**

All organisms have the same "machinery" for carrying out the reactions characteristic of living things. In the cells of such simple plants as bacteria and in the cells of man, there are organic catalysts which speed up reactions. These catalysts are called **enzymes.** All living things have a great number of enzymes, and many of them are the same in all living things.

Life is maintained only by the continuous output of energy. If this output ceases, the living system stalls and dies. The energy to maintain life comes from respiration, and this process, in essential features, is the same in plants and animals. Respiration also occurs in the same kinds of "powerhouses" in cells of all organisms. In our cells and those of plants there are minute granules, called **mitochondria,** in which respiration proceeds continuously. All organisms use the same "transmission." Some of the energy liberated in respiration produces high-energy phosphate bonds (see Chapter 7) which transmit the energy to various processes—growth, synthesis, uptake of minerals, and, in man, to the contraction of muscles, the smile, the winking of an eye, or walking.

Long, long ago there was developed a precise means of storing in microscopically small packets "blueprints" of organic forms and the methods for carrying out the plans. You, like plants and animals in general, developed from a microscopically small fertilized egg. Yet contained within this egg were the "blueprints" of what you are today, and in addition there was established in the egg a course of action leading to your development. Contained within the fertilized egg were two sets of chromosomes, one from the father and one from the mother. Genes are resident on the chromosomes, and they control the course of development. All plants and animals, from the simplest to the most complex, from the old models to the new ones, use this same mechanism for transmitting traits from parents to offspring. We do not have one set of laws of inheritance for man and other animals, still another for plants; instead, one set of laws is applicable to all living things.

Chemically, genes of plants and animals are nucleoproteins—huge molecules formed when a protein combines with nucleic acid, more precisely, **deoxyribonucleic acid,** abbreviated DNA. DNA, the chief carrier of genetic information, is formed by the joining together of one thousand or more smaller molecules, called nucleotides, of which there are just four kinds. Remarkably, genes of all animals, from amoeba to man, and genes of all plants, from molds to sunflowers, are formed from the same four kinds of nucleotides. Even though the building blocks for all DNA molecules are identical, the genes are not all alike. The sequence in which the nucleotides occur in the DNA molecule makes the difference between one kind of gene and another. To illustrate the great variety of molecules that could result with just four different kinds of nucleotides, suppose that beads of four colors, red, yellow, green, and blue, represent the four nucleotide types. Let us thread 250 of each color on a string as if we were making a necklace. By varying the color arrangement, we could make billions of different patterns. As we will see later (Chapter 15), DNA is not just a single strand, but instead is a double strand, with cross-linkages and a coiled configuration.

Not only are genes of all creatures constructed of the same building units, but they also operate in the same manner in all organisms. Genes control the synthesis of enzymes and other proteins, not directly, but by way of a messenger known as **ribonucleic acid** (RNA) which is packaged with protein to form submicroscopic granules called **ribosomes.** On the surface of the ribosomes, enzymes are synthesized.

Each step in a biochemical reaction is controlled by a definite enzyme, and each kind of enzyme is synthesized under the direction of a specific gene, that is, a certain DNA pattern. As we have seen, respiration occurs in all organisms. In respiration there are about fifty reactions, and each reaction is catalyzed by a definite enzyme. Therefore, about fifty different enzymes are involved in respiration, and so also are fifty different genes. Because the reactions are the same in plants and animals, it follows that the enzymes are also of the same kind, and so also are the genes. Here we show our kinship to plants by having certain genes that are identical to their genes.

When the earth's vegetation consisted only of simple plants and animals growing in water and when the land surfaces were barren of life, various systems of reproduction were being tried out. From these "pilot plants" there evolved a pattern of sexual reproduction which was incorporated into the new models of life. It may, at first, seem strange to you that in essential features the reproduction of squirrels is similar to that of the pine trees in which they live. Squirrels and pines produce sperms and eggs. Eggs become fertilized, and each fertilized egg develops into a new individual—

the fertilized egg of a squirrel into a squirrel, that of a pine into a pine tree. Another general feature of reproduction is a special type of division called **reduction division** or **meiosis,** a division which halves the number of chromosomes. As a result of reduction division the sperms and eggs have just half as many chromosomes as the body cells. For example, the body cells of a pea plant have fourteen chromosomes, but each sperm or egg has seven. When a sperm fertilizes an egg, the fertilized egg will have fourteen chromosomes, seven from the sperm and seven from the egg, and so also will the cells which develop from the fertilized egg. In molds and man, in pine trees and squirrels, and in nearly all plants and animals, this halving of the number of chromosomes is carried on in the same manner.

All living things grow. From certain substances secured from their environment, they make more of themselves. They form substances of enormous complexity from relatively simple compounds. We say they carry on **assimilation.** Growth involves not only assimilation, but cell division and enlargement. Cell division, the forming of two cells from one, is remarkably the same in plants and animals.

Primeval organisms, those of intermediate complexity, and the highly specialized plants and animals of today have the capacity to respond to stimuli. We say that all organisms are **irritable.** Plants, like animals, respond to such stimuli as gravity, light, injury, touch, and certain chemicals. If we put a plant on its side, the stem responds to gravity and grows upright. A plant placed near a window will curve toward the light. The leaves of a sensitive plant droop when the plant is touched. When other plants are touched or stroked, there may not be an external response, but an internal response occurs. Stroke a kitten and the response is obvious. Stroke the average leaf and there is no external expression, but there is an internal change. A stroked leaf respires more rapidly than an unstroked one.

We live in a colorful world of blue skies, red sunsets, green foliage, and variously colored flowers, fruits, homes, cars, and works of art. Fortunately, we can distinguish colors. Plants also sense colors. Thus moistened seeds of the saguaro cactus germinate promptly when exposed to those shades of red which are next to the orange part of the spectrum, but they do not germinate when exposed to deeper shades of red.

Plants can measure time, responding in one way when nights are long and in a different manner when the nights are short. For example, poinsettias flower when nights are 12 hours long, but they do not flower during nights of 11 hours. In a sense plants have built-in clocks.

Certain plants are extremely sensitive to temperature. Among the most sensitive is the slime mold, *Dictyostelium discoideum,* which glides toward warmer regions with a remarkable ability to detect differences of 0.0005° C.

The tendrils, structures which wrap around a support, of *Sicyos angulatus,* a climbing vine, are more sensitive to touch than our fingers. A microscopic thread weighing only eight-billionths of an ounce is enough to stimulate the tendril to curve, while human skin is unable to perceive this stimulus. The fact that most plants are fixed in one position may lead us to believe that they are not irritable. However, the examples cited clearly demonstrate that plants are irritable and respond to a variety of stimuli.

Life is orderly and precise. And life is diverse too; but even with diversity there is, as we have seen, a likeness among all living things, plant and animal (Fig. 1–1). All organisms have a cellular structure, and they assimilate, grow, reproduce, respire, and are irritable. Their development is controlled by genes, which function by controlling the synthesis of **enzymes,** organic catalysts that bring direction and control to biological processes. These collectively are properties which separate the living world from the nonliving one.

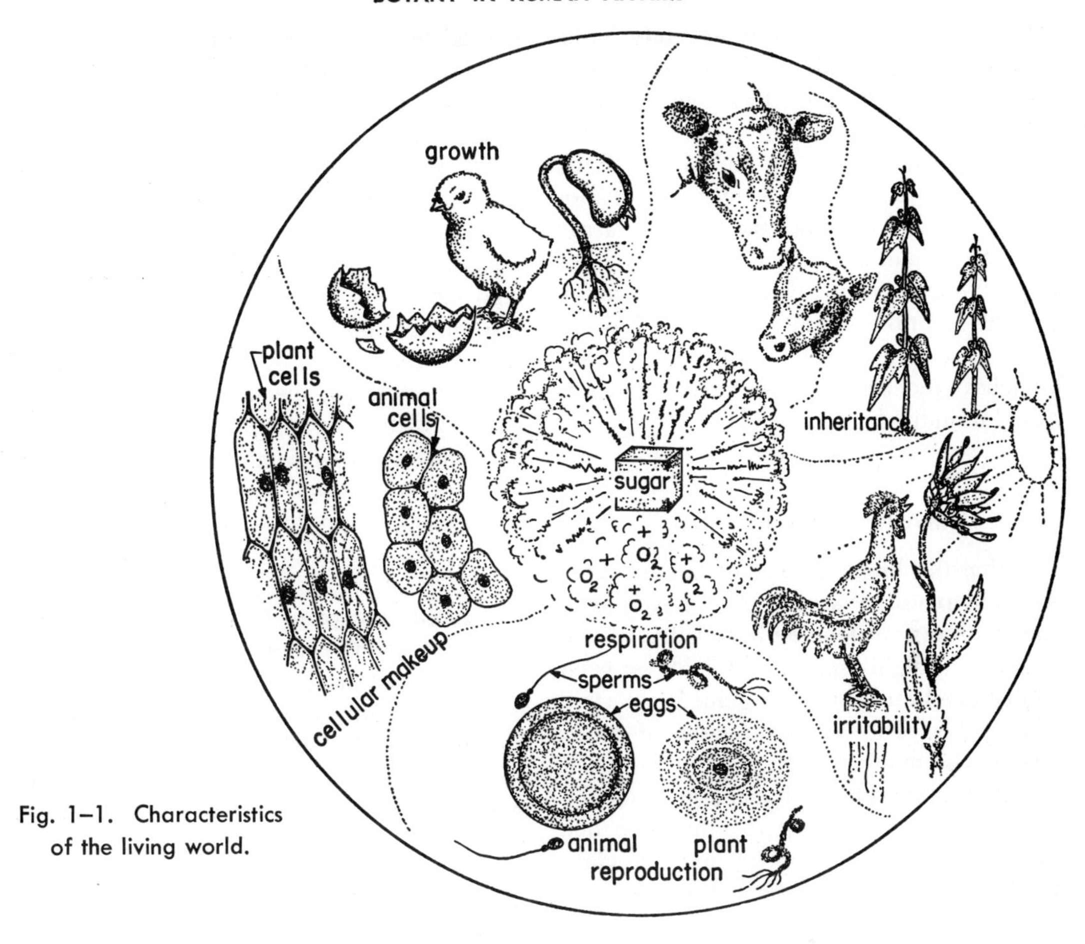

Fig. 1–1. Characteristics of the living world.

DIFFERENCES BETWEEN PLANTS AND ANIMALS

When we consider all plants and animals from the microscopic to the very large, only a few characteristics serve to distinguish plants from animals, and even here there are exceptions. Most animals are motile, and most plants remain in one place. However, there are exceptions. The sponge, an animal, is stationary. Some plants, certain algae and many bacteria, can swim.

Most plants have the green pigment, chlorophyll, and hence also the remarkable capacity to use light energy in the synthesis of food from simple inorganic substances. In contrast, animals secure their food ready made, feeding on plants or on animals which in turn have fed on plants. Here too there are exceptions; some plants are nongreen and cannot make their own food, bacteria and molds, for example.

The cells of plants have rigid cell walls usually constructed of cellulose. These cell walls form a somewhat rigid framework which is in harmony with a stationary mode of life. On the other hand, animal cells lack rigid cell walls of cellulose. Cellulose is not formed by any animal and hence is distinctive of plants.

Another feature, but one with many exceptions, which distinguishes plants from animals is the manner of growth. Many plants have unlimited growth and an indefinite number of parts, whereas the growth of animals is limited and the number of parts is definite. How tall is a man? How big is a mouse or an elephant? You could set fairly accurate limits because these animals attain a certain size and then stop growing. How big is a pine tree? Here you cannot give a definite answer because growth is unlimited. Each year the tree increases in height and diameter. In general, the older the tree, the larger it is. An animal has a characteristic number of parts—legs, arms, eyes, etc. In general, a plant has an indefinite number of parts. In the case of a tree the numbers of leaves, branches, and branch roots vary from individual to individual.

BRANCHES OF BOTANY

Within the broad science of botany there are many phases of study, which have developed with ever-expanding knowledge. Each comprises a highly specialized field, yet each is dependent on knowledge gained in certain other fields. The various phases of study concern the many aspects of plant life itself and its relationship to human life. As the knowledge gained is put together into one great story, a more complete understanding of plants will one day be the reward.

Taxonomy

Whether plants are being used for ornamental or commercial purposes or for study, it is essential to know their correct identity.

The identification and classification of plants is an exact study called **plant taxonomy.** The taxonomists study the parts of a plant, especially of the flower, and find out to which group the plant belongs. They then designate it with an official name, and that name is used for it throughout the world. People who grow nasturtiums should find the name *Tropaeolum majus* on the seed packet, whether the seed was produced and sold in England, India, South America, Germany, or the United States. The Mediterranean oak tree which provides the cork of commerce is *Quercus suber,* and those who wished to undertake cork growing on an experimental basis in southern United States had, of course, to obtain this species for the project.

Some taxonomists are explorers who search for, name, and classify new species. Each year plants heretofore unknown to man have been located. Many of the plants that you are familiar with—and take quite for granted—have been introduced from foreign lands.

Many of our garden plants are hybrids; that is, they have been created by crossbreeding different species. In placing plants in groups that are related to each other, the taxonomists have given the plant breeders essential information as to what kinds might be crossed to produce better varieties.

Genetics

A number of plant breeders who have relied on common sense and trial-and-error methods have become famous for their horticultural triumphs. However, much work has been wasted when plants did not respond as it was hoped they would. Genetics, a science of the twentieth century, is concerned with heredity and variation and provides a more exact basis for plant breeding. Plant geneticists study the ways in which plants transmit their own visible or invisible characteristics to their offspring, and the mechanism by which these characteristics are borne in the internal structure of the plants' own cells. The information that these geneticists have compiled gives a quicker route to the development of superior varieties of plants. Without this branch of botany, the world's food supply would be far more critical than it is. Even so, it is going to be a continual struggle to develop more and better crop plants to keep pace with an ever-increasing world population. In addition to practical benefits, genetics has contributed a great deal to our understanding of the living world, its origin, and evolution. Our understanding of genetics is developing rapidly. The basis of inheritance, the mechanism of gene action, the replication of genes, and the mutation of genes are fairly well understood. It has even been possible to alter the genetic makeup of certain plants, especially bacteria.

Morphology, Anatomy, and Cytology

Any specialized work with plants requires a knowledge of their form, structure, and the way in which they develop. Such information has been made available by the fields of morphology, anatomy, and cytology. If we study the outward forms of plants and their organs, their development, and their life histories, we are working in the field of plant morphology. If we center our attention on their internal structure, generally as revealed by a microscope, we are working in the field of plant anatomy. If we study the several intrinsic aspects of the smallest units of structure, the cells, we are working in the field of cytology. These three fields give evidence for plant relationships and for evolutionary trends.

Ecology

Living things seldom exist alone and independent of other living things. They grow in communities, made up of both plant and animal life, and each type of life contributes to and is influenced by the other organisms in the community. We are affected not only by the contributions plants make to our lives, but by the influences they exert on our mutual environment. For optimal benefit the community must be in a state of equilibrium. There must be a balance among the factors of soil, light, and water and the plants which use these; the organisms of the soil which contribute to its fertility; and the animals which use the plants. If the balance is upset, the community may be destroyed. Unwise treatment of plant communities can be destructive just as drought, fire, and flood are destructive. The phase of botany which deals with plant communities is known as **plant ecology.** The ecologist studies who lives where, who coexists with whom, and, considering the animals, who eats whom.

Conservation

Conservation is the preservation of natural resources through their wise use, and it involves the application of ecological principles. The decimation of our forests has led to destructive floods, and the wastage of land through overgrazing or unwise cultivation has brought about great damage by dust storms and erosion. It has been learned that man cannot wantonly destroy the plant cover of the earth without suffering dire consequences (Fig. 1–2). On the other hand, if we use our resources wisely, planning for their maintenance as we go along, we can have the benefit of their protection as well as the materials we need from them. The subject of conservation has become a major study in several universities and is one of the chief concerns of some government agencies such as the Soil Conservation Service. The American Museum of Natural History in New York has created a special department for the study of conservation. These moves have been made because it is increasingly recognized by leaders that the future welfare of America will be in large part determined by the acts of conservation of the present era.

Phycology

Plants of simple form, the algae (singular: alga), inhabit the lakes, streams, and oceans, and the botanists who study them are **phycologists.** The algae range from microscopic size, through sizes that can just be seen with the unaided eye, up to "giant" forms several hundred feet long. The small algae are food for the small creatures of the water, which in turn are eaten by larger animals and so on up to the fish, clams, oysters, shrimps, and lobsters that man consumes. The nutritional potentialities of the algae are just beginning to be realized. As the world's food problems become more and more acute, the possibility of producing cattle feed from the algae is being explored. Algae have long been used in industry, however, and the list of products made from them or utilizing them is most imposing.

Fig. 1–2. Mono Dam and reservoir filled with silt. To retard the silting-in of reservoirs, plant cover must be maintained on the headwaters of streams. (U.S. Forest Service.)

Mycology

Penicillin is effective in controlling many diseases: pneumonia, sore throat, tonsillitis, etc. This drug is obtained from a fungus commonly called blue mold. A number of other drugs are also derived from nongreen plants, fungi; among them are aureomycin, streptomycin, terramycin, and actinomycin. That branch of botany which deals with fungi is called **mycology.** Some mycologists are employed by large drug companies especially to breed molds for drug production. These scientists select the strains of fungi for quality and yield as carefully as crop plants are selected. Just any variety will not do. For instance, through strain selection and induced mutation, the yield of streptomycin from the fungus *Streptomyces griseus* has been increased threefold.

Many fungi are parasites and cause diseases of man, animals, and plants. Certain fungi bring about the decay of wood. Mycologists have studied these and have discovered a number of chemical agents to prevent the decay and prolong the life of lumber, railroad ties, and posts. Better control of such fungi may be reflected in lower railroad and telephone rates than would otherwise be possible.

Fungi are being increasingly used to produce valuable substances out of materials formerly wasted. Some, for example, can utilize sawdust for food and in their activities produce industrial alcohol. Many fungi are used in the manufacture of foods, such as bakery products, cheese, and beverages.

Physiology

A plant is a living entity. From its environment it takes raw materials and makes them into organic chemicals suitable for its own use. With these chemicals the plant makes more of its own living substance, makes and uses its own food, carries on digestion and assimilation, and grows and reproduces, according to the pattern laid down by its inheritance. **Plant physiology** is the field of study that observes, analyzes, and experiments with these life processes. What minerals does a plant need, how much of each, what does each mineral do in the plant, how does it transport the minerals and the chemicals it makes, and what happens if some mineral is lacking? What goes on in a plant during the day, and during the night, and how do variations in length of day and night or in the amount of light affect it? What chemicals within the plant stimulate it to grow, or to flower, or to respond to gravity, or to lose its leaves or shed its fruit? In solving some of these problems of plant nutrition and plant behavior, the physiologists have given us much practical information. For example, mineral deficiencies in the soil produce symptoms in the plants growing upon it. A study of the symptoms and an analysis of the soil will show what kind of fertilizer to apply to make up for the lack of certain minerals. There is hardly a farmer or home gardener who does not make use in one way or another of knowledge gained by the physiologists, from such simple things as what time of the year to sow seed and how much sun or shade to provide, to the more advanced use of plant hormones and even to growing plants without soil.

Molecular Biology

Over the centuries, the frontiers of botanical investigation have moved into ever-smaller realms—from the plant community to individual plants, and later to a study of tissues and cells. Now a new frontier is being explored, a frontier where attention is focused on the molecules making up the cell. Scientists are toiling to gain an understanding of the genetic code, to reveal how the sequence of nucleotides in DNA determines the sequence of amino acids in enzymes and other proteins. The knowledge gained from molecular studies is profoundly important in other botanical fields—in genetics, cytology, taxonomy, and physiology. As our knowledge of genes, and the manner of transmission of genetic information, grows, we may gain insights that might enable us to cure cancer and virus diseases, reduce the number of birth defects, and even permit us to modify traits of plants and animals. Indeed, the field of molecular biology promises many profound discoveries.

In our enthusiasm for this new field of biology, however, we should not neglect the traditional approaches which have contributed immensely to our understanding of the living world and which will continue to do so in the future. The application of knowledge gained from traditional studies has provided us with bountiful harvests and has enabled us to control many diseases of man, plants, and animals. Without this knowledge and application, hunger and disease would plague us today as they still do in underdeveloped countries. Many biologists are anxious and willing to help the underdeveloped countries to control disease and obtain higher yields. In other words, biology can be and should be applied on a worldwide scale.

INVESTIGATION AND APPLICATION

Some practical applications of botany have been suggested here. It is important to emphasize, however, that most plant scientists are primarily seeking an understanding of life and life processes. Their investigations may or may not be of immediate use to the general public. But the bits and pieces of information gathered by scientists through the years finally form a tremendous mass of knowledge. Discoveries that seem sudden and spectacular usually have as a background the patient work of many people. Work done for the sake of "pure" science may eventually lead to a product of unexpectedly practical use. For example, the early work on the classification of the blue molds, along with studies on how they reproduce and what conditions are required for their growth, was performed out of pure scientific curiosity. When it was discovered that an important drug could be obtained from the blue molds, this work suddenly became of great practical value.

Some people believe that we are entering a period of chemical agriculture. In this new era, chemicals will be increasingly used to perform farm work. Weed killers already supplement much of the labor of hoeing and cultivating; instead of hand thinning fruits, chemicals now are often applied. Chemicals also can prevent the preharvest "drop" of fruits and can hasten the rooting of cuttings.

These techniques were not originally developed by persons deliberately seeking weed killers or fruit-drop inhibitors. Rather, they were the culmination of a long series of experiments, the earliest ones being conducted merely to satisfy scientific curiosity. They began with the discovery by H. Söding in 1925 that some chemical or chemicals which influenced growth were produced in the tips of oat seedlings. Three years later, F. W. Went succeeded in extracting this material. In the early 1930's the chemical nature of these growth substances was determined. Growth-promoting chemicals were then synthesized (that is, they were artificially produced in the laboratory) and thus became readily available to investigators. Plants were treated in one way or another with these synthetic growth substances, and the effects were recorded. It was discovered that high concentrations did not accelerate growth, but actually retarded it. Further work demonstrated that at still greater strength such growth regulators would kill plants. A number of chemicals were found to act similarly to the naturally occurring growth substances. One such synthetic chemical was 2,4-D (2,4-dichlorophenoxyacetic acid), which is now widely used to control weeds. Thus the fact that home owners no longer need laboriously to dig dandelions from the lawn, but instead may eradicate them with 2,4-D, is the result of botanical study.

Inventions have played an important role in the development of biology. The microscope, invented in 1590 by Zacharias Jansen, permitted man to explore more deeply into the realm of the plant and animal kingdoms and to see minute structures and organisms he had not dreamed of. The electron microscope is a tool that makes it possible to investigate objects which are not visible with the ordinary microscope. Whereas the ordinary microscope gives a useful magnification of 1500 diameters, the electron microscope permits magnifications of 100,000 diameters. With the invention of the phase-contrast microscope, we can distinguish structures in living cells that are not clearly evident with the ordinary microscope. A microtome enables us to cut very thin sections of plants for microscopic examination.

Nuclear energy research has led to methods for studying life processes in plants. With such energy, radioactive elements can be produced on a large scale. These elements differ from the stable elements in giving off radiations which can be readily detected with a Geiger-Mueller counter or a photographic plate. There are almost unlimited possibilities for research using radioactive chemicals. A few examples are studies of movement of material in plants, of transformation of substances leading to the formation of proteins, fats, vitamins, antibiotics, and flavorings. Radioactive carbon is being used to unlock the secret of photosynthesis, the process by which green plants make sugar out of water and carbon dioxide. From these materials, free in the air and soil, plus soil minerals, plants produce all the foods which keep the animal and human world alive and functioning. Perhaps some day this process will be fully understood, and, with understanding, man may be able to synthesize food from the same inexpensive and readily available materials.

In many ways all sciences are interrelated, and advances in one science often lead to progress in others. For example, new knowledge and techniques in chemistry and physics may lead to a better understanding of plants. Reciprocally, advances in botany may aid the chemist and physicist.

Not all botanical investigations have required elaborate instruments. Significant theories have been developed from observations in the field and from simple experiments. Instruments certainly played a minor part in the monumental discoveries of Linnaeus, Darwin, and Mendel. From studies of plants Linnaeus gave us a precise system for naming plants, the binomial system. We associate the theory of evolution with Darwin. Mendel developed the laws of inheritance from studies of pea plants growing in a monastery garden.

We often look without seeing. Thousands of botany students and hundreds of botany teachers could have discovered penicillin long before Fleming reported on it in 1929. Each year they looked at bacterial plates contaminated with blue mold but did not see that the mold prevented bacteria from growing near it. Now that we know what to look for, thanks to Fleming, we can readily see the inhibition of bacterial growth by substances secreted by contaminating blue molds. Penicillin and more recently discovered antibiotics have alleviated human suffering. Furthermore, the discovery of antibiotics led to a huge industry, one unknown a couple of decades ago. In 1959 over 2,000,000 pounds of antibiotics were produced that had a value of one-third of a billion dollars.

The scientist approaches problems with an open mind, records data accurately and objectively, attempts to correlate his findings with those of other workers, and in many instances formulates a hypothesis, a tentative explanation, which can explain his observed facts and those of others. From the hypothesis one may be able to predict that specific results will be obtained when certain experiments are carried out.

If the predictions prove true, the hypothesis is then advanced to the rank of a theory. If the theory explains a number of seemingly unrelated phenomena and if there are no exceptions, the theory becomes a law, a careful statement about general uniformities which have been observed in natural phenomena. We have very few laws in biology; most of our generalizations stop at the theory stage. At present there are vast accumulations of facts awaiting a unifying theory to relate them to one another. For example, we do not have a clear understanding of how the hundreds of chemical reactions which occur in an organism are coordinated in the development of the plant or animal body.

CAREERS IN BOTANY

There are many opportunities for employment in botanical fields and in fields where botanical training is necessary. Commercial enterprises are finding more and more the need for trained plant scientists on their staffs; the research, explorations, and advice of botanists aid in developing better methods and in expanding their potentialities. Among these enterprises are the large food-growing and food-processing companies, pharmaceutical companies, and laboratories that produce various chemical products such as insecticides, fertilizers, and growth substances. State and federal research organizations also offer many opportunities to botanists in various fields.

Do not overlook teaching, a satisfying, interesting, and challenging vocation. A student who specializes in botany and obtains an advanced degree, preferably the doctor's degree, can expect excellent opportunities in college teaching. Twice as many college teachers of botany will be needed in 1970 as were employed in 1958 because of the increased numbers of college-age persons and the greater percentage of young persons who will go to college. High school teachers will also be scarce and the opportunities will be great. For both college and high school teaching the student needs a sound liberal education, a broad grasp of the communication arts, the humanities, and the social sciences, in order to relate science to the student's world. In addition, the prospective high school or college teacher needs training in related sciences, such as zoology, chemistry, physics, and mathematics. The high school teacher will generally teach biology instead of botany and hence should be well trained in zoology as well as botany. Secondary school teachers must likewise have certain professional courses in education. The education of biology teachers for secondary schools is best carried out as a joint venture by education and biology departments.

QUESTIONS

1. List the various ways that plants are important in our culture.

2. _____ Which of the following are not biologists? (A) Agronomist, (B) forester, (C) coal miner, (D) pharmacist, (E) glassblower, (F) horticulturalist, (G) florist, (H) machinist, (I) physician.

3. In what ways are plants like animals? How do they differ?

4. What are the distinguishing characteristics of life?

5. _____ In cells of both plants and animals, respiration goes on in minute granules known as (A) mitochondria, (B) ribosomes, (C) enzymes, (D) dictysomes.

6. _____ DNA, the chief carrier of genetic information, (A) is formed by linking together nucleotides of four kinds, (B) in plants is formed from nucleotides which differ from those which combine to form the DNA of animals, (C) operates by controlling the synthesis of carbohydrates, (D) directly controls respiration.

7. _____ Genes, large nucleoprotein molecules, (A) directly synthesize enzymes, (B) control the synthesis of enzymes and other proteins by way of an RNA intermediary, (C) are identical in all plants and animals.

8. _____ Plants do not (A) have internal clocks, (B) discriminate between shades of red, (C) respond to mechanical stimuli, (D) respond to gravity, (E) have limited growth, (F) have built-in calendars.

9. _____ Which one of the following is not characteristic of both plant and animal cells? (A) Ribosomes, (B) mitochondria, (C) DNA, (D) RNA, (E) cell walls of cellulose, (F) enzymes.

10. _____ Jones was working on a stream survey for the fish and game commission. He was studying a sample of water and noted one-celled, green organisms swimming around which had rigid walls. The organisms were probably (A) animals, (B) algae, (C) fungi, (D) insect eggs.

11. _____ A mycologist studies (A) algae, (B) fungi, (C) mosses, (D) seed plants.

12. _____ A botanist who identifies and classifies species of wheat is working in the field of (A) morphology, (B) anatomy, (C) physiology, (D) ecology, (E) taxonomy.

13. _____ Scientific experiments which seemed useless when they were performed may later prove to be very valuable and practical. (A) True, (B) false.

14. _____ Which one of the following requires the least evidence? (A) Law, (B) hypothesis, (C) theory.

15. Outline the procedure you would use to determine whether or not light favors the development of the red pigment in apple fruits.

16. Suggest a problem that might be solved through using radioactive chemicals.

17. Which branch, or branches, of botany most appeals to you and why?

18. List the advantages and disadvantages of college teaching as a career. Of high school teaching.

19. The public image of the botanist has not changed as much as the botanist himself in the past few decades. (A) True, (B) false.

---

20. To become familiar with units used in botany, match the following:

______	1. One pound.	1. One-thousandth of a gram.
______	2. One meter.	2. One-hundredth of a meter.
______	3. One centimeter (cm.).	3. 454 grams.
______	4. One milligram (mg.).	4. 39.37 inches.
______	5. One microgram (μg.).	5. One-millionth of a gram.
______	6. One decimeter (dm.).	6. Ten centimeters.
______	7. One liter (l.).	7. 1.06 quarts.
______	8. One milliliter (ml.).	8. One-thousandth of a liter.
______	9. One cubic centimeter (cc.).	9. One milliliter.
______	10. One micron (μ).	10. One-thousandth of a millimeter or one-millionth of a meter.

2

Kinds of Plants

We usually think of a plant as having leaves, roots, stems, flowers, and seeds. It is natural to do so, for plants that produce seeds are readily visible with the unaided eye and are of great significance. They clothe the prairies, forests, and cultivated fields. They furnish food for man, wildlife, and livestock, and they adorn our gardens, parks, and natural landscapes.

Yet there are plants which cannot be seen with the unaided eye. This microscopic world is a fascinating and important one. Microscopic plants are everywhere—in soil, water, and air, and on the bodies of animals and larger plants.

Between these relatively simple microscopic plants and the well-developed, complex, and usually conspicuous seed plants, there are plants of intermediate complexity—mosses, horsetails, and ferns, for example.

In our immediate environment there exist all grades of plants from the simplest to the most complex. Considering the whole plant kingdom, at least 340,000 species are now known, and many more await discovery.

Every kind of plant has been given two names, the combination of which is called a **binomial.** For example, the western yellow pine is *Pinus ponderosa. Pinus* is the genus and *ponderosa* the species. In the genus *Pinus* there are many species, each different from the others and yet all having sufficient similarity to place them in the same genus. The white pine of the northeastern states is *Pinus strobus,* while the southern yellow pine is *Pinus palustris.* Fir and pine have some characteristics in common. Both are evergreen trees which produce cones, but they differ in several ways. For example, the leaves (called needles in these plants) on all but one species of pine are borne in clusters, and there is always a sheath at the base, whereas the leaves of the fir are borne singly, and there is never a sheath at the base (Fig. 2–1). Because the differences between pines and firs are quite great, they are placed in different genera (plural of genus); pines in the genus *Pinus,* firs in the genus *Abies.* However, because pines and firs are alike in some ways, they are included in the same family. The family is called Pinaceae. Families are grouped into orders, orders into classes, and classes into phyla (singular: phylum), the largest divisions of the plant kingdom. For a complete classification of the plant kingdom see Chapter 29.

Plants occur in a myriad of forms and many sizes, but all plants can be placed in one or another of twelve groups, called phyla, of the plant kingdom which consists

Fig. 2–1. *Left,* needles of limber pine (*Pinus flexilis*). *Center,* needles of lodgepole pine (*Pinus contorta*). *Right,* a needle of concolor fir (*Abies concolor*). Species of *Pinus* have needles in clusters, with a sheath at the base. The needles of fir (*Abies*) are borne singly and are flat. Both genera are in the family Pinaceae.

of two subkingdoms, namely, **Thallophyta** and **Embryophyta.** The ten most primitive phyla, seven of algae and three of fungi, are classified in the subkingdom Thallophyta and are commonly called thallophytes. The thallophytes have a simple plant body lacking roots, stems, and leaves; the sex cells develop in unicellular organs, and the fertilized egg develops into an embryo after it has been liberated from the female sex organ. The embryo is on its own and is not nourished and sheltered by the parent.

Two phyla, **Bryophyta** and **Tracheophyta,** are classified in the subkingdom Embryophyta. In all members of this subkingdom the sex organs are multicellular, and the fertilized egg develops into a multicellular embryo while it is still enclosed in the female sex organ. The phylum Bryophyta includes the mosses and liverworts. The Tracheophyta, commonly called vascular plants, includes the horsetails, ferns, and seed plants.

THALLOPHYTA

The ten phyla of Thallophyta range in size from minute bacteria, some of which have a diameter less than 1/100,000 of an inch, to giant seaweeds which may attain lengths of 600 feet. All thallophytes have a simple body structure. A plant body which lacks roots, stems, and leaves is called a **thallus. Phyte** means plant. The earliest plants on earth were thallophytes, and many have remained unchanged for very long periods of geologic time. There are two main groups of thallophytes: algae and fungi. In certain instances an alga and a fungus grow together, forming a plant with a characteristic form. Such plants are called lichens.

Algae

The green coatings on flower pots and fountains are algae. So are the floating masses on ponds, which are made up of many fine threadlike strands or filaments.

The slippery coating on the rocks in streams is composed of algae, and a good fisherman knows that where algae are abundant so also are fish. Fresh-water algae make up the greater part of the pastures of lakes and streams. Most freshwater fishes, such as trout and bass, do not feed directly on algae but eat the small animals that depend on algae for food.

Some of the larger algae can be seen near the shores of oceans. We call them seaweeds or kelp, and many of them contribute to the odors we call "salt air." Far from shore the ocean water which seems so clear is teeming with microscopic life, both plant and animal. Many of the plants are one-celled algae, called **diatoms,** and they are the food for many animals, including one of the largest, the blue whale.

Fig. 2–2. A closed system, known as a Microterella, in which a mouse lived 6 weeks. (William J. Oswald.)

All algae contain the green pigment chlorophyll, and hence can manufacture food, as explained in Chapter 6. Certain algae have pigments in addition to chlorophyll, and these blend with the green, forming colors other than green. In a simple way we can classify algae on the basis of color. Thus we recognize blue-green algae, green algae, brown algae, and red algae. The blue-green and green algae are for the most part freshwater forms, whereas the brown and red algae are marine plants.

In the manufacture of food, a process called photosynthesis, algae use carbon dioxide and water and produce oxygen and food. Certain one-celled green algae, *Chlorella* for example, reproduce rapidly and photosynthesize energetically. Many botanists are experimenting with *Chlorella,* some to unravel the details of photosynthesis, others to design closed systems that could provide man with food, water, and oxygen on future orbiting space stations, at manned lunar bases, and on space ships headed for other planets. In the closed system illustrated in Figure 2–2, a mouse lived 6 weeks. The main features of this system are the mouse and a mixture of water, algae, and bacteria. When the system is balanced, the bacteria decay the animal wastes, providing the algae with minerals and some carbon dioxide. The algae are continuously lighted by a bank of lights and hence are constantly absorbing carbon dioxide and emitting the oxygen which is needed to support the system's animal occupant. The carbon dioxide given off by the mouse is used in photosynthesis. Each day a portion of the algal mixture is drained off automatically. The water is returned to the system, and the algae are mixed with other food and then injected by a grease gun fitting to the mouse's food dish. Scientists believe that a system such as described here can be adapted to a manned space vehicle.

Fungi

Beneficial bacteria, harmful bacteria, yeasts, blue molds, bread molds, mushrooms, truffles, and organisms which cause such plant diseases as smuts, rusts, blights, and powdery mildews are examples of this significant group of thallophytes. Like the algae, the bodies of fungi are simple, but they differ from algae in one important characteristic. Fungi lack chlorophyll; hence, they cannot make their own food. They obtain their food already elaborated either from dead organic matter or from living organisms. If they obtain their food from living organisms, they are known as **parasites;** if from dead organic matter, as **saprophytes.** Both parasites and saprophytes affect our daily lives. The parasitic fungi cause a great variety of diseases of man, of domestic and wild animals, and of plants.

The saprophytes bring about decay and thus return elements to the soil and atmosphere. Nothing is ever really lost in this economical world; instead, all things circulate. In their development plants use water, carbon dioxide, and certain minerals. Plants may be eaten by animals. When the plants and animals die, their remains are food for bacteria and fungi which bring about the return of minerals to the soil and carbon dioxide to the air. Some saprophytic fungi make products such as alcohol, glycerine, and lactic acid, and such antibiotics as penicillin, streptomycin, and aureomycin. Certain saprophytes destroy commercial products such as flour, bread, butter, meat, fabrics, glue, paint, electrical insulation, leather, camera lenses, wood in all its forms, and a variety of other things.

Three phyla are referred to as fungi; these are commonly known as bacteria, slime molds, and true fungi. **Bacteria** are the simplest and most ancient fungi; fossil bacteria have been found in gunflint chert which was 2 billion years old. They are minute, one-celled organisms that reproduce by dividing in two, a process called fission. We recognize three major shapes of bacteria: the **coccus, bacillus,** and **spirillum** forms. Bacteria which are spherical in shape are said to have a coccus form; those which are rod-shaped, a bacillus form; and those shaped like a corkscrew, a spirillum form (Fig. 2–3). Most

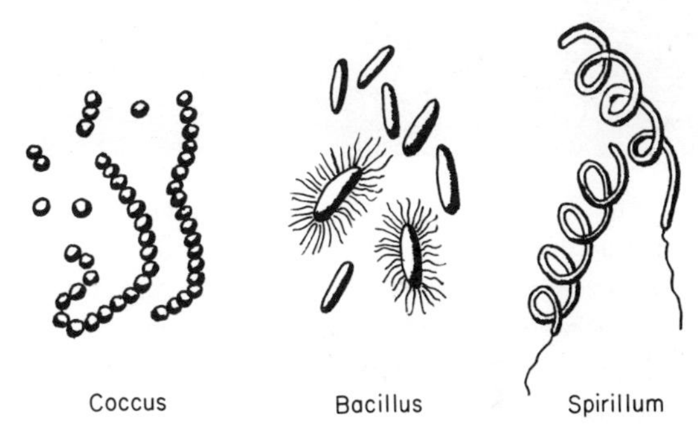

Fig. 2–3. The three general shapes of bacteria.

bacteria are saprophytes, but some are parasites. Anthrax, lobar pneumonia, typhoid fever, tuberculosis, diphtheria, tetanus, and syphilis are bacterial diseases that cause human suffering.

Slime molds are a relatively small group of fungi. The body of a slime mold is a gelatinous mass that glides over the substrate, which may be a fallen log or a decaying stump. At certain seasons the slime mold stops moving and produces spore cases filled with spores.

Most true fungi, a group that includes about 75,000 species, have a cottony body known as a **mycelium,** and most reproduce by aerially dispersed spores (Fig. 2–4). Included in the true fungi are such saprophytes as yeasts, blue molds, mushrooms, and puffballs. Many serious plant diseases, such as smuts, rusts, mildews, and blights, are caused by true fungi. One blight disease, the late blight of potatoes, is of historical interest. In 1845 this disease completely destroyed the potato crop in Ireland, and as a result more than one-quarter of a mil-

Fig. 2–4. A glass model of *Penicillium notatum*, a true fungus. The cobweb-like plant body is known as a mycelium, and the fungus reproduces by spores which are borne in chains at the tips of the erect branches. (Chicago Natural History Museum.)

lion Irish people died of starvation and one and one-half million emigrated to other countries. This catastrophe stimulated an interest in plant diseases, their cause and cure. Since 1845 much progress has been made in controlling plant diseases.

Lichens

Lichens are worldwide in distribution, growing from the arctic and the summits of high mountains to the tropics, where they clothe the branches and trunks of trees. In temperate regions lichens grow on soil, on logs, on branches of trees, and even on rocks where drought is frequent and prolonged. On the bare rock surface lichens are pioneers that in time build up enough soil to support the growth of other plants.

Although a lichen appears to be just one kind of plant, actually two different plants, one an alga and the other a fungus, associate together to form a lichen with a distinctive form. Both the alga and the fungus profit from this cooperative relationship known as **symbiosis.** From the alga the fungus obtains food, and in turn the fungus provides a home for the alga.

We can recognize three major kinds of lichens: **crustose** lichens, **foliose** ones, and **fruticose** kinds (Fig. 2–5). Crustose lichens often grow on rocks as do some foliose ones, those with a leaflike form. However, other foliose lichens may grow on soil and on standing or fallen trees. Fruticose lichens are often stringlike in form, and many grow suspended from tree branches.

BRYOPHYTA

The forest floor, rotting logs, cliffs, and stream banks are favorite haunts of mosses and liverworts, the interesting and beautifully formed members of the phylum Bryophyta. The moss that is usually noticed by the casual observer consists of a clump of small green plants, each with structures resembling a stem and leaves. Hairlike structures, called rhizoids, which anchor the moss and absorb water and minerals, radiate from the base of the stem. Wiry, brown stalks terminated by a capsule may extend upward from some leafy plants (Fig. 2–6). Spores, one-celled reproductive structures, are formed in the capsule, which at this stage is usually orange, brown, or red, contrasting prettily with the green leaves. Liverworts are commonly found growing on stream

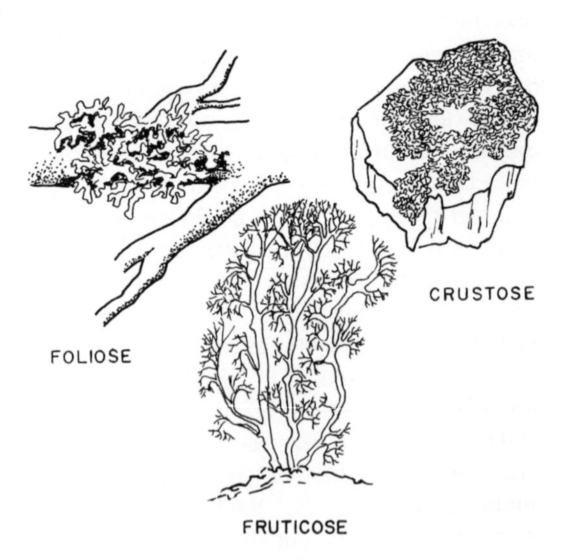

Fig. 2–5. Three kinds of lichens.

Fig. 2–6. Bryophytes. *Left, Marchantia,* a liverwort. *Right,* a moss. (About natural size.)

banks. Characteristically, they are green, ribbon-like forked plants which grow flat on the soil. In the liverwort, *Marchantia,* upright branches bearing sex organs are produced at certain seasons of the year (Fig. 2–6). The bryophytes descended from the algae and were among the first plants to invade the land. The bryophytes have more highly developed sex structures and more advanced methods of reproduction than the thallophytes.

TRACHEOPHYTA

Plants with a well-developed tissue system for the rapid conduction of water, minerals, and food are placed in the phylum Tracheophyta of the plant kindgom. The tracheophytes (vascular plants) are primarily land-inhabiting plants with well-developed roots, stems, and leaves. Familiar representatives of the phylum are club mosses, horsetails, ferns, and seed plants.

Club Mosses

Lycopodium and *Selaginella* are familiar club mosses, which are small plants, less than 2 feet tall, and have small leaves and reproduce by spores. In many species the spore-producing structures are aggregated into small cones that are located at the tips of the branches. *Lycopodium* and *Selaginella* are survivors of a large and diverse group which was abundant on earth during the great period of coal formation. Some of the ancient club mosses, known only as fossils, were large trees.

Horsetails

Horsetails, in the genus *Equisetum,* are the only living survivors of a great group of plants that occupied much of the land during the Devonian period, about 300,000,000 years ago (Fig. 2–7). At that time some horsetails were trees, but the present-day horsetails are small plants, at

Fig. 2–7. Fertile cones of horsetails (*Equisetum arvense*). The tip of a vegetative branch shows at the right. (About natural size.) (Woody Williams.)

most a few feet tall, with ridged, jointed, and hollow stems that bear small leaves in a whorled arrangement. Reproduction is by spores formed by structures which are aggregated to form a cone.

Ferns

Ferns grow in a variety of habitats and in practically all parts of the world. We generally associate ferns with moist, shady woods and ravines, but some are adapted to other habitats, such as swamps, marshes, rocky hillsides, unshaded fields, and crevices in cliffs. Most of our native ferns have an underground horizontal stem (known as a rhizome) that bears leaves (often called fronds) and an abundance of fine roots. Ferns, like other tracheophytes, have well-developed conducting systems. Ferns reproduce by spores formed in spore cases which generally develop on the undersurfaces of the leaves.

Seed Plants

The seed plants are the most recent additions to the earth's flora, and they now occupy areas which in the distant past were populated by ferns, horsetails, and related plants. The success of the seed plants largely is due to their production of seeds, in which a new individual, the embryo, is contained. Furthermore, within a seed there is a supply of food which is used by the developing seedling.

There are two major classes of seed plants, the **gymnosperms** and the **angiosperms.** The gymnosperms, our familiar evergreens such as pine, fir, and spruce, produce seeds in a cone, and the seeds are not enclosed in a fleshy or dry structure. Angiosperms are popularly called "flowering plants," and they have their seeds enclosed. The word gymnosperm means literally "naked seed"; and angiosperm, "covered seed."

Fig. 2–8. A fern growing in a crevice on a rocky cliff. Lichens are evident on the left.

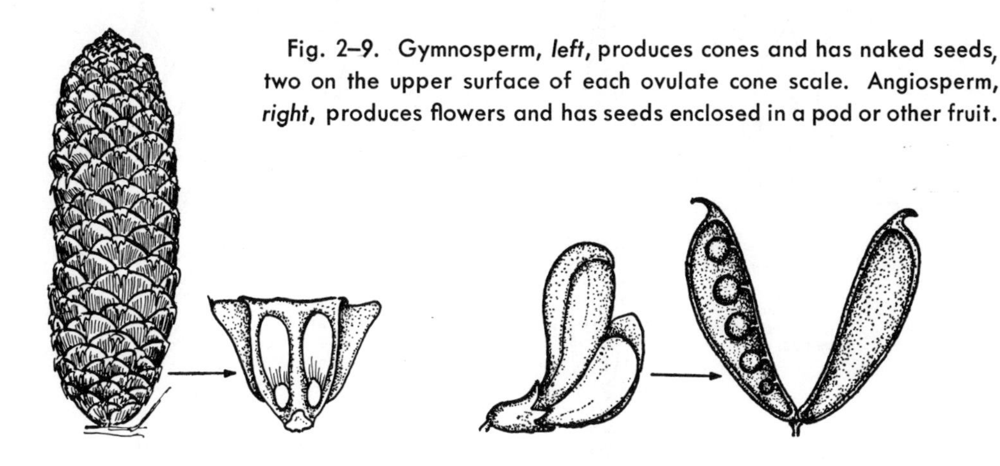

Fig. 2–9. Gymnosperm, *left*, produces cones and has naked seeds, two on the upper surface of each ovulate cone scale. Angiosperm, *right*, produces flowers and has seeds enclosed in a pod or other fruit.

Angiosperms are more advanced and diversified than the gymnosperms. All gymnosperms are either trees or shrubs. As you know, trees and shrubs are woody plants that continue to grow year after year. Trees are generally tall and have a single trunk, whereas shrubs are shorter plants with several stems arising near the ground. Included in the angiosperms are not only trees and shrubs, but also herbs. Herbs are plants with relatively soft stems. Among the herbaceous angiosperms are wheat, corn, pea, sunflower, marigold, and pansy.

All gymnosperms are perennials, which means they live year after year, but in angiosperms we find annuals and biennials as well as perennials. Plants which live during only one growing season are annuals; during one season the seed germinates, the plant develops and reproduces, following which the plant dies. Biennials live for 2 years and then die. Generally, during the first year, a biennial, such as sugar beet, stores large amounts of food, and in the second year it flowers and produces seeds, after which the plant dies. Perennials are plants that live longer than 2 years, often living indefinitely, and flowering repeatedly. Perennials may be either woody plants or herbaceous ones. Many herbaceous perennials, tulips and delphinium, for example, die down to ground level each year but persist from year to year as rhizomes, bulbs, or other underground parts.

Gymnosperms. Pines, firs, spruces, junipers, hemlocks, and Douglas-fir are representative gymnosperms. All of them produce two kinds of cones: male (staminate) and female (ovulate). Pollen is produced in the small, short-lived staminate cones. The pollen is distributed by wind, and some of it lands on young female cones, which then enlarge and produce seeds. Seeds develop on the upper surfaces of the scales of the ovulate cones as if on a series of shelves. Generally, two seeds rest on the upper surface of each scale. The seeds are actually naked and not enclosed in any structure.

Forests of gymnosperms are beautiful, inspiring, and of great economic importance. From them we obtain lumber, ties, poles, posts, and wood pulp which is used to make paper, rayon, and cellophane. The pitch collected from pine trees in the South is a source of turpentine and rosin.

Angiosperms. Angiosperms bear characteristic flowers and have the seeds enclosed in a fruit, using here the botanist's concept of fruit, which includes not only fleshy fruits but dry pods or podlike structures. According to this concept, a tomato and a pea pod are fruits just as are apples, cherries, and grapes. Seeds of tomato, apple, cherry, and grape are enclosed in a fleshy covering; those of pea, in a dry one, the pea pod.

Two subclasses of angiosperms may be recognized, the Monocotyledoneae and the

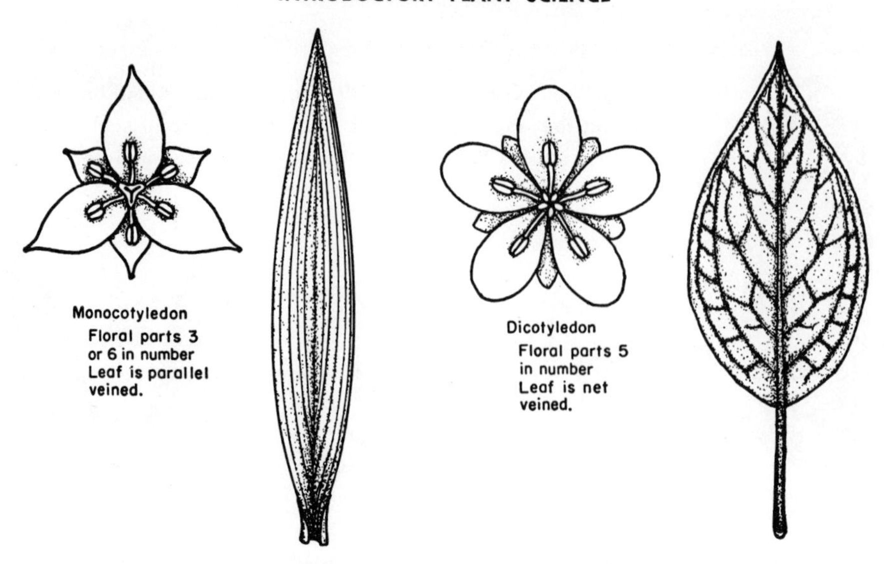

Fig. 2–10. Monocots and dicots are two subclasses of angiosperms.

Dicotyledoneae. The lily, grass, amaryllis, iris, and orchid families are important families of monocotyledons, popularly called monocots. All of the monocots have their flowers built on a pattern of three, or some simple multiple of three. Many monocot flowers have three petals and three sepals. Three or six stamens are typically present in each flower. The veins in a monocot leaf run parallel to each other. Furthermore, within a monocot seed there is one specialized leaf known as a cotyledon.

The mustard, potato, pea, rose, birch, oak, carrot, and sunflower families are important families of dicotyledons (dicots). The dicotyledons have two cotyledons in each seed. The leaves are net-veined and not parallel-veined. The flowers of dicots are built on a pattern of four or five or a multiple of four or five instead of three, as in the monocots.

ECONOMIC IMPORTANCE OF ANGIOSPERMS

Angiosperms are the foundation for our civilization. They provide us with food, spices, beverages, fibers, rubber, certain medicines, and some wood, and many are used for landscaping—among them, lawn grasses; shade trees such as the elm, maple, oak, and cottonwood; and flowers, for example, tulips, lilies, orchids, pansies, and geraniums.

Much of our food comes directly or indirectly from cereals, such as wheat, rice, corn, oats, barley, and rye. In temperate regions wheat is the most important cereal. From a worldwide viewpoint, however, rice is the most important; more than half the people of the earth rely on rice for a substantial part of their calories. The seeds of many legumes, for example, pea, bean, soybean, and lentil, are also much used.

Some widely used foods are obtained from vegetative organs of angiosperms. Of major importance are the Irish potato, sugar cane, sugar beets, sweet potatoes, yams, and the cassava. Shoots of many plants are forage for cattle, sheep, goats, and many game animals, which are sources of meat, leather, and other products.

Fruits are tasty, nutritious, and a source of several vitamins. Among the major fruits are apples, pears, peaches, citrus fruits, grapes, bananas, pineapples, and dates.

The angiosperms provide us with a number of spices. The most widely used is pepper, derived from the seeds and fruits of

Piper nigrum, a vine native to southeastern Asia. Cinnamon is obtained from the bark of *Cinnamomum zeylanicum,* an evergreen shrub or small tree that is native to Ceylon. Other flavoring substances include nutmeg from the seeds of *Myristica fragrans;* cloves, the flower buds of *Syzygium aromaticum;* and vanilla from the unripe, cured fruits of *Vanilla planifolia,* a climbing orchid that grows in tropical America.

Coffee, cocoa, and cola drinks are among the beverages prepared from seeds of angiosperms. Coffee is made from the ground seeds of several species of *Coffea;* cocoa and chocolate, from the seeds of *Theobroma cacao;* and cola drinks, from the seeds of *Cola nitida.* Tea, a beverage relished by more than half the people of the world, is made from the dried leaves of *Camellia sinensis,* a small tree or shrub that is native to China and northeastern India.

Important fibers that come from angiosperms include cotton, linen, sisal hemp, and Manila hemp. Cotton, the dominant textile fiber of the world, is obtained from hairs attached to the seed coats of seeds of various species of *Gossypium*—some of Old World origin; others, such as *Gossypium hirsutum* and *Gossypium barbadense,* of New World origin.

Angiosperms are also a source of various lumbers, such as oak, maple, birch, mahogany, and other hardwoods; of rubber; of some tannins; and of certain medicines such as quinine, digitalis, codeine, and belladonna.

QUESTIONS

1. _____ Every described plant has been given two names, the combination of which is called a binomial. In the name for corn, *Zea mays, Zea* is (A) the genus, (B) the species, (C) the family.

2. Arrange the following categories in the proper order, beginning with the largest: class, phylum, genus, order, species, family.

3. _____ With respect to algae, which one of the following statements is false? (A) Four groups of algae are blue-green algae, green algae, brown algae, and red algae. (B) All algae are so small that you need a microscope to see them. (C) Algae produce large amounts of the food used by aquatic animals. (D) During the day, algae release oxygen into the water. (E) Algae are thallophytes.

4. _____ Algae are in the same major group of plants as are the (A) mosses, (B) liverworts, (C) fungi, (D) seed plants.

5. _____ A closed ecological system in a spaceship would not provide (A) a means for the disposal of excrement, (B) a supply of food, (C) energy to keep the spaceship in motion, (D) a supply of oxygen.

6. _____ Three fishermen of equal skill went trout fishing. They fished different stretches of the same mountain stream. At the end of the day they compared their catches. The fisherman who caught the most fish was the one who fished (A) where the water was extremely clear and where the stream bottom was of pure sand; (B) where the stream went through a canyon and where the sides and bottom of the stream were of solid, clean rock; (C) where the bottom of the stream consisted of slippery, slimy boulders ranging in size from 6 inches to a foot or more (so-called rubble bottom).

7. _____ One-celled fungi, reproducing by fission, and of a coccus, bacillus, or spirillum shape, are (A) bacteria, (B) slime molds, (C) true fungi, (D) close relatives of lichens.

8. _____ Those fungi with a cottony body, known as a mycelium, and which generally reproduce by aerially dispersed spores are (A) bacteria, (B) true fungi, (C) slime molds, (D) lichens.

9. _____ The Irish famine of 1845 resulted because the potato plants were killed by a (A) parasitic true fungus, (B) a saprophytic true fungus, (C) a parasitic bacterium, (D) a saprophytic bacterium, (E) a virus.

10. _____ Which statement about lichens is false? (A) Lichens are closely related to bryophytes, which they resemble. (B) The body of a lichen consists of an alga and a fungus growing together. (C) The major kinds of lichens are crustose lichens, foliose lichens, and fruticose lichens. (D) Some lichens grow on rocks where they play a role in soil formation.

11. _____ Unlike the thallophytes, the mosses and liverworts (A) have complex vascular tissue, (B) have multicellular sex organs and embryos, (C) produce spores on the undersurfaces of leaves, (D) grow to lengths of a foot or more.

12. _____ Mosses and liverworts (A) are in the phylum Tracheophyta, (B) were among the first plants to invade the land, (C) are primarily aquatic plants, (D) are in the subkingdom Thallophyta.

13. _____ The phylum of the plant kingdom known as Tracheophyta includes (A) ferns, horsetails, and seed plants, (B) mosses and liverworts, (C) algae and fungi, (D) only the seed plants.

14. _____ Pine, fir, and spruce (A) are angiosperms with naked seeds, (B) produce pollen in staminate cones, and seeds on the upper surfaces of the scales of ovulate cones, (C) are seed plants that are on the rise, not declining, (D) are gymnosperms that evolved from angiosperms.

15. _____ If a certain plant has parallel-veined leaves and flowers built on a pattern of three, it is a (A) monocotyledon, (B) dicotyledon, (C) polycotyledon.

16. Indicate whether each of the following is an annual (A), a biennial (B), or a perennial (P):

_____ Geranium.	_____ Tulip.
_____ Bean.	_____ Blue grass.
_____ Corn.	_____ Maple.
_____ Carrot.	_____ Sugar beet.

17. Distinguish between trees, shrubs, and herbs.

18. _____ The plants that are most harmful to man are (A) fungi, (B) algae, (C) lichens, (D) mosses, (E) gymnosperms, (F) angiosperms.

19. _____ Most of our food comes directly or indirectly from (A) algae, (B) mosses, (C) gymnosperms, (D) monocots, (E) dicots.

---

20. To become familiar with units of measurement, match the following:

_____ 1. One meter.	1. 2.54 centimeters.
_____ 2. One inch.	2. 0.001 gram.
_____ 3. One liter.	3. 0.001 millimeter.
_____ 4. One milligram.	4. Conversion degrees Centigrade to Fahrenheit.
_____ 5. One micron.	5. Conversion degrees Fahrenheit to Centigrade.
_____ 6. (°C. × 9/5) +32	6. 453.6 grams.
_____ 7. (°F. −32) × 5/9	7. 1.06 quarts.
_____ 8. One pound.	8. 39.37 inches.

3

From Seed to Plant

The story of civilization is often said to have its origin in seeds. When man found that he could gather the seeds of wild plants, plant them in ground near his dwelling, and harvest a supply of food for which he formerly had to roam great distances, there developed a more stable kind of existence. With stability came the time and the interest to develop arts and skills. As he harvested his crop, he laid aside the best seed to be planted next season. Whether he had any idea that the better seed might produce a better crop, we do not know. Instinctively, he was doing what plant breeders do today.

The innovation of agriculture is known as the food-producing revolution and is one of the two principal economic revolutions. The other is the industrial revolution which began about 175 years ago and which is continuing today. Agriculture was started about 8000 years ago in the Near East. For 500,000 years before this momentous innovation small groups of men lived mostly in caves and spent much of their time gathering wild plants and in hunting and fishing. After the food-producing revolution men lived in villages, and then cities were built. Figure 3–1 shows some of the principal events in human history together with some indication of the size of the population.

Man in the New World invented agriculture independently, but later than in the Near East. The prehistoric Indians of our country grew many kinds of plants. Their development of maize is shrouded with mystery, except that we know they must have taken advantage of mutations as they occurred. Through thousands of years of selection they developed a large-eared, meaty-kerneled corn which cannot be traced back with certainty to any wild plants known today. This legacy from the Indians is our most valuable crop—one worth, in the United States, $4.5 billion annually. We are indebted to them for many other plants. Of course, some of our crop plants originated in the Old World. A comparison of what man was using in the Old World and in the New World in 500 B.C. is shown on the following page.

The two lists are almost completely different. Only one major item—cotton—is common to both areas and that is a great puzzle. Cotton was cultivated in Peru in 2000 B.C., but not in the Nile Valley until about 500 B.C. The different crops in the Old and New Worlds suggest that man in America developed his own culture without contact with the other continents. After the discovery of the Americas the white man was the international distributor. He brought

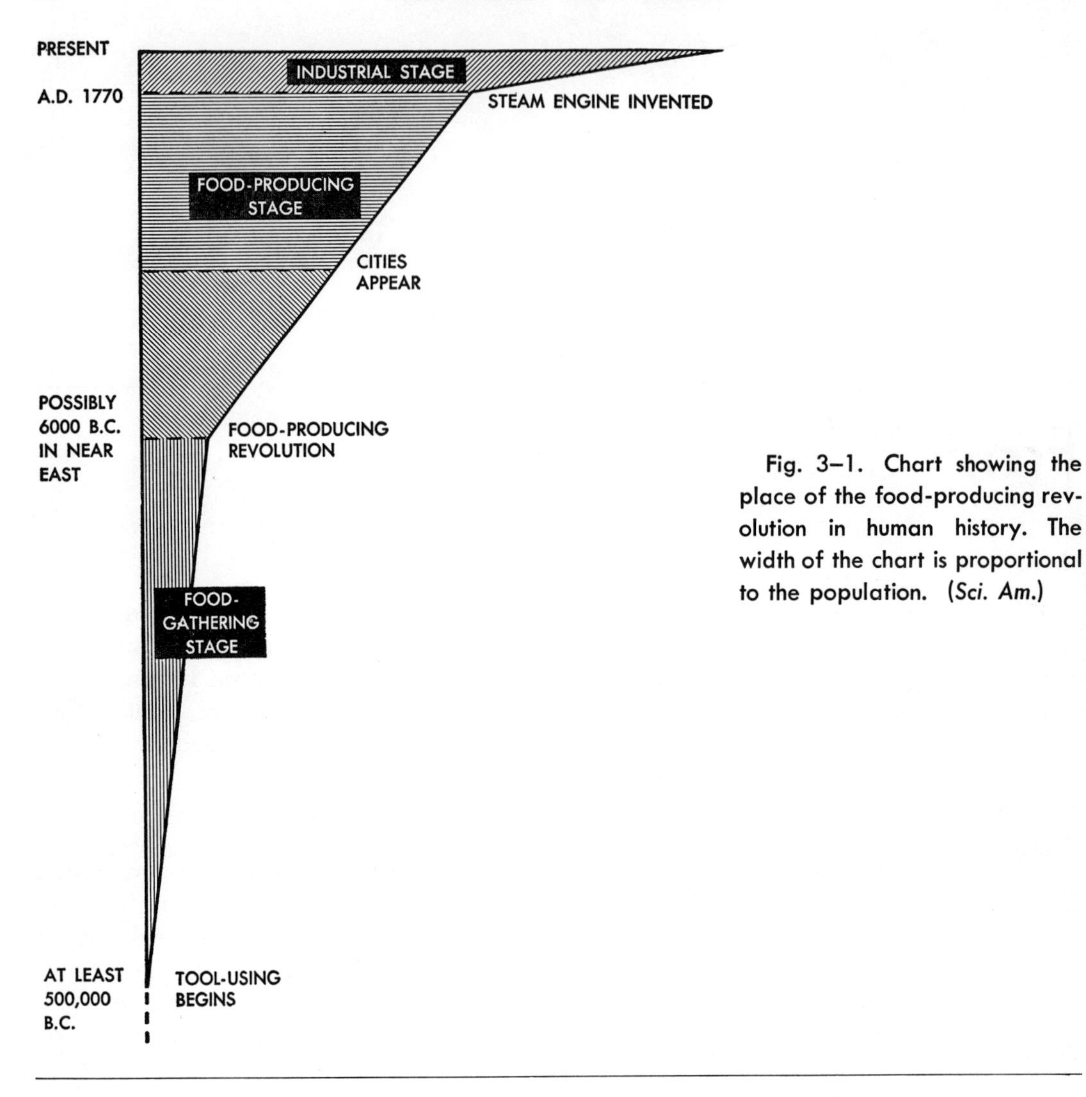

Fig. 3–1. Chart showing the place of the food-producing revolution in human history. The width of the chart is proportional to the population. (*Sci. Am.*)

OLD WORLD

1. Vegetables: artichoke, asparagus, cabbage, cucumber, garlic, lettuce, onion, spinach.
2. Fruits: apple, cherry, fig, grape, lemon, pear, plum.
3. Nuts and oilseeds: linseed, olive, poppyseed, walnut.
4. Roots: beet, carrot, parsnip, radish.
5. Legumes: broad bean, lentils, peas, soybean.
6. Cereals: barley, millet, oats, rice, wheat.
7. Condiments: cane sugar, mustard.
8. Industrial plants: gourds, flax, hemp, cotton.

NEW WORLD

1. Vegetables: cabbage palm, chayote.
2. Fruits: avocado, chirimoya, papaya, pineapple, raspberry, strawberry, tomato.
3. Nuts: brazil, cashew, hickory, peanut.
4. Roots or tubers: camote, manioc, potatoes (many varieties).
5. Legumes: beans (all varieties except the little-known broad bean, *Vicia faba*) and the soybean.
6. Cereals: corn, quinoa.
7. Condiments: peppers.
8. Beverages: chocolate, guayrisa, mate.
9. Industrial plants: gourds, cabuya, cotton, rubber.

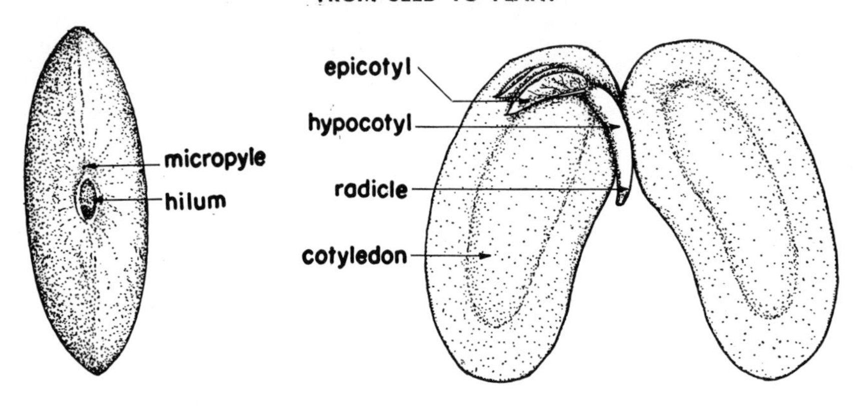

Fig. 3–2. Bean seed. *Left,* external view. *Right,* seed with seed coat removed.

chocolate, strawberries, vanilla, squash, potatoes, tomatoes, corn, pineapples, and other plants to be cultivated in Europe and elsewhere. The white man brought the banana, coffee, sugar cane, rice, wheat, and many others to the New World. Even diseases were interchanged. The white man brought smallpox to the New World; the Indian gave him syphilis.

STRUCTURE OF A TYPICAL SEED—A BEAN

A bean seed is large enough so that its parts can be seen with ease, and its germination and growth into a young seedling can be watched step by step. The bean seed consists of a seed coat, technically called **testa,** and an **embryo.** If you examine the surface of a bean seed, you will see the scar left where the seed separated from the stalk which supported it during development. This scar is called the **hilum.** Near the hilum there is a minute opening known as the **micropyle** (Fig. 3–2).

The seed coat can be removed easily from a soaked seed, thus revealing the embryo. The two large halves of the embryo, which contain the nutrient material sought by human beings, supply food for the developing seedling. They are called **cotyledons.** The other parts of the embryo lie between the cotyledons as you first see them opened (Fig. 3–2). Attached below the cotyledons is a stemlike structure (the **hypocotyl**) and a root (**radicle**). There is no evident demarcation between the hypocotyl and the radicle. The lower part is the radicle; the upper, the hypocotyl. Above where the cotyledons are attached is a stem tip (called the **epicotyl**) bearing a pair of miniature leaves. After the seed is planted, the two minute leaves grow to full size, and the stem tip (also called **shoot apex**) forms additional stem tissue, leaves, and, at maturity, flowers.

In a bean seed food is stored in the cotyledons. In a corn seed, or seed of some other cereal, some food is stored in the single cotyledon, but most is stored in a special tissue known as the **endosperm** (see Chapter 21). In seeds of certain dicotyledons, castor bean for example, food is also stored in the endosperm.

GERMINATION

A seed contains an embryonic plant in a resting condition. The resumption of growth is called germination. When the bean seed is planted and given conditions favorable for germination (water, oxygen, and a proper temperature), water is absorbed, the seed coat is ruptured, and the radicle grows down into the soil, developing into a root which anchors the seedling and absorbs water and minerals. The hypocotyl then elongates, lifting the cotyledons and epicotyl above the ground. The first leaves unfold and expand, and the stem tip grows

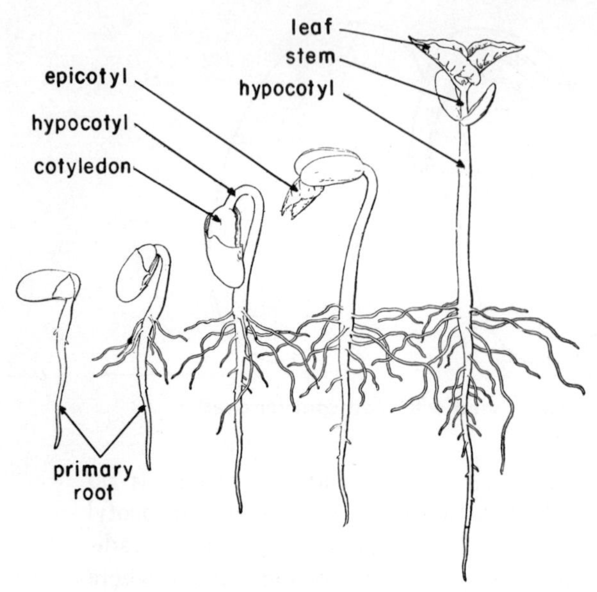

Fig. 3–3. The development of a bean seedling. (After Bergen.)

longer, soon to give rise to more leaves, and later to flowers (Fig. 3–3).

In the germination of pea and cereal seeds, the hypocotyl does not elongate and lift the cotyledons above the soil (see Chapter 21). In these plants the shoot which appears above ground develops entirely from the epicotyl, and the cotyledons remain in the soil.

The transmutation from a living embryo in a dormant condition to a growing young plant involves rapid changes within the seed. During these few days of germination, physiological activities are initiated and accelerated. Respiration becomes rapid, and the food stored within the cotyledons is digested. The digested food is transported to the growing parts of the embryo, and the cotyledons shrink and eventually wither away.

Regardless of how a seed is oriented in the ground, the root grows downward and the shoot upward. The directional growth is a response to gravity, to which the root responds positively and the shoot negatively. The controlling factor is an unequal distribution of hormones within the plant (see Chapter 17).

The seedling is on its own as soon as it has used up the food stored in the cotyledons. Its primary root grows longer, and branch roots develop, giving the plant a greater absorptive area. As the first leaves unfold and expand, they begin to make food from carbon dioxide and water. The cells of the stem tip divide. Some of the resulting cells remain stem cells, and thus the stem increases in length. Other cells formed at the stem tip develop into leaves.

The plant has two main parts: the root system below ground, and the shoot system, collectively the stem, leaves, and flowers, above ground. The organs of a flowering plant are conveniently grouped into vegetative and reproductive. The vegetative organs are the root, stem, and leaf; the reproductive ones are the flower, fruit, and seed. The organization of a seed plant is shown diagrammatically in Fig. 3–4.

VEGETATIVE ORGANS OF A PLANT

A man-made machine, even though it is designed to work automatically, is subject to all sorts of failure. A switch may break, a line may become disconnected, the bearings may wear out, or any one of a number of troubles may interfere with its working order. Power failure will put it completely out of operation. A plant puts to shame any such creation of man. It becomes located in the midst of raw materials, builds and repairs its own machine, and makes its own fuel. It can never suffer from a power shortage as long as the sun is in the sky, for it operates on solar energy. This wonderfully efficient mechanism is held in the deceptively simple structures of roots, stems, and leaves.

The roots absorb water and minerals from the soil, and they anchor the plant. The root system of a plant is very extensive. A painstaking examination of the roots of a rye plant was once made in order to learn how much absorbing and anchoring surface it possessed. The figures are impressive,

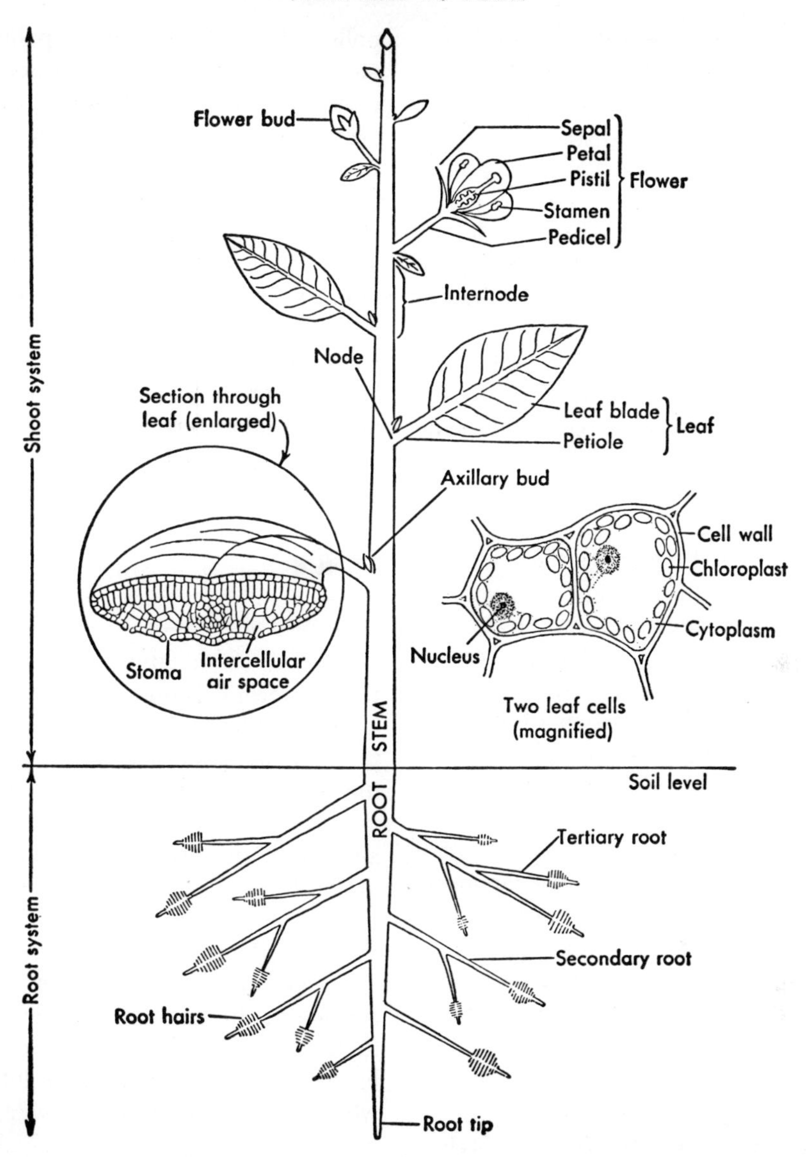

Fig. 3–4. Organization of a seed plant (diagrammatic). (Reproduced by permission from *The Essentials of Plant Biology* by Frank D. Kern. Copyright, 1947, by Harper & Row, Inc., New York.)

particularly since most of us have never really seen the whole underground structure of a plant. The number of roots was found to be 13,815,762, and their total length end to end was 387 miles. Remember, this was one rye plant. You might like to take pencil and paper and calculate the number of roots per acre. The absorbed water and minerals are conducted from the small roots to the larger ones and so on into the stem and leaves of the plant through a tissue called **xylem.**

The stem, in addition to supporting leaves and flowers, conducts water and minerals and **translocates** food. The water that moves up from the roots, carrying with it soil minerals, travels in the xylem. In a soft stem the amount of xylem is relatively small, but the accumulation of xylem through the years in a tree or shrub leads to the rigid structure we call wood. The sugar which is made in the leaves is distributed to all parts of the plant by long thin-walled sieve tubes contained in a tissue called **phloem.** In a tree the inner bark is the phloem. Leaves are attached to the stem at **nodes.** The region of the stem between two adjacent nodes is an **internode.** In other words, a stem consists of nodes and internodes.

A leaf generally has an expanded **blade** and a stalk, known as a **petiole,** which bears at the base two small appendages called **stipules.** Leaves are food-making organs, and here chlorophyll activates the manufacture of sugar (and the liberation of oxygen) from carbon dioxide and water in the presence of light. This is the process of **photosynthesis,** the most significant chemical reaction in the world. Carbon dioxide enters the leaf through minute pores called **stomata,** and water is distributed to the leaf cells by veins. Only a small portion of the water reaching the leaf cells is used in photosynthesis. A greater amount is evaporated from cells within the leaf and escapes through the leaf pores as water vapor. Loss of water vapor is called **transpiration.** On a hot, dry day transpiration is more rapid than on a cool, damp day.

REPRODUCTIVE ORGANS OF A PLANT

Flowers, fruits, and seeds are the reproductive organs of a seed plant. Although many seeds and fruits are essential for human and animal nutrition, they are primarily important to the plant world for the perpetuation of the species. A flower consists of certain fundamental parts, the interaction of which is necessary to produce new plants. The standard pattern of a flower consists of **sepals, petals, stamens,** and one or more **pistils.** Each stamen has a stalk, called a **filament,** and an **anther. Pollen** is produced in the anther, which opens when the pollen is mature. Dusty pollen is then picked up by the wind; sticky pollen, by visiting insects. The pistil, which is in the center of the flower, consists of three parts—a stigma, a style, and an ovary. The **stigma** is the uppermost part of the pistil. It receives pollen and has the capacity to allow only the appropriate pollen to function. The **style,** which is often columnar, connects the stigma to the enlarged **ovary** at the base of the pistil. One or more **ovules** are present in the ovary. After pollination and resultant fertilization, each ovule develops into a **seed,** and the ovary develops into a **fruit.**

BUDS

Terminal Buds

Careful examination of a plant reveals that a terminal bud is present at the tip of every stem. A terminal bud forms additional stem tissue and leaves or flowers. If the terminal bud is removed from a stem, that stem will no longer increase in length, and additional leaves will not be formed at the tip.

Axillary Buds

An axillary bud is present on the stem just above the point where a leaf is attached to the stem. Axillary buds are also called lateral buds. They develop into branches which may bear leaves, flowers, or both.

INTERRELATIONSHIPS BETWEEN PLANT ORGANS

In following chapters each organ of a flowering plant will be discussed separately in more detail. At this time, however, it should be emphasized that the organs of a plant are interrelated. Activities occurring in leaves affect stems and roots. Root processes influence what goes on in the other organs of a plant. For example, leaves depend on roots for a supply of water and minerals. Roots rely on leaves for a supply of food, without which the roots could not grow or absorb water and minerals, or even survive.

Chemicals produced in certain parts of a plant move to other organs and influence them. For example, the terminal bud produces a chemical called **auxin,** which moves downward and inhibits the development of lateral buds. If the terminal bud is removed, the supply of auxin is cut off, and then one or more lateral buds develop. Gardeners make use of this principle to shape plants and to secure plants with many branches.

GROWTH OF SEPARATED PLANT PARTS

Most flowering plants have the capacity to regenerate one or more missing parts. If a portion of a stem bearing leaves and buds is removed from a plant and placed in moist sand or water, roots will regenerate at the base. In some plants—begonias, African violets, and others—stem tissue as well as roots form at the base of the petiole (leaf stalk) when the petiole of a severed leaf is placed in moist sand or vermiculite, or in water.

Tissue Culture

Small portions of a plant may be grown in test tubes or flasks containing the requisite nutrients. The culture of portions of a plant in containers is called tissue culture. This is a field of research which is used to solve a number of biological problems.

The nutrient requirements of various tissues may be determined through the use of this technique. For example, roots grown in tissue culture require vitamin B_1. Intact plants in a field or garden, however, do not make better growth when vitamin B_1 is added to the soil. In intact plants the vitamin requirements of roots are met by vitamins synthesized by leaves and translocated to roots.

Embryos from mature seeds and developing embryos which are only one-twentieth of mature size grow well in flasks containing the appropriate nutrients. Embryos too weak to develop in soil will develop in culture solution. This opens new possibilities

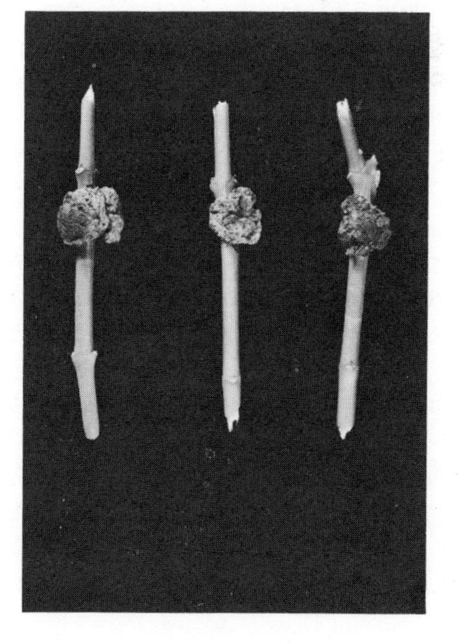

Fig. 3–5. *Left,* tumors free of bacteria growing in a flask. (Wisconsin Agricultural Experiment Station.) *Right,* tumors developing on plant stems which were inoculated with small pieces of bacteria-free tumor tissue. (Philip R. White.)

to the plant breeder because, with the tissue culture technique, he may obtain mature plants from hybrids which otherwise could not be grown.

Tumors develop on some plants when they are infected with *Agrobacterium tumefaciens,* a bacterium. The bacterium introduces a tumor-inducing substance into the host cells, and as a result a large primary tumor, containing bacterial cells, develops at the site of infection. Secondary tumors, free of bacteria, may arise some distance from the site of inoculation. Bits of these bacterial-free secondary tumors may be grown in tissue culture (Fig. 3–5). If pieces of the bacterial-free tumors are inoculated into healthy plants, the tumors will develop on the healthy plants. Although bacterial infection is necessary to induce the formation of the first tumor, once the tumor is formed, it can maintain itself in the absence of bacteria. Information gained from studies of plant tumors, or galls as they are sometimes called, may ultimately contribute to our understanding of the causes and control of tumors in man.

It is possible to raise tomatoes from flowers which have been removed from a tomato plant. After the flowers are removed, they are disinfected and then aseptically transferred to a flask containing nutrient solution. The tomatoes will develop in about 35 days, by which time they are red and taste like vine-ripened fruits (Fig. 3–6).

Fig. 3–6. From plant to test tube. Excised tomato flowers are able to produce fruits on synthetic media. (J. P. Nitsch.)

Frederick C. Steward and his collaborators have grown a complete carrot plant from a single cell obtained from a carrot root, an experiment which indicates that each plant cell has the capacity to develop into an entire plant. If a single cell is to develop into a whole plant, many nutrients must be present in the flask—carbohydrates, minerals, vitamins, hormones, amino acids, and unknown chemicals from coconut milk. From the single cell an unorganized mass of cells first develops. Within the mass, tissues differentiate and an embryo becomes organized. After the embryo is well formed, it is transferred to a flask containing a nutrient medium made solid with agar. Here the embryo grows into a carrot plant with its characteristic root and shoot. The development of an entire plant from a single vegetative cell, not a fertilized egg, is indeed a remarkable achievement. Whether or not zoologists will be able to duplicate this feat with animal and human cells remains to be seen.

QUESTIONS

1. _____ Which one of the following is false? (A) Agriculture was innovated about 8000 years ago. (B) Our present civilization would be impossible without agriculture. (C) Man in the New World invented agriculture independently and before it was practiced in the Near East. (D) Corn was developed by the American Indians.

2. Indicate whether each of the following is of Old World origin (O) or of New World origin (N):

_____	Potatoes.	_____	Squash.
_____	Rice.	_____	Rubber.
_____	Wheat.	_____	Sugar cane.
_____	Corn.	_____	Bananas.
_____	Tobacco.	_____	Alfalfa.
_____	Tomatoes.	_____	Tea.

3. Draw a bean seed in external and internal views and label fully.

4. _____ In bean seeds most of the food is stored in the (A) endosperm, (B) cotyledons, (C) epicotyl, (D) testa, (E) hypocotyl.

5. _____ The radicle of a bean seed develops into (A) a hypocotyl, (B) a root, (C) a stem.

6. _____ The epicotyl of a seed develops into (A) the root, (B) the hypocotyl, (C) a stem bearing leaves.

7. _____ Which one of the following is not required for the germination of bean seeds? (A) Light, (B) water, (C) oxygen, (D) proper temperature.

8. Name the vegetative organs of a plant and list their special functions.

9. _____ If the roots of one rye plant were placed end to end, they would extend (A) across the classroom, (B) across the campus, (C) from New York to Montreal, Canada.

10. _____ Seeds develop from (A) ovules, (B) ovaries, (C) anthers, (D) pistils.

11. _____ Pollen is produced in the (A) ovary, (B) anther, (C) filament, (D) stigma, (E) style.

12. _____ An axillary bud is located (A) just below a leaf, (B) just above where a leaf is attached to the stem, (C) in the center of an internode, (D) just above the terminal bud.

13. _____ Axillary buds develop into (A) leaves, (B) flowers, (C) branches.

14. Suggest a problem that might be solved by tissue culture.

15. Do you think that in the future zoologists will succeed in having rabbits develop from single rabbit cells cultured in flasks? Justify your statement. Will it be possible to have human babies develop from fertilized eggs transplanted to flasks containing appropriate nutrients?

---

16. The terms on the left are frequently used in the construction of botanical terminology. Indicate their meaning by matching the proper translation from the list on the right.

_____	1. Epi-	1. Between.
_____	2. Hypo-	2. Opening.
_____	3. Inter-	3. Naked.
_____	4. Photo-	4. On, upon.
_____	5. -pyle	5. Plant.
_____	6. Angio-	6. Vessel.
_____	7. Gymno-	7. Under, beneath.
_____	8. Morpho-	8. Seed.
_____	9. Phyto-, -phyte.	9. Form.
_____	10. Spermato-, -sperm.	10. Light.

4

Units of Protoplasm—Cells

Even though whales and human beings, rattlesnakes and dogs, carrots and lilacs, roses and bacteria, redwood trees and microscopic fungi appear to be entirely different, nevertheless they all have a certain uniformity of fundamental structure. A comparison of portions of any one with any other under the microscope would reveal this unity of form. It would be seen that both redwood and whale or dog and rose are built up of microscopic structural and functional units called **cells.** Some contents of the cells of a whale and the cells of a redwood would appear similar, the nucleus and cytoplasm, for example. The cells of the whale and the redwood would not be identical in shape, but neither would all the whale cells be alike in shape or all the redwood cells. Muscle cells of the whale would be unlike nerve cells or red blood cells. Some of the redwood cells would be boxlike; others spherical; still others, tubular. In any organism the cells are suited in size, shape, and function to the job they perform.

This similarity of form led T. H. Huxley to state that there is a hidden bond which connects the flower a girl wears in her hair to the blood that courses though her youthful veins. One of the great generalizations of biology is that all organisms, both plants and animals, are composed of three-dimensional cells. Within the cells the process of living takes place. Cells reproduce, assimilate, respire, respond to changes in the environment (that is, they are irritable), and absorb water and other materials from their internal or external environment. In other words, the activities of an organism result from cell actions.

If cells are units, it should be possible to culture them apart from a body. In the previous chapter the tissue culture technique for growing plant tissues was discussed. Animal cells can also be removed from the body and cultured in flasks. The cultivation of animal cells in flasks was first successfully carried out in 1906 by Ross G. Harrison, using nerve tissue from frog embryos. This was the beginning of the modern technique of tissue culture. This great tool has yielded richer benefits for mankind than Harrison and other pioneers could have foreseen and demonstrates once again that from theoretical studies practical applications may follow. The culture of animal cells in flasks made possible the Salk vaccine for polio control. The virus causing polio is grown in monkey kidney cells cultured in flasks. The virus is then treated to render it harmless, following which it is used as a vaccine.

Living heart cells, taken from a chick embryo, have been growing for more than 20 years in flasks, far longer than the life expectancy of a chicken. We generally assume that a piece of raw meat is dead. Certainly, the animal from which it came is dead, but some of the cells are alive. Dr. T. Strangeways, in England, used to demonstrate to his students the difference between the death of the organism and the death of the individual cells by culturing cells taken from sausage purchased in the market. Some of the cells grew. With the appropriate technique, cells grown in flasks are immortal.

The concept that living organisms are made up of cells has developed during the past 300 years. The discovery of the microscope in 1590 by Zacharias Jansen, a spectacle maker of Holland, permitted man to explore the microscopic world of life and the units which make up the complex bodies of multicellular plants and animals. The microscope was later improved by Robert Hooke, an Englishman, who was interested not only in optics, but in the structure of objects made evident with his microscope. In 1665 he published his observations in a book, *Micrographia*. Among other things, he described the boxlike structures which made up cork, and these units of structure he called cells. Because the cork cells that he examined were devoid of the living substance, protoplasm, his descriptions were based on only the cell walls. It did not occur to Hooke or to his immediate successors that both plants and animals were cellular in nature. Between 1808 and 1839, the cell theory was elaborated by a number of individuals, including Mirbel, Dutrochet, Turpin, Meyen, Von Mohl, Schleiden, and Schwann. According to this theory, the cell is the unit of structure and of function in both plants and animals. All the activities of the body are the result of cellular processes. The intact plant or animal is not a chaotic aggregation of cells, but a coordinated entity in which each cell contributes to the welfare of the whole organism. Each cell affects other cells and in turn is influenced by them. Furthermore, the modern cell theory states that cells originate from pre-existing cells.

The living substance in plant and animal cells is called **protoplasm.** All the metabolic and physiological processes known collectively as vital activity occur in protoplasm, which is indeed the physical basis of life. Both plant and animal cells take nonliving ingredients from their environment and organize them into living substance, that is, into protoplasm. Protoplasm is chemically very much alike in both plants and animals; yet the differences, though slight, are enough to make one kind characteristic of an oak, another kind characteristic of wheat, and another of a human being. Later in this chapter we shall consider the composition and structure of protoplasm. Before taking up this topic, let us consider the structure of cells.

STRUCTURE OF A PLANT CELL

The three-dimensional character of a green plant cell, as seen with a light microscope, is shown diagrammatically in Figure 4–1*A*. Leaf cells of *Elodea* are favorites for introducing cell structure, and the parts evident with a light microscope are shown in Figure 4–1*B*. The electron microscope reveals several cell structures that are not evident with the ordinary microscope and shows fine details of cellular structure. The use of this instrument has enhanced our knowledge of cells to a remarkable degree during the past decade. Figures 4–1*C* and 4–1*D* show the structure of a cell as established by studies with the electron microscope.

The living unit of protoplasm that makes up a cell is called the protoplast, and in plants it is enclosed by a rigid **cell wall,** a freely permeable membrane produced by

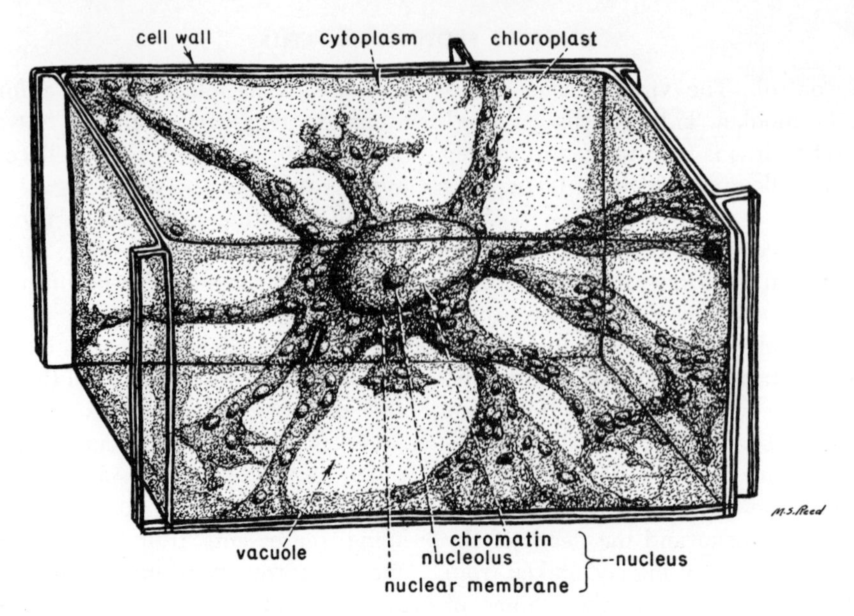

Fig. 4–1*A*. Diagrammatic drawing of a green plant cell.

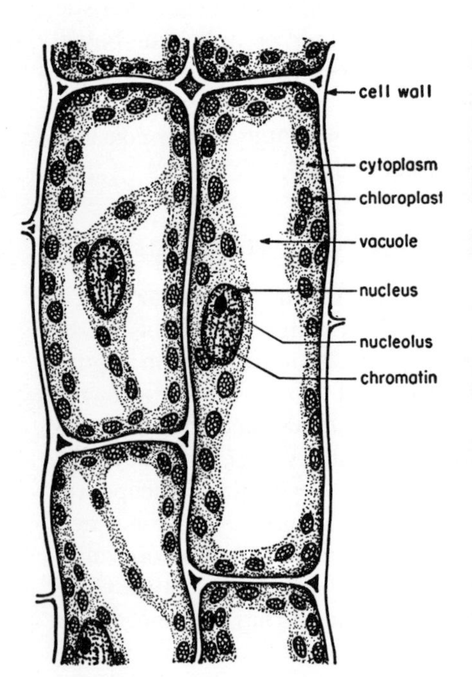

Fig. 4–1*B*. Cells of *Elodea* leaf.

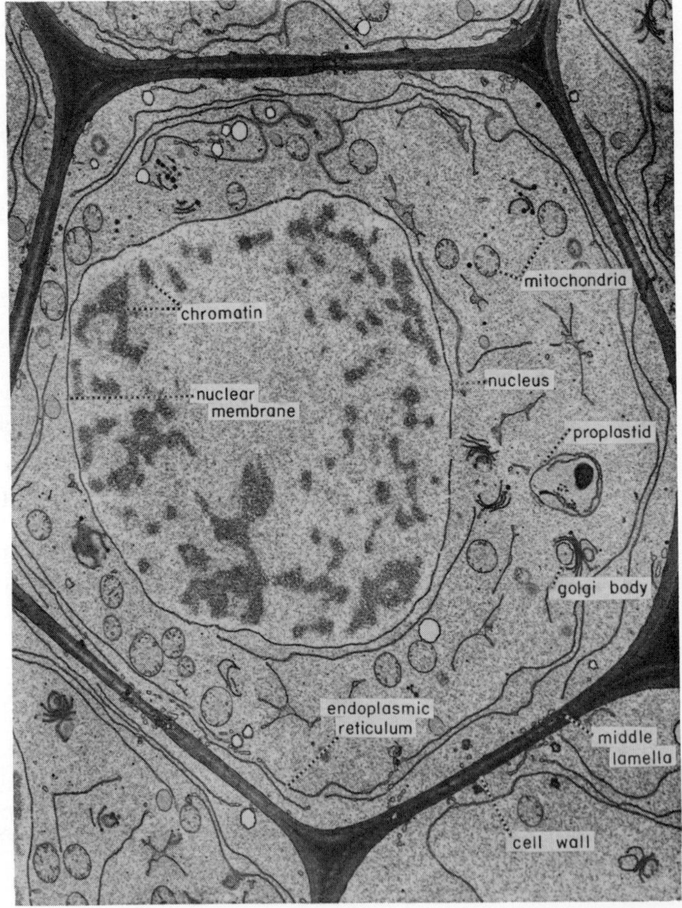

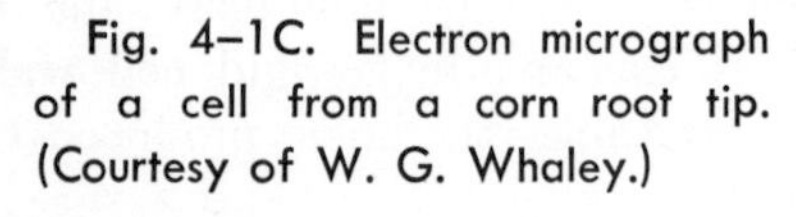

Fig. 4–1C. Electron micrograph of a cell from a corn root tip. (Courtesy of W. G. Whaley.)

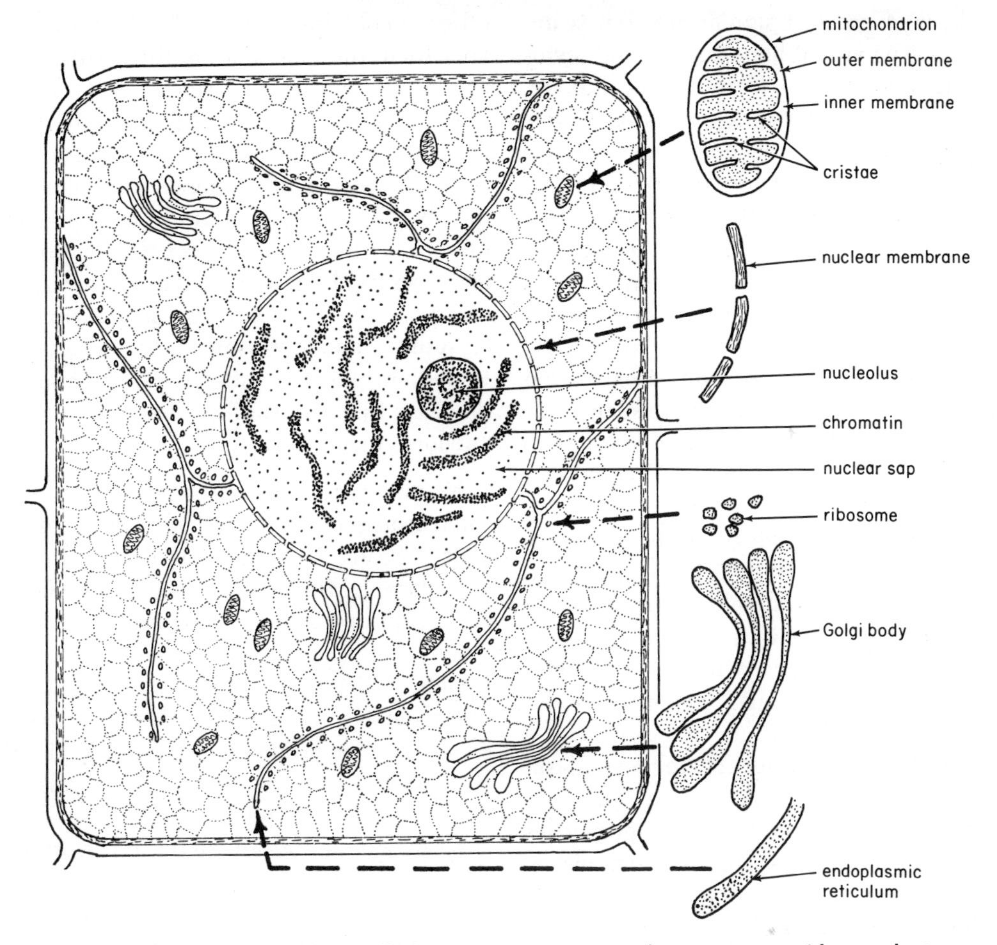

Fig. 4–1*D*. Diagram of a cell showing structures as they appear with an electron microscope.

the protoplast and made chiefly of cellulose. In a protoplast the living material, the protoplasm, is differentiated into **cytoplasm,** and certain structures contained within it, called **organelles.** Some nonliving materials—**cell inclusions,** such as starch grains, **vacuoles,** and crystals of salts or proteins—frequently occur within the protoplast and are referred to as **ergastic substances.**

Parts of a Cell Formed of Protoplasm

The protoplasm in a cell is not uniform throughout. Instead, there is a division of labor, with some structures performing one function and others different ones. The **nucleus** is the control center of the cell, and its directions are carried out by the cytoplasm and the organelles suspended in the cytoplasm. The organelles, each with a characteristic form and a specific action to carry on, are the **endoplasmic reticulum, plastids, mitochondria, ribosomes, microtubules,** and the **Golgi bodies** (also known as **dictyosomes**).

Cytoplasm. In young, actively dividing cells, the cytoplasm occupies most of the volume of a cell. In mature cells the cytoplasm usually occurs as a layer inside the cell wall, but strands of cytoplasm may traverse the cell.

The simple, grayish, and almost structureless appearance of the cytoplasm under the microscope belies the complicated processes of which it is capable. The submicroscopic structure, the arrangement of the compounds, is of greatest significance but is imperfectly known. When active, the cytoplasm is fluid; but in a dormant condition, such as in seeds, it is in a gel state. Under changing conditions the consistency of the cytoplasm may vary from the fluid to the gel state. When fluid, the cytoplasm often circulates around and around inside the cell. The streaming of cytoplasm is a fascinating phenomenon to observe under the microscope. It led Huxley to remark, "If such be the case [streaming in all living cells] the wonderful noonday silence of a tropical forest is, after all, due only to the dullness of our hearing, and could our ears catch the murmur of these tiny maelstroms, as they whirl in the innumerable myriads of living cells which constitute each tree we should be stunned, as with the roar of a great city." Many biological reactions occur within the cytoplasm. Numerous cellular enzymes are located in the cytoplasm.

The outermost surface of the cytoplasm forms the **plasma membrane,** a submicroscopic membrane with some properties unlike the main mass of the cytoplasm. Membranes, called **vacuolar membranes,** are formed where the cytoplasm contacts vacuoles in the cell. Both the plasma and vacuolar membranes are composed of proteins and fatty substances, and the electron microscope reveals that each membrane consists of two layers. The plasma and vacuolar membranes are **differentially permeable** and are the principal barriers to the unrestricted movement of materials into and out of cells. When alive, they permit some substances to enter or leave the cell, but restrict the entrance or exit of other compounds. For example, in a sugar beet the differentially permeable membranes prevent the sugar from diffusing out of the cells into the soil. If this were not true, the farmer would have the difficult job of harvesting the sugar from the soil instead of harvesting the beets. In a garden beet the membranes restrict the exit of the red pigment. When the membranes are killed by heat or by some other agent, they become freely permeable, and then the pigment in the garden beet or the sugar in the sugar beet diffuses out. In a living plant cell the differential permeability of membranes is not constant, but variable. Under some conditions the membranes are more permeable than under others. The permeability of the membranes increases with a rise in temperature, with injury to adjacent cells, and in the presence of sodium and potassium ions. In contrast to sodium and potassium ions, calcium ions decrease the permeability of the plasma and vacuolar membranes.

Mitochondria. As observed with the light microscope, mitochondria appear as minute rods, threads, and granules dispersed in the cytoplasm. Mitochondria, about 1000 in each cell, are formed from proteins, deoxyribonucleic acid, ribonucleic acid, and lipoids. They arise only from pre-existing mitochondria by division. They are passed on from generation to generation via the eggs. The electron microscope reveals that the surface of a mitochondrion consists of two layers: a uniform outer layer, and an inner one that bulges inward in many places to form plates known as **cristae** (Fig. 4–1*D*), which are studded with many minute spheres. Mitochondria, present in cells of all plants and animals, are the "powerhouses" of the cells. Within them, especially on the cristae, energy is released from food by respiration.

Plastids. In many plant cells some of the protoplasm is organized into microscopic bodies called plastids. Three kinds of plastids may be encountered in plant cells: **chloroplasts** (green bodies), **chromoplasts** (yellow to red bodies), and **leucoplasts** (colorless plastids). Plastids originate from minute granules known as **proplastids** which are transmitted from generation to generation by the reproductive cells, especially the egg. In growing leaves, fully developed chloroplasts multiply by division, in which process the plastids elongate and then constrict in the middle.

Green plant cells do not have green cytoplasm. Instead, the green pigments, **chlorophyll a** and **chlorophyll b,** are localized in chloroplasts, which are the food-manufacturing centers. In bryophytes and tracheophytes the chlorophylls occur in a ratio of three parts of chlorophyll a to one of chlorophyll b. It is within each chloroplast that carbon dioxide and water react in the presence of light to form sugar and oxygen.

Yellow pigments, **carotenes** and **xanthophylls,** are also present in a chloroplast. We do not see these colors in a green leaf because the more intense green pigments mask them. However, if the green pigments disappear, as they do in the leaves of many plants in the autumn, the yellow pigments become evident. It is then that the foliage of the birches, poplars, elms, oaks, and other trees gives beautiful gold highlights to the landscape. The yellow and green pigments in chloroplasts are not soluble in water, but they are soluble in alcohol, ether, acetone, petroleum ether, and other organic solvents. Chlorophyll loses its ability to make food when separated from a chloroplast.

The pigments present in a chloroplast can be separated as follows: The chloroplast pigments are extracted from fresh leaves by boiling in 80 per cent alcohol over a water bath. A drop of the extract is then placed, ½ inch from the bottom, on a V-shaped piece of filter paper about 8 inches long and 1 inch across at the wide end. When the drop is about dry, another drop of the extract is placed over the first. This is repeated until twenty-five drops have been placed at the same spot. The strip of paper is then suspended in a test tube containing 5 milliliters of a mixture of 90 per cent petroleum ether and 10 per cent acetone. The tip of the paper, but not the spot of pigments, is immersed in the mixture (Fig. 4–2). As the mixture ascends by capillarity, the pigments move with it for a certain distance, but not indefinitely because the pigments are held by the cellulose of the paper—some strongly, others weakly. The cellulose attracts chlorophyll b most strongly, then chlorophyll a, then xanthophyll, and carotene is least attracted and held by the cellulose. Hence an hour after the paper has been dipped in the petroleum ether–acetone mixture, four bands will be evident on the filter paper. At the greatest height is a band of orange carotene; below, a band of yellow xanthophyll; next below, a blue-green band of chlorophyll a; and lowermost is a green band of chlorophyll b. With this simple technique (called paper chromatography) we have separated the pigments in a mixture. Paper chromatography can also be used to separate the constituents in a mixture of different kinds of sugars, or amino acids or antibiotics, and is in general a most useful tool in biology.

The minute features of a chloroplast can be detected with an electron microscope. A mature chloroplast has a double-layered surface membrane which is differentially per-

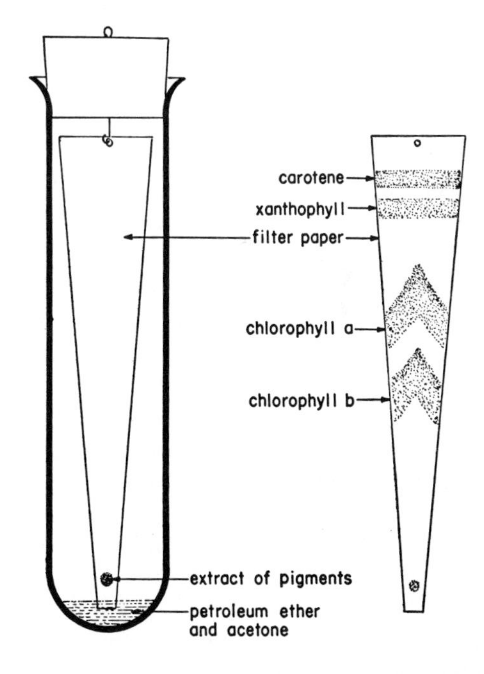

Fig. 4–2. Separation of chloroplast pigments by paper chromatography. *Left,* filter paper with spot of pigments in test tube containing a mixture of petroleum ether and acetone. *Right,* after 1 hour the strip of paper has been removed from the test tube. Notice that the four pigments have become separated.

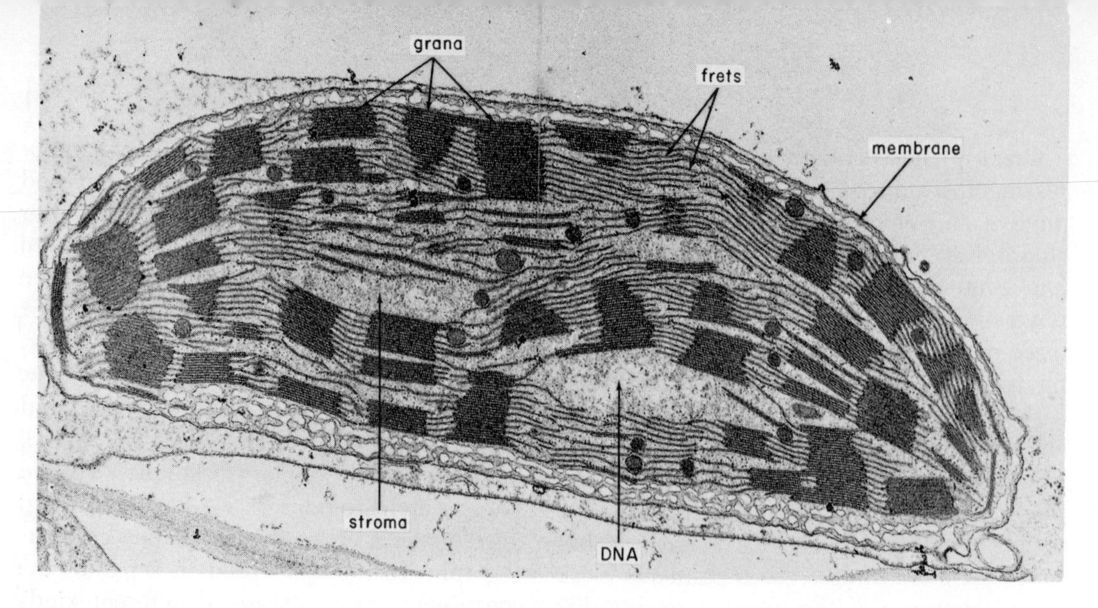

Fig. 4–3. Structure of a chloroplast of corn as revealed with an electron microscope. The stroma has many ribosomes, and the large clear region is probably an area of DNA. Magnification is about 30,000 times. (Courtesy of L. K. Shumwary.)

meable, a nongreen **stroma,** and green **grana.** About fifty disk-shaped grana are embedded in the stroma of each chloroplast (Fig. 4–3). The grana are interconnected by frets. Each granum has a framework of densely stacked protein layers between which are nucleoproteins, DNA, RNA, enzymes, chlorophylls, carotene, and xanthophyll—indeed, the complete apparatus for photosynthesis. Even when chloroplasts are removed from cells, they retain the capacity to carry out all stages of photosynthesis.

The yellow to orange or red colors of some flowers, fruits, and seeds are due to the presence of chromoplasts in the cells. These bodies occur, for example, in the cells of a carrot root, in nasturtium petals, and in cells of tomato, red pepper, lemon and orange fruits. Carotene is present in the chromoplasts of a carrot root, and the related pigment, **lycopene,** occurs in the chromoplasts of a tomato fruit (Fig. 4–4*A*). You should not assume, however, that all red colors of plants are due to chromoplasts. The red color of an apple or a garden beet is caused by an **anthocyanin** pigment. The anthocyanin pigments are soluble in water, whereas the pigments present in plastids are not. If you boil a beet, the pigment dissolves in the water; but if you boil a tomato, the pigment in the pulp does not go into solution. Try filtering tomato juice, and you will find that the filtrate is clear. The red pigment does not go into solution, but remains in the suspended pulp.

Leucoplasts (Fig. 4–4*B*) are colorless plastids in which starch is frequently stored. They are most abundant in roots, tubers, seeds, and other storage organs. They are not responsible for the white color of flowers or other organs. Such whiteness results because the air between the cells reflects light and gives the appearance which we call white.

Plastids may change from one type to another under certain external and internal conditions. When exposed to light, some leucoplasts in a potato tuber become transformed into chloroplasts. As certain fruits ripen, tomato for example, the chloroplasts change into chromoplasts, and the fruits become red.

Ribosomes. These minute granules, about 0.02 micron in diameter and hence only evident with the electron microscope, are assemblers of enzymes and other proteins. Many ribosomes are associated with the endoplasmic reticulum, but others are free in the cytoplasm. Evidence has indi-

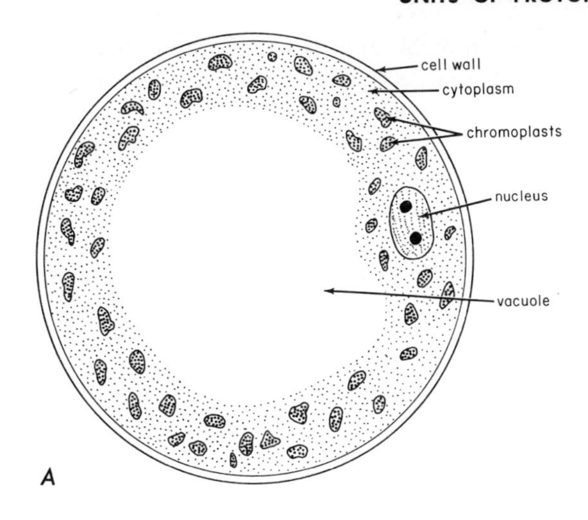

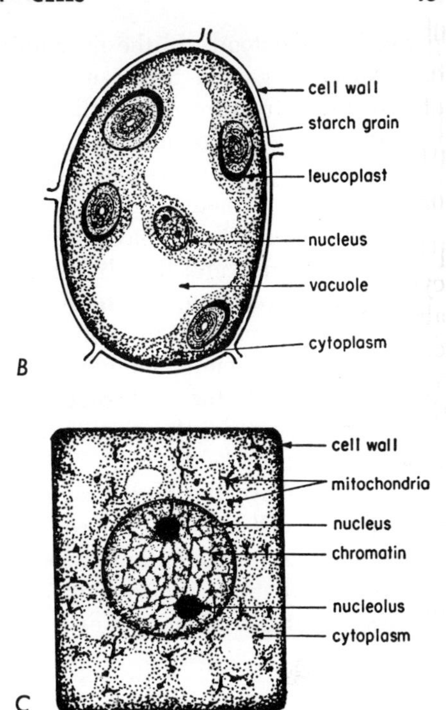

Fig. 4–4*A*. A cell from a tomato fruit. *B*. Cell from a bean cotyledon showing leucoplasts in which starch has been deposited. C. Fixed and stained cell from a root tip.

cated that several ribosomes may be linked to a strand of RNA. Such a collection of ribosomes is known as a **polyribosome** or **polysome.**

In the nucleus, ribonucleic acid (RNA) is packaged with protein to form a ribosome which then leaves the nucleus and enters the cytoplasm. The pattern of the RNA in a ribosome is controlled by the deoxyribonucleic acid (DNA) of a gene. All ribosomes in a cell appear similar, but some ribosomes contain one RNA pattern, whereas others have a different kind. One type of RNA synthesizes one kind of enzyme; a different type produces a different enzyme. Even when ribosomes are isolated from cells, they can synthesize proteins from amino acids provided the necessary enzymes and adenosine triphosphate, the energy source, are present.

Golgi bodies. These bodies, evident with the light microscope but more clearly revealed with the electron microscope, appear as droplets, platelike layers, or other configurations. Because Golgi bodies are abundant and well developed in actively secreting cells, their main role is probably secretion.

Endoplasmic reticulum. Traversing the cytoplasm from the plasma membrane to the nuclear membrane is a network of exceedingly fine double membranes known as the endoplasmic reticulum (Fig. 4–1*D*) which is evident only with the electron microscope. Some investigators believe that the plasma membrane folds inward to form the endoplasmic reticulum. The membranes, nucleus, mitochondria and ribosomes are linked to the endoplasmic reticulum which thus integrates the organelles. Within the fine tubes of the endoplasmic reticulum, substances may be moved from one part of a cell to another, thus facilitating intracellular transport. Furthermore, the endoplasmic reticulum of a certain cell may be joined to those of neighboring cells, thus expediting an exchange of materials between cells and serving as a pathway for the transmission of stimuli.

Microtubules. The electron microscope reveals certain structures which are still imperfectly understood. In the near future we can expect to gain a better understanding of the ultra structure of a cell. The most recently discovered organelles in the cytoplasm are microtubules—fine tubes about 0.02 micron in diameter and several microns in length. Microtubules may alternately expand and contract and thereby bring about cytoplasmic streaming. Microtubules may also participate in cell wall formation; when cell walls are thickening, many microtubules are aligned next to the wall, suggesting that they are playing a role in the deposition of cellulose fibrils in the wall. The spindle fibers which move the chromosomes to the poles during cell division are composed of a large number of microtubules; some scientists believe that the contraction of microtubules moves the chromosomes toward the poles.

Nucleus. Usually, one spherical or ovoid nucleus is embedded in the cytoplasm of each cell. In living cells details of nuclear structure are not clearly evident, the nucleus appearing as a grayish body. In cells which have been killed with appropriate reagents and stained, certain details of nuclear structure are visible. The nucleus is surrounded by a **nuclear membrane.** Within the nucleus there are one or more dense spherical **nucleoli** and an irregular network of **chromatin** (Fig. 4–4*C*). The remaining part of the nucleus is the **nuclear sap** or **karyolymph.** The fine details of these parts are not evident with the light microscope, but can be seen with the electron microscope.

The nuclear membrane, formed of protein and fatty substances, regulates the movement of materials between the nucleus and cytoplasm. The electron microscope shows that the nuclear membrane is a double layer, pierced by tiny pores just large enough to permit the ribosomes to move from the nucleus into the cytoplasm. The outer layer of the nuclear membrane is joined with the endoplasmic reticulum.

A nucleolus, generally spherical in shape, consists largely of nucleoproteins formed from protein and RNA. Because nucleoli contain large amounts of RNA, they are regarded as storage sites for this substance. Nucleoli are formed by certain chromosomes. Specific chromosomes bud off material which collects into nucleoli.

The nucleus, by means of the genes in it controls and coordinates the activities of a cell and functions in the transmission of genes from one generation to the next. As we learned earlier, the DNA of a gene controls the synthesis of a certain kind of RNA, specifically messenger RNA, which combines with protein to form a ribosome.

Genes are resident on the chromatin which, accordingly, is of prime significance in heredity and in directing cell activities. In a nondividing cell the chromatin consists of very thin filaments that are not easily identifiable. When a cell divides, the fine threads become thickly coated with additional nucleoprotein to form rod-shaped chromosomes which are easily seen with the light microscope.

The number of chromosomes formed from the chromatin is constant in a given species. From the chromatin in the nucleus of a dividing cell of corn, twenty chromosomes are formed; and from that of a lily, twenty-four chromosomes. Although the vegetative cells of corn have twenty chromosomes, the sex cells have ten chromosomes. Prior to the formation of sperms and eggs, a reduction division (meiosis) occurs which halves the number of chromosomes. The quantity of DNA is proportional to the number of chromosomes; the sex cells, sperms and eggs, have just half as much DNA as the body cells. Chromosomes bear many smaller units, the genes. The genes are submicroscopic structures whose existence has been demonstrated by breeding work. In any plant or animal there are thousands of them. Each gene has its particular place on a chromosome, like beads on a string, and each controls, or works with other genes to control, a certain characteristic or functional process. The genes do not all function simultaneously. Some do their work while the embryo is forming in the seed; others, while the plant is developing vegetatively. Some may function only seasonally—those which control flowering or leaf fall, for example.

Parts of a Cell Made of Nonliving Material

Not all parts of a cell are made up of protoplasm. The cell wall and cell inclusions are composed of nonliving substances.

The cell wall. In mature cells the cell walls are nonliving, fairly rigid structures. The rigid walls are secreted by the cytoplasm, and they give strength and form to plants.

In a young dividing cell the first thin wall that is formed is called the middle lamella, and it is composed largely of calcium pectate. As the cell matures, a primary wall, mostly of cellulose, is formed by the protoplast against the middle lamella. Later, secondary walls (Fig. 4–5), chiefly of cellulose but often containing lignin as well, may be deposited against the primary wall, thus resulting in a relatively thick cell wall which adds rigidity and strength. Wood is strong because the cell walls are thick and lignified. Wood consists of 50 to 60 per cent cellulose, a carbohydrate formed from glucose, and 25 to 30 per cent lignin, a complex organic chemical. Lignin contributes to the rigidity of wood, but does not markedly increase the tensile strength. When plants are grown rapidly, the cell walls remain relatively thin. In such crops as asparagus, lettuce, and celery, thin walls are desired because they make for tenderness.

In cells with thick walls, thin areas called pits are usually present. The pits are areas where no secondary wall material is deposited, and they may be simple pits or bordered pits. Simple pits are merely cylinders where no secondary wall has been deposited. In a bordered pit the secondary wall material arches over part of the pit. Through the thin areas of the wall, materials are exchanged between cells.

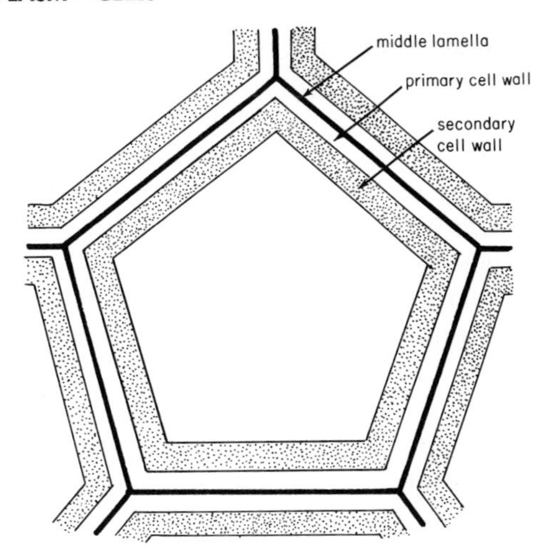

Fig. 4–5. The structure of the cell wall.

Materials may also be exchanged from cell to cell by **plasmodesmata.** Plasmodesmata are fine strands of protoplasm which extend through the cell walls and connect the cytoplasm of one cell with the cytoplasm of adjacent cells. Plasmodesmata are pathways for the exchange of materials between cells. All living cells in a tissue are linked by plasmodesmata.

Cell inclusions. A cell is not alive in every part. The nonliving structures in the protoplast are called cell inclusions. Starch grains, needle-like or star-shaped crystals of calcium oxalate or calcium sulfate, and vacuoles are typical cell inclusions. Vacuoles are spaces in a cell which are filled with cell sap. The cell sap is a solution of sugars, salts, organic acids, alkaloids, and other materials dissolved in water. In a young cell many small vacuoles are present, each surrounded by a vacuolar membrane. As the cell ages, the vacuoles fuse and increase in size until in a mature cell there is usually only one large central vacuole.

Anthocyanin pigments dissolved in the cell sap are responsible for blue, purple, and some red colors found in plants. The colors resulting from anthocyanin pigments delight the eye in apples, flowers, and in the autumn coloring of some trees. In most leaves the anthocyanins are not present during the summer but are formed during autumn. It is a pleasure to drive through the woods in autumn and watch the waves of color splashing in the wind. On your trip through the hills or mountains you will see that trees of the same species may be colored quite differently in separate locations. In one place they will have magnificent crimson leaves; elsewhere, under different conditions, the leaves may almost lack anthocyanin pigments. Bright light, drought, low temperatures, and soil that is low in nitrogen favor the formation of anthocyanin pigments, so in habitats where such conditions are fulfilled, the leaves are the deepest red.

KINDS OF CELLS

The cells at the tips of stems and roots, as well as those in certain other regions, are thin-walled cells with dense cytoplasm and a large nucleus. During the growing season, they divide, forming additional cells, which in turn may divide, thus bringing about growth. Cells having the capacity to divide are known as **meristematic cells.** The cell shown in Figure 4–4*C* is a characteristic meristematic cell.

In time the cells derived from meristematic cells take on characteristic features; they become specialized to carry on one function or another. The mature plant body is made up of many kinds of cells, differing in shape, size, and other characteristics. The differences are related to the physiological functions which each kind performs. Among the types of cells making up the plant are **epidermal cells, parenchyma cells, collenchyma cells, sclerenchyma cells, tracheids, vessel cells, sieve tube cells,** and **companion cells** (Fig. 4–6). In subsequent chapters we will see how these are organized into tissues.

Epidermal, parenchyma, and collenchyma cells are moderately specialized living cell types. Epidermal cells provide a protective covering, generally just one cell thick, over the plant. Epidermal cells of leaves and stems usually have thicker walls on the exposed side, which is also coated with a wax that reduces transpiration. Parenchyma cells form the bulk of the softer plant parts. They are thin-walled, and their dimensions are about equal. Parenchyma cells are generally subjected to the pressure of neighboring cells and under such pressure, as seen in a three-dimensional view, have fourteen sides (Fig. 4–7). Those parenchyma cells having chloroplasts manufacture food; others, such as those in a potato tuber, store food. Parenchyma cells are so unspecialized that they retain the capacity to develop into other cell types or even into entire organs if a suitable stimulus is applied.

In contrast to parenchyma cells, the walls of collenchyma cells are thickened, characteristically in the corners, and the cells are more elongate. They are generally present toward the outside of a stem where they furnish support against bending strains.

Sclerenchyma cells, vessel cells, tracheids, sieve tube cells, and companion cells represent highly specialized cell types. **Fibers** and **sclereids** are the two kinds of sclerenchyma cells—those with very thick cell walls which provide strength or hardness to plant parts. Fibers are long, thick-walled cells, with a tensile strength about equal to steel, and they have dovetailing end walls. At maturity the walls of fibers are so thick that only a small lifeless cavity remains. Sclereids, like fibers, have thick walls, but unlike fibers the cells are not elongate. Certain sclereids may be star-shaped; others, such as the stone cells which provide the gritty texture of pears, are nearly spherical. Stone cells make up the hard structures of plants such as the walls of nuts and the pit of a peach.

Vessels are long tubes which efficiently conduct water and minerals. Each tube is made up of cells joined end to end. During the development of a vessel, the end walls of adjoining cells are partially or completely digested, forming a pipelike structure through which water can move with a mini-

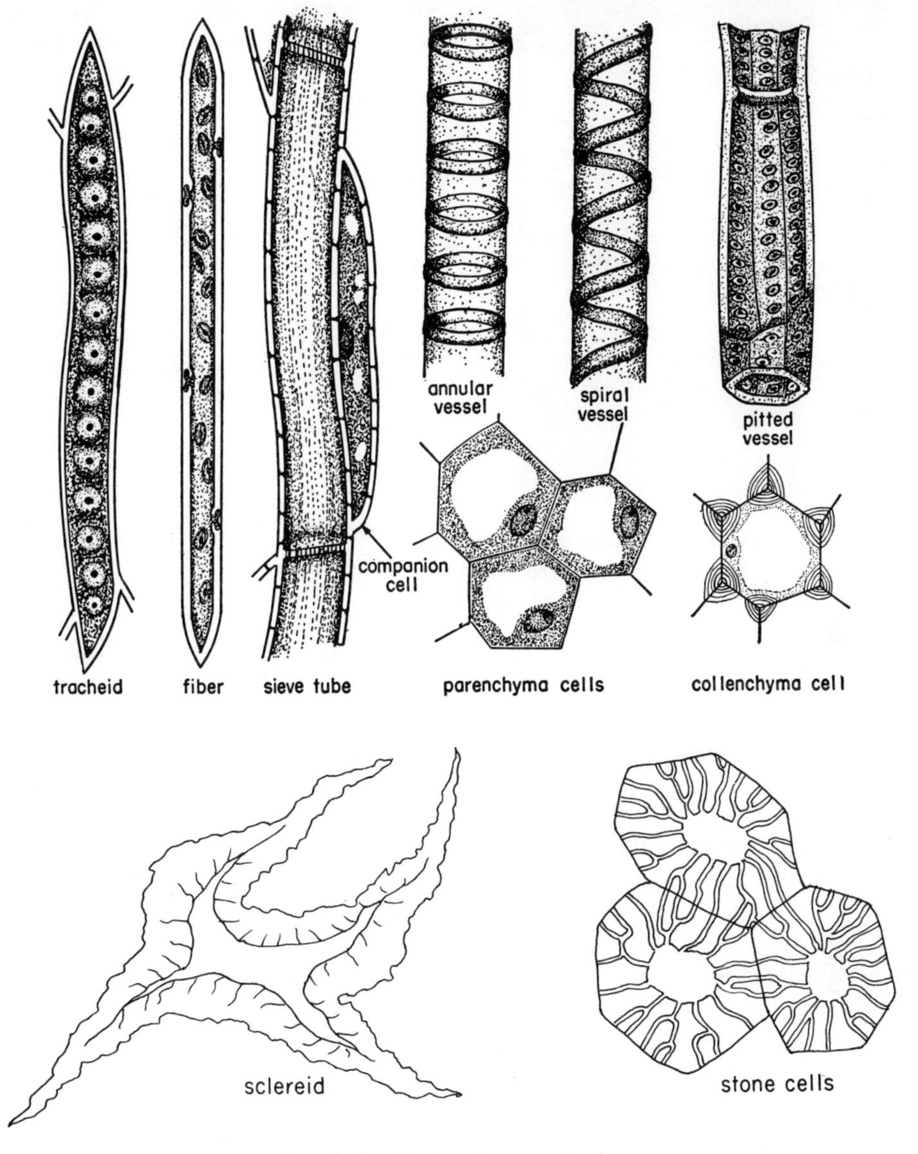

Fig. 4–6. Various types of cells.

mum of obstruction. As a vessel develops, its walls are thickened, and thereby strengthened, by the deposition of a chemical, called **lignocellulose.** In some vessels, **pitted vessels,** the walls are thickened uniformly except where thin areas, called pits, are located. In **annular vessels,** thickenings are deposited as rings; and in **spiral vessels,** as spirals. A mature vessel consists of dead cells; that is, the cells lack protoplasm.

Tracheids, like vessels, conduct water and minerals. Unlike vessels, the end walls of tracheids do not disappear. Tracheids are long cylindrical cells with tapering end walls. The length of tracheids varies between species and with age. In one experiment a scientist found that the tracheids of conifer seedlings averaged 1 millimeter in length, whereas those in a 120-year-old tree averaged 4 millimeters. As with vessels

Fig. 4–7. Three-dimensional structure of cells. Typically, each cell has fourteen facets. The above wax-plate reconstruction shows pith cells of *Eupatorium dubium.* (F. T. Lewis, *Am. J. Botany.*)

their walls may be strengthened by rings or spirals, or may be uniformly thickened except for bordered pits which appear as doughnut-like structures on the walls. Mature tracheids lack protoplasm. Tracheids and vessels along with parenchyma cells and fibers are grouped together, forming a complex tissue called xylem.

Tracheids are less perfectly adapted for moving water and minerals than are vessels. Primitive vascular plants, including most gymnosperms, have only tracheids for conduction, whereas more advanced plants, the angiosperms, have vessels in addition to tracheids. Vessels probably evolved from tracheids through the gradual disappearance of end walls, through a tendency for the end walls to become less oblique, and by a progressive shortening and widening of the cells.

Sieve tube cells are elongated cells with perforated end walls known as **sieve plates.** They are joined together end to end to form a long duct, called a **sieve tube,** through which sugar and other foods are moved. Mature sieve tube cells lack a nucleus but have cytoplasm, fine strands of which extend through the perforations from one cell to the cells above and below. This continuity of cytoplasm from cell to cell favors conduction of food. In the higher plants a smaller cell known as a **companion cell** is associated with a sieve tube cell. A companion cell has a nucleus as well as cytoplasm, and it has been suggested that this nucleus controls the activity of the neighboring sieve tube cell. Many sieve tubes, companion cells, and parenchyma cells are grouped together in a plant, forming a tissue called **phloem.**

Composition and Structure of Protoplasm

We have learned that all plants are built up of cells—units of protoplasm surrounded by walls. The phenomena of life are the result of the activities of protoplasm. Protoplasm is difficult to study because techniques so useful in chemistry and physics often result in the death of protoplasm. With death, changes occur in protoplasm so that what is studied by some methods is protoplasm which was alive rather than that which is alive. There are some techniques, however, for studying living protoplasm.

The microscope used to observe untreated protoplasm reveals little about its structure. Its microscopic appearance is very simple, resembling a minute drop of egg white. The all-important submicroscopic organization is not evident and can only be inferred.

Much greater magnifications can be obtained with an electron microscope than with the usual optical microscope. However, living cells cannot be studied with the electron microscope. Materials to be examined must be dehydrated, a process which usually results in death. The electron microscope is of great value, however, in studying the structure of cell parts—plastids, chromosomes, mitochondria, ribosomes, etc.

Even though cells may be $\frac{1}{200}$th of an inch or less across, it is possible to perform surgery on them with the aid of a micromanipulator (Fig. 4–8). This instrument enables the operator to maneuver microneedles, microscalpels, micropipettes, and even thermocouples into cells. If the nucleus of an amoeba, a one-celled animal, is removed with a microneedle, the amoeba dies within 20 days, whereas an amoeba with a nucleus lives indefinitely. Although an amoeba without a nucleus may live for a short time, the enucleated amoeba cannot divide and reproduce. From these studies we learn that the nucleus is necessary for survival and for cell division. The consistency of the cytoplasm and the nucleus has been estimated by moving a needle back and forth in them. In active cells both the cytoplasm and the nuclear sap are quite fluid.

The compounds present in protoplasm can be determined by chemical analysis. The composition and average molecular weight of each protoplasmic constituent are given in Table 4–1. Protoplasm from different organisms is made up of the same general ingredients, a fact which has led some investigators to state that there is a unity of substantial but not identical composition among all living things. In active cells protoplasm is about 85 per cent water. In dormant cells, such as in seeds, the water content is much less, about 8 per cent.

The "other organic substances" include carbohydrates, vitamins, hormones, and many other compounds. The inorganic material consists of salts of such elements as phosphorus, potassium, nitrogen, sulfur, calcium, iron, magnesium, manganese, copper, and zinc.

Fig. 4–8. Gamma "Chambers" micromanipulator. (Gamma Instrument Co., Inc.)

Table 4–1

Constituents of Protoplasm

Substance	Percentage of Fresh Weight	Average Molecular Weight
Water	85	18
Protein	10	36,000
Fats and related compounds	2	700
RNA	0.7	40,000
DNA	0.4	1,000,000
Other organic substances	0.4	250
Inorganic material	1.5	55

A mixture of the ingredients in appropriate amounts would not yield protoplasm, no more than would a mixture of glass. steel, and jewels form a watch. To make a watch, the materials must be fashioned in a definite way and be put together in a special pattern. The same is true in protoplasm, but here submicroscopic parts are involved. In protoplasm, molecules are arranged in intricate designs and connections. When properly fitted, the resulting protoplasm becomes a self-repairing, self-constructing, dynamic system. The most complex machine man has devised, say an electronic brain, is simple compared to the complexity of protoplasm.

Proteins and Nucleoproteins

The distinctive structure and activity of protoplasm reside, for the most part, in the proteins and nucleoproteins. Protein molecules are very complex, changeable, and large. Enzymes, organic substances which catalyze biological reactions, are complicated protein molecules. In any living cell there are many enzymes, each speeding up a particular biological reaction.

Many protein molecules in cells or test tubes can aggregate (clump) and disaggregate reversibly. When they are disaggregated, the protoplasm is fluid (a **sol**); when markedly aggregated, the protoplasm attains the viscosity of a **gel.** Intermediate degrees of aggregation may also exist. We are familiar with this phenomenon. In the first stage of the preparation of gelatin desserts, the gelatin molecules, which are proteins, are disaggregated. As the dessert cools, the molecules aggregate and form a gel.

Although the proteins are of prime significance in forming the architecture of protoplasm, it should be emphasized that they are only part of the living system. The other compounds also play a significant role in protoplasmic structure. It is the arrangement and distribution of all which makes the living system.

A **nucleoprotein** is made up of a protein combined with nucleic acid. Nucleic acid molecules, like those of proteins, are enormous. We recognize two major kinds of nucleic acid: deoxyribonucleic acid (DNA) and ribonucleic acid (RNA). As we have seen earlier, the hereditary units, genes, are nucleoproteins formed from DNA and protein, and they act by forming certain patterns of RNA which control the formation of enzymes and other proteins. We will have more to say about DNA and RNA in Chapter 15.

Let us digress for a moment and consider some nucleoprotein particles which by themselves have some living traits, namely, the viruses. Viruses exhibit characteristics generally associated with living things; that is, they can reproduce and mutate. Viruses are nucleoprotein particles that reproduce within the cells of a host plant or animal; all viruses are parasites. These submicroscopic infective agents cause such diseases as the common cold, influenza, measles, mumps, smallpox, and polio; and many

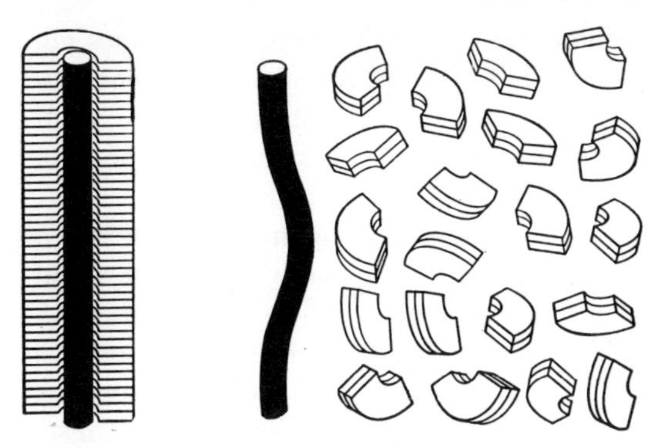

Fig. 4–9. A tobacco mosaic virus consists of a core of nucleic acid surrounded by protein (*left*). With appropriate treatment the virus can be separated into its constituents, namely, nucleic acid and protein (*right*). (*Sci. Am.*)

plant diseases, for example, tobacco mosaic, curly top of sugar beets, and aster yellows. A particle of tobacco mosaic virus has a cylinder of protein and a rodlike RNA core (Fig. 4–9). The protein makes up 94 per cent of the virus substance; and nucleic acid, 6 per cent. Recently, the tobacco mosaic virus has been split into its components—protein and nucleic acid. The protein component could not infect a tobacco plant, and the nucleic acid component was only slightly infective, about 0.1 per cent as infective as the intact virus. Even more remarkable than the splitting was the recombination under suitable conditions of the inactive protein and the slightly active nucleic acid to form a virus having the original capacity to bring about infection. This discovery may lead in time to new techniques for controlling certain virus diseases of man. It may now be possible to break down such viruses and reconstruct them in a form that could provide immunity to that particular virus disease without producing the disease itself. We may learn something about genes—their structure, reproduction, and mode of action—from studies of viruses, because genes, like viruses, are nucleoprotein particles.

Colloids

In protoplasm, protein molecules, aggregates of protein molecules, oil droplets, and some carbohydrates are dispersed in an aqueous solution of salts, sugars, and other soluble substances. The oil droplets and protein particles do not settle out because they are very small, being in the range of 1/1000th of a micron to $\frac{1}{10}$th of a micron in size (a micron is 1/1000th of a millimeter or about 1/25,000th of an inch). Particles which are in this range are in the **colloidal state,** and they remain dispersed when suspended in water or in a solution.

Colloids are one of three types of **dispersions**. The other two are true solutions and coarse suspensions. Many substances, for example, sugar and salt, dissolve in water to form **true solutions** in which the dissolved particles are less than about 1/1000th of a micron in size (the limit is somewhat arbitrary). In **coarse suspensions** the suspended particles are greater than $\frac{1}{10}$th of a micron, the upper limit of colloidal size, and such particles settle out. Sand settles to the bottom of a glass of water, but colloidal clay remains in suspension. Raindrops fall to earth, but colloidal droplets of water remain suspended and do not settle out until they are caused to unite by some natural or artificial means to form larger drops. Even gold, heavy though it is, will remain suspended in water if the gold particles are of colloidal dimensions. We can summarize the three types of disperse systems as follows:

Disperse Systems

Coarse Suspensions	Colloids	True Solution
Particles larger than 0.1 micron	0.1 micron to 0.001 micron	Particles less than 0.001 micron

In a colloidal system there are two phases: the **dispersed phase** and the **dispersion medium.** The colloidal particles are the dispersed phase, and the medium in which they are suspended is the dispersion medium. Fog is a colloidal system having water droplets as the dispersed phase and air as the dispersion medium. In protoplasm, protein particles and oil droplets constitute the dispersed phase; and a solution, the dispersion medium.

The total surface area of material in the colloidal state is tremendous. For example, if just 1 cubic centimeter of water is sprayed into droplets $\frac{1}{100}$th of a micron in diameter, the total area will be 6,000,000 square centimeters. Similarly, if proteins, fats, and certain carbohydrates are in the colloidal state, they will have an enormous total surface area. Many biological reactions occur at surfaces. Hence the large surface area provided by protoplasmic colloids enables biological reactions to go on at a rapid rate.

In protoplasm the colloidal particles may be arranged in a **spatial pattern.** Some biological reactions may take place at the surfaces of certain colloidal particles; and others, at the surfaces of distant ones. Hence a number of different reactions may go on simultaneously without one reaction interfering with another.

As far as we know, additional protoplasm can be formed only by protoplasm now in existence. In some unknown manner protoplasm organizes compounds, which by themselves are not living, into a living system. This process of forming new protoplasm is called **assimilation.** All the protoplasm now on earth has come from pre-existing protoplasm, and that in turn from other protoplasm, and so on back for millions of years. How the first protoplasm on earth came into being is not known.

ORIGIN OF LIFE

Before discussing the origin of protoplasm, or life if you please, we should remind ourselves that just because we cannot demonstrate a certain phenomenon does not prove that the phenomenon is nonexistent. We cannot demonstrate how life originated, but we do know that it came into being.

Some persons believe that life originated spontaneously out of nonliving matter about 3 billion years ago. Our earth is about 5 billion years old. No plant or animal on earth today appears spontaneously; from the simplest to the most complex, each organism reproduces its own kind and arises in no other way. Many ancient peoples believed in spontaneous generation, but this was disproved by a number of investigators. For centuries people believed that maggots originated from spoiled meat. In the seventeenth century Francesco Redi showed that meat placed under a screen, so that flies could not lay their eggs on it, never developed maggots. In the eighteenth century and on into the nineteenth some persons believed that microorganisms arose spontaneously. Lazarro Spalanzani in the eighteenth century and more conclusively Louis Pasteur in the nineteenth demonstrated that microbes do not originate spontaneously. Spalanzani demonstrated that microorganisms would not develop in a sealed flask containing a boiled broth. His experiments did not convince all scientists who believed that heating the air and sealing the flasks made conditions unfavorable for the spontaneous development of microorganisms. In 1860 Louis Pasteur proved to the satisfaction of all scientists that microorganisms do not arise spontaneously. Pasteur used a flask containing boiled broth, but instead of sealing the flask, he drew the neck out in a long, S-shaped curve with its end open to air. Air could flow in and out of the flask, but bacteria and molds were trapped on the walls of the curved neck. Except for occasional contaminations, no organisms developed in the sterile broth.

Could life have originated spontaneously 3 billion years ago when the earth was much different from what it is today? Here we enter the realm of speculation, for we have

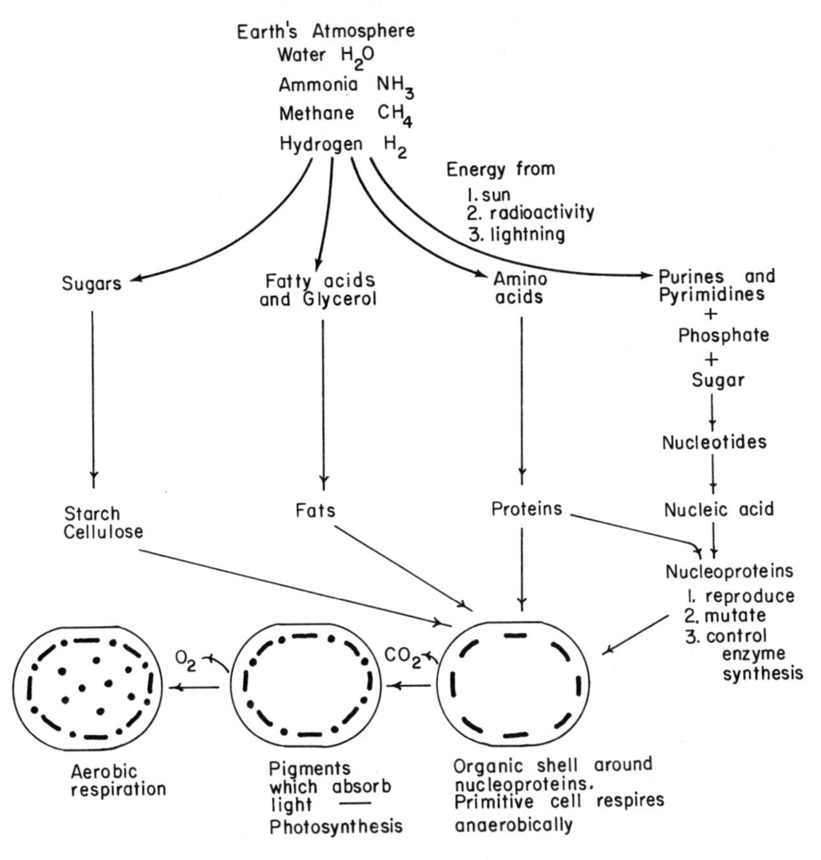

Fig. 4–10. Stages in the origin of cells.

no direct evidence as to where, or how, or when life originated. Certain planets may now have an atmosphere like that of the earth when life originated. Spectroscopic observations show that their atmospheres lack free oxygen and consist of hydrogen, ammonia, water, and methane. With energy from the sun, from ultraviolet radiations, from lightning or radioactivity, these gases may react to form sugars, fatty acids, glycerol, amino acids, purines, and pyrimidines. In the laboratory amino acids and proteins have been synthesized from ammonia, water, hydrogen, and methane, contained in an evacuated tube, and exposed to electric discharges (simulating lightning). Perhaps lightning discharges in the early days of the earth produced from atmospheric gases the materials out of which life arose (Fig. 4–10).

Not only could amino acids and proteins be produced from the earth's primordial gases, but so also could sugars, fatty acids, glycerol, purines, and pyrimidines. Starch, cellulose, and other carbohydrates could be synthesized from sugar; and fats, from glycerol and fatty acids. Most significantly, certain biochemists believe that purines and pyrimidines combined with phosphate and sugar to yield nucleotides; and these polymerized to produce nucleic acids which, in the earth's primordial water, gave direction and control to the synthesis of proteins from amino acids. Certain arrangements of amino acids, controlled by nucleic acid, yielded enzymes which catalyzed various reactions. Nucleic acid molecules, in turn, may have combined with protein molecules to form nucleoproteins which, like present-day viruses and genes, had the capacity to reproduce themselves and to mutate. The creation of the first nucleoproteins may have required hundreds of millions of years because it had to occur by chance. There was no blueprint to follow. Genes are the modern descendents of the first nucleoproteins.

It is assumed that the nucleoproteins duplicated themselves by selecting appropriate molecules from the primeval broth, thus forming more of their type. In time nucleoproteins became surrounded by a membrane that enclosed not only the nucleoproteins, but also water and organic and inorganic molecules. Thus the first cells came into being. According to this biochemical concept of genesis in the early seas, or more likely tide pools, progressive aggregation occurred. Simple molecules combined into complex ones which in turn aggregated into those of still greater complexity, and the huge molecules eventually aggregated into cells.

The events leading to the development of the first cell required many millions of years and depended, as we have seen, on the accumulation of many kinds of organic molecules. The molecules that formed persisted because at that time the atmosphere was devoid of oxygen, and hence no spontaneous oxidation of the organic substances could occur. Furthermore, prior to the formation of the first cell no organism was present to metabolize the organic substances for energy. Any organic molecules synthesized nonbiologically in nature today would soon either be consumed by organisms or oxidized.

Along with the development of the first cells, there evolved a system for the release of energy from some of the organic substances that the primitive cells absorbed from the surrounding water. Thus **anaerobic respiration,** a type of respiration in which oxygen is not used, came into being. In anaerobic respiration carbon dioxide is given off, and thus the waters and atmosphere became enriched with this indispensable gas. The accumulation of carbon dioxide in the air was followed by the evolution of the capacity to make food, that is, to carry on photosynthesis.

The primordial organisms survived at the expense of organic molecules that had accumulated during millions of years. As cells increased in number, the food supply dwindled and ultimately would have become exhausted; then life would have become extinct. Life persisted, however, because **photosynthesis** evolved. Certain mutant organisms produced pigments which absorbed light that supplied the energy for the manufacture of food from such simple raw materials as carbon dioxide and water. This development was an enormous step forward. Living organisms no longer depended upon the accumulation of organic matter from past ages; they could make their own. The evolution of photosynthesis not only provided organisms with food, but profoundly altered the atmosphere and also the earth's oceans. In photosynthesis oxygen is produced. The addition of oxygen to the air markedly changed the earth's atmosphere. The primordial gases of hydrogen, ammonia, and methane became oxidized to other compounds and were removed from the air. The nonbiologically synthesized organic substances in the water were also oxidized into simpler compounds. With their disappearance life became forever dependent on photosynthesis for food.

After oxygen became abundant on earth, certain primitive organisms evolved a mechanism for respiring aerobically, that is, for using oxygen to release the energy from foods. **Aerobic respiration** is many times more effective than anaerobic respiration—releasing about thirty times more energy from a given amount of food.

From these earliest forms of life which could carry on photosynthesis and respiration, there emerged present-day life in all its magnificent array. Living matter, seemingly so transitory, soft, and fragile, is more durable than the hardest stones and metals. Mountains, oceans, and even continents have come and gone a number of times during the past 2 billion years, but life has persisted and indeed has become progressively more abundant and varied.

The ideas previously discussed are the ones in vogue today, and some scientists believe that if other planets are chemically and physically similar to the earth, they should also be populated with organisms. Two planets of our system, Mars and Venus, are like the earth in some ways. The information relayed to earth by Mariner II indicates that Venus lacks a protective magnetic field, and that the temperature is too high, about 800° F, to support life. The photographs and other data taken by Mariner IV indicate that Mars is an inhospitable planet for life. The photographs reveal a moonlike landscape speckled with more than 10,000 craters. Mars lacks a magnetic field. The atmosphere is very thin and has little or no water. During its long history, about 5×10^9 years, Mars has never had sufficient water to form streams, to bring about erosion, or to fill oceans.

Table 4–2 summarizes the environments of the planets. Temperatures are not compatible with life as we know it on Venus, Mercury, Jupiter, Saturn, Uranus, and Neptune.

Table 4–2

Environments of the Planets

Planet	Surface Temperature	Atmosphere	Other Features
Earth	−71° to +57° C.	79% N_2; 20% O_2; 0.03% CO_2	Water present
Mars	−101° to +30° C.	Rarefied; CO_2 and N_2 present; O_2 assumed to be present	Length of day nearly the same as on earth
Venus	In excess of +400° C.	High CO_2; little O_2	Surface hidden by heavy cloud cover
Mercury	Side toward sun, +277° C. Side away from sun, −234° C.	None	One side always toward the sun
Jupiter, Saturn, Uranus, Neptune	Less than −150° C.	Hydrogen (H_2), Helium (He_2), Methane (CH_4), Ammonia (NH_3)	

QUESTIONS

1. _____ Small portions of plants or animals may be grown in test tubes or flasks containing the requisite nutrients. (A) True, (B) false.

2. _____ Which one of the following features of the cell theory is the basis for modern surgery and canning of foods? (A) All plants and animals are composed of cells or the products of cells. (B) All activities of organisms are activities of cells. (C) Cells originate from pre-existing cells.

3. _____ Robert Hooke is given credit for (A) developing the cell theory, (B) the discovery of cells, (C) the discovery of the nucleus, (D) stating that cells come only from pre-existing cells.

4. Diagram a green plant cell and label its parts.

5. _____ The plasma and vacuolar membranes are (A) impermeable, (B) permeable, (C) pseudopermeable, (D) differentially permeable.

6. _____ A membrane which lets water through readily but restricts the entrance or exit of salt or sugar is referred to as being (A) permeable, (B) impermeable, (C) imbibitory, (D) differentially permeable.

7. _____ Which one of the following is *not* a kind of plastid? (A) Chloroplast, (B) anthoplast, (C) leucoplast, (D) chromoplast.

8. Indicate which of the following are evident only with the electron microscope (E).

_____ 1. Plastids.
_____ 2. Mitochondria.
_____ 3. Ribosomes.
_____ 4. Endoplasmic reticulum.
_____ 5. Gene.
_____ 6. Parallel lamellae in a granum.
_____ 7. Pores in the nuclear membrane.
_____ 8. Double nature of membranes.
_____ 9. Cell wall.
_____10. Golgi bodies.
_____11. Chromosomes.
_____12. Plasmodesmata.

9. _____ The anthocyanin pigments are (A) found in plastids, (B) formed in darkness, (C) red or blue in color, (D) insoluble in water.

10. _____ Which of the following would most favor the development of red coloration of leaves in the autumn? (A) Cloudy weather, poor soil, abundance of water; (B) soil low in nitrogen and moisture, bright days; (C) soil high in nitrogen and moisture, bright days; (D) high temperature in autumn, bright days, soil high in water.

11. _____ Which one of the following is not dead at maturity and when functioning? (A) Tracheid, (B) vessel, (C) sieve tube, (D) fiber, (E) sclereid.

12. _____ Which one of the following statements about colloids is false? (A) Colloids are in the range of 0.1 micron to 0.001 micron. (B) Colloids are molecularly dispersed systems. (C) Colloids have a very large surface area. (D) Colloidal particles remain dispersed and do not settle out. (E) Many biological reactions occur at surfaces.

13. Distinguish between a true solution, a colloidal system, and a coarse suspension.

14. Discuss briefly the nature of viruses and their significance to mankind.

15. The following events are those which some scientists believe led to the origin of life. Arrange the events in proper sequence. Number the first event 1, the second 2, and so on.

_____ Primordial gases.
_____ Polymerization into complex molecules.
_____ Formation of simple molecules.
_____ Nucleoproteins.
_____ Membrane enclosing nucleoproteins.
_____ Aerobic respiration.
_____ Anaerobic respiration.
_____ Photosynthesis.

16. Do other planets support life and, if so, what type?

17. _____ Ribosomes are (A) large enough to be seen with the optical microscope, (B) granules which synthesize enzymes and other proteins, (C) the carriers of genetic information, (D) sites where photosynthesis occurs.

---

18. Match the following:

_____ 1. Mitochondria.	1. Colorless plastid in which starch is stored.
_____ 2. Chloroplast.	2. Assemblers of enzymes.
_____ 3. Leucoplast.	3. Repository of a unit of genetic information.
_____ 4. Ribosomes.	4. Storage site of RNA.
_____ 5. Golgi bodies.	5. Photosynthesis.
_____ 6. Endoplasmic reticulum.	6. Chief carrier of genetic information.
_____ 7. Nucleolus.	7. Controls sequence of amino acids in a protein.
_____ 8. DNA.	8. Facilitates intracellular and intercellular communication; provides structural integration of a cell.
_____ 9. RNA.	9. Site of respiration.
_____10. Gene.	10. Secretion.
_____11. Plasmodesmata.	11. Cytoplasmic strands between neighboring cells.
_____12. Nucleus.	12. Executive center of a cell.

19. Match the following:

_____	1. Sieve tube.	1. A tube which conducts water.
_____	2. Vessel.	2. Consists of cells, each with perforated end walls, placed end to end and serving to conduct food.
_____	3. Tracheid.	3. A thick-walled cell with tapering end walls which conducts water.
_____	4. Parenchyma cell.	4. A living isodiametric, thin-walled cell.
_____	5. Collenchyma cell.	5. A long cell with a heavy wall.
_____	6. Fiber.	6. A cell with thickenings in the corners.

20. The terms on the left are frequently used in the construction of botanical terminology. Indicate their meaning by matching the proper translation from the list on the right.

_____	1. Chloro-	1. Tissue.
_____	2. Chromo-	2. White.
_____	3. -enchyma.	3. Leaf.
_____	4. Leuco-	4. Green.
_____	5. -phyll.	5. Yellow.
_____	6. Scler-	6. Color.
_____	7. Xantho-	7. Hard.

5

Absorption of Materials by Cells

Experience has taught us something about the diffusion of gas molecules. For example, if a bottle of ether is opened for a short time in a room, soon the odor is detected throughout the room. In time the molecules become equally distributed. It is evident that molecules move.

Molecules, of which matter consists, are discrete submicroscopic bodies surrounded by space. The ether molecules move in space among the molecules of nitrogen, oxygen, and carbon dioxide of the air. Their movement is not unimpeded, however; they are constantly colliding with other ether molecules and with oxygen, nitrogen, and carbon dioxide molecules. The collisions with other like molecules are fewest in the direction of lesser concentration; this is the reason that molecules diffuse away from regions of greater concentration, like the ether molecules diffusing away from the bottle. The movement of molecules from where their concentration is greater to where it is lesser is known as **diffusion.**

Molecules diffuse independently of each other. For example, during the day, carbon dioxide molecules diffuse into a leaf, and oxygen molecules diffuse out.

Substances dissolved in water also diffuse, because water and other liquids are made up of molecules and space. Dissolved substances diffuse through the spaces between the molecules of the liquid. This can be demonstrated by dropping a crystal of a soluble dye in water. At first the dye molecules are more numerous in the vicinity of the crystal, and here the concentration of dye molecules is high. The dye molecules are in constant motion. Their movement is less interfered with by collisions in the direction of fewer dye molecules. As a consequence they diffuse from the region of higher concentration to the region of lesser concentration, ultimately becoming equally distributed. Their motion continues after they have become equally distributed; but as one molecule leaves a region, another enters it on the basis of chance.

CELLULAR ABSORPTION OF WATER BY OSMOSIS

Osmosis is the diffusion of water molecules across a differentially permeable membrane from regions of greater to regions of lesser concentration of water. Like gas and

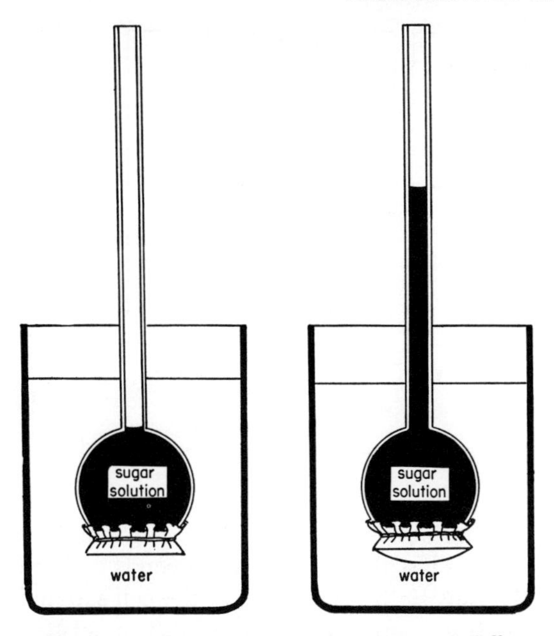

Fig. 5–1. Demonstration of osmosis. A differentially permeable membrane is fastened over the lower end of a thistle tube, which is then partially filled with a sugar solution (*left*). As water diffuses across the membrane from the beaker into the thistle tube, the column of sugar solution rises (*right*).

dye molecules, water molecules are in constant motion, diffusing from regions where they are numerous to where they are fewer in number. Different concentrations of water may be obtained by using the setup illustrated in Figure 5–1. In this setup a cellophane membrane is fastened over the enlarged end of a thistle tube. The thistle tube with membrane is called an **osmometer,** a device to demonstrate osmosis. The cellophane is **differentially permeable,** meaning that molecules of some substances can pass through it, whereas others cannot. Water molecules can pass through it readily, but sugar molecules cannot. The cellophane can be pictured as having submicroscopic pores, large enough for water molecules to get through, but not large enough to permit the passage of the larger sugar molecules. If the swollen basal part of the thistle tube is filled with a concentrated sugar solution and then placed in a beaker of water, the sugar solution rises in the tube (Fig. 5–1).

The column rises because water enters the osmometer. Water diffuses across the membrane from where the concentration of water is greater in the beaker to where the concentration of water is lesser inside the thistle tube. You should not assume that water molecules are diffusing only inward. Actually, some water molecules are also diffusing outward. However, because more water molecules diffuse into the osmometer than diffuse out, the net diffusion is from the beaker across the membrane into the osmometer (Fig. 5–2).

If a living cell is immersed in water, it takes up water by osmosis until the cell walls become so taut that further entrance of water is restricted (Fig. 5–3, left). The cell wall is freely permeable, permitting both dissolved substances and water to pass through. However, in a living cell the plasma and vacuolar membranes are differentially permeable and are comparable to

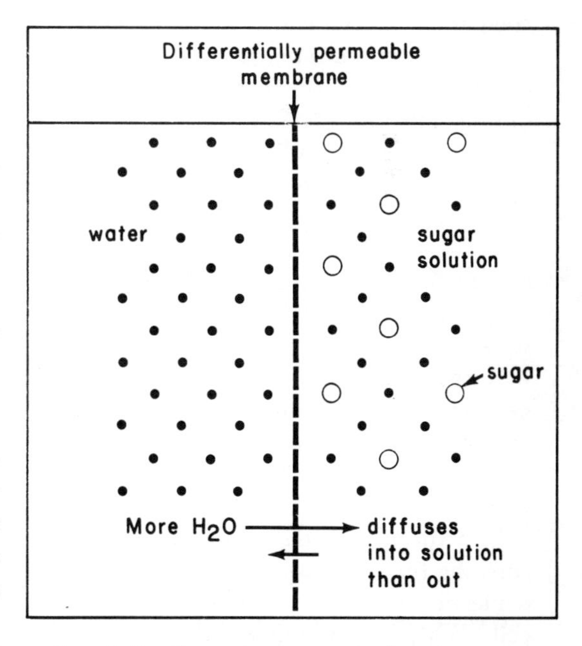

Fig. 5–2. The net movement of water across the differentially permeable membrane is from where the water concentration is greater to where it is lesser, a process called osmosis.

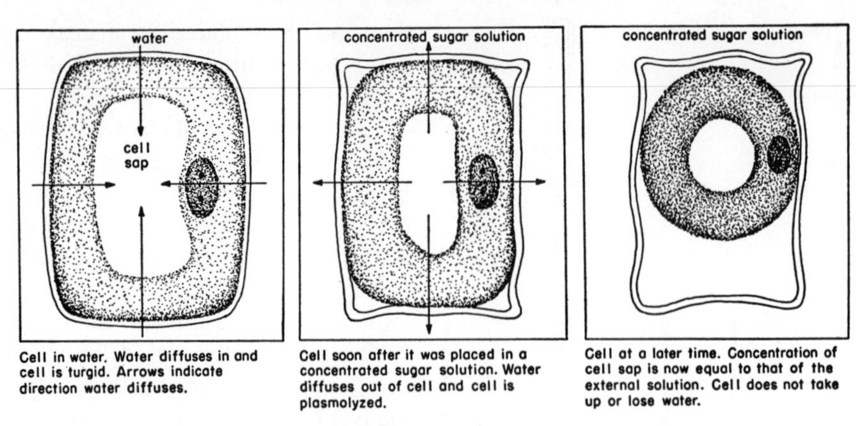

Fig. 5–3. Cell mounted in water (*left*) becomes turgid. Cell mounted in a concentrated sugar solution (*center and right*) becomes plasmolyzed.

the cellophane membrane of the osmometer. Like the cellophane, the plasma and vacuolar membranes permit small water molecules to pass through quickly but restrict the outward diffusion of substances dissolved in the cell sap. The cell sap is a solution of sugar, salts, and other materials dissolved in water. It is comparable to the sugar solution in the thistle tube, and the water in which the cells are immersed is comparable to the water in the beaker. Because the concentration of water is greater outside the cell than in the cell sap, water diffuses into the vacuole.

Celery, carrots, radishes, and other vegetables may be made crisp by soaking them in water. Water enters the cells by osmosis. The cells become filled with water, with a consequent pressure, called **turgor pressure,** on the cell walls. The cells are in a sense inflated with water and are said to be **turgid.** As water enters a cell, the turgor pressure brings about a stretching of the cell wall, which then exerts a counteracting pressure known as the **wall pressure** (Fig. 5–3).

In a turgid cell the water is under pressure, the turgor pressure, which increases the tendency for water to diffuse out of the cell. When a cell is immersed in water, the turgor pressure increases and in time attains a value equivalent to the osmotic concentration of the cell sap. At this point the cell will neither absorb nor lose water; that is, equilibrium is reached. Hence whether or not a cell will absorb water depends on both the osmotic concentration of the cell sap and the turgor pressure of the cell. The net tendency of water to enter a cell, known as the **diffusion pressure deficit** (DPD), is equal to the osmotic concentration (OC) minus the turgor pressure (TP) as shown below:

$$\text{DPD} = \text{OC} - \text{TP}$$

In a wilted cell, with a turgor pressure of zero, the net tendency for water to enter the cell when immersed in pure water is equal only to the osmotic concentration of the cell sap as follows:

$$\text{DPD} = \text{OC} - 0$$

In a partially turgid cell which has an osmotic concentration of 0.4 molar and a turgor pressure of 0.1 molar, the net tendency of water to enter the cell will be 0.3 molar as can be seen below:

$$\text{DPD} = 0.4 - 0.1 = 0.3$$

When a cell is fully turgid, such as one immersed in water, the net tendency for water to enter the cell will be zero. For example, the DPD of a cell with an osmotic concentration of 0.4 molar and a turgor pressure of 0.4 molar will be as follows:

$$\text{DPD} = 0.4 - 0.4 = 0$$

We can use molarity instead of actual pressures, because pressure is proportional to molarity. Theoretically, a 1-molar solution of a nonelectrolyte has a potential osmotic pressure of 22.4 atmospheres; a 0.5-molar solution, a pressure of 11.2 atmospheres; etc.

The turgor of cells makes for rigidity in leaves and nonwoody stems. The loss of turgor in such parts results in wilting. When the stem and leaf cells of a seedling, bean for example, are turgid, the seedling stands erect. If the stem and leaf cells lose more water than they absorb, the plant droops and the leaves wilt, a consequence of a loss of turgor in the cells.

The cell sap in root cells has a greater concentration of dissolved materials than does the soil solution. Since there are more water molecules per unit volume in the soil solution than in the cell sap, water enters root cells by osmosis.

PLASMOLYSIS

The situation in the osmometer may be reversed by filling the osmometer with water and immersing it in a beaker containing a concentrated sugar solution. With this setup water diffuses from the osmometer into the beaker, and the membrane collapses inward instead of being distended as in the previous experiment.

If a cell is immersed in a concentrated solution of salt or sugar, water leaves the cell. This osmotic loss of water with a concomitant shrinkage of the protoplast from the cell wall is called **plasmolysis** (Fig. 5–3). In this system the concentration of water is greater inside the vacuole (where the concentration of dissolved materials is lesser) than in the surrounding solution. As water diffuses out of the cell, the volume of the vacuole diminishes, and the cytoplasm shrinks away from the cell wall. The concentration of dissolved substances in the cell sap increases as water diffuses out. In time the concentration of dissolved substances in the cell sap becomes equal to that in the external solution. Then there are the same number of water molecules per unit volume in the cell sap and in the solution, at which time water will neither enter nor leave the cell.

If plants in pots are furnished with a great amount of chemical fertilizer (a mixture of salts of essential minerals), the concentration of dissolved materials in the soil may be greater than the concentration of dissolved materials in the cell sap. Under such a circumstance roots lose water instead of absorbing it. Weeds on paths, driveways, etc., may be killed by sprinkling common salt around them. Some soils in the West have a high content of chemical salts, and the roots of many species cannot absorb water from the concentrated soil solution. However, the cell sap of such native plants as glasswort, salt grass, and greasewood has a greater concentration of dissolved materials in the cell sap than is present in the soil solution, and hence they can thrive in soils high in salts, so-called alkali soils.

With continued use, some irrigated soils become increasingly salty. The river water applied may contain some salts which are not absorbed by plants and which accumulate in the soil. Certain rivers carry tremendous amounts of salt. The Colorado, for example, carries between 6 and 10 million tons of salt annually, and most is left on irrigated fields. With continued irrigation, the salt content may become so great that roots of crop plants cannot absorb water. When the salt content exceeds 3.5 atmospheres, the growth of many crops is depressed (Fig. 5–4).

The preservation of foods by sugaring or salting is based on principles of plasmolysis. Bacteria and fungi which cause food spoilage cannot absorb water from foods covered with salt or sugar and hence they cannot develop.

ABSORPTION OF WATER BY IMBIBITION

Colloidal substances such as starch, cellulose, and proteins take up water and swell by a process known as **imbibition.** The submicroscopic particles, either huge molecules

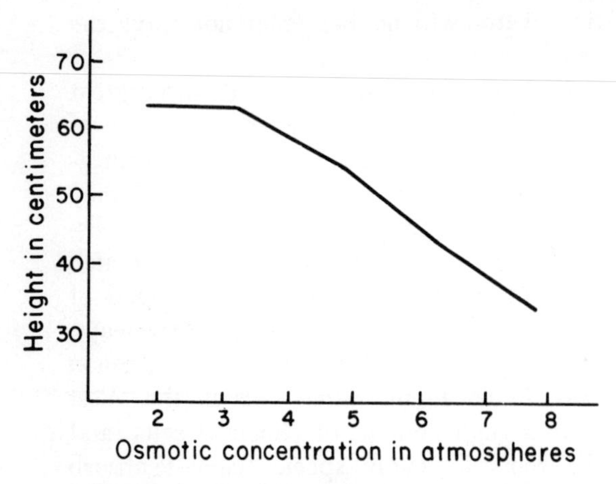

Fig. 5–4. The growth of tomato plants is depressed by high concentrations of sodium salts. The osmotic concentration of the nutrient solution which was used to water the plants was varied by adding sodium salts to the solution. (Data of Hayward and Long, *Plant Physiol.*)

or aggregates of molecules, attract water to their surfaces with considerable force. For example, the cellulose in wood, when wetted, swells with such force that wooden wedges can be used to split rocks. As colloids imbibe water, heat is liberated. When dry seeds are planted in moist soil, the cellulose of the cell walls, the starch, and the proteins imbibe considerable volumes of water. The initial uptake of water by seeds is largely due to imbibition, but as germination proceeds, osmotic forces also come into play.

INTAKE AND EXIT OF GASES FROM CELLS

Gases, such as carbon dioxide and oxygen, enter and leave cells primarily by diffusion. Before gases enter or leave cells, they must be in solution. Consider a cell immersed in water. If the concentration of a certain dissolved gas is greater in the cell than in the surrounding water, the gas will diffuse out of the cell. On the other hand, if the concentration is greater in the water, the gas will diffuse into the cell.

ABSORPTION OF MINERALS

As water enters root cells, minerals are not swept into the cells with the water. Water diffuses into cells independently of minerals. Often the concentration of a given mineral is greater inside a cell than in the solution which bathes it. According to the law of diffusion, the mineral in question should diffuse out of the cell. Actually, in many instances the cell continues to take up the mineral. The data obtained by D. R. Hoagland and A. R. Davis from studies of cells of *Nitella* (an alga) clearly illustrate this fact (Table 5–1).

In all instances the ions entered the cells even though their concentration was higher in the cell sap than in the pond water. The ions moved from where they were less concentrated to where they were more concentrated. Moving a substance from a low to a high concentration is comparable to mov-

Table 5–1

Analysis of the Cell Sap of *Nitella* and of the Pond Water in Which it Was Growing

Ion	Concentration in Sap*	Concentration in Pond Water*
Calcium	13.0	1.3
Magnesium	10.8	3.0
Sodium	49.9	1.2
Potassium	49.3	0.51

* Milliequivalents per liter. A milliequivalent of an ion is one-thousandth its gram ionic weight divided by its valence.

ing a weight from a low to a high position. Both involve an expenditure of energy. In an accumulating cell the energy comes from respiration. The ions are first bound (adsorbed) at the protoplasmic surface, and then they are transferred into the cytoplasm at the expense of respiratory energy. Adsorbed ions or molecules enter the cytoplasm when the plasma membrane invaginates and pinches off a vesicle in the cytoplasm. The minute vesicle then breaks down, leaving the molecules and ions in the cytoplasm. This mechanism for the active uptake of substances is known as **pinocytosis.** The ions and molecules may then be moved from the cytoplasm into the vacuole.

In order for cells of many plants to absorb minerals, oxygen must be available and respiration must occur. The uptake of minerals by root cells or other cells is checked when respiration diminishes. In soils which have been flooded, the oxygen concentration is low and respiration proceeds slowly. Under such circumstances minerals are absorbed with difficulty, and the plants show symptoms of mineral deficiency. If an oxygen deficit is maintained for too long, the roots of many plants die and later the tops succumb.

Species differ in the degree to which they accumulate ions and the kinds of ions they accumulate. Thus the marine alga *Valonia* accumulates large amounts of potassium, but does not accumulate sodium. The potassium concentration in the cells is forty times greater than in sea water, while the sodium concentration is only one-fifth as great as in sea water. Likewise, some crop plants will accumulate a certain ion to a greater extent than others. For example, pasture grasses accumulate radioactive zinc from irrigation water to a much greater extent than vegetables such as peas and corn. In Washington the water from the Columbia River that is used to cool the Hanford nuclear reactors becomes contaminated with radioactive zinc and other minerals. From the reactors the water flows back into the river. Downstream, farmers irrigate their fields, and the radioactive zinc accumulates in the hay and other crops in amounts as indicated in the table below.

Of the plants studied, grapes and corn did not accumulate zinc in concentrations above that of the water, whereas the other crops did so, often in striking amounts. Pasture grasses contained a concentration 440 times that in the water. When cattle fed on the pasture grasses, the concentration of radioactive zinc in their flesh and milk became relatively high.

The water still containing some radioactive zinc flows into the vast Pacific Ocean. Even though diluted, radioactive zinc still becomes incorporated in the bodies of many marine organisms—algae, mussels, clams, and oysters, for example. The amount in their bodies varies with the distance from the river's mouth, but even so, oysters and clams in Coos Bay, 200 miles south of the river's mouth, show radioactive zinc in their bodies.

Table 5–2

Sample	Concentration of Radioactive Zinc, $\mu\mu$c./g.	Concentration Factor, Produce/Water
Irrigation water	0.188	–
Grapes	0.089	0.47
Corn	0.16	0.83
String beans	0.29	1.50
Tomatoes	0.46	2.4
Peas	0.55	2.9
Milk, cow	4.88	26.0
Beef, flesh	5.23	28.0
Pasture	82.9	440.0

When people drink the milk, and eat the beef, fruits, vegetables, or sea foods, radioactive zinc enters their cells. Clearly, when we dump radioactive wastes into our waters, we have not gotten rid of them. Sooner or later they may become incorporated into our own bodies. In other words, as we pollute our water, so also do we become polluted. Neither man nor animals and plants can change radioactive minerals into harmless ones. Only time will lessen their radioactivity and harmful effects.

Plants and animals may accumulate organic substances as well as inorganic ones. For example, the Mississippi River is polluted with Endrin, a widely used insecticide. The concentration in the water varies from 0.054 to 0.134 parts per billion, but the flesh of fish contains seven parts per million, a concentration which kills them. Massive kills of fish in the lower Mississippi River have resulted because of pollution with Endrin and other insecticides. Moreover, the pesticides are found in shrimp from the Gulf of Mexico. The pesticides are, of course, being ingested by people, not only through fish and shrimp, but through the drinking water. The pesticides in the fish and shrimp threaten the fisheries industry on the Gulf area. Moreover, the Mississippi case may be an omen of more to come unless we are more careful with the use of pesticides. Of course, in agriculture and in public health the control of insects is essential. According to the World Health Organization, insect pests are responsible for half of all human deaths and deformities due to disease. Insects consume or destroy a high proportion of the food which man raises. Because of the hazards of insecticides scientists are searching for other methods of control. One method that shows promise involves the use of odorous chemicals to lure insects into traps. Some attractants have been isolated from female insects; others have been synthesized. The natural odorous sex lures are specific for the species and remarkably potent. For example, the odor emitted by one caged female pine sawfly attracted more than 11,000 males.

QUESTIONS

1. _____ The tendency of any substance when it occurs as a gas or in solution to become evenly distributed throughout the whole space available to it by moving from points of greater to points of lesser concentration is called (A) diffusion, (B) osmosis, (C) plasmolysis, (D) turgor.

2. _____ The diffusion of water through a differentially permeable membrane from where the concentration of water is greater to where it is lesser is called (A) turgor pressure, (B) osmosis, (C) diffusion pressure deficit, (D) plasmolysis.

3. _____ If a sac made of bladder or a similar osmotic membrane is filled with water and then placed in a vessel of molasses, water will (A) leave the sac by osmosis, (B) leave the sac by imbibition, (C) enter the sac by osmosis, (D) enter the sac by plasmolysis.

4. _____ The rounding up of protoplasm within a cell because of loss of water by osmosis is called (A) turgidity, (B) plasmoptysis, (C) plasmolysis, (D) reverse osmosis.

5. Three cells, A, B, and C, have an osmotic concentration of 0.5 molar. Cell A is mounted in water; cell B, in a 0.5 molar solution of sucrose; and cell C, in a 1.0 molar sucrose solution. Describe the appearance of the cells after 20 minutes immersion in their respective solutions.

6. _____ Cell A is in contact with cell B. Cell A has an osmotic concentration of 0.5 molar and a turgor pressure of 0.3 molar. Cell B has an osmotic concentration of 0.4 molar and a turgor pressure of 0.1 molar. Water will (A) diffuse from cell A into cell B, (B) diffuse from cell B into cell A, (C) will not diffuse in either direction.

7. _____ The turgor pressure (A) is equal in magnitude but opposite in direction to the wall pressure, (B) is always equal to the osmotic concentration, (C) has a high value in plasmolyzed cells.

8. _____ A cell was immersed in a-0.5-molar solution of glycerol which plasmolyzed the cell in 5 minutes. The cell, still in the glycerol solution, recovered from plasmolysis in 30 minutes. The data indicate (A) that the plasma and vacuolar membranes are permeable to glycerol, (B) that the membranes are not permeable to glycerol.

9. _____ Which one of the following practices is *not* based on osmotic phenomena? (A) The preservation of vegetables, meat, and fish by salting. (B) The splitting of rocks in quarrying by the insertion of dry wooden wedges into cracks and later wetting the wedges. (C) Making vegetables crisp by immersing in water. (D) The practice of bathing wounds with a salt solution instead of pure water.

10. _____ Which one of the following is false? (A) If a cell is immersed in a concentrated solution of salt or sugar, water leaves the cell. (B) Weeds on paths may be killed by sprinkling salt around them. (C) With continued use, some irrigated soils become increasingly salty. (D) Bacteria and fungi have the remarkable capacity to absorb water from foods covered with salt or sugar. (E) As water enters a cell, the turgor pressure brings about a stretching of the cell wall which then exerts a counteracting pressure.

11. _____ The capacity of some colloids to take up water and swell is called (A) absorption, (B) osmosis, (C) imbibition, (D) turgor, (E) suction tension.

12. _____ The accumulation of ions by cells (A) requires energy from respiration, (B) occurs equally well in the absence of oxygen, (C) occurs in algal cells but not in root cells.

13. _____ Which of the following is false? (A) As we pollute the air and water, so also do we become polluted. (B) Insecticides used for the protection of crops may injure fish and other aquatic organisms as well as birds and mammals. (C) The deleterious effect of excess salts in the soil may be lessened by flooding and draining—practices which will not harm farmers downstream. (D) Certain plants may accumulate radioactive zinc if it is present in the irrigation water.

14. _____ Which one of the following is false? (A) The movement of water into and out of cells is governed to a large measure by the principles of osmosis. (B) Water tends to diffuse across a differentially permeable membrane from regions of high water concentration to regions of lower water concentration. (C) The net tendency of water to enter the cell, the diffusion pressure deficit, is equal to the difference between the osmotic concentration of the cell and the turgor pressure exerted by the cell contents on the cell wall. (D) Ions enter cells primarily by diffusion, and the concentration in the cell is equal to that in the medium bathing the cell.

6

Food Manufacture—Photosynthesis

The productivity of gigantic steel mills and huge, noisy automobile factories pales in significance compared to the quiet, sustained activity of the earth's green plants: so many hundred million tons of steel, so many automobiles–but an estimated 375 billion tons of sugar are produced each year by the green plants of land and sea. Aquatic plants make 90 per cent of the total; and land plants, 10 per cent (Table 6–1). Plants of the oceans make more food than land plants because the ocean area is 2.7 times the land area and because environmental conditions are more uniform in the oceans than on land. A study of Table 6–1 reveals that forests make more food than crop plants. Not all forest areas are equally productive; a tropical rainforest makes far more food than does a thorn woodland of a semiarid region. Certain crops make more food per unit area than do others; a thrifty acre of sugar cane may produce twenty times more food than an average acre of corn.

Table 6–1

Amount of Food Made by Plants of Land and Sea

Habitat	Area (sq. miles)	Food Made/ Year/Sq. Mile (tons)	Total Food Made (tons)
Oceans	140 million	2410	337.5 billion
Land			
Forests	17	1488	25.3
Farms	10	920	9.2
Grasslands	12	208	2.5
Desert	13	38	0.5
Total for land plants			37.5 billion

Fig. 6–1. Food chains. *Left,* when plant products are eaten directly by man, there are only two links in the food chain, and man makes maximum use of the green plants. *Right,* in the ocean there are at least four links in the food chain—green plants, small animals, fish, and man; only a small part of the food made by plants is available as food for man. (*Sci. Am.*)

About 98 per cent of the food made by green plants is used by the green plants themselves and by nongreen plants, the bacteria and fungi. Approximately 0.2 per cent of the food is used directly or indirectly by the human race for nourishment, and all animals of the earth, great and small, use only 2 per cent. People who have no insight into the plant world give the advice "don't vegetate," not realizing how active green plants are. Green plants create; they manufacture food and products which we use as fibers, drugs, beverages, perfumes, spices, and vitamins from such readily available raw materials as water, carbon dioxide, and certain minerals. More noise and excitement go with the marketing of food than in its production. The supermarket hums and rings with activity. Few shoppers pause to think that the silent plants made all the food displayed on the shelves. Independent are green plants; dependent are we.

We may eat plant parts directly, corn flakes for example, or indirectly in the form of beef, pork, lamb, poultry, or fish. All livestock feed directly on plants; that is, they are herbivores, and they in turn are eaten by man. Here there are three links in the food chain—plants, animals, man. When we eat fish, there are at least four links in the food chain (Fig. 6–1): aquatic green plants — small animals — fish — man. The fewer the links in the food chain, the more efficiently man uses green plants for food. For example, if man eats corn directly, the food value for him is greater than if he feeds it to a steer and then consumes the beef. Not all corn fed to a steer is turned into beef; much is used by the animal itself. Under natural conditions about 90 per cent of the food eaten by a growing animal is respired, and only 10 per cent puts weight on the animal. One food chain in the ocean is as follows: algae—small animals—herring—tuna. The production of 1 pound of tuna requires 10 pounds of herring, 100 pounds of small animals, and 1000 pounds of algae. A boy would gain 1 pound by eating 10 pounds of tuna whose production would require 10,000 pounds of algae. If he were to consume algae directly, he would gain 1 pound by consuming only 10 pounds of algae.

Photosynthesis is the most important chemical reaction in the world. Let us analyze the process and learn what is used, what is formed, and what requirements must be fulfilled for the process.

MATERIALS USED AND MATERIALS FORMED IN PHOTOSYNTHESIS

Water and carbon dioxide are the raw materials used in photosynthesis, and glucose (a sugar) and oxygen are formed. Photosynthesis may be summarized as follows:

In words:
Carbon dioxide + water
→ glucose + oxygen

In symbols:

$$6CO_2 + 6H_2O \rightarrow C_6H_{12}O_6 + 6O_2$$

The balanced equation tells us that six molecules of carbon dioxide combine with six of water to form one glucose molecule and six oxygen molecules. Translated into grams and Calories needed to form 180 grams of glucose, 180 being the molecular weight of glucose, the equation would be:

$$264 \text{ grams } CO_2 + 108 \text{ grams } H_2O + 673 \text{ Calories} \rightarrow 180 \text{ grams } C_6H_{12}O_6 + 192 \text{ grams } O_2$$

The 673 Calories of energy from light are stored in the 180 grams of glucose. Photosynthesis not only produces energy-rich foods, but also produces oxygen which is essential for respiration. Now, as well as in the past, photosynthesis is the only source of oxygen for the earth's atmosphere and water. Green plants produce annually about 400 billion tons of oxygen. If it were not for photosynthesis, the atmosphere, oceans, lakes, and streams would be devoid of oxygen. As we will see later, the oxygen released in photosynthesis comes from water. For each molecule of glucose formed, six molecules of oxygen are released. To get six molecules of oxygen, twelve molecules of water are required. Accordingly, a more complete general formula for photosynthesis is as follows:

$$6CO_2 + 12H_2O \rightarrow C_6H_{12}O_6 + 6H_2O + 6O_2$$

Elodea, an aquatic plant, is often used to demonstrate that oxygen is released in photosynthesis. Oxygen bubbles are given off when an *Elodea* sprig is exposed to light, but not when the plant is kept in darkness. Rates of photosynthesis may be determined by measuring the amount of oxygen which plants produce. A simple experiment is to place an *Elodea* sprig in water and count the number of bubbles given off in a certain period, or the oxygen may be collected and measured with the apparatus illustrated in Figure 6–2. The stem of the *Elodea* is inserted in an inverted 1-milliliter pipette which is filled with water. As oxygen is liberated, the water column drops, and the change in the height of the water column represents the amount of oxygen released. More accurate measurements of the amount of oxygen released by the *Elodea* or other plants can be obtained with a manometer such as the one illustrated on page 85. Rates of photosynthesis can also be measured by determining the amounts of carbon dioxide used, and by determining the increase in dry weight of a leaf or part of a leaf.

ENERGY STORED BY PHOTOSYNTHESIS

The sugar made in photosynthesis can be changed by green plants into other organic compounds—starch, cellulose, fats, and also, with the addition of such elements as nitrogen, sulfur, and phosphorus, into proteins. The sugar, or organic compounds made from the sugar, can be oxidized either by burning or by respiration. Energy is liberated during oxidation. For example, a fire could be built with sugar, and energy as heat would be liberated. A fire could not be made from water or carbon dioxide, the materials used in photosynthesis. No energy is stored in such materials, but energy is stored in the sugar. Where does this energy come from?

The energy in the sugar comes from light. Approximately 3 per cent of the light energy received by a leaf is stored in the sugar or in foods made from sugar. Some of the remaining 97 per cent is reflected, some transmitted, some used to evaporate water, and some to increase the temperature of the leaf. It is that small portion of the

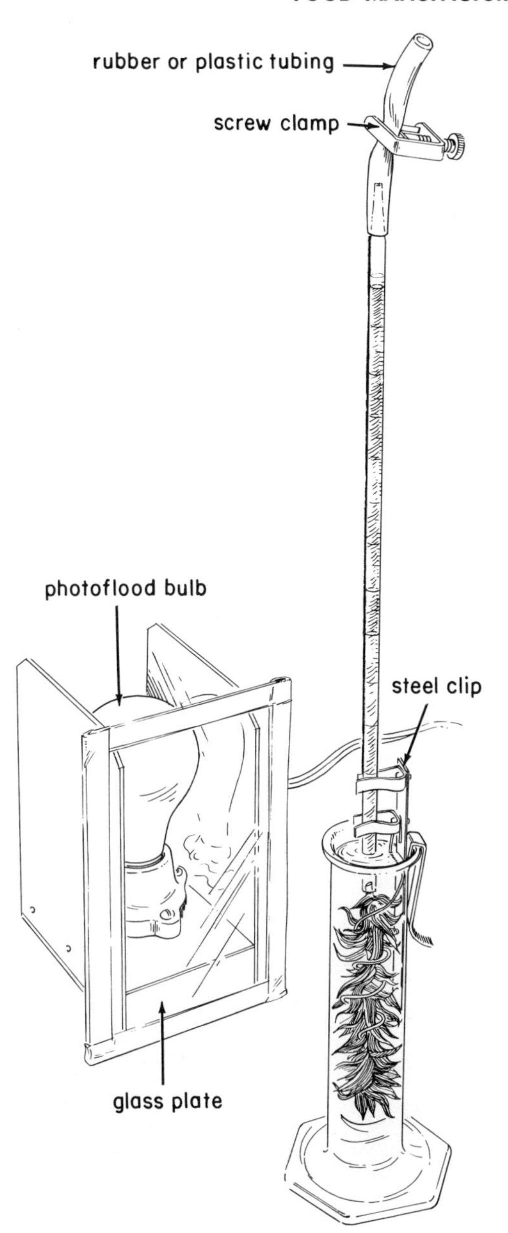

Fig. 6–2. Apparatus for measuring the production of oxygen by green plants. When the *Elodea* plants are illuminated, oxygen collects in the pipette. (From *Plants in Perspective* by E. H. Newcomb, G. C. Gerloff, and W. F. Whittingham. Copyright, 1964, by W. H. Freeman and Co., San Francisco.)

sun's energy stored in foods by green plants which keeps the plant and animal worlds going, most of our vehicles and machines moving, and our houses warm. Without the energy stored by green plants, no plant could grow, no animal could move, no brain could think.

It is a simple matter to demonstrate that light is necessary for photosynthesis. In many plants the glucose formed in photosynthesis is quickly changed into starch which then accumulates as starch grains in the chloroplasts. Hence in these plants we can tell whether or not photosynthesis is going on by testing leaves for starch. Starch can be detected by first heating leaves in alcohol to decolorize them, that is, to extract the chlorophyll, and then placing them in an iodine solution. If starch is present, the decolorized leaves turn blue. If a geranium plant is kept in the dark for 48 hours, the leaves will not give a positive starch test. While the plant is in the dark, the starch in the leaves is changed into sugar; some of the sugar is respired, and much of it is moved out of the leaves to other plant parts. A leaf depleted of starch may then be covered with a mask so that only a portion of the leaf receives light. After exposure to light, starch is present only where light struck the leaf (Fig. 6–3*A*).

How do green plants store the sun's energy and make it effective in photosynthesis? If you bubble carbon dioxide through plain water in the presence of light, no sugar forms and no energy is stored. Before light can be effective in the manufacture of food, it must be absorbed. In green leaves chlorophylls absorb visible light and are thereby converted to a higher energy, or "excited," state. The "excited" chlorophyll molecules have had some of their electrons raised from their normal energy state to a higher energy level. Later we will learn that the energy of excited chlorophylls, after several electron transfers, provides chemical energy in the form of ATP (adenosine triphosphate) and reducing power in the form of $TPN \cdot H_2$ (reduced triphosphopyridine nucleotide).

A simple experiment to demonstrate that chlorophyll is necessary for photosynthesis can be performed with variegated leaves—leaves which are green in places and colorless elsewhere. If the chlorophyll in such a leaf is extracted with alcohol and the leaf is then placed in iodine, the parts that were green will now turn blue. The blue areas coincide exactly with the original green pattern of the leaf (Fig. 6–3*B*).

Chlorophyll carries out its complete function of absorbing light and making food only when present in chloroplasts. Food would not be made in a test tube containing chlorophyll, water, and carbon dioxide. Hence photosynthesis is bound to the structure of the chloroplast and cannot be duplicated apart from that structure. However, with suitable supplements chloroplasts that have been isolated from cells will carry out all stages of photosynthesis, generally at a rate that is 15 per cent of that which occurs *in vivo*.

The radiant energy spectrum, shown in Figure 6–4, includes many types of radiations, ranging from cosmic rays with very short wave lengths to electric waves of extremely long wave lengths. Not all of these radiations come from the sun. Radiations from the sun include the longer ultraviolet, visible light, and infra-red. Of these, only the visible radiations are effective in photosynthesis.

Visible light from the sun is a composite of many light rays, varying in wave length from 390 millimicrons for the shortest violet to 760 millimicrons for the longest red rays (a millimicron is $\frac{1}{1000}$ of a micron, approximately one 25-millionth of an inch). Visible light may be separated into its constituent colors by passing it through water droplets or a prism. The spectral components are obvious in a rainbow where they are evident as violet, indigo, blue, blue-green, green, yellow, orange, and red light. Some colors are absorbed more completely by chlorophyll than others, and those most strongly absorbed are most effective in photosynthesis. Chlorophyll absorbs much of the red, blue, and violet light, but little of the green light. Hence a plant will make more food

Fig. 6–3*A*. Light is necessary for photosynthesis. A strip of black paper with the letter W cut out was placed over a leaf from a geranium plant which had been kept in the dark for 48 hours. The leaf was exposed to light for 12 hours, and then tested for starch with iodine. Starch is present only where light struck the leaf. *B*. Chlorophyll is necessary for photosynthesis. The *Coleus* leaf was green along the margin and white in the center. The chlorophyll was extracted with alcohol, and then the leaf was placed in iodine. Starch was present only in areas where chlorophyll had been present.

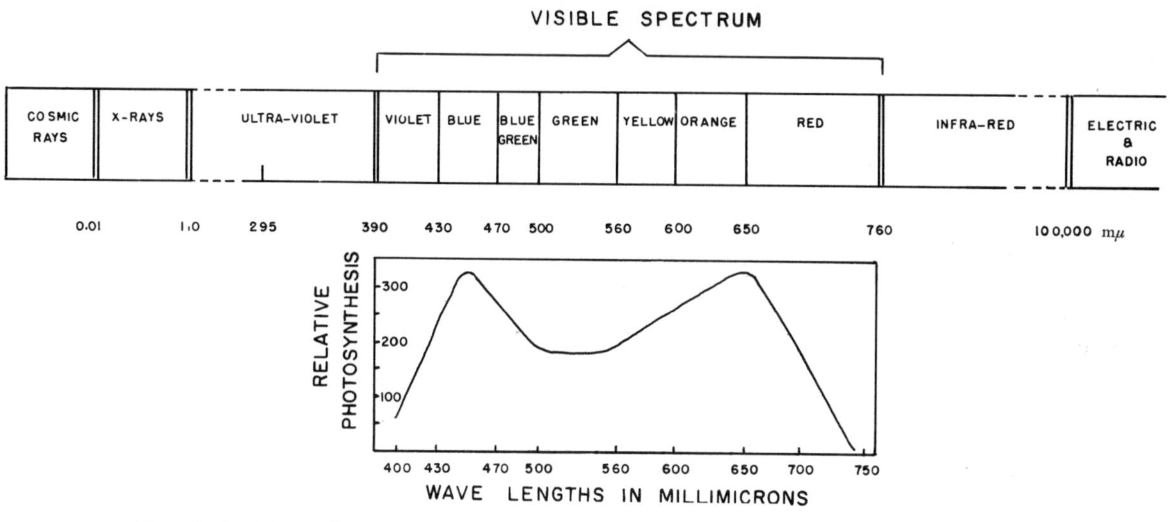

Fig. 6–4. The radiant energy spectrum and the relative rates of photosynthesis of wheat in different wave lengths of light. Notice that red and blue lights are more effective than green light.

when exposed to red or blue light than when exposed to green light of equal intensity. However, Figure 6–4 reveals that green light is effective in photosynthesis.

Not all of the energy stored by plants living now is released in our day. Huge masses of only partially decayed plant material may accumulate, later to be covered by rock. Coal, oil, and peat represent remains of plants of the past, and when such fuels are burned, the energy stored long ago by those plants is released.

Only a small percentage of the sun's energy is stored by green plants—about $1/2000$ of the 5.5×10^{23} small calories that the earth receives annually. Of course, much of the radiant energy falls where there are no plants. If we only consider the energy absorbed by the green-plant surface of the earth, the efficiency of photosynthesis in storing solar energy is 1 to several per cent. The sun's energy is the result of a nuclear reaction in which four hydrogen atoms, each with a mass of 1.008, are fused into one of helium with a mass of 4.003. In this reaction, 0.029 units of mass (the difference between four times 1.008, or 4.032, and 4.003) is converted into energy according to the Einstein equation of $E = mc^2$ in which E is the energy produced (in ergs), m is the mass of matter transformed (in grams), and c is the velocity of light (3×10^{10} cm./sec.). Within the sun about 120 million tons of matter are converted into energy every minute.

FACTORS AFFECTING THE RATE OF PHOTOSYNTHESIS

To obtain a maximum yield, the farmer, forester, and gardener should maintain conditions favorable for a high rate of photosynthesis.

Other things being equal, a plant with a large leaf area makes more food than one with a small leaf area. Healthy plants with expansive leaves develop in soils properly fertilized and watered. When fields are fertile and ample water is present, crops may be planted closer than in fields where water and minerals are limited. With closer planting the leaf area is greater and yields are higher. Photosynthesis occurs at a slow rate in plants suffering from mineral deficiencies. The removal or destruction of leaves or portions of leaves diminishes the amount of food made. Pest control and the control of disease prevent the loss of food-making units and also save food for the plants instead of allowing it to be used to nourish unwanted organisms.

Of prime importance to a good lawn is an extensive root system. The lawn should not be mown too closely. Enough leaf area should remain to make ample food for the development of a good root system and to tide the plant through the winter. In general, the lawn mower should be set for not less than 1½ inches through the season. Similarly, too close and frequent grazing of range plants may reduce the leaf area so that the native plants are weakened and the range deteriorates.

A number of environmental factors influence the rate of photosynthesis. Light intensity, light duration, temperature, and the amount of water and carbon dioxide influence the rate. We shall take up the factors separately and assume that all environmental conditions, except the one under discussion, are optimal. Under field conditions the several environmental factors interact. We will see shortly that as light intensity increases, photosynthesis speeds up, provided, of course, that the temperature is favorable and that there is an ample supply of water and carbon dioxide. Obviously, without a supply of carbon dioxide, the rate would not increase as the light intensity went up, nor would it if water were lacking or the temperature were 0° F. When light, water, and carbon dioxide are optimal, an increase in temperature from, say 50° to 70° F. brings about an increased rate, but photosynthesis would not be enhanced by an increase in temperature if the plant were kept in the dark. The optimal light intensity and temperature vary considerably with different species. Hence it is impossible to say that for all species a certain temperature and light intensity are optimal. The following discussion shows general trends only.

Light

As the light intensity increases, the rate of food manufacture increases almost directly until a certain light intensity is reached, above which the rate does not increase appreciably. The optimum light intensity varies with the species. For plants which grow in nature in shaded habitats (shade plants, such as African violets and wood sorrel), the optimum light intensity for photosynthesis is one-tenth to one-fifth of full sunlight (between 1000 and 2000 foot-candles). Frequently, higher light intensities are not beneficial, but detrimental. For plants that grow natively in full sun (nasturtiums, cereals, geraniums), the more light they receive up to full sunlight, the more food they make. That is true when considering the whole plant, but if isolated leaves are considered, food making increases in sun plants up to about one-third or one-fourth of full sunlight, and with higher intensities the rate does not increase. This difference in behavior between an intact plant and a leaf isolated from the plant can be readily explained. When a whole plant is exposed to full sunlight, only part of the leaves actually receive full sunlight, for some leaves are partially or wholly shaded by others. On the other hand, all portions of an isolated leaf receive equal light intensity.

The duration of light influences the amount of food made. Plants in greenhouses grow slowly during December, January, and February because there are only 9 or 10 hours of light each day during which photosynthesis can occur. During the long days of summer, more food is made and growth is rapid.

Plants can use artificial light as well as sunlight in photosynthesis. Orchids, African violets, gloxinias, wheat, radish, and many other plants grow and flower with fluorescent light. Cuttings may be rooted in a cabinet which is lighted with fluorescent light of the daylight or white-light types or by cold cathode lighting. Many botanical laboratories are now equipped with growth chambers (Fig. 6–5) lighted with fluorescent tubes and tungsten bulbs. The growth chambers are designed so that temperature, light intensity and duration, and humidity

Fig. 6–5. A growth chamber. (Percival Refrigeration and Manufacturing Co.)

can be controlled precisely. They are used for experiments designed to determine the effects of various environmental factors on the growth and flowering of plants.

Temperature

When light intensity is optimal and other conditions favorable, the rate of photosynthesis is influenced by temperature. At low temperatures photosynthesis proceeds slowly or not at all. The temperature at which photosynthesis becomes nearly zero varies with the species, being lower for plants from cold regions than for plants from tropical ones. Some plants, Norway spruce, for example, make a small amount of food when the temperature is as low as —6° C. (about 21° F.). For many temperate zone plants, as the temperature increases from about 0° C. (32° F.) to 25° C. (77° F.), the rate of photosynthesis increases, and within these limits on bright days it usually doubles for each 10° C. rise. At higher temperatures the rate becomes progressively less. The optimum temperature for tropical plants is higher than that for temperate plants.

Water

A deficiency of water in the soil markedly decreases the rate of photosynthesis as does an excessive amount such as is present in waterlogged soils. When the soil moisture is so low that apple leaves wilt, the photosynthetic rate of the wilted leaves is only 13 per cent of that of turgid leaves. After the tree is watered, the leaves soon regain turgidity, but photosynthesis remains somewhat less than normal for more than 2 days. Thus the retarding effect of drought on photosynthesis lingers even after the plants are again provided with water. Apple trees growing in a waterlogged soil, one with an excess of water and hence a deficiency of oxygen, carry on photosynthesis at a much reduced rate.

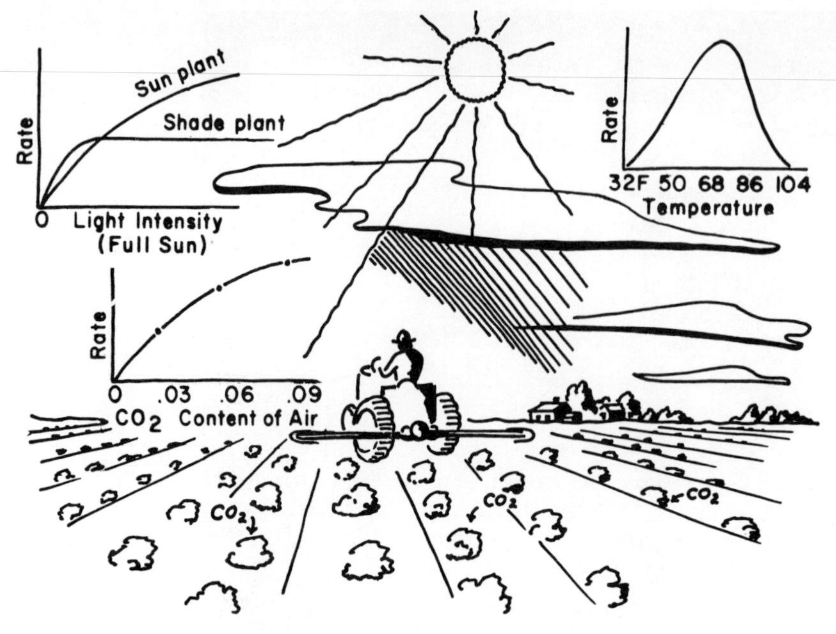

Fig. 6–6. Factors affecting photosynthesis.

Carbon Dioxide

When water, temperature, and light are favorable for photosynthesis, the rate of photosynthesis is limited by the percentage of carbon dioxide in the air (Fig. 6–6). Seventy-eight per cent of air is nitrogen, 21 per cent is oxygen, and only 0.03 per cent is carbon dioxide. In other words, only three parts by volume of carbon dioxide are present in 10,000 parts of air. On that relatively meager percentage depends all life on earth. Without carbon dioxide not a green plant would grow, and all animals would starve to death. Even though the percentage of carbon dioxide in the air is small, the total amount in the vast atmosphere of the earth is large, about 2200 billion tons. The oceans contain about one hundred times this quantity as dissolved gas and as carbonates. The great reserve of carbon dioxide in the air and in the oceans would last for 250 years even if none were returned to the air.

However, the carbon dioxide used in photosynthesis is in equilibrium with the return of carbon dioxide to the atmosphere by combustion of wood, coal, oil and gas; by the respiration of green plants, nongreen plants, and animals; volcanic activity; and diffusion from the ocean. The great reserve of carbon dioxide in the ocean water is a stabilizing factor; it maintains the average carbon dioxide content of the atmosphere at a constant level.

When light, temperature, and water supply are optimal, the rate of photosynthesis can be augmented by increasing the concentration of carbon dioxide in the air. With each added amount of carbon dioxide, the rate increases until the concentration is about ten times the normal amount. With further increase, the rate of food manufacture does not increase appreciably. In a greenhouse the fertilization of air may be practical for augmenting yields, but in the out-of-doors it would not be feasible because the added carbon dioxide soon would diffuse throughout the atmosphere. Furthermore, care must be exercised in fertilizing plants with carbon dioxide, because high concentrations may be injurious. For example, tomato plants grown with 0.3 per cent carbon dioxide in the air showed dead spots on the leaves within 2 weeks.

Although the carbon dioxide content of air averages 0.03 per cent, it may be greater or less than this under some circumstances. The carbon dioxide in the air above soil high in organic matter, where microbial activity is intense, is likely to be greater than 0.03 per cent. In a closed greenhouse and in a field of corn on a still day, the concentration will be less than 0.03 per cent. Opening the ventilators in the greenhouse will result in an increased carbon dioxide concentration. A slight breeze in a corn field will increase the supply of carbon dioxide. When no water deficit exists, the photosynthetic rate of plants in moving air is likely to be 20 per cent greater than in still air.

CHEMISTRY OF PHOTOSYNTHESIS

The equations previously given show only the beginning and the end of the process of photosynthesis and do not illustrate the intermediate stages and the complexity of the process. In recent years scientists have unraveled some of the many intermediate steps in photosynthesis, and these are summarized below.

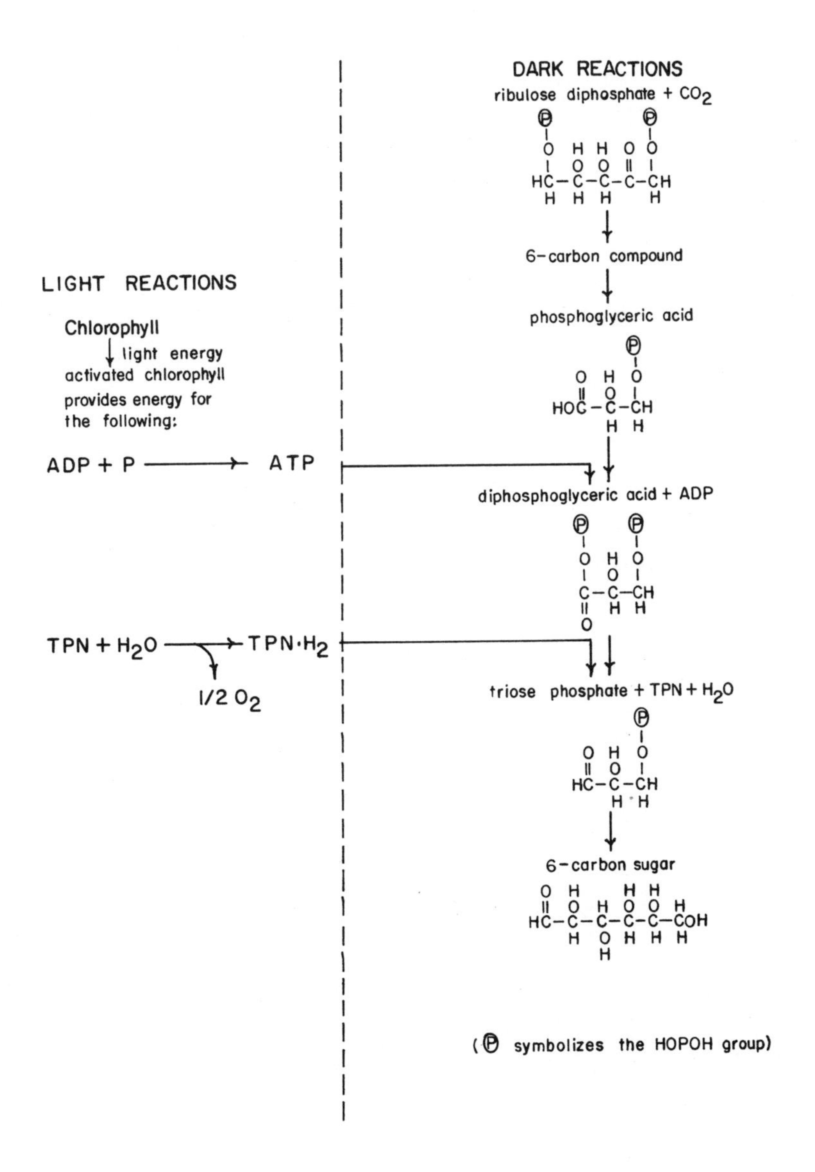

The steps in photosynthesis can be grouped in two series of reactions: light reactions, which require energy from light, and dark reactions, which do not directly require light energy, but which can proceed in light, as they usually do, or in the dark if **adenosine triphosphate** (ATP) and reducing power in the form of reduced triphosphopyridine nucleotide ($TPN \cdot H_2$) are present. $TPN \cdot H_2$ is formed when hydrogen ions from water combine with triphosphopyridine nucleotide (TPN). Adenosine triphosphate is the chief energy carrier in a cell. The light reactions occur in the green grana of chloroplasts, whereas the dark reactions proceed in the surrounding stroma. The light reactions are not greatly influenced by temperature, whereas the dark reactions are temperature sensitive, proceeding slowly at low temperatures, such as 0° C., and much more rapidly at moderate temperatures, 20° C. for example.

The light reactions are more involved than is indicated by the summary of photosynthesis, which shows only the end results of the reactions. Actually, the light reactions occur in two series, one called **cyclic photophosphorylation** and the other termed **noncyclic photophosphorylation.** Cyclic photophosphorylation produces ATP from ADP (adenosine diphosphate) and inorganic phosphate in the manner illustrated in Figure 6–7. As a molecule of chlorophyll a absorbs light, an electron is raised from its normal energy level to a higher one. The electron is then transferred from one compound to another, each at a successively lower energy level. (In Fig. 6–7, the gray circles represent the compounds.) As the electron is passed from one acceptor compound to one at a lower level, energy is released, and this energy is used to synthesize ATP from ADP and inorganic phosphate. After its energy is spent, the electron returns to chlorophyll a and the process is repeated.

Noncyclic photophosphorylation yields ATP and reducing power in the form of $TPN \cdot H_2$ (Fig. 6–8). As chlorophyll a absorbs light, an electron is raised to a higher energy level. The electron is then trans-

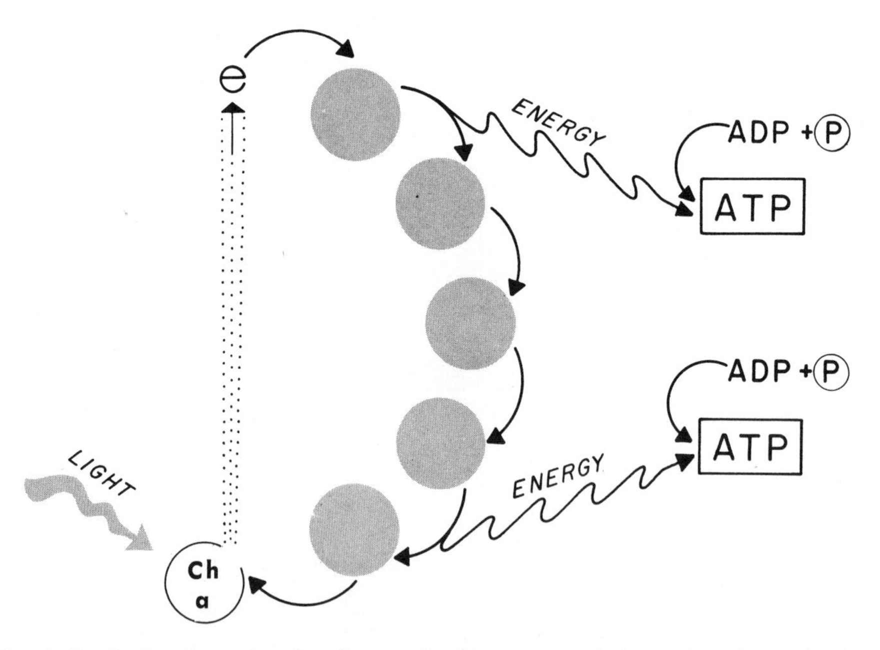

Fig. 6–7. Cyclic photophosphorylation. In this process, light energy is used to synthesize ATP from ADP and inorganic phosphate. For details, see text. (Reproduced by permission from *Biological Science* by William T. Keeton. Copyright, 1967, by W. W. Norton & Company, Inc., New York.)

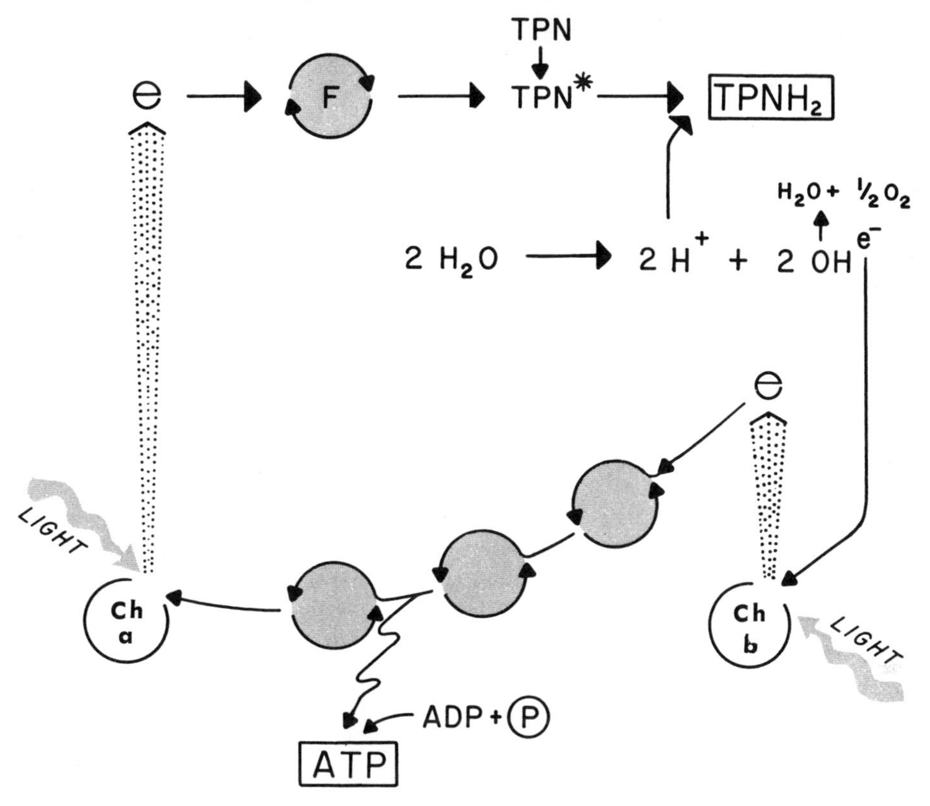

Fig. 6–8. Noncyclic photophosphorylation. In this process, light energy separates the hydrogen from water and results in the formation of $TPN \cdot H_2$ and ATP. For details, see text. (Reproduced by permission from *Biological Science* by William T. Keeton. Copyright, 1967, by W. W. Norton & Company, Inc., New York.)

ferred to ferredoxin (F), which passes it on to TPN, which becomes activated (indicated by an asterisk). The activated TPN* separates hydrogen ions (H^+) from water; these ions combine with TPN to yield $TPN \cdot H_2$. A second light reaction raises an electron from chlorophyll b to a higher energy level (lower right, Fig. 6–8). This electron is transferred from one acceptor compound to another down an energy gradient. The released energy is used to synthesize ATP. Finally, the electron reaches chlorophyll a, where it replaces the electron lost from chlorophyll a as a result of the first light event. An electron from the hydroxyl ions of water replaces that emitted from chlorophyll b. The remaining OH radicals combine to form water and oxygen.

In summary, in noncyclic photophosphorylation, energy from light separates the hydrogen from the oxygen of water and produces ATP from ADP and inorganic phosphate. The hydrogen from water combines with TPN, triphosphopyridine nucleotide, forming $TPN \cdot H_2$, reduced triphosphopyridine nucleotide. (Just as botanists sometimes change plant names, so also do biochemists change the names of chemicals. Thus, some biochemists refer to TPN as NADP, nicotinamide adenine dinucleotide phosphate. Both TPN and NADP refer to the same compound.)

As we have just seen, the oxygen liberated in photosynthesis comes from water, a reaction requiring light energy. This was discovered by furnishing plants with water containing ordinary hydrogen but a rare form of oxygen, one with an atomic weight of 18 instead of 16. When the evolved oxygen was tested, it was found to have an atomic weight of 18, indicating that the liberated oxygen came from water. Of course, in this experiment the oxygen in the CO_2 was the usual one with an atomic weight of 16.

In the dark reactions, carbon dioxide combines with **ribulose diphosphate,** a 5-carbon sugar, to yield a 6-carbon compound that splits into two molecules of **phosphoglyceric acid,** a 3-carbon substance. (In the outline only one of the two phosphoglyceric acid molecules is shown.)

Phosphoglyceric acid receives a phosphate group and energy from ATP and is thereby converted into **diphosphoglyceric acid** which then reacts with the hydrogen of $TPN \cdot H_2$ to yield **triose phosphate,** a phosphorylated sugar with three carbons; water, phosphate, and TPN are also formed. Two triose phosphate molecules then combine to form a 6-carbon sugar, such as glucose ($C_6H_{12}O_6$). If the reaction sequence from ribulose diphosphate to glucose is to continue, ribulose diphosphate must be continuously formed. A fraction of the triose phosphate and of the 6-carbon sugar produced in photosynthesis is sidetracked to form ribulose diphosphate. To be more specific, for every six ribulose diphosphate molecules and six carbon dioxide molecules that go through the reaction series, a net of one sugar molecule will remain, and the rest will be used to synthesize, by several reactions, six molecules of ribulose diphosphate.

The intermediate substances formed in photosynthesis were detected by providing plants with radioactive carbon dioxide ($C^{14}O_2$). At intervals, some as short as ½ second, after providing the plants with the labeled carbon, the plants were analyzed and the order in which the radioactive compounds appeared was determined (Fig. 6–9). The sequence is that which we have considered. Of further significance, some of

Fig. 6–9. Apparatus used to study the intermediate substances formed in photosynthesis. The algae in the flat flask are illuminated by the two lamps. Radioactive carbon dioxide enters the flask through the tube. At intervals the algae are removed by opening the stopcock. An extract of the algae is then analyzed by paper chromatography. (Melvin Calvin.)

the radioactive carbon was soon detected in certain amino acids, such as alanine and serine, indicating that in at least some species amino acid synthesis is intermeshed with photosynthesis.

FOOD AND MAN

Hunger, starvation, and famine have stalked the footsteps of man since the dawn of history. This may seem strange to you, for in the United States we eat better today than ever before. Scientists have shown us how to raise more and better food and how to preserve it so that we may feast on most anything every day in the year. But people elsewhere are not so fortunate. On an average, people in the United States have a daily Caloric intake of 3200 Calories, whereas in Egypt it is 2300 Calories, and in the Far East it is only about 1800 Calories. Only people in Australia and New Zealand have a greater protein intake than those in the United States; people of New Zealand and Australia consume about 65 grams per person per day and those in the United States, 60 grams. Persons in other countries are not so fortunate. In Turkey, Egypt, and the Far East, the average protein intake is less than 15 grams per person per day. At present nearly half the people in the world suffer from malnutrition to a greater or lesser degree. Even if one does not accept all the implications of the Malthusian theory, there can be no doubt that the population has outrun food supplies in many areas. The world's population, now at 3 billion, is increasing at a rate of 2 per cent per year—faster than at any other period in human history. If the same rate continues, the population will be 6 billion by the year 2000 and over 25 billion by the year 2075. However, there is evidence that the rate is beginning to decline in many countries, among them, China, Japan, India, Pakistan, Taiwan, and Korea; the decline is the result of social changes, family planning, and better education. It is predicted that the population of the United States will increase from the present 198 million to 300 million by the end of the century. The world's agriculture must meet the challenge of an increasing population. There are two different but entirely complementary ways of tackling this large problem that confronts mankind, namely, enhance productivity and increase crop acreage.

Productivity may be increased through breeding better plants, fertilizing and conserving the soil, improving cultural practices, and controlling plant diseases, insects, and weeds. Increased animal production will also play a role. Better breeds of livestock, proper feeding, and improved veterinary care can increase the efficiency of farm animals as producers of human food.

In most countries the better soils are already being cultivated. The task before us is to increase the productivity of poorer lands—deserts, for example. In the United States some desert areas have been reclaimed by irrigation and are very productive, but in arid regions most of the water is now being used. Many cities and farms now need additional water. Growth of cities and industries calls for increasing water, and water formerly used for irrigation may be diverted to their use. Experiments are now being directed toward making sea water suitable for irrigation. Some leads seem quite promising.

With increased population mankind will be forced to shift more of his caloric reliance to cereal grains and other plant products. Unsupplemented, these diets may be deficient in proteins and in certain vitamins and minerals. Even at the present time, about three-fourths of the people subsist to a large extent on such diets. In the future mankind may supplement his diet to a greater extent than at present with fish. It has been estimated that fish harvests could be doubled without serious risk to future fish resources. Various vitamins and amino acids have now been synthesized and could be used to supplement deficient diets. Perhaps man will make increasing use of lower plant forms. Yeast foods and alga cells are theoretically vast new sources of food; however, at present these forms have low palatability.

There is of course a possibility that scientists will unravel the secrets of photosynthesis and be able to make food artificially. As yet, scientists have not been able to imitate photosynthesis.

QUESTIONS

1. Describe two ways that photosynthesis is important to man.

2. Explain why five to ten times more land is required to produce food in the form of meat than is required to produce food, of similar caloric value, in the form of grains or vegetables.

3. _____ Assume that at 8 A.M. you punched out, from a number of leaves, 100 leaf disks with a total area of 100 $cm.^2$ After drying in an oven at 100° C., the disks weighed 5 grams. At noon you removed a similar sample and dried the disks. They weighed 6 grams. In the 4-hour period, 100 $cm.^2$ of leaf area had made (A) 1 gram of food, (B) slightly less than 1 gram of food, (C) somewhat more than 1 gram of food.

4. _____ In the presence of light, photosynthesis would proceed in (A) a test tube containing chlorophyll, water, and carbon dioxide; (B) isolated chloroplasts provided with water and carbon dioxide.

5. _____ Which of the following statements about photosynthesis is false? (A) Water and carbon dioxide are used. (B) Sugar and oxygen are formed. (C) Energy is liberated. (D) The oxygen liberated comes from water. (E) Light is necessary for photosynthesis. (F) Chlorophyll is necessary for photosynthesis.

6. _____ The dry weight of an acre of corn plants at the end of the season is approximately 6 tons. Most of this dry weight came from (A) minerals from the soil, (B) carbon dioxide from the air and water from the soil, (C) carbon dioxide from the air and minerals from the soil, (D) water from the soil, oxygen from the air, and minerals from the soil.

7. _____ Which one of the following statements with respect to photosynthesis is false? (A) Repeatedly cutting off the tops of perennial weeds will finally kill them. (B) Except for sea foods, such as clams, oysters, crabs, and various fish, all of our food comes directly or indirectly from photosynthesis. (C) A newly seeded lawn should not be cut as closely as an old lawn. (D) In cities where soft coal is burned, evergreen trees may do poorly because the stomata may become clogged.

8. List the principal factors affecting the photosynthetic rate.

9. What are the sources of the carbon dioxide in the atmosphere?

10. _____ The photosynthetic rate of sugar beet plants in an irrigated field on a bright summer day is limited by (A) water, (B) light intensity, (C) temperature, (D) the amount of carbon dioxide.

11. How may the world's food problem be solved?

12. _____ Which one of the following reactions requires light energy?

(A) Ribulose diphosphate + carbon dioxide $\rightarrow$ 6-carbon compound.
(B) Phosphoglyceric acid + ATP $\rightarrow$ diphosphoglyceric acid + ADP.
(C) $TPN + H_2O \rightarrow TPN \cdot H_2 + \frac{1}{2}O_2$.
(D) 2 triose phosphate $\rightarrow$ glucose.

13. _____ Which one of the following would be least effective in photosynthesis? (A) Red light, (B) infra-red radiations, (C) green light, (D) blue light.

14. _____ The following reactions (A) are known as light reactions and proceed only in light, (B) are known as dark reactions and proceed slowly at low temperatures and more rapidly at moderate temperatures, (C) are known as dark reactions which are not influenced by temperature, (D) directly obtain energy from chlorophyll which has been activated by light.

1. Ribulose diphosphate + carbon dioxide → 6-carbon compound.
2. 6-carbon compound → 2 phosphoglyceric acid.
3. Phosphoglyceric acid + ATP → diphosphoglyceric acid + ADP.
4. Diphosphoglyceric acid + TPN · H_2 → triose phosphate + TPN + H_2O.
5. 2 triose phosphate → 6-carbon sugar.

15. _____ Which one of the following is not formed as an intermediate substance or as an end product of photosynthesis? (A) Adenosine triphosphate, (B) alanine, (C) serine, (D) starch, (E) sugar.

16. The oxygen given off in photosynthesis (A) comes primarily from carbon dioxide, (B) is now and has been in the past the source of the oxygen in the earth's atmosphere, (C) is of no significance for aquatic animals.

17. _____ Organisms which make their own food are known as autotrophs, whereas those which cannot make their own food are referred to as heterotrophs. The heterotrophs depend on autotrophs for food. Some are parasites; others are saprophytes. Indicate which of the following are autotrophs (A) and which are heterotrophs (H).

_____ 1. Cattleya orchid.
_____ 2. Spanish moss.
_____ 3. Powdery mildew fungus.
_____ 4. Pitcher plant.
_____ 5. Rust fungus.
_____ 6. Mushroom.
_____ 7. Moss.
_____ 8. Seaweed.
_____ 9. Tomato plant.
_____ 10. Mistletoe.

7

Food Utilization—Respiration

AEROBIC RESPIRATION

Many a stalled motorist has discovered to his inconvenience that a car will not run on oxygen alone. Belatedly, he realizes that fuel is also necessary. When oxidized, the fuel furnishes the energy necessary to keep the car active. Living cells are also active, and hence require fuel, and usually oxygen. (We will see later that a few organisms can obtain energy from foods in the absence of oxygen.) Foods are the fuels (the containers of energy) of plant and animal cells. Glucose ($C_6H_{12}O_6$) is the food most frequently used in respiration. When it is oxidized, energy is released and made available for cellular activities, and carbon dioxide and water are formed. The process utilized by cells to release the energy in foods is known as **respiration.** The materials used in respiration and the substances formed are summarized below:

In words:

Glucose + oxygen → carbon dioxide + water + energy

In symbols:

$$C_6H_{12}O_6 + 6O_2 \rightarrow 6CO_2 + 6H_2O + \text{energy}$$

The balanced equation tells us that one molecule of glucose reacts with six molecules of oxygen to produce six molecules of carbon dioxide and six of water with the release of energy. Translated into grams and Calories, the equation would be:

$$180 \text{ grams } C_6H_{12}O_6 + 192 \text{ grams } O_2 \rightarrow 264 \text{ grams } CO_2 + 108 \text{ grams } H_2O + 673 \text{ Calories of energy}$$

Both day and night, respiration goes on continuously in every living cell, or more specifically, in or on the mitochondria present in all cells. The raw materials used in respiration flow to the mitochondria, and from them usable energy emerges. Plants use oxygen in respiration during the day as well as during the night. Carbon dioxide is one of the end products of respiration in plants, just as it is in animals. During the day, while the plant is using carbon dioxide in photosynthesis, it is not obvious that its cells are also producing it. At night, however, carbon dioxide is given off in easily measurable quantities. You will recall that the energy for photosynthesis comes from the sun. The energy for other plant processes and activities comes from respira-

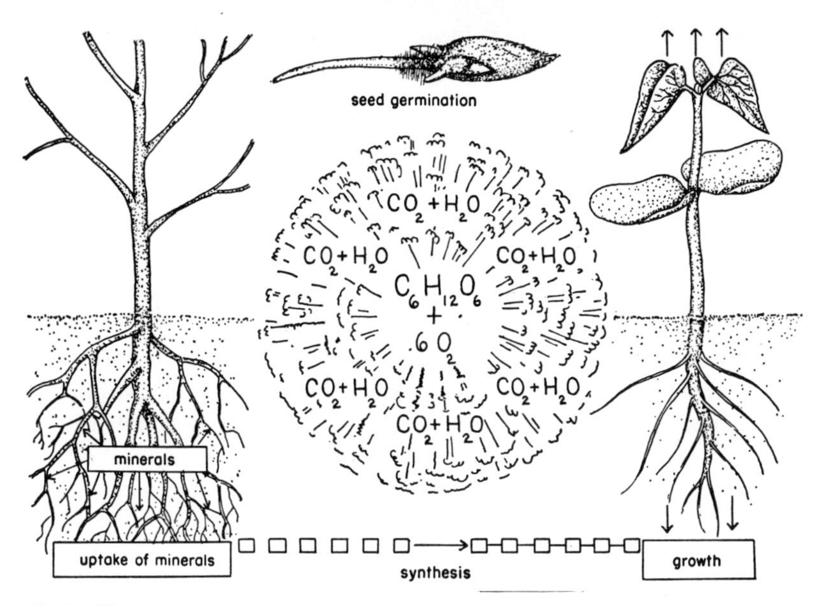

Fig. 7–1. The energy released in respiration is used in seed germination, growth, synthesis, uptake of minerals, and in other ways. Sugar and oxygen are used, and carbon dioxide and water are formed.

tion. In photosynthesis 673 large Calories are required to form 180 grams of glucose, and this amount of energy is liberated when 180 grams of glucose are respired. A large Calorie, the unit generally used when referring to energy values of foods, is that quantity of heat energy which raises the temperature of 1 kilogram of water 1°C.

Liberation of Energy

Respiration resembles burning. When sugar is burned in a flame, oxygen is used and carbon dioxide and water are produced. However, there are differences between burning and respiration. Burning occurs at high temperatures that would destroy protoplasm, but respiration occurs at temperatures compatible with life. In burning, all the energy is liberated as heat. Only a fraction of the energy released during respiration is liberated as heat. A significant amount is released without heat, and this energy is used in various ways. The ceaseless, persistent activities of plants require a large amount of energy. A tree seed germinates, sending its first root into the soil and lifting its leaves to the light (Fig. 7–1). Through the years, as that young plant develops into a tree, its great roots move a large volume of soil. Within the cells of a plant invisible, internal activities are constantly going on. Energy is required for each activity. It takes energy for cells to divide and enlarge, and so increase the length and girth of roots and stems, and the number of leaves. Energy is required for the roots to accumulate minerals from the soil. The increase in size from a developing embryo to a mature plant represents the division of a few original cells into an astronomical number of new cells, and it takes energy to synthesize the protoplasm of which they are made. The myriad chemical reactions that go on within a plant require energy: the synthesis of chlorophyll, other pigments, hormones, fats, vitamins, waxes, to name but a few. A few organisms liberate energy in the form of light.

The heat energy liberated during plant respiration is soon dissipated to the environment. However, if plant parts are confined in a thermos bottle, the temperature increases appreciably. The heat liberated by the respiration of bacteria and fungi growing in organic matter is the underlying principle of a hotbed, in which it is used to warm a volume of air.

Carbon Dioxide Is Formed

A simple experiment to demonstrate that germinating seeds produce carbon dioxide is shown in Figure 7–2*A*. Two tubes are partially filled with a barium hydroxide solution. Then a pad of cotton is placed above the solution in each test tube. Germinating seeds are placed on the pad in one test tube, but not in the other. The tubes are then stoppered. In a short time a precipitate of barium carbonate forms in the tube containing germinating seeds but not in the tube lacking seeds. Barium hydroxide combines with carbon dioxide to form an insoluble white precipitate of barium carbonate whose presence is evidence that carbon dioxide has been produced. If the experiment is carried out with other plant parts, the same result is obtained.

Rates of respiration can be obtained by determining the amount of carbon dioxide given off by a definite amount of plant material in a certain time. The apparatus illustrated in Figure 7–2*B* shows how this is done. The flask on the right is connected to a vacuum pump which pulls air through the system. As air bubbles through the barium hydroxide, contained in the left flask, carbon dioxide is removed from the entering air. The air free of carbon dioxide enters the flask containing the plant tissue and then bubbles through the 200 ml. of 0.01 normal potassium hydroxide contained in the flask on the right where the carbon dioxide evolved by the tissue is absorbed. After the desired time the residual potassium hydroxide in the right flask is titrated with 0.1 normal hydrochloric acid using phenolphthalein as an indicator. Then a 200 ml. aliquot of 0.01 normal potassium hydroxide that has not had air bubbled through it is titrated with 0.1 normal hydrochloric acid. From these two titrations the amount of carbon dioxide released in respiration can be determined by the following formula:

$$\text{Milligrams } CO_2 = V \times N \times 22$$

where V is the difference between the blank and experimental titrations in milliliters, and N is the normality of the acid used for titration.

Oxygen Is Used

Put some germinating seeds in a jar and then seal it. A day later open the jar and insert a burning match. The match will go out immediately, demonstrating that the germinating seeds used all the oxygen. Because oxygen is necessary for respiration and because energy is required for germination, most seeds do not sprout when oxygen is lacking. A striking, but simple, experiment to demonstrate that oxygen is necessary for germination can be carried out by wrapping one lot of corn grains, which are sandwiched between pieces of wet filter paper, in polyethylene film and a different lot in Saran wrap. When the packages are unwrapped 3 days later, it will be seen that the seeds wrapped in polyethylene film germinated, whereas those enclosed in Saran wrap did not. From other experiments we know that polyethylene film is permeable to oxygen, whereas Saran wrap is impermeable to oxygen.

Rates of respiration can be obtained by determining the amount of oxygen used by a tissue. The manometer illustrated in Figure 7–2*C* makes such determinations easy. Attached to the manometer is a flask containing a tissue and a potassium hydroxide solution. The carbon dioxide released in respiration is absorbed by the potassium hydroxide. As the tissue uses oxygen, the volume of air in the flask diminishes, and the column of fluid in the manometer rises in the right arm. After the experimental run the fluid in the right arm is returned to its starting position by rotating the dial which pushes a plunger down the right arm. When the manometer fluid is at the starting mark on the right arm, the amount of oxygen utilized can be read directly from the dial. The manometer illustrated, as well

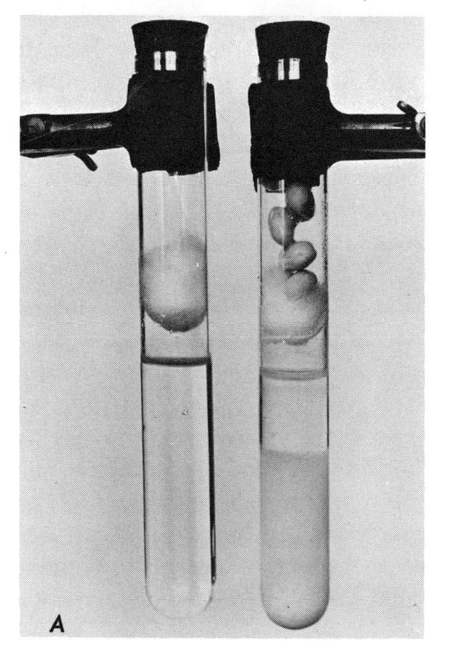

Fig. 7–2A. Germinating seeds produce carbon dioxide. B. The amount of carbon dioxide liberated in respiration can be determined with the gas train illustrated here. A suction pump is connected to the flask on the right. C. The amount of oxygen used in respiration can be determined with the manometer illustrated here. (Courtesy of Roger Gilmont Instruments, Inc.)

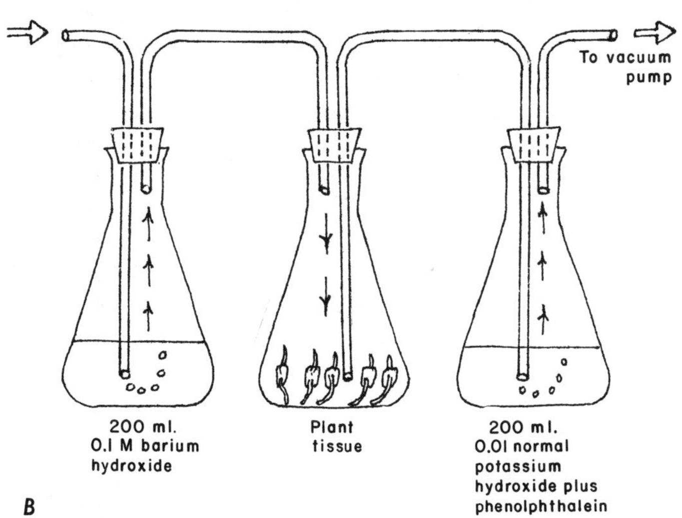

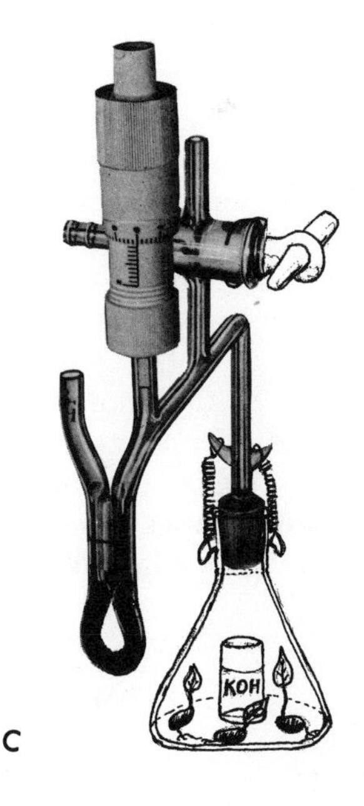

as similar types, may also be used for determining rates of photosynthesis. For example, the flask may contain algae suspended in a dilute carbonate solution which provides them with carbon dioxide. As photosynthesis proceeds, the released oxygen will increase the pressure in the flask, and the fluid in the manometer will rise in the left arm and descend in the right one. After the experimental run the column is returned to its starting position on the right arm, and the amount of oxygen evolved is read on the dial.

All parts of a plant require oxygen. Stems, leaves, flowers, and fruits obtain oxygen by diffusion from the surrounding atmosphere. Roots may or may not secure sufficient oxygen. Compaction of soil or a surplus of water in the soil decreases the oxygen available to roots and results in poor plant growth. Respiration must proceed normally if roots are to grow and function efficiently. Plants in a flooded field, or in waterlogged pots, will sicken and turn yellow. If the excess water is not removed to make way for oxygen, they drown, just as a man will drown, for lack of oxygen. Compact soils may be made more productive by cultivation, which loosens the soil and aerates it.

Food Is Used

The typical food respired is glucose. Because food is used in respiration, the dry weight of a plant (the weight after the water has been evaporated) tends to decrease unless offset by photosynthesis. If seeds are germinated in the dark, the resulting seedlings have a lower dry weight than the seeds had when they were planted. During the day, food is made in leaves at a faster rate than it is used in respiration. Hence, during the day, the dry weight of leaves increases. During the night, however, only respiration occurs, and accordingly the dry weight decreases from sundown to sunup.

When potatoes, apples, pears, beets, carrots, and other products are stored, respiration continues and their food value diminishes. For example, a carrot eaten the day it is pulled has more food value than one eaten a week afterward. In the 1-week interval some food is respired.

The growth of plants and their yield are dependent to some extent on the balance between the food-making process (photosynthesis) and the food-using process (respiration). Under favorable conditions the average amount of sugar synthesized by a corn plant during its life is about 9 grams per day, while the average amount respired is 2 grams per 24 hours. Hence, the average net daily gain is 7 grams. When, as in corn, the amount of food made is greater than the amount used, there is a surplus for growth and storage. When the amount made equals the amount used, there is no surplus for growth. To illustrate this balance concept, let us maintain plants under various conditions. In Table 7–1 hypothetical values of respiration and photosynthesis are recorded under different environmental conditions.

Under condition 1 plants make rapid growth. Under condition 2 less food is available for growth because, with a higher night temperature than in condition 1, more food is used in respiration. Under condition 3 (low light intensity) all of the food that is made is respired, and the plant does not grow. In forests some trees in the understory may be subject to such a condition. Deeply shaded trees may survive for a time, but as the dominant trees increase in size, the shade becomes greater and the

Table 7–1

Hypothetical Balance Sheet for Plants Under Different Conditions

Environmental Condition	Grams of Sugar Made by Plant per Day	Grams of Sugar Used by Plant per Day	Grams of Sugar Available for Growth and Storage
1. Light intensity and day temperature optimum for photosynthesis. Night temperature of 60° F.	100	20	80
2. Conditions same as above except night temperature is 80° F.	100	40	60
3. Light intensity very low, but other conditions favorable for photosynthesis	20	20	0
4. Short days of December and January, but plants indoors where other factors are favorable for photosynthesis	50	20	30

rate of food-making less, with the consequent death of such trees. Under condition 4 plants grow slowly even though temperatures, water, and soil factors are favorable.

Rates of Respiration

As we have seen, the rate of respiration can be determined by measuring the amount of carbon dioxide released or the amount of oxygen consumed. Let us compare respiratory rates in various plants and animals. In Table 7–2 the respiratory rates are recorded as cubic centimeters of oxygen used, per gram of tissue, per hour.

Table 7–2
Respiratory Rates of Certain Plants and Animals

Plants		Animals	
Organism	Oxygen Consumed (cc./gm./hr.)	Organism	Oxygen Consumed (cc./gm./hr.)
Azotobacter (a bacterium)	500.00	Paramecium	1.00
Carrot, mature leaves	0.44	Earthworm	0.06
Carrot, young leaves	1.13	Spider	1.40
Wheat—roots, young	0.70	Trout	0.22
Wheat—roots, older	0.85	Man	0.16–0.33
		Mouse, resting	2.50
		Mouse, running	20.00

The data in the table may surprise you. The lack of locomotion in plants leads people to believe that plants are torpid. The table reveals that plants respire more rapidly than we do. In terms of energy expended, they work harder than man. Of course, energy is not used for thought and locomotion, but instead, for growth, absorption, synthesis, and other productive processes.

Many factors influence the rate of respiration. Let us consider these next.

Factors Affecting Rate of Respiration

Water content of the tissue, temperature, oxygen supply, and the amount of carbon dioxide in the atmosphere influence the rate of respiration. Light does not directly affect the rate of respiration, but it may affect the food supply and hence indirectly affect the rate of respiration. In starving tissues respiration proceeds slowly. However, when food is available and when other conditions are the same, respiration goes on as rapidly in darkness as in light.

Hydration of tissues. The relationship between water content of cells and respiration is most evident in germinating seeds. Air-dry seeds have a respiratory rate so low that it is not measurable with ordinary techniques. When wheat seeds, as well as others, are planted in moist soil, the respiratory rate increases slightly as the water content goes up to 16 per cent. With a further increase in water content, a precipitous rise in the rate occurs. Certainly, before wheat seeds are placed in storage, their water content should be less than 16 per cent. If it is above this value, an excessive amount of food will be respired.

Temperatures. Within limits of about 0° to 35° C. (32° to 95° F.) an increase in temperature accelerates the rate of respiration. In the storage of such living plant parts as avocados, apples, bananas, pears, potatoes, carrots, flowers, etc., it is desirable to reduce the rate but at the same time keep the tissues alive. The respiratory rate can be kept at a relatively low value by storage at low temperatures. The temperature, however, should be high enough to prevent low-temperature injury. Certain fruits, vegetables, and flowers of tropical origin may be injured by temperatures of 10° C. (50° F.). The low-temperature injury is characterized by susceptibility to decay, discoloration, and failure to ripen properly.

Many plants grow better when the night temperature is about 10° F. lower than the day temperature. Operators of greenhouses generally run night temperatures 10° F. lower than day temperatures. During the day, the greenhouse is maintained at a temperature most favorable for photosynthesis. The night temperature is lowered to reduce the amount of food respired and thereby provide the plant with a greater supply for growth. Of course, the night temperature is not reduced to such an extent that it interferes with growth and flowering.

Oxygen. When all other environmental conditions are favorable, the rate of respiration diminishes as the oxygen concentration decreases below 21 per cent. However, augmenting the oxygen supply of the air to values above 21 per cent does not increase the respiratory rate. The influence of the oxygen content of the air on rates of respiration of avocados is shown in Figure 7–3. Except when temperature limits the rate, at 5° C., the rate of respiration diminishes as the oxygen concentration decreases. Similarly, in poorly aerated soils the respiration of roots decreases as the oxygen content of the soil goes down.

Carbon dioxide. It is a general rule in chemistry that rates of reactions decrease as the end products accumulate. One of the end products of respiration is carbon dioxide, and we might anticipate that as carbon dioxide accumulates, the rate would diminish. For many plants the rate does decrease as the carbon dioxide concentration of the air increases, but there are some exceptions. An increase in the carbon dioxide concentration decreases the rate of respiration of apples and sugar beets. For example, increased concentrations of carbon dioxide in the storage atmosphere (up to 12 per

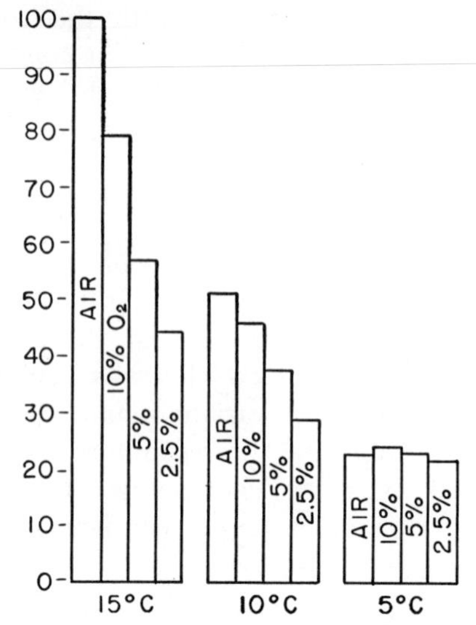

Fig. 7–3. Relative respiration rates (on ordinates) of Fuerte avocados with different oxygen concentrations in relation to temperature. The rate of respiration of avocados in air diminishes as the temperature drops. At 15° C. and at 10° C. the rate diminishes as the concentration of O_2 is lowered. At 5° C. the respiration rate is not influenced by oxygen concentration. (J. B. Biale, *Am. J. Botany.*)

cent) reduce the rate of respiration of sugar beets. For some unknown reason potato tubers do not follow the rule, because their respiratory rate actually increases as the carbon dioxide content increases.

In the storage of apples the three factors affecting respiration are often controlled. To slow down respiration and to keep the apples in prime condition, they are stored in a cold room in which the oxygen content has been much reduced and the carbon dioxide increased, an environment which keeps McIntosh apples crisp and fresh from September through May. Reducing the temperature without modifying the oxygen and carbon dioxide supply is also effective in reducing respiration and thereby prolonging the storage life of many fruits and vegetables. For example, apples respire as much food in 1 week at 70° F. as they do in 10 weeks at 32° F. Table 7–3 reviews the differences between photosynthesis and aerobic respiration.

Table 7–3
Photosynthesis and Aerobic Respiration

Photosynthesis	Respiration
1. Occurs only in green plants.	1. Occurs in all plants and animals.
2. Occurs only in green plant cells.	2. Occurs in all living cells.
3. Proceeds only in light.	3. Proceeds in both light and darkness.
4. Food is made.	4. Food is used.
5. Increases dry weight.	5. Lessens dry weight.
6. Carbon dioxide and water are used.	6. Sugar and oxygen are used.
7. Produces sugar and oxygen.	7. Produces carbon dioxide and water.
8. Stores energy in sugar.	8. Releases energy, about 34 per cent as heat, 66 per cent in a directly utilizable form.

CHEMISTRY OF AEROBIC RESPIRATION

We have learned that sugar and oxygen are used in aerobic respiration, and carbon dioxide and water are produced. This merely shows the beginning and the end of the process and not the intermediate steps. A sugar molecule is very complex in contrast to the simplicity of carbon dioxide and water. The sugar molecule is **degraded** (broken down) in steps—not all at once. In certain steps of respiration, energy is released. Respiration, unlike burning, is oxidation controlled by enzymes, and the energy is released bit by bit, rather than suddenly and explosively (Fig. 7–4). In aerobic respiration there are about fifty sequential reactions which can be divided into three main phases: **glycolysis,** the **citric acid cycle,** and **terminal oxidation.** Glycolysis occurs in the cytoplasm, whereas the citric acid cycle and terminal oxidation occur in mitochondria.

Glycolysis

The first steps of aerobic respiration, collectively known as glycolysis, occur in the absence of oxygen and lead to the formation of pyruvic acid, a 3-carbon acid with a formula of $CH_3COCOOH$.

Just as energy, in the form of heat, must be applied to wood before it will burst into flame, so also must energy be added to glucose to initiate respiration. A phosphate radical transferred from adenosine triphosphate (ATP) to glucose serves as the energy source to initiate respiration. As ATP gives up a phosphate radical to glucose, it becomes changed into adenosine diphosphate (ADP). The glucose phosphate undergoes internal rearrangement to yield fructose phosphate which then receives a phosphate radical from ATP to form fructose diphosphate. If we consider just the beginning and the end of these processes, we can summarize them as follows:

$$\text{Glucose} + 2\text{ATP} \rightarrow \text{fructose diphosphate} + 2\text{ADP}$$

Next, fructose diphosphate, a phosphorylated 6-carbon sugar, splits into two mole-

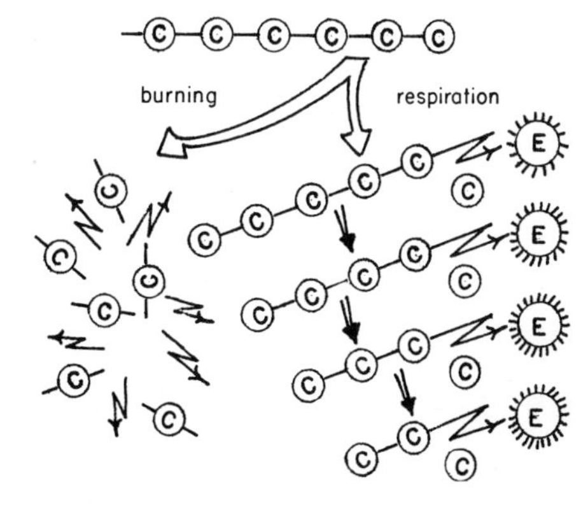

Fig. 7–4. When organic substances are burned, energy is released suddenly (*left*), whereas in respiration energy is released gradually because the carbon bonds are broken one at a time.

cules of phosphoglyceraldehyde (a 3-carbon compound) which in several steps is changed into pyruvic acid. As these reactions proceed, some energy is released, and this energy is packaged directly into new chemical energy, the energy of a high-energy phosphate bond formed when phosphate is linked to ADP. Four molecules of ATP are formed from four molecules of ADP and four molecules of phosphate, and four free radicals of hydrogen are liberated. The hydrogen is not liberated as a gas, but instead it is combined in pairs with a compound that accepts hydrogen, which we will call a **hydrogen acceptor** and designate by the letter A. Recall that two molecules of ATP were used to kindle the respiratory process and that four were formed in the reactions which change phosphoglyceraldehyde into pyruvic acid. This leaves a net gain of two ATP in glycolysis.

The two major steps in glycolysis may be summarized as follows:

Step 1:
Glucose + 2ATP → intermediates
→ fructose diphosphate + 2ADP

Step 2:
Fructose diphosphate + 2A + 4ADP
→ intermediates → 2 pyruvic acid
$+ 4ATP + 2AH_2$

If we consider just the beginning and the end of glycolysis, we may summarize the process as follows:

Glucose + 2 adenosine diphosphate + 2 phosphate + 2A → 2 pyruvic acid
$+ 2AH_2$ + 2 adenosine triphosphate

$$C_6H_{12}O_6 + 2ADP + 2P + 2A \rightarrow 2CH_3COCOOH + 2AH_2 + 2ATP$$

Citric Acid Cycle

In the citric acid cycle pyruvic acid is changed stepwise into carbon dioxide and hydrogen as shown in Figure 7–5. The cycle shown here is a simplified one and leaves out some of the acids formed, some of the reactants, and the enzymes involved.

You will notice that carbon dioxide is removed from pyruvic acid to form a 2-carbon fragment that combines with oxaloacetic

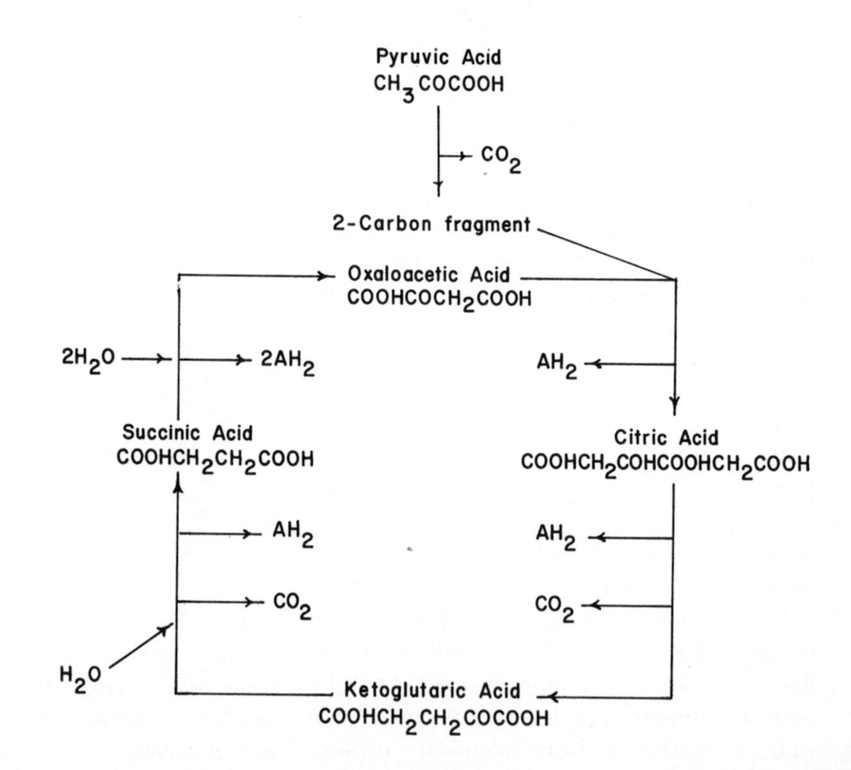

Fig. 7–5. The citric acid cycle.

acid, a 4-carbon acid, to form citric acid, an acid constructed of 6 carbons. Citric acid is the starting point for the synthesis of three other 6-carbon acids, formed in the following order: cis-aconitic acid, isocitric acid, and oxalosuccinic acid. One molecule of carbon dioxide is removed from oxalosuccinic acid to form ketoglutaric acid, a 5-carbon acid. (In the diagram substances given off are shown inside the cycle and those used on the outside.) Then a molecule of carbon dioxide is separated from ketoglutaric acid to form succinic acid, a 4-carbon acid, which is transformed through other 4-carbon intermediates, specifically fumaric acid and malic acid, into oxaloacetic acid, the starting point of the citric acid cycle. In the cycle the 2-carbon fragment has been broken down into carbon dioxide and hydrogen. In one revolution of the cycle three molecules of water are used, and two molecules of carbon dioxide and five molecules of hydrogen are formed. The hydrogen is not free gas; instead, it is combined with a hydrogen acceptor to form AH_2. (Four of the five hydrogen molecules are combined with the acceptor, technically known as diphosphopyridine nucleotide, and one is joined with triphosphopyridine nucleotide. For simplicity we designate these compounds as hydrogen acceptors, A.)

In aerobic respiration the hydrogen formed in the citric acid cycle later combines with oxygen to form water and to release energy. The oxidation of the hydrogen is known as terminal oxidation.

Terminal Oxidation

In glycolysis, as one glucose is changed into two molecules of pyruvic acid, two molecules of AH_2 are formed. We have just seen that for each molecule of pyruvic acid that goes through the citric acid cycle, five molecules of AH_2 are formed. Hence for two molecules of pyruvic acid (the number formed from one glucose molecule) that go through the citric acid cycle, ten molecules of AH_2 will result. Thus for each glucose molecule respired, twelve AH_2 molecules are formed. The hydrogen of AH_2 is now transferred to oxygen yielding water and liberating energy which is used to hook phosphate radicals, by means of high-energy phosphate bonds, to ADP, forming ATP. The transfer of hydrogen to oxygen is catalyzed by several enzymes and with the aid of several intermediate substances; among them are FAD (flavine adenine dinucleotide) and the cytochromes. For each AH_2 that combines with oxygen, three molecules of ATP are formed. From twelve molecules of AH_2, thirty-six molecules of ATP are produced. We can summarize the terminal oxidation as follows:

$$12AH_2 + 6O_2 \rightarrow 12A + 12H_2O$$

The energy released in oxidation is used to synthesize ATP from ADP and inorganic phosphate as follows:

$$36ADP + 36H_3PO_4 \rightarrow 36ATP$$

To the thirty-six molecules of ATP formed in terminal oxidation we must add the two that are formed in glycolysis to get the total number of ATP molecules formed as one glucose molecule is respired. Thirty-eight molecules of ATP are formed for each molecule of glucose respired.

The citric acid cycle and the terminal oxidation may be summarized as follows:

Pyruvic acid + oxygen → carbon dioxide + water

$$CH_3COCOOH + 2\tfrac{1}{2}O_2 \rightarrow 3CO_2 + 2H_2O$$

In the oxidation of two molecules of pyruvic acid, five molecules of oxygen would be used, and six of carbon dioxide and four of water would be produced. The oxidation of the two molecules of AH_2 produced in glycolysis would require one oxygen molecule, and two molecules of water would be formed. Thus for each glucose molecule respired, six molecules of oxygen are used, and six molecules of carbon dioxide and of water are formed.

Role of ATP

As we have just seen, much of the energy released in respiration is used to hook phosphate radicals (designated as P in the following formula) by means of high-energy phosphate bonds to ADP. The ADP is thereby converted into ATP as summarized in the diagram below:

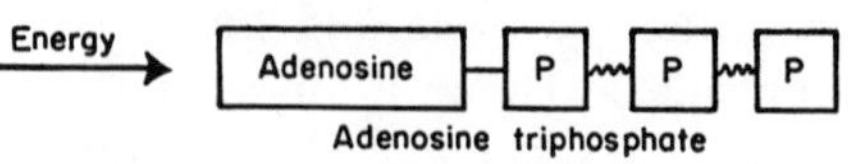

About 66 per cent of the energy liberated in respiration is stored in the terminal phosphate bond of ATP, which is known as a high-energy phosphate bond and is represented by a wavy line. The other 34 per cent is liberated as heat. The ATP emerges from mitochondria and diffuses to sites where energy must be utilized. Some of the energy stored in ATP can be transferred to some other compound simply by transferring the terminal phosphate radical with its energy supply to that compound. After receiving the energy, the recipient may then react. For example, glucose molecules will not spontaneously combine to form starch, but glucose molecules, each with a phosphate group attached thereto, will link together to form starch in the presence of the enzyme starch phosphorylase. The following summarizes the steps in the formation of starch from glucose:

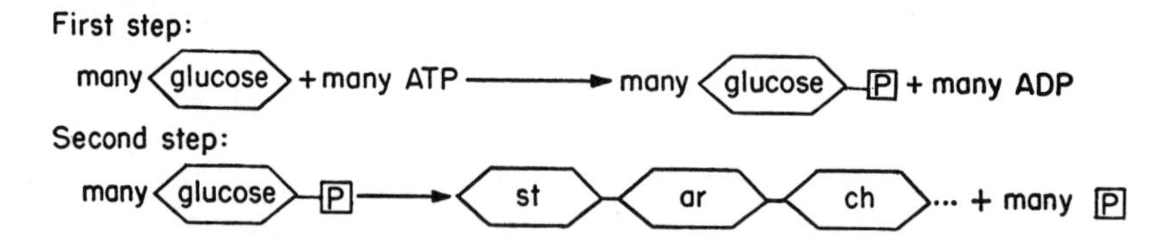

In the transfer of the phosphate radical from ATP to glucose, some energy of the high-energy phosphate bond is dissipated as heat; hence the phosphate bond of glucose phosphate is one of lesser energy than the terminal phosphate bond of ATP. This bond of lesser energy is represented by a dash. Nevertheless, the energy of the phosphate bond of glucose phosphate is ample for starch synthesis. When the phosphate bonds are broken in the second step, energy is released, starch is synthesized, and phosphate is formed. The ADP released in the first step goes back to the respiratory cycle where energy again hooks on a phosphate radical. ATP furnishes energy not only for starch synthesis, but also for the synthesis of proteins, fats, and other compounds. The energy released when the terminal phosphate bond is broken is also used for mechanical work such as cytoplasmic streaming, movement of chromosomes, and in animals for muscular contraction. The energy may also be used for the accumulation of ions (Fig. 7–6).

Respiration is not only significant because it releases energy, but also because it is intimately related to the synthesis of cell constituents. In any living cell great numbers of glucose molecules are respired each day. Many are completely oxidized to carbon dioxide and water, but some are not. Instead, the intermediate substances are diverted to the synthesis of fats and proteins (Fig. 7–7). Molecules of the 2 carbon compound formed from pyruvic acid may combine to form fatty acids which then combine with glycerol, formed from a 3-carbon intermediate, to yield a fat. For example, nine molecules of the 2-carbon compound may be joined to form stearic acid, a fatty acid containing eighteen carbons ($C_{17}H_{35}COOH$). Three molecules of stearic acid combine with one of glycerol ($CH_2OHCHOHCH_2OH$) to form the fat known as tristearin.

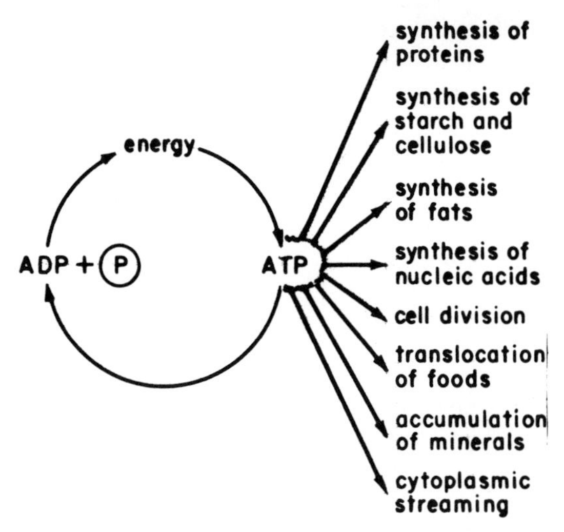

Fig. 7–6. The ATP produced in respiration does biological work. It provides energy for synthesis, cell division, translocation, ion accumulation, and cytoplasmic streaming.

Some molecules of ketoglutaric acid, produced in the citric acid cycle, react with ammonia to form glutamic acid ($COOHCH_2$-CH_2CHNH_2COOH), an amino acid with five carbons. Oxaloacetic acid may combine with ammonia, yielding aspartic acid ($COOHCHNH_2CH_2COOH$). Other amino acids may be formed. Many amino acid molecules are joined together to form a protein. We will have more to say about fat synthesis and protein synthesis in Chapter 15.

ANAEROBIC RESPIRATION

Our earth's earliest inhabitants lived in an environment devoid of oxygen. They released energy from food by a type of respiration in which atmospheric oxygen was not used—**anaerobic respiration.** This primordial process of releasing some energy from food in the absence of oxygen is retained by organisms now living. As we have just seen, glycolysis releases some energy from food in the absence of oxygen. Even those organisms which use oxygen in respiration still carry on glycolysis, thus showing their kinship to primordial creatures.

Among present-day plants, some obtain all their energy from anaerobic respiration; for example, yeast plants and certain bacteria. As yeast plants respire anaerobically, they transform glucose into carbon dioxide and ethyl alcohol as follows:

Glucose → carbon dioxide + alcohol + 21 Calories

$$C_6H_{12}O_6 \rightarrow 2CO_2 + 2C_2H_5OH + 21 \text{ Calories}$$

In the making of bread and similar products, the carbon dioxide produced by yeast leavens the bread, and hence it is the desired product. In the brewing industry alcohol is the wanted substance. Twenty-one Calories of energy are released for each 180 grams of glucose respired anaerobically; this is a small amount compared to the 673 Calories which are released when 180 grams of glucose are respired aerobically. In anaerobic respiration of yeast some energy is liberated as heat and another part, that stored in ATP, is available for the activities of yeast plants.

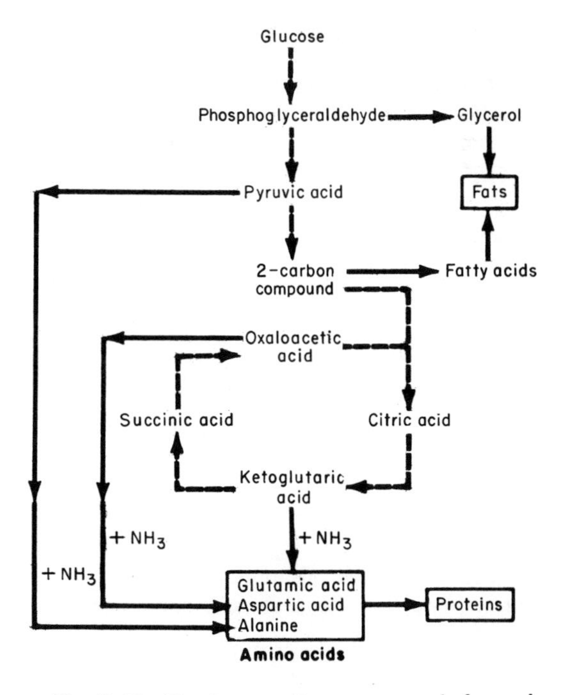

Fig. 7–7. The intermediate compounds formed in respiration are the starting points for the synthesis of fats and protein.

Certain bacteria also respire anaerobically. Some produce carbon dioxide and ethyl alcohol as do yeast plants. Anaerobic respiration in different kinds may produce butyric acid, as in the spoiling of butter, or lactic acid, as in the souring of milk. In other instances, still different products may be formed during anaerobic respiration. Some of these may be poisonous to human beings if consumed—for example, certain **ptomaines.** Food spoilage in sealed cans is brought about by microorganisms which respire anaerobically. The production of carbon dioxide during anaerobic respiration causes the ends of the cans to bulge. Bulging end walls, therefore, are a sign that the food is inedible.

Higher plants may survive for a relatively short time where oxygen is lacking by obtaining energy through anaerobic respiration. Usually, however, the plants will not survive for very long under anaerobic conditions because the amount of energy liberated per unit of food used is small and because toxic substances are produced.

CHEMISTRY OF ANAEROBIC RESPIRATION

In anaerobic respiration, as in aerobic respiration, glucose is changed into pyruvic acid by glycolysis. Each molecule of glucose yields two molecules of pyruvic acid, two of ATP, and two of AH_2.

As yeast respires anaerobically, the pyruvic acid is changed into alcohol and carbon dioxide as follows:

$$2 \text{ pyruvic acid} + 2\,AH_2 \rightarrow 2 \text{ ethyl alcohol} + 2 \text{ carbon dioxide} + 2A$$

$$2CH_3COCOOH + 2AH_2 \rightarrow 2C_2H_5OH + 2CO_2 + 2A$$

Certain bacteria transform pyruvic acid into lactic acid as follows:

$$2 \text{ pyruvic acid} + 2AH_2 \rightarrow 2 \text{ lactic acid} + 2A$$

$$2CH_3COCOOH + 2AH_2 \rightarrow 2CH_3CHOHCOOH + 2A$$

The transformation of pyruvic acid into ethyl alcohol or lactic acid yields no energy. The only energy useful to organisms respiring anaerobically is that stored in the two molecules of ATP which are formed during glycolysis.

Comparison of Aerobic and Anaerobic Respiration

The following diagram shows the beginning product and the end products of aerobic and anaerobic respiration.

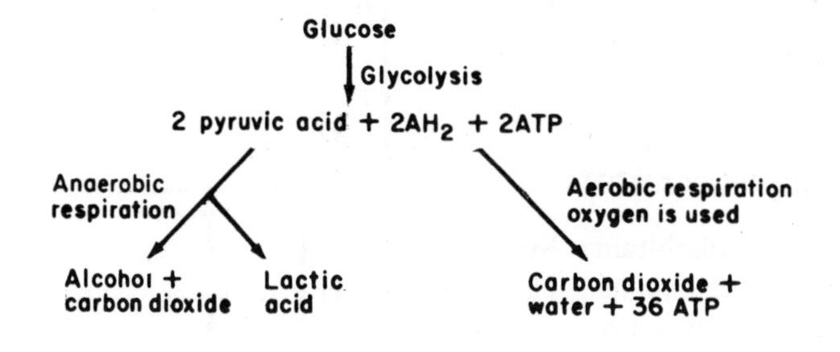

QUESTIONS

1. ______ Which one of the following is the correct equation for aerobic respiration?

(A) $C_6H_{12}O_6$ (glucose) + $6O_2$ (oxygen) → $6CO_2$ (carbon dioxide) + $6H_2O$ (water)

(B) $6CO_2$ (carbon dioxide) + $6H_2O$ (water) → $C_6H_{12}O_6$ (glucose) + $6O_2$ (oxygen)

(C) $C_6H_{12}O_6$ (glucose) → $2C_2H_5OH$ (alcohol) + $2CO_2$ (carbon dioxide)

2. A grain elevator employee was working alone in a grain elevator filled to within 7 feet of the top with flaxseed which contained about

9.0 per cent moisture. The employee did not leave the elevator at his usual time. An investigation was then made. The employee was found dead in the elevator. How could you account for his death?

3. _____ Which one of the following statements about aerobic respiration is false? (A) The energy released in respiration changes adenosine triphosphate into adenosine diphosphate. (B) Sugar and oxygen are used, and carbon dioxide and water are produced. (C) In respiration, energy is released, and this energy is used for synthesis, accumulation of ions, growth, etc. (D) As a result of respiration the dry weight of a plant tends to decrease.

4. _____ If a plant respires more food than it makes (such as may occur in deep shade), the plant will (A) survive indefinitely, (B) increase in dry weight, (C) lose in dry weight and may die (when stored food is used up), (D) grow very rapidly in order to get out of the shade.

5. Are the growth and yield of plants related to the balance between respiration and photosynthesis? Explain.

6. _____ With respect to respiration, which one of the following statements is *not* correct? (A) When bread dough is "raised" with yeast, the yeast plants respire and produce carbon dioxide and alcohol. (B) If a plant were to be grown in air which had been freed of oxygen, it would live longer in darkness than in light. (C) Pebbles, bits of broken pottery, or similar coarse material should be placed in the bottom of a flower pot in which a plant is to be grown. (D) In aerobic respiration, oxygen and glucose are used, and carbon dioxide and water are produced.

7. _____ The food value of a potato will (A) not be affected by storage at a low temperature, (B) increase with storage, (C) decrease with storage because of the loss of water, (D) decrease with storage because part of the food is respired.

8. What factors affect the rate of respiration? How would a greenhouse operator use this information? A fruit storage man?

9. Contrast respiration and photosynthesis.

10. _____ Which one of the following represents the correct equation for anaerobic respiration as carried out by yeast plants?

(A) Glucose → ethyl alcohol + carbon dioxide.

(B) Glucose + oxygen → carbon dioxide + water.

(C) Carbon dioxide + water → glucose + oxygen.

(D) Ethyl alcohol + oxygen → acetic acid.

11. _____ Which one of the following correctly summarizes glycolysis?

(A) Glucose + 2ADP + 2 phosphate + 2A → 2 pyruvic acid + $2AH_2$ + 2ATP.

(B) Pyruvic acid + oxygen → carbon dioxide + water.

(C) $12AH_2 + 6O_2 \rightarrow 12A + 12H_2O$.

12. _____ The citric acid cycle occurs (A) only in aerobic respiration, (B) only in anaerobic respiration, (C) in both aerobic and anaerobic respiration.

13. Outline the essential features of the citric acid cycle.

14. _____ For each molecule of glucose that is respired aerobically, (A) 38 ATP molecules are formed, (B) 2 ATP molecules are synthesized, (C) 18 ATP molecules result, (D) 36 molecules of ADP are produced.

15. _____ Which one of the following is false? (A) The energy released when the terminal phosphate bond of ATP is broken may be used for the accumulation of ions and for mechanical work. (B) Some of the energy stored in ATP may be transferred to some other compound by transferring the terminal phosphate with its energy supply to that compound. (C) Although ATP is the chief energy carrier in plant cells, AMP (adenosine monophosphate) is primarily used by animal cells. (D) About 66 per cent of the energy liberated in respiration is stored in the terminal phosphate bond of ATP.

16. _____ Which one of the following is not a step in the anaerobic respiration of yeast plants?

(A) $2CH_3COCOOH + 2AH_2 \rightarrow 2CH_3CHOHCOOH + 2A$.

(B) $C_6H_{12}O_6 + 2ADP + 2P + 2A \rightarrow 2CH_3COCOOH + 2AH_2 + 2ATP$.

(C) $2CH_3COCOOH + 2AH_2 \rightarrow 2C_2H_5OH + 2CO_2 + 2A$.

17. _____ Judged by the amount of energy expended per unit weight, which of the following is least active? (A) Mouse, (B) man, (C) young carrot leaves, (D) *Azotobacter*, a bacterium.

8

Structure of Leaves

EXTERNAL FEATURES OF LEAVES

The primary function of leaves is photosynthesis, for which they are specially adapted. Typically, the blade (**lamina**) of a leaf is a flat, thin, much expanded structure that exposes a large surface to light and air. The number of leaves on trees and their total area is immense. For example, an orange tree 29 years old bears 170,000 leaves having an area of 2,000,000 square centimeters. Even herbaceous plants expose a large area to light. Thus just one stem of an alfalfa plant may support eighty-eight leaves with a combined area of 16,000 square centimeters.

Leaves of many plants have three parts; a **blade,** a **petiole,** and, at the junction of the petiole and stem, two **stipules** (Fig. 8–1). Leaves of certain species may lack one or more of these structures. In asters, whose leaves lack a petiole, the blade is attached directly to the stem, and the leaf is said to be **sessile.** However, in most plants the blade is supported by a stalk, the **petiole,** which transports materials between the blade and stem and which makes adjustments to wind and light. Frequently, the petioles bend and elongate in such a way that the blades become favorably oriented to light, even to the extent of producing a regular flat pattern known as a leaf-mosaic, as in ivy. In addition to the blade and petiole some leaves have two appendages, known as **stipules,** at the base of the petiole. In roses and many other plants the stipules are small and functionless, but in peas and pansies the stipules are an important part of the leaf surface.

Although foliage leaves are uniform in their primary function, they are remarkably diverse in size, shape, texture, leaf margins, arrangement of veins, presence or absence of hairs, and duration of life. We identify many plants by their foliage. Through the use of leaf characters we can recognize an oak, rose, maple, lilac, holly, geranium, African violet, spinach, pea, and other plants.

In many plants the foliage lasts only one season, but in pines, holly, and other plants the leaves live for several years and the trees are said to be evergreen. Leaves range in size from banana leaves which may be 10 feet long to the minute leaves of desert and alpine plants. In shape, leaves vary from narrow, elongated structures to the circular ones of the nasturtium and water lily, with many intermediate shapes. Some common

Fig. 8–1. A simple leaf is made up of an expanded blade, a stalklike petiole, and basal stipules.

leaf shapes and their names are given in Figure 8–2*A*. Leaves of maple and oak are **lobed,** whereas those of lilac are not lobed; neither do they have teeth. A leaf margin which lacks teeth is said to be **entire.** The leaf margins of some plants have pointed teeth extending outward (**dentate**), while those of others have teeth directed toward the tip of the leaf (**serrate**). Certain leaves have rounded teeth on their margins, and the margins are then said to be **crenate.** Figure 8–2*B* illustrates the kinds of leaf margins, leaf apices, and leaf bases and gives the terms that have been coined to describe these external features of leaves.

Hairs are present on the foliage of many plants but are lacking on that of different species. The hairs may be soft, stiff, or glandular. The glandular hairs of some species of wild geraniums, as well as other plants, produce a sticky secretion. The hairs of nettles are sharp, and the tips are readily broken off when touched. The sharp broken end pierces the skin, and irritating contents flow out of the hair. A leaf hair develops from a single epidermal cell or a cluster of epidermal cells.

In grasses, irises, and most other monocots the main veins of a leaf run parallel to each other and the leaf is said to be **parallel-veined.** The more conspicuous veins of a geranium, foxglove, melon, or other dicot leaf form a network, and this type is called **net venation.** If a leaf is net-veined, the arrangement may be either **palmate** or **pinnate.** In a foxglove leaf the main branch veins originate from a prominent midrib, and the leaf is **pinnately net-veined.** A melon leaf has several principal veins starting from the base of the blade, and it is therefore **palmately net-veined** (Fig. 8–3).

Leaves are either **simple** or **compound.** The blade of a simple leaf is in one piece, whereas the blade of a compound leaf consists of two or more pieces called **leaflets.** Leaves may be either pinnately compound or palmately compound (Fig. 8–4). In rose and other pinnately compound leaves the leaflets are attached at intervals on the sides of a stalk, known as the **rachis,** which is an extension of the petiole. In Virginia creeper and other palmately compound leaves all of the leaflets are attached at the end of the petiole. At times it may be difficult to determine whether a certain structure is a leaf or a leaflet. If there is a bud above where the structure is attached, it is a leaf. If a bud is lacking, it is a leaflet. A bud is always present just above where a leaf is attached to a stem.

Leaves are arranged on stems in an **alternate, opposite,** or **whorled** manner (Fig. 8–5). If one leaf is present at a node, the arrangement is alternate. When a pair of leaves occurs at a node, the distribution is

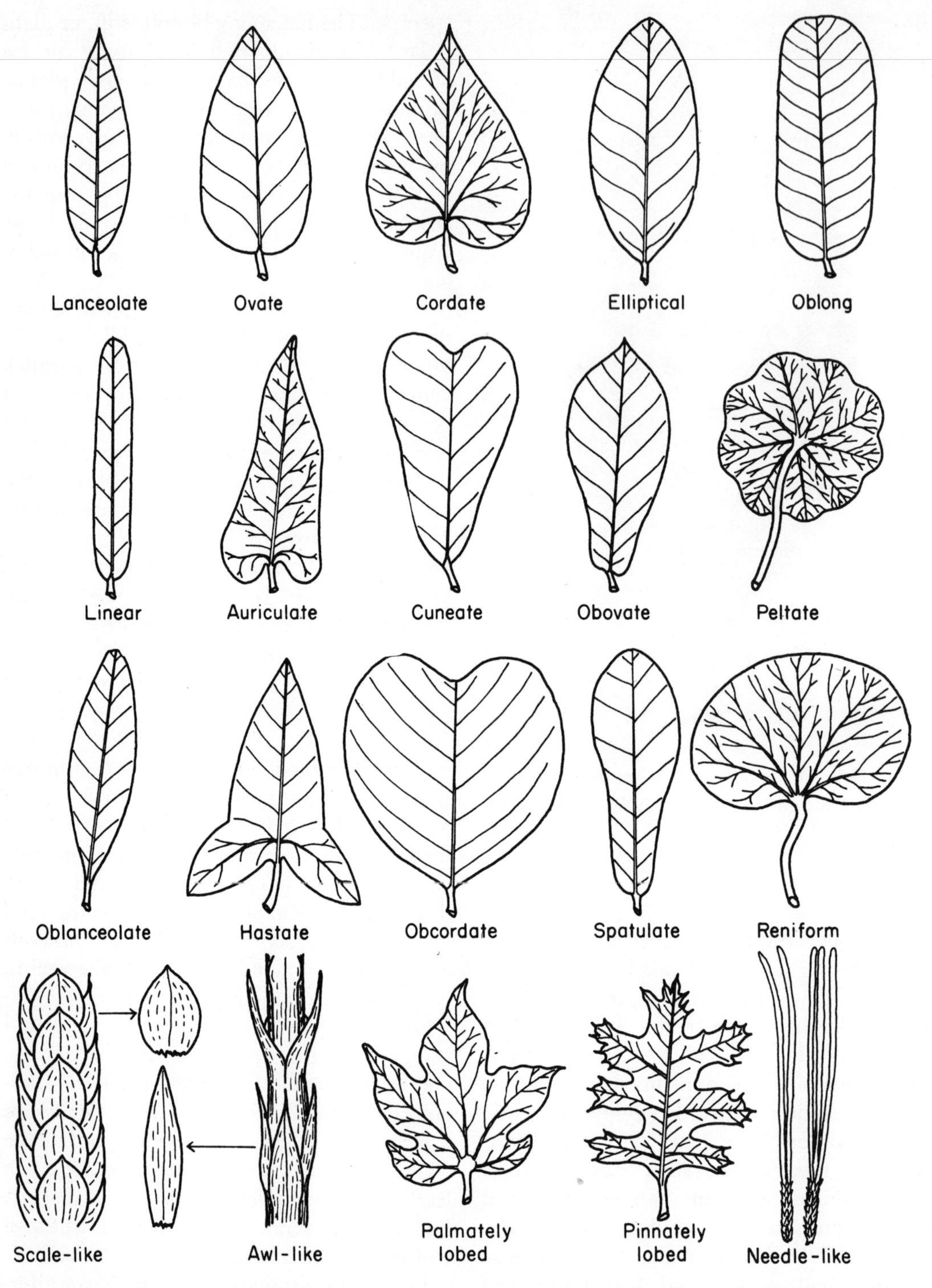

Fig. 8–2*A*. Variety of shapes found in leaves. (William M. Carlton, *Laboratory Studies in General Botany*. Copyright © 1961, The Ronald Press Company, New York.)

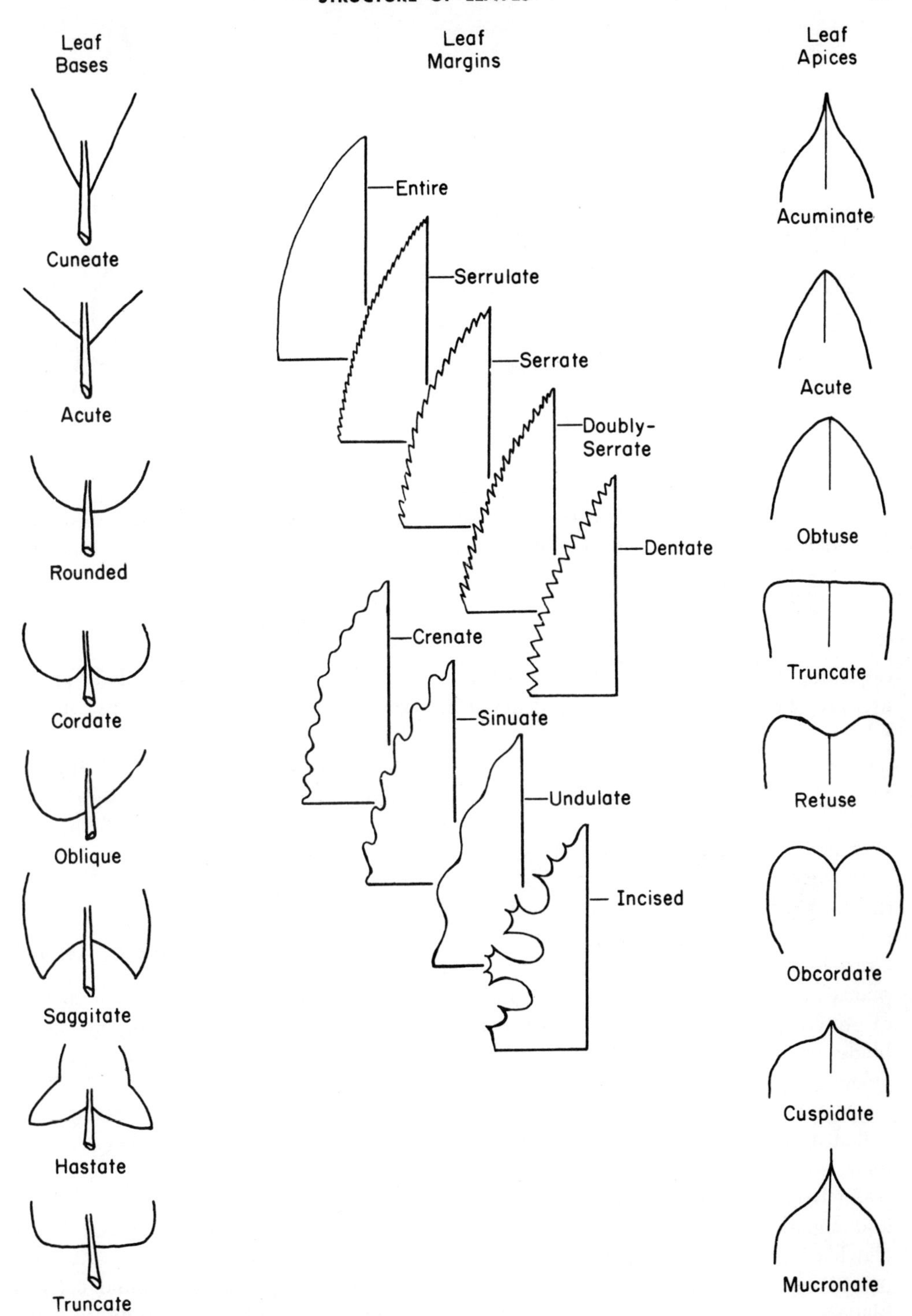

Fig. 8–2*B*. External features of leaves. (William M. Carlton, *Laboratory Studies in General Botany*. Copyright © 1961, The Ronald Press Company, New York.)

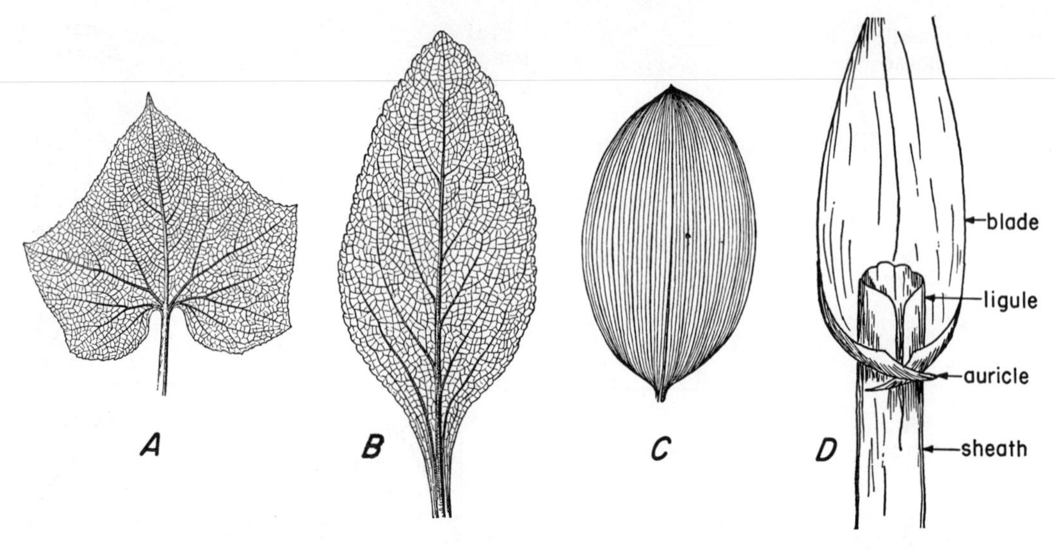

Fig. 8–3*A*. Palmately net-veined leaf of melon. *B*. Pinnately net-veined leaf of foxglove. C. Parallel-veined leaf of Solomon's seal. *D*. A portion of a grass leaf. (*A*, *B*, and C after Bergen.)

opposite; and when more than two leaves are present at a node, the arrangement is whorled.

GRASS LEAVES

Grasses are the major plants of range and pasture, and their foliage comes to man in the form of mutton, lamb, and beef. Furthermore, all of the cereals are grasses. Grass leaves are unlike those of many other plants in external parts and in their manner of growth. A grass leaf consists of a narrow **blade** and a cylindrical **sheath** which envelops the stem from one node to near the one just above. A collar-like structure, called a **ligule,** is present where the blade joins the sheath, and it may prevent rain water from flowing down between the stem and sheath. In certain grasses two small clawlike projections, called **auricles,** are present on the sheath where it joins the blade.

Grasses withstand grazing or mowing better than most other plants because leaf growth occurs near the base of the blade. If the terminal portion of a grass leaf is eaten or mowed off, the blade grows from the base and soon attains its original length. On the other hand, if a clover leaf is eaten, that is the end of that leaf. A young clover leaflet grows throughout, not just at the base. After reaching a certain size, growth ceases and is not renewed if the terminal part is removed.

INTERNAL STRUCTURE OF LEAVES

Leaves are adapted for food manufacture not only in external form, but in internal structure as well. The anatomy of a grape leaf is illustrated in Figure 8–6. The three tissues which make up a foliage leaf are **epidermis, mesophyll,** and **veins.**

The Epidermis

Each leaf surface is covered by an epidermis made up of closely fitting, interlocked, transparent **epidermal cells** and, at least on one surface, green **guard cells.** The outer walls of the transparent epidermal cells and the guard cells are coated with a waxy material—**cutin.** This wax layer is called the **cuticle,** and it is secreted by the epidermal cells. The epidermis with the cuticle protects the underlying cells from injury and prevents an excessive loss of water. In some plants a portion of the leaf surface may not be completely impervious to water. It has been demonstrated that certain plants can take up water through their leaves. When these leaves were sprayed with water or covered with rain or dew, some water entered. The lower or upper epidermis or both, depending on the species, is perforated by numerous lens-shaped pores (called **stomata;** singular, **stoma**) of which there may be as many as 62,500 per square centimeter. An entire corn plant has about 200 million stomata. The guard cells open and close the stomata. Guard cells contain chloroplasts, are often sausage-shaped, and lie one on each side of a stoma. In other words, guard cells always occur in pairs. The walls of the guard cells are thicker on the side toward the opening. Generally, stomata are closed at night and open during the day. As the sun comes up in the morning, the guard cells use carbon dioxide and water, which when combined is carbonic acid, in photosynthesis, and thus the cell sap becomes less acid. With decreasing acidity the enzyme starch phosphorylase converts starch into glucose with a resultant increase in the osmotic concentration of the cell sap. A pair of guard cells then absorbs water from the ordinary epidermal cells which, lacking chloroplasts, do not carry on photosynthesis. As a pair of guard cells absorbs water, the thin outer walls bulge into the surrounding epidermal cells, and the thicker inner walls follow, thus leaving a space, a stoma, between them. After sundown the cell sap in the guard cells becomes more acid as carbon dioxide and water are formed in respiration. Then the enzyme starch phosphorylase converts sugar into starch with a resulting decrease in the osmotic concentration of the guard cells. In time the osmotic concentration of the guard

Fig. 8–4. Compounding in leaves. *Left,* pinnately compound leaf of rose. *Right,* palmately compound leaf of Virginia creeper.

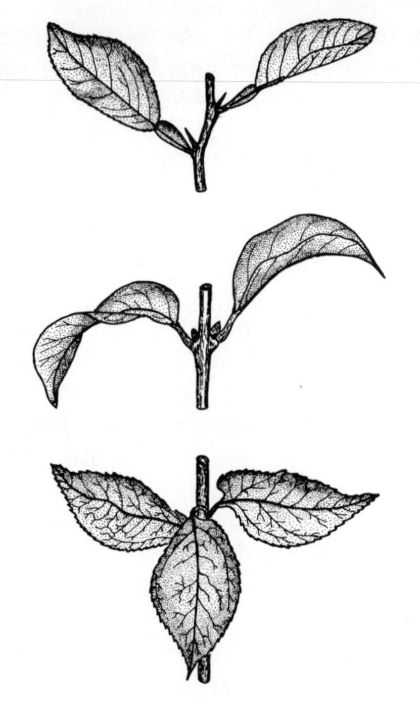

Fig. 8–5. Alternate (*upper*), opposite (*center*), and whorled (*lower*) arrangement of leaves. (William M. Carlton, *Laboratory Studies in General Botany*. Copyright © 1961, The Ronald Press Company, New York.)

cells becomes less than that of the ordinary epidermal cells, and then water diffuses from the guard cells into the surrounding epidermal cells. As the guard cells collapse, the two thick inner walls come in contact and the stoma is closed. According to this mechanism, you would expect stomata to be open during the day and closed at night and during drought. In most species the expected results actually occur (Fig. 8–7).

In addition to light and water the carbon dioxide concentration of the substomatal cavity (the space inside a stoma) regulates the opening and closing of a stoma. When the carbon dioxide concentration in the cavity is greater than 0.03 per cent, the guard cells close the stoma; when less than 0.03 per cent, they open the stoma. During night, the carbon dioxide concentration becomes greater than 0.03 per cent as carbon dioxide, formed in respiration, accumulates. When the leaf is lighted, carbon dioxide is used, and the concentration falls below 0.03 per cent. The guard cells then take up water, become turgid, and thus bring about an opening of the stoma.

Gases diffuse into and out of leaves through the stomata. When a leaf is making food, carbon dioxide diffuses through the stomata into the leaf and oxygen diffuses out. Although stomata are the chief portals for the entrance of carbon dioxide, they are not the only means by which carbon dioxide may enter a leaf. When scientists provided plants with carbon dioxide containing radioactive carbon (C^{14}), they discovered that some carbon dioxide entered leaves even when the stomata were closed. Apparently, the carbon dioxide diffused directly through the cuticle and ordinary epidermal cells. Water vapor also diffuses out through stomata. The fact that stomata are very numerous may be seen from Table 8–1.

Table 8–1

Number of Stomata per Square Centimeter

Plant	Upper Surface	Under Surface
Olea europaea (olive)	0	62,500
Acer pseudoplatanus (Norway maple)	0	40,000
Cucurbita pepo (pumpkin)	2800	26,900
Lycopersicum esculentum (tomato)	1200	13,000
Phaseolus vulgaris (bean)	4000	28,100
Zea mays (corn)	9000	15,800
Nymphaea alba (water lily)	46,000	0

The water lily, whose leaves float on the water surface, have stomata only on the upper surface. These stomata permit a ready exchange of gases between the leaf and the air above. The olive and maple have stomata only on the lower surface, which is less exposed than the upper one. Although pumpkin, tomato, bean, and corn have stomata on both the upper and lower leaf surfaces, they are more numerous on the lower epidermis.

The Mesophyll

Except for the veins, the space between the upper and lower epidermis is occupied by mesophyll cells. These are food-making cells with thin walls and numerous chloroplasts. Between the mesophyll cells there are air spaces through which carbon dioxide and other gases diffuse. The intercellular spaces form a continuous system of air passages throughout the leaf. In the leaves of many plants the mesophyll cells toward the upper surface are cylindrical in shape and are known as **palisade cells,** whereas those toward the lower surface are rounded and are called **spongy cells** (Fig. 8–6). In the leaves of some other plants a palisade layer is present next to the lower epidermis as well as next to the upper, with spongy tissue in between. In corn leaves only spongy cells make up the mesophyll. Certain upper epidermal cells, the **motor cells** (also called **bulliform cells**), of corn are very large. When these cells lose water, the blade rolls up. When they become turgid, the blade unrolls and becomes flat (Fig. 8–8).

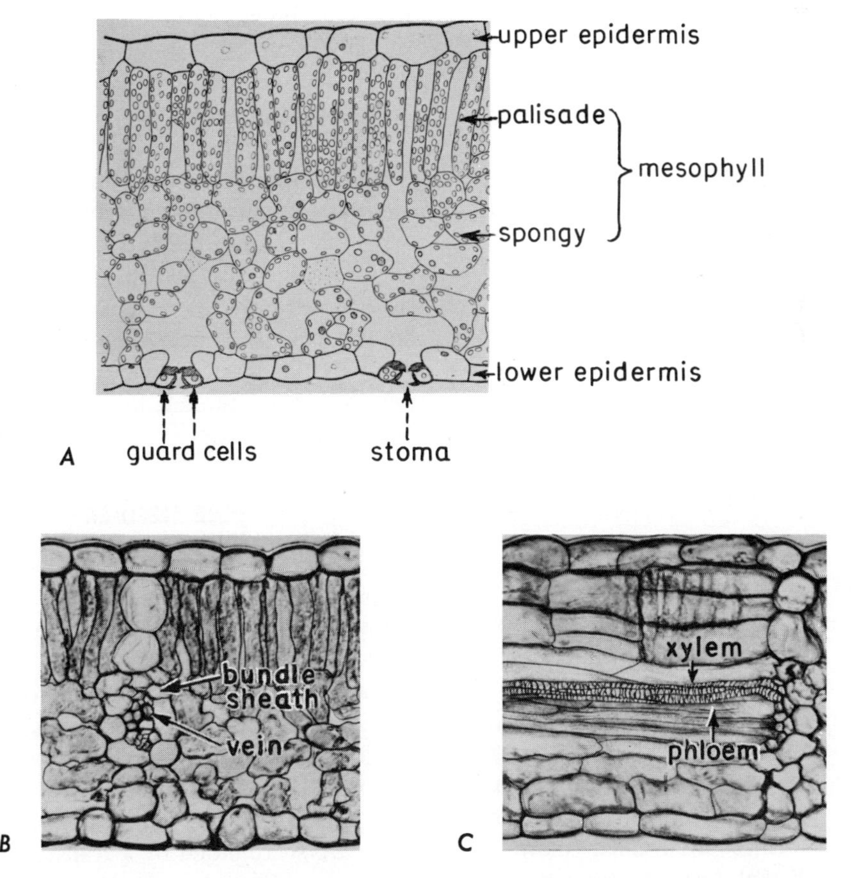

Fig. 8–6. Sections of grape leaves. Sections *A* and *B* were cut at right angles to a vein; *C* was cut parallel to a vein. (Robert B. Wylie.)

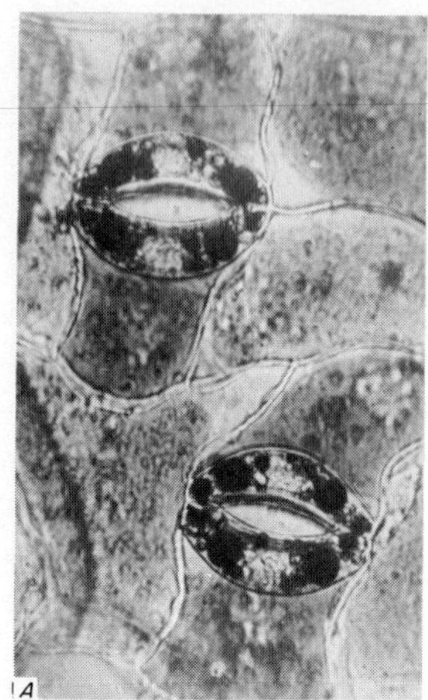

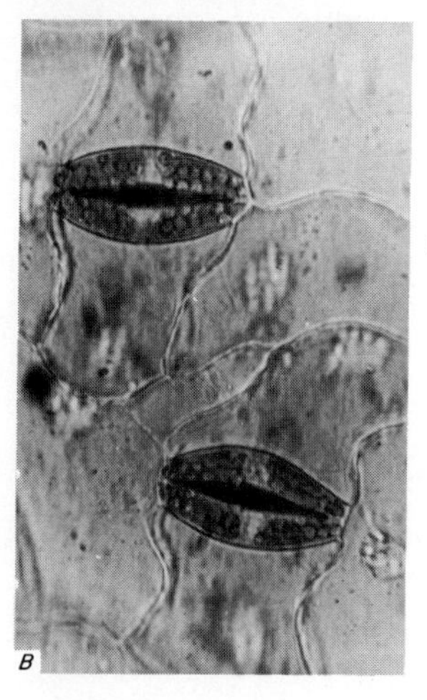

Fig. 8–7. Lower surfaces of leaves. *Top,* Open stomata. *Bottom,* Closed stomata. Note that two guard cells enclose a stoma and that chloroplasts are present in the guard cells, but that they are absent in the ordinary epidermal cells. (Paulo deTarso Alvim.)

The Veins

The veins form the structural framework, and they conduct water, minerals, and food. The midrib and the larger veins are visible to the unaided eye, but in addition to these there are innumerable smaller veins which can be seen only with a hand lens. The system of veins is so efficiently spaced that no mesophyll cell is more than a few cells distant from a vein or veinlet (Fig. 8–9). Each vein consists of **conducting tissue** surrounded by a **bundle sheath.** The sheath cells are thin-walled, with or without chloroplasts; they are elongated and parallel with the vein. They are in contact with the palisade and spongy cells. Substances moving from or into the vascular tissues must pass through the bundle sheath. Two conducting tissues are present in a vein: **xylem** toward the upper epidermis, and **phloem** toward the lower. The xylem consists of **vessels** and **tracheids** which conduct water and minerals. The characteristic cells of the phloem are sieve tube cells, long cells with perforated end walls, that conduct food.

Fibers, thick-walled elongated cells, are usually associated with the larger veins. They may surround the vein or be located above and below the vein, resulting in a girder-like support which resembles an I-beam. The strengthening tissues, the veins and fibers, are so located that the leaf will easily support its own weight and that of rain, dew, and snow.

The vascular tissue in a leaf is continuous with that in a stem, thus forming a continuous passageway for the movement of water and minerals into the leaf and for the translocation of foods out of the leaf. One to several strands of vascular tissue, each known as a leaf trace, extend outward from the vascular tissue in a stem, enter the petiole, and then branch to form the veins in the blade.

PINE NEEDLES

Like leaves of other plants, the needles of pine have an epidermis, mesophyll, and veins (Fig. 8–10). The epidermal cells have thick walls and a heavy cuticle. The stomata are sunken in pits, and the mesophyll cells are markedly lobed and compactly arranged. **Resin ducts** are present in the mesophyll. Leaves of pine have two veins, each consisting of xylem and phloem. The veins are surrounded by a **transfusion tissue,** which may store water, that in turn is enclosed by a well-developed endodermis.

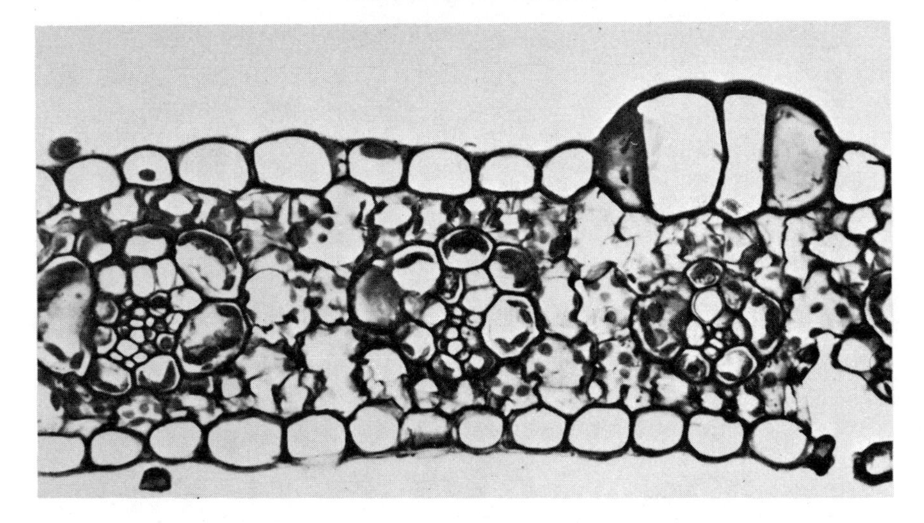

Fig. 8–8. Cross-section of a corn leaf showing upper epidermis, mesophyll, veins, and lower epidermis. The large cells of the upper epidermis are motor cells. When they lose water, the leaf rolls up and thus lessens transpiration. Note the stoma in the lower right-hand corner. (Leo W. Mericle.)

THE PLANT AS A FOOD-MAKING FACTORY

The plant as a food-making factory is so well designed that it could hardly be improved by a skillful engineer. Instead of one great concentrated unit all under one roof, the plant has billions of units (the cells) held in many thin, separate stories (the leaves), all exposed to light. The mesophyll cells of a leaf communicate with the outside air through the stomata. Carbon dioxide enters a leaf through the open stomata, diffuses through the **intercellular spaces,** becomes dissolved in the water of cell walls, and enters the mesophyll cells in solution. Vessels and tracheids of the xylem bring to the mesophyll cells water absorbed from the soil by the roots. Within the cells the two raw materials, carbon dioxide and water, come in contact with chlorophyll, the agent that acts to combine them into sugar with the release of oxygen. The oxygen diffuses out through the stomata. The sugar moves out of the leaf through the sieve tubes of the phloem and is distributed to other parts of the plant.

LEAF STRUCTURE AND ENVIRONMENTAL FACTORS

Leaf structure may be markedly altered by such environmental factors as water content of the soil, humidity, temperature, and light intensity. Leaves developing in humid air generally have a greater leaf area, more chloroplasts, and fewer vascular bundles than those which form in drier air. Leaves which develop in bright light are thicker, have a heavier cuticle, more veins, and smaller intercellular air spaces than those which grow in shade (Fig. 8–11).

LEAF FALL

Throughout the summer the leaf has lived an intense life, forming from water and carbon dioxide with the energy of sunlight a large quantity of sugar, some of which was moved to more permanent plant parts. Before leaves fall, they generally surrender to the plant that bears them the remainder of their nutrients. Prior to actual leaf fall there develops in most trees and shrubs an **abscis-**

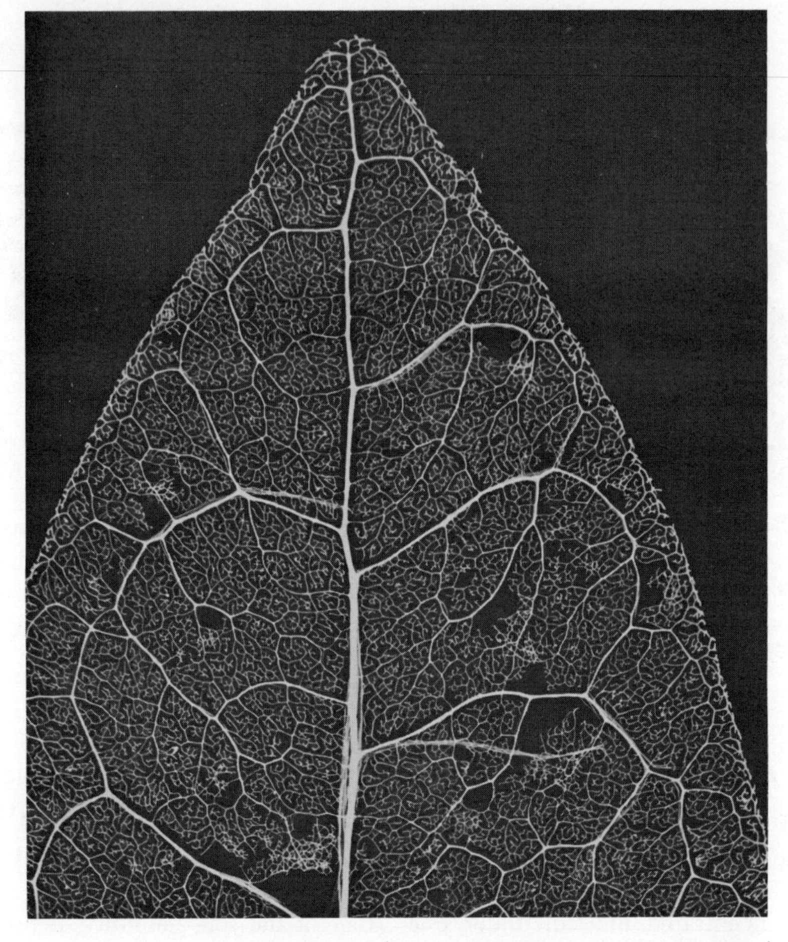

Fig. 8–9. Veins of a privet leaf. (R. T. Whittenberger and J. Naghski.)

sion layer (also called **separation layer**), several cells wide, at the base of the petiole (Fig. 8–12). In autumn the walls of these cells undergo chemical changes which result in their separation. The cell wall material breaks down into a gelatinous mass which no longer holds adjacent cells of the abscission layer together. The leaf is then held on only by the vascular strands, an attachment so delicate that a gust of wind will snap the strands and the leaf will fall. In a number of plants a layer of cork forms prior to leaf fall. The cork covers the scar left when a leaf drops; hence there is no open wound. This is certainly splendid surgery, for the scar is ready before the operation is completed. However, in other plants the cork does not form until after the leaf has been separated. In a few plants leaves are separated from the stem without the formation of a separation layer. In such instances the walls of ordinary cells at the base of the petiole weaken and separate.

In early autumn this author can observe from his office window a cottonwood which is green on one side and golden on the other. One side bears leaves with the green of summer; the other, leaves of autumnal color. The green is near the street light; the gold is away from it. Because the days are artificially prolonged by the light, the leaves near it are green; where the light in-

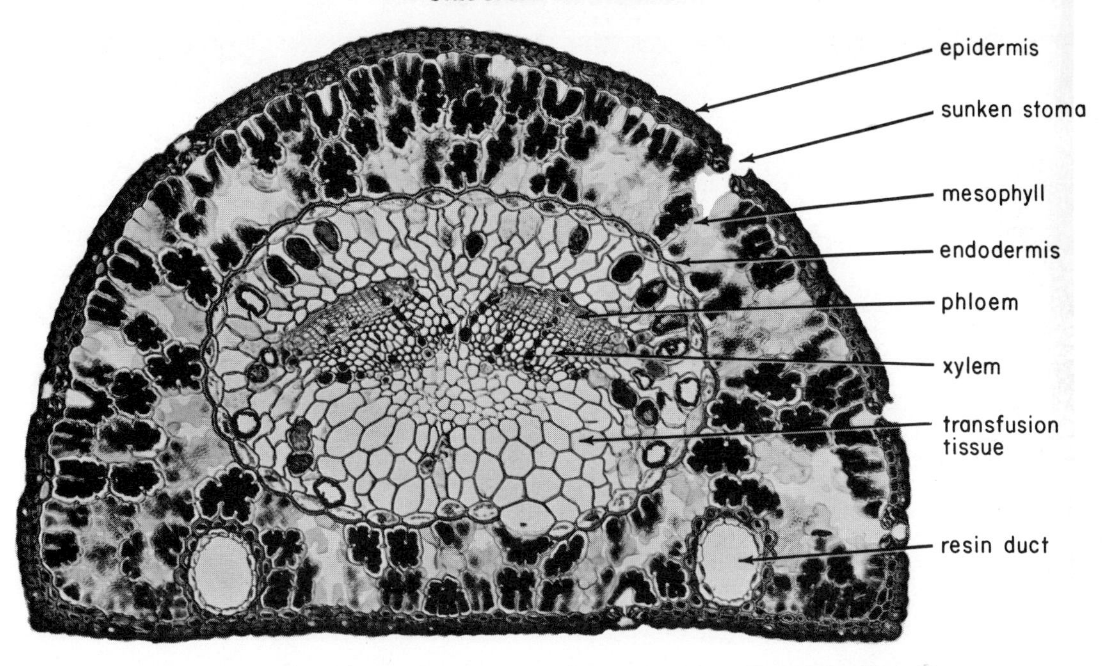

Fig. 8–10. Cross-section of a pine needle. (J. Limbach, Ripon Microslides.)

tensity is too low to influence the branches, they are yellow. Later in autumn the branches away from the light are bare of leaves, while leaves still cling to the branches near the light. This simple observation suggests that leaf fall is conditioned to a large extent by the length of day. When the leaves begin to fall, other events are taking place in nature—those which are also initiated by the short days of autumn. The brook trout in the streams color brilliantly, especially the males, and migrate to gravel beds and bars where nests are made and eggs are laid and fertilized. Here where trout were so scarce during the summer months, they are now so abundant that they literally churn the water when disturbed. Soon there will be a flight of northern ducks, and the junco as well as other birds will migrate southward. The ptarmigan in the high country above timberline will change from a mottled brown and black

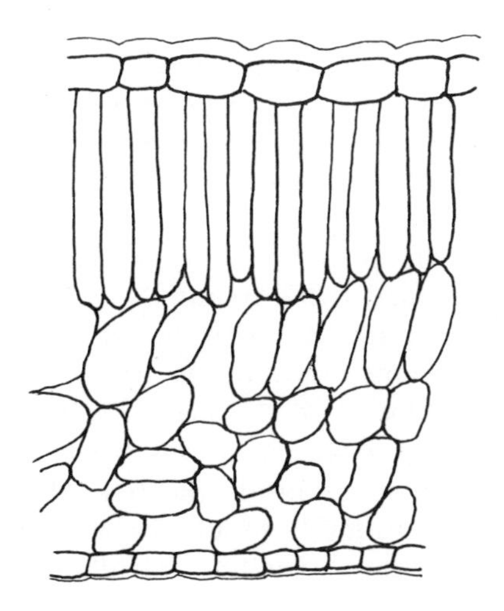

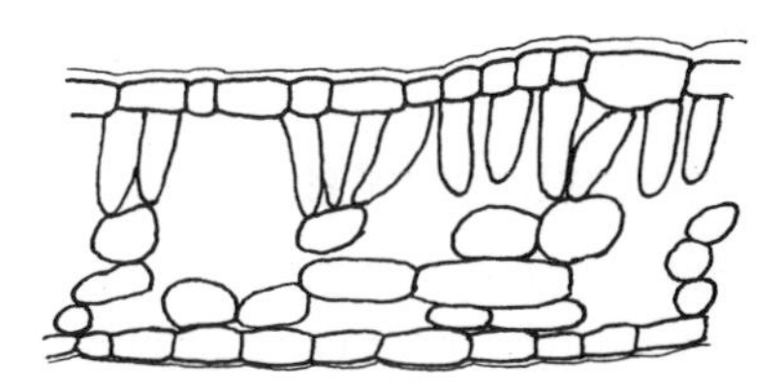

Fig. 8–11. Effect of light intensity on leaf structure. *Left*, leaf from the south periphery of an isolated maple tree (*Acer saccharum*). *Above*, leaf from the base of a maple tree which was growing in a forest. (Modified from H. C. Hanson, *Am. J. Botany*.)

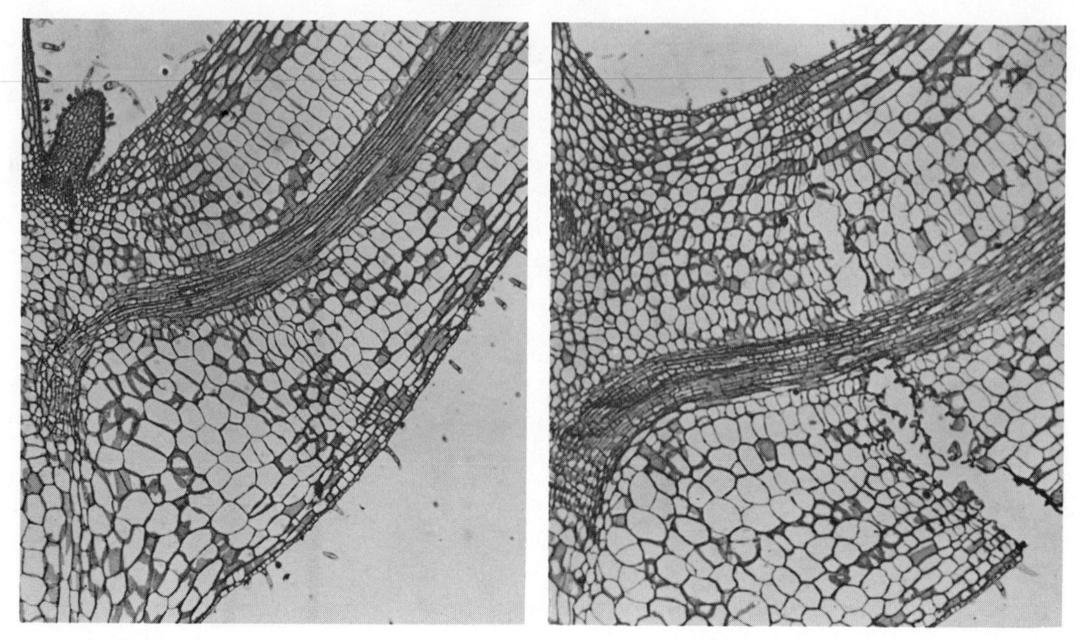

Fig. 8–12. Longitudinal sections of the base of the petioles of two *Coleus* leaves. *Left*, the abscission layer is formed. *Right*, abscission is nearly completed. (Richard A. Popham and T. J. Johnson.)

to white, and the reddish brown coat of the bloodthirsty weasel will change to white. In the garden aphids and spider mites lay their dormant eggs. These are all responses to the short autumnal days and not primarily to diminished temperatures. A striking feature of these adaptations is that the response is made before severe winter actually comes. The blasts of winter do not catch our native plants and animals unprepared. The short days of fall bring about changes having survival value.

SPECIALIZED LEAVES

Leaves are not always flat expanded structures which carry on photosynthesis. Some leaves are modified in form and function, carrying on such activities as support, food storage, or protection against drought, and some even trap insects for supplementary nutrients. From their shapes and functions you could hardly recognize some of these specialized structures as leaves, but their position on the stem indicates that they are leaves. Each is located just below an axillary bud, and a structure occupying this position is a leaf regardless of its shape. **Bud scales** are small modified leaves which protect the delicate inner tissues of buds of trees and shrubs from desiccation.

In barberry and certain species of cacti the **spines** are specialized leaves. In the black locust, acacias, and crown-of-thorns the spines are modified stipules. The hornlike spines of certain Mexican and Central American acacias are large and hollow and are tenanted by fierce stinging ants that make a small hole for their entrance and exit near one end of the spine (Fig. 8–13). Glands on the petiole secrete abundant nectar, and the tips of many leaflets bear pear-shaped bodies (called **Belt's corpuscles**), about $\frac{1}{12}$ inch long, that are rich in oil and protein. The ants forage on the honey and nutritious globules. If the tree is molested by man or if predacious insects, such as the leaf-cutting ants, invade the tree, the ants swarm out furiously to attack the intruders with their stings. The tree provides the ants

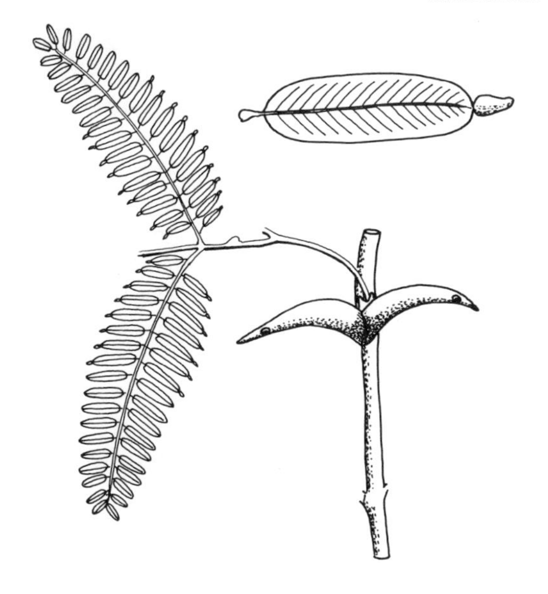

Fig. 8–13. In *Acacia* the hollow spines, modified stipules, house an army of ants which protect the tree from predacious insects. The ants feed on nectar and Belt's corpuscles which are produced at the tips of the leaflets.

with room and board, and the ant army in turn protects the tree from invaders.

In some plants, leaves or parts of leaves are specialized as **tendrils** which twine about objects and thus support the shoot. In clematis and nasturtium the petioles of ordinary foliage leaves act as tendrils; in sweet peas and garden peas the leaflets are modified into twining structures (Fig. 8–14). The thick fleshy structures of onion, tulip, and other bulbs are modified leaves which store food. In *Sarracenia* and *Nepenthes* the leaves are attractive pitcher-like structures which attract insects by their color and odor. In the bottom of each pitcher there is a small pool of fluid which contains enzymes that digest the trapped insects. From the bodies of insects these **carnivorous plants** secure minerals and supplementary food, but the plants are not wholly dependent on insects for minerals and food. Some minerals are secured from the soil in which they are rooted, and the leaves have

Fig. 8–14. The tendrils of sweet pea are modified leaflets.

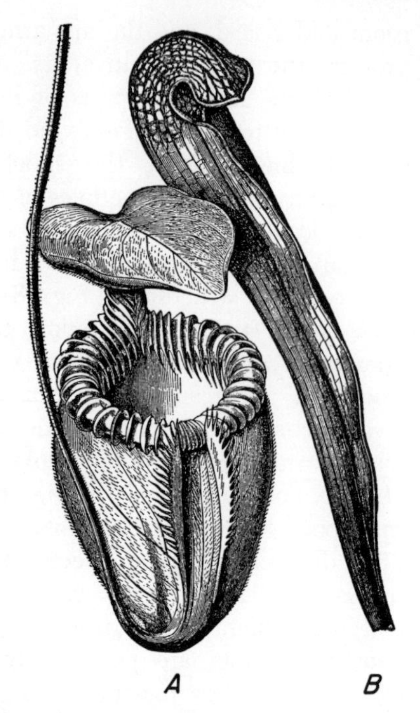

Fig. 8–15. The pitcher-shaped leaves of *Nepenthes villosa* (*A*) and *Sarracenia variolaris* (*B*) capture insects. (After Kerner.)

chlorophyll and hence carry on photosynthesis (Fig. 8–15).

USES OF LEAVES

Man is not primarily a leaf-eater. He prefers to get his food in a more palatable and concentrated form than it occurs in leaves of grass and other forage plants. He relies on cattle and sheep to convert the forage into meat. Certain leaves are consumed directly by man. The leaves of lettuce, chard, cabbage, Brussels sprouts, watercress, and spinach, and the petioles of celery and rhubarb are palatable, and they are excellent sources of minerals and some vitamins which are essential in our diets. Man eats for pleasure as well as for nutrition and prefers to have his food attractively seasoned. The leaves, or products secured from leaves, of some plants are used for flavoring. Sage, bay, marjoram, parsley, thyme, peppermint, and spearmint are leaves or leaf products. You are certainly familiar with the use of tea leaves and tobacco leaves.

Some fibers, drugs, and waxes come from leaves. The leaves of certain species of *Agave* yield sisal hemp, a fiber used in making rope and twine. Drugs obtained from leaves include cocaine, to relieve pain; digitalis, a heart stimulant; belladonna, to dilate the pupil of the eye; and such ones as witch hazel, senna, and pennyroyal. Most commercial waxes, such as Carnauba wax, come from the cuticles of leaves and are obtained by simply immersing the leaves in hot water. The wax melts and floats to the surface. Carnauba wax is an ingredient in many products, such as floor and automobile waxes, cosmetics, carbon paper, and phonograph records.

QUESTIONS

1. _____ The three parts of a leaf are (A) leaf, petiole, stipules; (B) blade, stem, stipules; (C) blade, petiole, stipules; (D) blade, petiole, axillary bud.

2. _____ The leaves of corn and wheat are (A) parallel-veined, (B) palmately net-veined, (C) pinnately net-veined.

3. Why are the leaflets of a rose not considered separate leaves?

4. _____ If there are two leaves at a node, the arrangement is (A) alternate, (B) opposite, (C) whorled.

5. Diagram a cross-section of a leaf and label the parts (one part is the upper epidermis).

6. _____ Chloroplasts are not present in (A) spongy mesophyll cells, (B) palisade mesophyll cells, (C) ordinary epidermal cells, (D) guard cells.

7. _____ If the leaves on a certain plant are oppositely arranged, the axillary buds will be (A) opposite, (B) alternate, (C) whorled, (D) unknown.

8. _____ Transpiration, loss of water as vapor, from leaves of the olive would be more reduced (A) if the upper surface of the leaf were coated with Vaseline, (B) if the lower surface were coated with Vaseline.

9. ____ If the following events occur in a pair of guard cells, the stoma will (A) open, (B) close, (C) neither open nor close.

1. Carbon dioxide and water used in photosynthesis.
2. Decreased acidity of cell sap.
3. Starch → sugar.
4. Increase in the osmotic concentration.
5. Uptake of water from surrounding cells.
6. Greater turgor in guard cells than in ordinary epidermal cells.

10. ____ Which one of the following statements about stomata is false? (A) Stomata are usually open during the day. (B) Stomata generally close during drought. (C) Stomata open when the CO_2 content of the substomatal cavity is greater than 0.03 per cent.

11. ____ Which one of the following statements about a vein is false? (A) The veins form the structural framework of a leaf. (B) Xylem and phloem are present in a vein. (C) Xylem conducts water and phloem conducts food. (D) The phloem occupies the upper half of a vein, that is, the part toward the upper epidermis.

12. Explain fully how the structure of a leaf is adapted to photosynthesis.

13. ____ Leaves readily fall off deciduous trees in the autumn because (A) the leaves are brittle, (B) an abscission layer is formed, (C) the temperatures are low, (D) the soil is dry.

14. ____ Which one of the following autumnal factors is most important in bringing about leaf fall, dormancy, bird migration, and spawning of brook trout? (A) Short days of autumn, (B) cool weather in autumn, (C) bright light of clear autumnal days.

15. Have leaves of some plants been modified to perform special functions? Give examples.

16. List some economically important uses of leaves and leaf products.

17. ____ Which one of the following does not come from leaves? (A) Sage, (B) peppermint, (C) cocaine, (D) opium, (E) digitalis, (F) belladonna.

18. Indicate whether the following terms are singular (S), plural (P), or both (B).

____ 1. Stoma.
____ 2. Stomata.
____ 3. Algae.
____ 4. Alga.
____ 5. Nuclei.
____ 6. Nucleus.
____ 7. Fungus.
____ 8. Fungi.
____ 9. Genera.
____ 10. Leaf.
____ 11. Species.
____ 12. Bacteria.
____ 13. Primordium.
____ 14. Stimuli.

---

19. Place the number of the appropriate term from the column on the right in the space to the left of each statement.

____ 1. The layer of waxlike substance overlaying the epidermis.
____ 2. The elongated cylindrical cells below the upper epidermis.
____ 3. The loosely arranged somewhat round cells toward the lower leaf surface.
____ 4. A pore surrounded by two guard cells.
____ 5. One of the divisions of a compound leaf.

1. Leaflet.
2. Cuticle.
3. Spongy mesophyll.
4. Palisade mesophyll.
5. Stoma.
6. Vein.
7. Bundle sheath.
8. Stipule.

20. Complete the following table.

Leaf Structure	*Principal Function*
1. Cuticle.	1. ____
2. Epidermis.	2. ____
3. Chloroplasts.	3. ____
4. Spongy mesophyll.	4. ____
5. Palisade mesophyll.	5. ____
6. Xylem.	6. ____
7. Phloem.	7. ____
8. Guard cells.	8. ____

9

Water

There is something restful and appealing about water—a brook, a lake, an ocean. But water can be frightening too—a driving blizzard; a cloudburst, washing away soil, tearing out bridges, flooding towns; a storm at sea. Man struggles to control excess water, to prevent floods, and to drain land so that the soil will contain both air and water. Man also struggles to compensate for a deficiency of water. As the world population grows, as more lands are irrigated, as industry spreads into new regions, and as more homes are built, the world's thirst becomes greater and greater.

In many areas of the world water is now scarce and becoming scarcer. In arid regions the water table is dropping, and the underground water supply is dwindling. The sea is a tremendous source, and our scientists have made progress in converting sea water into fresh water. The cost is still too high for commercial use, but costs are likely to become less as new methods for removing the salt are developed. In the future, atomic energy may be used to pump water through extensive pipeline systems from regions of surplus water to areas having a scarcity.

All land life depends on the water cycle—the hydrological cycle. The sun desalts ocean water effectively. Water vapor from the oceans, the major source of the land's water, is carried by winds over the land. As the air ascends mountains or is lifted in frontal disturbances or rises in unstable air masses, the air expands, it is cooled, clouds form, small water droplets become larger, and rain or snow falls on the land.

Most of the world's water is present in the oceans and inland seas—a lesser amount in icecaps and glaciers, a still smaller volume as liquid fresh water, and least as water vapor in the atmosphere (Table 9–1). Some of the liquid fresh water is in rivers, streams, and lakes; another fraction is in the soil; and a third is groundwater which may be shallow or deep, even ½ mile below the surface.

Table 9–1

The World's Water

Source	Per Cent
Oceans and inland seas	97.2
Icecaps and glaciers	2.15
Liquid fresh water	0.63
Atmosphere	0.001

The total income of water as rain, snow, sleet, and hail for the continental United States, excluding Alaska, is estimated to be 1430 cubic miles of water each year. Of this amount about 1000 cubic miles are

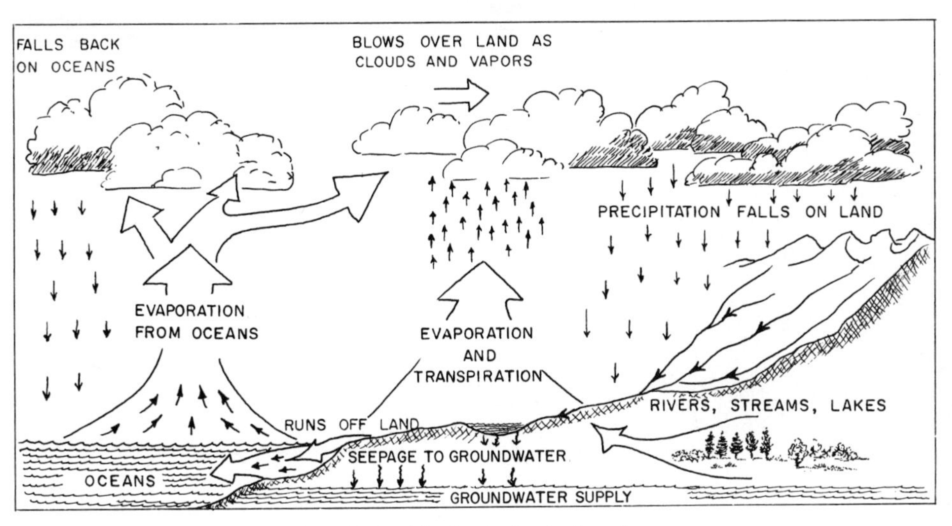

Fig. 9–1. The hydrological cycle.

returned to the air through evaporation and transpiration. A small fraction of the water evaporated and transpired may later fall on the land as rain or snow. However, much does not. The air made moister by transpiration and evaporation is carried back to the ocean where it again becomes charged with moisture and awaits circulation back to the land. In other words, the great bulk of the precipitation on land comes from water vapor originating from oceans. The rivers return about 390 cubic miles of water to the oceans yearly. The remaining 40 cubic miles go underground; some of it seeps into rivers and lakes and ultimately returns to the oceans (Fig. 9–1).

Even though an understanding of hydrology is vital to people everywhere, the science has been a neglected one. To stimulate interest in hydrology and to promote research and understanding in this field, an International Hydrological Decade began throughout the world on January 1, 1965.

Long, long ago life originated in a water habitat; and since the beginning, water and life have been inextricably bound together. We have seen that active protoplasm contains 85 to 90 per cent water, and that there are more water molecules in it than all others combined. There are about 18,000 water molecules in active protoplasm for every one of protein and for every 100 molecules of inorganic substances. Resting cells, those of seeds for example, have a much lower percentage of water. If reactions are to go on rapidly, cells must be nearly saturated with water. Water plays many roles in the life of green plants. We have already learned that water is a raw material for photosynthesis, and that water intake by cells makes them turgid. Water is necessary for growth and for the formation, solution, and transportation of many substances. Digestion of starch, proteins, and other compounds occurs only in the presence of water. Water is the medium in which all cellular reactions occur. Minerals enter roots only when dissolved in water and are conducted upward in this medium. The sugar formed in leaves moves out of them in an aqueous medium. Land plants retain and use some of the water taken up through their roots, but they lose a much greater amount through evaporation from their aerial parts, especially the leaves.

TRANSPIRATION

The wet walls of mesophyll cells are constantly exposed to air in intercellular spaces. Accordingly, water evaporates from the wet walls, accumulates as vapor in the

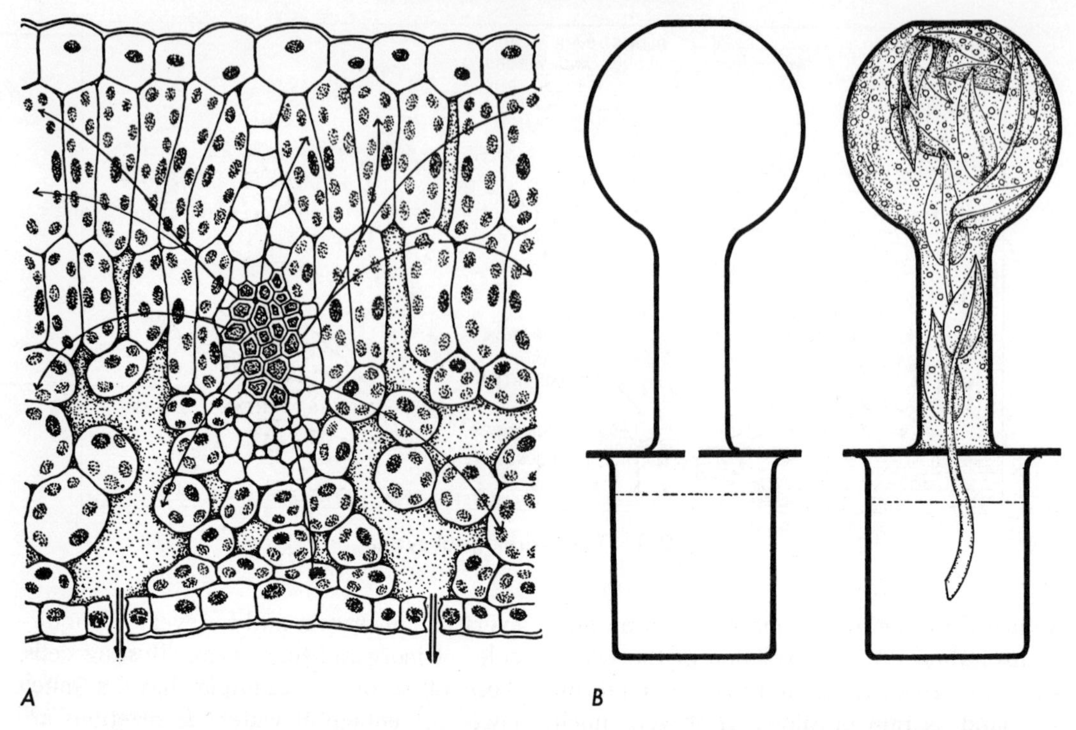

Fig. 9–2*A*. Transpiration from a leaf. Water passes from the vein to the leaf cells. Water evaporates from the cell walls and accumulates as vapor (shown as dots) in the spaces between the cells. In stomatal transpiration the vapor escapes through open stomata. *B*. Demonstration of transpiration. The transpired water condenses on the inner surface of the flask (*right*). No water droplets form on the control flask.

air spaces, and diffuses out of the leaf. This loss of water as vapor is referred to as **transpiration** (Fig. 9–2*A*). Most of the vapor escapes through open stomata, but some may diffuse through the cuticle. If the water escapes through stomata, the process is known as **stomatal transpiration,** whereas if it diffuses through the cuticle, it is referred to as **cuticular transpiration.** Even leaves with a thick cuticle will have some cuticular transpiration, possibly through minute breaks in the cuticle. In most species less than 10 per cent of the water vapor escapes through the cuticle, the remainder escaping through open stomata. Transpiration is not limited to leaves; twigs, fruits, flowers, tubers, and other plant parts also transpire.

Transpiration can be easily demonstrated by placing an inverted flask over a plant shoot with the cut end extending through a hole in a piece of cardboard into a beaker of water (Fig. 9–2*B*). A similar setup without a shoot serves as a control. In a short time drops of water condense on the inner wall of the flask which surrounds the plant, but not on the control flask. The water is given off as vapor from the leaves and then condenses on the flask.

The amount of water transpired by a plant may be obtained by weighing the plant and the container in which it is growing at intervals. The container and the soil surface are covered with foil to prevent evaporation. Assume that a plant and its container weigh 1000 grams at 8 A.M. and 992 grams at 12 noon. In the 4-hour period the plant transpired 8 grams of water, or 2 grams per hour. Next the leaf area should be determined so that the results can be expressed as units of water transpired per unit time

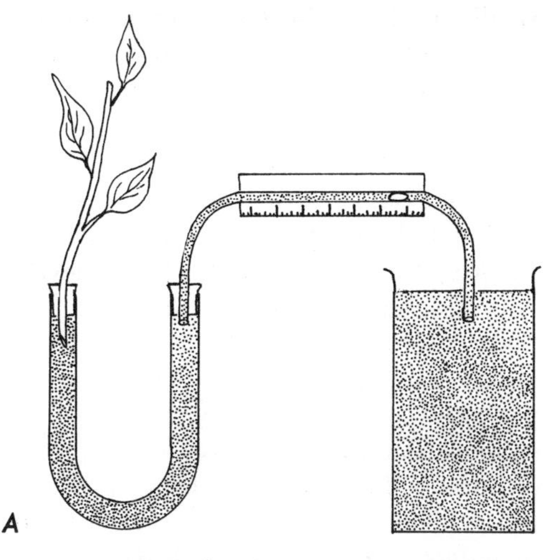

Fig. 9–3*A*. The rate of transpiration may be determined by using a potometer. *B*. Light, temperature, relative humidity, and wind influence the rate of transpiration.

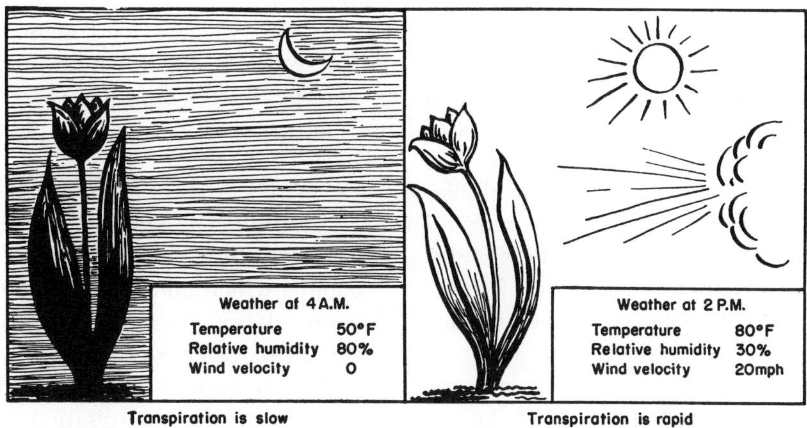

per unit area. Usual rates of transpiration vary between 0.5 and 2.5 grams per hour per square decimeter of leaf area.

A **potometer** such as the one illustrated in Figure 9–3*A* is useful for determining the effects of environmental factors on rates of transpiration. The potometer is filled with water. As water is transpired, the column of water in the capillary tube moves along, and air enters the end of the tube. After air has ascended about ¼ inch in the vertical part of the capillary, the end of the tube is immersed in water. As transpiration continues, the water column, now broken by an air bubble, continues to move up and along the capillary tube. The rate of transpiration is determined by noting the time required for the air bubble to move a certain distance.

An amazing amount of water passes through a plant in a season; it is absorbed by the roots, and much of it is lost as vapor through the leaves. A tree may transpire 50 gallons in a day. One large apple tree transpires 1800 gallons (close to 7 tons) in its growing season. In nature it sometimes happens that springs tapped by the roots of large trees dry up when the leaves come out. A single corn plant may transpire 400 pounds of water during the growing season, and an acre of corn plants, 1200 tons. This vast amount of water is absorbed by roots and moved through the stem into the leaves.

Weeds, like crop plants, transpire rapidly. One of the major reasons for cultivation is to eliminate the competition for water between weeds and crop plants. In some areas the annual precipitation is not sufficient for crop production. However, crops may be raised in such regions by **summer fallowing** and eliminating weeds. When the land lacks plants, water accumulates in the soil. In this system crop plants have 1 full year's precipitation plus that stored in the soil while the land was fallow.

The lavish use of water by plants is further evident from a consideration of the number of pounds of water used to form 1 pound of dry weight. Sorghum plants use 250 pounds of water for every pound of dry matter accumulated. The amount of water used by some other plants to accumulate 1 pound of dry matter follows: corn, 350 pounds; wheat, 500; potatoes, 636; alfalfa, 900. The units of water (in pounds, grams, tons, etc.) required to form a unit weight of dry matter (in the same system of measurements) is known as the **water requirement** or the **transpiration ratio.** For example, the water requirement of alfalfa, as seen above, is 900. The water requirement varies to some extent with environmental conditions because the rate of transpiration is influenced by a number of surrounding factors.

Let us see if the plant benefits from transpiration. The evaporation of water lowers the temperature of leaves 2° to 6° C. This may be of some value on very hot days. When leaves are exposed to bright sun, their temperature may be 10° C., or even more, above that of the surrounding air in spite of transpiration. Although transpiration plays a role in lowering leaf temperatures, other factors are far more important. When the temperature of a leaf is above that of the surrounding air, the leaf loses heat by conduction, convection, and radiation. If a leaf exposed to full sun did not give off heat by these means, its temperature would be at the boiling point in a few minutes.

A leaf receiving 2.1 calories of radiant energy per square centimeter per minute removes about 35 per cent of the heat by transpiration, 60 per cent by radiation, and about 5 per cent by convection. About 600 small calories of heat will be dissipated for each gram of water transpired. If transpiration is checked by coating the leaves with Vaseline, the leaf temperature rises 2° to 6° C., and then the dissipation of heat by radiation and convection increases. Such augmented radiation and convection keeps the leaf from rising to a still higher level.

It has been suggested that transpiration plays a role in bringing minerals and water to the leaves, but the evidence for this is not clear-cut. It may, at first thought, seem strange that the process of transpiration is of little value to the plant, but it should be recalled that leaves are primarily food-manufacturing organs. If leaves are constructed for efficient food manufacture, the loss of water is inevitable because the walls of mesophyll cells must be wet, and stomata must be present in order for photosynthesis to occur.

Environmental Factors Affecting Transpiration

Factors which affect evaporation influence the rate of transpiration. Clothes on a line dry rapidly when the relative humidity is low, the temperature high, and the day windy. Likewise, transpiration in plants is rapid under such conditions (Fig. 9–3*B*). The rate of transpiration is also influenced by light and soil factors.

Relative humidity. When air is dry, the relative humidity is low; when moist, it is high. Relative humidity is the amount of water vapor the air actually has in it compared to the amount it could hold at that temperature. The result is given as a percentage. If the relative humidity is 100 per cent, the air is saturated; and if a leaf is at the same temperature as the air, transpiration does not occur. If other environmental conditions remain constant, the rate of transpiration is inversely proportional to the relative humidity. Transpiration is rapid at low relative humidities, and slow at high ones.

Temperature. Because evaporation occurs more rapidly at high temperatures than at low ones, transpiration increases as the temperature rises. Furthermore, air at high temperatures can hold more water vapor than air at low temperatures. For instance, air at 90° F. can hold about twice as much water vapor as air at 70° F. If the air at 70° F. is completely saturated, the relative humidity is 100 per cent. But if the temperature rises to 90° F., that amount of water is no longer enough to saturate the air; in fact, it will result in a relative humidity of approximately 50 per cent. For every rise of 20° F. the relative humidity is about halved, provided no moisture is added to or subtracted from the air. Conversely, as the temperature goes down, the relative humidity increases. During the day when temperatures are high, the relative humidity is low and transpiration is rapid. As the temperature decreases during the night, the relative humidity increases and transpiration diminishes. If the drop in temperature is great enough, the relative humidity becomes 100 per cent; fog or dew will then form, and transpiration will cease.

Vapor pressure deficit. When other conditions are constant, light and wind, for example, transpiration is proportional to the vapor pressure deficit of the air, which is the difference between the saturation vapor pressure and the actual vapor pressure. The saturation vapor pressure is the amount of vapor the air can hold, and this value varies, as we have seen, with temperature. The saturation vapor pressure of air at 20° C. is 17.54 mm. of mercury; and at 30° C., 31.82 mm. of mercury. Assume that on a certain day the temperature is 20° C. and the relative humidity is 25 per cent. The actual vapor pressure is 25 per cent of 17.54, or 4.38, and the vapor pressure deficit is 17.54 minus 4.38, or 13.16. Assume that on a second day the temperature is 30° C. and the relative humidity is 25 per cent. Our problem is to determine how much faster transpiration would be on the second day than on the first. On the second day the actual vapor pressure is 25 per cent of 31.82, or 7.95, and the vapor pressure deficit is 31.82 minus 7.95, or 23.87. On the second day transpiration would occur 23.87/13.16, a little less than two times faster than on the first day when the temperature was 20° C.

Wind. If the air is still, the relative humidity of the air immediately outside the stomata is higher than the surrounding air. If the day is windy, the moister pockets of air are replaced by the drier air of the atmosphere, and transpiration is augmented. The wind velocity in an orchard or other place may be reduced by planting trees along the border, that is, by a windbreak. The windbreak not only lessens transpiration, but also reduces the injurious effects of wind action on growth.

Although wind generally increases transpiration, it may, under certain conditions, result in a decrease in transpiration. Hot dry winds may bring about the closure of stomata with a consequent decrease in transpiration. Furthermore, movement of air over a leaf surface may carry heat from the leaf; the lowered leaf temperature may result in a decreased rate of transpiration.

Light. Light brings about the opening of stomata and furnishes the energy for evaporation. Furthermore, light increases leaf temperature, which results in a greater diffusion gradient. Hence, transpiration is more rapid during the day than during the night as can be seen in Figure 9-4. The graph also shows the amount of water absorbed throughout the day. You will notice that during the night, absorption exceeds transpiration; whereas during the day, transpiration exceeds absorption.

Soil factors. Transpiration is reduced when water is not available to the plant, as during periods of drought and when the soil is frozen or at a temperature so low that water is not absorbed by roots. Both absorption and transpiration go on slowly in plants growing in saline soil and in soils which are poorly aerated. The diminished

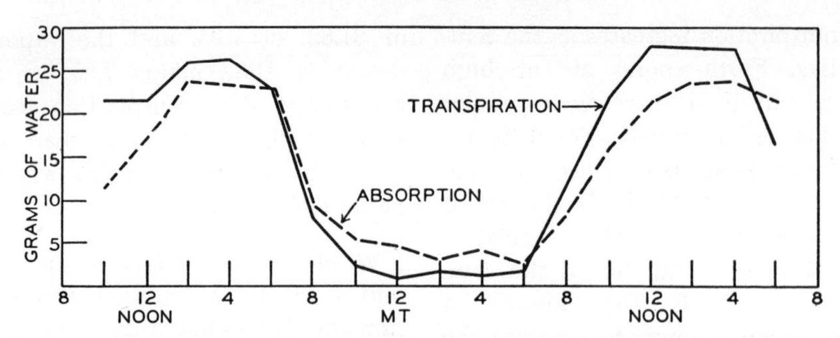

Fig. 9–4. Comparative daily periodicities of transpiration and absorption of water in the loblolly pine (*Pinus taeda*). (From *Introduction to Plant Physiology*, by B. S. Meyer, D. B. Anderson, and R. H. Böhning. Copyright © 1960, D. Van Nostrand Co., Inc., Princeton, N.J. Data of Kramer, *Am. J. Botany*.)

transpiration resulting from decreased absorption results, at least in part, from stomatal closure.

Physiological and Structural Features of Leaves Which Affect Transpiration

Some plants have a number of structural and physiological characteristics which enable them to survive drought.

Leaf area small. Fir, pine, and spruce have a relatively small leaf area, and accordingly the amount of water transpired is low. Cacti and other desert plants lack foliage leaves, and therefore the amount of water transpired is very small. Although photosynthesis occurs in the stems, the area of these is relatively small; therefore, little food is made and growth is very slow.

Periodic reduction of leaf area. The annual shoots of many perennial plants die at the end of the growing season, leaving tubers, bulbs, or similar living parts in the soil. In temperate latitudes most trees and shrubs drop their leaves in the autumn. In tropical regions where a dry season alternates with a moist one, the trees and shrubs shed their leaves at the onset of the dry season, resulting in a landscape that resembles autumn in temperate zones. Even though temperatures are warm, there are more golds in the landscape than greens. Such adaptations markedly reduce the transpiring area and hence reduce water loss. In northerly latitudes winter is a period of severe drought. The soil water may be frozen and not available to roots, or the soil temperatures may be so low that even liquid water is absorbed with difficulty.

Temporary reduction of leaf area. When exposed to air and illumination, the leaves of some plants turn their edges toward bright sun. The leaves of others, corn and many grasses, roll up during drought.

Structural modifications. A thick cuticle, the development of more than one layer of epidermal cells, stomata sunken in pits or grooves, and a compact arrangement of mesophyll cells are features which reduce the rate of transpiration. These modifications are found in plants, such as the oleander (Fig. 9–5), that live natively where drought occurs frequently or at definite seasons.

Water storage tissue. Plants such as cacti, agave, and sedum have tissues in which water is stored. Mucilaginous substances which retain water avidly are present in the cells of such plants. Plants with fleshy stems and/or leaves in which water is stored are commonly called succulents. The strange and varied forms of the succulents, the rare beauty of some of their flowers, and their adaptations for desert life make them appealing plants. The plants are exposed to scorching sun interrupted only oc-

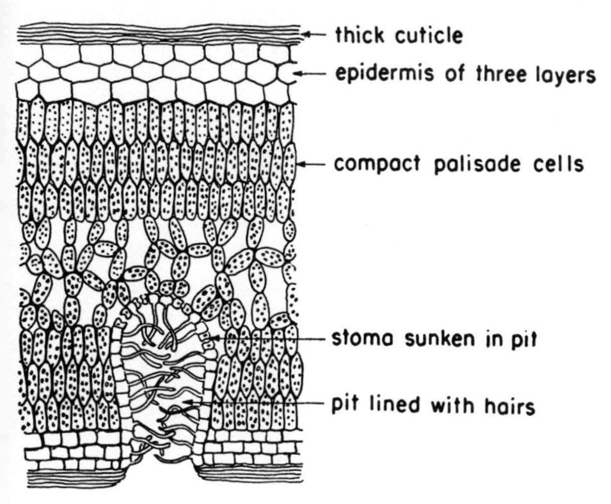

Fig. 9–5. An oleander leaf has many features which reduce transpiration.

casionally by heavy rains. After a rain the extensive roots absorb water, which is then stored in fleshy stems or fleshy leaves or both. Cacti are the most conspicuous succulents of our deserts. But there are plants other than cacti which have the succulent habit, and in the deserts of India and Africa they are the dominant plants.

Distantly related plants have evolved similar forms and structures which store water; thus we find plants of a succulent habit in the cactus, lily, amaryllis, crassula, milkweed, and spurge families. Such evolution is referred to as **convergent evolution,** and in this example has led to similar adaptations to a xerophytic way of life.

Resistant protoplasm. The protoplasm of some plants, the creosote bush (*Larrea tridentata*), for example, is able to survive considerable water loss without being killed. During drought periods, the water content of the leaves on a fresh-weight basis may be reduced to 33 per cent without injury to the plants. In contrast, the water content of mesophyte leaves varies from 66 to 76 per cent. Leaves of many mesophytes, bean for example, are killed when drought is prolonged. The ability of protoplasm to survive prolonged desiccation is probably the major feature of drought-enduring plants.

BALANCE BETWEEN ABSORPTION AND TRANSPIRATION

When more water is taken up by roots than is transpired, water is available for photosynthesis, growth, and other processes. With this favorable water balance plants thrive, provided, of course, that other environmental factors are favorable. On the other hand, when transpiration exceeds absorption, growth ceases and certain plants, sunflower for example, wilt. Furthermore, prolonged wilting may bring about the dropping of leaves, flowers, and fruits, with a consequent decline in yield, in such plants as apple, almond, cotton, cucumber, orange, tomato, and watermelon. If the unfavorable balance between uptake and loss continues for too long, drought-susceptible plants die.

Evergreen trees may be injured during winter because water is transpired faster than it is absorbed from frozen soil. Trees near timberline are injured year after year. Injury is especially severe in the early spring when the tops that extend above snow level are exposed to the warming winds, the dry air, and the bright sun. Transpiration goes on relatively rapidly through the tops, but the roots can absorb very little water from the still frozen soil. The deep snowdrifts that cover the lower parts of the tree protect them from drying. Each year the tops are killed back, while the lower parts survive, and the dwarfed and matted trees that result are what we call wind timber.

When trees, shrubs, and herbs are transplanted, transpiration may exceed absorption until new roots are formed. The injury may be minimized by transplanting deciduous trees and shrubs in the fall or spring when the plants are leafless, by leaving a ball of earth around the roots, by transplanting on cloudy or rainy days, and by covering the newly set plants. Another method is to spray plants with a latex plastic.

In the summer on hot, bright, windy days, plants may temporarily wilt even though the soil is abundantly supplied with water, be-

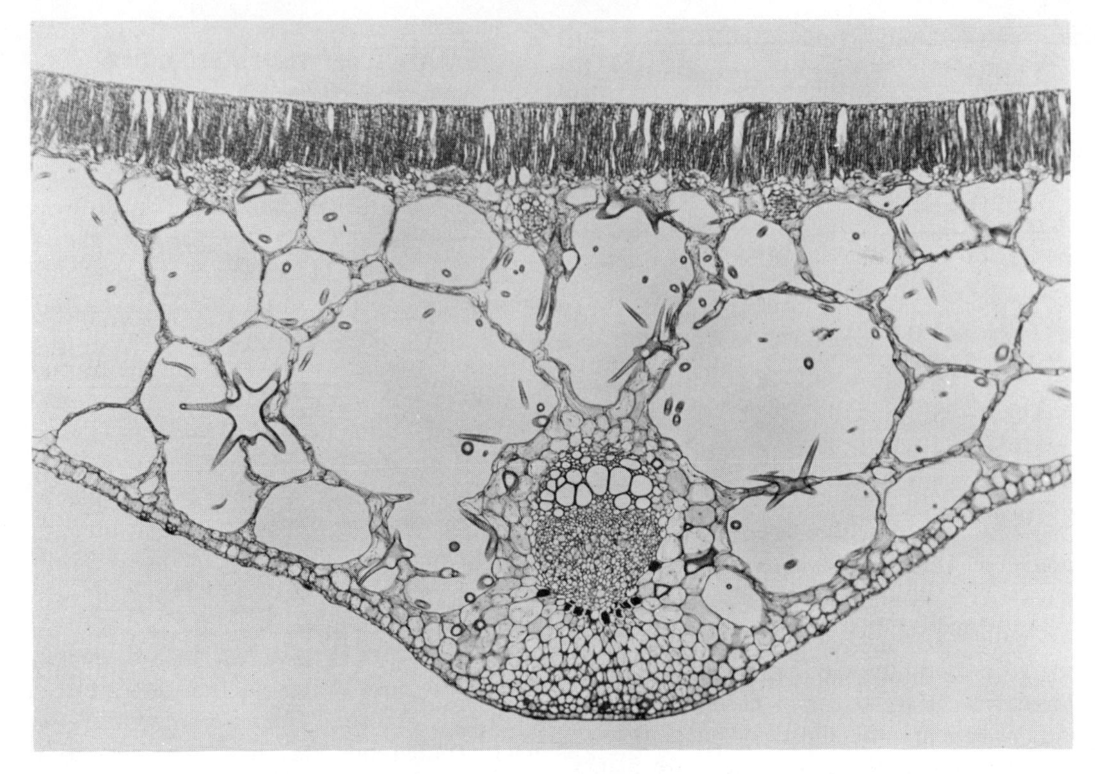

Fig. 9–6. A leaf of a water lily. Note the aerenchyma tissue with its large air spaces. (J. Limbach, Ripon Microslides.)

cause absorption and conduction cannot keep pace with transpiration. Such plants usually recover at night when transpiration is reduced. In greenhouses, growers maintain a favorable water balance by shading the glass and by wetting walks, stagings, etc., to increase the relative humidity and thereby lower the rate of transpiration.

CLASSIFICATION OF PLANTS ON BASIS OF WATER SUPPLY

The language or accent of a man reveals his geographical origin. The structure of a plant reveals, if not the country, at least the kind of water supply it had in its original home. You know that some plants grow submerged in water, others in deserts, and others where the soil is uniformly moist during the growing season. Plants that grow submerged, partly submerged, or floating in lakes, ponds, streams, or marshes are called **hydrophytes.** Water lilies and pickerelweed are typical examples. Hydrophytes have a scanty root system, a small amount of water-conducting tissue and supporting tissue, and an abundance of tissue for conducting and storing air, so-called **aerenchyma.** A leaf of a water lily (Fig. 9–6) illustrates typical features of many hydrophytes. In this floating leaf stomata are present only on the upper epidermis, below which there is a layer of palisade cells. Large air spaces make up most of the chambered spongy tissue (the aerenchyma). The star-shaped cells scattered in the spongy tissue are sclereids.

At the other extreme are the plants that live where drought is frequent, for example, cacti, sedums, sagebrush, oleander, and yucca. These are called **xerophytes.** Such plants have a small leaf area, a thick cuticle, sunken stomata, and resistant protoplasm. A number of them have water-storage tissues. Strangely, some xerophytes are found in tropical forests where precipitation is high and uniform throughout the year. These xerophytes do not grow in the soil, but instead they are perched on trunks and branches of trees where they are cut off from soil moisture. These xerophytes are known as **epiphytes,** and they are not parasitic. Their roots do not penetrate the branch, but instead ramify in the humus that accumulates on the branches. The familiar orchids seen in florists' shops are epiphytes, as are many members of the pineapple family, commonly called **bromeliads.** The leaves of bromeliads are arranged to form a reservoir in which rain water accumulates. Such water is used by the bromeliads and provides a home for many aquatic animals such as mosquito larvae and polliwogs of tree frogs. In many regions bromeliads form the aerial swamps of the tropics. It may seem strange to speak of aerial swamps because we usually associate swamps with flooded land surfaces and not with trunks and branches of trees. In certain regions the control of mosquitos with insecticides has been difficult because the bromeliads were out of reach of the sprays.

Plants which grow where the habitat is neither extremely wet nor very dry are called **mesophytes,** "middle plants." You can tell them by their profusion of leaves, which are usually thin and expanded and subject to wilting. They develop in regions where the water supply is fairly regular and where the soil is uniformly moist during the growing season. Sweet peas, roses, begonias, beans, cotton, and tobacco are representative mesophytes.

For a more detailed discussion of hydrophytes, mesophytes, and xerophytes, see Chapter 26.

WATER DETERMINES THE TYPE OF VEGETATION

Among all the factors necessary to plant life, you cannot point to water or to any other single factor and say, "This is the more important, or this, or this." Each factor has its specific relation to the plant, and often must be described by itself, but each factor acts in response to, and in accordance with, all the other factors. All are interdependent. A plant requires oxygen, carbon dioxide, light, warmth, minerals from the soil, and water, and cannot survive long if any one factor is removed. Oxygen and minerals are of no use if the plant cannot have light and carbon dioxide, and a plant can use none of these if water is absent. Certainly, water alone is of no use without the others. Even all these factors together cannot function if the temperature is too low or too high to allow the natural processes of life to be carried on.

Yet among the factors, water is a determining agent. The amount of water in the soil and its distribution through the year make the difference between desert and jungle, between sagebrush deserts and rich farming land.

Precipitation is unevenly distributed over the earth's surface. In extremely arid regions there are periods of a year or more in which there is no precipitation. If such lands are irrigated, often they are productive. For example, valleys along the Peruvian Coast which are irrigated with water from the high Andes produce large yields of sugar cane and other crops. Arid regions, where the annual precipitation is less than 15 inches a year, characteristically support desert vegetation and can be farmed only when the lands are irrigated. Grasslands occur in semiarid regions, where the annual precipitation is between 15 and 30 inches; in the moister grasslands, crops can be raised without irrigation. Forests develop in humid regions, where precipitation is in excess of 30 inches a year.

The precipitation limits for the development of deserts, grasslands, and forests represent average values. They vary with latitude and altitude. For example, in northerly latitudes or at high elevations forests may develop where the precipitation is 20 inches a year, and grasslands may oc-

Fig. 9–7. Guttation from a strawberry leaf. (Courtesy of J. Arthur Herrick, Kent State University.)

cur where the precipitation is 10 inches annually. In northerly latitudes and at high altitudes transpiration and evaporation are less than in tropical latitudes and at low elevations. Where transpiration and evaporation are low, plants use water more effectively than where these processes take place rapidly.

GUTTATION

Most water is lost from plants as vapor. However, some water may be given off in liquid form, a process called **guttation** (Fig. 9–7). Guttation occurs in some plants when absorption of water exceeds transpiration, the surplus water being exuded through **hydathodes.** Hydathodes are modified stomata, located at the ends of veins. Unlike the usual stomata, hydathodes remain open day and night. The liquid water appears as glistening drops at the tips of veins. Frequently, water of guttation is mistaken for dew, but it can be distinguished from dew by its regular arrangement on the leaves. Guttation can be readily demonstrated by covering a pot of corn or wheat seedlings with a bell jar. In time the air will become nearly saturated, and then absorption will exceed transpiration, resulting in the exudation of water at the vein tips.

QUESTIONS

1. How may the world's thirst for pure clean water be alleviated?

2. Outline the hydrologic cycle.

3. Discuss the importance of water in the life of a plant.

4. Indicate by an X which of the following practices are related to transpiration.

1. _____ Salting of nuts.
2. _____ Use of trees as windbreaks.
3. _____ Removal of weeds.
4. _____ Food preservation by salting or drying.
5. _____ Placing cellophane caps over newly planted seedlings.
6. _____ Hilling soil around rose shrubs in the autumn.
7. _____ Wrapping cut flowers in waxed paper.
8. _____ Fine spray of water over vegetables in a supermarket.
9. _____ Packing roots of nursery stock in moist sphagnum moss.

5. _____ The water requirement would be least for a plant growing in (A) moist soil of low fertility, (B) moist soil of high fertility, (C) soil so dry that little water is available to the plant.

6. _____ The units of water required to form a unit of dry matter are known as (A) field capacity, (B) water requirement or transpiration ratio, (C) photosynthetic efficiency, (D) wilting percentage.

7. Discuss the significance of transpiration in the life of a plant.

8. Discuss four environmental factors that influence the rate of transpiration.

9. _____ An increase in which one of the following usually results in a decrease in the rate of transpiration? (A) Temperature, (B) relative humidity, (C) wind velocity, (D) light intensity.

10. _____ Most water is lost from plants by (A) stomatal transpiration, (B) cuticular transpiration, (C) lenticular transpiration, (D) guttation.

11. Describe two methods for determining rates of transpiration.

12. _____ When light, wind, and soil moisture are constant, the rate of transpiration is (A) proportional to the relative humidity, (B) proportional to the barometric pressure, (C) proportional to the vapor pressure deficit of the air, (D) unaffected by other environmental variables.

13. _____ During a typical day, (A) absorption exceeds transpiration when the sun shines brightly, (B) transpiration exceeds absorption during the night, (C) transpiration is greater than absorption during the early afternoon, (D) the transpiration rate is about the same during both day and night.

14. _____ Transpiration of deciduous trees and shrubs ceases when they are leafless during the winter months. (A) True, (B) false.

15. Check only those leaf anatomical characteristics of a plant that would transpire at a very slow rate:

_____ Thick cuticle.
_____ Many guard cells.
_____ Sunken stomata.
_____ Large broad leaf.
_____ Little spongy mesophyll.
_____ Much mechanical (thick-walled) tissue.
_____ Few intercellular spaces.
_____ Stomata on both leaf surfaces.

16. _____ A plant which grows in a very dry region would probably not have one of the following characteristics. Which characteristic would it not have? (A) Resistant protoplasm, (B) large air spaces between the mesophyll cells and elevated stomata, (C) stomata which are located in pits or grooves which may be partly filled with hairs, (D) capacity of the leaves to roll.

17. What are some common gardening practices which are primarily intended to reduce transpiration?

18. _____ Which one of the following should be moved with a large ball of earth around the roots? (A) Apple, (B) spruce, (C) lilac, (D) rose.

19. Why are trees at timberline dwarfed?

20. How could you classify plants on the basis of water supply?

21. _____ The loss of water as liquid, usually at the tips of the veins, is called (A) guttation, (B) photosynthesis, (C) respiration, (D) transpiration.

22. _____ The plants that form the aerial swamps in tropical regions are (A) orchids, (B) bromeliads, (C) terrestrial plants of great variety, (D) hydrophytes.

23. _____ Most epiphytes are (A) xerophytes, (B) mesophytes, (C) hydrophytes.

24. _____ A leaf of a water lily, a typical hydrophyte, is characterized by (A) aerenchyma tissue, (B) absence of stomata, (C) water-storage tissue, (D) absence of a cuticle.

25. _____ Plants growing in salt marshes or on moist soil high in salt content are (A) xerophytes, (B) mesophytes, (C) hydrophytes.

26. _____ In 1748 Guettard enclosed a branch in a spherical glass jar. After a brief interval he noted droplets of water uniformly distributed over the inner surface. The result indicates that (A) the water came from guttation, (B) plants give off small droplets of water through open stomata, (C) the water vapor lost by the shoot condensed on the wall of the container, (D) the shoot was respiring rapidly.

27. A botanist harvested the shoots of a group of experimental plants. The fresh weight of the shoots was 200 grams. After drying the shoots at a temperature of 70° C., they weighed 40 grams. What was the percentage of water on a fresh-weight basis? On a dry-weight basis?

10

Roots

Seed plants are fixed in position, and in nature they become located by chance. Some seeds fall where the soil is deep and fertile; others, where the soil is sparse and poor. A seed of a tree may land in a crack in otherwise solid rock. Here the seed germinates, and the roots extend into the crevice where wind-blown soil and organic matter have accumulated. In this rugged and picturesque place the tree stands, perhaps for 2000 or more years, until it dies (Fig. 10–1). Cultivated plants are pretty much pampered. The soil is carefully prepared, and the nutrient and water requirements of the plants are looked after. In the cultivation of plants the environment of the root may be greatly modified, but the environment of the shoot cannot be altered to any great extent.

FUNCTIONS OF ROOTS

If seed plants are to thrive, their immediate environment must furnish them with the requisites for survival, growth, and development: oxygen and carbon dioxide from the atmosphere, water and minerals from the soil. Water and minerals are absorbed from the soil by roots which ramify extensively through the soil. You will recall that if the roots of just one rye plant were placed end to end, they would extend 387 miles. Certainly trees also have extensive root systems. The water and minerals absorbed from the soil by roots are conducted by the xylem to all parts of the plant. Injury to roots by parasitic fungi and animal parasites interferes with absorption and, hence, with plant growth and development. Food is stored in the roots of many plants, in ordinary roots or in modified roots, such as those of the sugar beet, carrot, sweet potato, and dahlia.

Roots anchor plants and, in areas undisturbed by man, maintain them upright even during severe storms. Violent storms have battered old trees repeatedly, and they have withstood all onslaughts. No storm, war, or other disaster has felled the Arkewyke yew at Runnymede, England. This tree now measures 31 feet around, and in its presence the English barons compelled King John to sign the Magna Charta in 1215. The enormous and ancient redwoods in Muir Woods, across the bay from San Francisco, have survived many storms and earthquakes, quakes which have crumpled man-built structures. Moreover, during this long period, the trees have kept the soil in place so that it is now as deep and fertile as it was 2000 years ago when the present trees were seedlings. How different are things where man

Fig. 10–1. Located by chance, this yellow pine survives in a rugged environment.

has exercised control. Many ancient cities and pueblos have been completely buried by soil eroded from areas disturbed by man.

When winds blow violently, the tensile strength of the root system is tested. As the crown sways, the roots are alternately taut and loose like the cables of a ship at anchor in a gale. Like cables, roots have high tensile strength and are moderately flexible.

Some people believe that a too sheltered and easy life is not good for a man. The unsheltered tree, one constantly buffeted by winds, is the sturdy one. A few years ago an Australian scientist compared free-swaying trees with trees which had been staked so that they could not sway in the wind. The trees buffeted by the wind had larger trunks and sturdier roots than the tethered ones. After being staked for 2 years, the trees were no longer stable in a normal environment. After the stakes were removed, the first strong winds laid them low.

CHARACTERISTICS OF ROOTS

Not all underground plant parts are roots. Thin underground stems, called **rhizomes,** may be mistaken for roots. Of the fleshy underground plant parts, some are roots and others are stems. A carrot is a root, but an Irish potato is a stem.

In weed control practice and in plant propagation, it is essential that roots be distinguished from stems. If the shoot of a plant is severed from the roots with a hoe or cultivator, generally the plant does not survive. On the other hand, if the aerial shoot is severed from an underground stem, usually the plant survives because soon buds on the underground stem develop into new aerial shoots. Most plants can be propagated from portions of a stem, but only a few from pieces of a root. For example, portions of a potato can be used to propagate potato plants, but parts of a carrot cannot be used to start carrot plants.

The absence or presence of nodes and internodes can be used to distinguish roots from stems. While stems have nodes and internodes, these structures do not occur on roots. Leaves are attached at nodes, and a bud is present on the stem above each leaf. Although the leaves of underground stems may be mere scales, their positions can be recognized and the buds are usually evident. Buds and leaves are lacking on roots.

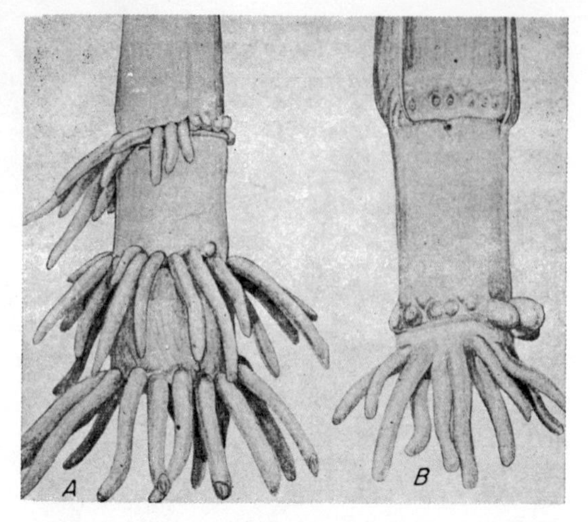

Fig. 10–2. Adventitious roots of a sorghum plant. *A.* Root crown with three whorls of adventitious roots in various stages of development. *B.* Base of young stalk with adventitious roots in early stage of development. (Ernst Artschwager, *U.S.D.A. Tech. Bull. 957.*)

ROOT SYSTEMS

When a seed germinates, the radicle develops into the primary root which soon produces branch roots that may in turn branch. The branches of the **primary root** as well as those developing from the branch roots are known as **secondary roots.** In certain species the primary root grows downward, and the secondary roots extend horizontally, thus forming a dense mat just below the soil surface. In other species the secondary roots grow horizontally for a relatively short distance and then angle downward. Sometimes roots originate from stems, less commonly from leaves, and such roots, that is, those originating from a structure other than a root or radicle, are called **adventitious roots** (Fig. 10–2). If a stem cutting of geranium, ivy, coleus, or other plant is placed in water or moist sand, adventitious roots develop at the base of the stem. From the petiole of an African violet leaf which is inserted in a rooting medium, adventitious roots originate. Roots that develop from modified stems such as bulbs and rhizomes are also adventitious roots.

The root system of a plant is the entire mass of roots. The depth of roots, the number of branch roots, the diameters of roots, and other features vary with species. We recognize two major kinds of root systems: **taproot systems** and **diffuse root systems** which are also known as **fibrous root systems.** Kentucky bluegrass (Fig. 10–3*A*), cereals, and many other plants have diffuse root systems, but other species, for example, alfalfa and sugar beet (Fig. 10–3*B*), have taproot systems. In a taproot system there is only one main vertical root which gives rise to branch roots of smaller diameter, whereas in a diffuse root system there are many main roots, all of about equal diameter, and all developing branch roots. In cereals, as well as in many other plants, most of the roots comprising the diffuse root system are adventitious roots. In other species, however, the root system consists of only primary and secondary roots.

Roots occupying the same volume of soil share the water and minerals it contains, and those that survive are those that acquire

Fig. 10–3*A*. Root system of Kentucky bluegrass (*Poa pratensis*). The distance from the top of the soil to the tip of the longest root is 3 feet. (J. E. Weaver and John W. Voight.) *B*. Taproot system of the sugar beet. (Great Western Sugar Co.)

a share large enough to meet their requirements. In developing forests, the amount of water in the soil may be insufficient for the best growth of all the trees. If the stand is thinned, the growth rate of the remaining trees is markedly accelerated. In planning a garden, it is always best to keep small plants out of the reach of tree roots. Even a lawn is difficult to maintain where the area is spotted with trees.

A knowledge of the depth and extent of root systems can be used to plan irrigation and fertilization practices. Many roots of lawn grasses go down at least a foot, and a more prosperous lawn will result when the soil is watered to this depth. For such deep-rooted plants as pumpkins, enough water should be given to wet the soil to a depth of about 6 feet. Such a supply will last a fairly long time. In contrast to pumpkins, onions are an extremely shallow-rooted crop, and they require frequent but light irrigations. Table 10–1 shows the depth of vegetable roots.

The root systems of trees cover wide areas. It is not unusual for an elm tree 36 feet tall to have roots radiating from the base for a distance of 60 feet, or almost double the height of the tree, and these go 4 feet deep. A spruce tree 30 feet tall may have roots 16 feet long and 4 feet deep. Some oaks have roots that extend to a depth of 15 or more feet. Table 10–2 gives the relative lengths of roots of a number of kinds of trees and shrubs. The information given in this table shows why a large area around a tree in your yard must be watered and fertilized if the tree is to thrive.

Table 10–1

Depth of Rooting of Truck Crops

Shallow-rooted (down to 2 feet)	Moderately Deep-rooted (down to 4 feet)	Deep-rooted (down to 6 feet)
Brussels sprouts	Beans, pole	Artichokes
Cabbage	Beans, snap, spring	Asparagus
Cauliflower	Beans, snap, fall	Cantaloupes
Celery	Beets	Lima beans
Lettuce, winter	Carrots	Parsnips
Lettuce, summer and fall	Chard	Pumpkins
Onions	Cucumber	Squash, winter
Potatoes	Eggplant	Sweet potatoes
Radishes	Peas	Tomatoes
Spinach	Peppers	Watermelons
Broccoli	Squash, summer	
Sweet corn	Turnips	

Table 10–2

Relative Length of Woody Roots

	Short Roots (root spread less than height of plant)	Intermediate Length Roots (root spread equal to or exceeding plant height)	Long Roots (root spread one and one-half times plant height, or more)	Extra Long Roots (root spread twice the plant height or more)
Evergreens	Colorado Spruce Black Hills Spruce Western Yellow Pine	Colorado Juniper Red Cedar	Jack Pine	
Shrubs	Tamarix	Tartarian Honeysuckle Caragana Buckthorn	Common Lilac Silver Buffaloberry	Chokecherry
Deciduous Trees	Basswood	Soft Maple Dwarf Asiatic Elm Northern Cottonwood American Plum Hackberry Green Ash Boxelder	Russian Olive Golden Willow Bronze Golden Willow Apple (Hibernal) Butternut Amur Maple American Elm Siberian Crabapple	Mossy Cup Oak Black Walnut

One of the early lessons western ranchers had to learn was that the number of cattle that can be grazed on an acre of land depends on the kind of feed the land offers. Two more steers per acre simply took that much food away from the rest, with the result that the rancher got not a single pound more of beef than he would if he had not pastured the two additional animals. The same principle applies to spacing garden plants in proportion to their requirements for water and minerals. In a unit of garden soil there is only so much mineral supply and so much water available. If you grow one plant in that unit, you get a large, robust, fruitful plant. If you grow two or three plants in the same unit, often each one is smaller, and so is their yield. To get the most profit from a piece of land, either in crops or in flowers, it is well to space the plants to their best advantage. The most favorable spacing may be determined by the amount of water likely to be present in the soil during the growing season. This may be modified if supplemental water can be given.

Nature automatically takes care of the spacing of plants. You may have noticed how few and far between are the plants in a desert. A cactus here, and then a number of feet away, another one. Those plants have survived because they got a head start over others near them, and finally their roots attained a monopoly on the water supply in their immediate vicinity.

ROOT TIPS

All roots terminate in delicate white tips which have usually a diameter of less than 1/16 inch. Water and minerals are absorbed by root tips, and growth of roots occurs at the tip. As root tips increase in length, previously untapped supplies of water and minerals are encountered.

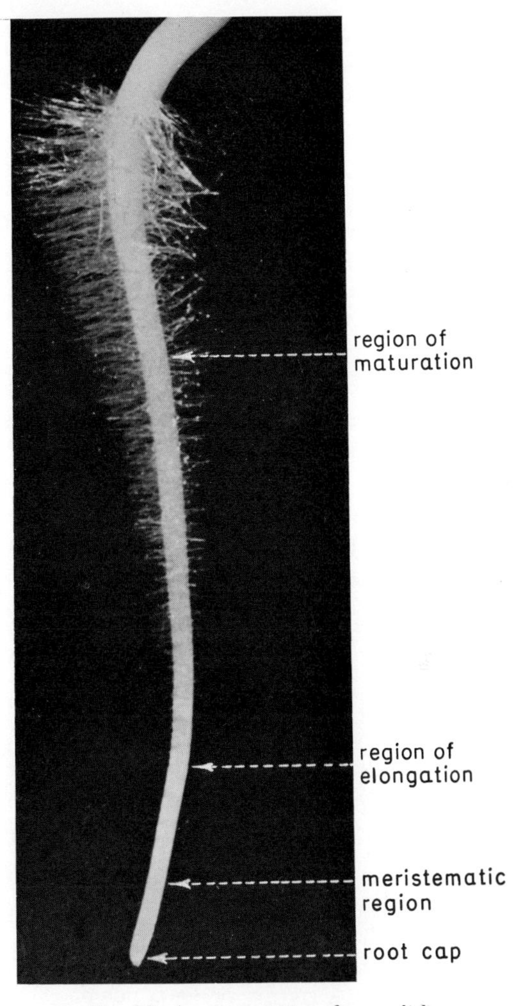

Fig. 10–4. A root tip of a radish.

Regions of a Root Tip

At the tip of a root, four regions can be recognized: the root cap, the meristematic region, the region of elongation, and the region of maturation (Fig. 10–4 and Fig. 10–5).

The root cap. The root cap is a covering which protects the tip of the root. As the root pushes its way through the soil, some root cap cells are crushed. The mucilaginous material released from the injured cells serves as a lubricant for the root tip. Even though root cap cells are constantly destroyed, their number remains fairly constant because new ones are continually formed by the meristematic region.

The meristematic region. The thin-walled cells of this region are alike, and they are constantly dividing, thus adding new cells to the root cap and to the main part of the root. Through division of cells and the subsequent enlargement of the newly formed cells, roots increase in length. If the meristematic region is cut off, or bitten off by an animal, additional root cells are not formed, and the root does not grow in length.

The region of elongation. The cells formed in the meristematic region become longer in the region of elongation. The increase in size results largely from an osmotic uptake of water and to a lesser extent by the formation of additional protoplasm. The cells formed in the meristematic region have, at first, many small vacuoles. As these cells take up water by osmosis, the vacuoles increase in size and fuse. In fully elongated cells, generally only one large central vacuole is present. As the cells elongate, additional cell wall material is deposited on the cell walls; hence the walls do not become thinner as they are stretched by the internal pressure resulting from the osmotic uptake of water. The cells in the elongation region are not uniform in size and shape as they are in the meristematic region. Some cells are long; others are short. In other words, differentiation has begun.

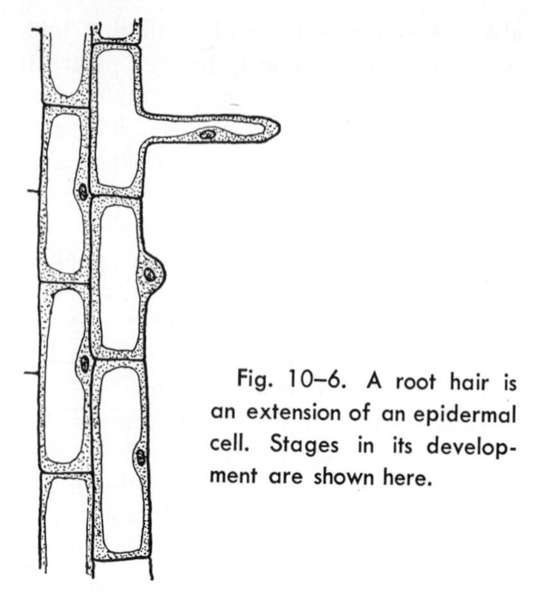

Fig. 10–6. A root hair is an extension of an epidermal cell. Stages in its development are shown here.

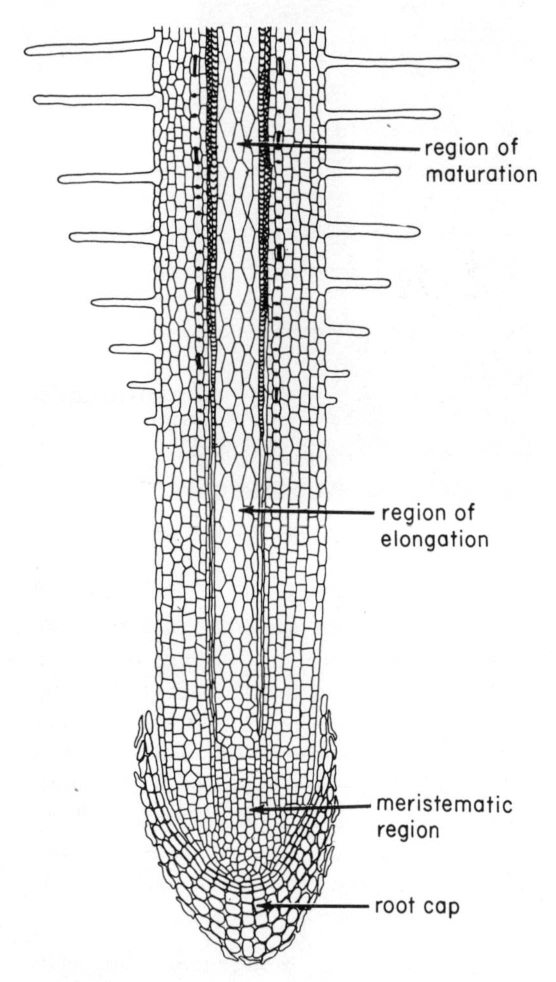

Fig. 10–5. Diagram of a longitudinal section of a root tip. (William M. Carlton, *Laboratory Studies in General Botany*. Copyright © 1961, The Ronald Press Company, New York.)

The region of maturation. In this region cells become specialized in structure and function. Some cells take on structural features which enable them to conduct water. Others become specialized for the conduction of food; and still others for food storage. Root hairs develop in the younger part of the maturation region as finger-like extensions of the epidermal cells. Once initiated, the root hair elongates rapidly and may attain full length, up to ½ inch, in a few hours. A root hair is simply part of an epidermal cell. The nucleus is near the tip, and the cytoplasm in the hair is continuous with that in the epidermal cell (Fig. 10–6). Root hairs adhere to soil particles and furnish a very large surface for the absorption of water and minerals; they increase the absorbing area about twentyfold. It has been estimated that one rye plant has 14 billion root hairs with a combined area of 4000 square feet. Maximum absorption of water occurs in the root hair region, although water is also absorbed in the zone of cell enlargement. Cells of the root cap and meristematic region absorb little water. Minerals are absorbed and accumulated by cells of the meristematic region as well as by those in the region of elongation and root hair zone.

Root hairs are short-lived structures, but as the older root hairs die, newer ones forming at the younger end of the region take over their function of absorbing water and minerals. There is a constantly changing supply of root hairs, which are ever advancing in the soil. As a root grows longer at the tip, new epidermal cells are formed which in turn form new root hairs. The cell sap of root hairs has a higher concentration of dissolved materials (sugars, salts, etc.) than does the soil solution. Therefore, water diffuses from the soil solution into the root hairs. The absorbed water diffuses through other root tissues and ultimately enters vessel and tracheids.

ANATOMY OF A DICOT ROOT

The anatomy of roots is linked closely with function. Within the root, vessels and tracheids of the xylem conduct the absorbed water and minerals through the small roots to the larger ones and so on into vessels and tracheids in the stems and leaves. Roots rely on leaves for food which is necessary for their growth. The food is conducted from the leaves to the roots by sieve tubes of the phloem. Xylem and phloem are the conducting, or vascular, tissues of a plant. The xylem, as we have seen, is the tissue which conducts water and minerals; the phloem is the food-conducting tissue. Let us locate the xylem and phloem and other tissues by studying a cross-section of a dicot root from the region just back of the root hairs. The cells in this region were derived from the root tip. The tissues differentiated from cells which originated at the root tip are known as **primary tissues.**

In a cross-section of a root three regions can be readily recognized. In the center there is a solid cylinder known as the **stele,** consisting of **xylem, cambium, phloem,** and **pericycle.** Surrounding the stele is the **cortex,** a region many cells wide. On the very outside is the **epidermis.** Let us now consider these regions in more detail. Cross-sections of roots are shown in Figure 10–7 and a longitudinal section is shown in Figure 10–8.

The Epidermis

The single layer of cells on the exterior of a root comprise the epidermis whose cells generally lack a cuticle. In the younger part of the maturation region root hairs extend outward from the epidermal cells. In a more mature part, as seen in Figure 10–7, the root hairs have shriveled and are no longer evident.

The Cortex

The innermost layer of the cortex region is known as the **endodermis,** a ring only one cell wide. The wall of a young endodermal cell is thin except for a thick band on the radial and transverse walls. This thickened strip is suberized or cutinized and is known as the **Casparian strip.** The Casparian strip may prevent water from diffusing through the radial and transverse walls and thus direct the water through the protoplasts of endodermal cells. As the cells get older, the endodermal cells between the xylem points develop thick walls. The other cells of the cortex, those from the endodermis to the epidermis, are **parenchyma cells.** You will recall that these are thin-walled, living cells, whose dimensions are about equal. Food in the form of starch grains is frequently stored in parenchyma cells of the cortex.

The Stele

The central cylinder of a root is the stele. It is bounded externally by the **pericycle.** Xylem and phloem occur inside the pericycle. In dicotyledons a **cambium layer** is often present between the xylem and phloem, but it is absent in monocot roots. Let us next consider the parts of the stele in more detail.

The pericycle. The pericycle is a continuous ring of cells, usually one cell wide. The pericycle is important because here **branch**

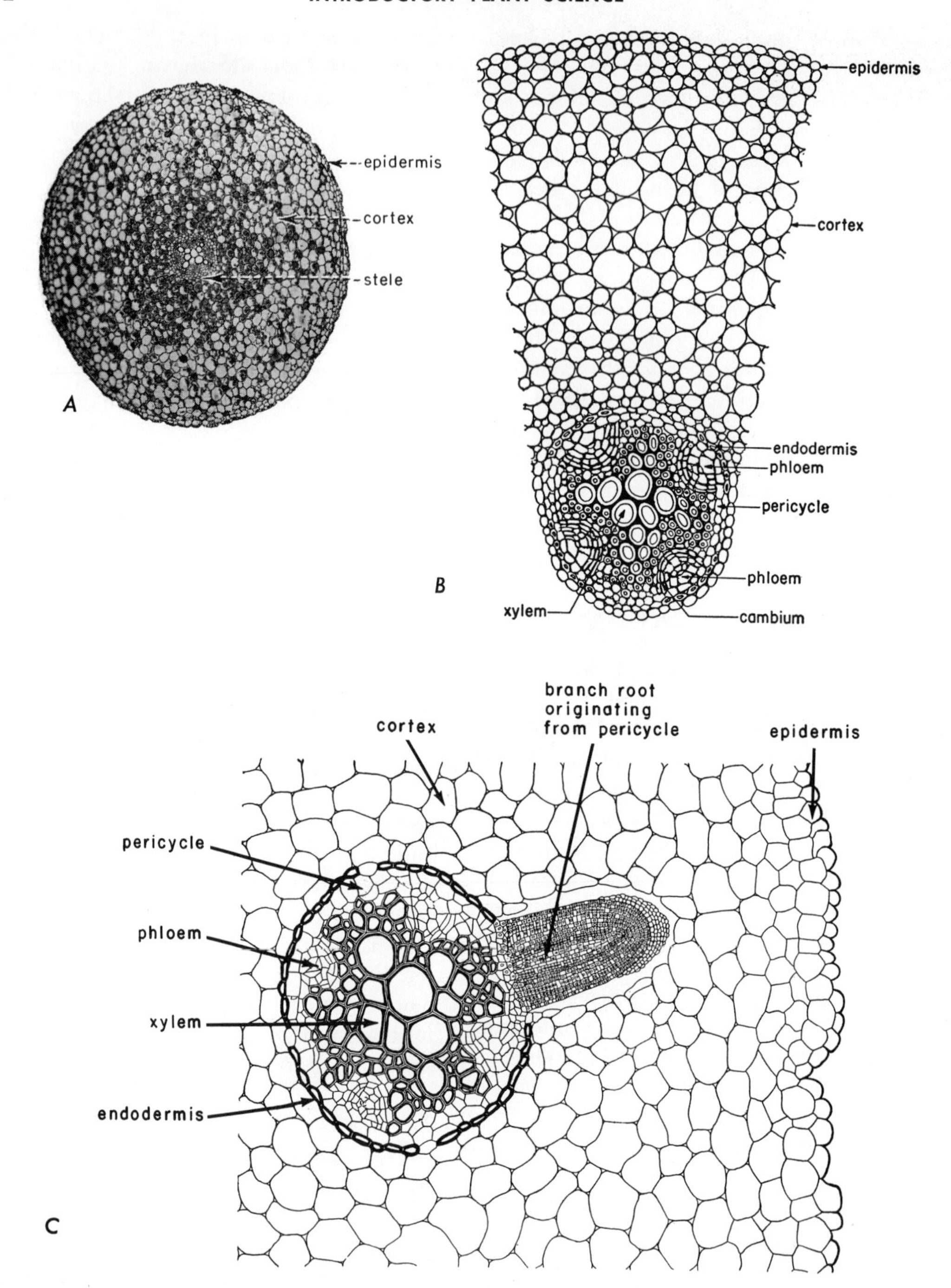

Fig. 10–7*A*. Cross-section of a root of a buttercup (*Ranunculus*). The section was taken just back of the region of root hairs. The stele is the central core; it consists of pericycle, phloem, cambium, and xylem. The xylem resembles a four-pointed star. (Copyright, General Biological Supply House, Inc.) *B*. Sector from a cross-section of a buttercup root. C. Portion of root cross-section showing the development of a lateral root. (From *The Plant World*, 4th ed., by Harry J. Fuller and Zane B. Carothers. Copyright © 1963, Holt, Rinehart & Winston, Inc., New York.)

roots arise and **cork** originates. Branch roots arise back of the root hair zone. A branch root originates through the division of pericycle cells which are opposite a xylem point. Here the pericycle cells divide in a plane parallel to the root surface. Shortly, a small conical mass of cells is developed, and these cells through division form a young branch root. The young root pushes the endodermis aside and grows outward through the cortex (Fig. 10–7*C*) and in time reaches the soil. By the time the branch root emerges into the soil, it has a root cap protecting its apical meristem, and its other tissues have begun to differentiate. Later the conducting tissues of the branch root establish connection with those of the main root. As more and more branch roots develop, an increasing soil area is tapped for water and minerals. Do not confuse a branch root with a root hair. A root hair is merely an extension of an epidermal cell. A branch root has a great many cells, and each root has a root cap, a meristematic region, a region of elongation, and a mature region.

A root may be induced to form an abundance of branch roots by removing the tip. During transplanting, many root tips are destroyed and hence many branch roots develop. Frequent transplanting results in the development of a compact root system which withstands transplanting better than a straggly root system.

The **outer bark,** technically called **cork,** of an old root originates in the pericycle. The cork is formed by the division of **cork cambium** cells which are formed in the pericycle. The epidermis and cortex are sloughed off as the cork develops. The root then consists of cork, a cork cambium, pericycle, phloem, cambium, and xylem, in that order, starting from the outside and working toward the center.

The xylem. The xylem occupies the center of the stele, appearing as a four-pointed star in Figure 10–7. The smaller cells of the xylem are tracheids; the larger ones are vessels. Both kinds conduct water and minerals, and when mature they lack protoplasm. You will recall that tracheids are

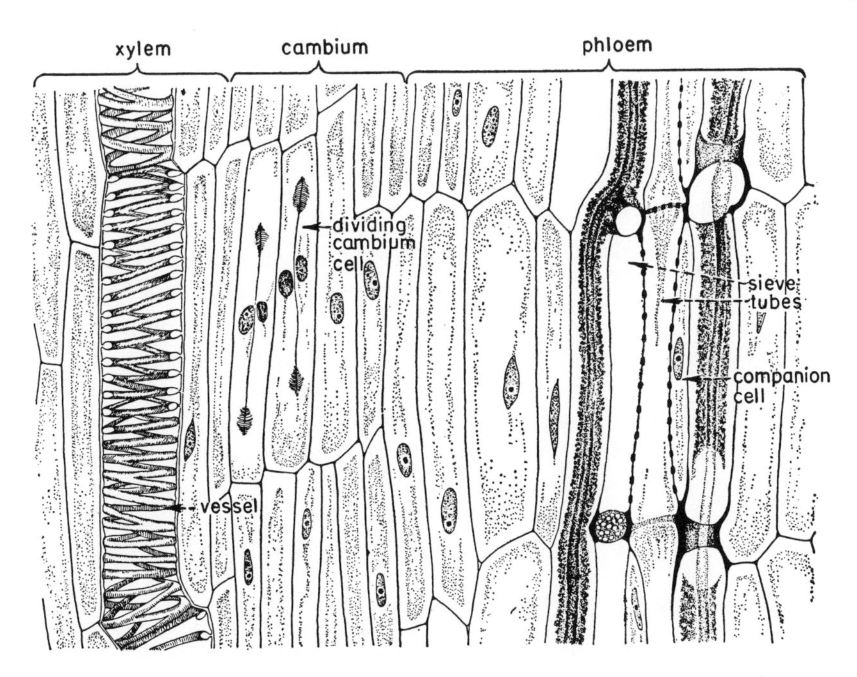

Fig. 10–8. A longitudinal section through a Russian dandelion root showing xylem, cambium, and phloem. (Ernst Artschwager, *U.S.D.A. Tech. Bull. 843.*)

long, cylindrical cells with tapering end walls. Vessels are long tubes made up of cells joined end to end with the cross-walls digested out. Some vessels and tracheids have pits in their thickened walls, but others have the thickenings in the form of rings or spirals. Such thickenings add strength to xylem tissue. The xylem tissue of a root forms a tough central core which provides high tensile strength with flexibility—properties leading to effective anchorage of the plant.

You will notice in Figure 10–7*A* that the outer xylem cells are smaller than the inner ones. This pattern results because the outer cells develop secondary walls, which prevent enlargement, before the inner cells. The small xylem cells, those first to mature, are called **protoxylem cells.** The secondary thickenings, generally in the form of rings or spirals, are deposited on protoxylem cells when they still have a narrow diameter. While cell walls of protoxylem cells are being thickened, the walls of the inner xylem cells, called **metaxylem,** remain pliable, and accordingly the cells enlarge. Pitted vessels and tracheids, as well as scalariform and reticulate ones, occur in the metaxylem. When the sequence of xylem development is from the outside toward the inside, the process is said to follow an **exarch** pattern. Roots have an exarch pattern. As we will see later, stems have a reverse pattern, an **endarch** one, in which development proceeds from inner cells to outer ones. Many botanists believe that the exarch pattern characteristic of roots is more primitive than the endarch one.

The phloem. The phloem is located between the arms, or points, of xylem. You will notice in Figure 10–7*B* that the xylem arms occupy radii different from the four groups of phloem cells. The **radial arrangement** (a xylem arm on one radius, a group of phloem cells on a different radius) is characteristic of roots, but not of stems. The radial arrangement in a root permits water and minerals absorbed from the soil to enter the xylem without first passing through the phloem. The water absorbed from the soil moves across the epidermis, cortex, and pericyle, and then enters the xylem. If the phloem and xylem were on the same radius, as they are in stems, water and minerals would pass through the phloem before entering the xylem.

Large, thin-walled sieve tube cells and smaller companion cells make up the phloem tissue. We have seen earlier (Chapter 4) that sieve tube cells are joined together end to end to form long ducts through which food is transported. The end walls of sieve tubes (called **sieve plates**) are perforated. Nuclei are absent in mature sieve tube cells, but cytoplasm is present. A companion cell has a thin wall, dense cytoplasm, and a large nucleus and is adjacent to a sieve tube cell.

The cambium. A cambium layer is present between the xylem and phloem in the roots of many dicotyledons. The cells of this layer divide to form additional phloem cells toward the outside and additional xylem cells toward the inside of the root.

Figure 10–9 illustrates the formation of xylem and phloem by the division of a cambial cell. The cambial cell divides, and the inner of the two cells differentiates into a xylem cell (X′), while the outer cell remains a cambial cell. The cambial cell divides again. The outer cell differentiates into a phloem cell (P′), and the inner one remains a cambial cell. Subsequent divisions of the cambial cell result in the formation of additional xylem and phloem. Usually, the cambium produces more xylem cells than phloem cells. Roots of monocotyledons do not have a cambium layer, and therefore their roots do not increase in diameter. Roots of corn, wheat, and other monocots remain slender.

SECONDARY GROWTH OF DICOT ROOTS

The phloem formed by the cambium is known as **secondary phloem,** and the xylem formed by the cambium is known as **secon-**

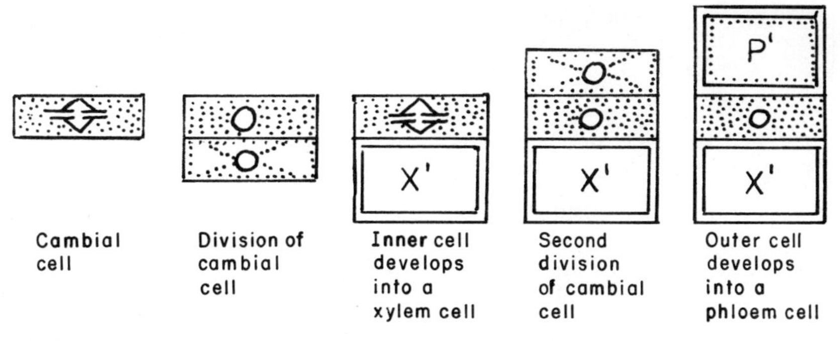

Fig. 10–9. Secondary xylem and phloem are formed by the division of cambial cells.

dary xylem. Secondary xylem is first formed in the area between the xylem points. As the secondary xylem accumulates, the xylem becomes circular. The xylem is then surrounded by a narrow layer of cambium, outside of which there is a cylinder of secondary phloem, formed, of course, by the cambium (Fig. 10–10). The isolated patches of primary phloem are external to the secondary phloem. As secondary xylem and phloem accumulate, the primary phloem is crushed by the pressure of the growing tissue within.

As the years go by, many secondary phloem cells in roots of trees and shrubs are crushed and hence do not accumulate to a great extent. The thick-walled xylem cells resist crushing and accumulate with each passing year. In old roots of trees and shrubs a large amount of xylem, commonly called wood, is present. The wood of an old root shows annual rings, as does that of the trunk (Fig. 10–11).

In the older root the pericycle remains intact. The cells of this layer have the capacity to divide. Certain cells form a cork

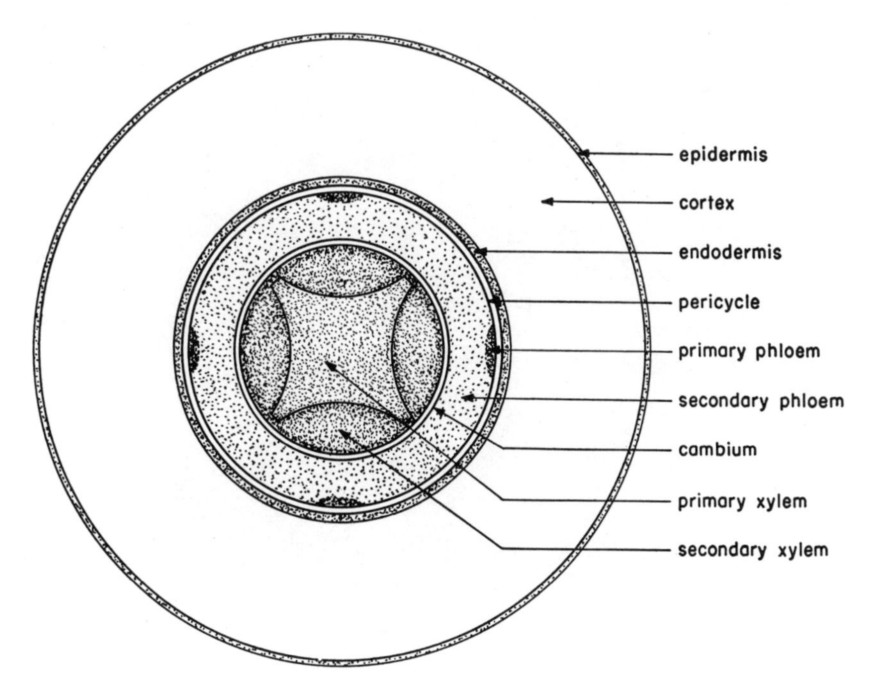

Fig. 10–10. Diagram of a buttercup root after secondary xylem and phloem have developed.

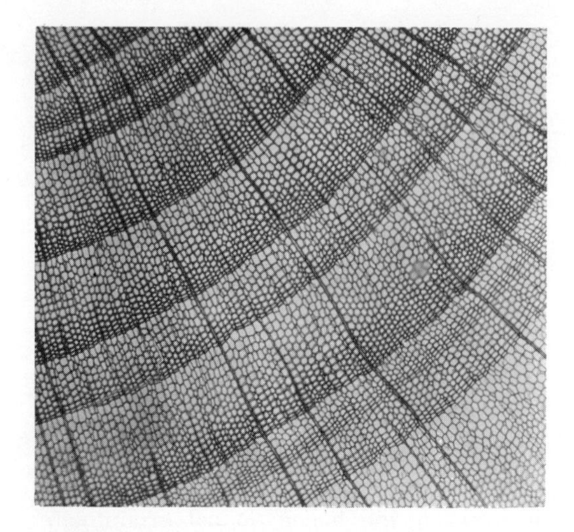

Fig. 10–11. Secondary xylem of a root of spruce (*Picea glauca*). Annual rings are evident in the xylem (wood). (M. W. Bannan.)

cambium whose cells divide to form cork (outer bark). In many dicot roots the cortex is sloughed off from the root, and then the outermost tissue is the cork.

ANATOMY OF A MONOCOT ROOT

Roots of monocotyledonous plants differ from those of dicotyledons in several ways as can be seen by comparing the buttercup root (Fig. 10–7*B*) with that of *Smilax* (Fig. 10–12), a monocot. Unlike the dicot roots, the monocot root has a central pith. Roots of dicots have few xylem points and groups of phloem cells, whereas monocots have many, usually more than ten in the *Smilax* root. Notice also that the *Smilax* root has several layers of pericycle cells, while the buttercup root has only one. Most dicots have a cambium layer in both roots and stems, whereas monocots lack a cambium layer.

SPECIALIZED ROOTS

The roots of some plants are modified to carry on special activities. Roots of some biennials and many perennials are enlarged organs which store considerable food, usu-

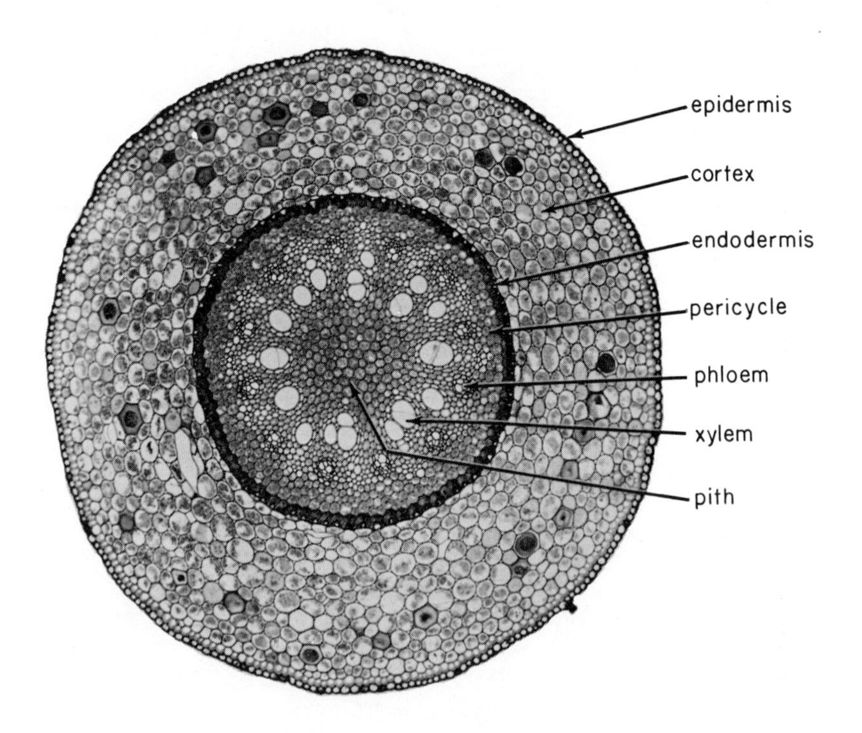

Fig. 10–12. Cross-section of a *Smilax* root showing features of a monocot root. (J. Limbach, Ripon Microslides.)

ally in the form of sugar or starch or both. A sugar beet contains 15 to 20 per cent sugar. Considerable starch as well as sugar is stored in sweet potatoes and parsnips. In carrots, turnips, and beets the taproot is modified for food storage, whereas in dahlias, peonies, and sweet potatoes it is the side roots which are tuberous. Because the tuberous roots of dahlias and peonies lack buds, those plants cannot be propagated from portions of a tuberous root unless a bud-bearing portion of the stem is included.

Cattleyas and other tropical orchids have **aerial roots** which anchor the plants to trees or other supports. These aerial roots have a special tissue on the outside, the **velamen,** which was formed by the division of young epidermal cells. The velamen absorbs water and dissolved minerals from the debris which collects on the branches or support—also from rain and dew. In some species of orchids, *Vanilla,* for example, the aerial roots are green, and hence they manufacture food as well as anchor the plant and absorb water and minerals.

From the shaded side of stems of ivy and some other vines adventitious roots develop which adhere at their tips to walls or other supports.

In corn and screw pine adventitious roots are formed on the stem some distance above the soil. They grow down into the soil at an angle and support the stem. Such roots are called **prop roots.**

Related to prop roots are roots of several species of fig. In the strangling fig the roots become the tree trunk. The life story of this fig, a tree that kills the host that supported it during youth, is a strange one. A fig seed is carried by a bird or mammal to a branch of some tree where the seed germinates. The young fig develops and grows as an epiphyte. As the decades go by, the roots grow downward, encircling the tree trunk. In time they reach the soil from which they then absorb water and minerals. The roots surrounding the trunk enlarge and fuse, thus forming a mantle around it. In the meantime the crown of the fig has become very large. The massive crown and the encircling roots ultimately kill the host tree, leaving the fig standing alone.

Along ocean shores and estuaries in subtropical and tropical regions mangroves may thrive on the beaches. The mangroves develop horizontal roots over the mud or just below the surface. From the horizontal roots, air roots, called **pneumatophores,** extend upward into the atmosphere. The vertical roots have large intercellular spaces where air is stored and conducted to plant parts below the water.

Bald cypress trees which grow in swamps develop root projections, known as "knees," which extend above the water. We are still uncertain as to their role. Formerly, it was believed that the "knees" conducted air to the submerged roots, but investigations have cast some doubt on this idea (Fig. 10–13).

The common dodder and the mistletoe are parasitic seed plants that produce modified roots known as **haustoria** which penetrate the stem of the host. The dodder (Fig. 10–14) is a nongreen plant that is entirely dependent on the host for food, water, and minerals. The mistletoe, on the other hand, is green and makes at least some of its own food, but it depends entirely on the host tree for water and minerals. The dwarf mistletoe is a serious pest of lodgepole pine, Douglas-fir, and black spruce and may reduce the lumber production of infested trees by 30 to 50 per cent. The haustoria of the mistletoe invade the bark and become imbedded in the sapwood of their hosts, from which they absorb nutrients and water. The dwarf mistletoe produces small sticky seeds which are explosively discharged when mature. Seeds striking a limb or carried there by birds or squirrels germinate and send their modified roots, the haustoria, on their parasitizing mission.

The shoots of some plants, dandelion for example, may be pulled closer to the soil by **contractile roots,** those which shorten as their parenchyma cells become shorter and

Fig. 10–13. "Knees" of bald cypress. (U.S. Forest Service.)

wider. Bulbs of many plants are often pulled deeper into the soil by the contraction of their roots. The contractile parts of roots may shorten to just one-third of their original length in a few weeks time.

USES OF ROOTS

The enlarged roots of a number of plants are used as food. You are familiar with roots of sweet potatoes, beets, carrots, parsnips, and turnips. Roots of certain plants yield flavoring materials—horseradish, licorice, and sarsaparilla, for example.

Rotenone (sometimes called derris), a valuable insecticide, is obtained from the roots of various species of *Lonchocarpus* and *Derris*. Rotenone has been used for centuries by primitive people as a fish poison. Only in recent years have wildlife specialists used it to kill suckers and other fish. A lake may have a great many rough fish and very few game fish. Rotenone is added to the water and kills all fish. Later the lake is stocked with game fish, which then do not have to share the food with unwanted kinds.

The medicine reserpine, a tranquilizing drug, is used in the treatment of high blood pressure and in calming excitable subjects. Sufferers of nervous and mental disorders at times have benefited from this drug. The drug is isolated from roots of various species of *Rauvolfia* (also spelled *Rauwolfia*), among them *Rauvolfia serpentina, R. vomitoria,* and *R. tetraphylla.* Cortisone is widely used in medicine, and this drug can be synthesized from chemicals isolated from certain Mexican yams. Other drugs obtained from roots are aconitine, gentian, ginseng, ipecac, licorice, and rhubarb.

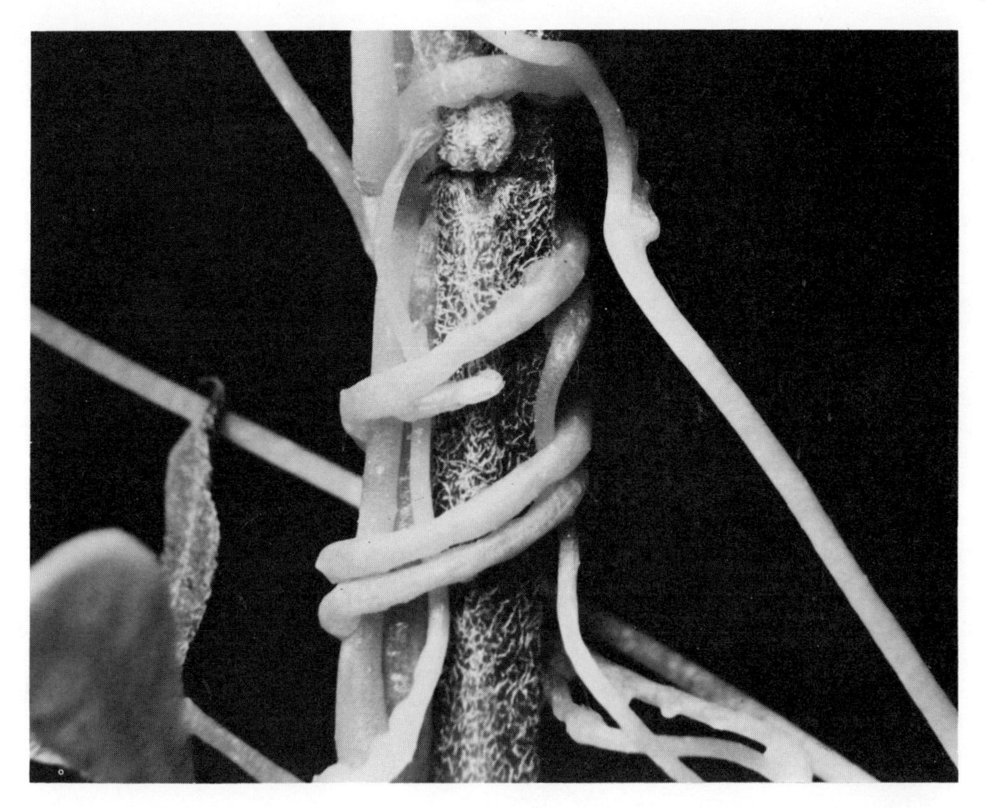

Fig. 10–14. Dodder, a nongreen parasitic seed plant, encircling a chrysanthemum stem. Haustoria penetrate the host and absorb food. (Nature in Pictures, Dr. Ross E. Hutchins.)

QUESTIONS

1. _____ Trees buffeted by the wind (A) do not make growth adjustments because they lack a nervous system, (B) are sturdier than staked trees, (C) are not more resistant to wind than previously staked trees.

2. _____ Roots of many plants may be likened to (A) cables, (B) reinforced concrete columns, (C) I-beams, (D) bridges.

3. List three functions of roots.

4. _____ Which one of the following plants does not store large quantities of food in its roots? (A) Sweet potato, (B) carrot, (C) parsnip, (D) Irish potato, (E) sugar beet.

5. _____ A certain underground plant part had nodes and internodes on it. It was a (A) root, (B) stem, (C) petiole, (D) leaf, (E) mycorrhiza.

6. _____ Alfalfa and sugar beets have (A) adventitious root systems, (B) fibrous root systems, (C) taproot systems.

7. _____ Roots produced at the end of a stem cutting are (A) primary roots, (B) secondary roots, (C) adventitious roots.

8. _____ Which one of the following requires light but frequent irrigations? (A) Onions, (B) tomatoes, (C) lima beans, (D) carrots.

9. _____ If the tip of a root is removed, (A) the root will continue to increase in length, (B) the root will no longer increase in length.

10. _____ Which of the following statements about a root tip is false? (A) The root cap is a covering which protects the tip of the root. (B) In the region of elongation, cells divide and thus increase the length of the root. (C) In the region of maturation, cells become specialized in structure and function. (D) Root hairs develop in the younger part of the maturation region as finger-like extensions of the epidermal cells.

11. _____ Each root hair is made up of two or more cells. (A) True, (B) false.

12. Name the tissues of a mature root in correct order, beginning at the outside of a cross-section. Use these terms: epidermis, pericycle, endodermis, xylem, phloem, cortex, cambium.

1. __________ 5. __________
2. __________ 6. __________
3. __________ 7. __________
4. __________

13. _____ Foods such as sugars are conducted in roots by (A) sieve tubes of the xylem, (B) sieve tubes of the phloem, (C) vessels of the xylem, (D) vessels of the phloem, (E) the pericycle, (F) the cambium, (G) the cortex.

14. Which one of the tissues mentioned in question 13 conducts water and minerals? __________.

15. _____ Both the cork (outer bark) of roots and branch roots originate by the division of (A) cambial cells, (B) pericycle cells, (C) cortex cells, (D) endodermal cells.

16. _____ When, as in roots, the developmental sequence of the xylem is from the outside toward the inside, the pattern is (A) endarch, a primitive type; (B) endarch, an advanced pattern; (C) exarch, a primitive type; (D) exarch, an advanced pattern.

17. _____ The first primary xylem cells to mature are called (A) metaxylem and the vessels are pitted, (B) metaxylem and the vessels have annular or spiral thickenings, (C) protoxylem and the vessels are pitted, (D) protoxylem and the vessels have thickenings generally in the form of rings or spirals.

18. With an × indicate the features which characterize the root of *Smilax*, a monocot.

_____ 1. Cambium present.
_____ 2. Branch roots formed by the pericycle.
_____ 3. Exarch development.
_____ 4. Pith present.
_____ 5. Fewer than five xylem points.
_____ 6. Radial arrangement of xylem and phloem.
_____ 7. Pericycle of several layers.
_____ 8. Cortex present.

19. _____ Which one of the following is most harmful to the host? (A) Orchid, (B) mistletoe, (C) dodder, (D) strangling fig.

20. List four roots which are of economic importance.

21. Match the following:

_____	1. Orchid.	1. Haustoria.
_____	2. Corn.	2. "Knees."
_____	3. Bald cypress.	3. Prop roots.
_____	4. Dodder.	4. Velamen.
_____	5. Sugar beet.	5. Food storage.

22. _____ The drug obtained from roots that aids the mentally ill is (A) reserpine, (B) cortisone, (C) rotenone, (D) aconitine, (E) ginseng, (F) ipecac.

23. List some methods for preventing soil erosion.

---

24. Place the number of the appropriate term from the column on the right in the space to the left of each statement.

_____	1. The region where cells divide to increase the length.	1. Adventitious root.
_____	2. A tubular outgrowth of an epidermal cell.	2. Root hair.
_____	3. The star-shaped tissue in the center of a buttercup root.	3. Xylem.
_____	4. The tissue between the xylem points.	4. Phloem.
_____	5. The ring of cells just inside the endodermis.	5. Meristematic region.
_____	6. A root which is produced at the node of a stem.	6. Pericycle.

11

Soils

Soil is more than just dirt. Soil is highly complex, constantly changing, affecting the growth and development of plants and in turn affected by plants, both large and microscopic. It has a structure much like that of a cake, in that particles are held together in crumbs, connected by films of moisture, with air spaces in between. In each crumb of soil that you can see with your unaided eye there are rock particles of various sizes, mineral salts dissolved in water held on the surface of the particles, bits of decaying plant and animal material, and thousands of microscopic plants and animals.

Many of the rock particles in this crumb are the size of sand, many are smaller than sand and are called **silt,** and still smaller particles are called **clay.** All of the factors that have fashioned the face of the earth have contributed to the making of soils. The uplift of the earth's crust formed mountains and offered the exposed rocks to the weathering effects of wind and rain, heat and cold. As the mountains have been leveled by these forces, the rocks have been broken up and washed into the valleys. Glaciers added their grinding force, to shear rocks from the parent stratum, to pulverize them as they carried them along in their frozen undersurface, and finally to deposit the resulting gravel when the ice melted. Ancient lakes and rivers and inland seas, few of which were seen by man, have left the pattern of their terraces and beaches and the deposits of their beds. Erosion and floods continuing into our time have moved quantities of soil from one area to another. The type of rock shown in the particles of soil may therefore be a mixture of native rock and rock that originated far away, and the soil in your backyard is the result of the changing topography and climates of yesterday, as well as the climate and topography of today.

The rock particles are only the beginning of the story of the soil. A bucket of wet sand is not a good soil, nor is a bucket of wet silt or clay. If you mix these three together you have a beginning, for a good soil (loam) has a mixture of various-sized particles. Other ingredients of equal importance would have to be present, however, before a farmer would approve of the mixture for growing plants. The other ingredients necessary to make the mixture a good soil are organic matter, air, and microorganisms. Organic matter makes the soil more porous and increases its air-holding and water-holding capacity. Furthermore, organic matter is a rich storehouse of nutrients for plants growing in the soil, but some of the nutrients would not be available to the green plants without the action of soil bacteria and fungi. These microorganisms feed on the dead material, digesting it and releasing into the soil mineral compounds that can be absorbed by the roots of higher plants. We call the process decay. It is part of the never-ending cycle of activity that gives back into the soil the materials used by green plants. And here time plays its role.

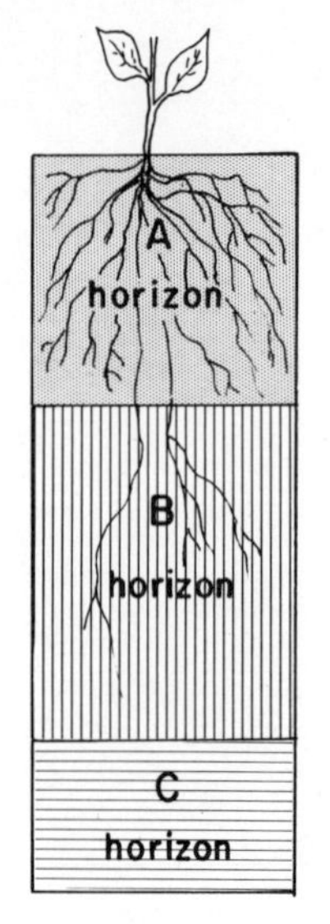

Fig. 11–1. A soil profile showing the three horizons. Starting from the surface and going down, the horizons are designated A, B, and C.

Climate, soil, and vegetation are intimately related. Black soils contain a large amount of organic matter, the result of long periods of heavy vegetation. Black soil, subhumid climate, and tall grass go together, and today these three combine to give the rich farmlands of our country. Red and gray soils have a sparse amount of organic matter. In nature, light-colored, leached soils, a cool, moist climate, and evergreen forests go together, as do brown soils, cool and semiarid climate, and short grass plains. Each soil has three major layers, one above the other, and these layers are referred to as **horizons** (Fig. 11–1).

If you examine a vertical section through soil, as exposed in a gully, road cut, or excavation, you will see the three horizons formed through the interplay of biological and chemical agents working on the parent material over long periods of time. In the A and B horizons there may be several subdivisions. Most of the biological activity goes on in the A horizon where organic matter mainly accumulates and where roots and soil organisms are most abundant. In many soils the A horizon is darker in color and sandier in texture than the B horizon below. Water percolating through the soil of the A horizon tends to leach out some of the soluble salts and clay particles. Hence the A horizon is the zone where leaching occurs. The minerals and clay carried down by water from the A horizon accumulate in the B horizon, the zone of concentration. At least in certain soils some roots penetrate the B horizon from which minerals are absorbed and translocated to the shoots. When the shoots die and decay, these minerals then become part of the A horizon. Although there is some microbial activity in the B horizon, it is less intense than in the horizon above. The soil in the B horizon is usually compact and bright in color. The B horizon grades into the C horizon, which is composed of the parent material from which the soil was derived. Little or no biological activity occurs in the C horizon. Only the general features of the soil profile have been considered. The depth, color, fertility, and texture of the three horizons varies greatly with climatic factors, the type of vegetation and other biological agents, and the nature of the parent rock. Thus soils in the great plains differ markedly from those of eastern forest regions.

COMPONENTS OF THE SOIL SYSTEM

We have seen that a number of components make up the soil system: rock particles, organic matter, air, water, soil organisms, and dissolved substances. Let us now consider these in more detail.

Rock Particles

The rock particles of soil range in size from sand to silt to clay. The diameters of the various types of rock particles are: coarse sand, 2.0 to 0.2 mm.; fine sand, 0.2 to 0.02 mm.; silt, 0.02 to 0.002 mm.; and clay, less than 0.002 mm. Most soils contain a mixture of rock particles of various sizes. Soils which have high percentages of all four types are known as loam soils; those which are predominantly made of clay particles, as clay soils; those with a high percentage of sand and a low percentage of silt and clay, as sandy soils, etc. Table 11–1 shows the percentage of coarse sand, fine sand, silt, and clay in three different soils—a sandy loam, a heavy loam, and a clay soil. The percentages of the various sizes present determine to some extent the texture of the soil, the capacity of the soil to hold water, and the amount of water in the soil which is available to plants. For example, a clay soil holds considerably more water than a sandy soil.

The clay particles are especially important in plant nutrition. They have a tremendous total surface area. It has been estimated that 1 pound of clay has a total surface area equal to 100 acres of land. Water is held on this large area. Moreover, the clay particles are negatively charged and hence attract and hold positively charged ions such as calcium, potassium, magnesium, and ammonium. These bound ions are not readily leached out of the soil, but they can be taken up by plant roots which are in contact with clay particles. Roots exchange hydrogen ions for the minerals held on the clay particles. You will recall that roots respire, forming carbon dioxide and water. The carbon dioxide and water combine to form carbonic acid. The hydrogen ions of this acid are exchanged for the positively charged ions held on clay surfaces. That is, the root cells release hydrogen ions which take the place of ammonium, potassium, or some other ion on the clay particles. The released ion then enters the root cell. Even though clay has many desirable properties, a soil made up of only clay particles is not an ideal agricultural soil. Clay soils often have poor physical properties, being sticky and hard to work when wet and being very hard when dry. The best agricultural soils are loams which are made up of sand and silt as well as clay particles.

Table 11–1

Percentages of Particles in Typical Soils

Name of Particle	Sandy Loam	Heavy Loam	Clay
Coarse sand	66	15	1
Fine sand	14	36	10
Silt	10	21	19
Clay	10	28	70

Organic Matter

The organic matter in soils consists of the partially decayed remains of plants and animals. Organic matter is vital for a good growing soil. It promotes granular soil structure, retains plant nutrients, and supplies food for microorganisms and soil animals which enrich the soil. As the organic matter decays, a balanced supply of nutrients is made available to plants. Soils rich in organic matter have a high water-holding capacity and usually they are well aerated. The addition of organic matter to soils in the form of peat, leaves, manure, or plowed-in cover crop increases the productivity of soils.

Air

Spaces are present between the soil particles, the rock particles, and organic matter. The total space between such particles is known as the **pore space,** which may make up 30 to 60 per cent of the soil volume. When soils are dry, the pore space is filled with air. When poorly drained soils are saturated with water, the space is occupied entirely with water, a condition unfavorable to the growth of most species. The principal exceptions to this rule are the hydrophytes, plants native to lakes, streams, and marshes. Most species grow best when both air and water are present in the soil. The oxygen of the soil air is utilized in the respiration of soil organisms and roots. The energy released in respiration is necessary for the accumulation of minerals and for root growth. Root growth, and also shoot growth, of many species is retarded when the oxygen concentration of the soil drops below 10 per cent. In poorly aerated soils, especially undrained and flooded ones, oxygen concentrations may drop below this value and may even approach zero. In fields drainage makes soils more productive. Plants in pots thrive better when ample drainage is provided. In naturally porous, well-drained soils, those with a good crumb structure, and in soils loosened by plowing or cultivation there is a free exchange of gases between the atmosphere and the soil. Consequently in such soils the soil air does not deviate greatly from the atmosphere. Oxygen diffuses into the soil and carbon dioxide diffuses out. In poorly aerated soils oxygen is low and carbon dioxide high, up to 15 per cent. Although most injury results from the low oxygen concentration, certain species may be harmed by the high carbon dioxide concentration (Fig. 11–2).

Water

We may gripe when it rains during a football game or picnic. We know, although we do not praise it, that the rain is essential for farms, forests, range lands, and gardens; that it helps fill reservoirs for human use; and that some of the storm's water goes into rivers and streams to benefit people too distant to be annoyed by it.

The immediate fate of the water that descends to earth is governed by the conditions it finds where it falls. On sloping lands, bare of vegetation, or only sparsely covered, much of the water runs off, carrying precious topsoil with it into flooding streams. In poorly drained fields, the water may stand, causing damage to crops. Well-managed farms, gardens, and lawns, also adequately covered natural areas, usually receive the full benefit of normal amounts of precipitation, for most of it penetrates the soil.

A light rain penetrates to a depth of 3 or 4 inches, a heavy rain 1 foot or more. After the water has had time to percolate through the soil, the moistened layer has a uniform water content. It is then said to be at its **field capacity.** The field capacity is the upper limit of the water storage in a drained soil. Another rain coming soon afterward moves through the moistened layer and wets another layer just below. So with each succeeding rain, the soil becomes moistened deeper and deeper. If rainfall is abundant in a region and if the water table is close to the surface, some of the water enters the water table. However, in many agricultural regions the precipitation never percolates to the water table. In these regions the upper layer of soil is moist, the lower layers are dry. When soils are at or below field capacity, water moves extremely slowly, or not at all, to adjacent dry areas. If you forget to set the sprinkler in one area of the lawn, the grass may shrivel in that spot whereas in the surrounding watered area it is green. The water did not move from the sprinkled area to the unwatered one. The continued growth of roots favors absorption of water. As roots increase in length, they may contact additional supplies of water and minerals.

Although water does not move at an appreciable rate through soils which are at or below field capacity, it does move up through soils from the water table by **capil-**

Fig. 11–2. Low oxygen concentration in the nutrient solution retards the growth of tomato plants. The plants were grown in separate solution cultures with different oxygen concentrations and were assembled in one container for the photograph. The per cent oxygen saturation of the solutions was as follows, from left to right: 1, 3, 5, 10, and 20. (Courtesy of L. C. Erickson, University of California, Riverside.)

larity. If the water table is close to the surface, the roots of plants may use this supply of moisture, which may have risen several feet. The height to which water ascends by capillarity varies with the soil type, being lower for sandy soils than for loam or clay soils. In a sandy soil water rises about 1 foot by capillarity whereas in a loam soil it ascends about 3 feet. In most agricultural regions, the water table is so far below the surface that it cannot be reached by roots of crop plants.

Ancient Indians in Peru knew that water would rise only so far in soil and in some areas the water was too far below the surface to be available to their crops. In areas where the water table was quite close to the surface, they removed the upper two feet of soil so that the roots of crop plants could reach the capillary water.

Plants use the reservoir of water in the soil. If drought occurs and no water is applied to the drying soil, the water available to the roots is used up, and the plants cease to grow. Shallow-rooted ones suffer first, while those with deep and extensive root systems are the last to show any symptoms, for they have a much larger reservoir upon which to draw.

Evaporation of water from the soil takes place in the upper few inches. Below that, most of the water is removed from the soil by roots of plants. If no plants are growing on the soil, as when summer fallowing is practiced, the water is stored for later use. Summer fallowing to conserve moisture one year for use the next year is an accepted practice in most of the Great Plains area and in other areas where precipitation is limited or where the rainfall pattern does not correspond with the growing season. Because of wind and water erosion the recommended method is to alternate strips of crops with strips of fallow land, the latter between 75 and 150 feet wide, depending on soil type and danger of erosion.

Not all of the water in the soil is available to roots. Some is so strongly held by surfaces of soil particles, with a tension of 15 or more atmospheres, that roots cannot absorb it. The percentage of water in soil that is not available to roots is called the **wilting percentage.** The wilting percentage (also called permanent wilting percentage) is the lower limit at which the soil reservoir may be considered empty as far as plants are concerned. When the soil moisture is reduced to the wilting percentage, plants **permanently wilt.** A wilted plant which will not recover when placed in a saturated atmosphere is said to be permanently wilted and will become turgid only if water is added to the soil. The percentage of unavailable water varies with the kind of soil. In sandy soils only a small percentage is unavailable to roots, in clay soils a high percentage. If the soil is of the same kind, the amount of unavailable water is the same, or nearly so, for a great variety of plants, both mesophytes and xerophytes. Table 11–2 shows that in a clay soil water becomes unavailable when the percentage of moisture is between 14 and 15 per cent.

Table 11–2

Wilting Percentages for Different Plants Growing in a Clay Soil

Plant	Wilting Percentage
Coleus	14.2
Corn	15.0
Hollyhock	14.2
Jimson-weed	15.0
Lettuce	14.6
Pepper	14.6
Pigweed	14.5
Sorghum	14.2
Sunflower	14.0
Thistle	15.0

Although clay soils have a higher percentage of unavailable water than sandy soils, the total available water, from field capacity down to the wilting percentage, is much greater in clay soils than in sandy ones. At field capacity clay soils hold much more water than sandy ones. For example, at field capacity 100 grams of dry silty clay loam soil holds 34 grams of water, whereas 100 grams of dry fine sand soil holds only 12 grams. In a silty clay loam, soil water is available to roots from 34 per cent water (the field capacity) down to 15 per cent water (the wilting percentage). In a fine sand soil the total water available ranges from 12 per cent (the field capacity) down to 4 per cent (the wilting percentage). A comparison of the total water available to plants growing in a fine sand soil and in a silty clay loam is shown in Figure 11–3. When soil is above field capacity for a considerable time, many plants grow poorly or die because in this range the soil is poorly aerated.

Soil Organisms

Most soils are teeming with animals and plants. Protozoa, nematodes, insects, earthworms, and burrowing animals make up the soil fauna. Algae, bacteria, and fungi comprise the flora of the soil. The number of organisms present in fertile soils is tremendous. Where there is an optimum supply of water, air, and organic matter, there are about 6 billion bacteria per ounce of soil. Bacteria and fungi cause decay of the organic matter which is thus returning compounds to the soil to be used by new generations of plants. Some soil fungi inhibit the growth of other fungi. For example, some strains of an Actinomycete produce a chemical (an antibiotic) which retards the growth of *Pythium arrhenomanes,* a fungus which causes a severe root rot of sugar cane.

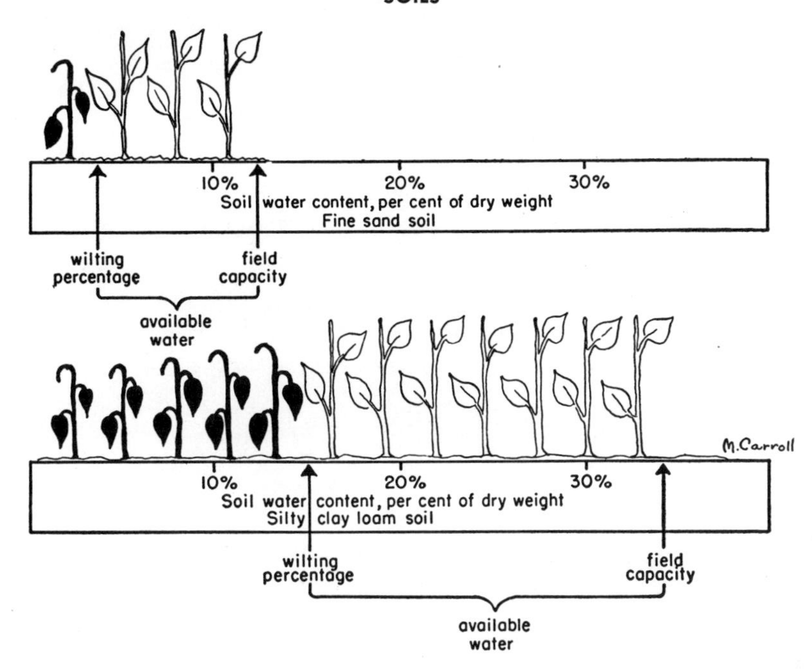

Fig. 11–3. Comparison of available water in a silty clay loam soil with that in a fine sand soil.

Nitrogen fixation. Certain soil bacteria change atmospheric nitrogen into nitrogen compounds which higher plants can use. This transformation of unavailable into available nitrogen is known as **nitrogen fixation.** *Clostridium* and *Azotobacter* are nitrogen-fixing bacteria which live free in the soil. Species of the genus *Rhizobium* live in nodules on roots of leguminous plants. They, too, fix nitrogen, and thus provide the leguminous plants with available nitrogen. Figure 11–4 shows the effect of inoculating red clover plants with *Rhizobium trifolium.* The plants were grown in quartz sand and were watered with a nutrient solution which lacked nitrogen. The inoculated plants, provided with available nitrogen by the bacteria, thrived whereas the uninoculated ones did not. For a more complete discussion of nitrogen fixation see Chapter 31.

Mycorrhiza. Some fungi grow in association with roots of certain species of plants, to their mutual benefit. In such instances the green plants make better growth when these fungi are present, and the fungi in turn obtain food from the green plants. In some instances the fungi cover the roots externally; in others the fungi are in the cells of the root. The symbiotic association of a root with a fungus is known as a **mycorrhiza** (Fig. 11–5). If the fungus is on the outside of a root, the mycorrhiza is known as an **ectotrophic mycorrhiza;** if the fungus is inside the cells, the mycorrhiza is **endotrophic.** The fungal threads of an ectotrophic mycorrhiza permeate the soil and absorb water and minerals, some of which become available to the roots. In soils high in organic matter, those of bogs and forests, the fungus brings about decay and thus makes nitrogen and other minerals available. The green plant benefits from the relationship by being provided with minerals. Mycorrhizal fungi are found on the roots of chestnut, beech, oak, and other trees, and on roots of members of the orchid and heath families. Not all associations of fungi with roots are beneficial to the green plant. In many instances fungi are parasitic on the roots and retard plant growth.

Fig. 11–4. Red clover plants whose roots were inoculated with *Rhizobium trifolium* grew vigorously, whereas those not inoculated grew poorly. (O. N. Allen, University of Wisconsin.)

Dissolved Substances

Salts of various minerals are in solution in the soil water and these are readily taken in by plant roots. However, not all the ions of salts are free in the soil water. We have seen that some ions are held on surfaces of soil particles; these ions are said to be **adsorbed.** The adsorbed ions can be taken up by roots which are in close contact with the soil particles. These roots exchange hydrogen ions for those of calcium, magnesium, potassium, ammonium, and other ions. Certain salts are only slightly soluble and at any one time there are just so many ions in solution. As these are taken up by roots, more go into solution. A fraction of the minerals present in soil are taken up by bacteria and other soil-dwelling organisms, and this fraction is then not available to roots. However, after death and decay of soil-dwelling organisms, the minerals are released and then are available to roots.

Under natural conditions, there is a continuous cycle of minerals out of the soil, into plants, and back to soil as a result of decay. If the cycle is broken by harvesting crops, the fertility of the soil diminishes unless fertilizers are added. The bodies of livestock sold off grazing country also represent a significant depletion of soil nutrients, especially phosphorus.

Roots are the primary absorbing organs of a plant and from the soil they take up both water and minerals. Roots absorb not only essential minerals but also nonessential ones. In other words, they do not discriminate absolutely between essential and nonessential minerals. An analysis of a plant often reveals fluorine, iodine, cobalt, sodium, silicon, aluminum, and sometimes even gold

Fig. 11–5. *Left,* an endotrophic mycorrhiza of hickory (*Carya ovata*). *Right,* an ectotrophic mycorrhiza of red maple (*Acer rubrum*). (W. B. McDougall, *Am. J. Botany.*)

and uranium, elements which play no role in plant metabolism. A high uranium content in shoots of deep-rooted trees and shrubs may give a prospector a clue that uranium deposits occur below the soil surface and a high content of gold may indicate a rich deposit. Selenium, a nonessential element, is abundant in certain shale soils that support a characteristic vegetation of woody aster, and certain species of *Astragalus* and *Stanleya*. The selenium present in the shoots of these plants is toxic to cattle and sheep which graze them. Fruits and vegetables growing on soils containing fluorine and iodine will contain these elements which, although not required by plants, are essential in human nutrition—fluorine to retard dental caries and iodine to prevent goiter.

Although under natural conditions minerals are absorbed primarily by roots, they can enter a plant through the leaves, as when a fertilizer solution is sprayed on them. Plants can take in nitrogen, phosphorus, potassium, iron, zinc, and other nutrients through the leaves. In alkaline soils, the iron in the soil may be insoluble and unavailable to plants. This deficiency may be overcome by spraying the foliage with a dilute solution of iron sulfate. In other areas zinc and copper deficiencies have been corrected by an application of salts of these elements to the foliage. Recently it has been demonstrated that woody plants can even take in minerals through the bark. When dormant trees and shrubs were sprayed with salts of nitrogen, phosphorus, and potassium, these entered the plants.

ESSENTIAL ELEMENTS

Plants require sixteen elements for growth. From the air and water they get carbon, hydrogen, and oxygen. From the soil they obtain nitrogen, phosphorus, potassium, calcium, magnesium, iron, sulfur, and trace amounts of manganese, boron, zinc, copper, molybdenum, and chlorine. The available nitrogen in the soil was derived originally from the atmosphere, whereas the remaining twelve minerals are contained in, and hence derived from, the parent rock material.

Micronutrients

Boron, copper, zinc, manganese, molybdenum, and chlorine are known as **micronutrients,** or **trace elements,** because only small amounts are required for plant growth and development. For example, a tablespoon or two of molybdenum per acre on some deficient soils makes a striking difference between healthy, vigorous plants and weak nonproductive ones. Of course, exact diagnosis is necessary before treatment. As a matter of fact, plants are injured by concentrations of trace elements only slightly higher than the concentrations that are beneficial. For normal plant growth they need not be present in amounts greater than one part per million. Most soils have an ample supply of trace elements, but in some areas micronutrients may be lacking. In several regions fruit and nut trees have benefited when zinc compounds were applied. In some areas the addition of manganese has been beneficial. In other localities the application of boron has resulted in better crops of celery, cauliflower, turnips, and sugar beets.

The micronutrients play essential roles in metabolism. Some micronutrients are activators of enzymes; others are constituents of enzymes. For example, copper is incorporated into ascorbic acid oxidase, a respiratory enzyme. Zinc is associated with an enzyme that synthesizes tryptophane from which indoleacetic acid, a plant hormone, is synthesized (Fig. 11–6*A*). Molybdenum is the active center of nitrate reductase, an enzyme necessary for the reduction of nitrate to ammonia. Ammonia combines with certain organic acids to form amino acids; the amino acids are then joined together to form a protein.

The information gained from studies of the micronutrient requirements of plants greatly benefits farmers. For example, in Florida the value of the increased yields resulting from the application of certain micronutrients amounts to about $90 million annually. This is three times as much as has been appropriated by the state of Florida for agricultural research during her entire history. Of course, farmers in other states and nations benefit from the application of this knowledge and rewards come in year after year.

Macronutrients

Nitrogen, phosphorus, potassium, calcium, magnesium, sulfur, and iron are used in greater amounts than are the trace elements. These elements are taken up by plants as inorganic ions in the form of nitrate (NO_3), ammonium (NH_4), phosphate (PO_4), sulfate (SO_4), potassium (K), calcium (Ca), magnesium (Mg), and iron (Fe). Most soils have large reserves of calcium, magnesium, sulfur, and iron, and hence such elements seldom need be added. Plants require large amounts of nitrogen, phosphorus, and potassium. Because the soil reserves of such elements are limited, it is usually necessary to add compounds of nitrogen, phosphorus, and potassium to cultivated soils. Manufacturers of fertilizers state on the labels the amounts of these three elements which the fertilizers contain by symbols such as 5–10–5, 4–12–4, etc. The numbers signify, respectively, the percentages by analysis of nitrogen as the pure element, phosphorus as P_2O_5, and potassium as K_2O.

Nitrogen, phosphorus, and potassium, as well as other minerals, are removed from the soil by cropping, grazing, runoff, and leaching. A ton of wheat grain contains about 40 pounds of nitrogen, 8 pounds of phosphorus, and 9 pounds of potassium, minerals which came from the soil. A ton of cattle corresponds to a soil depletion of 54 pounds of nitrogen, 15 pounds of phosphorus, and 3 pounds of potassium. Such rates of removal will quickly exhaust a typical soil unless the losses are made up by additions of fertilizer.

Of the short-range factors capable of increasing agricultural productivity, so desperately needed in developing countries, the largest yields and the most substantial returns on invested capital result from the addition of chemical fertilizer to the soil. The application of fertilizer to underfertilized soils has dramatic results: crop yields are often increased between 100 and 200 per cent.

A

B

Fig. 11–6*A*. *Left*, zinc deficiency, shown in a branch from an orange tree. The leaves are mottled, and the fruit is small and borne upright. *Right*, a branch from a tree which was treated with zinc. The foliage is healthy, and the fruit is large and pendant. (University of Florida, College of Agriculture, Citrus Experiment Station.) *B*. Plant growth is abnormal when essential minerals are deficient. The check (CK) tobacco plant was grown with all the essential minerals. Notice the weak growth when potassium (-K), calcium (-Ca), boron (-B), magnesium (-Mg), nitrogen (-N), phosphorous (-P), or sulfur (-S) was lacking. (W. Rei Robbins and the New Jersey Agricultural Experiment Station.)

A vital task for biologists is the education of farmers—particularly farmers in developing countries—in the effective use of fertilizers. Developing nations should establish credit plans so that impoverished farmers can buy adequate supplies of fertilizer. To aid developing nations produce enough food for their people, the developed nation might consider granting credit for the purchase of fertilizer or for the building of fertilizer plants. As the world's population increases, the need for chemical fertilizers will increase tremendously. The world's production of fertilizers containing nitrogen, phosphorus, and potassium is now 30 million tons. In the year 2000, when the population of the earth may be 6 billion, 120 million tons will be needed.

Roles of macronutrients. Macronutrients play many roles in plant nutrition. They are building materials, and they influence osmotic pressure, imbibition, and permeability. In the following discussion only a few of the specific roles that each macronutrient performs are discussed. In addition to these they have other functions. In the following discussion refer to Figure 11–6*B*.

Nitrogen is an essential part of molecules of proteins, nucleic acids, chlorophylls, and alkaloids. A deficiency of nitrogen results in stunted growth and a yellowing of foliage. Nitrogen can be added to the soil in organic form or as certain salts. Dried blood and manure are fairly rich in nitrogen. Sodium nitrate, calcium nitrate, ammonium sulfate, and ammonium nitrate are salts which furnish available nitrogen to plants. Where large-scale irrigation is practiced, ammonia may be dissolved in the irrigation water. With special equipment anhydrous ammonia gas can be injected directly into the soil.

Phosphorus is a constituent of nucleic acid; of phospholipids, some of which are incorporated in membranes; and of ATP, the major energy carrier. Phosphorus is also necessary for carbohydrate transformations and for respiration and cell division. When phosphorus is deficient, plants are stunted and the foliage is dark green.

Potassium catalyzes the formation of proteins, fats, and carbohydrates. It is also necessary for cell division. Plants deficient in potassium are dwarfed, and frequently the tips and edges of the leaves are dead.

Sulfur is incorporated into proteins. When sulfur is deficient, dead spots form on the leaves and the leaves are light green.

Calcium is an essential part of the middle lamella, where it is found as calcium pectate. It favors the translocation of carbohydrates and amino acids. Plants deficient in calcium are stunted and frequently the growing points die.

Magnesium is a constituent of chlorophyll. Leaves of plants deficient in magnesium are greenish-yellow in color.

Although iron is not a part of the chlorophyll molecule, it is necessary for the formation of chlorophyll. Iron plays a role in respiration and is necessary for growth, including the growth of isolated roots. The leaves of iron-deficient plants are yellow, as would be expected. While iron salts are usually present in the soil, they may be insoluble in certain soils, especially alkaline ones. Ferrous sulfate may be used to acidify soils and at the same time to furnish plants with available iron. Even in some acid soils there may not be enough available iron to meet the demands of certain plants. Iron deficiencies may often be overcome by an application of ferrous sulfate. Recently chelated iron compounds have been found to be very effective in preventing iron deficiency in citrus and many other plants.

That plants show characteristic symptoms when a certain nutrient is lacking is evident from the key on page 153, furnished by the Ohio Agricultural Experiment Station.

DETERMINATION OF ESSENTIAL ELEMENTS

The minerals essential for the growth and development of plants have been determined by culturing plants in water instead of in soil. Paraffined vessels are filled with

Key to Nutrient Deficiency Symptoms

1. Effects general on whole plant or localized on older, lower leaves.
   2. Effects usually general on whole plant, although often manifested by yellowing and dying of older leaves.
      3. Foliage light green. Growth stunted, stalks slender, and a few new breaks. Leaves small, lower ones lighter yellow than upper. Yellowing followed by a drying to a light brown color, usually little dropping. Minus nitrogen.
      3. Foliage dark green. Retarded growth. Lower leaves sometimes yellow between veins but more often purplish, particularly on petiole. Leaves dropping early. Minus phosphorus.
   2. Effects usually local on older, lower leaves.
      4. Lower leaves mottled, usually with necrotic areas near tip and margins. Yellowing beginning at margin and continuing toward center. Margins later becoming brown and curving under, and older leaves dropping. Minus potassium.
      4. Lower leaves chlorotic and usually necrotic in late stages. Chlorosis between the veins, veins normal green. Leaf margins curling upward or downward or developing a puckering effect. Necrosis developing between the veins very suddenly, usually within 24 hours. Minus magnesium.
1. Effects localized on new leaves.
   5. Terminal bud remaining alive.
      6. Leaves chlorotic between the veins; veins remaining green.
         7. Necrotic spots usually absent. In extreme cases necrosis of margins and tip of leaf, sometimes extending inward, developing large areas. Larger veins only remaining green. Minus iron.

            Note: Certain cultural factors, such as high *p*H, overwatering, low temperature, and nematodes on roots, may cause identical symptoms. However, the symptoms are still probably of iron deficiency in the plant due to unavailability of iron caused by these factors.
         7. Necrotic spots usually present and scattered over the leaf surface. Checkered or finely netted effect produced by even the smallest veins remaining green. Poor bloom, both size and color. Minus manganese.
      6. Leaves light green, veins lighter than adjoining interveinal areas. Some necrotic spots. Little or no drying of older leaves. Minus sulfur.
   5. Terminal bud usually dead.
      8. Necrosis at tips and margins of young leaves. Young leaves often definitely hooked at tip. Death of roots actually preceding all the above symptoms. Minus calcium.
      8. Breakdown at base of young leaves. Stems and petioles brittle. Death of roots, particularly the meristematic tips. Minus boron.

---

carefully distilled and redistilled water to which the desired minerals are added. Salts used in the experiments must first be meticulously purified. One group of plants is given all the minerals previously found to be essential plus the one about which information is sought, whereas another group is given all except the mineral in question. A comparison of the two groups gives a clue as to the role of that particular mineral. Perhaps this mineral has been present in soils or as an impurity in salts used in fertilizers, so that all along it has been playing an unsuspected role in plant growth. Determination of its specific contribution will aid in diagnosing an ailment that might otherwise go uncured.

HYDROPONICS OR NUTRICULTURE

At first, the raising of plants without soil was mainly of interest to students of plant nutrition, helping them to determine what kinds and amounts of nutrients are required. In recent years, it has been adopted on a limited scale for the commercial production of plants.

Soilless culture methods are practical for the production of some greenhouse crops and for the outdoor production of a few vegetables and floral crops in mild climates.

Three general methods may be used: (1) water culture, (2) sand culture, (3) subirrigation culture (gravel or cinder culture).

1. *Water Culture.* In the water culture method, plants are grown with their roots suspended in a nutrient solution contained in a shallow tank. The plants are supported above the solution by wire netting covered with straw, wood shavings, or rice hulls. Usually provision is made for the aeration of the culture solution.

2. *Sand Culture.* In sand culture, benches or beds are filled with sand which is watered with a nutrient solution. It is a simple and satisfactory method but is more expensive than the subirrigation method.

3. *Subirrigation.* In the subirrigation method, watertight benches or beds are filled with gravel, cinders, or haydite (sintered shale). Haydite, which is used in the preparation of low-density concrete, is inert, fairly porous, and has a high water-holding capacity. At intervals, the culture solution is pumped into the benches, thus wetting the aggregate. The culture solution drains back into the storage cistern. The solution is used over and over again. It must be tested at intervals, however, and chemicals that have been depleted must be added. Subirrigation can be made automatic by the installation of a time switch to turn on the pump.

Botanists have developed many nutrient solutions which support thrifty plant growth but which vary in the salts used and their ratio to one another. Among the favorite nutrient solutions is that formulated by Hoagland. The composition of Hoagland's nutrient solution is as follows:

Salt	Grams per Liter of Distilled Water
Calcium nitrate—$CaNO_3 \cdot 4H_2O$	1.18
Potassium nitrate—KNO_3	0.51
Potassium phosphate—KH_2PO_4	0.14
Magnesium sulfate—$MgSO_4 \cdot 7H_2O$	0.49
Ferric tartrate—$FeC_4H_4O_6$	0.005
Boric acid—H_3BO_3	0.0029
Manganese chloride—$MnCl_2 \cdot 4H_2O$	0.0018
Zinc sulfate—$ZnSO_4 \cdot 7H_2O$	0.00022
Cupric sulfate—$CuSO_4 \cdot 5H_2O$	0.00008
Molybdic acid—$H_2MoO_4 \cdot H_2O$	0.00002

CHARACTERISTICS OF PRODUCTIVE SOIL

A productive soil furnishes plants with the essential minerals in the proper ratio and with a continuous supply of water and oxygen. The fertility of soil may be determined by analyzing soil extracts for essential minerals or by analyzing plant parts, especially petioles. If soil is low in available phosphorus, nitrogen, and potassium, the petioles also will be low in these minerals. A more direct method is to compare the yields of plants growing in soil to which minerals have been added with those growing in unfertilized soil.

In addition, a good growing soil is free from such harmful factors as disease organisms, insect pests, and an excess of salts.

Some seed plants produce chemicals which are injurious to other species. For example, wormwood (*Artemisia absinthium*) depresses the growth of plants around it. Black walnut (*Juglans nigra*), guayule (*Parthenium argentatum*), and brittlebush (*Encelia farinosa*) produce chemicals detrimental to other species in the same soil. In *Eucalyptus* groves the observer is impressed by the bare soil, which may be partially, if not entirely, due to the influence of chemicals washed out from the fallen leaves and flower buds. The effects of such plants on others has been referred to as chemical warfare among the plants.

The reaction of the soil should be suited to the species which are grown. **Soil reaction** refers to the degree of acidity or alkalinity of the soil. The *p*H system is used to designate soil reactions. The *p*H is the logarithm of the reciprocal of the hydrogen ion concentration. At *p*H 7 the soil is neutral. If the *p*H is less than 7, the soil is acid in reaction; the lower the figure, the more acid the soil. Because *p*H values are logarithms, a soil with a *p*H of 5 is ten times as acid as one with a *p*H of 6. Soils with *p*H values above 7 are alkaline; and the greater the number, the more alkaline the soil.

Most plants make their best growth at a *p*H of 6 or 7, but such plants as blueberries, rhododendrons, azaleas, citrus fruits, and African violets prefer a more acid soil, one with a *p*H of about five.

In some regions, soils are too acid for the best development of crop and garden plants. To make the soils less acid, lime is added.

In many parts of the West, soils are too alkaline for the best growth of some crop plants. Such soils may be made neutral by the addition of sulfur or by the incorporation of leaf mold or peat into the soil. It is difficult and expensive to change alkaline soils to neutral or acid soils. In field practice the *p*H of the soils is not altered, but species and varieties adapted to alkaline soils are selected for culture. For small gardens and for the pot culture of plants, it is feasible to change the soil from an alkaline to a neutral or an acid soil.

SOILS, PLANTS, AND NUTRITION

Animal nutrition begins with the soil. Plants growing on infertile soils have a low mineral content. Animals feeding on such plants may show symptoms of one or another mineral deficiency. For example, phosphorus is often deficient in virgin soils throughout the Atlantic and Gulf Coastal Plains. The phosphorus content of the forage produced on such soils is not sufficient to meet the dietary requirements of cattle. Cobalt, although not required by plants, is essential for animals. If soils are deficient in cobalt, the cobalt content of the forage does not meet the requirements of cattle and sheep. In some areas, the copper requirements of livestock are not satisfied by forage. Mineral deficiencies of animals may be corrected either by fertilizing the soil or by supplying the animals with minerals.

Human nutrition is also related to soil fertility. Human beings obtain most of their minerals from plant and animal foods. If foods of human beings are deficient in minerals, dietary deficiencies become evident. Increasing attention is being directed to the production of fruits and vegetables to meet dietary requirements as well as to be tasty and attractive.

EFFECT OF PLANTS ON THE SOIL

So far we have been concerned with the contributions of soil to plants. Now let us consider the contributions that the plant cover makes to the soil. During hard rains plants break the fall of raindrops. Their impact is reduced, the soil is not compacted, and splash erosion is retarded. Soils with a good plant cover are porous; hence much of the water enters the ground instead of running off. Trees in a forest shade snow and thereby retard melting. When snow melts gradually, streams flow more evenly. Such evenness of stream flow is beneficial to farmers who depend upon irrigation in July and August, to owners of hydroelectric plants, to trout fishermen, and to cities depending upon rivers for water.

Plants play an important role in the development and maintenance of fertile soils. They retard wind and water erosion. Some minerals are taken from deeper layers of soil by roots and moved into the shoots. When roots and shoots decay, the minerals are left in the upper layer of soil. In tropical regions, where the rainfall may be several hundred inches a year, a dense forest is especially important in maintaining soil fertility. As fast as plant debris decays, the liberated minerals are taken up by roots and hence cannot be leached out by the heavy rains.

QUESTIONS

1. _____ If a soil has a dark color, it usually indicates that the soil (A) is rich in organic matter, (B) has many fine rock particles, (C) has large rock particles, (D) is rich in air and water, (E) has an abundance of air in it.

2. Discuss four components of soil.

3. Why is it necessary to drain wet land before ordinary crops can be grown thereon?

4. _____ The total available water, from field capacity to the wilting percentage, is greatest in a (A) sandy soil, (B) loam soil, (C) clay loam soil.

5. _____ A tomato seedling is planted in each of four pots containing a coarse sand, a sandy loam, a clay loam, and a clay soil, respectively, and the pots are sealed against water loss. Initially the soil was at field capacity in each pot. The water content of the soil when the plants permanently wilt would be greatest in the (A) coarse sand, (B) sandy loam, (C) clay loam, (D) clay.

6. _____ In the A horizon (A) leaching of minerals and clay occurs, (B) the leached minerals accumulate, (C) microbial activity is low, (D) the soil is lighter in color than in the B horizon.

7. _____ The productivity of soils for farm crops is related to the vegetation which the area supported prior to its cultivation. Which of the following would be most suitable for farming? (A) Desert, (B) short grass plains, (C) tall grass prairie, (D) tropical rainforest, (E) coniferous forest, (F) deciduous forest.

8. _____ Which one of the following is false? (A) Organic matter in soil promotes good aeration and increases the water-holding capacity. (B) Tropical soils are generally richer in organic matter than soils in Iowa. (C) Certain fungi in the soil may produce chemicals which inhibit the growth of other fungi. (D) As organic matter decays, minerals are made available to roots. (E) The symbiotic association of a root with a fungus is known as a mycorrhiza.

9. _____ Which one of the following is false? (A) The wilting percentage is the same, or nearly so, for xerophytes and mesophytes. (B) At field capacity both water and air are present in soil. (C) Water rises higher by capillarity from the water table in sandy soils than in clay soils. (D) Continuous growth of roots may result in additional supplies of water and minerals available to the plant.

10. _____ Which one of the following is false? (A) Animal nutrition begins with the soil. (B) A soil with a *p*H of 6 is alkaline. (C) Plants can be raised in water, sand, or gravel if a nutrient solution is used. (D) The minerals essential for plants have been determined by culturing plants in water containing the appropriate dissolved minerals.

11. _____ Peas and beans are often rotated with other crops like grains because (A) the peas and beans fix nitrogen, (B) bacteria which grow in nodules on peas and beans change atmospheric nitrogen into compounds of nitrogen which higher plants can use, (C) the bacteria which grow on peas and beans make phosphates available, (D) of custom and the practice is not desirable.

12. _____ The three elements that are likely to be deficient in many farm soils are (A) calcium, iron, and nitrogen; (B) iron, nitrogen, and potassium; (C) nitrogen, phosphorus, and potassium; (D) magnesium, sulfur, and phosphates.

13. _____ A soil with a *p*H of 8.6 could be made more productive by adding (A) potassium hydroxide, (B) sulfur, (C) lime.

14. Name one function for each of the following minerals: magnesium, nitrogen, sulfur, phosphorus, calcium.

15. _____ Which one of the following is false? (A) The determination of the exact nutrient requirements of a plant may be done with soil to which various nutrients are added or subtracted. (B) Nutrients required in very small amounts are called micronutrients or trace elements. (C) Plants take up small amounts of many of the mineral elements present in the soil in addition to the thirteen essential ones. (D) Certain elements taken up by some plants may be toxic to animals.

16. Describe a commercial setup for raising plants without soil.

17. _____ Which one of the following may accumulate in plants in amounts great enough to be injurious to livestock? (A) Gold, (B) selenium, (C) uranium, (D) aluminum.

18. _____ Cations, such as calcium, potassium, magnesium, and ammonium, which are adsorbed on clay colloids (A) may be absorbed by roots which are in close contact with the clay particles, (B) cannot be taken up by roots, (C) are easily leached from soils, (D) are generally more soluble in fat solvents than in water.

19. _____ Cultivation of soil is not beneficial in (A) destroying weeds which compete with crop plants for water and minerals, (B) loosening soil, thus facilitating gaseous exchange between the soil and atmosphere, (C) conserving moisture by retarding evaporation from soils which are at or below field capacity.

20. Does man today use the soil resource more or less wisely than in Pre-Columbian times?

21. In 1860, Julius Sachs grew plants without soil using a nutrient solution made from distilled water containing the following:

Salt	*Gm./liter*
KNO_3	1.0
$Ca_3(PO_4)_2$	0.5
$MgSO_4 \cdot 7H_2O$	0.5
$CaSO_4$	0.5
NaCl	0.25
$FeSO_4$	Trace

The plants grew vigorously because (A) the salts listed are the only ones required for plant growth and development, (B) impurities in the salts or water provided the plants with micronutrients, (C) all plants do not require micronutrients.

22. _____ In the early seventeenth century, van Helmont planted a willow weighing 5 pounds in a box containing 200 pounds of dry soil. Five years later the tree weighed 169 pounds. He weighed the soil and found that its weight had diminished (A) by about 164 pounds, (B) by only a few ounces, (C) not at all.

23. A plant physiologist allowed plants to permanently wilt. He then removed a sample of soil which weighed 110 grams. The soil was then placed in an oven maintained at 105° C. The dry soil weighed 100 grams. What is the wilting percentage?

12

Stems

The immense cathedral-like stems of forest trees, the long encircling stems of vines, the very short stem of the dandelion, the hard stem of ebony, the soft herbaceous stem of asparagus, the spiny stem of a cactus, the annual stem of a bean, and the perennial stem of a catalpa tree are certainly unlike in external appearance—but all have a similar origin, all have the same functions to perform, and all possess essentially the same tissues for carrying out the various functions. All stems bear leaves, either full-sized leaves or rudimentary ones. Indeed, a stem may be defined as the structure which bears leaves. Buds are also present on stems. You will recall that roots lack leaves and buds.

FUNCTIONS OF STEMS

Stems support and display flowers, fruits, and leaves. In nature there is intense competition among plants for light, and stems have evolved many and varied features for exposing the leaves to the sun. The stems of even the tallest trees support themselves, while those of vines and scramblers depend on other plants for their support.

The stem carries the conducting tissues between the roots, or intake centers for water and nutrients, and the leaves, or food-making organs. The water and minerals that move up from the roots travel in the vessels and tracheids of the xylem. The xylem continues through the roots and stems and into the leaves. In a soft stem, a herbaceous one, the amount of xylem is relatively small, but the accumulation of xylem through the years in a tree or shrub leads to the rigid tissue we call wood. The food that is made in the leaves is distributed to all parts of the plant by the sieve tubes of the phloem. The sieve tubes run from the leaves to the stem, through the stem into the roots, always parallel to the xylem.

Green stems manufacture food. In most plants only a small amount is made by stems, but in cacti and other plants which lack foliage leaves, all food is produced by green stems. Food is stored in stems, in ordinary ones or in such specialized stems as the tuber and corm; the latter is a short, upright underground stem (see Chapter 18). In sugar cane, a major commercial source of sugar, the sugar is stored in an ordinary stem. The potato tuber is a principal food source for man. The potato, although native to the Western Hemisphere, is now cultivated in most countries. In Polynesia the corms of taro are a major food.

SPECIALIZED STEMS

The stems of some plants are modified into specialized structures such as **tendrils, thorns, tubers, rhizomes, corms,** and **runners.** The tendrils of the grape, Virginia creeper (Fig. 12–1*A*), and Boston ivy are modified stems. Certain branches of the osage orange, honey locust, and buffaloberry are modified to form sharp-pointed structures known as thorns (Fig. 12–1*B*). Not all tendrils and pointed structures are modified stems. You will recall that leaves may be modified into tendrils and spines. The **prickles** on the stem of a rose plant are neither modified stems nor leaves, but instead they are outgrowths of the epidermis. For information about tubers, rhizomes, bulbs, stolons, and corms see Chapter 18.

A structure that develops from a bud and that bears leaves, either full-sized or rudimentary, is a stem. A plant part—tendril or spine, for example—that occurs at a node and has a bud just above it is a leaf. Structures such as rose prickles which are not located at nodes and which do not develop from buds are neither stems nor leaves.

BUDS AND SHOOTS

Nodes and Internodes

A shoot of a geranium plant, which consists of a stem and leaves, is illustrated in Figure 12–2. You will notice that the stem is made up of nodes and internodes. A node is the region of a stem where a leaf or leaves are attached. You will recall that one leaf is present at each node when the leaf arrangement is alternate, two at a node when opposite, and more than two when whorled.

Terminal Bud

A terminal bud is present at the tip of each stem. The terminal bud of a geranium stem consists of immature leaves and a **stem tip** or **shoot apex.** At the tip of the stem, cells are dividing. Some of the resulting cells become stem cells and thereby increase the length of the stem; others are organized into leaves.

Axillary Buds

Above each leaf, in the axil between the petiole and the stem, an axillary bud (also called a lateral bud) is present. Each axillary bud of a geranium plant, like the terminal bud, consists of a shoot apex enveloped in immature leaves. As long as the terminal bud is present, few of the axillary buds will develop. This is because in the terminal bud a chemical, called **auxin,** is produced which moves downward and inhibits the growth of the axillary buds. If the terminal bud is cut off or used up in the formation of flowers, the supply of auxin ceases and the axillary buds develop. An axillary bud, like the terminal bud, grows into a branch (Fig. 12–3).

Accessory Buds

In certain species, more than one bud is present in the nodal region. For example, in the apricot three buds are present just above where each leaf is attached to the stem. The central bud is the axillary bud and the other two are accessory buds. The axillary bud develops into a branch bearing leaves whereas each accessory bud develops into a branch bearing flowers.

Flower Buds, Leaf Buds, and Mixed Buds

In geraniums, some buds, the flower buds, develop into branches bearing only flowers; other buds, the leaf buds, develop into branches bearing leaves. In other plants, apple for example, some buds develop into branches which bear both leaves and flowers; these are known as mixed buds.

Covered and Naked Buds

Both the terminal buds and axillary buds of temperate zone trees and shrubs are usually encased in bud scales which are shed when, and if, the bud begins to develop; such buds are known as covered buds. Buds of herbaceous plants are usually de-

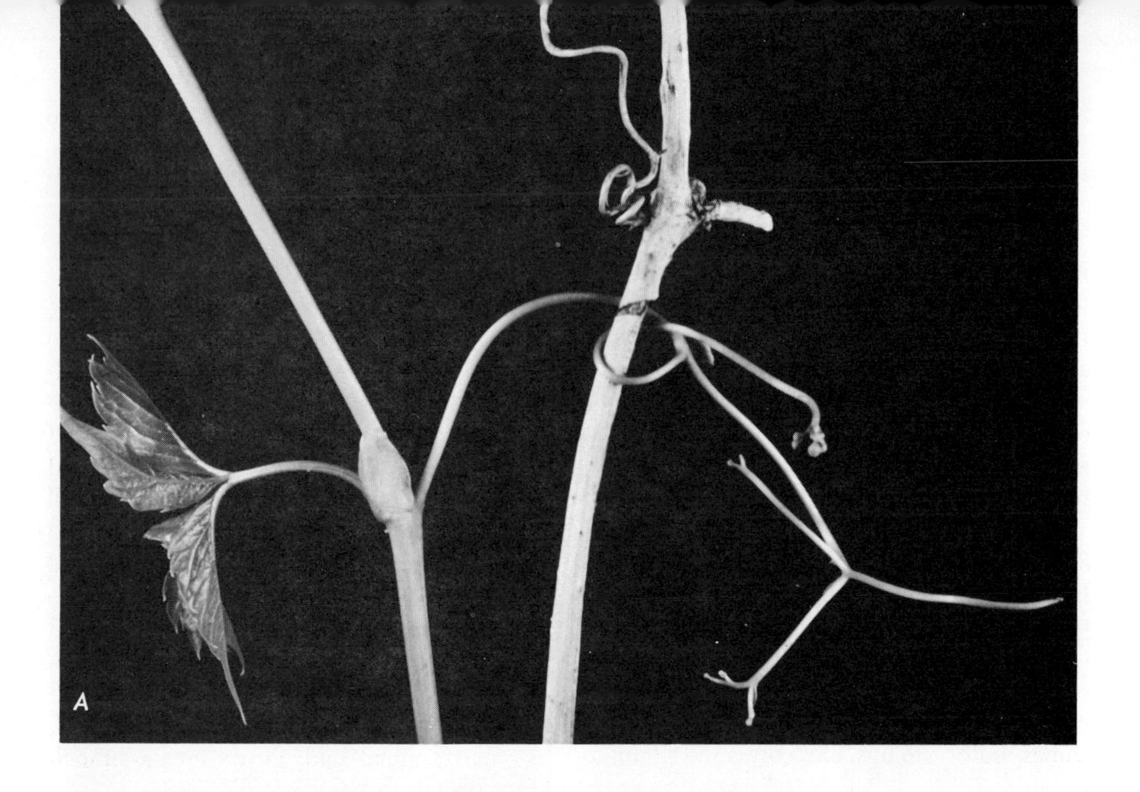

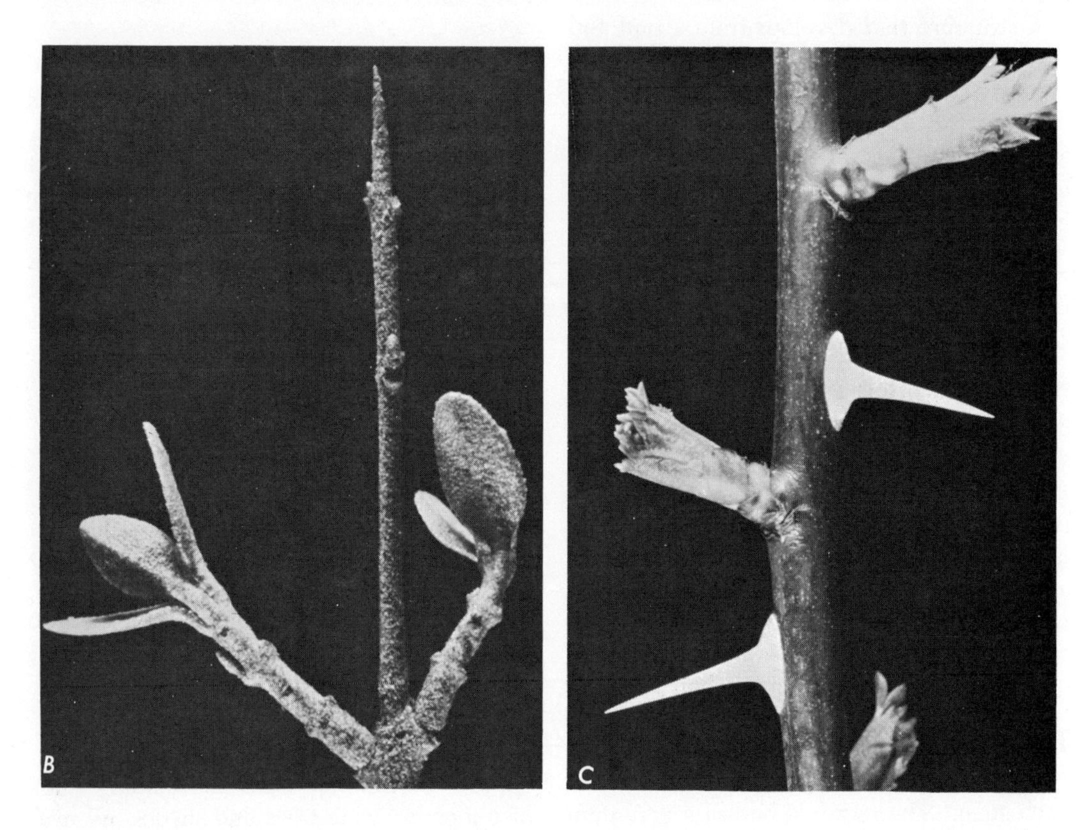

Fig. 12–1*A*. The tendril of a Virginia creeper is a modified stem. *B*. The thorn of a buffaloberry is a modified stem. C. The prickles of a rose are outgrowths of the epidermis.

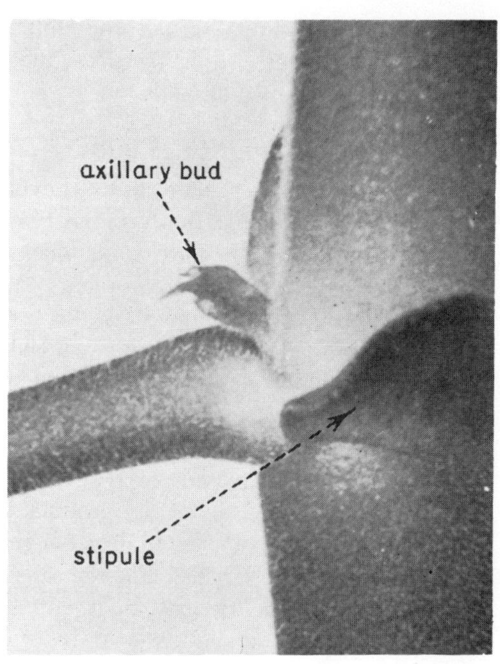

Fig. 12–2. Stems and their external structure. *Left*, part of a geranium plant. New leaves as well as additional stem cells are formed by the terminal bud. Leaves are attached to the stem at nodes. *Right*, details of a stem at a node. An axillary bud is present above the horizontal petiole. Stipules are also evident.

Fig. 12–3. Branch development. After the terminal bud of this Martha Washington geranium was removed, the axillary buds developed into branches. (Henry T. Northen and Rebecca T. Northen, *The Complete Book of Greenhouse Gardening*. Copyright © 1956, The Ronald Press Company, New York.)

void of bud scales and the delicate stem tip is protected only by the young leaves; we call these naked buds.

Pruning in Relation to Buds

A knowledge that buds develop into branches and that the terminal bud inhibits the growth of axillary buds may be used to control the shapes of plants. Branched plants may be obtained by the removal of the terminal bud, following which the axillary buds develop into branches. After the branches have grown to the desired length, their terminal buds may in turn be removed. This may be repeated until a much-branched plant is produced. Plants can be trained to one main stem by removing the axillary buds. Flowers of large size and good quality may be secured by **disbudding.** In disbudding, only one flower bud is permitted to develop on a stem; the other flower buds are removed. Carnations, chrysanthemums, peonies, calendulas, and other plants are frequently disbudded.

Commercial rose growers often time their crop by removing the terminal buds at certain dates. After the terminal bud of a rose stem is removed, the axillary buds develop into branches which bear leaves at first and ultimately flowers. The grower knows the number of days it takes for a bud to grow into a branch bearing a flower, and this many days before the desired flowering time he removes the terminal buds.

THE STEM TIP

Stems continue to increase in length as long as they live, a phenomenon especially obvious in trees and shrubs. Experience shows us, however, that not all parts of a branch or trunk of a tree continually increase in length. This is noticeable in the main trunks of trees. Branches on trees remain the same distance from the ground year after year, as do initials carved on tree trunks or nails driven into the trunks. Stems increase in length only at or near the stem

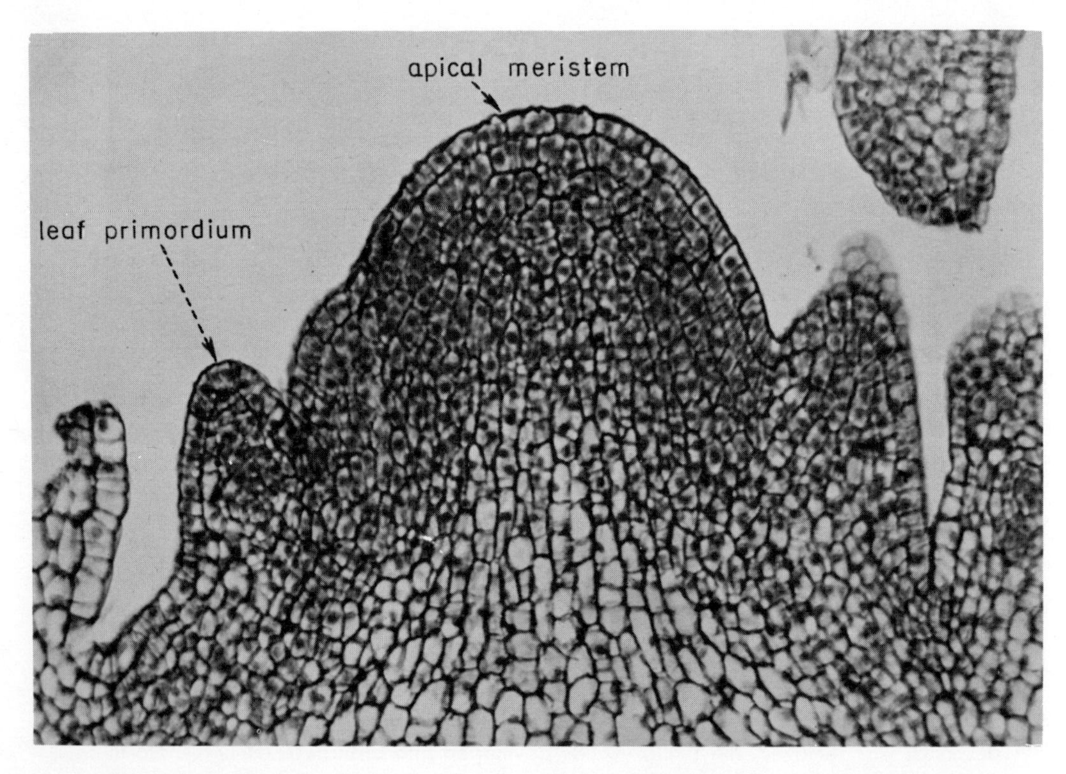

Fig. 12–4. Stem tip of a lupine (*Lupinus alba*), showing apical meristem and leaf primordia. The leaf primordia develop into leaves. (Ernest Ball.)

tips. A Christmas tree may show ten annual rings at the base of the tree, but only one will be evident at the top, the stem cells added during the past season.

Figure 12–4 is a photomicrograph of the stem tip or shoot apex of a lupine. The dome-shaped mass of small cells with dense protoplasm and small vacuoles, at the very tip, is the **apical meristem.** During the growing season, the cells of the apical meristem divide. Some of the resulting cells develop into stem cells which, after elongation, increase the length of the stem. Other cells formed by the division of cells at the apical meristem become organized into **leaf primordia** which ultimately, through cell division, enlargement, and differentiation, become leaves.

At the very tip of the stem, cell division is most active. Back from the tip cells elongate; often they become ten or more times longer than the meristematic cells. Cell elongation results from an uptake of water, an increase in the size of vacuoles, and the formation of additional cell-wall material and protoplasm. The zone of elongation may extend a considerable distance behind the apex, in some species for 6 inches, a distance which would include several internodes.

While the cells are elongating, differentiation of the specialized stem tissues begins and then three regions can be noted: The **protoderm** on the outside; strands of **procambium;** and the remainder, the **ground meristem** (Figs. 12–5 and 12–6). As differentiation progresses, the protoderm develops into the **epidermis,** a layer just one cell thick which covers and protects the underlying tissue. As in leaves, the epidermis allows, by means of stomata, an exchange of gases. The **cortex, pith,** and **pith rays** develop from the ground meristem. The cortex, usually made of collenchyma and parenchyma cells, extends from the epidermis to the vascular tissue. The pith, typically of parenchyma cells, occupies the center of the stem. The pith rays connect the pith with the cortex. The cells of the procambium are narrower, have denser cytoplasm, and are

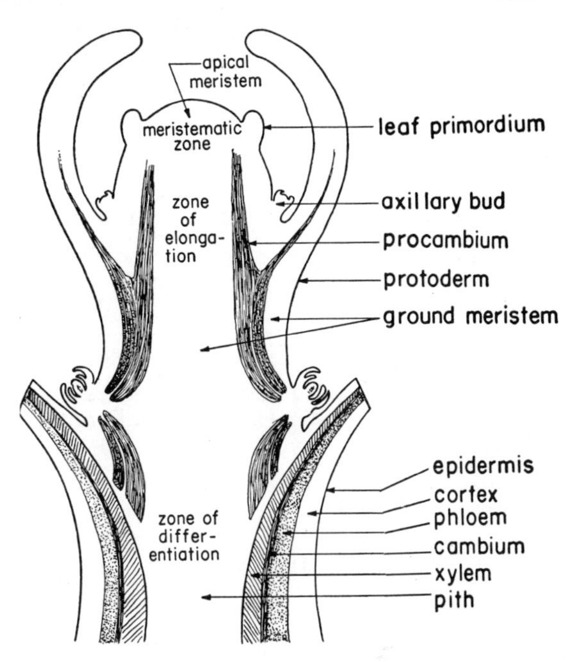

Fig. 12–5. Growth zones in the terminal part of a stem.

longer than those of the ground meristem. The procambium gives rise to the primary vascular tissues. The innermost cells of each procambium strand mature into **primary xylem,** the outer ones, into **primary phloem.** Between the primary xylem and phloem there remains in dicotyledons a single layer of cells, the **cambium** layer, whose function we will consider a little later. The mature tissues that originate from cells derived from the apical meristem are called primary tissues. The phloem derived from the procambium, which in turn originated from cells formed at the apical meristem, is known as primary phloem. The xylem derived from procambium cells is primary xylem.

The location of the primary tissues can be determined by microscopic study of a cross-section of a stem from the mature region.

TYPES OF STEMS

In the following discussion, we shall study two herbaceous stems, and in the next chap-

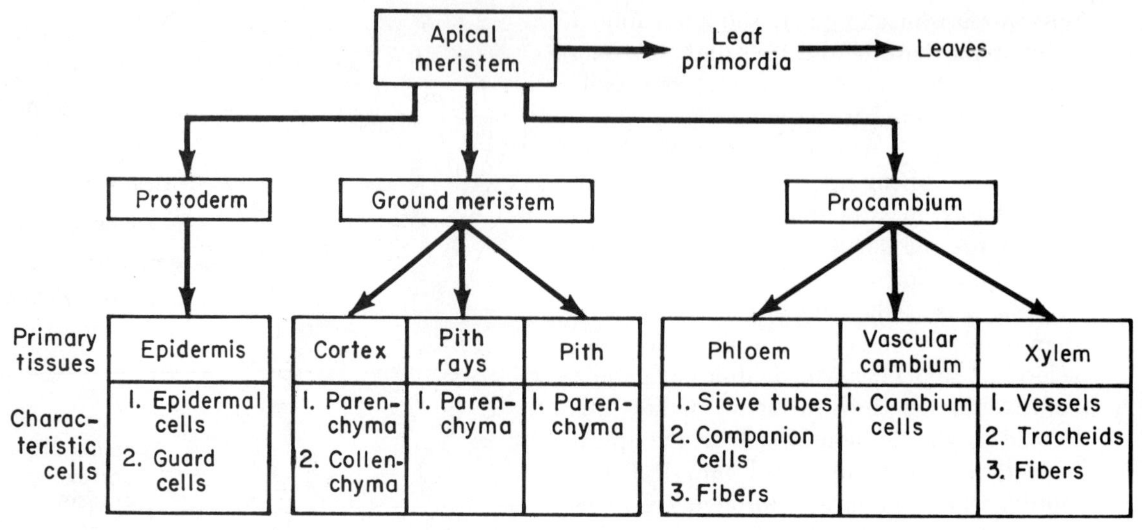

Fig. 12–6. Flow chart showing tissues produced from cells formed by the apical meristem.

ter a woody stem. The first herbaceous stem is that of corn, which is a representative monocotyledonous plant. The second is that of an alfalfa plant, a dicotyledonous plant. In a monocotyledonous stem the vascular bundles are scattered, whereas in a dicot stem they are arranged in a circle. Furthermore, a cambium layer is absent in stems of monocots, but is present in stems of many dicotyledons.

MONOCOTYLEDONOUS STEMS

A cross-section of a corn stem after tissues have differentiated is illustrated in Figure 12–7. The outermost layer of cells is the epidermis. The somewhat circular structures scattered throughout the stem (Fig. 12–7*A*) are **vascular bundles**—structures concerned with conduction. Each vascular bundle consists of xylem toward the inside and phloem toward the outside. In the accompanying figures we see the vascular bundles in cross-section. In lengthwise view the vascular bundles would appear as long strings extending upward in the stem.

Epidermis

The single layer of protective cells on the outside is the epidermis. The cells are small, fit together tightly, and are covered on the outside with a cuticle.

Ground Parenchyma

The ground parenchyma is the tissue between the vascular bundles. The cells in this region are large, thin-walled, and loosely arranged; they are parenchyma cells.

Vascular Bundles

In each vascular bundle **phloem** is present toward the outside and **xylem** toward the inside (Fig. 12–7*B*). The xylem is characterized by two large, pitted vessels (the **metaxylem**) and one to several smaller vessels (the **protoxylem**) with spiral or ring thickenings, below which there is generally an air space. You will recall that in stems the first xylem cells to mature are the inner ones which hence are smaller than the outer xylem cells which mature later. Between the two large vessels there are several tracheids. The vessels and tracheids conduct water and minerals and also furnish support.

Small, almost square companion cells with dark cytoplasm alternate with large, clear sieve tubes in the phloem region. In some

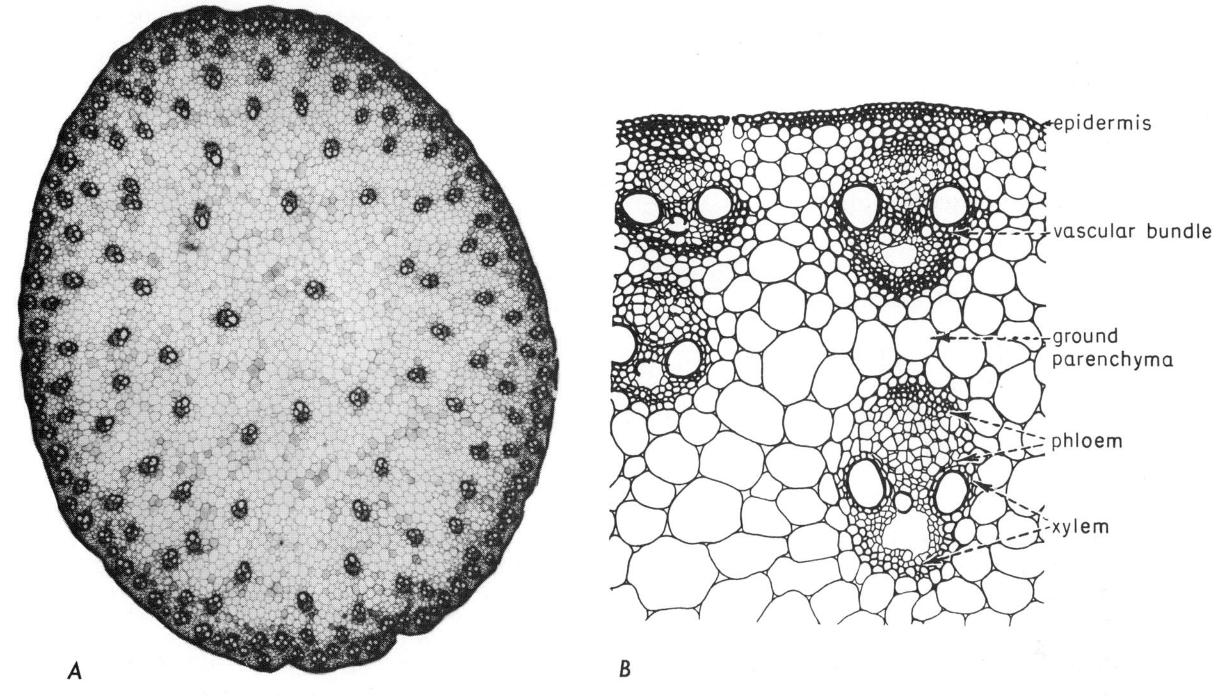

Fig. 12–7*A*. A cross-section of a corn stem. (George H. Conant.) *B*. A portion of a corn stem. The two arrows from the word "phloem" mark the inner and outer boundaries of the phloem. Similarly, the two arrows from the word "xylem" mark the limits of this tissue. The large cells of the phloem are sieve tubes; the small ones are companion cells. The large cells of the xylem are vessels. Notice the large, irregularly shaped air space in the xylem. (George S. Avery, Jr., *Am. J. Botany*.)

vascular bundles, several parenchyma cells are present on the outer margin of the group of sieve tubes and companion cells. The sieve tubes conduct food.

In an older corn stem, fibers surround each vascular bundle, being especially well developed on the outer and inner sides of each bundle. **Fibers** are long thick-walled cells with dovetailing end walls which add strength to the stem. When fully developed, each fiber has a very small cavity and no living contents. Varieties of corn with many well-developed fibers resist being blown down by wind and beaten down by hail better than those with few.

Because a cambium layer is lacking, no additional xylem and phloem can be formed. The increase in diameter of a corn or other herbaceous monocot stem results from an increase in the size of cells rather than from the formation of additional cells. At the nodes, the vascular bundles join one another and some vascular bundles extend into the leaves.

Intercalary Growth

In corn and other grasses the stem tissue just above a node remains soft and meristematic after the tissues in the rest of the internode have matured. Through division of cells just above a node, and their subsequent enlargement, the stem increases considerably in length. The growth resulting from the activity of these meristematic cells is called intercalary growth in contrast to the more usual apical growth which results from division and enlargement of cells at the stem tip. Grasses, of course, have apical growth as well as intercalary growth. Some stem cells and all leaves originate at the stem tip, as in other plants.

HERBACEOUS DICOTYLEDONOUS STEMS

In the following discussion an alfalfa stem will be used as a typical example of a herbaceous dicotyledonous stem.

Young Herbaceous Stem

Epidermis, cortex, vascular bundles, pith, and pith rays are evident in a cross-section of a young alfalfa stem (Fig. 12–8A).

Epidermis. The epidermis consists of a single layer of cells which have a cuticle on their outer walls. The epidermis protects the stem and retards transpiration.

Cortex. The region between the epidermis and the phloem is known as cortex. In an alfalfa stem chlorenchyma and collenchyma cells make up the cortex. **Chlorenchyma** cells are thin-walled, living cells containing chloroplasts. **Collenchyma** cells are cells whose walls are thickened in the corners, and they strengthen the stem. They are living cells with cytoplasm, a nucleus, and frequently chloroplasts. The collenchyma is most strongly developed beneath the ridges of the stem.

Vascular bundles. In alfalfa and other dicot stems, the vascular bundles are arranged in a circle in contrast to the scattered arrangement in a corn stem. The vascular bundles not only serve to conduct water, minerals, and food, but they also furnish strength to the stem. Xylem occupies the inner part of each vascular bundle; then comes the cambium, and toward the outside of the bundle, the phloem. The most distinctive cells of the xylem are the thick-walled vessels and tracheids which are nonliving and thus appear empty. Parenchyma cells, with their living contents, are also present in the xylem. The inner primary xylem cells are smaller than the outer ones because maturation proceeds in an outward direction, just the reverse of that occurring in roots. The developmental sequence in stems is said to be **endarch** in contrast to the **exarch** pattern in roots. The first xylem cells to mature, the **protoxylem** cells, are relatively small cells with secondary thickenings in the form of rings or spirals. The larger xylem cells, those maturing later, are known as **metaxylem** cells; their walls are thicker than those of the protoxylem and in some vessels the entire wall is thickened except for pits.

In the phloem we find **sieve tubes, companion cells,** and on the outer border of each bundle, a group of fibers called **phloem fibers.** The sieve tubes are relatively large, and adjacent to them are the smaller companion cells, each with a nucleus and dense cytoplasm.

At each node, the conducting tissues of the stem connect with those of the leaf and the axillary bud. The vascular strand leading to a leaf is known as a **leaf trace** and that extending into a bud or branch is called a **branch trace.** Just above each leaf and branch trace there is a gap in the vascular tissue of the stem which is usually filled with parenchymatous tissue. The gap above a leaf trace is known as a leaf gap, whereas that above a branch trace is called a branch gap.

The cambium layer is one cell wide and is situated between the xylem and phloem. The **cambium** cells divide. Of the two cells resulting from the division of each, one remains a cambial cell and the other differentiates into either a xylem cell or a phloem cell. Xylem cells are formed inside of the cambium and phloem cells outside of the cambium. The cambium forms more xylem cells than phloem cells, perhaps eight xylem cells for each phloem cell. Xylem formed by the cambium is known as **secondary xylem,** whereas that present prior to the beginning of cambial activity is known as **primary xylem.** The phloem formed by the cambium is called **secondary phloem,** and that previously present is known as **primary phloem.** You will recall that both primary xylem and primary phloem were differentiated from cells which originated at the stem tip. The conducting cells of the xylem, both primary and secondary, are vessels and

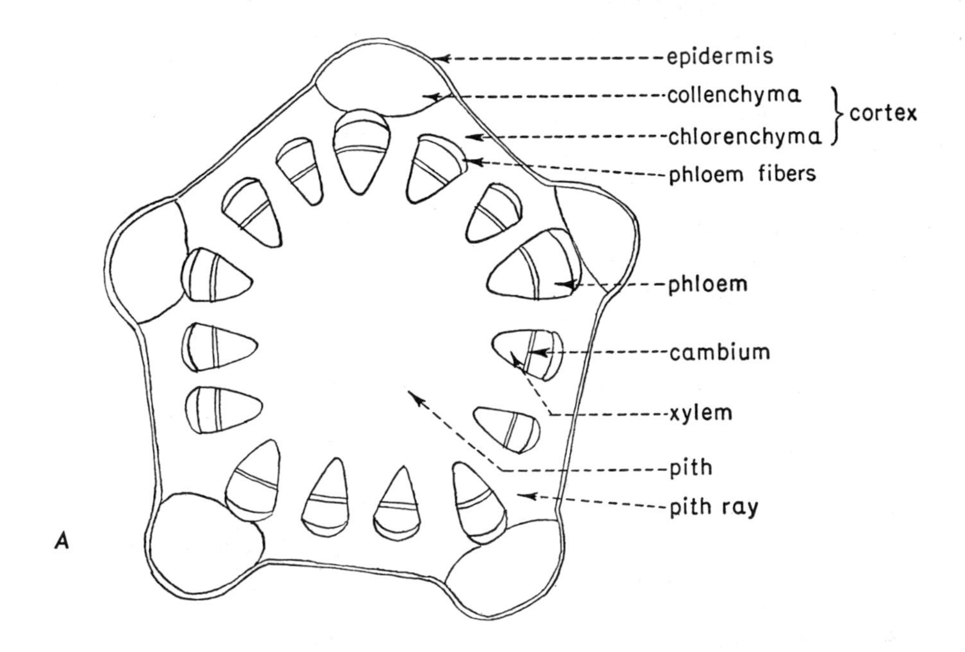

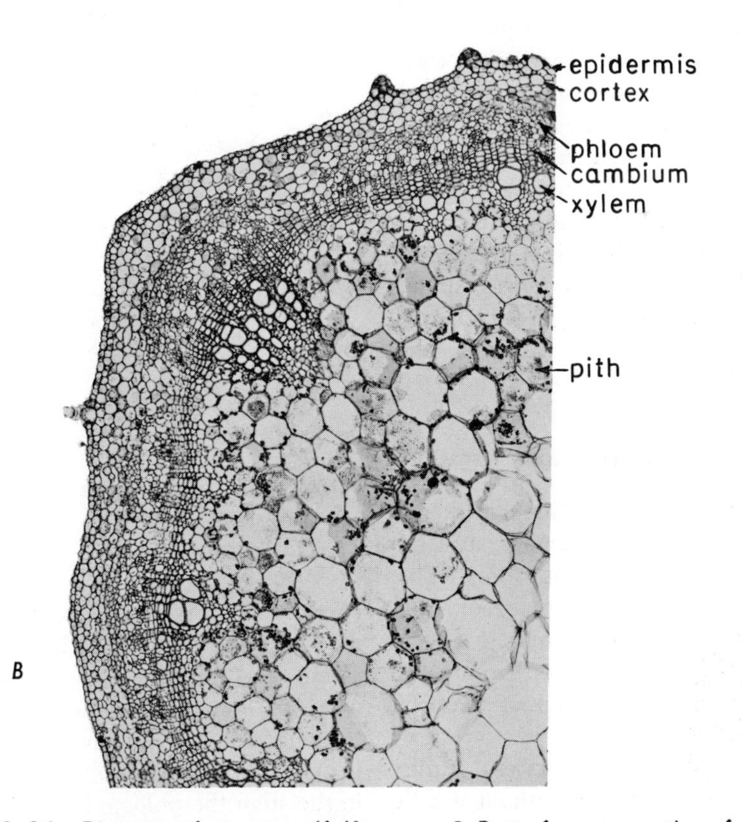

Fig. 12–8*A*. Diagram of a young alfalfa stem. *B*. Part of a cross-section of an alfalfa stem after the cambium has become continuous. (John E. Sass, *Botanical Microtechnique*.)

tracheids. Sieve tubes and companion cells occur in both the primary and secondary phloem.

Pith. The pith occupies the center of the stem and consists of parenchyma cells. The principal function of the pith is food storage.

Pith rays. Pith rays occur between the vascular bundles, and they extend from the pith to the cortex. They are radial extensions of the pith, and, like the pith, they are made up of parenchyma cells. Pith rays store food and conduct materials radially.

Older Herbaceous Stem

In an older alfalfa stem, the cambium layer is continuous and more xylem and phloem are present than in a younger stem. In a young stem, cambium is found only in the vascular bundles, and not between them, and this cambium is known as **fascicular cambium.** As the stem gets older, the fascicular cambiums of adjacent bundles become connected by strips of cambium cells, known as **interfascicular cambium,** formed by the division of pith ray cells. After the cambium has become continuous, secondary xylem and secondary phloem are produced both in the vascular bundles and between the bundles. In other words, in an older stem there is a continuous cylinder of xylem and a continuous cylinder of phloem (Fig. 12–8*B*). Figure 12–9 diagrammatically illustrates the stages in the growth of an herbaceous dicot stem.

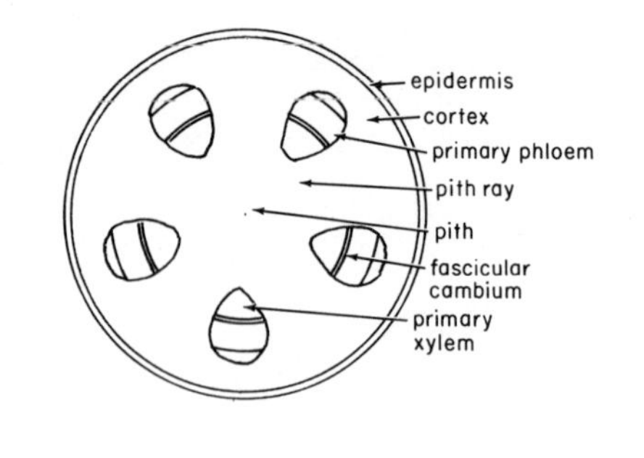

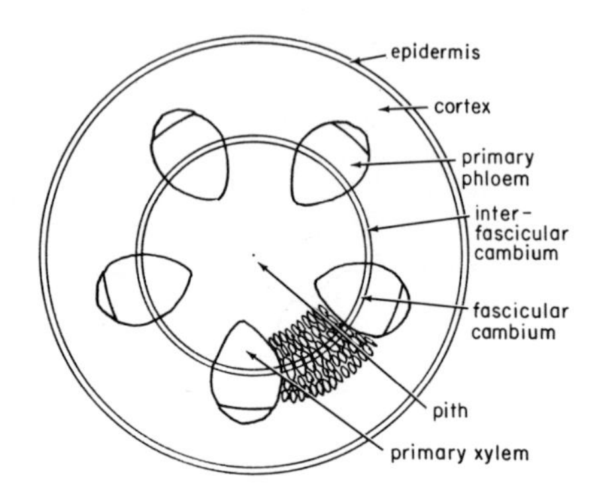

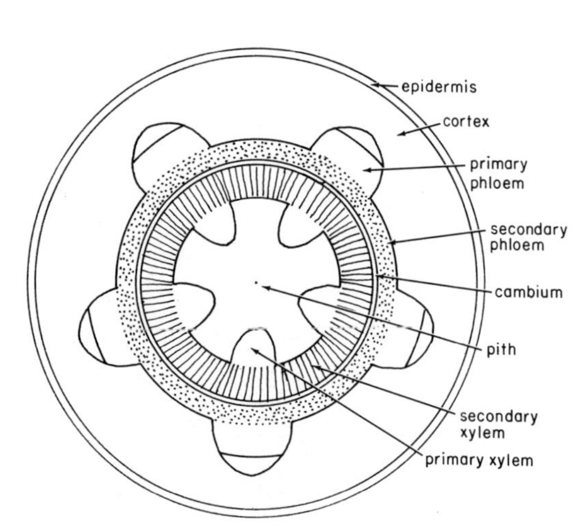

Fig. 12–9. Stages in the growth of a herbaceous dicot stem.

COMPARISON OF ROOTS AND STEMS

The stem and the root differ in many ways. The stem has leaves, nodes, and internodes, structures absent from roots. The tip of a root is protected by a root cap; that of a stem, only by overarching leaves. The branches of a stem develop from buds that originate superficially whereas those of the root originate internally in the pericycle. Stems have endarch xylem either with a pith, as in alfalfa, or without a pith but with scattered vascular bundles, as in corn. Roots have exarch xylem and in many roots a pith is lacking. In a root, primary xylem and phloem occupy different radii, whereas in a stem these tissues occur on the same radius. In other words, in the root primary xylem and phloem alternate, but in the stem the phloem is opposite the xylem (a pattern called **collateral arrangement**).

In many roots, but not all, the xylem is a central core, a pattern which results in high tensile strength with flexibility, but in low bending strength. Bending strain is best resisted with supporting tissue toward the outside. Collenchyma cells, fibers, and xylem cells have thickened walls and are strong. In a stem these are toward the outside (see Fig. 12–8) with a large weak pith occupying the center, an arrangement resulting in efficient support and resistance to bending even though a minimum amount of construction material is used. A root may be compared to a cable; the stem, to a reinforced concrete column—that is, a column consisting of steel rods scientifically located and embedded in concrete. In a stem the xylem and groups of fibers and collenchyma cells are comparable to the steel rods and the softer plant tissues to the concrete. The flexibility of the fibers and other strengthening tissues and their firm adhesion to the ground tissue enables the stem to bend without the risk of rupture. Just as the number of steel rods and their location determine the strength of a reinforced concrete column, so also the number and position of the strengthening strands govern the strength of a stem. However, as in structures designed by skillful engineers, more than one arrangement may be followed. An engineer may design several columns of equal strength and cost and yet follow a different plan in each instance. Similarly in the plant world several distinct plans result in stems of equal strength.

QUESTIONS

1. List four functions of stems.

2. Give the distinguishing characteristics of a root, a stem, and a leaf.

3. _____ Nodes, internodes, leaves, and buds are present on (A) stems, (B) leaves, (C) roots.

4. _____ Leaves are attached to the stem at (A) nodes, (B) internodes.

5. _____ Which one of the following is false? (A) Initials carved on tree trunks get higher from the ground with each passing year. (B) New leaves are formed at the tip of a stem. (C) If the tip of a stem is removed, that stem ceases to become longer and no additional leaves are formed on that stem. (D) The dome-shaped mass of small cells at the very tip of a stem is the apical meristem.

6. _____ Leaves originate (A) in the pericycle, (B) at the very tip of a stem, (C) at the base of the stem, (D) in the cambium layer.

7. _____ Primary xylem and phloem are differentiated from (A) procambium, (B) protoderm, (C) ground meristem.

8. _____ Buds which develop into branches bearing both leaves and flowers are called (A) leaf buds, (B) flower buds, (C) mixed buds.

9. _____ A plant with many branches can be produced by (A) removing the lateral buds, (B) removing the axillary buds, (C) removing the terminal bud.

10. _____ Snapdragon plants have opposite leaves (two at a node). When snapdragon plants are about 6 inches high, florists remove the upper part of the stem, including the terminal bud (pinch it off) so that the plant will branch. If the florist pinches the plant so that four leaves remain (two pairs of two) he would eventually obtain a plant with (A) four flowering branches (B) six flowering branches (C) no flowering branches, (D) one flowering branch.

11. Distinguish between primary and secondary tissues.

12. _____ The primary tissues in a stem develop from (A) cells produced by the apical meristem, (B) cells formed by division of interfascicular cambium cells, (C) cells formed by division of fascicular cambium cells, (D) cells of the ground meristem.

13. _____ Assume that cubes measuring one two-hundredths of an inch are removed from a plant and cultured in flasks containing appropriate nutrients. Which one of the following cubes is most likely to develop into an entire plant with roots, stems, and leaves? (A) Cube from apical meristem of a stem, (B) cube from meristematic region of a root tip, (C) cube from a leaf primordium.

14. Indicate whether the following are dead (D) or living (L) when carrying out their specific functions.

_____ 1. Sieve tube.
_____ 2. Companion cell.
_____ 3. Vessel.
_____ 4. Tracheid.
_____ 5. Guard cells.
_____ 6. Palisade cells.
_____ 7. Fibers.
_____ 8. Meristematic cells.

15. _____ Structurally, a stem most resembles (A) a cable, (B) a reinforced concrete column, (C) a bridge.

16. _____ The vascular bundles in a corn stem are regularly arranged in a circle. (A) True, (B) false.

17. _____ Stems of monocotyledonous plants (such as corn, wheat, lilies, tulips, etc.) never have a large diameter. The stems do not increase in diameter by forming secondary xylem and phloem because (A) a cambium layer is absent, (B) a pericycle is absent, (C) an endodermis is lacking, (D) a cortex is absent.

18. _____ If you were breeding asparagus plants for tenderness, which of the following would you select as parents? (A) Those with a large amount of xylem, (B) those with a small amount of xylem.

19. Arrange the following parts of a dicot stem in correct order, starting from the outside and working toward the center: epidermis, pith, cortex, xylem, cambium, phloem.

1. __________	4. __________
2. __________	5. __________
3. __________	6. __________

20. Distinguish between a monocot stem and a dicot stem.

21. List three differences between stems and roots.

	Stems	*Roots*
1.	__________	__________
2.	__________	__________
3.	__________	__________

22. In this and previous chapters we considered the three vegetative organs of a plant: the leaf, stem, and root. Label the diagrams on the opposite page.

---

23. By matching, indicate the types of cells which are present in the five regions of a stem. Also indicate (6–9) the functions of the cells listed in the right-hand column.

_____ 1. Epidermis.	1. Epidermal cells.
_____ 2. Cortex.	2. Guard cells.
_____ 3. Phloem.	3. Vessels.
_____ 4. Xylem.	4. Fibers.
_____ 5. Pith.	5. Collenchyma cells.
_____ 6. Support.	6. Sieve tubes.
_____ 7. Conduction of water.	7. Parenchyma cells.
_____ 8. Conduction of sugar.	8. Companion cells.
_____ 9. Retard transpiration.	9. Tracheids.

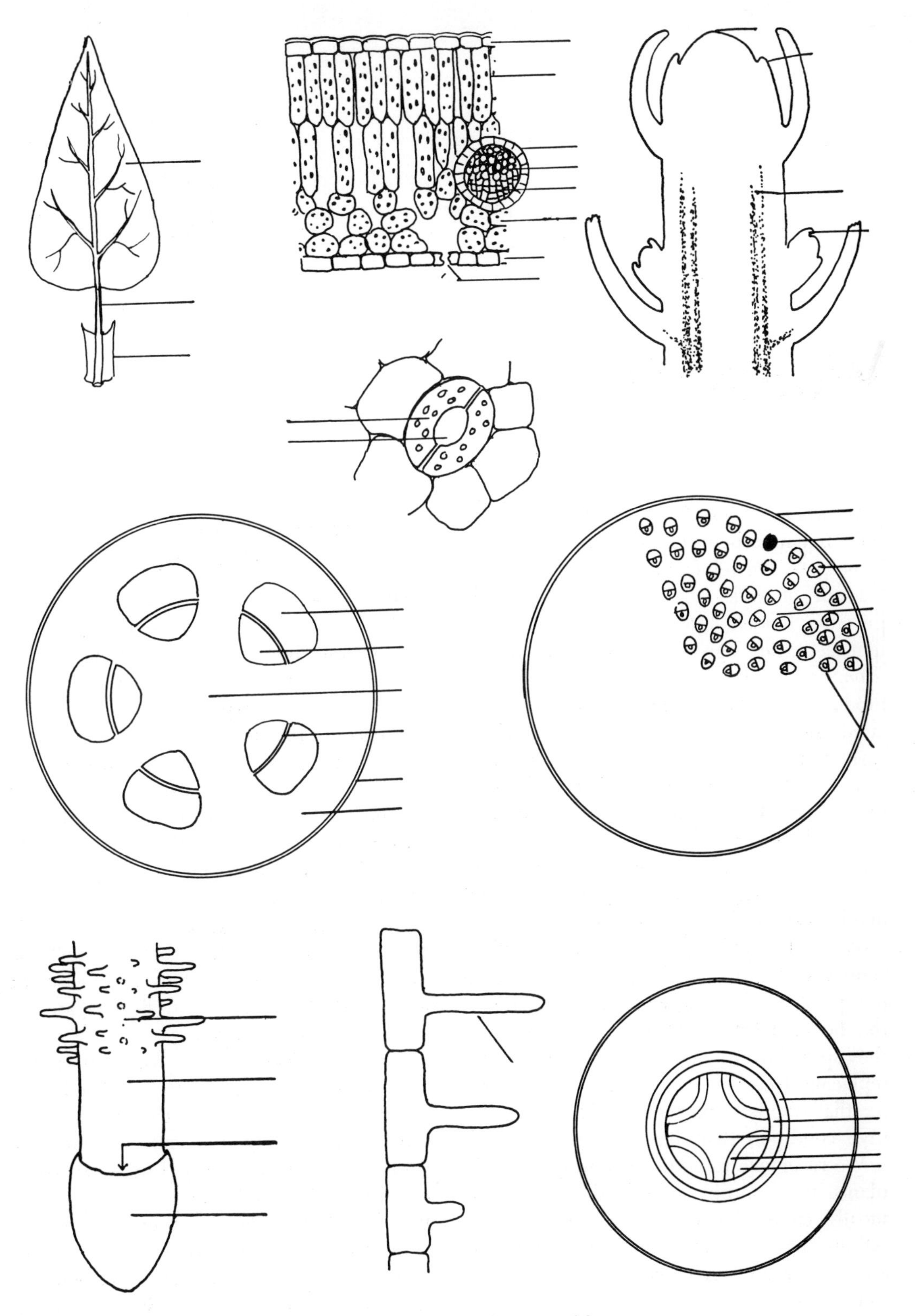

Diagrams for Question 22.

13

Woody Stems

We admire trees. Their giant stature, age, and magnificence endear them to us. Hence, we are reluctant to admit that they are out of style, that they are relatively primitive models; but so they are. Two million centuries ago *Calamites, Lepidodendron,* and *Sigillaria,* enormous trees that reproduced by spores and not seeds, populated much of the earth's surface. These early models gave way to more familiar trees—the pines, redwoods, firs, and other gymnosperms. For more than 50 million years these trees dominated the landscape. Then the flowering trees—magnolias, willows, cherries, elms, oaks, and maples—evolved and in certain areas they replaced the gymnosperms. About a million centuries ago bigness went out of style in the plant world and new smaller models, the herbaceous angiosperms, evolved; and they spread over much of the earth's surface, replacing trees in many areas. Hence we see that woody stems represent a more primitive condition than herbaceous ones.

We recognize three major kinds of woody plants: trees, shrubs, and woody vines, technically called **lianas.** Trees are generally tall and have a single trunk whereas shrubs are shorter and have more than one main stem.

SEASONAL PHENOMENA IN TREES AND SHRUBS

During winter the branches of lilacs, birches, and cottonwoods appear bare and lifeless. But how naked is a branch of a lilac shrub when it is bitterly cold outdoors? It might surprise you to learn that next season's full array of leaves and flowers is present on a lilac shrub during winter. You can prove this for yourself. Cut a branch of a lilac and from a number of buds remove the bud scales. These scales, you will recall, are modified leaves that protect the inner organs not so much from cold as from winter drought. You will notice that immature leaves are present inside some buds and immature flowers in other ones. Both the leaves and flowers are remarkably well developed.

When the days become warm in spring, the buds begin to grow—the leaf buds into branches bearing leaves, the flower buds into stems bearing blossoms. During the summer, in June or early July, next year's leaves and flowers begin to develop in the immature buds. The weather and internal conditions of the plant at this season determine the number of flowers which will appear the following spring. By the time of leaf fall in the autumn the leaf and flower buds are completely formed. It follows that the weather at flowering time in the spring cannot change the course of development of a particular bud of a lilac.

Not all trees and shrubs have both flower buds and leaf buds. In the snowball, rose, and hydrangea all the buds on a winter twig are leaf buds. In the spring these buds develop into branches bearing leaves. Later in the season flowers are produced on the current season's growth.

If you carefully examine winter twigs of a number of trees and shrubs, you will notice that the twigs of each species are distinctive. The axillary buds of the lilac occur in pairs at a node, whereas in cottonwood only one bud is present at a node (Fig. 13–1). Each bud of a willow is surrounded by a single bud scale, whereas several bud scales envelop each bud of a lilac or cottonwood. The shape and size of **leaf scars,** the number of **vascular bundle scars,** and the form and color of **lenticels** are other features that can be used to distinguish one kind of tree or shrub from another. Let us become familiar with these structures by examining a cottonwood twig.

EXTERNAL FEATURES OF A COTTONWOOD TWIG

Buds

Both the axillary and terminal buds of a cottonwood twig (Fig. 13–2) are surrounded by resinous bud scales which protect the delicate inner tissues from mechanical injury and drought. In the spring, the flower buds develop into branches bearing flowers before the leaf buds commence to grow. In addition to terminal buds and axillary ones, adventitious buds may form on cottonwood stems and roots. **Adventitious buds** are those which originate on leaves, roots, or stems at places other than nodes.

Bud Scale Scars

Following the swelling of the terminal bud in the spring, the bud scales fall off, leaving a ring of scars which marks the beginning of the new season's growth. The age of a twig can be determined by counting the rings of bud scale scars and the distance between the rings of scars gives a record of growing conditions in past years.

Leaf Scars

A leaf scar is a corky area left on the stem after a leaf is shed. The leaf scar marks the position where the petiole was attached to the stem. Within each leaf scar small **vascular bundle scars** are evident, remnants of vascular bundles which extended from the stem into the leaf. See if you can recognize the vascular bundle scars in the leaf scars of Figure 13-2.

Lenticels

Lenticels are elongated or circular areas which protrude above the surface of the bark, and they can be seen with the unaided eye. Lenticels are made up of loosely arranged cells with large intercellular air spaces through which gases diffuse into and out of the stem.

TREE TRUNKS

Consider for a moment the number of things which you own that were made from the main stems, the trunks, of trees. You would have to list your textbooks, some furniture, photographic paper and film, rayon clothing, your home, or at least a part of it, and many other objects and substances. These items are made from the wood of trees.

The forest products industry is a large one employing about 3.5 million people, or about 6 per cent of our nation's labor force. The value of the forest products produced annually is about $15 billion or somewhat more than 3 per cent of the total national income.

Fig. 13–1. Lilac buds in winter. *Upper,* a pair of leaf buds; the bud scales have been removed from the bud on the left, thus exposing next summer's leaves. *Lower,* the bud scales have been removed from one of the two flower buds.

Fig. 13–2. A cottonwood twig in the spring, when the flower buds are just opening.

The bark of many species is wasted but that of the cork oak is harvested and manufactured into corks, gaskets, and linoleum. Oak and hemlock barks yield a high percentage of tannin, which is used in the manufacture of leather. Quinine, for the treatment of malaria, is extracted from the bark of several species of *Cinchona.* Cinnamon, a valuable spice, comes from the bark of the cinnamon tree. In the inner bark of *Hevea brasiliensis* and other rubber trees there are **latex tubes** containing latex that exudes when the tree is tapped. From the crude latex, rubber is obtained, and this plays a vital role in our economy. Chicle, the latex of the sapodilla tree, is the characteristic ingredient of chewing gum. Turpentine is secured from the resin which drips from cut faces of pine trees.

If we examine the stump of a felled tree, we see bark, wood, and a minute amount of pith. The cambium layer is present between the bark and the wood.

Bark

The bark consists of two distinct tissues, **outer bark,** or **cork,** as it is technically called, and **inner bark.** The outer bark is hard, quite thick, and furnishes protection. Outer bark resists or prevents infection by fungi. Many fungi which attack tree trunks cannot enter through intact bark but may enter through wounds such as those formed when branches are broken off, by bears clawing the bark, and by deer feeding on inner bark or rubbing the velvet from their antlers. Outer bark (cork) is a poor conductor of heat and may insulate the interior tissues against extreme fluctuations of external temperatures. For example, trees with thick bark may not be injured by a mild forest fire whereas those with thin bark may be killed. As additional wood and inner bark are formed by the cambium, the outer bark is forced outward. Because the outer bark cannot expand, it cracks into ridges, plates, or scales, depending on the species, and some of it may slough off.

The inner bark is a much thinner layer than the outer bark. The inner bark is the phloem and conducts food. The characteristic cells of the phloem are living. On the other hand, the characteristic cells of the outer bark, cork cells, are dead.

Cambium

The cambium layer is situated between the inner bark (phloem) and the wood (xylem). Because the cambium layer is just one cell wide, it is not evident to the unaided eye. Additional xylem is formed by the cambium on its inner side and additional phloem on its outer side.

Wood

In the wood of a tree, **annual rings** appear, one for each year of the tree's life.

Fig. 13–3. A forester making growth studies with an increment borer. (U.S. Forest Service.)

An annual ring consists of a layer of spring wood and a layer of summer wood. The **spring wood** is made up of large cells formed in early season, and the **summer wood** of smaller, thicker-walled cells that were formed later in the season. (We will have more to say about spring wood and summer wood a little later on.)

In an old tree the innermost group of rings is frequently darker than the outermost group. The inner core of wood is called **heartwood,** and it adds support to the tree but does not conduct water. The sheath of younger wood is called **sapwood.** The outermost rings of the sapwood conduct water and minerals and, like the heartwood, add support. The darker color of the heartwood results from the infiltration of the cell walls with gums, tannins, resins, and oils. As the tree ages, the inner sapwood is transformed into heartwood. Hence, the heartwood increases with age whereas the sapwood remains about constant.

An examination of the wood of a stump reveals the past history of the tree—poor growing seasons result in narrow annual rings, favorable growing seasons in wide ones. Precipitation varies from year to year and cycles of drought alternate with moist periods. Annual rings are narrow when drought prevails and wide when moisture is ample. Removal or death of competing trees also is indicated by wider rings. The age of a tree and the width of annual rings may be determined by examining the stump of a felled tree or by using an increment borer on a standing tree. An increment borer consists of a hollow bit which is screwed into the trunk of a tree. A sleeve is then inserted into the bit. When the sleeve is removed, a core from the tree is obtained (Fig. 13–3).

Annual rings have been used by archaeologists to date Indian ruins of the Southwest. A "Tree-Ring Calendar" was developed by matching the inner annual rings from a very old living tree with the outer rings in the beam of an old house. These in turn were matched with annual rings of older beams and this was continued, resulting in a calendar which stretched back for more than 2000 years. Pueblo Bonito, a great thousand-room ruin in New Mexico, was dated by this calendar; the oldest timber was cut in the year 919. Most of the ruin was constructed from 1050 to 1085.

Vascular rays traverse the wood and inner bark. Rays, spring wood, and summer wood are features which give wood its grain. The grain of a board, or the appearance of a thin section under the microscope, depends on whether the section is a cross-section or a longitudinal one, and if longitudinal, whether the section has been cut at right angles to the rays or parallel to them (Fig. 13–4). If boards are cut from a log at right angles to the rays, the grain will be as illustrated in Figure 13–5, right. To a botanist such a section is known as a **tangential section.** To a lumberman, the board is plain-sawed. If the boards are cut parallel to the rays, the grain will have a different appearance (Fig. 13–5, center). This section is known as a **radial section,** or, as the lumberman says, the log has been quarter-sawed. It is called quarter-sawed because the log is first sawed into quarters and then boards are cut from each sector.

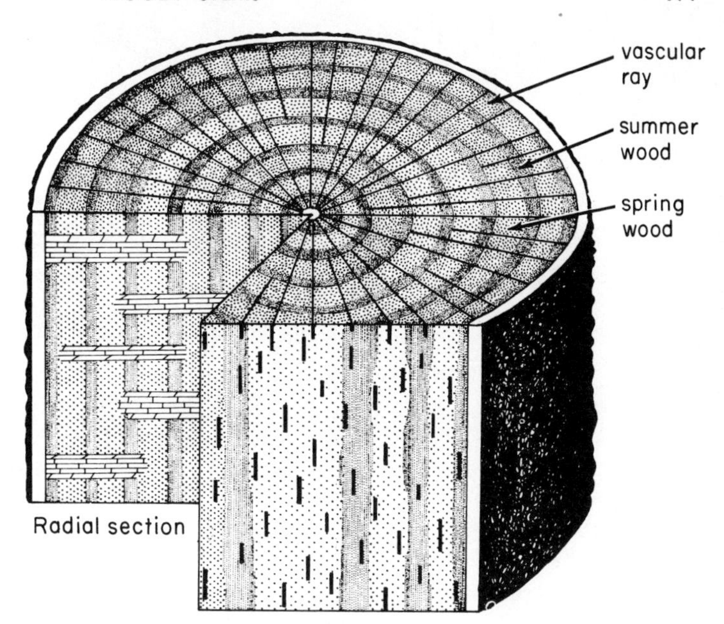

Fig. 13–4. Transverse, tangential, and radial sections of a hardwood stem. (William M. Carlton, *Laboratory Studies in General Botany*. Copyright © 1961, The Ronald Press Company, New York.)

A forester thinks of wood in terms of its structure, properties, and uses. He is interested in grain, color, durability, strength and bending strength, hardness, and the relationship of these properties to the uses of wood. He recognizes that the microscopic structure of wood determines many of its qualities and characteristics.

Tree species may be divided into **hardwoods** and **softwoods.** The softwood species are gymnosperms; the hardwoods, angiosperms. The wood of such gymnosperms as pine, Douglas-fir, spruce, and hemlock lack vessels, possessing only tracheids for water conduction. Woods which lack vessels are referred to as **nonporous** as contrasted to **porous wood** in which vessels are present (Fig. 13–6). Angiosperms have porous wood. In some hardwoods, oaks, for example, the vessels (pores) of the spring wood are very large, whereas those in the summer wood are much smaller and less numerous. Such wood is known as **ring porous.** When, as in maple, the vessels are more nearly uniform in size and more evenly distributed, the wood is referred to as **diffuse porous.**

Resin canals occur in the bark and wood of fir, pine, spruce, and other softwood species, but they are lacking in most hardwood species. Resin canals extend vertically in the bark and wood and branch at intervals to form horizontal canals. The resin canals are lined with living cells that secrete oleoresin, the substance from which turpentine and rosin are obtained commercially.

Knots

Knots are bases of branches which, through the formation of xylem around them, have become embedded in wood. When the base of a dead branch is buried in the xylem of the trunk, the knot is loose in contrast to the tight knot formed when a living branch is buried. If the lower branches are pruned from a young tree, the trunk at maturity will have small knots only in the center and most of the lumber sawed from it will be knot-free and hence of high quality.

MICROSCOPIC STRUCTURE OF A YOUNG WOODY STEM

A cross-section of a woody stem which is less than one year old is illustrated in

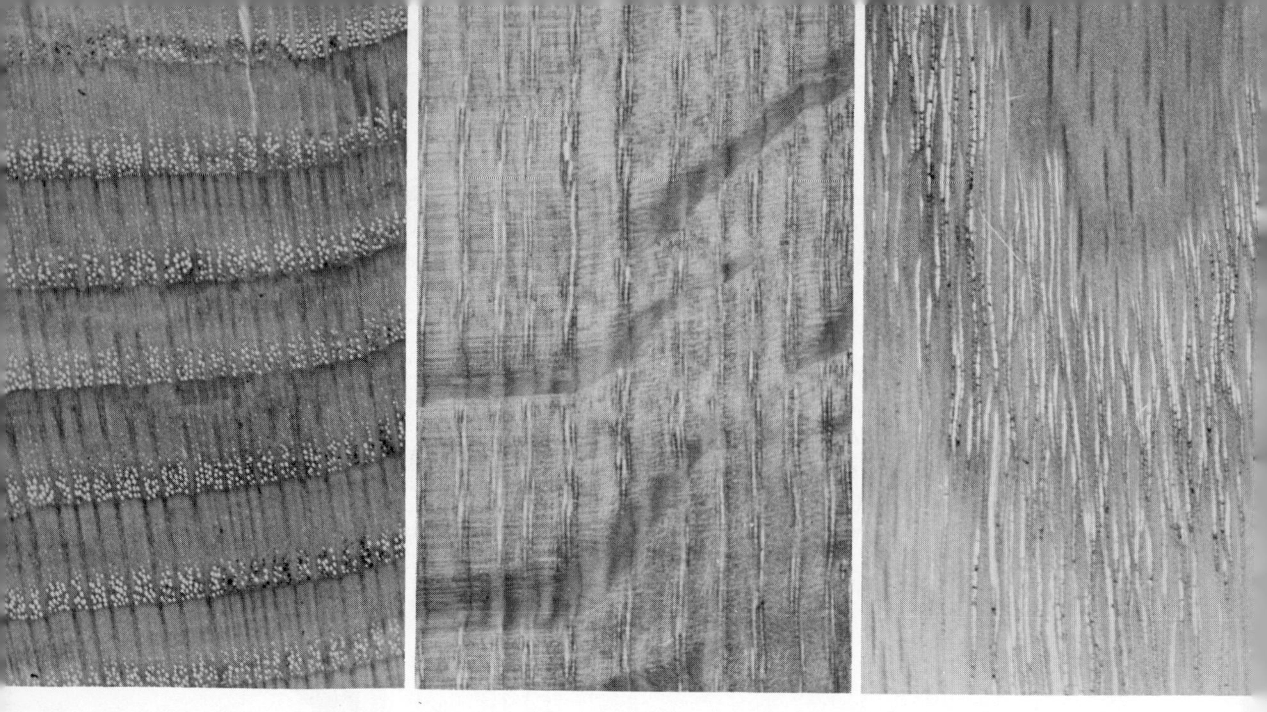

Fig. 13–5. The grain of wood is determined by the direction of the cut. From left to right, cross, radial, and tangential sections of red oak (*Quercus borealis*).

Figure 13–7. A young woody stem is essentially the same as a herbaceous dicotyledonous stem. However, with age the herbaceous stem and the woody stem become quite different. Much more cork, secondary xylem, and phloem are formed by a woody stem than by a herbaceous stem.

Epidermis

The epidermis is the outermost layer of cells. In a woody stem, one month or less old, the epidermis is the only protective tissue. Later, cork is formed. After cork is formed the epidermis dies and shrivels.

Cork

When a stem is about one month old, cork is not present. Before cork can be formed, a **cork cambium** must develop. The origin of the cork cambium from a ring of subepidermal cortical cells is illustrated in Figure 13–8. The cortical cells divide and the continuous ring of inner cells becomes the cork cambium, also known as the **phellogen.** (In a few plants, the cork cambium is formed by the division of epidermal cells and in others by the division of cells in the inner cortex or even in the secondary phloem.) During the growing season, the cork cambium cells divide, forming cork, also called **phellem,** toward the outside (Fig. 13–8). Occasionally, the cork cambium forms a few parenchyma-like cells, called **phelloderm** cells, toward the inside. The walls of the cork cells become impregnated with a waxy, waterproof material known as **suberin.** After impregnation, the cork cells die, but they still retard water loss and protect the stem.

Lenticels

The impervious cork is interrupted here and there by lenticels. At lenticels, the cork cambium produces large, loosely arranged cells, through which gaseous exchange occurs.

Cortex

The cortex is made up of thin-walled, living cells whose three dimensions are about equal. You will recall that such cells are known as parenchyma cells.

Phloem

Sieve tubes, companion cells, and parenchyma cells normally occur in the phloem. Some of the parenchyma cells are oriented to form **phloem rays,** which store food and

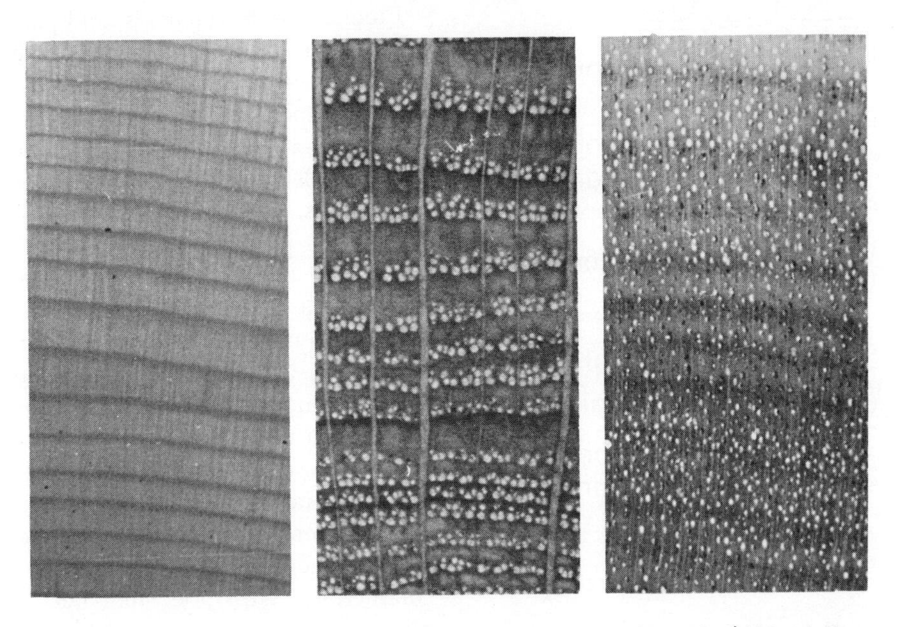

Fig. 13–6. Wood structure. *Left*, nonporous wood of balsam fir (*Abies balsamea*). *Center*, ring-porous wood of white oak (*Quercus alba*). *Right*, diffuse-porous wood of black walnut (*Juglans nigra*).

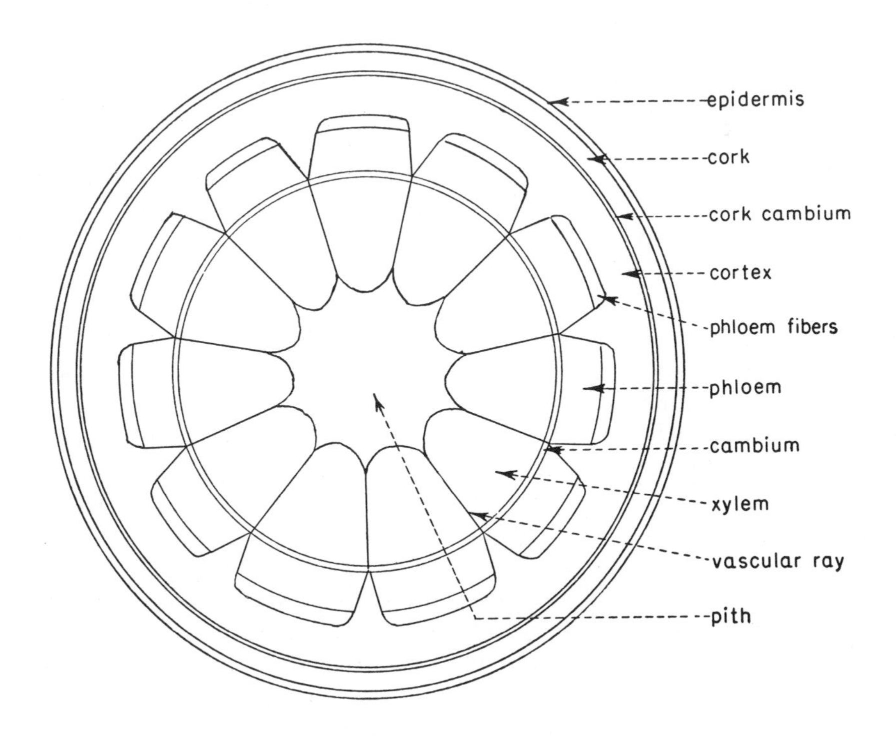

Fig. 13–7. Diagram of a cross-section of a young woody stem.

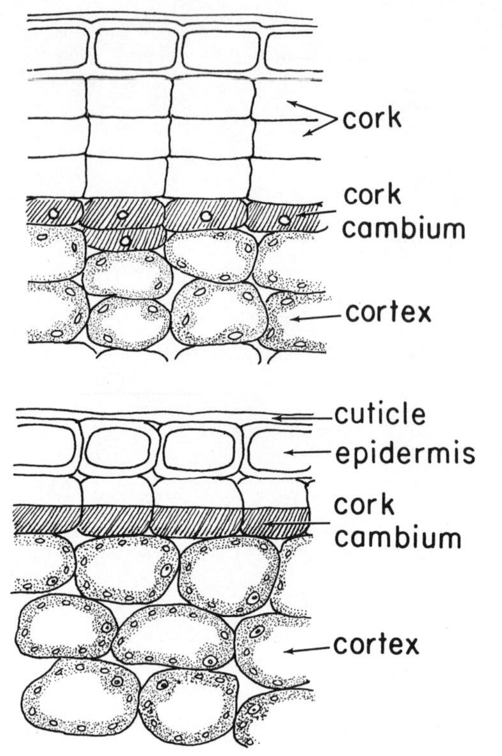

Fig. 13–8. *Upper,* cork formed by division of cork cambium cells. *Lower,* origin of the cork cambium from a ring of subepidermal cells.

conduct materials in a lateral direction. Fibers are present in the phloem of many woody plants.

Cambium

In cross-section (Fig. 13–9*A*) cambial cells have a brick shape. In lengthwise section (Fig. 13–9*B*) cambial cells which form sieve tubes of the phloem, and vessels, tracheids, and fibers of the xylem are long cells that taper at both ends (so-called **fusiform initials**), whereas those which form the cells that make up a ray (the **vascular ray initials**) are much shorter and are aggregated into lens-shaped clusters. As new xylem develops from the divisions of the cambium cells, the cambium and all of the cells external to it move outward and the girth of the cambium layer increases through the radial division of the cambium cells. As new phloem cells develop, the cells exterior to them are displaced outward.

Xylem (Wood)

Fibers, vessels, tracheids, and parenchyma cells occur in the xylem of woody angiosperms. Vessels are not present in the xylem of gymnosperms, but tracheids, fibers, and parenchyma cells are present. Some parenchyma cells of the xylem are oriented into xylem rays, which are continuous with the rays of the phloem. A **phloem ray** plus the **xylem ray** that joins it are together often called a **vascular ray.** In Figure 13–10, the vascular rays appear as lines radiating across the xylem and extending through the phloem. In the xylem all rays are narrow, but certain ones flare out in the phloem. A study of a radial or tangential section reveals that the rays are several to many cells high. Lateral movement of water, minerals, and food occurs in vascular rays—that is, conduction across the stem occurs in the rays.

Pith

The pith occupies the center of the stem and is made up of parenchyma cells.

AN OLDER WOODY STEM

In an old woody stem, considerably more secondary xylem, secondary phloem, and cork are present than in a young woody stem (Fig. 13–10). Pith and cortex do not increase in amount with age.

Xylem (Wood) of an Old Woody Stem

Each year the cambium produces an additional layer of wood. The most recently formed wood is the wood next to the cambium layer. The oldest wood is that next to the pith. The layer of wood produced during one growing season is known as an **annual ring.** In the spring of any one growing season, the cambium produces large xylem cells, but later in the season it produces smaller and thicker-walled cells. The wood made up of large cells is known as **spring wood,** whereas that made up of small cells is called **summer wood.** Each annual

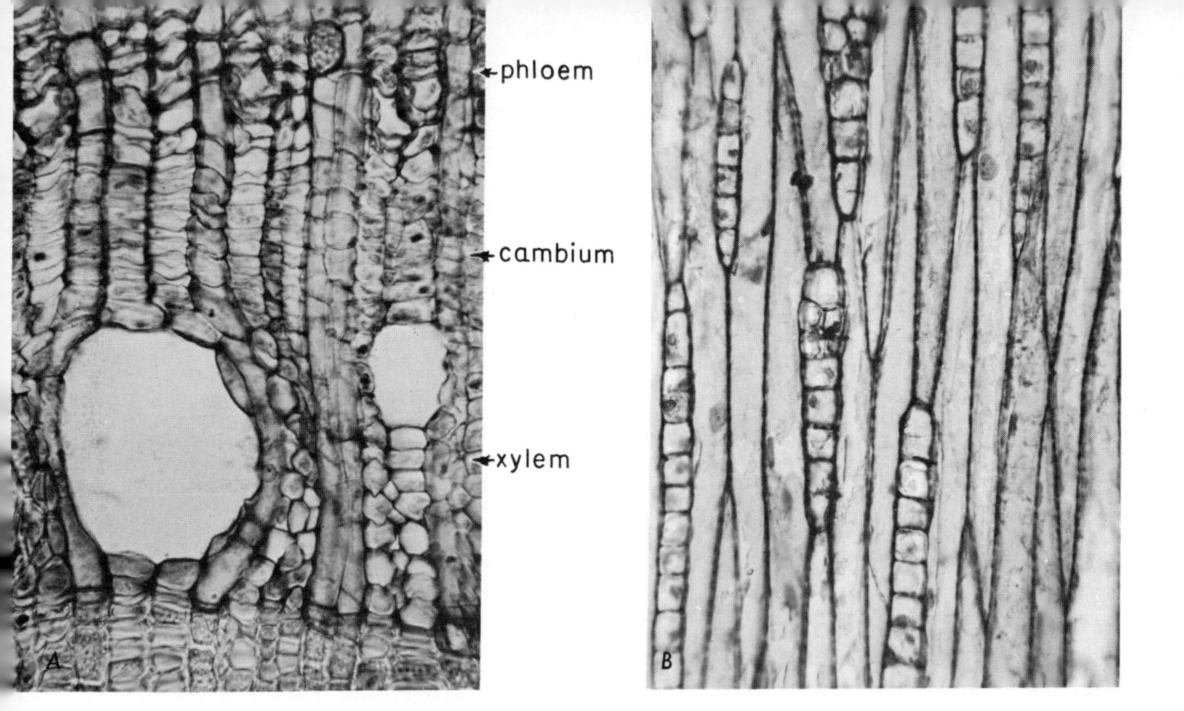

Fig. 13–9A. Growth within a stem. The cambium layer (shown in cross-section) of a pecan stem is forming secondary phloem toward the outside and secondary xylem toward the inside. Sieve plates are evident in some of the sieve tubes. Note the two large vessels in the xylem. *B*. Tangential section of cambium from a branch of pecan (*Carya illinoensis*). The short cambium cells divide to form parenchyma cells, whereas the long ones form sieve tubes, fibers, tracheids, or vessels. (Ernst Artschwager.)

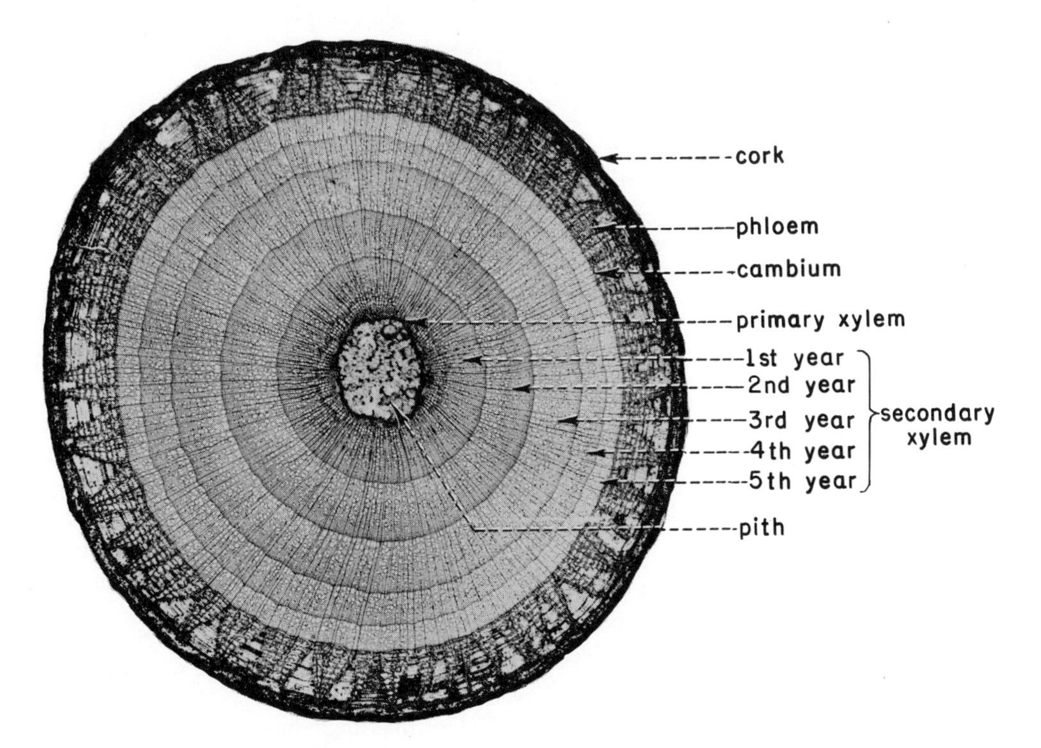

Fig. 13–10. A cross-section of a 5-year old *Tilia* stem. (Copyright, General Biological Supply House, Inc.)

ring consists of a layer of spring wood and a layer of summer wood. The difference in cell structure between these two makes annual rings evident (Fig. 13–11).

Phloem

During each growing season the cambium produces additional phloem cells. The phloem (inner bark) does not increase to the extent that the xylem does because the thin-walled phloem cells are crushed. Furthermore, the cambium produces more xylem cells than phloem cells.

Cork

Cork is continually formed by the cork cambium during the growing season. With each passing year the cork becomes thicker. In some trees—birch and beech, for example—the original cork cambium persists for the life of the trees. In other species, the original cork cambium is active for several years and then is replaced by a new one formed within the living tissues of the bark. Commercial cork is obtained at intervals of 8 to 10 years by stripping the bark from the cork oak tree, a native of the Mediterranean region. Removing the bark destroys the cork cambium, but a new one develops from living cells of the bark.

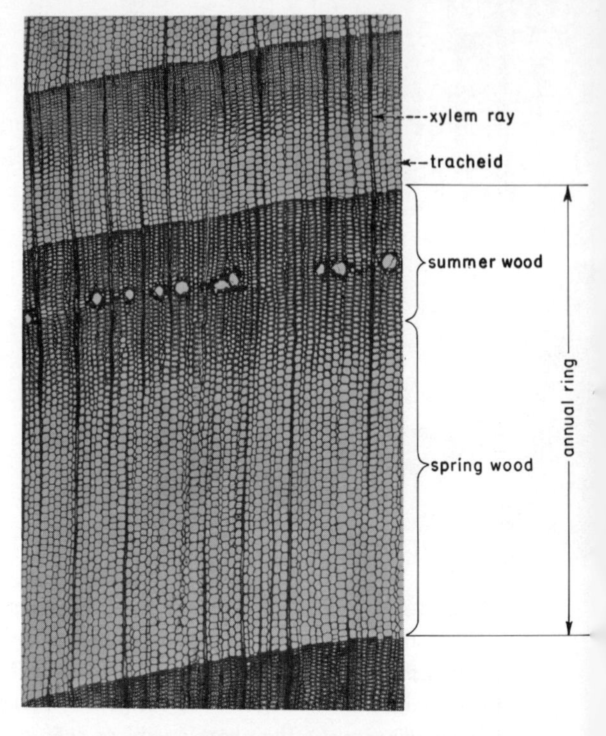

Fig. 13–11. Cross-section of a portion of the wood of Douglas fir, showing two complete annual rings and parts of two others. (U.S. Forest Service.)

HEALING OF WOUNDS

If a portion of a stem is removed down to the wood or if a branch is cut off, the exposed edges of the cambium produce cushions of meristematic cells which grow over the wounded area. Eventually the edges meet and grow together. Differentiation of xylem, phloem, and cambium occurs, and these tissues become continuous with the corresponding tissues of the uninjured parts of the stem.

QUESTIONS

1. Draw a cottonwood twig in winter condition. Label fully.

2. Describe two ways of determining the age of a twig.

3. List all of the items which you own which were made from wood.

4. _____ Starting from the outside of a tree and working toward the center, the various regions in correct order would be (A) outer bark, inner bark, cambium, sapwood, heartwood, pith; (B) outer bark, regular cambium, inner bark, sapwood, heartwood, pith; (C) outer bark, inner bark, sapwood, cambium, heartwood; (D) phloem, outer bark, sapwood, heartwood, cambium, pith.

5. Describe the gross structure of a tree trunk as seen in cross-section.

6. What contributions has the study of annual rings made to other sciences?

7. _____ The youngest wood in a tree is that which is (A) toward the outside of the tree, (B) toward the center of the tree.

8. _____ The Christmas tree which I cut in the adjacent forest showed ten annual rings at the base of the tree, each annual ring consisting of spring wood and summer wood. Had I cut a cross-section of the stem from the very top of the tree I would have noted (A) only one annual ring, (B) ten annual rings, (C) four annual rings, (D) nine annual rings.

9. _____ Vessels are present in (A) hardwoods, (B) softwoods, (C) wood of conifers.

10. _____ Which one of the following statements about wood is false? (A) The grain of wood is chiefly due to the difference between the spring and summer wood and the rays. (B) Knots are embedded branches or portions of branches. (C) Knots are usually more abundant toward the outside of a tree than toward the inside. (D) A tree growing in the open will usually have more knots than a tree growing in a forest. (E) Paper, rayon, and cellophane are made from the cellulose of wood.

11. _____ A forester is interested in the cambium layer because (A) it forms more inner bark and wood, (B) it conducts the water and hence keeps the tree alive, (C) it conducts food and hence the roots survive, (D) it forms wood (xylem) but not inner bark (phloem).

12. _____ Which one of the following is false? (A) The xylem conducts water and minerals. (B) The phloem conducts food. (C) The cells of the cambium divide to form secondary xylem and phloem. (D) The spring wood of an annual ring is harder and less porous than the summer wood.

13. _____ Annual rings appear in wood because (A) the cambium is present as a layer between each annual ring, (B) the xylem cells formed in spring are larger than those formed in summer, (C) the wood cells formed in late summer are larger than those formed in the spring.

14. _____ In the trunk of a tree the cork (outer bark) is formed by the (A) pericycle, (B) regular cambium, (C) cork cambium, (D) division of the ray cells.

15. Place the number of the appropriate term from the column at the right in the space to the left of each statement.

_____	1. The parenchyma tissue at the center of a stem.	1. Cambium.
_____	2. The meristem which gives rise to secondary xylem and phloem.	2. Cork.
_____	3. An external layer of cells whose walls are impregnated with suberin.	3. Xylem.
_____	4. The vascular tissue which is primarily involved in water transport.	4. Pith.
_____	5. The tissue containing sieve tubes and companion cells.	5. Phloem.

16. _____ A certain stem had scattered vascular bundles and lacked a cambium. This stem is likely to be (A) a monocot stem, (B) a dicot herbaceous stem, (C) a gymnosperm stem, (D) a dicot woody stem.

17. _____ In a murder trial the prosecuting attorney claimed that a club made from a living tree branch was the weapon. The murder occurred in early June. The defense had the branch examined with a microscope. The cells just inside the cambium were small cells. The club (A) was the weapon, (B) was not the weapon.

18. _____ An investigator staked some trees. Other trees were allowed to sway with the wind. After several years he noted that the staked trees had a lesser diameter than the free-swaying trees. The experiment demonstrates that (A) swaying promotes absorption of water and minerals, (B) the free-swaying trees try to overcome the strain by becoming stronger, (C) swaying brings about internal changes which enhance cambial activity, (D) swaying lessens cambial activity.

19. _____ The highest quality lumber generally comes from (A) the wood toward the outside of a log, (B) the wood near the center of a log, (C) from trees that are growing in the open where they do not compete with neighboring trees.

20. _____ Which one of the following is false? (A) A tree grows in height as a result of the formation and enlargement of cells produced by the meristematic cells at the stem tip. (B) Woody stems are more advanced than herbaceous ones. (C) The thickness of annual rings is usually a reflection of climatic conditions that prevailed when the rings developed. (D) In an old tree the innermost annual rings become plugged up and undergo chemical changes to form heartwood.

21. _____ Which one of the following is false? (A) A vascular cambium originates from the procambium and adjacent parenchyma cells in the rays. (B) Fusiform initials of the cambium produce cells which differentiate into elongated conducting and strengthening cells. (C) Ray initials differentiate into vascular ray cells. (D) As a tree ages, periodically the vascular cambium forms phellogen cells.

14

Conduction

Life has followed many paths. Circulation in our bodies is precise and efficient. Our heart pumps many gallons of blood each hour through one hundred thousand miles of arteries and veins. Our blood carries food and oxygen to all cells. We are impressed with the efficiency and precision of the heart and circulatory system in our bodies, but we seldom are even aware of the movements of food, water, and minerals in a plant. There are no heart beats, no pulse throbs, and yet food is translocated to all living cells. In the primeval forests of the Pacific Coast, some Douglas-firs tower 300 feet into the sky. The food made in the needles moves to all parts of these woodland giants. A large tree has roots totaling many miles in length. How many we cannot say because no one has measured the total length of the roots of a large tree. Recall, however, that just one rye plant had roots 387 miles long. Food made in the needles moves to all the root tips. Food is translocated not only to roots but also to stems and developing seeds and fruits.

The upward movement of water and minerals in a tree is as impressive as the translocation of food. The water absorbed by the miles and miles of roots moves into the xylem through which it ascends, ultimately being distributed to all the leaves and other parts. In certain trees water may be lifted 300 feet, a formidable task.

Xylem and phloem have been seen to be the conducting tissues of higher plants. The xylem in a root is continuous with that in a stem. From the xylem in a stem strands of xylem extend into leaves, floral parts, and fruits. Likewise, the phloem forms a continuous conducting system in a plant. Water and minerals are moved through vessels and tracheids of the xylem; foods and other organic materials are conducted by sieve tubes of the phloem.

CONDUCTION OF WATER AND MINERALS

Perhaps you have observed that leaves of trees do not wilt for a long time after a tree has been girdled by rabbits, porcupines, or some other agent. The removal of the outer and inner bark does not immediately interfere with the conduction of water. This indicates that the wood, not the bark, is the water-conducting tissue. Ultimately, however, the girdled tree dies because the roots are deprived of food.

Rates of water conduction may be determined by injecting a dye solution into the xylem and then periodically sampling the stem at different heights. The results obtained by injecting water colored with acid fuchsin into the trunks of six tree species are recorded in Table 14–1.

Table 14–1

Rates of Water Movement for Six Tree Species

Tree	Feet per Hour
Balsam poplar	13–30
Yellow birch	3–16
Beech	12–14
American elm	68–141
Red maple	5–15
Sugar maple	6–13

At the time the experiment was carried out, the rate of water movement varied among species, being slowest in the yellow birch and most rapid in the American elm where in one tree water ascended 141 feet in an hour's time.

Experiments carried out with radioactive isotopes demonstrate that minerals are conducted upward in the xylem. We now know that a chemical element may exist in different forms called isotopes. Isotopes of certain elements give off radiations; these are known as radioactive isotopes, and they have the same chemical properties as the stable forms. The radiation given off enables us to detect the element and measure it quantitatively even when present in minute amounts. The radiation may be detected by a Geiger counter or by radiographs. Radiographs are prepared by placing a photographic film against a plant or section of a plant that previously has been treated with a radioactive isotope. Of course, the film is not exposed to light; instead the procedures are carried out in the dark. The film is then processed in the usual way (Fig. 14–1).

In one experiment a scientist carefully separated the bark, both outer and inner, from the wood for a distance of 9 inches.

Fig. 14–1. Radiograph of a bean plant Radioactive phosphorus was absorbed by the roots, translocated by the stem, and accumulated by the young leaves and terminal bud. The radiograph was made by placing the plant on photographic film in the dark and then developing the film. The radioactive phosphorus affects the film just as light does. Dark areas indicate high concentration of radioactive phosphorus. (Courtesy of O. Biddulph.)

He inserted a sheet of paraffined paper between the xylem and the phloem. The plant was then watered with a solution containing radioactive potassium. After an interval of time the xylem and phloem from the region where the tissues were separated were analyzed. The xylem had a high content of radioactive potassium, the phloem a very low concentration. When the same experiment is carried out with radioactive phosphorus or zinc, similar results are obtained indicating that minerals are conducted upward in the xylem. However, as they move upward in the xylem, living cells along the

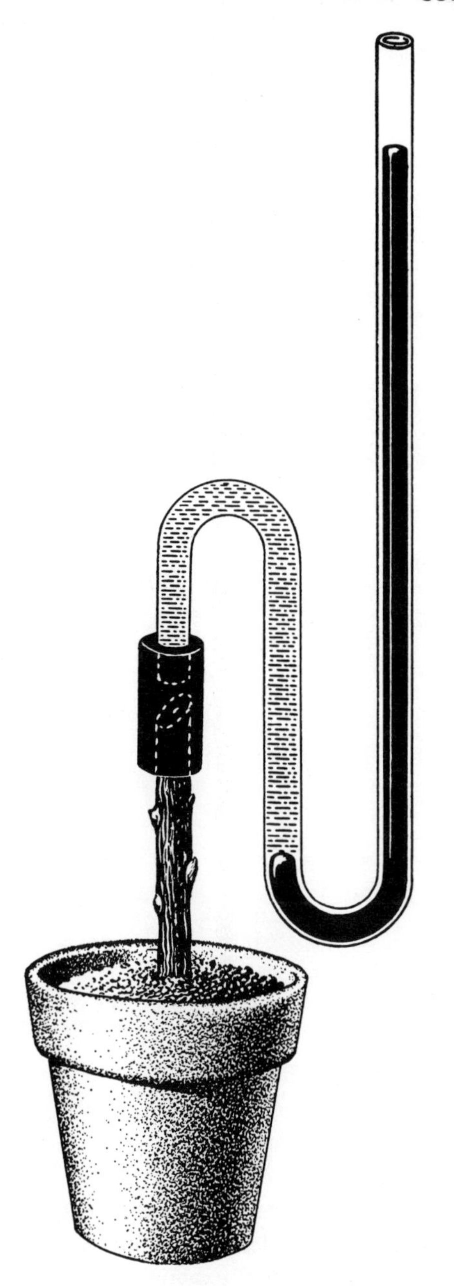

Fig. 14–2. Root pressure is demonstrated by fastening a mercury manometer to a cut stem. (*Sci. Am.*)

path may absorb some of the minerals. In other words, some lateral movement of minerals into the cambium and phloem does occur.

Minerals dissolved in water move upward in the xylem. The solution is called sap. In the spring, sugar which has been stored in roots and stems may ascend in the sap of some trees—sugar maple, for example. When sugar maple trees are tapped in the spring, the sugar solution exudes from xylem tissue. Although in the spring of the year sugar in certain trees ascends in the xylem, the xylem is not the main tissue which conducts food the year around. Throughout most of the year food is translocated by the phloem.

Mechanisms of Water Movement

Two theories are currently used to explain the rise of sap in plants: (1) root pressure and (2) transpiration pull and water cohesion.

Root pressure. Especially in the spring of the year, root pressure plays a part in the ascent of sap in some plants. If the leafy top of a plant is cut off, the stump frequently bleeds. The exudation is sap, and it flows for some time from the cut surface. The pressure causing exudation originates from osmotic forces in the roots and hence is called root pressure. As roots absorb water, the water is forced up the stem. The magnitude of the root pressure responsible for bleeding can be determined by attaching a manometer, in the manner illustrated in Figure 14–2, to a cut stem. As water is forced up through the stem, the mercury rises in the column, indicating that root pressure is operating. A root pressure of 1 atmosphere would cause the mercury to rise 76 centimeters. The values obtained with different species range from zero in some species up to 3 atmospheres in others. A pressure of 3 atmospheres would raise sap about 100 feet.

Many trees are taller than 100 feet; and in these, root pressure is probably insufficient to lift water to the uppermost leaves. For example, some Douglas-fir trees attain a height of 300 feet. To reach the upper needles on these trees, water must rise from roots many feet beneath the ground, a vertical distance of more than 300 feet. Even in small trees, shrubs, and herbaceous plants, root pressure is not the only means of raising sap. Many species do not bleed when the stems are cut, indicating that root pressure is not operating. In many plants, root pressure is negligible or nonexistent in summer when transpiration is highest. Furthermore, water moves upward in branches which have been removed from a plant, demonstrating that roots are not necessary for sap rise. For these reasons we must look for another mechanism to lift sap, one that can move water to great heights and that operates in all plants. Such a mechanism is transpiration pull and water cohesion.

Transpiration pull and water cohesion. The most generally accepted theory of sap ascension is the transpiration pull and water cohesion theory. According to this theory, water is pulled up the stem instead of being pushed up as in root pressure. The general features of this theory are as follows: As water is transpired from mesophyll cells, or used in photosynthesis, the concentration of solutes in the cells increases and they develop a water deficit. The cells then absorb water from vessels and tracheids by osmosis and imbibition. Cells adjacent to conducting elements will absorb water directly from vessels and tracheids; distant ones will absorb water from closer cells which in turn will absorb water from the conducting elements. As mesophyll cells absorb water from vessels and tracheids, they pull the intact columns of water upward. The long water columns, extending from the leaf through the stem into the roots, are not ruptured because water is highly cohesive. The pull on the water columns by transpiring or photosynthesizing cells is great enough to lift water at least as high as the tallest trees, and the cohesion of water molecules is great enough to keep the columns intact.

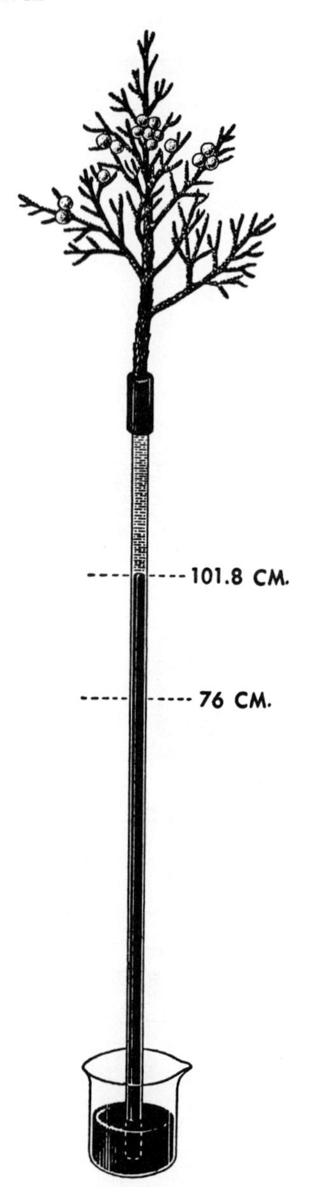

Fig. 14–3. As the cedar twig transpires, the water column is pulled up and the mercury rises in the tube; in this experiment the mercury ascended to 101.8 cm. Atmospheric pressure will hold the mercury at only 76 cm. (*Sci. Am.*)

We can demonstrate transpiration pull by attaching a branch to a long glass tube filled with air-free water and with the base of the tube placed in mercury (Fig. 14–3). As water is transpired or used by the shoot, the water column will be pulled up and the mercury will rise in the tube. In one experiment the tension created by the branch lifted mercury about 102 cm., 26 cm. higher than it could be forced by atmospheric pressure. A tension of 102 cm. (about 40 inches) would lift a water column about 46 feet. In the intact plant much greater tensions develop.

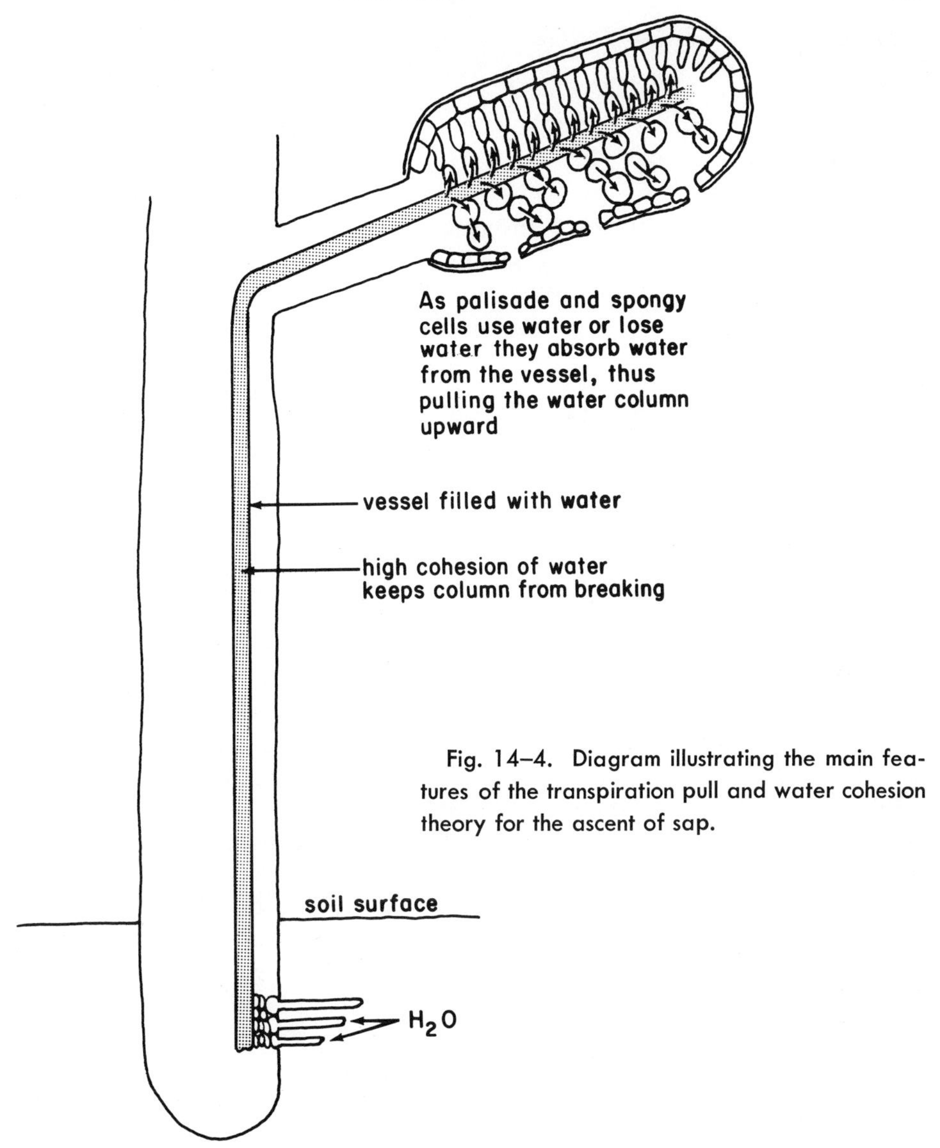

Fig. 14–4. Diagram illustrating the main features of the transpiration pull and water cohesion theory for the ascent of sap.

As water is pulled up in the trunk of a tree, the vessels and tracheids tend to become narrower. Actually, when transpiration is rapid, the diameter of a tree is slightly less than when transpiration is very low. Vessels and tracheids do not collapse as water is pulled through them because they are reinforced by rings, spirals, or heavy walls.

When a stem bearing leaves, or leaves and flowers, is cut from a plant, transpiration pull continues and the columns of water move upward. As they ascend, air enters the vessels and tracheids at the cut end. If the stem is later placed in water, water enters the vessels and tracheids, but such water is separated from the main columns by air bubbles and hence conduction is slow. If the basal end of the stem is cut off under water to remove the air bubbles, the columns become intact and water is conducted rapidly.

Water can move across a stem as well as in an upward direction. Thus, if the roots are removed from one side of a tree, the leaves above do not wilt. Some of the

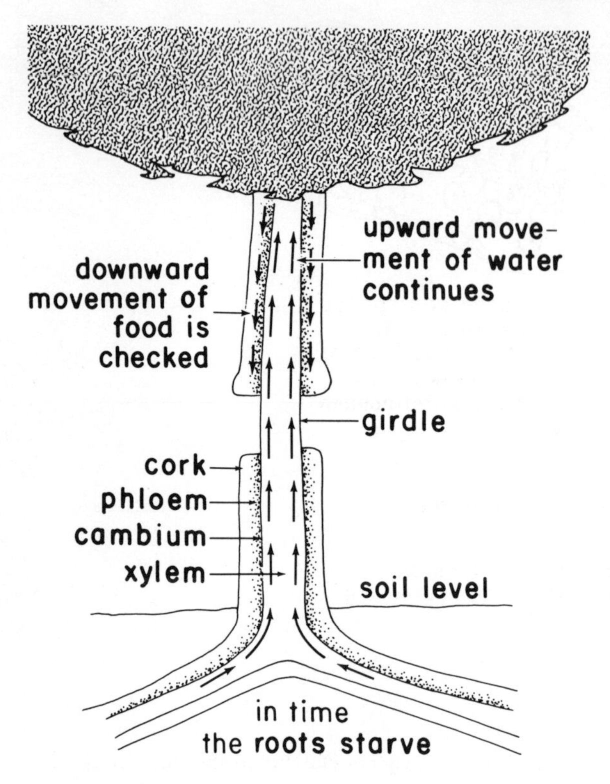

Fig. 14–5. A girdle checks the downward movement of food, but does not interrupt the upward path of water.

water absorbed by the roots moves across the stem and then ascends. Furthermore, if we saw half-way through the trunk of a tree, all of the leaves remain turgid. Even when two cuts, only 6 inches apart, are made on opposite sides of the trunk, water still ascends. Obviously water moves around the cuts, thus demonstrating that water moves laterally perhaps through the pits in the walls of tracheids and vessels. We can visualize the water as being intact not only in a vertical direction, but also in a crosswise direction.

CONDUCTION OF FOOD

Earlier we said a plant was deceptively simple. Although a plant appears to be quiescent, it is amazingly active and coordinated. Food is translocated to all plant parts, to every living cell, in amounts coordinate with their activities. The movement is orderly and is controlled in some way. At certain times large amounts of food are translocated to stem tips and root tips, as when growth commences in the spring. At other periods abundant food is translocated into developing seeds and fruits; and, at still other seasons, into storage sites such as roots, bulbs, and tubers. Thus, sometimes food is moving upward in the phloem, sometimes downward. At one time certain substances may even be ascending while others are descending. In other words, the movement of materials in the phloem is bidirectional. Only living cells translocate food. When the phloem is killed, food movement ceases. Temperature influences the rate of translocation. For example, translocation out of bean leaves is three times faster at 20° C. than at 10° C.

A number of observations indicate that foods, such as sugars, usually sucrose, and amino acids, are conducted in the phloem (inner bark). Perhaps you have seen girdled trees or trees which have had a tight band around the trunk. A swelling appears above the girdle or band (Fig. 14–5). The swelling results because food accumulates above the girdle or band and augments growth. The downward movement of food is checked by the band or halted by the girdle. If the main trunk of a tree is girdled, the tree ultimately dies because the roots starve. The tree also dies if the phloem is destroyed by a fungus, as it is in the chestnut tree blight, or by bark beetles.

If a small branch of a tree is girdled, the branch may live for a number of years. If the girdle is very narrow, the girdled region is healed over and the branch will live indefinitely. Girdling of branches may be used to increase the size of fruits and to initiate flowering. The sugar produced by the leaves on a girdled branch cannot be translocated to distant parts of the tree and it accumulates in the branch. The increased amount of sugar in the girdled branch favors the formation of large fruits. High sugar concentration often induces flowering. Hence under some circumstances girdled branches flower whereas intact ones do not.

Vertical movement of food occurs in the sieve tubes of the phloem and transverse movement of food in the vascular rays.

The mechanism of food movement is still uncertain. At one time it was believed that foods diffused through the sieve tubes of the phloem. However, recent data demonstrate that foods are conducted much more rapidly than could be accounted for by diffusion. Recent experiments, using a C^{14} tracer technique, demonstrate that in a bean plant sucrose is conducted at a rate of 84 centimeters per hour.

Some investigators suggest that protoplasmic streaming is involved in the transport of foods. According to this theory, sugar and other foods diffuse across the narrow intervening wall from one sieve tube cell into the one below. Within the sieve tube cell the sugar is carried rapidly from one end to the other in the streaming cytoplasm. Then the sugar diffuses across the wall into the next cell, where it enters the cytoplasmic stream. These steps are repeated until the sugar reaches its destination.

Another theory favored by many botanists to explain food movement is based on a mass flow of food dissolved in water. The features of this theory can be demonstrated with a model (Fig. 14–6). Two osmometers (A and B), made of cellophane, are connected by a piece of glass tubing as illustrated. Osmometer A is filled with a highly concentrated sugar solution that is colored with a dye, whereas osmometer B is filled with a dilute sugar solution. Both osmometers are immersed in a trough filled with water. As osmometer A absorbs water, a turgor pressure develops that is greater than the tendency for B to absorb water. The high turgor pressure forces water out of osmometer B into the vessel of water. The sugar solution from osmometer A then flows through the tube into osmometer B, replacing the water that is being forced out. The process continues as long as the sugar concentration in A exceeds that in B. Let us now compare the model to a plant. Osmometer A represents a leaf cell with a high sugar concentration. The tube is analogous to a sieve tube of the phloem. Osmometer B could be a root cell, a cambium cell, or some other consumer cell which is utilizing food or changing sugar into starch. The trough represents the xylem which circulates water to osmometer A.

The essential features of this mass flow theory as it actually goes on in a plant are illustrated in Figure 14–7. Water containing food passes from one mesophyll cell to an-

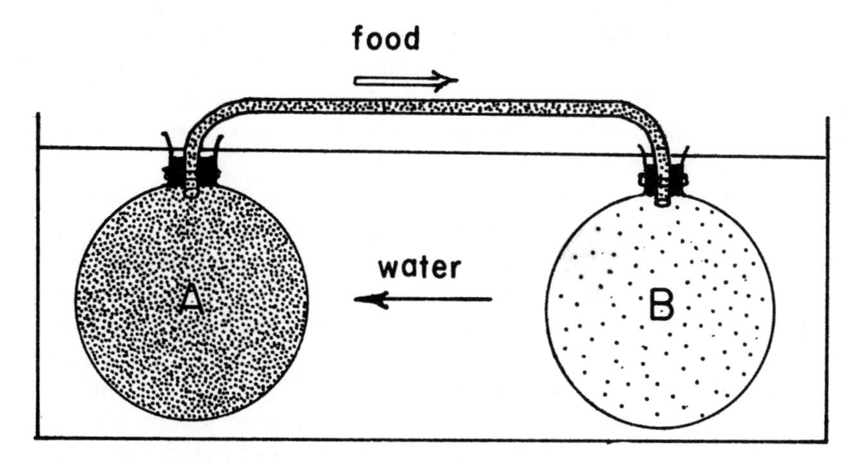

Fig. 14–6. A model to demonstrate the mass flow of sugar.

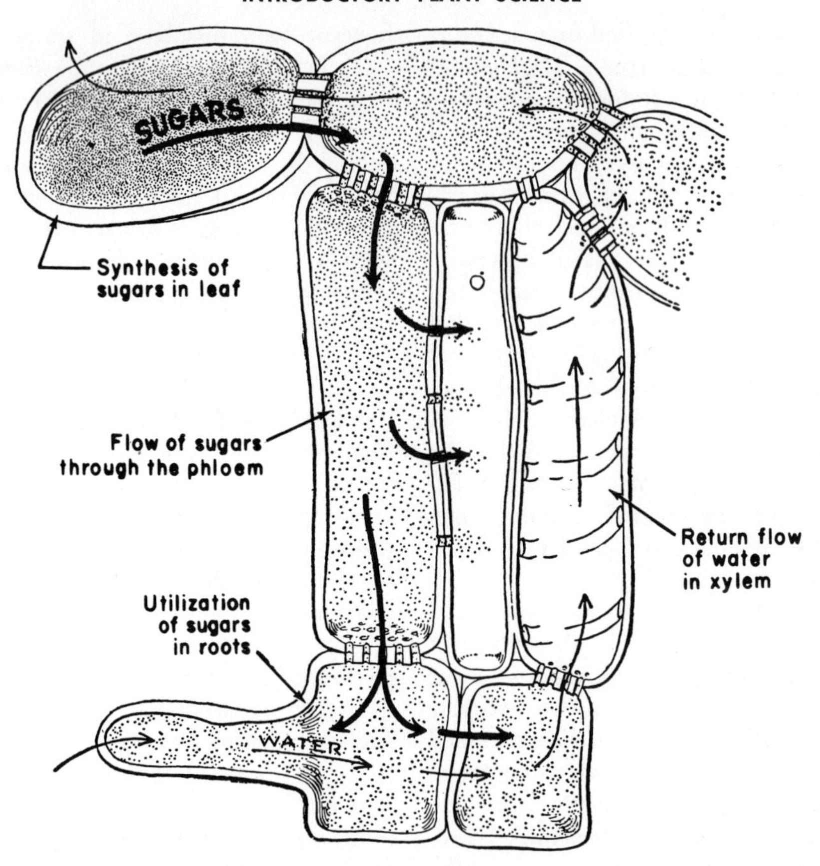

Fig. 14–7. Diagram illustrating the mass flow theory of food movement. (Reproduced by permission from *Principles of Plant Physiology* by James Bonner and Arthur W. Galston. Copyright, 1952, by W. H. Freeman & Co., San Francisco.)

other through fine strands of cytoplasm (plasmodesmata) which extend through pores in the cell walls. Similarly, cytoplasmic strands connect the cytoplasm in the cells adjacent to a sieve tube with the cytoplasm in the sieve tube. The sugar concentration is great in mesophyll cells and hence they take up water from the veins by osmosis. As the cells take up water, the turgor pressure increases and forces the cell solution via the cytoplasmic strands into and down the sieve tubes. If the flow of water containing sugar and other substances is to continue, water must be removed from the sieve tubes. The high pressure forces some water out of the sieve tubes, across the cambium, and into the xylem, which returns it to the leaves. The flow will continue as long as the sugar concentration in the mesophyll cells remains greater than in the root cells. The sugar content of mesophyll cells is kept high by photosynthesis. In the root, sugar is used in respiration and growth and some may be changed to starch; as a result, the sugar concentration is relatively low.

As the solution flows out of leaves into sieve tubes, it may carry other chemicals with it. When certain chemicals are sprayed on leaves, they may enter mesophyll cells and later be translocated, along with sugar, by the phloem. Among the chemicals which may move out in this manner are insecticides, weed killers, and minerals. Minerals are translocated from senescent leaves in the phloem.

QUESTIONS

1. _____ In 1675, Malpighi was engaged in a series of experiments on conduction. In one experiment, he removed a ring of bark from the trunk of a small tree. He probably noted that (A) the region above the girdle soon died, (B) the upward movement of water was influenced little or not at all by the removal of the bark, (C) the movement of food from shoots to roots continued as before the operation, (D) growth and swelling occurred below the girdle.

2. _____ In 1727, Stephen Hales (1677–1761) published a book called *Vegetable Statiks* in which he related some of his experiments. In one experiment he placed cut branches in colored water and discovered that (A) the phloem took up the dye, (B) the xylem became stained, (C) the leaves wilted in a short time.

3. _____ In 1892, Julius Sachs noted that a sunflower leaf, attached to a plant, lost 12 per cent of its dry weight during a 10-hour night. A detached sunflower leaf would lose during the same period (A) 12 per cent of its dry weight, (B) more than 12 per cent of its dry weight, (C) less than 12 per cent of its dry weight.

4. _____ Water is conducted by (A) the tracheids and, in angiosperms, vessels of the heartwood, (B) the tracheids and, in angiosperms, vessels of sapwood which is generally less than 24 years old, (C) the sieve tubes of the sapwood, (D) the sieve tubes of the heartwood.

5. _____ If one separates the bark from the wood in a part of a stem, waters the plant with a solution containing radioactive potassium, and then later tests the tissues for radioactivity, one would find that the most radioactive tissue is the (A) sapwood, (B) heartwood, (C) inner bark, (D) pith.

6. _____ In a stem (A) water and minerals move only upward, (B) water and minerals move laterally as well as upward, (C) sugar, especially sucrose, moves downward and laterally but not upward, (D) the minerals which are sprayed on leaves are conducted by the xylem.

7. How can it be demonstrated that minerals move in the xylem?

8. _____ Two processes for moving water upward in plants are (A) osmosis and capillarity, (B) root pressure and transpiration pull, (C) turgor pressure and diffusion, (D) plasmolysis and suction tension.

9. _____ In some plant diseases the bacteria or true fungi grow and develop in the vessels of the xylem. In such diseased plants (A) the leaves will wilt, (B) the conduction of sugars will be interfered with, (C) the roots will become plasmolyzed, (D) the plants will thrive.

10. _____ If both the inner and outer bark of a tree are removed, (A) the leaves will wilt, (B) the tree will survive, (C) the tree will die because the movement of food to the roots is checked, (D) the tree will die because the movement of water is prevented.

11. Trace the path of water from the soil to a palisade cell of a leaf.

12. Describe the transpiration pull and water cohesion theory for the rise of sap.

13. _____ Which one of the following is false? (A) Mesophyll cells absorb water from vessels and tracheids by osmosis and imbibition. (B) Water in vessels and tracheids moves up in intact columns. (C) As water is pulled up in the trunk of a tree, the vessels and tracheids tend to become narrower. (D) Water, being a liquid, has a very low cohesive strength.

14. _____ Which of the following theories can best account for the rise of water to the tops of very tall trees? (A) Root pressure, (B) transpiration pull and water cohesion, (C) capillarity, (D) pumping by living cells.

15. Explain how food is conducted in a plant.

16. _____ The diameter of a tree trunk would be least (A) when transpiration is rapid, (B) when transpiration is slow, (C) during the night, (D) when the leaf cells are turgid.

15

Metabolism

Starting with simple ingredients green plants manufacture such complex substances as carbohydrates, fats, waxes, sterols, proteins, pigments, hormones, and vitamins. In addition, a number of species synthesize latex, resins, perfumes, drugs, and alkaloids. Not all reactions occurring in plants result in the formation of complex substances from simpler compounds. In respiration and digestion, complex materials are broken down into simple ones. Both the constructive and destructive transformations which occur in organisms are collectively designated as **metabolism.** Metabolism is defined as the sum total of all the chemical activities associated with protoplasm. The constructive metabolic processes are called **anabolism** and the destructive ones **catabolism.**

Anabolic processes, such as photosynthesis and the synthesis of starch, cellulose, fats, and proteins, do not proceed spontaneously but instead require energy, from light for photosynthesis and from ATP for other reactions. In anabolic processes atoms are linked together and the energy utilized to join them is stored in the chemical bonds. When these bonds are broken, as they are in catabolic processes, energy is released. In other words, in catabolic processes complex molecules are changed into simpler ones and energy is released.

In an earlier chapter we considered photosynthesis, an anabolic process. You will recall that in photosynthesis carbon dioxide and water combine to yield glucose and oxygen. Energy from light is stored in the sugar formed.

The energy in the sugar is released in respiration, a catabolic process. In aerobic respiration, sugar and oxygen are used and water and carbon dioxide are formed. Of course, this shows only the beginning and end products. Many intermediate compounds are formed, and some of these are used in the formation of fats and proteins.

FOODS

The synthetic capacity of plants is amazing, far exceeding that of animals. From the sugar formed in photosynthesis other carbohydrates and also fats are produced. Plants make proteins and nucleoproteins from sugar and salts of nitrogen, sulfur, and phosphorus.

The foods of organisms fall mainly into three great classes: carbohydrates, fats, and proteins, the last including the nucleoproteins. These compounds contain energy and contribute materially to the growth and repair of tissues. Organic substances with these characteristics are technically called **foods.** When foods are defined in this way, the foods of plants and animals are the same. The difference between plants and animals is not in the kinds of food, but in their manner of obtaining food. Green plants synthesize food from inorganic materials such as water, carbon dioxide, and certain minerals, whereas animals cannot make food from these simple molecules but instead must secure their food ready made.

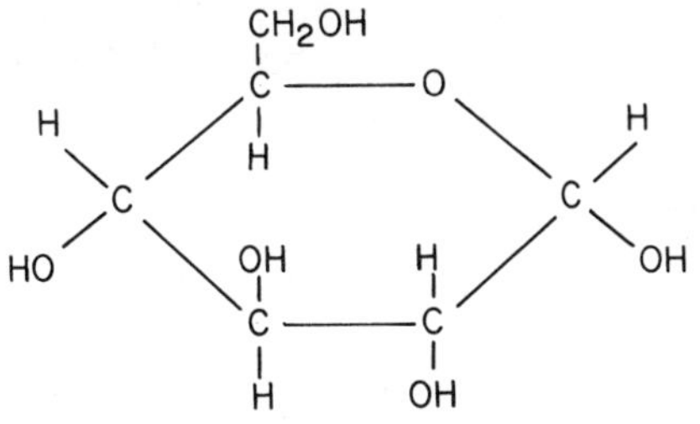

Glucose ($C_6H_{12}O_6$)

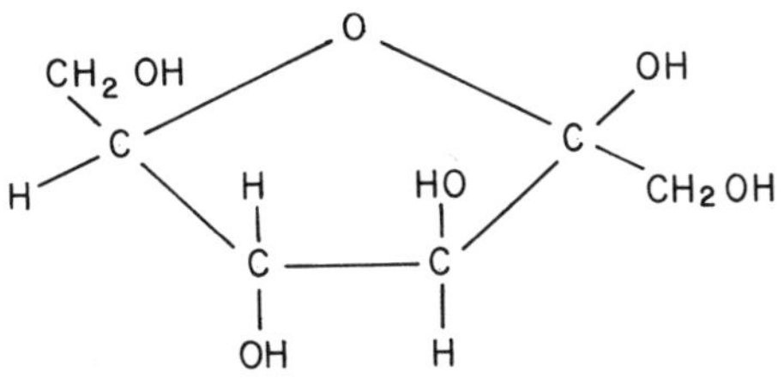

Fructose ($C_6H_{12}O_6$)

Carbohydrates

All carbohydrates contain carbon, hydrogen, and oxygen. The hydrogen and oxygen atoms are present in the same ratio as they are in water, two hydrogens for each oxygen. Some carbohydrates are sugars; others, such as starch and cellulose, are nonsugars.

Sugars. The sugars are the simplest carbohydrates. They are soluble in water, taste sweet, and the molecules are relatively small. Starting with the sugar **glucose,** the plant makes many different sugars—**ribose, fructose, sucrose,** and **maltose,** for example. Ribose, glucose, and fructose are monosaccharides. Sucrose and maltose are disaccharides—that is, sugars formed from the union of two monosaccharides.

Ribose is a **pentose** sugar, a sugar with five carbon atoms and a general formula of $C_5H_{10}O_5$. Related to ribose is **deoxyribose,** a 5-carbon sugarlike compound which has a hydrogen substituted for one of the OH groups of ribose. The formula for deoxyribose, is $C_5H_{10}O_4$. As we will see later, ribose is used in the synthesis of RNA and deoxyribose in the formation of DNA.

Glucose and fructose are **hexose** sugars, containing six carbons and having a formula of $C_6H_{12}O_6$; they occur in practically all plants. Although glucose and fructose have the same general formula, the groups within the molecule are arranged differently (see above) and, accordingly, glucose and fructose have unlike properties.

Sucrose and maltose are **disaccharides** with a formula of $C_{12}H_{22}O_{11}$. Sucrose is the main sugar of commerce and it is obtained commercially from the roots of sugar beets and the stalks of sugar cane. Plants synthesize sucrose from glucose and fructose as follows:

$$\text{Glucose} + \text{fructose} \rightarrow \text{sucrose} + \text{water}$$
$$C_6H_{12}O_6 + C_6H_{12}O_6 \rightarrow C_{12}H_{22}O_{11} + H_2O$$

Maltose is another sugar which is frequently found in plants. Two molecules of glucose combine to form one molecule of maltose and one of water as follows:

$$\text{Glucose} + \text{glucose} \rightarrow \text{maltose} + \text{water}$$
$$C_6H_{12}O_6 + C_6H_{12}O_6 \rightarrow C_{12}H_{22}O_{11} + H_2O$$

Nonsugars. Starch and cellulose are known as **polysaccharides** because they are formed by linking together many (*n* number) glucose molecules. When one glucose molecule is joined to another glucose molecule or a chain of glucose molecules, a molecule of water is split off. The figure at the top of page 196 shows how glucose molecules are joined together to form starch.

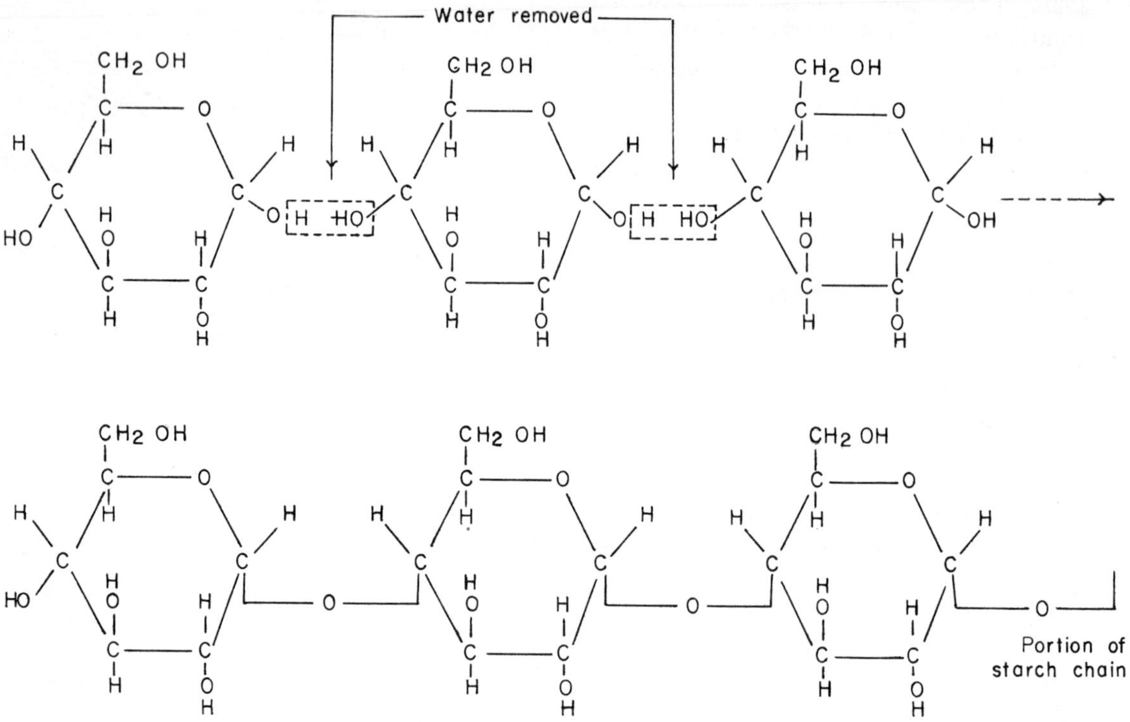

Water removed

Portion of starch chain

The general formula for the formation of starch and cellulose is as follows:

$$n \text{ glucose} \rightarrow \text{starch or cellulose} + nH_2O$$

$$nC_6H_{12}O_6 \rightarrow (C_6H_{10}O_5)_n + nH_2O$$

Although both starch and cellulose are built up from glucose, they differ in many ways and uses. A starch molecule is a coiled chain synthesized by linking together 300 to 1000 glucose units, whereas cellulose is a straight chain synthesized from about 2000 glucose molecules. Furthermore, the glucose that combines to form cellulose differs slightly from that which combines to form starch. Starch is synthesized from alpha-glucose while cellulose is formed from beta-glucose. Starch gives a blue color in the presence of iodine, but cellulose does not. Cellulose in cell walls forms the structural framework of a plant. Starch is a reserve food that can be digested and used. The seeds of many species contain considerable quantities of starch, as do tubers, fleshy roots, and woody twigs.

Starch is deposited as grains in leucoplasts and chloroplasts. Potato starch grains are single, but those of peas and oats are compound. Single grains result when only one starch grain is formed in a plastid; compound grains, when two or more grains develop within the same plastid. In a **starch grain** there is a darker region known as the **hilum** which is surrounded by **layers.** The layers are more prominent in some types of starch grains than in others, and they result because the starch deposited during the day is denser than that deposited during the night. Starch grains of one plant species are unlike those of any other species. The form and size of starch grains as well as the position of the hilum and clearness of the layers can be used to detect adulterants of foods and drugs and to identify commercially prepared products (Fig. 15–1).

Lipids

A second important class of foods is the lipids, fatty or oily substances that are generally insoluble in water, but soluble in such organic solvents as carbon tetrachloride, ether, and gasoline. The lipids include the fats and oils, the waxes, and the phospholipids.

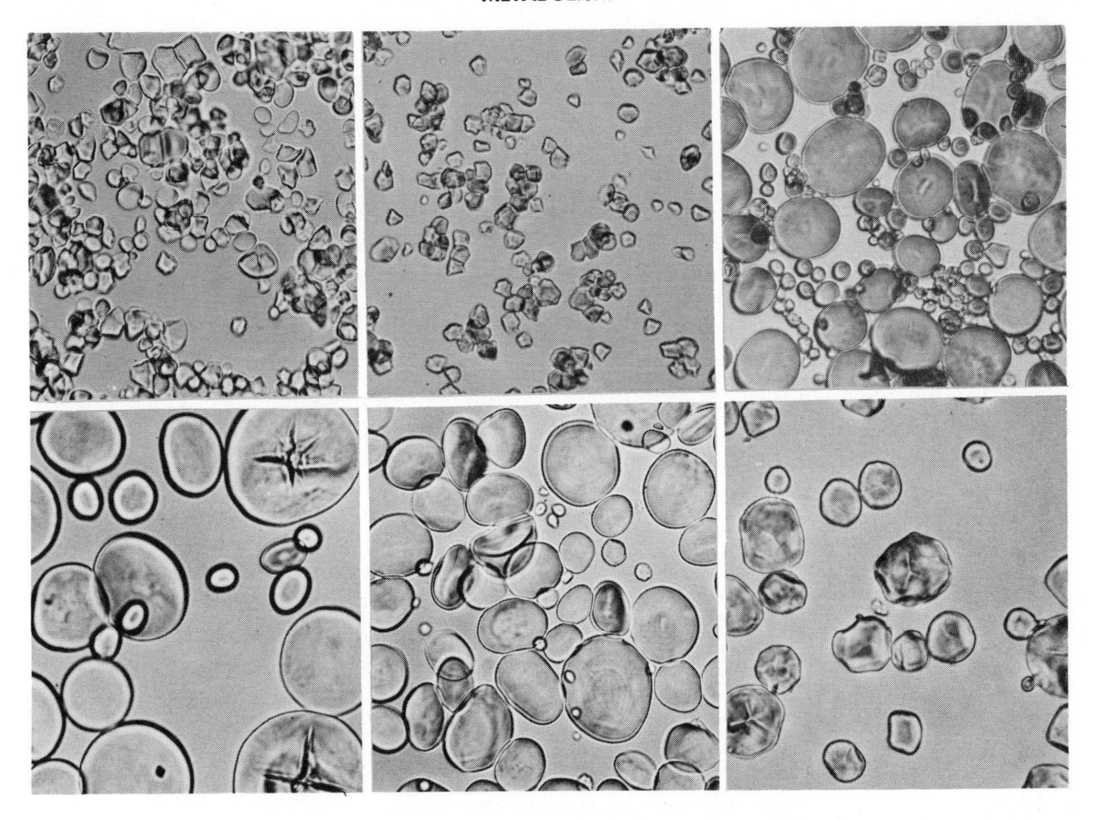

Fig. 15–1. Starch grains from various seeds. *Left to right, upper:* oats, rice, wheat; *lower:* rye, barley, grain sorghum. (M. M. MacMasters and C. E. Rist.)

The seeds of about 90 per cent of all species store food principally in the form of a fat or oil. We use the oil stored in seeds or fruits of the cacao, castor bean, coconut, corn, cotton, flax, olive, peanut, soya bean, and tung as food, as cooking fats, or in the manufacture of soaps, paints, and cosmetics. Fats which are fluid at room temperatures are called oils and those which are solid at these temperatures are called fats proper. The oils can be readily converted into fats by combining them with hydrogen, a process called hydrogenation.

Fats are efficient forms of food storage because they fill spaces completely and have a high caloric value. A pound of fat has as much potential energy as 2.25 pounds of sugar. Weight for weight more energy is stored in fats than in proteins. Animals, like many plants, store food as fat, and both can manufacture fats indirectly from glucose.

Fats and oils are formed from the combination of three molecules of a fatty acid with one of glycerol as illustrated below. All fatty acids, as well as other organic acids, are characterized by the carboxyl group (COOH). Glycerol is an alcohol with three hydroxyl (OH) radicals. The hydroxyl group occurs in all alcohols. In

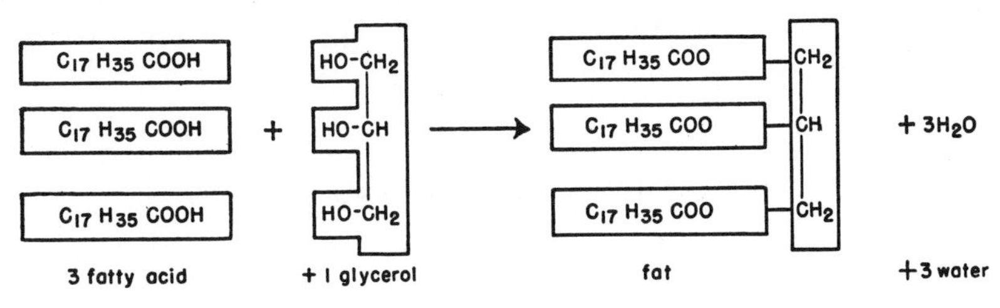

the example above, three molecules of stearic acid combine with glycerol to yield the fat tristearin. Fatty acids other than stearic may combine with glycerol to form a fat, among them lauric acid ($C_{11}H_{23}COOH$), palmitic acid ($C_{15}H_{31}COOH$), oleic acid ($C_{17}H_{33}COOH$), and linoleic acid ($C_{17}H_{31}$-$COOH$). Linoleic acid is the major fatty acid in corn oil, an oil much in demand. To increase the supply, plant breeders developed new varieties of corn which were higher in oil content. In two selection cycles the oil content was increased from 4.7 to 7.0 per cent. In certain fats, the three fatty acid molecules are the same. More often, three different fatty acids combine with one molecule of glycerol to form a fat.

The formation of fatty acids and glycerol is intimately related to respiration. The energy required for synthesis comes from respiration, as do the substances from which fatty acids and glycerol are formed. Fatty acids are formed by linking together substances containing two carbon atoms. For example, nine molecules of a 2-carbon compound are linked to form a molecule of stearic acid. Glycerol is formed from a compound containing three carbons. Both the 2-carbon and 3-carbon substances are intermediate materials formed when sugar is respired.

Waxes, important waterproofing chemicals, are the chief constituents of the cuticle which covers surfaces of leaves, young stems, flowers, and fruits, and of the suberin present in walls of cork cells. We make use of plant waxes as polishes for automobiles, floors, and shoes.

In the synthesis of a wax a straight-chain alcohol having twenty-four to thirty-six carbons and one hydroxyl group (OH) combines with one molecule of a fatty acid yielding one wax molecule and one of water. The formation of a wax from palmitic acid and octacosanol, an alcohol, is shown below:

Palmitic acid + octacosanol → wax + water

$$C_{15}H_{31}COOH + C_{28}H_{57}OH \rightarrow C_{15}H_{31}COOC_{28}H_{57} + H_2O$$

Alcohols other than octacosanol may combine with palmitic acid or some other fatty acid to form different waxes.

Like fats, phospholipids are synthesized in plant cells from fatty acids and glycerol, but in phospholipids only two fatty acid molecules combine with glycerol. The third hydroxyl group of glycerol combines with phosphoric acid. Hence a phospholipid molecule may be diagrammed as follows:

```
Fatty acid——CH2
             |
Fatty acid——CH
             |
Phosphate——CH2
```

In more complex phospholipids other groups, such as choline or ethanolamine, may be linked to the phosphate group.

Phospholipids are present in all cells. They are essential constituents of cell membranes, chloroplasts, and cytoplasm.

Proteins

Some proteins form the architecture of protoplasm, others serve as reserve food, and some are enzymes. Proteins are the largest and most complex molecules known, with molecular weights ranging from 13,000 to 10 million. Proteins are formed by the linking together of many amino acid molecules. There are about twenty-two different kinds of amino acids, but not every type of protein contains all of them. Amino acids are hooked together in chains hundreds to thousands of units long, with a great variety of branching and folding. In a certain protein, different kinds of amino acids are linked together, and each kind is present many times. A virtually infinite number of different proteins is possible. Organisms exploit this possibility, for no two species, animal or plant, possess exactly the same protein.

Amino acids are relatively small molecules with molecular weights in the range from 100 to 200, and they are made up of carbon, hydrogen, oxygen, and nitrogen. In addition to these elements, certain amino acids also contain sulfur. The simplest amino acid is glycine, which has a formula of CH_2NH_2COOH. Others are more complicated, but all amino acids have a carboxyl group (COOH) and an amino (NH_2) group. Among the amino acids containing sulfur is cysteine ($HSCH_2CHNH_2COOH$).

While plants can synthesize amino acids from carbohydrates and inorganic salts of nitrogen and sulfur, animals cannot. Animals are therefore dependent on plants, not only for a supply of carbohydrates, but also for their essential amino acids. Our diet need not include all of the amino acids which are incorporated in our proteins. Our diet must include eight essential amino acids. From these, other kinds may be synthesized. Not all foods furnish the eight essential amino acids. Wheat, rice, and corn are deficient in lysine. Adding lysine to products made from these cereals would make their proteins the equal of those in milk. Peas and beans are high in lysine but deficient in methionine and tryptophane, amino acids which are present in wheat. A diet of beans and wheat would provide us with all essential amino acids.

Plants synthesize amino acids from certain organic acids and ammonia. In some species amino acid synthesis is intermeshed with both photosynthesis and respiration. That amino acids are synthesized during photosynthesis was discovered by providing plants with carbon dioxide containing radioactive carbon (C^{14}). The radioactive carbon was soon detected in such amino acids as serine and alanine and in carbohydrates.

In nearly all plants the synthesis of some amino acids is coupled to respiration. The ammonia used in the synthesis of amino acids is produced by the reduction of nitrate. This reduction requires hydrogen and energy, both coming from respiration, and is summarized in the following equation:

Nitrate + hydrogen + energy → ammonia + water

An organic acid, formed as an intermediate in respiration, then combines with the ammonia to yield an amino acid as follows:

Ammonia + acid formed in respiration → amino acid + water

Because a variety of organic acids are formed in respiration, many kinds of amino acids may be synthesized. Amino acids are linked together by peptide bonds to form a protein. A peptide bond is a bond that connects carbon to nitrogen to carbon. The figure at the top of page 200 illustrates the formation of peptide bonds. Here alanine joins with glycine which in turn joins with another glycine molecule. The peptide bonds form as water is removed.

RNA located in a ribosome controls the sequence of amino acids in the protein chain thus synthesizing a specific protein which may be a certain enzyme. Let us illustrate the significance of RNA by showing how four amino acids, designated A, B, C, and D, might be fitted together in the absence of RNA and in its presence. We use four amino acids for simplicity; actually hundreds are linked together to form a protein.

PROTEIN FORMATION WITHOUT RNA CONTROL

Amino Acids	*Proteins*
A, B, C, D	A—B—C—D B—C—D—A D—A—C—B C—A—B—D

Without control many different proteins might result—for example, A—B—C—D and B—C—D—A.

PROTEIN FORMATION WITH RNA CONTROL

Amino Acids	*Protein*
A, B, C, D	A—B—C—D

With RNA control the amino acids are joined together in just one way, in the example to form the protein A—B—C—D. A different kind of RNA, formed under the

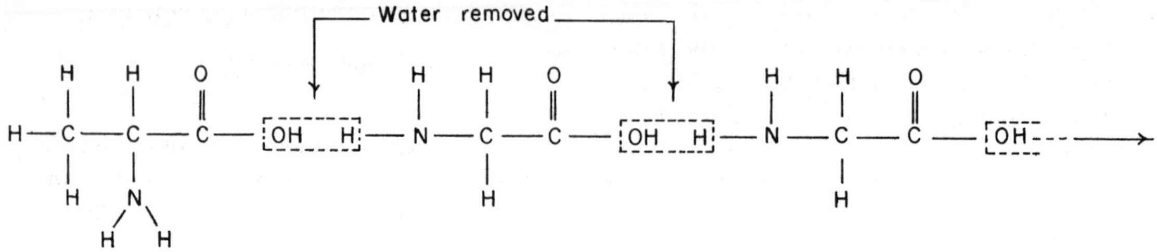

```
                         ┌──── Water removed ────────────────┐
                         ↓                                   ↓
     H    H    O                    H    H    O                        H    H    O
     |    |    ‖                    |    |    ‖                        |    |    ‖
H ── C ── C ── C ── [OH    H] ── N ── C ── C ── [OH    H] ── N ── C ── C ── [OH] - - ──→
     |    |                              |                                  |
     H    N                              H                                  H
         / \
        H   H

      alanine                          glycine                          glycine
```

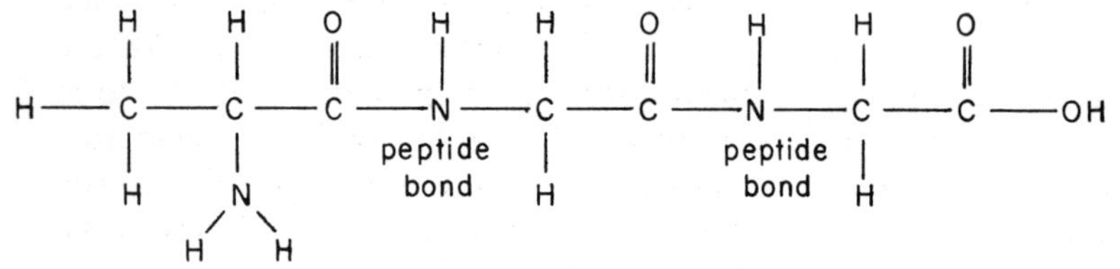

```
     H      H      O      H      H      O      H      H      O
     |      |      ‖      |      |      ‖      |      |      ‖
H ── C ──── C ──── C ──── N ──── C ──── C ──── N ──── C ──── C ──── OH     Portion of protein
     |      |          peptide   |          peptide   |
     H      N           bond     H           bond     H
           / \
          H   H
```

direction of a different gene, could produce a different arrangement—for example, D—B—C—A—but any one type of RNA always results in just one kind of protein.

The RNA incorporated in ribosomes is more precisely known as "messenger RNA" and it is this RNA that we have just considered. Another type of RNA, called "transfer RNA," also plays a role in protein synthesis. Transfer RNA selects amino acids from the cytoplasm and carries them to the ribosome where they are assembled into proteins under the direction of messenger RNA.

NUCLEIC ACIDS

Nucleic acid molecules, like protein molecules, are immense. The smaller molecules, the building blocks, that link together to form a nucleic acid are known as **nucleotides,** which come in two sets. One set of nucleotides is used to synthesize RNA, a different set to synthesize DNA.

The four kinds of nucleotides which join together to form RNA are known as ribose nucleotides and they are listed below:

Ribose Nucleotides
1. Adenine–ribose–phosphate
2. Guanine–ribose–phosphate
3. Cytosine–ribose–phosphate
4. Uracil–ribose–phosphate

All of the ribose nucleotides contain ribose linked to phosphate and to a nitrogenous base which may be adenine, guanine, cytosine, or uracil. Hundreds of these nucleotides are linked together by bonds connecting the phosphate of one nucleotide to the ribose of the adjacent one to form RNA. The sequence of the nucleotides in a "messenger RNA" molecule is controlled by a gene and the sequence in the "messenger RNA" controls the sequence with which amino acids are fitted together to form a protein. A molecule of RNA is a single strand and not a double helix as is DNA.

The nucleotides, called deoxyribonucleotides, that are joined to form DNA are also of four kinds, but they differ from those that link together to form RNA as can be seen below:

Deoxyribonucleotides
1. Adenine–deoxyribose–phosphate
2. Guanine–deoxyribose–phosphate
3. Cytosine–deoxyribose–phosphate
4. Thymine–deoxyribose–phosphate

All four deoxyribonucleotides contain the sugar deoxyribose and not ribose. The nitrogenous bases are the same as in the ribonucleotides except in No. 4 where thymine occurs instead of uracil. Adenine and guanine belong to a class of compounds called purines; cytosine and thymine are

Adenine

Thymine

Cytosine

Guanine

pyrimidines. The structural formulas for the nitrogenous bases are given above.

In the synthesis of DNA the nucleotides are joined together in such a manner that the DNA contains an enormous amount of coded information and so that it may replicate itself—that is, form a duplicate which is like the original molecule in every way.

The manner in which the nucleotides combine to form DNA may be visualized by referring to Figure 15–2, which schematically shows the four types of nucleotides that combine to form DNA. In the models D represents deoxyribose and P phosphate. A is adenine, T is thymine, C stands for cytosine, and G for guanine.

Parts of two molecules of DNA are illustrated in Figure 15–2. Notice that each molecule is a double chain with cross-linkages connecting the two chains. Each chain is formed as the nucleotides are connected end to end by bonds between the phosphate of one nucleotide and the deoxyribose of the one below. The cross-linkages between

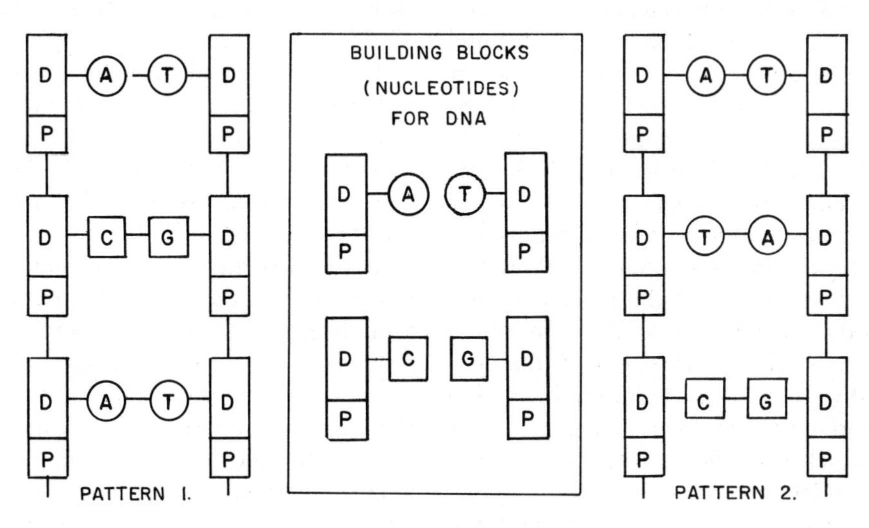

Fig. 15–2. Synthesis of DNA.

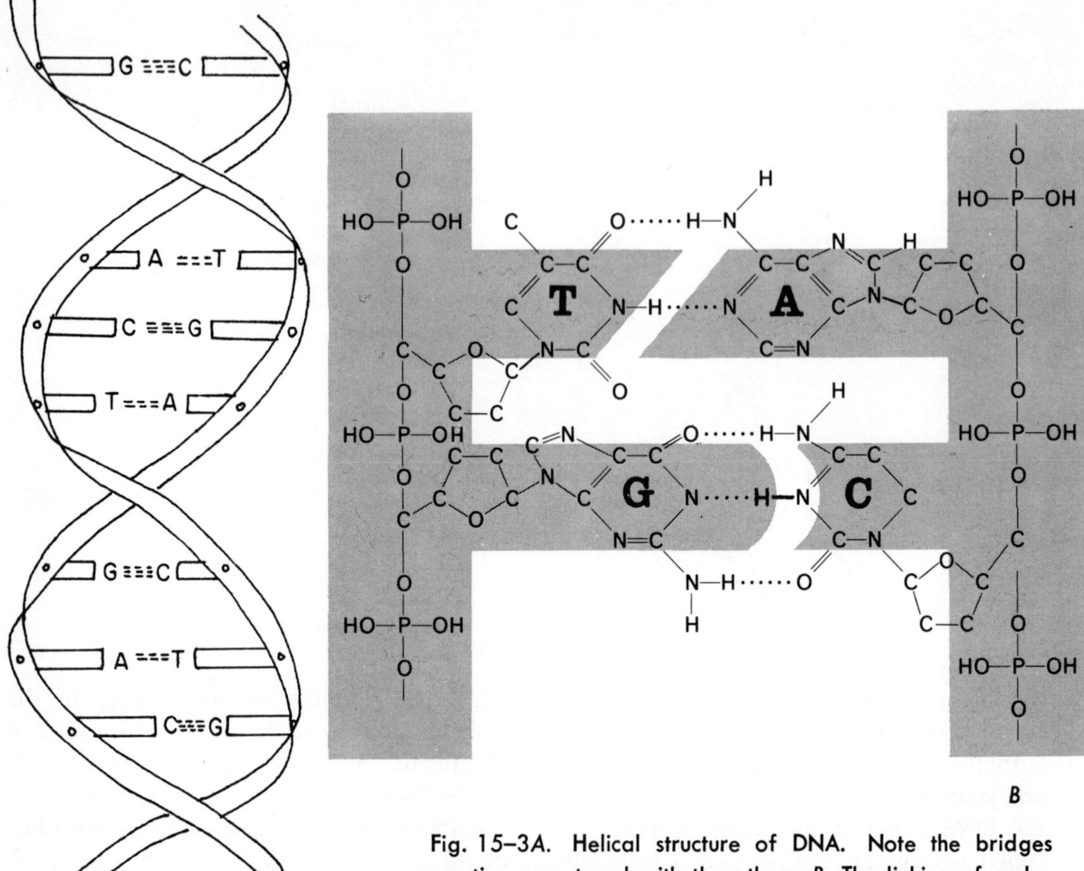

Fig. 15–3*A*. Helical structure of DNA. Note the bridges connecting one strand with the other. *B*. The linking of nucleotides to form DNA. Only four nucleotides are shown. Actually, thousands of nucleotides combine to form DNA. (From *Biology* by E. L. Cockrum, W. J. McCauley, and N. A. Younggren. Copyright © 1966, W. B. Saunders Co., Philadelphia.)

the two chains are between adenine and thymine and between cytosine and guanine and never between cytosine and adenine or guanine and thymine. The bridges connecting the two chains of a molecule seem to carry the coded information and the sequence differs with each kind of nucleic acid, of which there may be countless billions of types. In the diagram we show just two types. In pattern 1 the adenine-to-thymine bridge is followed by the cytosine-to-guanine one, whereas in pattern 2 the adenine-to-thymine link is followed by a thymine-to-adenine link. Our diagram is, of course, much simplified. Actually hundreds of nucleotides are joined together, not just six. Moreover, the nucleic acid molecule is coiled resembling a spiral staircase (Fig. 15–3*A*).

The fact that adenine always joins with thymine and guanine with cytosine ensures that when the molecule replicates itself, the two resulting molecules will be identical. The replication of a DNA molecule is illustrated in Figure 15–4.

In the first step of replication the DNA molecule separates into two chains; separation begins at one end and gradually progresses to the opposite end. A pool of nucleotides is present in the nucleus where the DNA is synthesized. A thymine nucleotide of the pool connects with an adenine nucleotide of the original chain and an adenine one of the pool with a thymine of an original chain. Similarly, cytosine always joins with guanine and guanine with cytosine. Thus, when replication is complete, two identical DNA molecules will result and each will be a perfect copy of the parent molecule.

We have studied the synthesis of DNA; now let us consider the synthesis of messenger RNA as it occurs next to a DNA strand in the nucleus (Fig. 15–5, *left*). The sequence of bases in the DNA strand governs the sequence of bases in the messenger RNA strand. The adenine nucleotide of DNA specifies the location of the uracil nucleotide of the messenger RNA; a uracil nucleotide is analogous to a thymine nucleotide of DNA and hence uracil always pairs with adenine. The guanine nucleotide of DNA governs the location of a cytosine nucleotide of the messenger RNA and the cytosine of DNA the position of the guanine nucleotide of messenger RNA. In other words, a base sequence of adenine (A), guanine (G), and cytosine (C) specifies a sequence of uracil (U), cytosine, and guanine in the RNA strand. After the nucleotides are aligned, they are joined together to form messenger RNA, which then leaves the nucleus. The messenger RNA coats the surface of a ribosome where it governs the sequence of amino acids in a protein.

The amino acids which will combine to form a protein are activated and then each kind combines with its particular transfer RNA molecule. Amino acid 1 joins with one type of transfer RNA, amino acid 2 with a different transfer RNA, and so on. The transfer RNA transports the amino acid to the messenger RNA. The messenger RNA specifies the position of attachment of the transfer RNA. The adenine of transfer RNA always pairs with uracil of messenger RNA and guanine always with cytosine. After the amino acids are aligned in the determined sequence, they are joined together to form a protein.

Recent evidence indicates that a sequence of three bases in DNA specifies a complimentary three-base sequence in messenger RNA, which then controls the location of

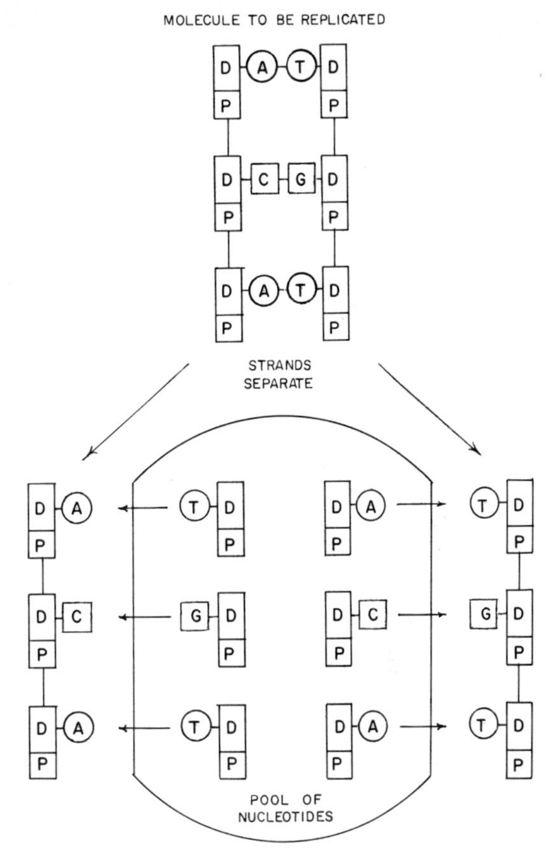

Fig. 15–4. Replication of DNA.

an amino acid via the three bases of its transfer RNA.

When three adenines (AAA) occur in sequence in messenger RNA, the amino acid lysine will be aligned. When three uracils (UUU) occur one after the other, phenylalanine will be the selected amino acid. The genetic code of messenger RNA for some amino acids is given in Table 15–1.

Table 15–1

Genetic Code for Messenger RNA and the Amino Acids

Amino Acid	Triplet Code
Alanine	CCG
Cysteine	UGU
Glycine	UGG
Lysine	AAA
Phenylalanine	UUU
Proline	CCC

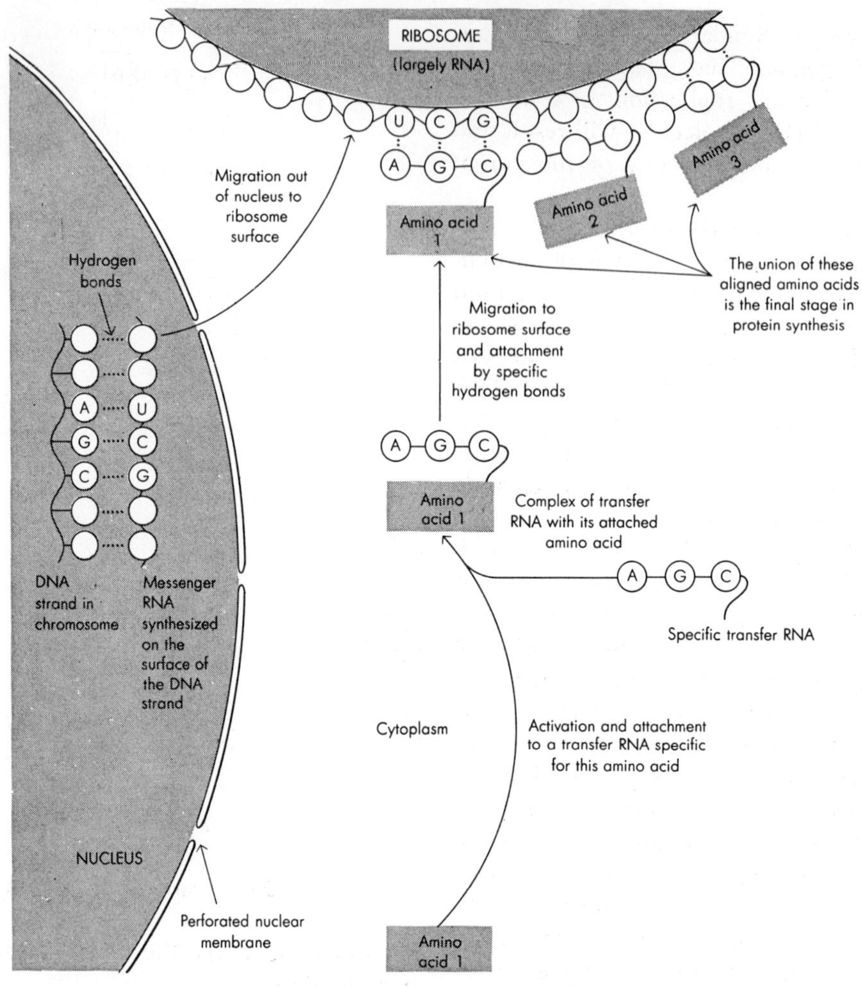

Fig. 15–5. The synthesis of messenger RNA in the nucleus and the control of protein synthesis by RNA. (Arthur W. Galston, *The Life of the Green Plant.* 1964, 2d ed. By permission of Prentice-Hall, Inc., Englewood Cliffs, N.J.)

If the sequence of bases in messenger RNA were as given in the table (CCG · UGU · UGG · AAA · UUU · CCC), amino acids would be linked as follows:

Alanine–cysteine–glycine–lysine–phenylalanine–proline

Incidentally, the sequence in the DNA strand which controlled the RNA strand which controlled the sequence of amino acids would be GGC · ACA · ACC · TTT · AAA · GGG.

Nucleic acids, both RNA and DNA, are joined to proteins, forming nucleoproteins. Nucleoproteins made of RNA are found throughout the cell, but DNA nucleoproteins occur chiefly in the nuclei of cells where they are the structural components of genes. Recent evidence indicates that chloroplasts and mitochondria contain a strand of DNA that may be the code for their included enzymes.

We can summarize the steps leading to the synthesis of nucleoproteins as follows:

1. Pyrimidine or purine + sugar + phosphate → nucleotide
2. Many nucleotides → nucleic acid
3. Nucleic acid + protein → nucleoprotein

VITAMINS

The higher green plants synthesize a number of vitamins, including thiamine (vitamin B_1), riboflavin (B_2), pyridoxin (B_6), niacin (another B vitamin), ascorbic acid (vitamin C), and vitamin K. The vitamins play essential roles in the metabolism of a plant as they do in animals. Many vitamins, especially those of the B complex, are prosthetic groups of enzymes. Some vitamins are formed only in certain plant parts from which they may be translocated to other sites. For example, Vitamin B_1 is synthesized in the leaves of plants, but not in the roots. Roots severed from a plant and grown in tissue culture will not grow unless vitamin B_1 is added to the nutrient medium. However, intact plants thrive equally well with or without an external supply of vitamin B_1.

Unlike green plants, not all bacteria and true fungi synthesize the vitamins required in their metabolism. If these organisms are to thrive, they must be provided with an external source of vitamins. Many animals, including man, cannot synthesize vitamins; accordingly, vitamins must be included in their diets.

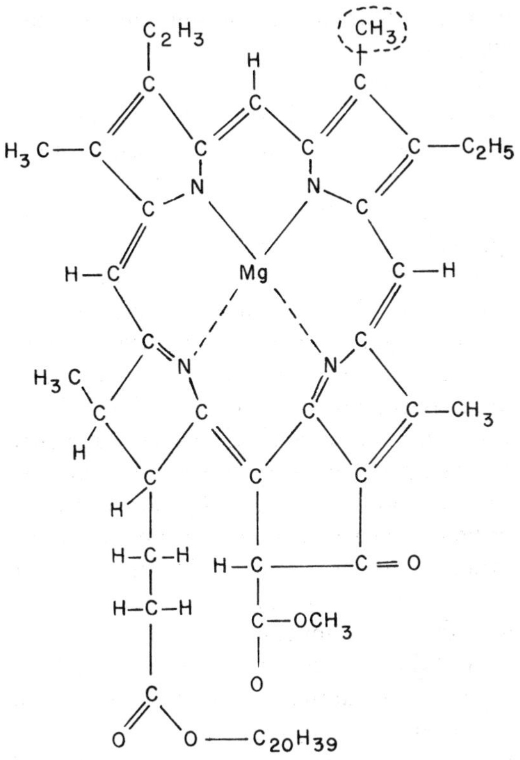

PIGMENTS

Chlorophyll a, chlorophyll b, carotene, and xanthophyll are pigments which all green plants synthesize. Chlorophyll a has the formula $C_{55}H_{72}O_5N_4Mg$, and chlorophyll b the formula of $C_{55}H_{70}O_6N_4Mg$. The structural formula of chlorophyll a is shown in Figure 15–6. The carbon, hydrogen, and oxygen in chlorophyll come from carbohydrates, and the nitrogen and magnesium from inorganic salts. As shown in Chapter 11, while iron is not a part of a chlorophyll molecule, it is essential for chlorophyll formation. In 1961 chlorophyll was synthesized in the laboratory.

Carotene, $C_{40}H_{56}$, and xanthophyll, $C_{40}H_{56}O_2$, are yellow to orange pigments. Carotene and xanthophyll occur with chlorophyll in chloroplasts. They also occur independently of chlorophyll in chromoplasts present in carrot roots, nasturtium petals, and many other yellow to orange plant parts.

Fig. 15–6. Structural formula of chlorophyll a. Chlorophyll b differs from chlorophyll a in having an HCO group instead of the CH_3 group which is circled.

Anthocyanin pigments which furnish red, blue, and purple colors to flowers, fruits, and leaves are synthesized by some plants. These pigments are soluble in water and occur in the cell sap. An anthocyanin pigment is formed by the combination of sugar with a colored substance known as **anthocyanidin,** of which there are three kinds—pelargonidin, cyanidin, and delphinidin. Pelargonidin occurs in flowers of geranium, cyanidin in cornflowers, and delphinidin in larkspur blossoms. The color of an anthocyanin varies with the acidity or alkalinity of the cell sap, being red in acid solution, violet in neutral, and blue in an alkaline medium. If you add a little ammonia to a glass of beet juice, the juice will turn from red to blue as the solution becomes alkaline.

The development of anthocyanin pigments is influenced by genetic factors and environmental conditions. If the appropriate genes are lacking, the pigments are not produced. In plants which have the required genes, the amount of pigment which forms is often influenced by environmental conditions. Environmental factors are of special significance in the development of anthocyanins in fruits, leaves, and stems. A high concentration of sugar in the cells favors the formation of these pigments. Bright light, low temperatures, and low nitrogen supply result in a high sugar concentration, hence in a large amount of anthocyanin. During autumn the nitrogen in leaves moves out of the leaves into the stem and the low temperatures which generally prevail favor the conversion of starch to sugar. If the autumn days are bright, sugar is formed by photosynthesis, and this accumulates in the leaves. The combination of low, but not freezing, temperatures with bright weather in the autumn brings about a brilliant coloring of the foliage of many species. The fact that stems and leaves of plants low in phosphorus become reddish or purplish in color indicates that low phosphorus favors the formation of anthocyanin pigments.

SECONDARY MATERIALS SYNTHESIZED BY SOME PLANTS

Certain species synthesize particular substances which do not seem to play an essential role in metabolism. However, these chemicals may play secondary roles, such as the attraction of insects to effect pollination or to make the plant unpalatable to various predators. Such chemicals are not produced by all species but occur in scattered species in a number of families. Among the secondary substances produced by some plants are alkaloids, essential oils, latex, and tannins, substances which may be useful to us.

Alkaloids

These are nitrogenous bases with a ring structure formed from carbon and nitrogen atoms. A familiar alkaloid is caffeine, which is not widely distributed in the plant world but instead is confined to leaves and seeds of a few species such as tea, coffee, and cola. The structural formula of caffeine is given below.

```
                O                    H
                ‖                    |
                C                    C—H
      H       /   \                 /|
      |      /     \               / H
    H—C—N            C————N
      |    |          ‖     \
      H    |          ‖       CH
           |          ‖      //
      O=C             C————N
           \         /
             \     /
               N
               |
             H—C—H
               |
               H
```

Caffeine

Other alkaloids produced by certain species are morphine, quinine, nicotine, and theobromine. In some families few or no species produce alkaloids, whereas in others, such as the Solanaceae—which includes tobacco, deadly nightshade, and henbane—many species do so.

Essential Oils

Essential oils are aromatic substances that are responsible for the characteristic flavors and odors of many plant parts, such as flowers, oranges, mint leaves, cinnamon bark, and juniper branches. Essential oils, isolated from plants by steam distillation, are economically important in food flavoring and in perfumes. Essential oils differ chemically from the true oils and fats. They are not formed from fatty acids and glycerol, but are synthesized from isoprene, a 5-carbon compound. They are included in a group of compounds known as the terpenes. In some plants the essential oils are produced in glands or glandular hairs. Less than 1 per cent of the seed plants synthesize essential oils and often these plants are more frequent in some families, such as the Pinaceae, Umbelliferae, and Myrtaceae, than in others.

The best-known essential oil is turpentine. Turpentine and rosin are obtained from pine resin, which is elaborated by glandular cells that line the resin canals present in the bark and wood of pine trees. To get a flow of resin a V-shaped strip of bark and ½ inch of wood is removed from the face of a pine tree. The crude resin is distilled; the distillate is turpentine and the finer part of the residue is rosin. The industry, called the naval stores industry, is centered in Georgia and Florida. About 25 million gallons of turpentine and 800 million pounds of rosin are produced each year.

Rubber

Like the essential oils, rubber is an isoprene derivative. Although 2000 species produce rubber, most of the rubber of commerce comes from the latex of *Hevea brasiliensis,* a tropical tree (Fig. 15–7). The rubber occurs as microscopic particles suspended in a liquid serum, called latex, contained in latex tubes of the bark. Because the latex is under pressure in these tubes, it flows out through incisions in the bark, a technique for collecting latex.

Fig. 15–7. Tapping a rubber tree (*Hevea*). The latex flows from the incision and is collected.

ENZYMES

Photosynthesis, respiration, digestion, and the synthesis of various products result from chemical activities in cells. Generally these reactions do not occur apart from cells at ordinary temperatures and pressures. What agents enable the many biological reactions occurring in cells to go on rapidly at these same temperatures and pressures?

Within cells there are organic catalysts known as enzymes which speed up reactions. **Catalysts** are substances which promote reactions without being used up, and they are effective in low concentrations. Most chemical activities in cells are brought about by specific enzymes. Without enzymes it would be impossible for reactions to proceed rapidly enough to maintain life. Enzymes are universally present in protoplasm of all plants and animals and are indeed the machinery of cells. In a cell there are many kinds of enzymes. Some catalyze reactions leading to the production of food, many play a role in respiration, different ones function in protein synthesis, etc. A given enzyme is specific; it can bring about only one kind of chemical action.

In just one cell ten thousand different kinds of enzymes control ten thousand different reactions. Some enzymes do their work in the cytoplasm, others in the nucleus. At least fifty kinds of enzymes are located in each mitochondrion, those enzymes which catalyze the fifty steps in respiration. A set of enzymes, complete for photosynthesis, is located in each chloroplast.

All enzymes are proteins which, as we have seen, are long molecules formed by the linking together of 150 or more amino acid molecules of twenty-two kinds. The sequence in which the different amino acids succeed one another down the length of the chain determines what kind of enzyme the molecule is. One sequence yields one type of enzyme, while a different arrangement results in a different enzyme, each catalyzing a different reaction.

If some enzymes are to function, they require a helper group called a prosthetic group. On this basis we can recognize two major groups of enzymes, those which are only protein and those which are protein-linked with a prosthetic group. The prosthetic group in certain enzymes is copper, zinc, or manganese, and in others it is a vitamin, such as thiamine or riboflavin. Many enzymes have been isolated from cells and have performed their work in test tubes. Enzymes work by combining with the **substrate** (the substance undergoing a reaction). When combined with an enzyme, the substrate is more reactive. After the reaction has occurred, the enzyme is free to combine with another substrate molecule and bring about its reaction. Hence we see that enzymes are not used up. The general features of enzyme action are shown in Figure 15–8.

At low temperatures enzymatic reactions slow down, but the enzymes are not destroyed. Enzymes are completely destroyed by boiling, however. We make use of this knowledge in storing certain foods. In ripening corn and pea seeds the enzyme **starch phosphorylase** changes sugar into starch, a reaction which proceeds when the freshly harvested seeds are kept in a warm room. As the sugar changes to starch, they lose their pleasant sweet flavor. The sweet taste can be retained for a longer time by storing the corn or peas in a refrigerator where the reaction will go on very slowly. The flavor can be retained indefinitely by destroying the enzyme with heat. The boiling of foods prior to quick freezing is designed to destroy enzymes and thereby maintain high quality of the frozen product.

Many enzymatic reactions are reversible. For example, in guard cells starch phosphorylase changes starch into sugar when the *p*H is between 6.0 and 7.4, and sugar into starch at a *p*H of 5 or less. Under some conditions, such as in germinating seeds, the enzyme lipase splits fat into glycerol and fatty acids. Under other conditions, as in ripening seeds, the same enzyme brings about the synthesis of fat from fatty acids and glycerol.

Rates of enzymatic reactions are affected by temperature, *p*H, hydration, and the concentrations of the enzyme, substrate, and end products. As temperature increases from 0° to about 40° C., the rate increases. Certain reactions go on more rapidly at one hydrogen ion concentration than at higher or lower ones; for example, the digestion of starch by diastase is most rapid at *p*H 5. Enzyme activity in dry seeds is extremely

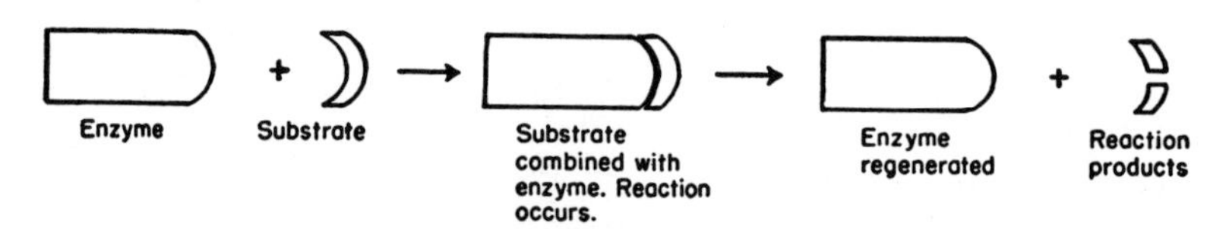

Fig. 15–8. Enzymes work by combining with the substrate, thereby lowering the activation energy. After the reaction is completed, the enzyme is regenerated.

low, but, as the seeds imbibe water, the enzymes become progressively more active. Within limits the rates of enzymatic reactions increase as the concentrations of the substrate and enzyme increase.

As the end products accumulate, the enzymatic reaction decreases or stops, an effective feedback mechanism that maintains the concentration of the end product within narrow limits. For example, in five steps, catalyzed by five different enzymes, the amino acid threonine is converted into isoleucine, also an amino acid. The presence of isoleucine in excess blocks the first step of the reaction series and thus effectively prevents an excessive accumulation of isoleucine. This method of self-regulation occurs in many other biological reactions and tends to maintain the end products in concentrations compatible with life.

Digestive enzymes are named by adding the suffix -ase to the root of the substance upon which the enzyme acts. Sucrase acts on sucrose, maltase on maltose, amylase on amylose (the more technical name for starch), etc. Enzymes concerned with respiration and photosynthesis are named from the type of reaction which they catalyze: the enzyme dehydrogenase, e.g., separates hydrogen from some compound.

DIGESTION

When a seed is planted, the starch, protein, and fat stored in the seed cannot be used directly by the developing embryo. Large molecules of starch, proteins, and fats cannot be moved as such to regions where growth occurs. Before they can be translocated and used, they must be changed into smaller molecules which are soluble in water. The process whereby large organic molecules are split into smaller, soluble ones is known as **digestion.**

Digestion of Carbohydrates

Starch, maltose, and sucrose are split into smaller molecules by particular enzymes when they are digested within cells of green plants. Most of the higher plants cannot digest cellulose, but many fungi can digest this compound.

Starch. A molecule of starch is very large. It can be digested in three ways by three different enzymes. **Beta-amylase** digests starch into the soluble sugar maltose. **Alpha-amylase** splits starch into molecules of intermediate size called dextrins. Dextrins in turn are changed to maltose by the enzyme **dextrinase,** and finally the maltose is changed to glucose by **maltase.** This series of reactions and the enzymes used may be summarized as follows:

$$1 \text{ starch} + n \text{ water} \xrightarrow{\text{alpha amylase}} \text{dextrins} \xrightarrow{\text{dextrinase}} \text{maltose} \xrightarrow{\text{maltase}} n \text{ glucose}$$

The above collection of enzymes which digest starch is commonly known as **diastase.** Starch can also be digested by the enzyme **starch phosphorylase.** When starch is digested by this enzyme, phosphoric acid enters into the reaction, instead of the water that is used when starch is digested by alpha- or beta-amylase. The end product of the digestion of starch by starch phosphorylase is glucose phosphate, which has a higher energy content than glucose.

Cellulose. The enzyme **cellulase** splits the long cellulose molecule into many molecules of the disaccharide **cellobiose,** which in turn are cleaved by **cellobiase** into glucose molecules. In these reactions water is used.

$$1 \text{ cellulose} + \text{water} \xrightarrow{\text{cellulase}} \text{cellobiose} \xrightarrow{\text{cellobiase}} n \text{ glucose}$$

Mammals, including man, lack the enzyme cellulase, and hence they cannot digest cellulose. Cattle and sheep do not produce cellulase, but this enzyme is formed by certain bacteria that normally live in their alimentary canals. Because of these bacteria, cattle and sheep can utilize cellulose for food.

Wood-destroying fungi secrete cellulase and cellobiase, thus changing the cellulose of wood into sugar and thereby enabling the threadlike bodies of these fungi to penetrate the hardest wood. They use the sugar as food. Some of the sugar is respired and some is assimilated. Fungi can destroy so much of the cellulose of a living tree that the wood becomes hollow or punky. Other fungi use the cellulose in fallen leaves and twigs for food and bring about the decay of these materials.

Sucrose. **Sucrase** splits sucrose into glucose and fructose, as follows:

$$\text{Sucrose} + \text{water} \xrightarrow{\text{sucrase}} \text{glucose} + \text{fructose}$$

$$C_{12}H_{22}O_{11} + H_2O \longrightarrow C_6H_{12}O_6 + C_6H_{12}O_6$$

Digestion of Proteins

Proteins are digested by the enzyme **protease.** Protease splits large protein molecules into units of smaller size and ultimately into a variety of amino acids. There may be as many as twenty-two different amino acids. Previously it was stated that enzymes are specific; for example, sucrase acts on sucrose but not on maltose. There are countless different kinds of proteins, but they can all be digested by the same protease. The different proteins have like chemical bonds. In any protein, carbon is linked to nitrogen, which in turn is linked to another carbon. This carbon-to-nitrogen-to-carbon bond is called a **peptide bond.** Proteases break such bonds. The protease isolated from the latex which exudes from incisions made into green fruits of the papaya plant is known as papain and this enzyme digests both plant and animal proteins. Papain is widely used to tenderize meats (that is, partially digest them). If we soaked meat long enough in papain, we could drink the "meat" because in time the protein would be changed into soluble amino acids. Papain is also used in the brewing industry to digest proteins that would otherwise precipitate when the beer is cooled.

Plants of an especially interesting group are the insectivorous plants. Venus flytrap is illustrated in Figure 15–9. The blade of each leaf is made of two parts hinged together. The two parts snap together when an insect touches the sensitive inner bristles. The leaf then secretes protease and the resulting amino acids diffuse into the leaf where they are utilized.

The insectivorous habit evolved independently in different families. About 440 species in five families have one or another device for capturing insects and also mechanisms for their digestion. Among the insectivorous plants are sundew, pitcher plants, and the bladderwort. Venus flytrap, sundew, and bladderwort have active traps, whereas the pitcher plants have passive pitfalls. Insectivorous plants grow in rather swampy soil with a low inorganic nitrogen content. From the captured insects the plants secure an organic nitrogen supply to supplement the inorganic nitrogen from the soil. The organic supply improves growth as compared with plants not furnished a supply of insects. All insectivorous plants have chlorophyll and hence carry on photosynthesis.

Digestion of Fats

Lipase digests fat into glycerol and fatty acids as follows:

$$1 \text{ fat} + 3 \text{ water} \underset{}{\overset{\text{lipase}}{\rightleftarrows}} 3 \text{ fatty acids} + 1 \text{ glycerol}$$

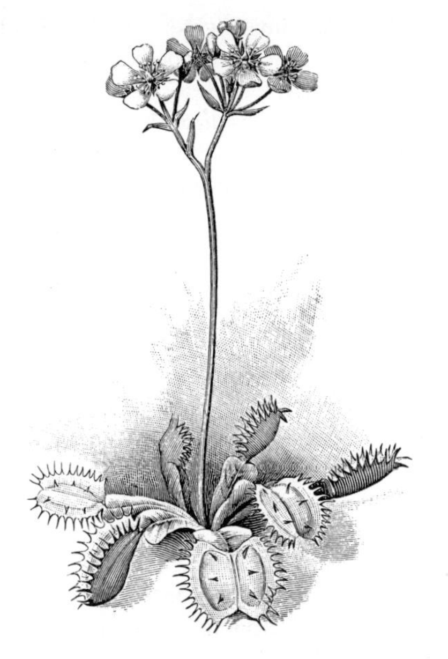

Fig. 15–9. Venus flytrap. (After Bergen.)

In the digestion of a fat, each fat molecule takes up three molecules of water. Three fatty acid molecules and one of glycerol are formed. The glycerol and fatty acids are not used directly in respiration and growth. Instead, they are converted into sugar which is then utilized. The steps in the conversion are unknown, but when seeds rich in fats germinate, the concentration of sugar increases as the fat content diminishes. During ripening of fatty seeds, sugar is translocated to the seeds where it is transformed into glycerol and fatty acids, which combine to form fat.

Summary of Digestion

The digestive reactions going on in plants are summarized below:

Substrate	Enzyme	End Product
Starch	Beta-amylase	Maltose
Starch	Alpha-amylase	Dextrins
Starch	Starch phosphorylase	Glucose phosphate
Dextrin	Dextrinase	Maltose
Cellulose	Cellulase	Cellobiose
Cellobiose	Cellobiase	Glucose
Maltose	Maltase	Glucose
Sucrose	Sucrase	Glucose and fructose
Protein	Protease	Amino acids
Fat	Lipase	Glycerol and fatty acids

QUESTIONS

1. Classify the following processes as catabolic (C), anabolic (A), or neither (N) catabolic nor anabolic.

_____ 1. Photosynthesis.
_____ 2. Respiration.
_____ 3. Transpiration.
_____ 4. Assimilation.
_____ 5 Digestion.
_____ 6. Protein synthesis.

2. Define food. Name the three classes of foods.

3. _____ In aerobic respiration a variety of organic acids are formed. Certain ones may combine (A) with nitrates to form amino acids, (B) with ammonia and sulfur in the reduced form to yield amino acids, (C) directly with glycerol to form fats, (D) with an anthocyanidin to form an anthocyanin pigment.

4. Match the process on the left with the organelle on the right.

_____	1. Photosynthesis.	1. Ribosome.
_____	2. Respiration.	2. Nucleus.
_____	3. DNA synthesis.	3. Grana.
_____	4. Starch synthesis.	4. Mitochondria.
_____	5. Enzyme synthesis.	5. Leucoplast.

5. _____ Which one of the following is a pentose sugar with a formula of $C_5H_{10}O_5$? (A) Ribose, (B) deoxyribose, (C) fructose, (D) mannose, (E) glucose.

6. _____ Both starch and cellulose are (A) formed from amino acids, (B) polysaccharides synthesized from glucose, (C) polysaccharides formed from fructose, (D) water-soluble carbohydrates.

7. _____ Which one of the following does not have a general formula of $C_6H_{12}O_6$? (A) Glucose, (B) fructose, (C) sucrose, (D) mannose.

8. _____ Which one of the following is false? (A) Fats and oils are formed from the combination of three molecules of a fatty acid with one of glycerol. (B) It is common knowledge that, weight for weight, more energy is stored in sugar than in fats. (C) The energy required for fat synthesis comes from respiration. (D) Animals, like plants, can manufacture fats indirectly from glucose.

9. _____ Which one of the following is not a lipid? (A) Fat, (B) phospholipid, (C) essential oil, (D) wax.

10. _____ Waxes are waterproofing compounds which are (A) synthesized from glycerol and fatty acids, (B) synthesized from fatty acids and long-chain alcohols having one hydroxyl group, (C) formed when a glycerol molecule combines with two molecules of a fatty acid and one of phosphoric acid, (D) often incorporated into the plasma and vacuolar membranes.

11. _____ Which one of the following is false? (A) All amino acids contain carbon, hydrogen, oxygen, and nitrogen. (B) Amino acids are made by plants from certain organic acids and ammonia. (C) Animals, like plants, can synthesize amino acids from carbohydrates and inorganic salts of nitrogen and sulfur. (D) Amino acids combine to form a protein molecule. (E) No two species possess exactly the same protein.

12. _____ An investigator provided illuminated algal cells with carbon dioxide containing radioactive carbon. A few minutes later he analyzed the cells. Radioactive carbon would not be detected in (A) amino acids, (B) monosaccharides, (C) isoprene derivatives, (D) phosphoglyceric acid.

13. _____ The two groups which are present in all amino acids are (A) COOH and NH_2, (B) COOH and SH, (C) SH and NH_2, (D) CHO and OH.

14. _____ The sequence of fitting amino acids together to form a protein which may be a specific enzyme is controlled (A) by messenger RNA, (B) directly by DNA, (C) by hormones of diverse kinds, (D) by protein templates.

15. _____ The substance formed under the direction of DNA which controls the synthesis of enzymes is (A) ribonucleic acid, (B) deoxyribonucleic acid, (C) mitochondrial centers.

16. _____ Which one of the following nucleotides is not used in the synthesis of DNA? (A) Uracil–ribose–phosphate, (B) adenine–deoxyribose–phosphate (C) guanine–deoxyribose–phosphate, (D) cytosine–deoxyribose–phosphate, (E) thymine-deoxyribose–phosphate.

17. _____ Which one of the following cross-linkages is never found in DNA? (A) Adenine-thymine, (B) thymine-adenine, (C) cytosine-guanine, (D) guanine-cytosine, (E) guanine-thymine.

18. Explain the replication of DNA.

19. _____ Which one of the following is false? (A) The RNA incorporated in ribosomes is "messenger RNA." (B) "Transfer RNA" carries specific amino acids to the ribosomes. (C) The building blocks for nucleic acid, a long molecule, are nucleotides. (D) RNA—like DNA—is a double-chain molecule with cross-linkages between the two strands.

20. _____ Which one of the following is false? (A) Green plants synthesize a number of vitamins such as thiamine, riboflavin, pyridoxin, niacin, ascorbic acid, and vitamin K. (B) Several vitamins are prosthetic groups or coenzymes of essential enzymes. (C) Roots severed from a plant and grown in tissue culture will not grow unless the nutrient medium contains thiamine (vitamin B_1). (D) Many green plants grow better when thiamine is incorporated into the soil.

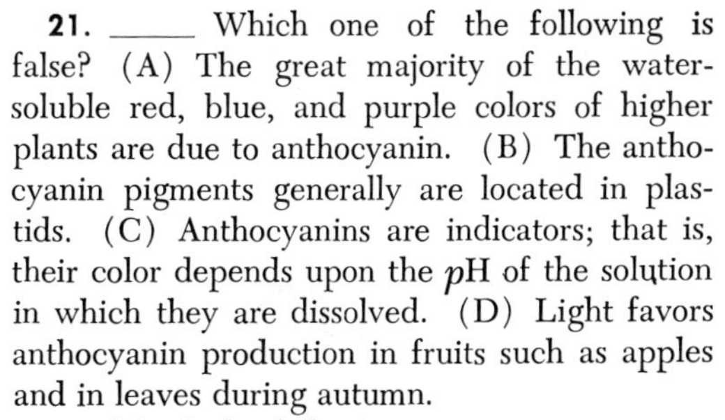

21. _____ Which one of the following is false? (A) The great majority of the water-soluble red, blue, and purple colors of higher plants are due to anthocyanin. (B) The anthocyanin pigments generally are located in plastids. (C) Anthocyanins are indicators; that is, their color depends upon the *p*H of the solution in which they are dissolved. (D) Light favors anthocyanin production in fruits such as apples and in leaves during autumn.

22. Match the following:

_____	1. Chlorophyll a.	1. $C_{40}H_{56}O_2$.
_____	2. Chlorophyll b.	2. $C_{55}H_{72}O_5N_4Mg$.
_____	3. Carotene.	3. $C_{55}H_{70}O_6N_4Mg$.
_____	4. Xanthophyll.	4. $C_{40}H_{56}$.

23. _____ Which one of the following is not a part of a chlorophyll molecule? (A) Carbon, (B) hydrogen, (C) oxygen, (D) nitrogen, (E) magnesium, (F) iron.

24. _____ The pigments which are responsible for blue, purple, and some red colors of plants are: (A) chlorophyll, (B) anthocyanin, (C) xanthophyll, (D) carotene.

25. _____ In order to bleach celery plants, gardeners cover them with earth or wrap them in paper. This practice is effective because (A) light is necessary for the formation of carotene, (B) the darkness destroys the chlorophyll, (C) light is necessary for chlorophyll formation, (D) light is necessary for photosynthesis.

26. What roles do alkaloids and essential oils play in the life of a plant?

27. _____ Which one of the following is false? (A) DNA makes the RNA. (B) RNA makes enzymes. (C) Enzymes convert food into the building blocks which will be used for the synthesis of DNA, RNA, and enzymes. (D) One gene contains the information required for the synthesis of several different kinds of enzymes.

28. _____ Which one of the following is false? (A) Each enzyme is specific to one reaction or to a group of closely related reactions. (B) Many enzymes, when extracted from a cell, are able to carry out their action quite apart from organized living material. (C) Enzymes are proteins, but they may have a prosthetic group. (D) Enzymes are used up in the reactions which they catalyze. (E) Enzymes can be destroyed by boiling.

29. List some factors that influence the rates of enzymatic reactions.

30. _____ Insectivorous plants (A) are non-green plants, (B) secure an organic nitrogen supply from captured insects, (C) are classified together in one family, (D) grow in sandy soil.

31. _____ Some fungi can grow through wood because they secrete (A) lipase, (B) papain, (C) cellulase, (D) sucrase, (E) diastase.

32. _____ Which one of the following is false? (A) The enzyme which digests cellulose is called diastase. (B) Upon digestion, fats are changed into fatty acids and glycerol. (C) Upon digestion, proteins yield amino acids. (D) Starch can be digested to form glucose. (E) Alpha-amylase splits starch into dextrins.

33. _____ Insoluble foods such as starch and fats (A) can be moved as such from one part of the plant to another, (B) must be digested before they can be translocated from one part of a plant to another, (C) must be assimilated before they can be transported.

34. _____ Proteases are present in pineapples. Such proteases can be used to tenderize meats. Which one of the following statements about the tenderizing of meat with pineapple juice is correct? (A) Canned pineapple juice should be used. (B) Juice from fresh pineapples should be used. (C) Either fresh or canned pineapple juice can be used to tenderize meat. (D) In the tenderizing of meat, the protein is changed to sugar.

16

Growth

All organisms grow. They transmute material into more of themselves. From such ingredients as water, minerals, proteins, carbohydrates, fats, vitamins, hormones, etc., organisms form additional protoplasm. The formation of protoplasm is called **assimilation.** Existing protoplasm forms more of itself—a remarkable phenomenon and a unique characteristic of living systems. Certainly an existing automobile could not form a duplicate of itself from a heap of glass, steel, copper, rubber, and other materials.

Growth involves an irreversible increase in size which is usually, but not necessarily, accompanied by an increase in dry weight. In the practical measurement of growth, an increase in length, area, volume, or in fresh or dry weight is regarded as growth. Plants and animals do not grow at a uniform rate through their entire lives. Growth is slow when they are young, then gradually becomes more rapid until a maximum is reached, following which the rate diminishes (Fig. 16–1). For example, a corn plant grows slowly for the first 4 weeks, then rapidly for the next 8 or 9 weeks, then slowly again, and finally growth stops. In trees the time intervals are years instead of weeks. In trees the first few years of slow growth are followed by many decades of rapid growth, succeeded in turn by a long period when growth is very slow.

Growth is more than just an increasing amount of the plant. Differential growth of plant parts results in a characteristic shape (Fig. 16–2). Each plant species has a distinctive form, developed by growth patterns. In a certain species cells may grow outward from the leaf epidermis, forming hairs over the surface, and cells along the margin of a developing leaf may grow outward in the form of teeth. In a different species, such growth may not occur and hence their leaves lack hairs and teeth. The shapes of leaves, flowers, fruits, seeds, roots, and stems are the result of genetically controlled growth patterns.

A number of interesting problems are related to growth. What is the stimulus which initiates growth? Cells within a body may grow for a while and, when mature, cease to grow. Why do they stop growing? Cells which have ceased growing may resume growth, sometimes producing normal tissues but under some circumstances producing tumors. In a sense, the cancer problem is a growth problem.

STAGES OF GROWTH

Growth occurs in three stages: cell division, cell enlargement, and cell differentiation.

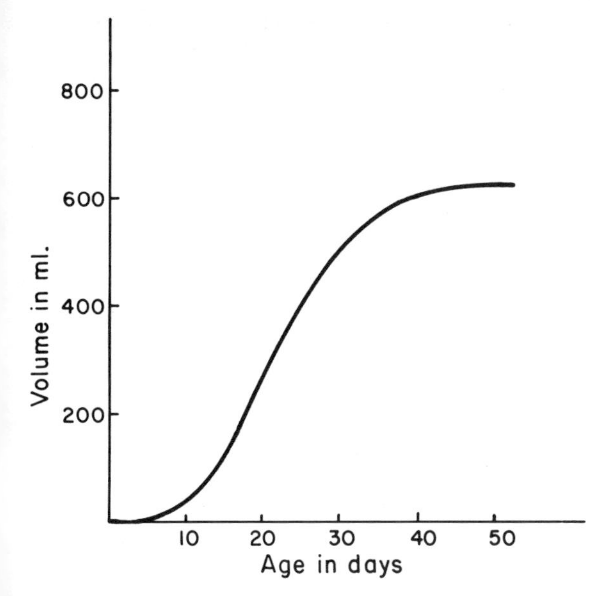

Fig. 16–1. A characteristic growth curve illustrated by a cucumber fruit. (Felix G. Gustafson, *Plant Physiol.*)

Cell Division

Cells increase in number by an extremely exact process. A cell does not just split into two. A consideration of the development of a plant or animal shows why the process must be precise. Plants and animals usually start life from a single cell, the fertilized egg. **Genes** located on **chromosomes** are present within the fertilized egg. They determine to a large extent the form and physiology of the organism which develops from this fertilized egg. When the egg divides and when the newly formed cells divide, none of the genes can be lost if development and reproduction are to be normal. After the body is fully formed, reproduction occurs, and the genes are transmitted to the new generation. The cells concerned with reproduction must have, at first, all of the genes which were present in the fertilized egg. Because genes are on chromosomes, a precise distribution of the chromosomes with each division insures an exact distribution of genes. Every time a cell divides, each of the resulting cells has the same number and kind of chromosomes as the parent cell.

Cell division involves two phases: the division of the nucleus which contains the chromosomes, and the partitioning of the cell. The process of nuclear division is known as **mitosis.** The partitioning of the cell is called **cytokinesis.**

Mitosis. The period of metabolic activity when cells are not dividing is called the **interphase** (Fig. 16–3*A* and *B*), a time when the chromatin in the nucleus is in the form of threads so thin that they are difficult to recognize. During late interphase the DNA content of the nucleus is doubled. For example, if ten units of DNA are present in a nucleus in early interphase, twenty units will be present in late interphase. The first visible sign that a cell is dividing is seen in the nucleus. When mitosis begins, the chromatin threads thicken and become more evident. Each thickened chromatin thread is a chromosome. At this stage the double nature of each chromosome is not seen; each chromosome appears as a single strand. As mitosis proceeds, the double nature becomes

Fig. 16–2. A flower of the orchid *Stanhopea*. Growth patterns produce the distinctive form.

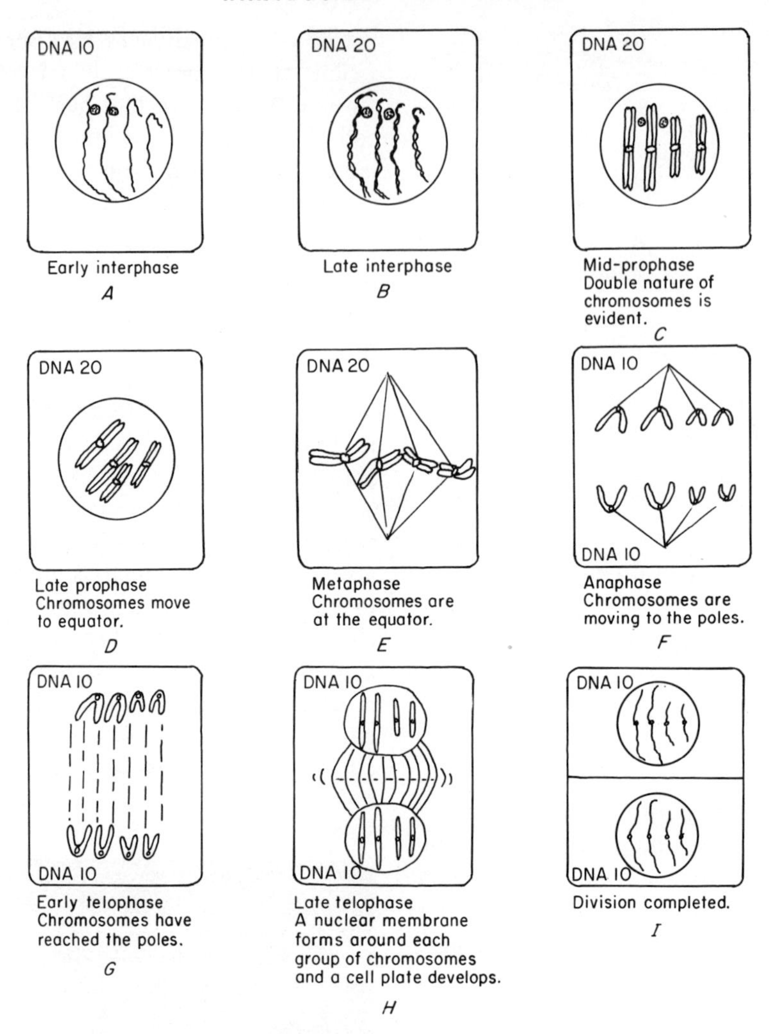

Fig. 16–3. Diagram of cell division.

evident and then each chromosome appears as a double filament. The two members of each doubled chromosome lie closely parallel and twined about each other, and they are joined together at a single point known as the **centromere.** The location of the centromere varies from one kind of chromosome to another. In one chromosome it may be midway between the ends; in a different one, closer to one end than the other. While the two chromosomes are still joined by one centromere, each chromosome is known as a **chromatid** (Fig. 16–3*C*).

The chromosomes shorten and thicken, becoming easily recognizable. After the chromosomes have become distinctly shorter, the nucleoli fade and disappear, as does the nuclear membrane. The electron microscope reveals that the nuclear membrane fragments into segments which resemble the units of the endoplasmic reticulum. Some segments move toward one pole, others to the opposite one. As the nuclear membrane breaks down, the chromosomes move to the center of the cell—to the **equator** as the center is called (Fig. 16–3*D*). While the chromosomes are moving to and becoming arranged at the equator, **microtubules** become oriented to form a **spindle.** Each spindle fiber is composed of many microtubules. Chemically, the spindle consists of protein, lipoids, and ribonucleic acid. In stained cells, the spindle appears as a series of fibers radiating from the two poles of the cell toward the equator. Certain spindle fibers, but not all, become attached to the centromeres and such fibers play a role in the subsequent movement of the chromosomes toward the poles. This completes the first stage of mitosis, which is called the **prophase** stage. All the events occurring from interphase until the chromosomes are at the equator constitute the prophase stage.

When the chromosomes are at the equator, the cell is in the **metaphase** stage of mitosis (Fig. 16–3*E*). The chromosomes are arranged so that at least one part of each chromosome is at the equator. The portion located at the equator is the centromere, also known as the **spindle attachment region;** and its position is constant for any one chromosome. The spindle fibers which move the chromosomes to the poles are attached at these regions. When the chromosomes are first arranged at the equator, the centromere still holds each pair of chromatids together. The metaphase stage terminates when each centromere duplicates itself. When duplication is complete, every chromosome has a centromere and then the chromosomes begin to move to the poles.

One member of each chromosomal pair moves toward one pole; the other goes to the opposite pole. When the chromosomes are moving to the poles, the stage of mitosis is called **anaphase** (Fig. 16–3*F*). When energized with ATP, the microtubules of each spindle fiber contract and thereby pull one member of each duplicated chromosome to one pole and the other member to the opposite pole. When the chromosomes are moving to the poles, it is clearly evident that the same number and kind are going to each pole and that the number and kind going to each pole are exactly like those of the original cell.

When the chromosomes reach the poles and stop moving, the **telophase** stage begins. At first the chromosomes are clumped together, but later they separate and lengthen into long threads, which in time become indistinct (Fig. 16–3*G*, *H*, and *I*). While the chromosomes are lengthening, a nuclear membrane forms around each group of chromosomes and the nucleoli reappear. The nuclear membrane is formed from units of the original membrane and perhaps certain segments of the endoplasmic reticulum. These units become oriented around the chromosomes and then fuse to form the new nuclear membrane.

Cytokinesis. In early telophase the spindle fibers still extend between the two groups of chromosomes or the reorganizing nuclei. As telophase continues, the spindle begins to disappear near the nuclei and to widen at the equator where in time it extends across the cell. At the equatorial plane of the spindle a line of minute droplets of pectic substances is formed. These droplets become transformed into a thin partition, called the **cell plate,** which begins to form in the center of the spindle and then progresses to the sides of the cell, thereby dividing the cytoplasm into two parts, each with its newly formed nucleus. Later cellulose walls are deposited by each new cell against the cell plate. The first cellulose wall is known as the **primary wall.** After the primary wall is formed, the cell plate is called the **middle lamella.** Cell division is now complete and the two new cells enter the interphase stage.

Interphase. During the interphase stage events occur which make the next division possible. During interphase the DNA content of a nucleus becomes doubled and only after such doubling can the cell again divide. In this stage each DNA molecule forms a replica of itself in the manner discussed in Chapter 15. The DNA content of the late interphase nucleus remains constant during the prophase and metaphase stages. During anaphase each group of chromosomes going to the poles has just one-half the DNA content of the late interphase nucleus, as does each reorganizing nucleus during the telophase stage. Thus, if the late interphase nucleus has twenty units of DNA, so also will the prophase nucleus; however, each telophase nucleus will have just ten units.

Summary of cell division and its stages. The division of a cell is a continuous process. At first, the chromosomes become clearly evident and later their double nature is seen. Then the nuclear membrane and nucleoli disappear and concomitantly the spindle develops and the chromosomes move to the equator. Later one member of each doubled chromosome moves to one pole, the other member to the opposite pole. At the poles, the nuclei reorganize and concomitantly a cell plate forms between the nuclei, thus dividing the cell in two. It is often convenient to divide the process into four stages: prophase, metaphase, anaphase, and telophase. All of the stages prior to the time when the chromosomes are at the equator are known as prophase stages. When the chromosomes are at the equator, the cell is in the metaphase stage of division. When the chromosomes are moving to the poles, the stage is anaphase; the reorganization stage is the telophase. During the interphase each DNA molecule replicates itself.

Cell Enlargement

A newly formed cell soon enlarges. The surface area of the cell wall increases as additional cell-wall material is formed and deposited. At the same time the cell absorbs water which enters the vacuoles thus increasing their volume and also the cell volume. In a recently formed cell there are many small vacuoles which coalesce as they increase in size. Accordingly, the number of vacuoles decreases as the cell enlarges. When enlargement is complete, the cell may have just one large central vacuole bordered by a thin layer of cytoplasm. Even though there appears to be

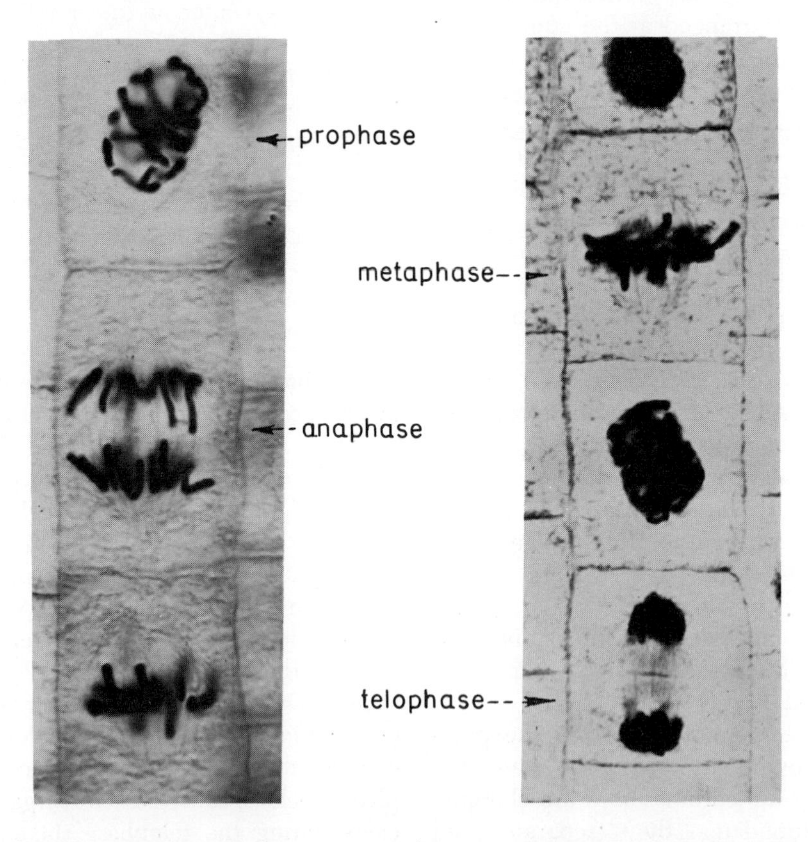

Fig. 16–4. Mitosis in an onion root tip. (Carolina Biological Supply Company.)

less cytoplasm in an enlarged cell than in a newly formed one, actually there is more. Hence, we learn that cell enlargement involves an increase in the area of the cell wall, an uptake of water, the fusion of vacuoles, and an increase in the amount of cytoplasm.

Differentiation

All living cells in a plant contain the same genetic information and hence each cell should have the capacity to develop into an entire plant. Earlier we learned that a single cell from a carrot did indeed develop into an entire plant, just as would a fertilized egg. Although all plant cells have the same genetic information, they are not all alike structurally or functionally. Differentiation of cells with a resulting characteristic shape and function occurs during plant development.

At the tip of a stem, cells divide. Some of the resulting cells develop into leaves, others with the same genes into stem cells. The cells which will become stem cells originally have the same appearance; but, as development proceeds, some become vessels, others tracheids, fibers, sieve tube cells, or parenchyma cells. What brings about this variation in cell types from cells which were originally alike and which have the same genes? Molecular biologists have discovered that cells developing into one tissue have different kinds and amounts of enzymes from those which are developing into a different tissue. Thus a cell which is developing into a fiber has an enzyme system that forms and deposits secondary wall materials, especially lignin, whereas one forming a parenchyma cell may lack this system or have it in a lesser amount. Furthermore, in the maturation region of a root tip only some epidermal cells develop root hairs. Those which form root hairs have certain enzymes which are lacking in cells which do not produce root hairs. We know that in an indirect way genes synthesize enzymes. Therefore you might assume that, because all cells have the same genes, they will also have the same enzymes. Since all cells do not have the same array of enzymes, there must be a "programming center" in the nucleus which instructs some genes to act at some times and in some cells but to remain quiescent in other cells and at other stages of development. In other words, the "programming center" instructs genes when to act and when not to act. For example in a red rose, the leaf cells have genes carrying the information that is required for the synthesis of enzymes that produce anthocyanin pigments, but in the leaf these genes do not perform. On the other hand, in the petals they do act. Many other genes that are quiescent during the vegetative development of a plant become active at flowering time—for example, those which control the structure of a flower, and the synthesis of nectar and essential oils. We have used the term "programming center" to include many factors, both internal and external, that may activate or inactivate genes. Repression of gene action seems to involve a chemical combination between DNA and a **histone protein;** activation results when the DNA is freed from the histone. The activity of genes may be controlled by location, environment, and hormones. When callus tissue, the tissue which develops adjacent to an injury, is grown in flasks containing the appropriate nutrients, only certain cells differentiate into vascular tissue. Certain internal cells of the mass differentiate into vascular elements whereas the peripheral ones do not. In the intact plant, the outer layers of a stem exert pressure on the cambium and its derivatives. This pressure controls the differentiation of xylem and phloem from cells formed by the cambium. When the pressure is relieved, the cells do not differentiate into xylem and phloem, but instead into callus tissue. The pressure on the cells can be relieved by bending the cambium and adjacent cells out on a flap cut from the stem. The cells on the flap differentiate into callus tissue and not into xylem and phloem. When the flap is bound back into the stem, normal differentiation occurs.

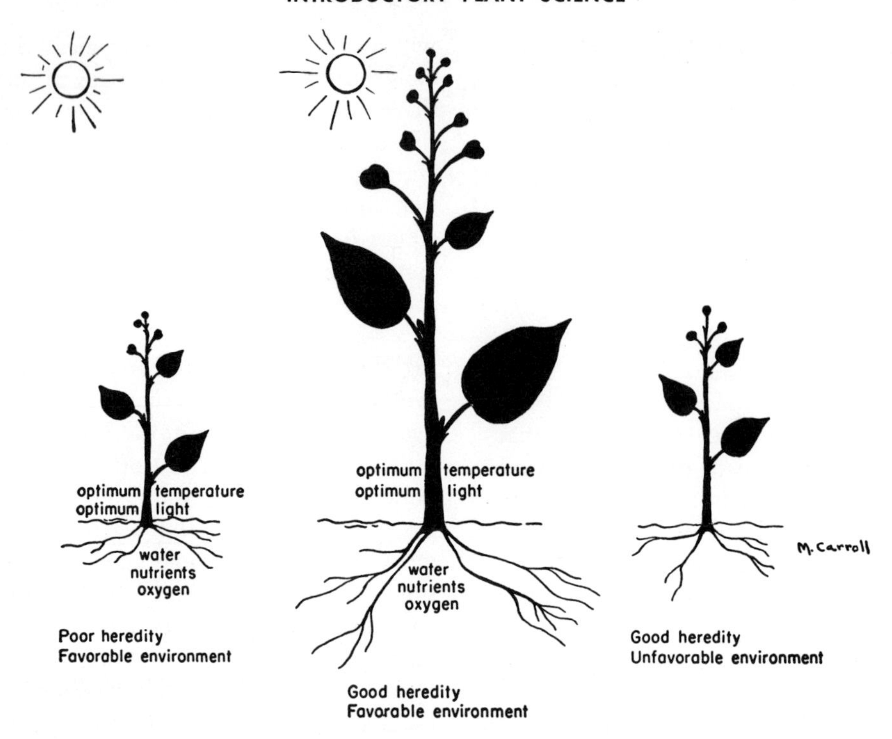

Fig. 16–5. A plant is the end result of both heredity and environment.

A familiar example of the environment controlling gene action is the synthesis of chlorophyll. In the dark the genes are not active whereas in the light they are active, synthesizing the RNA required for the production of the enzymes which are essential for chlorophyll formation.

ENVIRONMENTAL FACTORS AFFECTING GROWTH

Most plants start life from a fertilized egg which contains the potentiality of a particular plant, a corn plant, an oak tree, a geranium. If this inheritance is to be developed, there must be an appropriate environment. Without this environment, growth and development cannot occur. In other words, a plant is the end result of both its heredity and environment (Fig. 16–5). In the culture of plants we must consider both factors. A plant with superior heredity will not give choice flowers or fruits or a high yield if grown under poor environmental conditions. Similarly, a plant with poor heredity will yield only inferior products when cultivated under the best environmental conditions. Hence to get superior plant products and high yields, we must start with the best varieties and grow them in a favorable environment. A variety which may be superior under one set of environmental conditions may not necessarily be superior when the environment is changed. For example, a variety ideal for a humid climate is not necessarily the best for an arid region.

Among the environmental factors that influence growth are temperature, light, oxygen, water, and nutrients. Plants selected to be grown in a given locality should be those which are adapted to the prevailing environment. To some extent the climate in our cities is unlike that of the surrounding rural areas. Wind velocity and light intensity may be less in our backyards and temperature somewhat higher than in the open country. Even within the borders of your yard, different climates, microclimates, may be found. Some plants may thrive in certain areas of your yard whereas others might not. An evergreen growing on one side of a house or building may be sheltered from strong winds which would injure a tree growing in an exposed situation. Temperature and light intensities are higher on the south side of a building than on the north side.

The pollution of air in cities and in the surrounding country may retard the growth of plants, or kill them, and of course it is disagreeable to us. Sulfur dioxide and fluorine from smelters, and illuminating gas from leaky mains interfere with plant growth, as does smog. Smoke particles may clog stomata and thereby lower photosynthesis and growth.

Temperature

At low temperatures, plants grow very slowly or not at all. The lower limit for growth varies considerably with the species. For example, winter cereals make some growth at temperatures of 34° to 40° F., whereas in that temperature range pumpkins and melons do not grow. The minimum temperature for growth of the latter is about 50° F.

As the temperature increases above the minimum, growth is accelerated until a certain temperature is attained, above which growth becomes slower. When a certain temperature is exceeded, growth ceases. The temperature at which growth is most rapid varies greatly with the species, being higher for tropical plants than for those of temperate regions. The optimum temperature may also vary with the age of the plants and with the duration of the experiment. Plants may make rapid growth at a temperature of 86° F. if kept at this temperature for a short time, but if maintained for several days growth may cease. For these and other reasons it is difficult to state that growth will be most rapid at a certain temperature. But in general, a number of temperate zone species make most rapid growth when the temperature is between 68° and 77° F., above which growth occurs more slowly, ultimately ceasing at about 86° F., at least when this temperature is maintained for many days. So far we have been considering air temperatures. Soil temperatures also influence the growth of roots and shoots. For example, when the soil temperature is 60° F., shoots and roots of timothy grow faster than at a lower or higher soil temperature.

Fig. 16–6. Effect of temperature on *Browallia*. *Left*, grown with a night temperature of 50° F.; *right*, 65° F. (Kenneth Post.)

Night temperatures markedly influence the rate of growth. Some plants make their best growth at a night temperature of 50°, others at 60° or 65° F. With few exceptions plants grow better when night temperatures are lower than day temperatures, which is, of course, the usual condition in nature. Greenhouse growers generally run the night temperature 10° or 15° F. lower than the day temperature. In a carefully controlled experiment tomatoes grown with a day temperature of about 80° F. grew best when the night temperature was 64° F.

Peas grow better when night temperatures are lower than day temperatures. Pea plants grow poorly when both day and night temperatures are 50° F., 57° F., or 68° F. On the other hand, they grow vigorously at a day temperature of 68° F. and a night temperature of 57° F. When peas are grown for several generations at a constant temperature, their growth diminishes with each generation. Plants grown for five generations at a constant temperature, 50° F. for example, are tiny weak plants that produce few seeds. When plants from these seeds are grown with a day temperature of 68° F. and a night temperature of 57° F., the mature plants are smaller and less vigorous than those which develop from seeds collected from plants grown with a day temperature of 68° F. and a night temperature of 57° F. In other words, the parental environment affected the vigor of the offspring.

In many plants, the optimal night temperature varies with the age of the plant. The growth of young tomato seedlings is most rapid at a night temperature of 80° F., but as the plants grow older the optimal night temperature becomes progressively lower, finally reaching values of 55°—64° F., depending on the variety. Orchids, like tomatoes, have a higher optimal night temperature in their early stage of development than later on.

It should not be inferred, however, that all plants have a higher optimal temperature in early stages of development than in later ones. Many species develop better when temperatures are lower in early stages of development than in later ones. The sequence from a lower to a higher temperature optimum coincides with the natural variations of temperature from spring to summer. High temperatures after seed sowing frequently produce weak, low-yielding plants. If seeds of many species are sown late in the spring, the plants do poorly because of high temperatures. If sown early, more vigorous plants are produced. Peas, beets, sweet peas, larkspur, and many others should be planted early.

Light

Light is necessary for photosynthesis and hence indirectly for growth. Without food, plants do not grow. However, if food is available, plants grow in the absence of light, although the growth is not normal. Plants which develop entirely in darkness lack chlorophyll, have excessively elongated weak stems, and small, poorly developed leaves. Those with such characteristics are said to be **etiolated.** Although light is not essential for growth, it is necessary for normal development of plant parts and their tissues.

The intensity of light influences plant growth. Most crop and ornamental plants —for example, tobacco, wheat, corn, peas, carnations, and snapdragons—make stocky, vigorous growth and flower profusely with full sun. These and similar plants are called "sun plants." When grown with intermediate light intensities, sun plants are taller and have larger, thinner leaves, but fewer flowers, than when grown with full sunlight. Large thin tobacco leaves make choice cigar wrappers and sell at premium prices. Hence, tobacco for wrappers is often grown under light shade provided by cheesecloth.

Plants grow best when they are grown with the full visible spectrum of sunlight or with artificial illumination which has a spectral distribution similar to sunlight. Botanists have studied the effects of light quality, or color, on plant growth. Early botanists noted that plants grown with blue or violet light tended to be dwarf; those in red light, tall and spindly. In recent years we have learned that all hues of red do not affect plant growth in the same manner. Red light with wavelengths ranging from 650 millimicrons to 700 millimicrons promotes the germination of lettuce seeds (see Chapter 21), whereas red light with wavelengths ranging from 700 millimicrons to 760 millimicrons prevents germination. The red light of shorter wavelengths is referred to as "red light" and that of longer wave lengths is known as "far-red."

The growth of etiolated bean seedlings is affected in different ways by these two hues of red (Fig. 16–7). The leaves of bean seedlings grown in the dark are small and folded together and the stem has a characteristic hook just below the leaves. If a bean seedling which has been grown without light is exposed to red light for just 2 minutes before being returned to complete darkness, the leaves later expand and the stem straightens. Exposure to far-red light does not bring about this response; instead the plant resembles one kept in continuous darkness. Furthermore, if a bean seedling is exposed to red light for 2 minutes and then to far-red light for 5 minutes, the plant resembles one kept in continuous darkness. In this instance the promoting effect of the red light has been nullified by the subsequent exposure to far-red light. It is characteristic of the red, far-red effects that the plant response is determined only by the kind of light they last receive.

Before light can affect any process, whether growth, photoperiodism, or photosynthesis, the light must be absorbed by some pigment. Chlorophylls absorb the light that is used in photosynthesis. Phytochrome is the pigment that absorbs the light which influences photoperiodism and some aspects of growth. Phytochrome exists in two forms, one which absorbs red light, with an absorption peak at 660 millimicrons, and the second form which has absorption peaks at 710 and 730 millimicrons in the far-red range.

Radiations outside of the visible range may influence growth. The ultraviolet radiation from the sun does not promote growth. The glass in a greenhouse does not transmit ultraviolet radiations and yet the plants thrive as well in the greenhouse as outdoors. High dosages of ultraviolet and of X rays and gamma radiations often result in abnormal growth and even death (Fig. 16–8*B*). "Heavy nuclei radiation" in space results in abnormal growth (Fig. 16–8*A*).

Fig. 16–7. Bean seedlings after being kept in continuous darkness (*left*) and after being given 2 minutes of red, 2 minutes of red and 5 minutes of far-red, and 5 minutes of far-red only (*from left to right*). Both before and after the exposure to red or far-red the plants were maintained in the dark. (Agricultural Research Service, Plant Industry Station, U.S.D.A.)

Prolonged growth of green plants is intimately related to food manufacture for which light is required. Hence, plants grown in light of sufficient intensity to promote a maximum rate of photosynthesis usually make better growth than those grown with a lesser intensity. Because the light intensity at which photosynthesis is maximal varies with the species, a light intensity sufficient for good growth of some plants may be insufficient for the best growth of others. Greenhouse operators recognize this by growing some plants under shade, Afri-

Fig. 16–8A. Plants grown from corn seeds exposed to radiations in space are abnormal; notice the two longitudinal slits in the leaf on the right. The seed from which this plant developed was flown to an altitude of 130,000 feet by balloon. The effects are unlike those obtained with X rays or alpha particles and may be the result of "heavy nuclei radiation." (U.S. Atomic Energy Commission.) *B.* Gamma rays from radioactive cobalt resulted in abnormal growth of the dahlia leaf on the right. The control leaf on the left shows normal development, while the leaf treated each day for 2 months with 250 Roentgens of gamma rays from radioactive cobalt shows almost complete suppression of the leaf blade. (Brookhaven National Laboratory.)

can violets, for example, and others, such as carnations, with a maximum amount of light. Around homes, sun plants make poorer growth on the north side than on the south side of the house because of lesser light intensity. On the other hand, shade plants develop better on the north side.

Length of day not only influences photosynthesis but also has a marked influence on dormancy and flowering. Length of day influences these phenomena directly and not indirectly through photosynthesis. The short days of autumn are a stimulus which brings about a cessation of growth in many plants, a phenomenon not related to food supply. For example, when the days are 12 hours long or less and the night temperature 60° F. or above, growth and flowering of tuberous begonias cease. Concomitantly, the tuber below ground increases in size as food is translocated to it from the shoot. Catalpa, elm, birch, red maple, dogwood, and other trees respond to the short days of autumn by ceasing to grow and becoming dormant. Because the autumnal responses which lead to dormancy and hardening are initiated by short days instead of low temperatures, the plants are resistant to cold before severe weather comes.

The length of day has a marked influence on flowering. Plants are grouped according to their response to day length into what are called short-day plants, long-day plants, and day-neutral plants. The **short-day plants** in general flower when the days are less than 13 or 14 hours long. Among these are some species that will flower only when the day is less than 12 hours, and others for which 14 hours represent a short day. The **long-day plants** produce flower buds when the days are longer than 13 or 14 hours. The **day-neutral** or **indeterminate** species flower when the days are either long or short. (The topic of length of day and flowering is more fully discussed in Chapter 22.)

Length of day also influences the development of tubers, bulbs, and other underground plant parts. When day lengths are in a certain range, a hormone produced in the leaves is translocated to underground organs where it brings about a characteristic growth pattern. Many varieties of potatoes develop tubers when the days are short and not when they are longer than some critical length. On the other hand, onion bulbs develop only when the days are long, longer than some critical duration. The sweet Spanish onion forms bulbs only when days are longer than 12 hours, whereas the "Zittau yellow" variety develops bulbs only when days are longer than 13 hours.

Biological clocks. The fact that plant development is influenced by day length indicates that plants measure time. In photoperiodic responses—that is, the effects of day length on flowering, onset of dormancy, and development of underground plant parts—the plants may use the cues of alternate light and dark to measure time. But even when there are no cues of light and dark, certain plants can measure time with remarkable accuracy. For example, *Gonyaulax polyedra,* a marine alga, luminesces most brightly at midnight. Even when the alga is maintained day after day in continuous darkness, it glows most brilliantly at the same time with an error of no more than 2 or 3 minutes. Without a watch could you tell time as accurately? The flowers of many plants close shortly after sundown and remain closed until sunrise. We used to believe that their responses were controlled by light, but now we know that the rhythm of opening and closing of some species continues even when the plants are maintained day after day in light or darkness. For example, if *Kalanchoë blossfeldiana,* a familiar house plant, is kept in continuous darkness, the flowers will follow the rhythm shown in the following table:

Hours in Continuous Darkness	Flower Behavior
0	Closed
12	Open
24	Closed
36	Open
48	Closed
60	Open

Even in a uniform environment, time is measured with great precision, the flowers alternately opening and closing at 12-hour intervals.

The leaflets of bean and *Oxalis* exhibit sleep movements. During the day the leaflets are expanded whereas during the night they are folded together. However, these responses are not conditioned by the alternation between light and dark periods, because they continue even when the plants are maintained in continuous light or darkness. For example, when bean plants are grown with continuous light, the leaflets still fold together at a specific time each day.

Our understanding of these internal rhythms, or endogenous rhythms, and the nature of the biological clock remains incomplete. We know that the biological clock is not affected by changes in temperature and that it can be reset.

Oxygen

Growth is markedly retarded when plants are grown where oxygen is deficient. Usually the shoots of plants receive an ample supply of oxygen, but the roots may or may not secure sufficient oxygen to grow and function normally.

Water

During years of drought farmers and ranchers suffer considerable financial loss. With an inadequate water supply growth is poor and yields low. Plants grow well when ample but not excessive moisture is available. If the soil is continuously above field capacity, as it may be in poorly drained fields, plants grow slowly because roots are deprived of oxygen. In the range from field capacity to just above the wilting percentage plants make good growth. When the soil is at or below the wilting percentage, water is not available to roots and growth ceases.

Soil Nutrients

The quantity and nature of soil nutrients influence growth. When any of the macro- or micronutrients are lacking, growth is retarded. Growth and yield may be increased by proper fertilization of the soil.

Mechanical Stimuli and Growth

Plants make growth adjustments to various environmental stresses—adjustments made, of course, without the aid of sight, hearing, and feeling and without a nervous system. In Chapter 10, we learned that free-swaying trees had larger trunks and were sturdier than trees which had been staked. Similar results were noted when sunflowers which had been grown on a vibrating machine were compared to unshaken plants.

Pumpkins normally develop and ripen on the ground. If the vine is staked in an upright position so that the developing pumpkins hang freely in the air, the stalks will be called on to support a weight that increases with each passing day. Growth adjustments are made; the stalks supporting the fruits are thicker and sturdier than those of pumpkins developing on the ground. Normally, a petiole of black hellebore can bear a load of 12 ounces, but it may support a weight of 120 ounces if the load is gradually increased. Similarly, tendrils make growth adjustments. Those which are supporting a plant are larger than those which are unemployed—that is, those not anchored to a support.

THE SEASONS

Summer

Although we associate summer with vacation time, it is otherwise in nature. In summer plants and animals accumulate stores of energy for another year. Bear and deer are getting fat; bees are storing honey. Photosynthesis goes on rapidly, and much of the food is stored. Annuals store food in seeds; perennials in stems, roots, and perhaps in bulbs, tubers, and similar underground parts. The predominant green of the summer landscape is accented by the bright colors of flowers. In summer many plants come into flower, and seeds and fruits develop. The buds of trees and shrubs which will be carried through the forthcoming cold season have formed, and even though weather is still warm they remain dormant.

Autumn

The short days of autumn bring about autumnal coloration, leaf fall, and dormancy in many native plants. Frost or no frost, the leaves of our native plants color brilliantly and with dramatic suddenness. Later the leaves are shed. At the same time changes occur in the protoplasm of the persistent buds and stems, leading to dormancy and to hardiness. A November cold wave does not catch them unprepared. The changes are referred to as **hardening.** Hardened and dormant plants may survive very low temperatures, whereas unhardened plants would be killed. For example, needles of northern pines are killed by temperatures of 18° F. above zero in summer, but in the hardened winter state, they withstand temperatures of at least 40° F. below zero. Before leaves fall from trees and shrubs and before shoots of herbaceous perennials die, the nutrients are translocated from the perishable parts into the persistent ones.

Winter

In winter when trees and shrubs are decorated with snow and the temperature is below freezing, the plants are not completely inactive. The winter buds, by means of invisible processes occurring during winter, acquire the capacity for further develop-

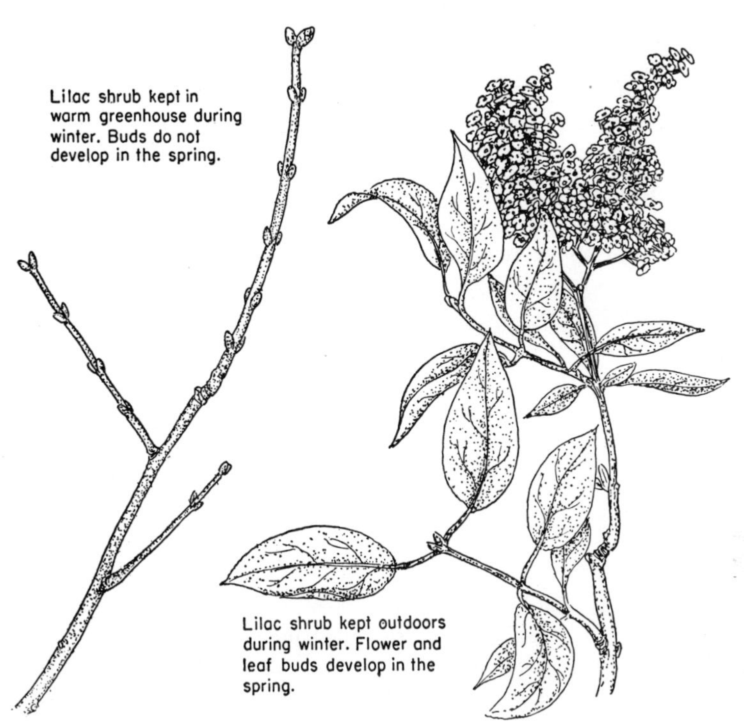

Fig. 16–9. Low temperatures during winter break the rest periods of buds. In the figure above are shown two lilac shrubs as they appear in late spring.

ment which was lacking in them during the previous autumn. Without an exposure to low temperatures the buds of many plants will not grow even though they are later exposed to favorable temperature and moisture conditions (Fig. 16–9). With respect to trees and shrubs of temperate regions, only temperatures below 46° F. are effective in breaking the bud dormancy. Each variety of plant requires a certain minimum exposure to cold. Plants native to cold climates require a longer chilling period than those from regions with warmer winters. Northern plants when grown in the South frequently show a delayed and erratic opening of the buds in spring. Southern plants may be killed farther north when their chilling requirements have been met before the danger of late frosts is over.

Earlier we learned that plants have "clocks." In a sense they also have calendars that record the passing of the winter months. The calendars are based on an inhibiting chemical which is formed during the growing season. The substance accumulates in the buds and prevents their development. During the winter months, the concentration gradually diminishes and in time reaches such a low level that growth is possible. Then, if the buds are provided with warmth and water, they develop. In summary, we note that plants record the passage of the winter months by means of a mechanism that progressively destroys the inhibitory chemical.

Low temperatures are the major natural means for breaking the rest periods of buds. However, scientists have discovered chemicals which will induce the same result. One of the most effective is ethylene chlorohydrin. This chemical will break the rest periods of buds of trees, shrubs, and potatoes. Freshly harvested potatoes will normally not sprout, but if the tubers are exposed to the vapors of ethylene chlorohydrin, the buds develop.

Spring

The ancients saw in spring the promise of fruitfulness and fertility in all living things. To many it was a miracle that all nature, which was apparently dead, should begin life anew. At springtime, in temperate latitudes, the period of rest is over, the changes of winter are completed, and the buds on the boughs are ready to expand their leaves and flowers. Dormancy of many seeds has been broken, buds on rhizomes and corms, and shoots within bulbs are ready to grow, awaiting only warmth and moisture.

QUESTIONS

1. _____ Which one of the following is false? (A) Growth is an irreversible increase in size. (B) New cells are formed at regions called meristems. (C) The rate of growth of a plant or plant organ varies with age, being initially rapid, then slower, then slower still, and finally zero. (D) The three phases or stages of growth are cell division, cell enlargement, and differentiation.

2. What is the biological significance of mitosis?

3. Diagram the process of mitosis.

4. _____ Which one of the following statements about mitosis is false? (A) In the prophase stage each chromosome produces an exact duplicate of itself. (B) In the prophase stage, the DNA content of the nucleus becomes doubled. (C) When the chromosomes are at the equator and before the centromeres become duplicated, the cell is in the metaphase stage. (D) Following duplication of the centromeres the chromosomes move toward the poles, a stage known as anaphase. (E) The reorganization stage is the telophase.

5. _____ In higher plants the partitioning of the cell in two (A) is called mitosis, (B) is brought about by the formation of a cell plate, (C) occurs when new nuclear membranes, formed from units of the original membrane and segments of the endoplasmic reticulum, develop, (D) is brought about by invagination of the cell walls.

6. _____ Chromosomal numbers are often used when studying plant relationships. The chromosomal number could be determined most easily during the (A) interphase stage, (B) prophase stage, (C) metaphase stage, (D) anaphase stage, (E) telophase stage.

7. Where do roots increase in length? Stems?

8. Name five meristematic regions in a flowering plant.

9. _____ Cells with identical genes (A) will have the same enzymes, (B) will be alike structurally and functionally, (C) may be different in appearance.

10. Discuss four environmental factors that affect growth.

11. _____ Which one of the following is false? (A) Most species grow best when the temperature is constant during the day and night. (B) The temperature at which growth is most rapid is higher for tropical plants than for those of temperate zones. (C) Tomatoes grow better when the night temperature is lower than the day temperature. (D) The optimum temperature may vary with the age of the plant.

12. _____ Which one of the following is false? (A) Plants which develop in light are said to be etiolated. (B) If food is available, plants will grow in the dark. (C) Plants grown in light shade have larger leaves than those grown in bright sun. (D) Plants which flower in greenhouses at Christmas are short-day plants. (E) Light is necessary for the normal development of plant parts and their tissues.

13. _____ Which of the following tobacco plants would be taller and have larger, thinner leaves? (A) One grown in full sun, (B) one grown under one or two layers of cheesecloth, (C) one grown with very low light intensity of about 100 foot-candles.

14. _____ Plants grow well when the soil moisture is (A) continuously above field capacity, (B) in the range from field capacity to just above the wilting percentage, (C) below the wilting percentage.

15. _____ Which one of the following colors does not promote germination of lettuce seeds and the expansion of bean leaves? (A) White light, (B) far-red light, (C) red light.

16. _____ Which one of the following is false? (A) Before light can be effective in any process, the light must be absorbed. (B) Phytochrome is the pigment that absorbs the light which influences photoperiodism and some aspects of growth. (C) Within plant cells, phytochrome exists in three forms: one form absorbs light with wavelengths of 660 millimicrons, a second light of 700 millimicrons, and a third light of 730 millimicrons. (D) It is characteristic of the red, far-red effects that the plant response is determined only by the color of light they last receive.

17. _____ As far as is known, length of day has little or no effect on (A) the development of tubers, bulbs, and other underground plant parts, (B) respiration, (C) photosynthesis, (D) flowering, (E) onset of dormancy.

18. Plants can measure time without the cues of day and night. (A) True. (B) false.

19. _____ Which one of the following is false? (A) Although the environment may affect the replication of genes, it has little effect on the action of genes. (B) Under some environmental conditions, certain genes may be quiescent whereas under different environmental conditions, the genes may be active. (C) Both heredity and environment interact to control the size of a plant and many of its features. (D) The environment may affect the action of genes as well as such processes as photosynthesis, respiration, and transpiration.

20. _____ Ethylene chlorohydrin can be used to (A) break rest periods, (B) hasten the rooting of cuttings, (C) hasten healing of graft unions, (D) retard the fall of apples from trees.

21. Describe the activities of plants during the four seasons.

22. _____ Which lilac shrub would leaf out and flower first in the spring? (A) A shrub maintained in a warm greenhouse during winter, (B) a shrub growing outdoors.

17

Plant Hormones

Gardeners and farmers for generations increased plant growth and yield by breeding and by giving careful attention to such environmental factors as light, temperature, soils, and irrigation. Now we are learning something about the internal factors which control the growth and development of a plant, and with this knowledge, greater control of plant development is possible.

The organs of a plant are coordinated, interrelated, and interdependent. Each part of a plant is influenced by other parts and in turn affects them. For example, roots depend upon leaves for food and vitamin B_1, leaves rely on roots for water and minerals; both are dependent upon the stem for conduction. In addition to coordinating mechanisms based on nutrition, there are correlations between plant parts which are controlled by chemical messengers known as hormones. The hormones of plants, like those of animals, are produced in one part of the body and travel to other sites where they produce special effects. Very small amounts of a hormone produce marked responses.

A flowering hormone seems to be produced in leaves under certain environmental conditions. Apparently it is translocated to a stem tip where it induces the formation of flower parts. The flowering hormone, called **florigen,** has not as yet been isolated from plants.

A hormone is produced in terminal buds which moves downward and inhibits the development of axillary buds. If the terminal bud is removed, the stream of hormone ceases, and the axillary buds develop into branches.

A similar hormonal mechanism may be responsible for apical dominance in gymnosperms. In pines, spruces, and others, the lateral branches are nearly horizontal as long as the terminal shoot (leader) is present. If the terminal shoot is removed, one or more of the lateral branches will grow upright. Apparently the leader produces a hormone which controls the direction of growth of the branches. After the leader is removed, the hormonal control is no longer present.

The end responses brought about by hormones are readily detected. Thus we can see the effects of hormones on growth, cell enlargement, cell division, bud development, and leaf fall; we can measure the effects of hormones on respiration and digestion. Currently, scientists are striving to understand the molecular changes, induced by hormones, which bring about the evident responses. Hormones may affect biological processes in several ways. Certain hormones may alter the structure of the cell wall or the plasma and vacuolar membranes. Other hormones may activate enzymes such as dehydrogenase and phosphorylase. Some hormones may operate at the gene level by changing inactive genes into active ones which then synthesize messenger RNA which soon brings about the synthesis of enzymes. We still have much to learn before we will thoroughly understand the interrelationships between hormones, DNA, RNA, enzyme synthesis, metabolic processes, and plant development.

Many aspects of growth are controlled by a single hormone called **auxin,** chemically known as **indoleacetic acid** ($C_{10}H_9O_2N$). Indoleacetic acid is synthesized by plants from tryptophane, an amino acid. Because zinc is necessary for tryptophane synthesis, a zinc deficiency is associated with a low concentration of indoleacetic acid in the plant. The structural formulas of tryptophane and indoleacetic acid are shown below.

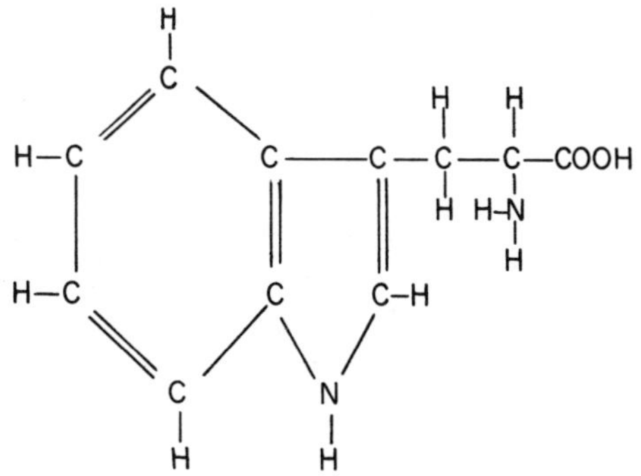

Tryptophane

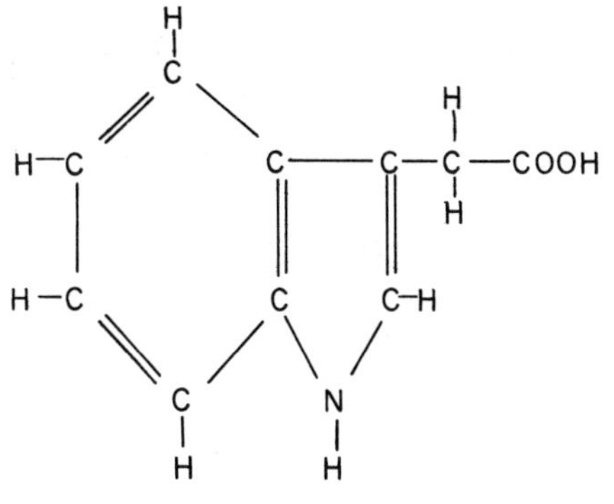

Indoleacetic acid

Auxin is not produced in glands but is produced in leaves, growing buds, root tips, and certain other places. From these sites auxin is translocated to other tissues generally via the phloem. In stems and coleoptiles auxin transport is polar; that is, auxin moves from tip to base, but not in the opposite direction. The velocity of polar translocation varies between 0.5 and 1.5 centimeters per hour. Auxin is concentrated in the tips of shoots and roots and decreases in concentration as distances from the tips increase. Auxin is used up in the processes which it controls. Hence, for continued coordination of plant parts there must be continued auxin formation. Indoleacetic acid can be synthesized by chemists, as can a number of other chemicals which exert similar physiological effects: indolepropionic acid, indolebutyric acid, alpha-naphthaleneacetic acid, 2,4-dichlorophenoxyacetic acid (2,4-D), and others. These substances, like the naturally occurring auxin, influence plant growth.

The synthetic chemicals which produce growth reactions similar to those brought about by the naturally occurring indoleacetic acid are also frequently called auxins. Auxins are also called growth substances and growth regulators. Strictly speaking, the term plant hormone is reserved for the naturally occurring substance, because by definition plant hormones are chemicals produced in one part of the plant, not by chemists, and moved elsewhere where they bring about characteristic responses when present in only trace amounts.

The effects obtained with growth substances depend on the concentration used. At relatively low concentrations, growth is accelerated. Higher concentrations retard growth and still higher ones kill plants. Furthermore, various organs of a plant respond differently to the same concentration of growth substance. For example, concentrations of growth substances which accelerate stem growth are high enough to retard the growth of roots. Root

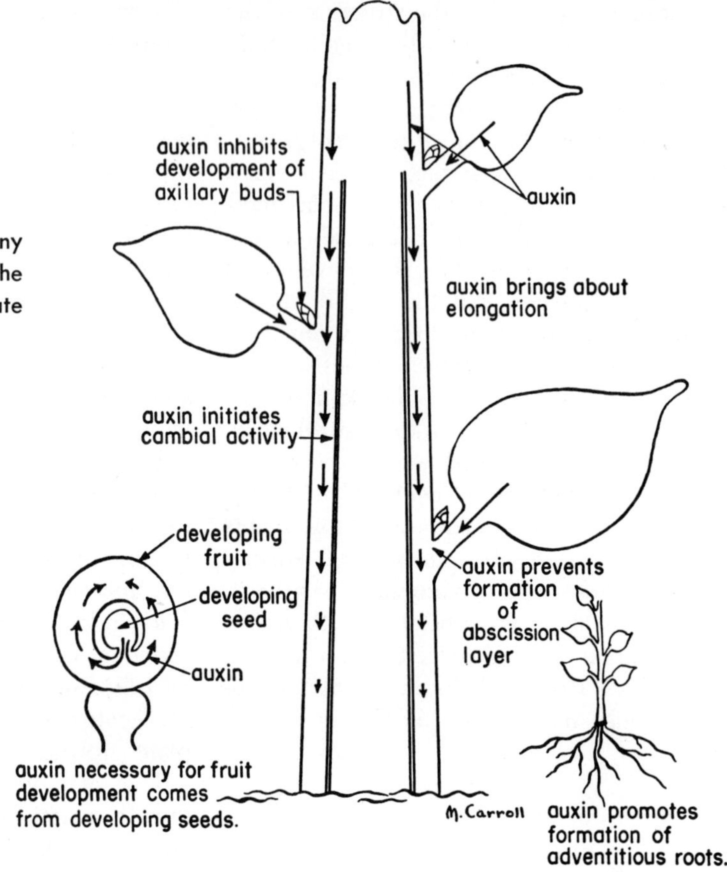

Fig. 17–1. Auxin plays many roles in the life of a plant. The arrows in the plant indicate auxin.

growth is augmented only by extremely low concentrations.

Auxin plays many roles in the life of a plant. It controls development of axillary buds, cell division and elongation, leaf and fruit fall, fruit development, and formation of adventitious roots. These roles are illustrated in Figure 17–1.

EFFECTS OF AUXIN ON GROWTH

Effects of Auxin on Cell Enlargement

Coleoptiles of oat seedlings are frequently used to demonstrate the effect of auxin on cell enlargement. The embryonic leaves of an oat embryo are enclosed by the coleoptile. When an oat seed germinates, the coleoptile is the first structure to appear above the soil. Later in the germination process, the leaves emerge from the coleoptile. However, for studying the effect of a plant hormone, the coleoptile is used before the leaves emerge, when it has a length of 25–40 mm. Further growth is brought about by cell enlargement and not by cell division. The cells at the mid-part of the coleoptile elongate, but those at the tip or extreme base do not.

Even though growth occurs only in the center portion, removal of the tip causes growth to cease (Fig. 17–2). This indicates the production of a plant hormone by the tip which moves downward and causes the cells of the mid-region to elongate. The removal of the tip eliminates the supply of this plant hormone.

If a severed tip is cemented with gelatin to the stump of a cut coleoptile, the coleoptile resumes growth because the cells of the center part are again supplied with auxin.

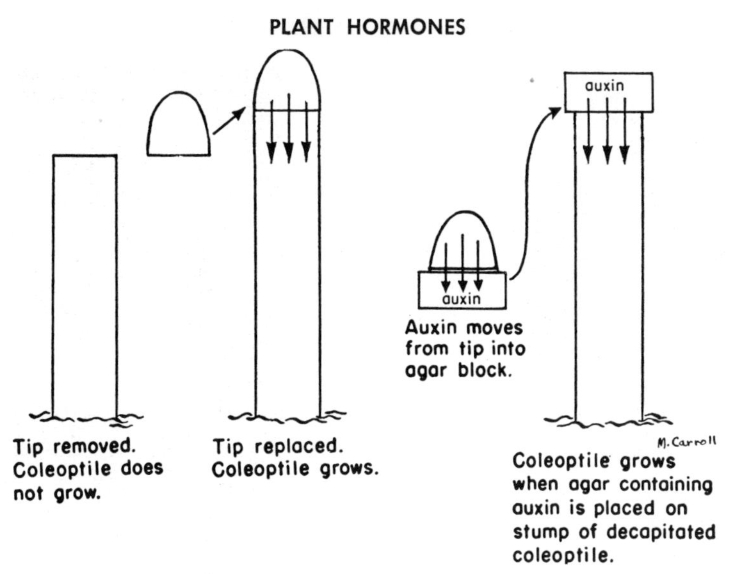

Fig. 17–2. Auxin produced in the coleoptile tip is necessary for growth.

The hormone present in the tip of an oat coleoptile may be isolated by placing the tip on a block of agar. The hormone diffuses from the tip into the agar. If a block of agar containing the hormone is placed on a decapitated coleoptile, the coleoptile grows (Fig. 17–2). If such a block is placed on one side of a coleoptile, the coleoptile bends because growth is more rapid on the treated side than on the opposite side. Auxin increases the plasticity of cell walls and enhances water absorption, thus bringing about cell elongation.

Auxin causes stem cells as well as coleoptile cells to elongate. In a stem much auxin is produced in the terminal bud and moves downward. If the stem tip is removed, the stem will not increase in length because the dividing cells at the apex are removed and because the supply of hormone to the elongating region is shut off. Without a supply of hormone, cell elongation does not occur. If a mixture of lanolin (wool fat) and auxin is smeared on one side of a stem, the smeared side grows more rapidly than the untreated side, and the stem bends (Fig. 17–3A).

Effect of Auxin on Cell Division

Auxin controls cell division as well as cell enlargement. For example, cell division in the cambial layer is activated by auxin which originates in developing buds and young leaves. The first demonstration of auxin control of cambial growth was carried out with sunflower seedlings. When the top of a seedling was removed, no cambial growth occurred in the region below the cut, but when the decapitated seedling was provided with an artificial supply, cambial activity was initiated. In tissue culture, cells of the Jerusalem artichoke do not divide when auxin is lacking, but they do divide when a minute amount is present in the culture solution.

OTHER EFFECTS OF AUXIN

When the auxin supply is ample, leaves and fruits remain on stems. When the supply is reduced, leaves and fruits fall from the branches. During summer the leaf blade produces auxin which moves down the petiole and prevents leaf fall. In autumn the waning leaf produces an inadequate supply of auxin and the leaf is shed.

Fruit development is also controlled by auxin. The initial supply of auxin comes from the pollen tube, but later the supply comes to a large extent from the seeds developing in the fruit.

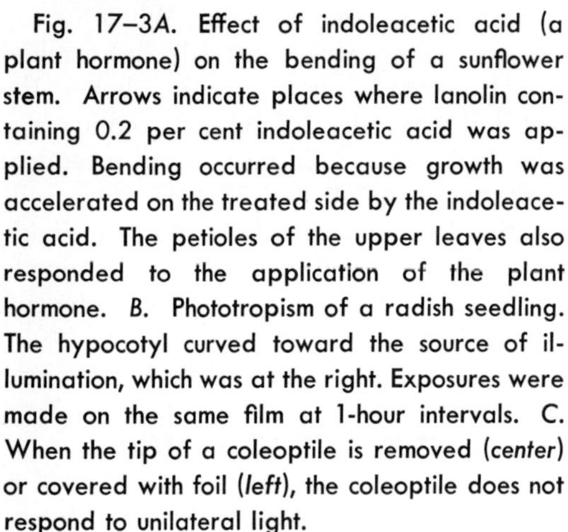

Fig. 17–3*A*. Effect of indoleacetic acid (a plant hormone) on the bending of a sunflower stem. Arrows indicate places where lanolin containing 0.2 per cent indoleacetic acid was applied. Bending occurred because growth was accelerated on the treated side by the indoleacetic acid. The petioles of the upper leaves also responded to the application of the plant hormone. *B*. Phototropism of a radish seedling. The hypocotyl curved toward the source of illumination, which was at the right. Exposures were made on the same film at 1-hour intervals. C. When the tip of a coleoptile is removed (*center*) or covered with foil (*left*), the coleoptile does not respond to unilateral light.

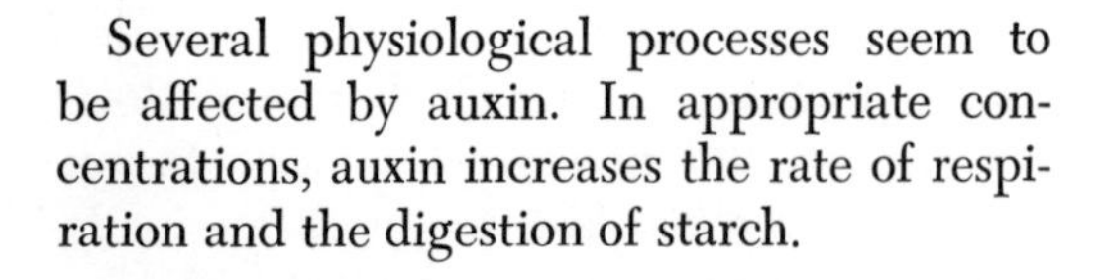

Several physiological processes seem to be affected by auxin. In appropriate concentrations, auxin increases the rate of respiration and the digestion of starch.

RELATIONSHIP OF AUXIN TO TROPISMS

A plant organ bends because the growth rate on one side of the organ is greater than on the opposite side. Curvatures which result from unequal growth are known as **tropisms.** Light, gravity, contact, and unequal distribution of water and chemicals are stimuli which cause unequal growth. The response of a plant to unilateral light is known as **phototropism,** to gravity as **geotropism,** to unequal distribution of water as **hydrotropism,** to chemicals as **chemotropism,** and to contact as **thigmotropism.** Plant movements toward or away from such stimuli are probably controlled by plant hormones. The hormonal basis of phototropism and geotropism are well established and these will be discussed in detail. The hormonal bases of other plant responses are not fully understood.

Phototropism

Interesting experiments to reveal features of phototropism can be performed with oat seedlings, specifically with the coleoptile. A coleoptile curves toward light, a response brought about by an accumulation of auxin in cells on the shaded side. An analysis of the auxin concentration on the two sides reveals that two-thirds of the auxin is located in cells on the shaded side and one-third in cells on the lighted side. The cells provided with the greater supply of auxin elongate more than those with lesser amount. The greater growth on the shaded side causes the coleoptile to curve toward the light (Fig. 17–3*C*). Even though the greater growth occurs some distance from the coleoptile tip, the removal of the tip or covering of the tip with foil prevents a phototropic response (Fig. 17–3*C*). In other words, the coleoptile tip perceives the stimulus, but the response occurs in cells below the tip.

Plants do not respond to all colors of light. The quality of light determines whether or not a phototropic response will occur. Coleoptiles curve toward blue-green, blue, and violet light, but not toward green, yellow, and red light. In other words, visible radiations with wave lengths shorter than 550 millimicrons elicit a phototropic response whereas those of greater wave lengths do not. In an earlier chapter we learned that if light is to be effective it must be absorbed. Two pigments that absorb light toward the blue end of the spectrum are carotene and riboflavin, and these may be the sensitizing pigments.

Stems, as well as coleoptiles, usually bend toward light. Here, as in coleoptiles, the auxin accumulates in cells on the shaded side where growth becomes greater than on the lighted side. Leaves also exhibit phototropism, but the mechanism bringing about their orientation to light is not well known.

Geotropism

Regardless of how a seed is planted, typically the root grows downward and the shoot upward (Fig. 17–4). The layman generally believes that roots grow down to search for water and the shoots upward to obtain light. He believes that plants do things consciously—that they have intelligence. However, this cannot be the correct explanation, because plants are not endowed with intelligence. Moreover, the shoot grows upward and the root downward, even when the seeds are planted in a dark room. The resulting curvatures are responses to gravity—geotropisms. Roots respond positively to gravity; shoots negatively. If a stem is placed horizontally, the plant hormone accumulates in cells toward the lower surface, stimulating growth of the lower cells and thereby bringing about an upward curvature. If a root is placed horizontally, the auxin also accumulates in cells toward the lower surface. However, the concentration becomes so high in root cells toward the lower surface that growth is retarded. (You will recall that roots are extremely sensitive to plant hormones.) The upper surface of a root with a small amount grows faster than the lower surface where a high concentration is present. Hence, the root bends downward.

Hydrotropism

Some roots, such as the aerial roots of orchids, curve toward a moist surface, a response known as hydrotropism. However, you should not assume that roots sense water at a distance and go in search of it. The response merely indicates that if the air, or for other plants the soil, is moister on one side of a root than on the opposite side the root will curve toward the moister air or soil. We can demonstrate that roots cannot sense water at a distance by growing a plant in soil made of three layers, an upper layer of moist soil, a middle layer of dry

Fig. 17–4. Geotropism. Three corn grains were oriented in different ways. The roots all responded positively to gravity; the shoots responded negatively. The grains were germinated in the dark.

soil, and a lower layer of moist soil. A subsequent study of the roots shows that they are confined to the upper moist layer. They do not penetrate the dry soil and, of course, never reach the moist lowest layer.

TURGOR MOVEMENTS

Not all plant movements result from the unequal growth of opposite sides of an organ. Some movements are produced by turgor changes in cells or groups of cells. Unlike tropisms, in which the movements are slow and the effects lasting, turgor movements are rapid, temporary, and reversible. The opening and closing of stomata are turgor movements. An increase in the turgor pressure of guard cells opens the stomata; a decrease results in their closure.

In some plants, turgor movements occur in response to a mechanical stimulus such as touch. Among the most dramatic movements are those occurring in the pinnately compound leaves of *Mimosa*, the sensitive plant. Touch a *Mimosa* leaf and in less than one-tenth of a second the leaf will drop down from an erect position to a pendant one. The movement is brought about by changes in the turgor of cells in the pulvinus, a swollen area at the base of the petiole, which consists of large thin-walled cells with spaces between them. When a *Mimosa* leaf is unstimulated and in an erect position, all cells of the pulvinus are turgid. But when the leaf is stimulated by touch or heat, a signal is transmitted to the pulvinus whose lowermost cells then lose water rapidly. As water moves from the lower cells into the intercellular spaces, the cells become flaccid and the greater turgor pressure of the upper cells forces the leaf to droop. The signal transmitted from the stimulated region to the pulvinus may involve auxin, but our information about the transmission mechanism is incomplete. As the leaf droops, the leaflets fold upward, a response brought about by the loss of turgor in the upper cells of the pulvini which are located at the bases of the leaflets. After stimulation the flaccid cells of the pulvinus at the base of the petiole and the pulvini at the bases of the leaflets gradually absorb water, become turgid, and thus restore the leaf and its leaflets to their normal position. Often recovery will be complete in about 10 minutes.

Other turgor movements occur in some species. For example, when stamens in a barberry flower are bumped by an insect, they fold inward and thus dust the insect with pollen.

The so-called sleep movements of leaves of many plants, especially legumes, result from turgor changes in certain cells. As we have previously noted (Chapter 16), bean leaflets are expanded during the day and folded together during the night.

PRACTICAL BENEFITS GAINED FROM RESEARCH WITH AUXIN

At first, the discovery of plant hormones was chiefly of theoretical interest, but since then a number of practical uses for growth substances have been found. They are used to retard the preharvest drop of fruit, for the production of seedless fruits without pollination, for hastening the rooting of cuttings, for retarding the development of buds, and for the control of weeds.

Prevention of Preharvest Drop of Fruits

Each year the premature drop of fruits from trees results in large losses to growers. The fruits fall before they are ripe and they are bruised as they hit the ground. If apple and pear trees are sprayed with a growth substance, the fruits do not fall prematurely. Many commercial preparations of growth substances are now widely used by orchardists to prevent preharvest drop.

Production of Fruit Without Pollination

You will recall that the auxin necessary for fruit development comes from the pollen tube and to a greater extent from the developing seeds. In the absence of pollination, seeds do not develop and hence there can be no flow of auxin into the developing fruit. In other words, under natural conditions pollination is generally necessary for fruit development. However, synthetic growth substances can be substituted for the natural supply and will bring about fruit development in the absence of pollination when sprayed on the flowers.

Rooting of Cuttings

Auxin has root-forming activity. If the base of a shoot is inserted in moist sand, roots will generally form at the cut end, the necessary auxin coming from the young leaves and terminal bud. The amount of naturally produced auxin may not be optimal for the quickest and best root development. Hence the artificial application of auxin to the basal end generally hastens root formation and increases the number formed (Fig. 17–5).

Retarding the Development of Buds or Eyes

The sprouting of potatoes during storage lessens their market value. This sprouting may be prevented by treating the tubers with a growth regulator. Growth regulators are also used to retard the opening of buds during the storage of roses and other nursery stock.

Fig. 17–5. Cuttings of variegated holly showing effect of indolebutyric acid on induction of roots. Photograph taken 36 days after treatment. *Left*, untreated cuttings; *right*, treated ones. (Boyce Thompson Institute for Plant Research.)

Destruction of Weeds

The chemicals which accelerate growth in low concentrations kill plants when used in high concentrations. Plants vary in their sensitivity to growth-regulating chemicals. Hence, concentrations can be used which will kill some species, but not others. Most grasses are resistant to the killing effects of growth-regulating chemicals, whereas many species of broad-leaved weeds are sensitive. Accordingly, these chemicals may be used for the selective killing of some weeds. Minute quantities are absorbed and translocated to all parts of the plant. In susceptible plants the first effects are bending, twisting, thickening and curling of leaves with gradual color change, usually to yellow. Later the plants die.

The most generally used growth-regulating chemicals for killing weeds are sodium or ammonium salts of 2,4-dichlorophenoxyacetic acid (2,4-D) or the amines or esters of it. Tables 17–1 and 17–2 show the relative sensitivity of various species to 2,4-D. The esters are more volatile than the salts and hence require greater care in their use. Their vapors may injure sensitive plants as much as one-half mile away. The selective toxicity of weed killers may result from differences in rates of penetration into the plants and the rates of degradation or inactivation within the plants. In resistant plants the herbicide would enter slowly and that which does enter would be broken down or inactivated.

At first, 2,4-D was used primarily by homeowners to eliminate lawn weeds, then by farmers to control weeds in fields of cereals. In more recent years, vast acreages clothed with native vegetation have been sprayed with 2,4-D. In many states, trees and shrubs along power lines and telephone lines have been killed with 2,4-D, leaving a seared path through the once beautiful forest. In western states sagebrush over wide areas has been killed to improve forage for cattle. After the sagebrush is killed, the grasses, freed from competition, make greater growth and hence the sprayed area

Table 17–1

Sensitivity of Crops, Ornamentals, and Woody Plants to 2,4-D

Easily Killed or Severely Injured	Less Easily Killed or Injured	Generally Not Easily Injured
Sugar beets	Flax	Oats
Soy beans	Corn	Wheat
Alfalfa	Sudangrass	Barley
Sweet clover	Sorghum	Rye
Other legumes	Potatoes	Bromegrass
Peas	Buckwheat	Wheatgrass
Beans	Millets	Bluegrass
Tomatoes	Elm	Redtop
Most vegetables	Strawberries	Timothy
Flowering ornamentals	Asparagus	Reed canarygrass
Creeping bent grasses	Rhubarb	Fescue
Broad-leaved shrubs	Sumac	Wild rose
All fruit trees	Grapes	Raspberries
Most bush fruits	Gladiolus	Buckbrush
Willows (species variable)	Iris	Silverberry
Poison ivy	Tulips	Spruce
Caragana		Cedars
Boxelder		Pines
Cottonwood		Ash

Table 17–2

Sensitivity of Weeds to 2,4-D

Easily Killed or Severely Injured	Less Easily Killed or Injured	Generally Not Easily Injured
Beggar's tick	Canada thistle	Asters
Burdock	Dock	Barnyard grass
Chickweed	Field bindweed (creeping jenny)	Buffalo bur
Cocklebur	Horsetail	Catchfly
Dandelion	Goat's beard	Corncockle
Dragonhead mint	Lamb's quarters	Foxtail
False flax	Peppergrass (annual)	Goldenrod
Frenchweed	Peppergrass (perennial)	Ground cherry
Gumweed	Plantain (buckhorn)	Horse nettle
Kochia	Russian thistle	Knotweeds
Locoweed	Shepherd's purse	Leafy spurge
Mallows	Sow thistle (perennial)	Milkweed
Marsh elder	Wild buckwheat	Pigeon grass
Morning glory	Wild carrot	Purslane
Mustards	Wormwood	Quackgrass
Pigweeds		Russian knapweed
Plantain (common)		Sandbur
Ragweed		Toadflax
Smartweed (lady's thumb)		White cockle
Sow thistle (annual)		Wild oats
Sunflower		Witch grass
Wild lettuce		
Wild licorice		

supports a larger number of cattle. In the past, the spraying has been done without regard to its effect on other creatures, such as antelope and sage grouse. When sagebrush is killed, the food supply of these beautiful and interesting animals diminishes. In other areas willows have been destroyed with 2,4-D to improve forage. Once attractive areas become blackened and ugly. With the loss of willows, the food for beaver diminishes and they move out of the area. Their ponds, so important as fisheries and for water conservation, are no longer kept in repair and they wash out. Before embarking on any spraying program, all values must be taken into account.

OTHER PLANT HORMONES

In addition to auxin, the best-known plant hormone, other hormones have been isolated. Still others have not been isolated, but their presence is inferred from various physiological responses. Among the other plant hormones are the gibberellins, the kinins, and the wound hormone.

Gibberellins

For decades farmers in Japan noted that some rice seedlings grew more rapidly than others, but were weaker and often died prematurely. Such seedlings had a disease, known as the "foolish seedling disease," caused by a fungus, *Gibberella fujikuroi.* Japanese scientists became interested in the factors which brought about the enhanced growth. In 1926 a Japanese plant pathologist discovered that filtrates from cultures of the fungus produced characteristic symptoms of the "foolish seedling disease" when applied to rice seedlings. The fungus secreted a chemical or chemicals which promoted elongation of rice seedlings. In 1935 crystals of gibberellic acid were isolated from extracts of the fungus. Appropriate concentrations of gibberellic acid promote the growth of many plants.

Fig. 17–6. Cabbage plants treated with gibberellic acid (*right*) flowered the first year, whereas the untreated plants on the left did not. The treated plants were taller than control plants, which flowered during the second year. (S. H. Wittwer, Michigan State University.)

Gibberellic acid ($C_{19}H_{22}O_6$), a tetracyclic dihydroxy lactonic acid, is one form of a group of compounds known as gibberellins which promote cell division and cell elongation. Gibberellic acid markedly enhances the growth of many plants, among them pea, bean, tulip, poplar, maple, geranium, petunia, and aster. Certain varieties of asters tripled in height 3 to 4 weeks after treatment. Dwarf varieties of peas which had been sprayed with a dilute solution grew three times their normal height and exceeded in height certain tall varieties of peas. Treated bush beans developed into tall twining plants resembling pole beans. Very small amounts of the acid bring about large growth responses. One millionth of a gram often enhances the growth of a plant.

In recent years gibberellins have been isolated from healthy seed plants and it seems likely that they are naturally occurring plant hormones. The amount present in plants varies with the species and the variety, being low in dwarf varieties. The growth of genetically dwarf varieties of certain plants, corn and peas for example, may be restricted by the small amount of gibberellic acid which they synthesize. When dwarf pea plants are provided with an external supply, their growth is greatly enhanced and in concentrations of 1/10,000 to $\frac{1}{10}$ of a microgram per plant growth is proportional to the amount of gibberellic acid provided. (A microgram is one-millionth of a gram.)

Gibberellic acid also affects flowering. It hastens flowering of lettuce, snap beans, and early tomatoes, but retards blossoming of peppers. Gibberellic acid induces flowering of long-day plants even when they are grown with short days. However, it will not induce flowering in short-day plants which are grown under long-day conditions. Many biennial plants, for example, cabbage (Fig. 17–6), complete their life cycle in one year when treated with gibberellic acid; such treated plants do not require the cold of winter to set the stage for flowering. Instead they flower during the first year. The

Table 17–3

Comparison of Gibberellin and Auxin

Activity	Gibberellin	Auxin
Curvature of oat coleoptile in standard test	No	Yes
Retards abscission of leaves	No	Yes
Initiation of adventitious roots	No	Yes
Moderate concentrations inhibit root elongation	No	Yes
Development of axillary buds inhibited	No	Yes
Induces development of callus tissue	No	Yes
External applications promote growth, especially of dwarf varieties	Yes	No
Promotes seed germination and breaks dormancy of seeds	Yes	No
Induces flowering of biennials during the first year	Yes	No
Induces flowering of long-day plants when they are grown with short days	Yes	No
Promotes parthenocarpic fruit development	Yes	Yes

application of gibberellic acid to seedless grape clusters results in the development of a greater number of grapes and fruit larger than normal. Gibberellic acid hastens the germination of seeds of many plants—among them pea, bean, cotton, and sugar beet—and may overcome some types of seed dormancy. The rest periods of buds of trees, shrubs, and potato tubers may be overcome by treatment with gibberellic acid.

Gibberellins can be distinguished from auxins not only by their chemical nature, but also by their effects on plants. Table 17–3 summarizes the effects of gibberellin and auxin on various aspects of plant development.

Kinins

Kinins, like gibberellins, promote cell division and cell enlargement. Kinins are required for DNA replication, an essential feature of mitosis. Although they have not been isolated from plants, experimental evidence indicates that they do occur naturally and that they represent a third class of hormones, a class distinct from the auxins and gibberellins. The first kinin to be studied was kinetin, 6-furfurylaminopurine, which was isolated from old degraded preparations of DNA obtained from herring sperm. Since then several related compounds with similar biological activity have been isolated; collectively these compounds as well as kinetin are referred to as kinins. Most experiments have been carried out with kinetin, which is readily available.

Tissue-culture studies demonstrate that certain aspects of plant development are controlled by the proportion of kinetin to auxin. When a specific ratio of kinetin to auxin is present in the nutrient medium, callus tissue grows as an undifferentiated mass; cell division is unrestrained and organs do not develop. If the ratio of kinetin to auxin is increased by augmenting the amount of kinetin or decreasing the amount of auxin, a dramatic developmental change occurs. With the higher ratio, buds develop from the previously undifferentiated mass and in time an entire plant develops. Such studies have enhanced our knowledge about unrestrained growth of cells and about the factors which control differentiation and development (Fig. 17–7).

Kinetin may direct food and other metabolites to specific sites. In an untreated plant metabolites move from older, senescent leaves into the younger ones near the stem tip. However, if kinetin is applied to an old waning leaf, the flow may be reversed and, as metabolites move from the younger leaf into the older one, the younger one may senesce and the older regain vigor. Application of kinetin to senescing leaves stimulates RNA synthesis.

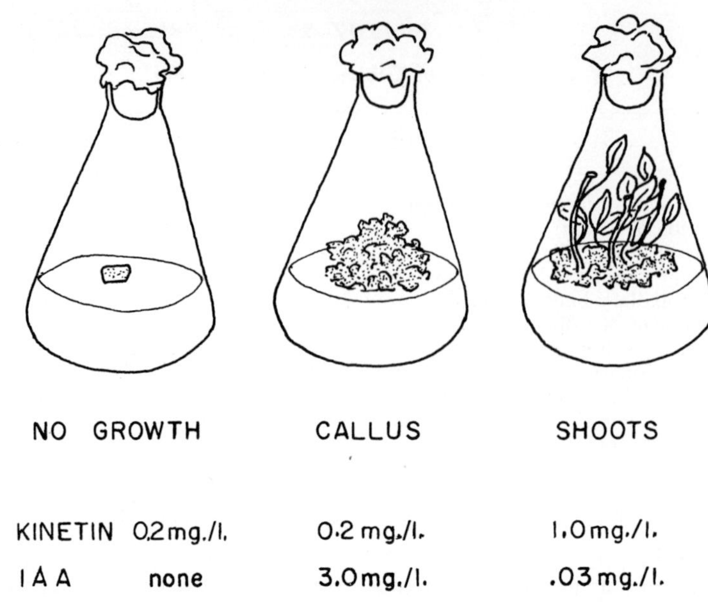

Fig. 17–7. Plant development is influenced by kinetin and indoleacetic acid. Without IAA the callus tissue, grown from the pith of a tobacco stem, does not grow. When the ratio of kinetin to IAA is low, the callus grows in an unrestrained manner. When the ratio of kinetin to IAA is high, shoots develop from the callus.

Wound Hormones

If freshly cut plant tissue is immediately and thoroughly rinsed with water, few cell divisions occur in the cells adjacent to the injury; whereas, if the tissue is not rinsed, the adjacent cells divide and the wound heals. Furthermore, if extracts of macerated tissue are applied to a thoroughly rinsed wound, the adjacent cells divide. Experiments such as these indicate that injured cells produce hormones which promote cell division in the intact cells adjacent to the injury.

QUESTIONS

1. In what ways do the roots, stems, and leaves affect the development of each other as well as themselves?

2. Describe an experiment which indicates that a growth hormone is produced in the tip of a coleoptile.

3. What plant activities are controlled by auxin?

4. _____ Which one of the following statements is false? (A) Auxin is primarily formed in apical buds and young leaves. (B) Auxin is responsible for the control and promotion of stem elongation. (C) Auxin inhibits the growth of axillary buds. (D) Auxin plays a role in the initiation and maintenance of cambial activity. (E) All organs—leaves, stems, roots, and buds—are equally sensitive to auxin.

5. _____ The blades were removed from two coleus leaves. The end of the petiole of one was smeared with auxin in lanolin; the petiole of the other with lanolin. Which petiole would absciss sooner? (A) The one smeared with auxin in lanolin, (B) the one smeared with just lanolin.

6. _____ An investigator had three sunflower plants. He removed the tops from two plants and smeared the stump of one with lanolin and the other with auxin in lanolin. The other plant he left intact. On which plant would the axillary buds (lateral buds) develop the soonest? (A) On the intact plant, (B) on the plant smeared with lanolin, (C) on the plant smeared with auxin and lanolin.

7. Explain fully the mechanism which is responsible for a primary root growing down and the primary stem growing up.

8. _____ If a plant is illuminated from one side, the auxin accumulates on the shaded side and thus causes that side to grow faster, with a resulting curvature toward the light. This response of stems to unequal lighting is known as (A) geotropism, (B) thigmotropism, (C) hydrotropism, (D) phototropism.

9. _____ Which one of the following is false? (A) The response of a plant to gravity is called geotropism. (B) The concentration of auxin which stimulates the growth of a stem also stimulates root growth. (C) When a root is placed in a horizontal position the auxin accumulates in cells toward the lower surface. (D) The downward bending (positive curvature) of a root results because growth is faster on the upper root surface than on the lower surface.

10. _____ Which one of the following coleoptiles would bend toward the light source? (A) One with a tinfoil cap placed over the tip of the coleoptile, (B) one with the tip cut off, (C) one with the tip cut off and then cemented back on with gelatin.

11. _____ Auxin does not (A) increase the plasticity of cell walls, (B) increase the rate of photosynthesis, (C) increase the uptake of water by cells, (D) increase the rate of respiration.

12. _____ The drooping of leaves of *Mimosa,* as well as those of other plants, is brought about (A) by greater growth on the upper side of a petiole than on the lower, (B) by turgor changes in cells of the pulvini, (C) by an accumulation of auxin in upper cells of the pulvini, (D) only by mechanical stimuli such as touch.

13. _____ An investigator sprayed plants of a dwarf variety of pea with dilute solutions of indoleacetic acid, gibberellic acid, naphthalene acetic acid, and 2, 4-dichlorophenoxyacetic acid, respectively. Two weeks later he measured the plants and probably found that the tallest plants were those sprayed with (A) indoleacetic acid, (B) gibberellic acid, (C) naphthaleneacetic acid, (D) 2, 4-dichlorophenoxyacetic acid.

14. _____ Which one of the following is false? (A) Gibberellic acid induces flowering of long-day plants even when they are grown with short days. (B) Gibberellic acid induces flowering of short-day plants which are grown with long days. (C) Biennials sprayed with an appropriate solution of gibberellic acid will flower during the first year. (D) Biennials treated with gibberellic acid do not require the cold of winter to set the stage for flowering.

15. _____ Growth substances are used commercially in several ways. However they are not used for (A) preventing apples from falling off trees, (B) hastening the rooting of cuttings, (C) killing weeds, (D) hastening the flowering of pineapples, (E) preventing wood from warping.

16. _____ Which one of the following growth substances is most frequently used to kill weeds? (A) Indoleacetic acid, (B) indolebutyric acid, (C) naphthaleneacetic acid, (D) 2–4 dichlorophenoxyacetic acid.

17. _____ If freshly harvested potatoes are treated with a growth substances such as 2,4-D or the methyl ester of naphthaleneacetic acid, their rest period will be broken. (A) True, (B) false.

18

Reproduction of Seed Plants—Asexual Methods

Many plants can be propagated (increased in number) without using true seed. Seeds are the end result of sexual reproduction. If plant parts other than seeds are used to increase plants, the methods are known as asexual methods.

Asexual methods are commonly used to propagate fruit trees, small fruits, some ornamental plants, sugar cane, pineapples, and such vegetables as onions, asparagus, and horseradish. Many standard varieties of plants have combinations of qualities which man wants to perpetuate. If asexual methods of reproduction are used, all of the offspring are exactly like the parents. If the plants are started from seeds, the offspring may not be uniform. Some of the seedlings may have all of the desired qualities, but others may be inferior. Of course, there is always the possibility that a few seedlings out of many will be superior to the parents.

Some cultivated plants have lost the capacity to develop seeds—among them, the banana, and seedless grapes and oranges. Such plants can be increased only by asexual methods. When plants are propagated by asexual methods, large plants can be produced in less time than when they are started from seeds. For example, a tulip started from a bulb will flower a year or two before one grown from seed. Likewise geraniums or roses started from cuttings or potatoes from tubers will mature in less time than those started from seeds.

Plants may be propagated asexually by specialized stems, root tubers, stem cuttings, leaf cuttings, root cuttings, and budding and grafting.

Under natural conditions, many plants reproduce vegetatively by bulbs, rhizomes, runners, tubers, and corms. Plants with runners and rhizomes invade surrounding areas. For example, Marram grass rhizomes extend outward as much as 15 feet a year. Rhizomes of other plants advance more slowly, some only a few inches a year.

SPECIALIZED OR MODIFIED STEMS

Rhizomes, runners, tubers, corms, bulbs, and bulbils are modified stems that can be used to propagate plants. Considerable

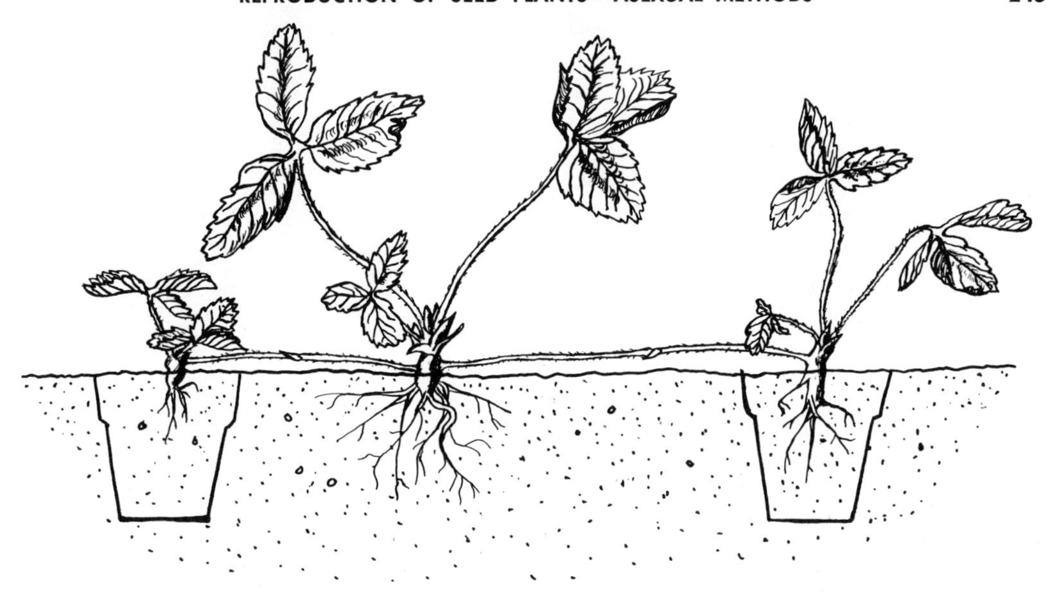

Fig. 18–1. Propagation of strawberry plants by runners. Pots of soil were placed under the runners where new plants were developing.

food is stored in all modified stems except runners.

Rhizomes

Rhizomes are horizontal stems which grow below or at the surface of the soil. Like all stems, they have buds which develop into branches. In addition, adventitious roots are present. From one rhizome several plants can be obtained by dividing it into sections, each with at least one bud. The division of rhizomes is used to propagate certain orchids, also asparagus, bearded iris, canna, and Solomon's seal, among other plants. Many weeds—quackgrass, for example—have rhizomes; if they are cut into sections during soil cultivation each portion becomes potentially a new plant.

Tubers

These are much thickened underground stems, of which the potato is an example. Potatoes are usually started from pieces of tubers which have at least one bud on them. After planting, the bud develops into a shoot which soon appears above ground. Adventitious roots develop on the shoot and permeate the soil. Later rhizomes develop from buds at underground nodes. The tip of each rhizome grows into a tuber replete with stored food and bearing spirally arranged minute scalelike leaves with buds in their axils.

Runners or Stolons

Runners, also called stolons, are relatively thin stems that grow horizontally above the soil surface. New plants develop at nodes where the runner is in contact with soil. In the strawberry young plants develop at every other node (Fig. 18–1). In agriculture, the young plants are separated from the parent and transplanted to the desired location. In nature, the young plants become independent only after the death of the connecting stolons.

Corms

Short, upright, underground stems are known as corms. Superficially they resemble bulbs, but in corms the bulk of the tissue is stem tissue, whereas in bulbs it consists of fleshy scalelike leaves. Gladiolus and crocus are characterized by corms, which are most frequently used for their propagation. In the gladiolus the parent corm dies and shrivels before the end of the season,

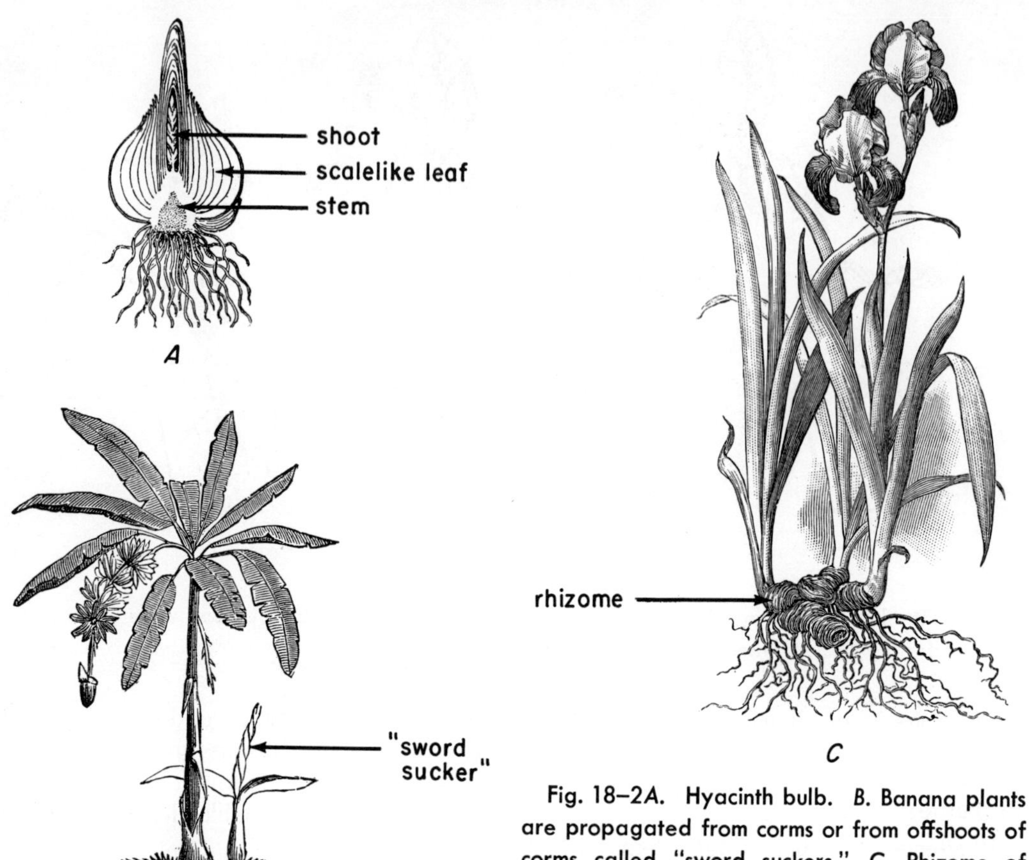

Fig. 18–2A. Hyacinth bulb. *B.* Banana plants are propagated from corms or from offshoots of corms called "sword suckers." C. Rhizome of iris. (After Bergen.)

but it is replaced by a new corm of flowering size. Several to many small corms, called **cormels,** are produced each year. These can be planted separately and thereby the population may be increased. Plants developing from cormels do not flower until the second or third year.

Compared to crocus and gladiolus, the corms of a banana plant are immense, measuring 1 foot in diameter. A new banana plantation may be started from offshoots produced by the corms or by dividing the corms into sections, each with a bud. With a favorable environment, the plants develop rapidly, often attaining a height of 30 feet and bearing fruit when 15 months old. Because a shoot produces fruit just once, the plant is cut down after harvest. A shoot from the old corm is then allowed to grow and bear the next crop.

Bulbs

Lilies, tulips, hyacinths, daffodils, and onions are usually started from bulbs, structures which are modified buds that consist of many fleshy scalelike leaves attached to a short stem which also bears a shoot (Fig. 18–2). After the bulb is planted the shoot emerges. The initial growth is made at the expense of the food stored in the scalelike leaves. The shoot within a mature tulip bulb consists of a stem bearing small leaves and a flower bud. Flower buds are not present in bulbs of all plants. The shoot within an onion is made up of a stem bearing leaves, but no flowers. In the onion the flowers form after the emerged plant has made considerable growth. The tulip bulb which is planted shrivels up almost completely before the end of the growing season, but a new large bulb and from one

Fig. 18–3. Preparing sugar cane cuttings. (A. J. Mangelsdorf.)

to four small flat ones develop each year. The large one will flower the next season, but the small ones will not flower until another year or later. In daffodils, the planted bulb does not die at the end of the season but continues to grow, each year producing one or more new bulbs beside it.

Bulbils

These are small bulblike structures which serve to propagate some plants. In several species of lilies, bulbils are formed in the axils of the upper leaves. These may be removed and planted. In about two years the plants will flower. In onions, bulbils may form in the place of flowers. They may be detached and planted.

ROOT TUBERS

Dahlias and peonies are among the plants which have thick fleshy roots known as **root tubers.** These alone cannot be used for propagation, because buds are not present on them. Therefore, when propagating plants with root tubers a piece of the stem with a bud on it should be left attached to one or more of the fleshy roots.

CUTTINGS

Stem Cuttings

Sugar cane, carnations, chrysanthemums, and geraniums are examples of nonwoody plants which are usually propagated by stem cuttings. Stem cuttings are made by cutting off the terminal portion of a stem just below a node. The cutting should have about four leaves. The lowest leaf is removed and the cutting is then placed in moist sand, vermiculite, or peat, to a depth of about 1 inch. The cuttings are kept in a humid, well-lighted place until roots form, after which they are planted in soil. Sugar cane cuttings are made by cutting the stem into segments, and these are planted directly in the field (Fig. 18–3). Trees and shrubs are also propagated from stem cuttings.

A foolproof way to propagate some trees and shrubs is by **layerage.** In layerage, a branch is bent to the ground and partially covered with soil. Roots form where the branch is in contact with the soil. Later, the branch with its roots is separated from the parent.

Air layerage (Fig. 18–4) can be used to increase many plants, trees, shrubs, and house plants. A cut is made half-way through the stem and the stem is slit up the middle for a distance of 3 inches, care being taken not to break it. The cut surface is then dusted with a rooting powder. Some moist sphagnum moss is then inserted in the slit stem so that it will not grow together again. Next, sphagnum moss is packed around the area. The moss is then covered with a sheet of polyethylene film, which is tied tightly to the stem. After roots form, the branch is severed from the parent.

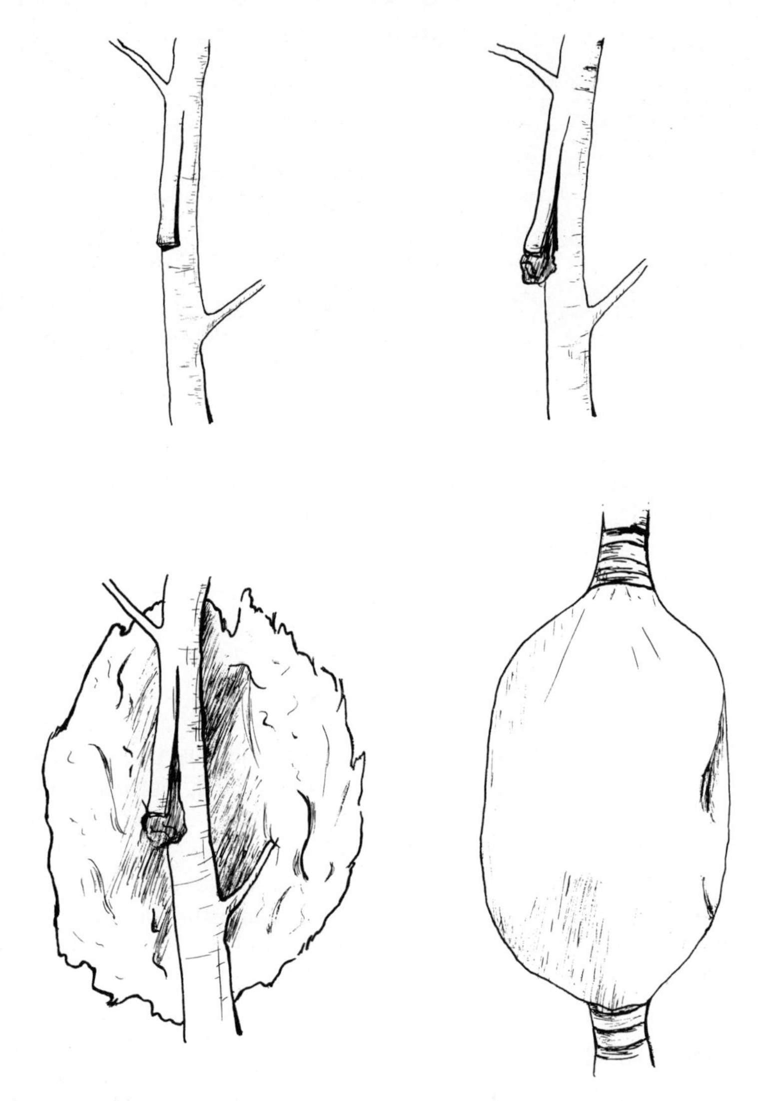

Fig. 18–4. Stages in the propagation of a shrub by air layerage. *Upper left,* the slit stem, which is dusted with a rooting powder. *Upper right,* sphagnum moss is inserted in the slit. *Lower left,* moist sphagnum moss is wrapped around the cut area. *Lower right,* the moss has been covered with a sheet of polyethylene film. The roots will form inside without further attention.

Fig. 18–5. New plants from leaves. All five of these young plants formed at the base of the petiole of this African violet leaf. Four of the plants have been separated from the petiole; the fifth one is still attached. The petiole was inserted in moist vermiculite.

Leaf Cuttings

African violets, begonias, and gloxinias are plants that can be increased by leaf cuttings. If the petiole of an African violet leaf is placed in moist sand or vermiculite, one or more stems bearing leaves and roots regenerate at the base of the petiole. The small plants formed at this point are later potted (Fig. 18–5). Rex begonias may be propagated by placing the blade on sand. The major veins are then cut. A new plant develops at each cut.

The most striking example of propagation from leaves occurs in the genus *Kalanchoë* (Fig. 18–6). In some species, tiny plants develop in the notches of the leaf margin even while the leaves are attached to the plant. These plants drop to the ground where they continue their development.

Root Cuttings

Typically, roots lack the capacity to regenerate shoots. However, there are exceptions to this rule. Roots of blackberry, raspberry, cottonwood, and horseradish regenerate shoots. Hence, such plants can be propagated by root cuttings. Such weeds as sow thistle and field bindweed reproduce naturally from shoots that originate from roots. The shoots develop from buds which originate in the pericycle.

The sweet potato may be quickly propagated from its fleshy roots. If a root is partially submerged in water, adventitious shoots soon develop. In commercial practice, sweet potato roots are planted in moist soil or sand. After shoots bearing roots have developed, the shoots are removed and planted in the field.

GRAFTING AND BUDDING

New varieties of plants cannot be produced by budding and grafting. These are not techniques for hybridizing plants. Instead, budding and grafting are methods of

Fig. 18–6. A leaf of *Kalanchoë*, showing young plants forming along the margin.

propagating existing varieties. They are resorted to because many varieties of lilacs, roses, apples, peaches, plums, grapefruits, grapes, etc., do not come true from seed. These plants can be increased by stem cuttings. However, budding and grafting have some advantages over stem cuttings. For example, in grafted stock, roots may be used which are resistant to insects, disease, drought, and other factors.

Grafting

Grafting consists of joining together the parts of two different plants. It is done in such a way that the cambium layers of the two parts are at least in partial contact. The top part, which will be the new shoot, is known as the **scion,** whereas the basal part is known as the **stock.** Flowers, fruits, and leaves which develop from above the graft union will be exactly like the scion. They will not have any characteristics of the stock. Plant parts which develop below the graft union will be like the stock. On occasion, a rose shrub that produces beautiful roses one year produces less handsome and different roses the following year. It is

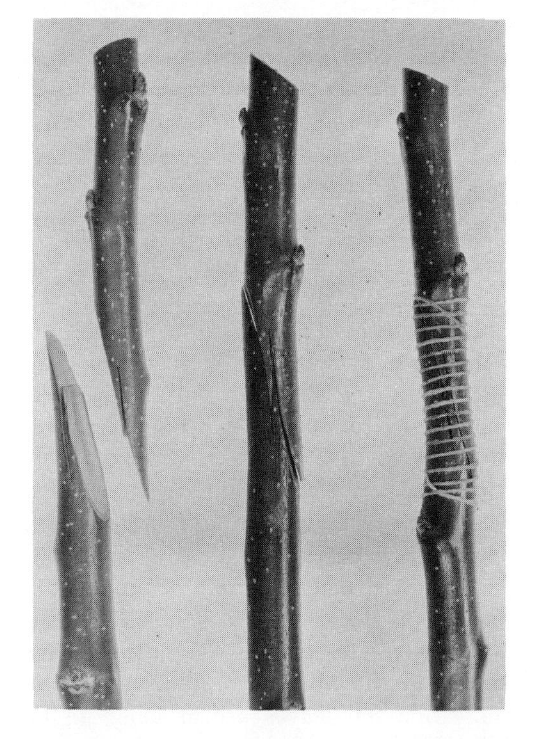

Fig. 18–7. Steps in whip or tongue grafting. *Left to right,* stock and scion prepared, parts fitted together, graft wrapped with waxed light twine. (C. J. Hansen and E. R. Eggers, *Calif. Agr. Ext. Serv. Circ. 96*).

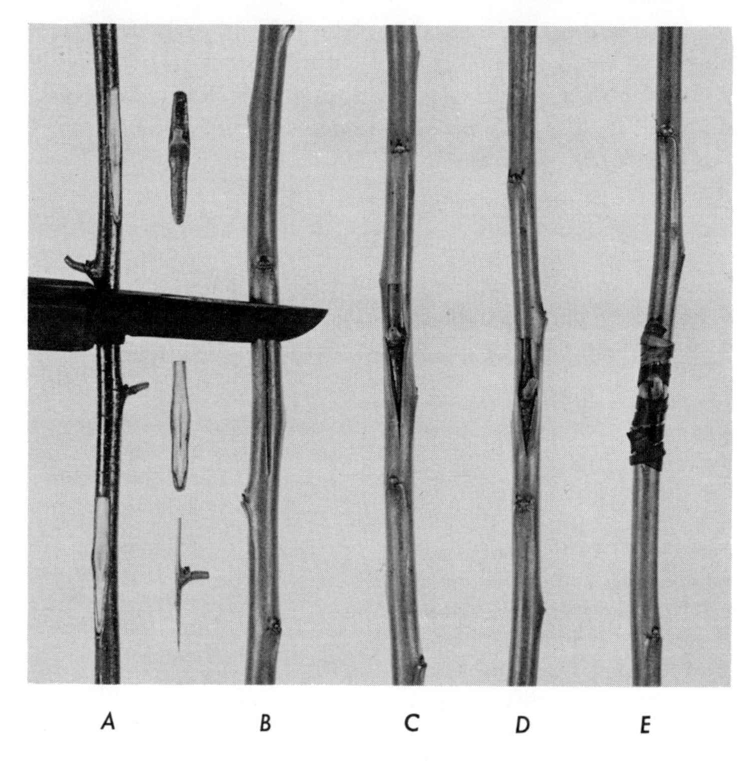

Fig. 18–8. T- or shield budding. *A*, bud stick with some buds removed (cuts should be started below buds); *B*, making cross-cut at top of vertical slit; *C*, bud partly inserted; *D*, bud in place; *E*, bud tied with rubber band. (C. J. Hansen and E. R. Eggers, *Calif. Agr. Ext. Serv. Circ. 96.*)

likely that the part above the graft union was winterkilled and that the flowering branches developed from buds below the graft.

There are many techniques for joining stock to scion. One of the most widely used is known as **whip** or **tongue grafting.** The grafting is done from January through March when the plants are dormant. The stock may be a piece of root or a stem with roots. The stock and scion should have about the same diameter. They are joined together as shown in Figure 18–7, making sure that the cambium layers are in contact. After the stock and scion are fitted together, the grafted portions are wrapped with No. 18 cotton string that has been dipped in paraffin or wax and then dried. After wrapping, the grafts are stored in moist peat, sawdust, or sand, at a temperature of 40° F. In the spring they are planted in the field.

The scion and stock should be of the same or closely related species. Apple is usually grafted on apple and pear on pear rootstock. However, pear may be grafted on quince, in which case the tree will be dwarf. Peach is usually grafted on plum stock. Among ornamentals, lilac is generally grafted on California privet stock; plants grown with such roots have fewer suckers than if grafted on lilac stock.

Budding

Budding is more frequently used by nurserymen than grafting in the propagation of trees and shrubs. Budding is essentially the same as grafting except that a single bud instead of a branch is grafted on the stock. The cambium of the bud should be in contact with the cambium of the stock. **Shield** or **T-budding** (Fig. 18–8) is usually used. Budding is generally performed in late July or early August. The bud unites with the stock in two weeks. The following spring the stock is cut off just above the bud. Suckers coming from the stock should be removed.

QUESTIONS

1. _____ If plants are propagated with asexual methods, the offspring (A) are uniform, (B) exhibit great variability, some being superior to the parents.

2. Distinguish between vegetative reproduction and seed reproduction, and describe several types of vegetative reproduction.

3. Define or describe a rhizome, tuber, runner, corm, bulb.

4. _____ Rhizomes are (A) horizontal stems which grow below or at the surface of the soil, (B) much thickened and swollen underground stems, (C) short upright underground stems, (D) short stems surrounded by fleshy leaves.

5. _____ With respect to modified organs, which one of the following statements is false? (A) A horizontal underground stem such as that of quackgrass which continues year after year to grow at one end is known as a rhizome. (B) When a portion of a rhizome becomes thickened, such as in the Irish potato, it is known as a tuber. (C) The sweet potato, carrot, dahlia, and peony have tubers. (D) A bulb, as in lily, tulip, or onion, is a modified bud.

6. _____ Which one or more of the following grafts are likely to be successful? (A) Apple on pear, (B) lilac on privet, (C) wheat on corn, (D) tobacco on tomato, (E) cherry on orange.

7. Gardeners have noted that rose shrubs which produce attractive hybrid tea roses one year produce during the following year roses of less attractive form and coloring. Explain.

8. _____ Assume that you want to perpetuate a mutant albino sunflower plant. This colorless mutant might be perpetuated by (A) leaf cuttings, (B) stem cuttings, (C) grafting.

9. Match the usual method of propagating the plants listed in the left-hand column:

_____	1. Poinsettia.	1. Corms.
_____	2. Chrysanthemum	2. Tubers.
_____	3. Gladiolus.	3. Fleshy roots.
_____	4. Tulip.	4. Bulbs.
_____	5. African violet.	5. Stem cuttings.
_____	6. Banana.	6. Leaf cuttings.
_____	7. Sugar cane.	7. Budding or grafting.
_____	8. Potato.	
_____	9. Sweet potato.	
_____	10. Apple.	

10. _____ Why do horticulturalists propagate many plants asexually instead of using seeds?

11. _____ Which one of the following is false? (A) Grafting is a technique for hybridizing (crossing) plants. (B) Grafting consists of joining together the parts of two different plants. (C) The top part, that which will be the new shoot, is known as the scion. (D) The part which furnishes the roots is the stock. (E) The cambium of the stock must be in contact with the cambium of the scion to insure a successful graft.

12. _____ If a branch from a Jonathan apple tree is grafted to the roots of a Malling tree, the crown of the resulting tree will bear (A) Malling apples, (B) Jonathan apples, (C) apples intermediate between Malling and Jonathan.

19

Sexual Reproduction—Flower Structure

There is more in plants than outward beauty, although it is the gross beauty that attracts our attention. In the fields and forests the blue masses of delphinium or lupine, the golden blanket of mustard, buttercup, or California poppy, or the flaming azaleas add beauty to the landscape. Many who cherish the colors and myriad shapes never see the most intriguing, most ingenious parts of flowers. Those who are curious, and who have the patience to look for the smaller things, the inside parts of a flower, the shape of a seed or fruit, have a new world of interest in plants spread before them.

Flowers are not only of decorative value, but, of more significance, they lead to the development of seeds and fruits so essential for human and animal nutrition and for perpetuation of the various races of cultivated and native plants. A flower consists of certain fundamental parts, the interaction of which is necessary to produce new plants. Fruits and the seeds contained therein are the end-products of sexual reproduction.

GENERAL FEATURES OF SEXUAL REPRODUCTION

In practically all plants and animals sexual reproduction involves the formation of sperms and eggs, the union of one sperm with each egg, and the development of the fertilized egg into a new individual. Another general feature of sexual reproduction is **reduction division** (also called meiosis) in cells, a type of division which results in cells with just half the number of chromosomes that the parent cell had.

PARTS OF A COMPLETE FLOWER

Before learning the details of sexual reproduction as it occurs in flowering plants, we must become familiar with the various parts of a flower. A flower is a reproductive branch made up of a stem with leaves which are altered to carry on reproduction. A complete flower consists of a stem tip (called the **receptacle**), upon or around which are the **sepals, petals, stamens,** and one or more **pistils**, all of which are modified leaves. Those which make up the pistil are known as **carpels.** Like foliage leaves, the floral parts are produced at the tip of the main stem, at the tips of branch stems, or both (Fig. 19–1). In flowers of hybrid tea roses, carnations, and other plants with double flowers, some stamens have metamorphosed into petals. In certain flowers we can see transition forms between stamens and petals. It is likely that long ago petals of all plants originated from stamens.

Accessory Parts of a Flower

Sepals and petals are accessory parts of a flower because they do not play a direct role in reproduction. Together the sepals and petals make up the **perianth** of a flower.

Sepals. The sepals, collectively called the **calyx,** are the outer-most whorl of floral parts. They are frequently green and leaf-like in appearance. In the bud stage, they protect the other floral parts from rain, mechanical injury, and insects.

Petals. The petals, collectively referred to as the **corolla,** are inside the sepals and are usually conspicuous and beautiful. They attract insects by their display and fragrance and often reward them with nectar, produced in glands called **nectaries.** Nectaries are frequently located at the base of the petals.

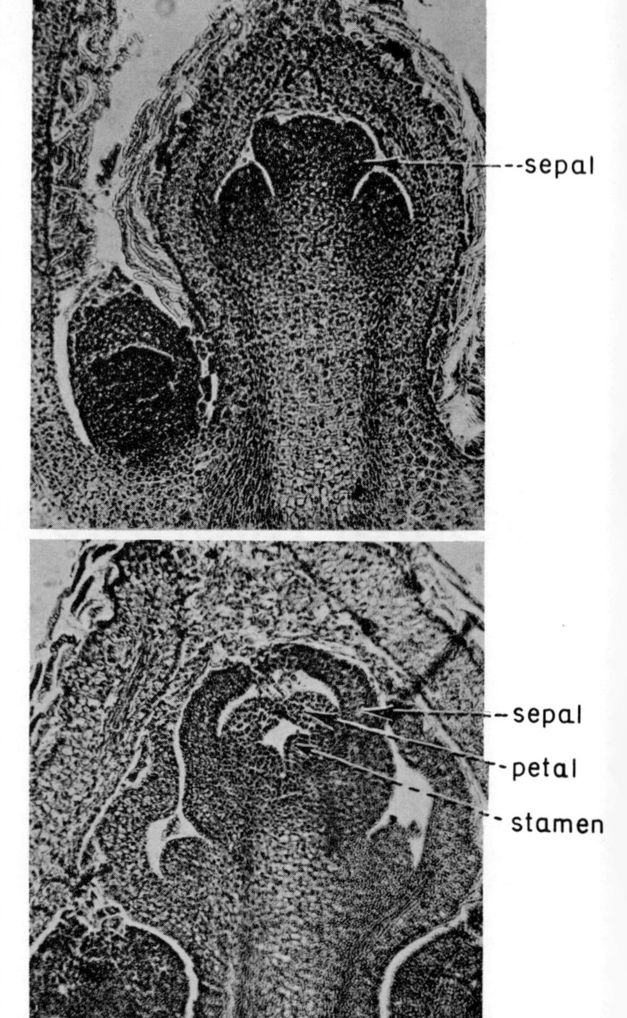

Fig. 19–1. Development of floral parts in an olive bud. The floral parts, like foliage leaves, are produced at the tip of a stem. *Upper,* a young bud showing the appearance of sepal primordia. *Lower,* an older bud showing sepal, petal, and stamen primordia. (H. T. Hartmann.)

Essential Parts of a Flower

Stamens and pistils are the essential parts of a flower (Fig. 19–2). They are directly related to sexual reproduction.

Stamens. Collectively the stamens in a flower are known as the **androecium.** The number of stamens in a flower ranges from one, as in certain orchids, to a hundred or more as in certain species of the buttercup family. Primitive species have many stamens in each flower whereas more advanced ones have few stamens. Each stamen has a stalk, the **filament,** with an **anther** at its tip. Pollen is produced in the anther, which opens when the pollen is ripe. Dusty pollen floats away on the wind and sticky pollen is picked up by visiting insects or birds.

Pistil. The pistil is in the center of the flower and consists of 3 parts, a **stigma,**

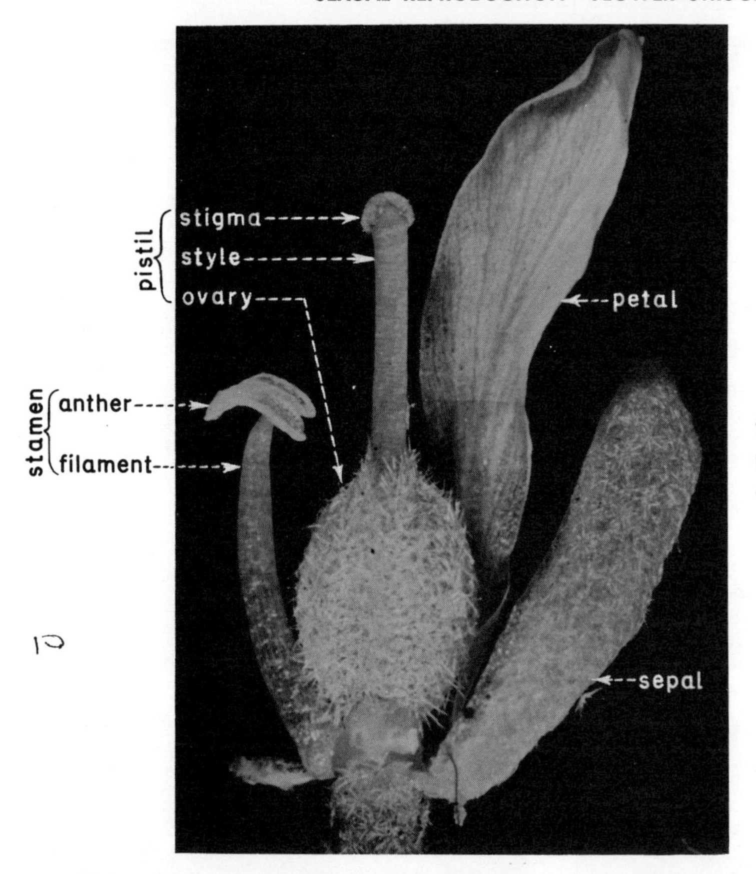

Fig. 19–2. Parts of a flower from which all but one sepal, petal, and stamen have been cut away.

a **style,** and an **ovary.** The uppermost part, the stigma, receives pollen and permits the appropriate pollen to function. The style connects the stigma to the enlarged ovary, the lower part of the pistil. The ovary is a hollow structure which contains one or more **ovules.** After pollination and fertilization, each ovule develops into a seed. Each ovule is borne on a short stalk known as a **funiculus.** The ovules arise from special regions within the ovary known as **placentae.** The location of the placentae is referred to as placentation, of which there may be three types, namely: **parietal, axial,** and **central** (Fig. 19–3). If the placentae are on the ovary wall, the placentation is parietal. If they are on a central axis in an ovary having more than one cavity (**locule**), the placentation is **axial.** The placentation is **central** if the placentae are borne on a stalk which extends from the base of the ovary upward into the cavity of a unilocular ovary.

A pistil is formed from one or more modified leaves known as **carpels.** A carpel is a minute leaf which instead of remaining flat folds together so that the edges come into contact, thus forming a pistil. A simple pistil, such as that of a pea flower, is composed of one carpel. A compound pistil—for example, that of tomato or hollyhock—is made up of two or more united carpels (Fig. 19–4). In flowers of certain primitive species, several to many pistils are present in each flower; whereas, in more advanced species, one pistil occurs in each flower. The pistil or pistils in one flower are called the **gynoecium.**

VARIETY IN FLOWERS

Flowers exhibit great diversity. What makes them so fascinating is the many ways in which the theme has been varied, from species to species, from genus to genus, from family to family. Flowers of unlike species

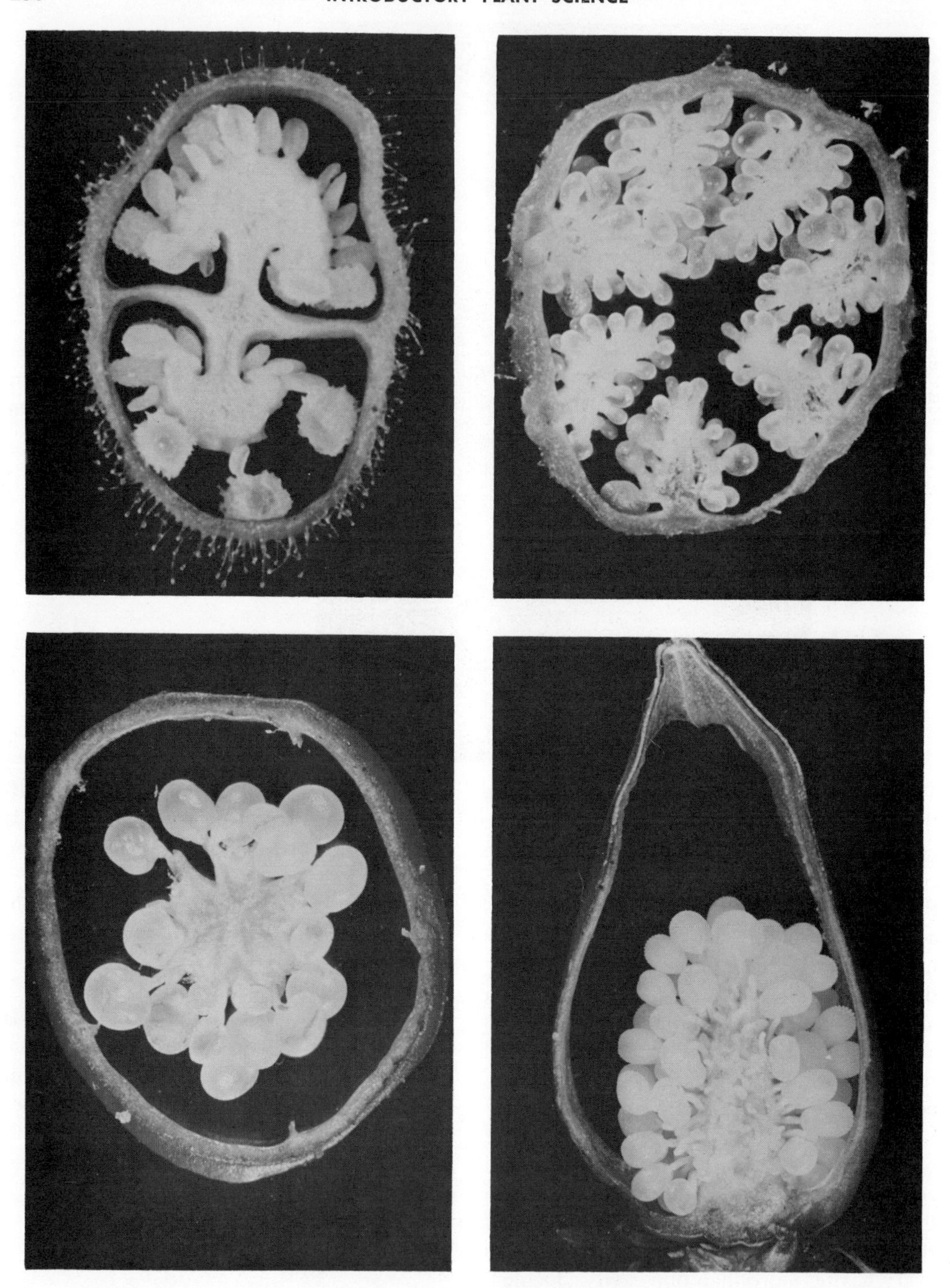

Fig. 19–3. Types of placentation. *Upper left*, cross-section of an ovary of a snapdragon showing axial placentation. *Upper right*, parietal placentation in a cross-section of an ovary of poppy. *Lower left*, cross-section, and *lower right*, longitudinal section, of a *Lychnis* ovary showing central placentation. Notice in all figures that each potential seed (ovule) has a short stalk (funiculus) which is attached to the placenta.

Fig. 19–4. Compound pistil of a hollyhock. This pistil is made up of many united carpels.

differ in size, color, number of parts, union or separateness of floral parts, and myriad other ways.

Some flowers are a foot across, others are so tiny that their parts can scarcely be seen except with a hand lens. Some have a hundred or more petals, others but three, and some lack petals altogether, in which instance they are said to be **apetalous.** The flowers of some species have petals, but lack sepals, the **asepalous** condition. Some flowers lack both sepals and petals. Flowers which have all four kinds of floral parts—sepals, petals, stamens, and a pistil or pistils—are called **complete flowers,** whereas those which lack one or more floral whorls are said to be **incomplete.**

In many species stamens and a pistil are present in the same flower and the flower is known as a **perfect flower.** In others, stamens and a pistil do not occur in the same flower and the flowers are said to be **imperfect.** In corn the flowers bearing stamens, the **staminate flowers,** and those containing a pistil, the **pistillate flowers,** are on the same plant. The staminate flowers are on the tassel, the pistillate ones, of which the silk is the stigma and style, on the ear. The tuberous begonia is another that bears staminate and pistillate flowers on the same plant (Fig. 19–5). Such a plant is called **monoecious.** In other words, in a monoecious plant both unisexual male and female flowers are borne on one and the same plant.

Willow, date, cottonwood, holly, and bittersweet bear staminate and pistillate flowers on different plants. They are said to be **dioecious.** In dioecious plants the sexes are separated as they are among the higher animals. The male plants produce staminate flowers; the female ones, pistillate flowers (Fig. 19–6). Four thousand years ago the Assyrians knew that there were two kinds of date trees and generally they planted one tree bearing staminate flowers for each fifty bearing pistillate ones. They hand-pollinated the pistillate flowers by shaking branches bearing staminate flowers over them and thus obtained an excellent crop. Many a gardener has planted just one holly tree or bittersweet vine only to discover later that the tree or vine did not produce berries. To obtain berries on such plants both sexes must be in close proximity, or the pistillate plant must have a branch from a staminate plant grafted on it.

In some flowers the petals are all alike in size and form, and the corolla resembles a star. These flowers are radially symmetrical and they have what is called a **regular corolla.** In others, the petals are not alike and the corolla is bilaterally symmetrical. Such a corolla is said to be **irregular.** Among the flowers with an irregular corolla is the sweet pea, which has five unlike petals. The largest petal, the one toward the back, is known as the standard. Two petals, known as wings, are situated one on each side of the standard. Two smaller petals are united to form a keel, which surrounds the ten stamens and single pistil.

Fig. 19–5. The tuberous begonia is monoecious. *Left,* a staminate flower. *Right,* a pistillate flower.

Fig. 19–6. Willows are dioecious. *Left,* staminate flowers which were produced on one shrub. *Right,* pistillate flowers which developed on a different shrub.

The stamens of a sweet pea flower are distinctive. The filaments of nine stamens are united and that of the tenth is not united (Fig. 19–7). Hence in the sweet pea, there are two groups of stamens, nine in one group and only one in the other. When the pattern of union forms two groups, the stamens are **diadelphous.**

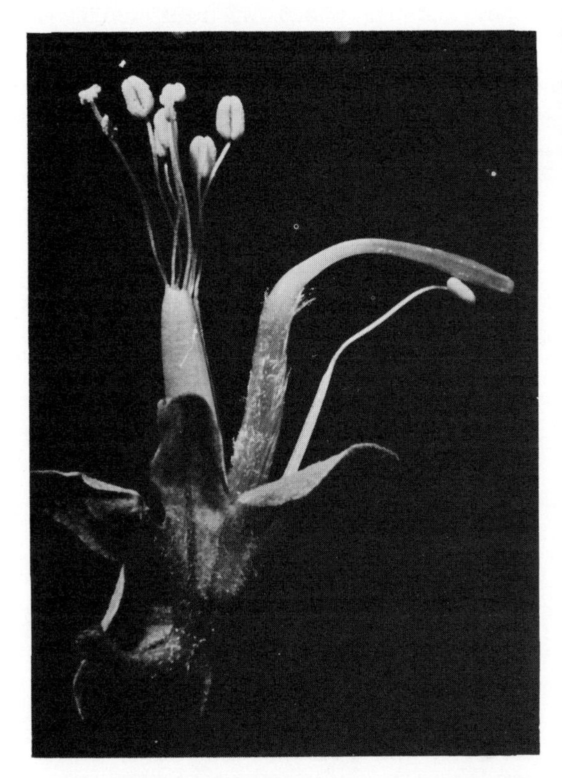

Fig. 19–7. Diadelphous stamens of a sweet pea flower. Note that the calyx is synsepalous. The petals have been removed to show these parts and also the pistil.

In most species the stamens are separate, but in others they are united by their filaments or anthers. If, as in hollyhock, the stamens are united to form a single tube, the stamens are said to be **monadelphous,** in contrast to the diadelphous condition as found in sweet pea.

In many species, and again the sweet pea is an example, one carpel is present in a flower, while others possess more than one carpel. In the latter instance the carpels may be separate or united to form a compound pistil. In one strawberry flower there are many pistils, each made of only one carpel. In a hollyhock flower there is one pistil (Fig. 19–4) made up of many united carpels.

Fig. 19–8. Hypogynous flower of death camass.

In some plants the calyx, corolla, and stamens originate below the ovary, and the flower is then said to be **hypogynous** (Fig. 19–8). The ovary of a hypogynous flower is **superior,** meaning that it is above the place where the other floral parts originate. In cherry (Fig. 19–9) the sepals, petals, and stamens are inserted on an urnlike rim, the **floral tube** (also called **hypan-**

Fig. 19–9. Perigynous flower of a Nanking cherry.

Fig. 19–10. Epigynous flower of a crabapple.

thium), which surrounds the ovary, but which is not fused to the ovary. Such flowers are **perigynous** and the ovary is still considered to be superior. In apple flowers (Fig. 19–10) the floral tube is fused to the ovary, and the calyx, corolla, and stamens appear to arise from the top of the ovary. When, as in apple, the floral parts appear to be situated on top of the ovary, the flower is **epigynous.** Because the ovary is below the sepals, petals, and stamens, the ovary is said to be **inferior.**

Stamens and pistils show great diversity in number and form. In some plants it is even difficult to recognize the stamens and the parts of a pistil. In breeding plants, it is necessary to transfer pollen from the anther to the stigma. Certainly before this can be done, you must know where the anthers are and where the stigma is located. In an orchid the anther and stigma are somewhat obscure, but with the aid of Figure 19–11 you should have no trouble locating these essential parts. In an orchid flower, the three stamens and the pistil are fused together, forming a structure known as the **column.** The column bears an anther at its tip, below which the stigma is located. The pollen occurs in masses called **pollinia.** The corolla of an orchid is irregular; two petals are alike in shape and size, but the third is modified to form a **lip,** also called **labellum.** Figure 19–12 summarizes flower variation.

Structure of the Composite Flower

The composite family (Compositae), also known as the sunflower family, includes many familiar plants, such as sunflower, aster, chrysanthemum, dandelion, and Transvaal daisy. What appears to be one flower of these plants is really a cluster of flowers, known as a head. What looks like three flowers in Figure 19–13 is actually three clusters of flowers, each containing perhaps 100 flowers. Below each cluster there are a number of bracts making up the involucre. In the dandelion all of the flowers in the head are alike, and all have a strap-shaped corolla, known as a **ligulate corolla,** consisting of five united petals. In sunflower (Fig. 19–14), also in Transvaal daisy, the marginal flowers, called **ray flowers,** have ligulate corollas, whereas the central ones, called **disk flowers,** have **tubular corollas.** Both the ray and disk flowers have a corolla consisting of five united petals. The calyx consists of scales, collectively called the **pappus.** In sunflower the ray flowers are sterile; they lack stamens and a functional pistil. Within each disk flower a pistil is present, consisting of an ovary, a style, and a stigma with two branches. Five stamens are present in each disk flower. The filaments of the stamens are not united, but the anthers are united into a ring which surrounds the style.

Structure of a Grass Flower

Grasses are of prime significance in human economy. Among the important ones are lawn grasses, sugar cane, bamboo, forage grasses, corn, wheat, rice, oats, and barley. Their flowers are quite inconspicuous and are seldom noticed by most people. One or more flowers usually grow together in a small cluster called a **spikelet,** more or less enclosed by two scales called

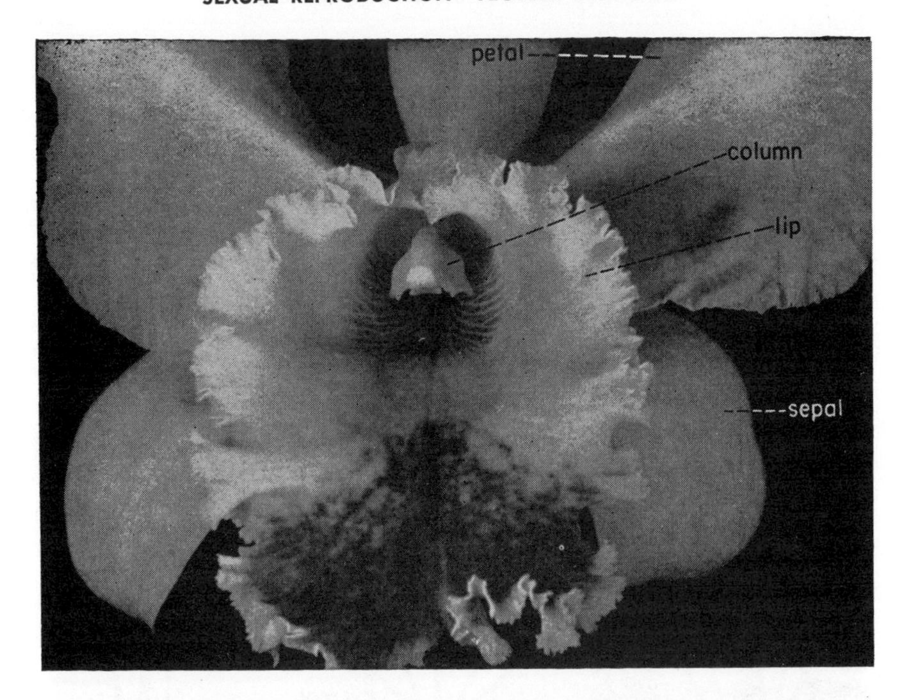

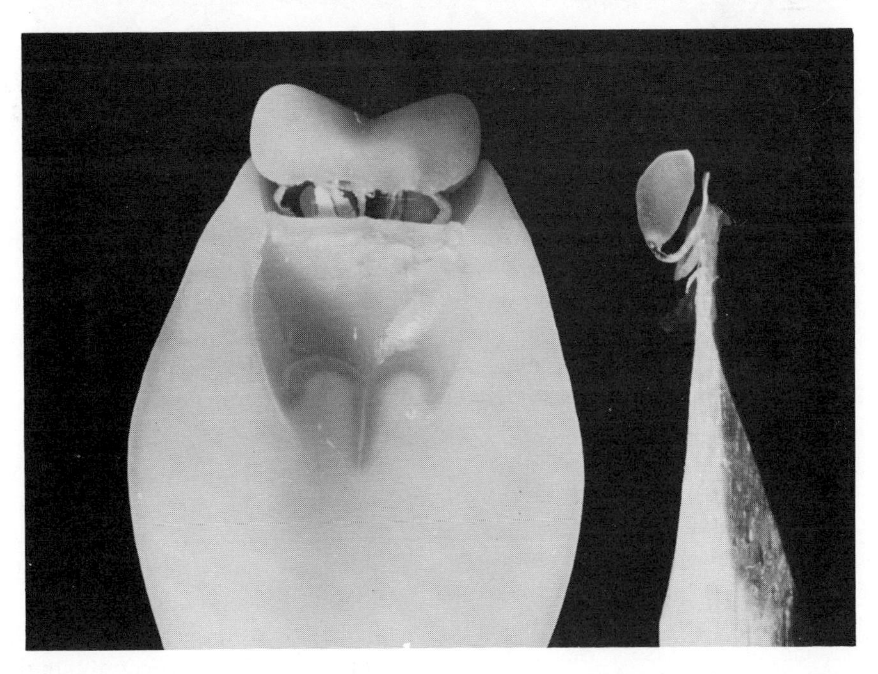

Fig. 19–11. A closeup of a cattleya flower. Notice the large lip (labellum) that surrounds the column, the two other petals, and the three narrower sepals. *Lower*, details of a column. The hinged structure at the top is the anther; it contains four pollen masses, called *pollinia*. The sunken area below the anther is the stigma. A pollinium is stuck to a sharpened matchstick at the right. (Henry T. Northen and Rebecca T. Northen, *The Complete Book of Greenhouse Gardening*. Copyright © 1956, The Ronald Press Company, New York.)

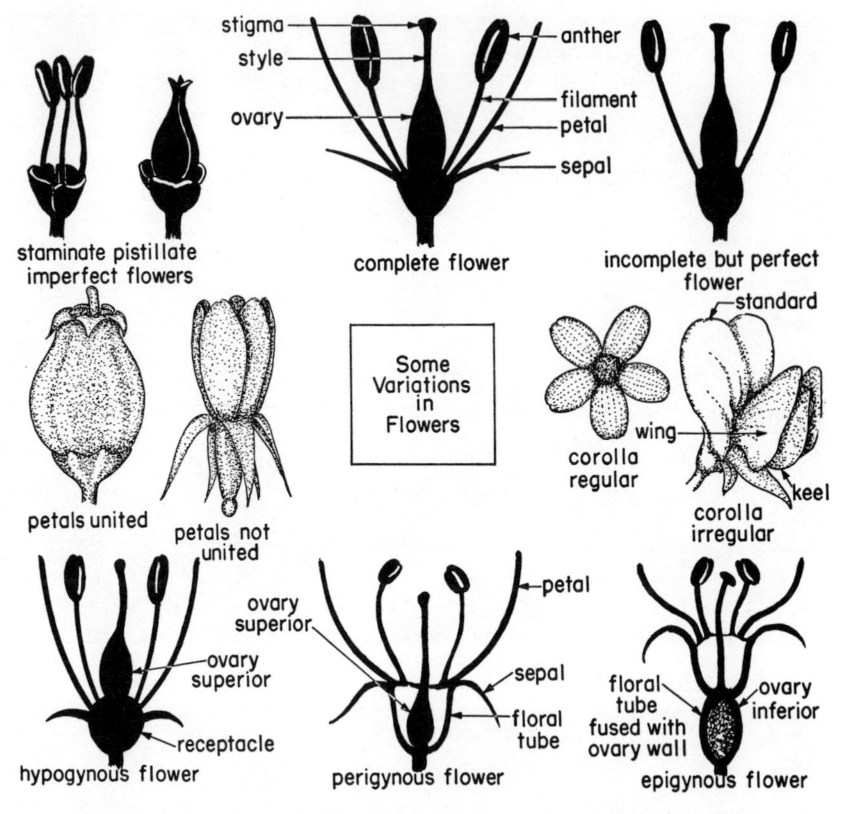

Fig. 19–12. Summary of variations in flowers.

glumes (Fig. 19–15). The shortened axis of the spikelet which bears the glumes and flowers is called the **rachilla.** Each flower is enclosed between two bracts; the outer bract is the **lemma** and the inner one, the **palea.** Each flower consists of a pistil, three stamens, and two small scalelike structures known as **lodicules.** The pistil consists of an ovary (containing one ovule) and two feathery stigmas. In wheat, barley, and certain other grasses, the spikelets are attached directly to the main stem (known as a rachis) and the inflorescence is a spike. In oats and some other grasses the spikelets grow at ends of branches of the rachis, forming an inflorescence known as a **panicle.**

INFLORESCENCES

In some plants the individual flowers are large and showy. In lilacs and other plants, the individual flowers are small, but together they form a conspicuous cluster. A cluster of flowers is called an **inflorescence.** There are two major types, indeterminate inflorescences and determinate ones.

Indeterminate Inflorescences

In indeterminate inflorescences the lower or outer buds open first. The terminal growing point of the main axis of the inflorescence grows continuously and produces additional flower buds. Seeds may be ripe in fruits at the base of the stem, but at the tip of the inflorescence only buds may be present. Some types of indeterminate inflorescences are described below.

Raceme. The individual flowers of a raceme are borne on short branches called **pedicels** which are about equal in length. The main axis (the peduncle) is elongated

Fig. 19–13. The Transvaal daisy. (Bodger Photo.)

and the flowers are spaced at regular intervals. Each flower develops from a bud in the axil of a modified leaf (bract). Larkspur, currant, and snapdragon have racemes (Fig. 19–16).

Spike. A spike is like a raceme, except that the flowers are not borne on pedicels but are attached directly to the main axis. Plantain illustrates the spike.

Catkin. Catkins are spikes made up of staminate or pistillate flowers. Willow, cottonwood, birch, and alder have catkins.

Panicle. A panicle is a branched raceme. In a panicle each branch of the main axis bears more than one flower, whereas in a raceme each branch (pedicel) is terminated by a single flower. Lilac and grape have panicles.

Umbel. Onions, celery, carrots, and dill are typical of plants which have umbels. The pedicels of an umbel are attached at approximately the same place at the top of the main axis instead of being spaced at regular intervals along the axis as they are in racemes and spikes.

Head. A head resembles a very short and dense spike. The crowded flowers lack

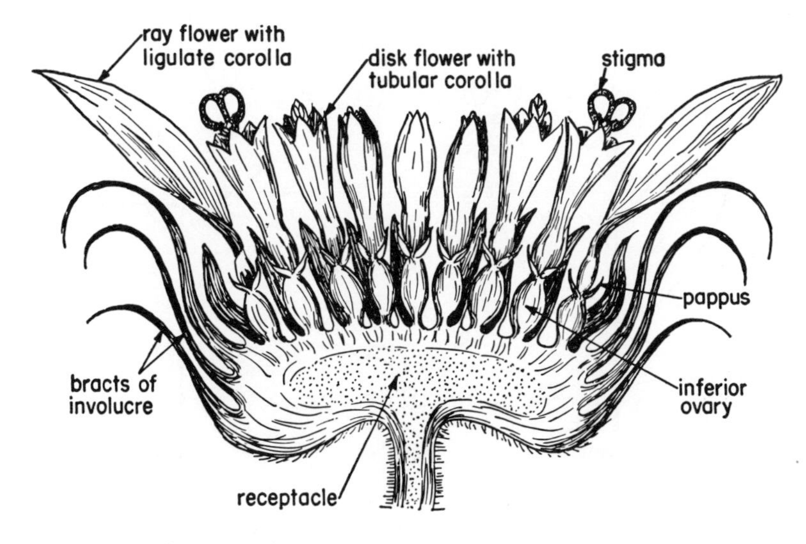

Fig. 19–14. Section through a sunflower head.

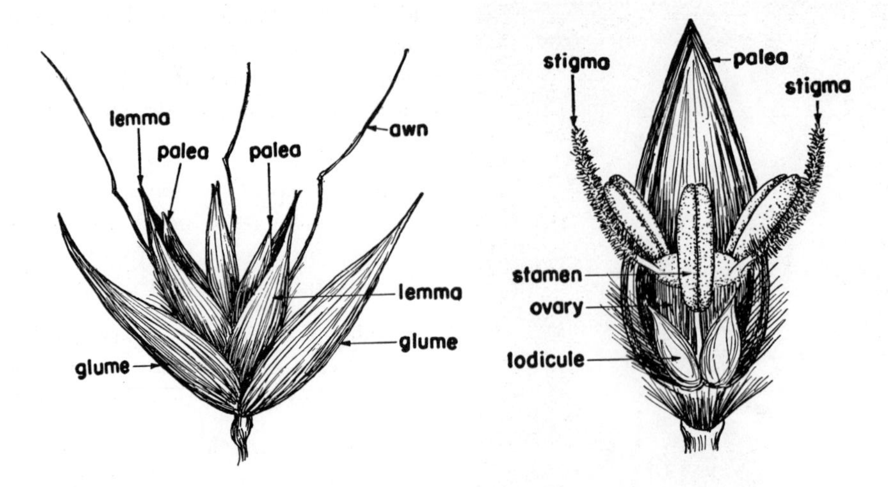

Fig. 19–15. *Left*, an oat spikelet. *Right*, a single oat flower.

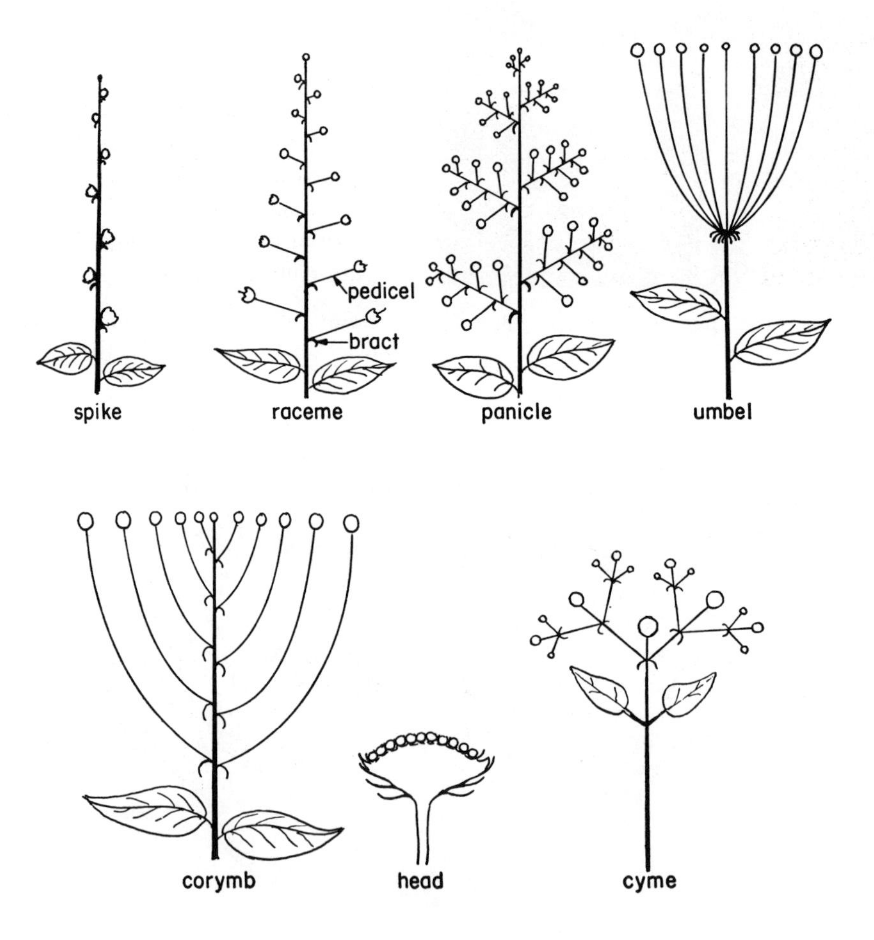

Fig. 19–16. Diagram showing the arrangement of flowers in common inflorescence types.

pedicels, and they are borne on a disklike expansion at the tip of the stem. Sunflower, aster, chrysanthemum, and dandelion have heads.

Corymb. A corymb is like a raceme, but the lower flowers have longer pedicels than the upper flowers. Hence the corymb has a more or less flat-topped appearance. The cultivated cherry and candytuft are typical examples.

Determinate Inflorescences

The terminal growing point of the main axis of a determinate inflorescence develops into a flower bud and hence the growth of the main axis is arrested. Later flowers develop on side branches below the terminal flower. The youngest flowers are those most distant from the apex of the inflorescence. Determinate inflorescences are known as **cymes.** Phlox and crabapple have determinate inflorescences.

QUESTIONS

1. Name and describe the floral organs of a complete flower, and state their functions.

2. _____ The essential parts of a flower are (A) sepals and petals, (B) stamens and pistil, (C) stigma and style, (D) filament and ovary.

3. Name in order, from the base upward, the parts of a pistil; of a stamen. What are the ovules? What is the part of the ovary to which the ovules are attached? What is the funiculus?

4. What are carpels? Of how many carpels is a simple pistil composed? A compound pistil? Name a flower having a simple pistil and one having a compound pistil.

5. _____ The structures located in ovaries which develop into seeds are called (A) integuments, (B) seeds, (C) ovules, (D) funiculus.

6. _____ The three parts of a pistil are (A) stigma, filament, and ovary; (B) stigma, style, and ovary; (C) anther and filament; (D) anther, style, and ovary.

7. _____ A certain flower had stamens and a pistil, but lacked sepals and petals. This flower is (A) imperfect, (B) perfect, (C) complete.

8. Match the collective term that is used for each of the floral parts:

_____	1. Sepals.	1. Corolla.
_____	2. Petals.	2. Gynoecium.
_____	3. Stamens.	3. Androecium.
_____	4. Pistil or pistils.	4. Calyx.

9. _____ When, as in cottonwood, the staminate flowers are produced by one plant and the pistillate by a different plant, the species is (A) monoecious, (B) hermaphroditic, (C) dioecious, (D) perfect.

10. _____ In the Christmas holly some trees produce flowers which have pistils in them and other trees produce flowers having only stamens. If you wanted branches bearing red berries you should (A) plant female trees, (B) plant male trees, (C) plant one of each.

11. _____ Which one of the following plants is dioecious? (A) Corn, (B) wheat, (C) willow, (D) sweet pea.

12. _____ When the petals are all alike and the corolla somewhat resembles a star, the corolla is (A) regular, (B) irregular, (C) diadelphous, (D) monodelphous.

13. _____ Which one of the following has an irregular corolla? (A) Pea flower, (B) a lily flower, (C) a tulip, (D) an apple blossom.

14. _____ One "sunflower" (A) is an inflorescence made up of only tubular flowers, (B) is an inflorescence having ray flowers with ligulate corollas and disk flowers with tubular corollas, (C) is a group of flowers subtended by involucral bracts which make up the calyx.

15. _____ The two scales which more or less enclose a grass spikelet are known as (A) glumes, (B) lemmas, (C) paleas, (D) lodicules.

16. _____ Which one of the following is false? (A) Each grass flower is enclosed between two bracts, the lemma and the palea. (B) Each spikelet is more or less enclosed by two bracts known as glumes. (C) Each grass flower consists of a pistil, three stamens, and two lodicules. (D) In most species of the grass family several to many ovules are borne in each ovary.

17. Describe the structure of a composite inflorescence (head).

18. Describe the structure of a grass spikelet and of a grass flower.

19. Match the following:

______	1. Sepals, petals, and stamens below the ovary.	1. Epigynous flower.
______	2. Sepals, petals, and stamens inserted on floral tube which is not fused to ovary.	2. Hypogynous flower.
______	3. Sepals, petals, and stamens appear to arise from the top of the ovary.	3. Perigynous flower.

20. Match the following:

______	1. Raceme.	1. Sessile flowers upon an elongated axis.
______	2. Spike.	2. Flowers on unbranched pedicels and occurring at intervals along an axis.
______	3. Panicle.	3. Pedicels bearing flowers radiating from about the same place.
______	4. Umbel.	4. A dense cluster of flowers on a receptacle.
______	5. Head.	5. An inflorescence having repeated branching, with each branch bearing a flower.
______	6. Cyme.	6. A determinate inflorescence.

21. What is the major difference between the flowers of monocotyledons and those of dicotyledons?

20

Sexual Reproduction of Seed Plants

A farmer sows grains of wheat or corn. The seeds absorb materials from the soil and air; they germinate; food is made and used in various ways; the plants develop, flower, and in due time produce seeds. All of these phases from seed germination to harvest occur with such regularity that we take them for granted and seldom consider how neatly timed and correlated they are. In previous chapters we considered the vegetative activities of a plant. Let us now center our attention on the reproductive process as it occurs in seed plants.

There are many phases in the sexual reproduction of plants, and they must occur regularly and be timed precisely if seeds are to be produced. The pollen formed in the anthers must be transferred to the stigma when it is receptive. This process is known as pollination. On the stigma a pollen grain germinates and sends a tube through the tissues of the stigma and style into an ovule, where the tube releases two sperms. One of the sperms unites with an egg which has been produced in the ovule. After the egg is fertilized it develops into a new plant. The second sperm unites with a pair of polar nuclei to form an endosperm nucleus which develops into a tissue called endosperm. After the egg and polar nuclei are fertilized, the ovule develops into a seed and the ovary into a fruit. So much for the general features; now let us consider the details.

POLLEN FORMATION

In the developing anther four **pollen sacs** are differentiated, each surrounded by a layer of cells known as the **tapetum.** Many **pollen mother cells** are present in each sac. Keep in mind that an anther is a three-dimensional structure and that pollen mother cells are found throughout the length of the pollen sacs. Each pollen mother cell undergoes **reduction division** (also called **meiosis**) to form four **pollen grains** which contain only one set of chromosomes apiece, half of the **diploid** number (called **haploid**). We shall have more to say about reduction division in Chapter 23. Let us now outline the significant features of reduction division. Two divisions are required to halve the number of chromo-

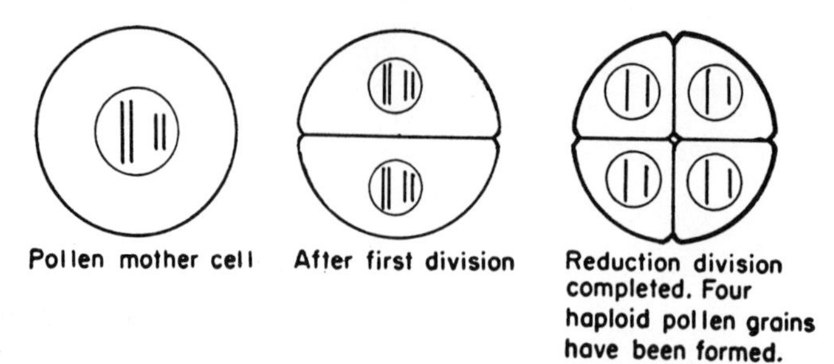

Fig. 20–1. Diagram showing the halving of the chromosome number by reduction division.

somes. Hence four haploid pollen grains result from the reduction division of each pollen mother cell. Figure 20–1 shows a pollen mother cell after the first and second divisions. You will notice that the pollen mother cell has four chromosomes, two long ones and two short. When reduction division is completed, each cell has just two chromosomes, one long chromosome and a short one; the chromosomal number has been halved. To take a specific example, the nucleus of a pollen mother cell of pea has fourteen chromosomes. The nucleus in each pollen grain has just half of fourteen, that is, seven.

Next the nucleus in each pollen grain divides by mitosis so that each grain now has two haploid nuclei. (In a pea each of the two nuclei would have seven chromosomes.) One of the nuclei is the **tube nucleus** and the other is the **generative nucleus.** The pollen is now mature (Fig. 20–2). The wall of each pollen grain consists of a thick outer wall, known as the **exine,** and a thin inner one, called the **intine.** One or more pores are usually present in the exine through which the intine bulges out to form a pollen tube when the pollen grain germinates. Pollen grains of most monocots have just one pore and are said to be **monocolpate,** whereas those of most dicots have three pores and are therefore **tricolpate** (see pollen of red oak and sagebrush in Fig. 20–3). After the pollen is mature, the tissue on each side of the anther between the two pollen sacs breaks down, and thereby the anther is opened and the pollen freed.

Pollen grains produced by one species are different from those of any other species. They vary in color, size, and shape, and they may be plain or decorated with a fascinating pattern of crests, ridges, or knobs. Species can be identified by their pollen. A physician who specializes in treatments for hay fever exposes to the air glass plates that are covered with Vaseline. From an examination of the plates he determines the kinds of pollen in the air. Knowing this, he plans appropriate treatment for his patients.

POLLINATION

The transfer of pollen from an anther to a stigma is called pollination. If the pollen is transferred to stigmas of flowers on the same plant, **self-pollination** results; if it is transferred to flowers on a different plant, **cross-pollination** is effected. Cross-pollination brings about a combination of genes from two parents, resulting in a greater variety of offspring than with self-pollination. The increased variability of offspring may result in a greater adaptability to different environments, a feature of evolutionary advantage to a species.

Self-Pollination

In such plants as garden peas, wheat, oats, tobacco, and cotton, self-pollination is the rule. In other species, self-pollination does not occur under natural conditions, but it can be effected by man. Even in self-pollinated plants cross-pollination occurs in about 1 to 3 per cent of the flowers.

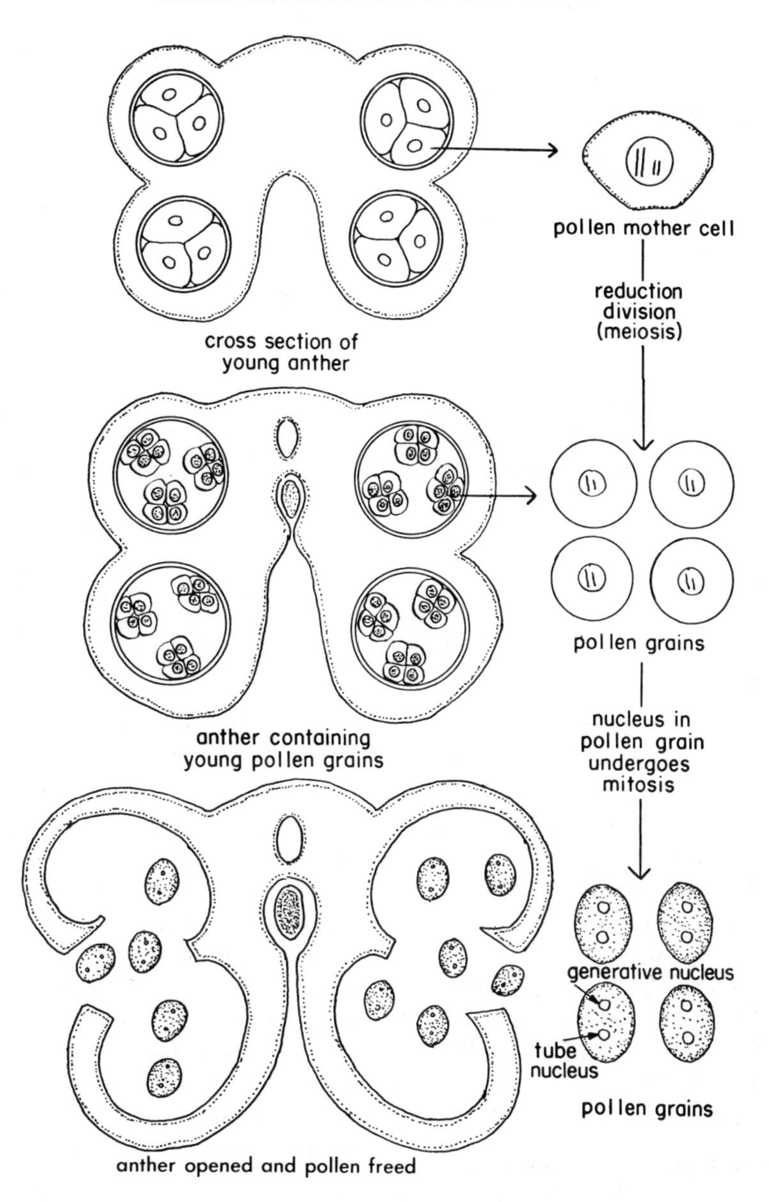

Fig. 20–2. Formation of pollen in anther.

Cross-Pollination

Many flowers have curious and interesting arrangements of floral parts which insure cross-pollination. In such dioecious plants as the oak, willow, and cottonwood, all pollination is necessarily cross-pollination. In others, cross-pollination results because the stamens in a flower mature before the stigma is receptive, or the stigma matures before the anthers. In orchids the pollen is attached to insects as they leave a flower rather than when they enter it, and the attached pollen is deposited on the stigma of the next flower visited. In rye the pollen cannot germinate on the stigma of the flower that produces it. In other species the pollen tubes which develop from pollen produced by the same plant grow more slowly than those produced by pollen from a different plant. The faster growing tubes provide the sperms that fertilize the eggs.

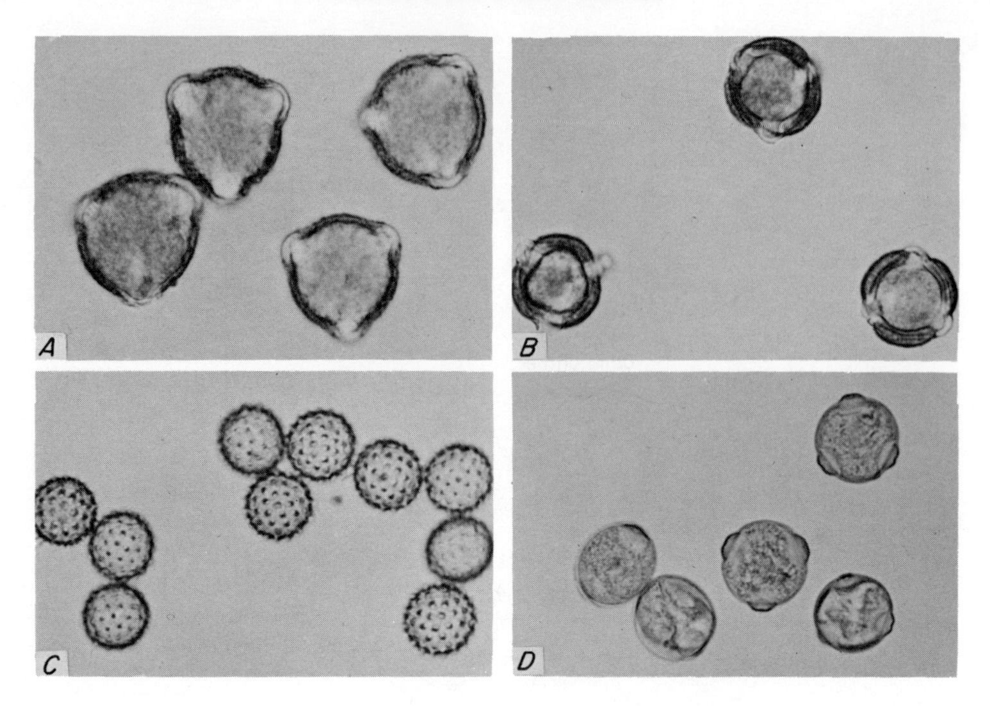

Fig. 20–3. Distinctive patterns in pollen grains, shown with photomicrographs. *A.* Red oak ×450. *B.* Sagebrush ×450. C. Short ragweed ×315. *D.* Hemp ×450. (Abbott Laboratories.)

In some species self-pollination occurs if cross-pollination has not been achieved. In *Epilobium* the style continues to grow as the flower ages and, if cross-pollination does not occur, self-pollination is accomplished when the stigma contacts neighboring anthers. In *Digitalis* the old corolla with its attached stamens is shed and as it falls from the flower the anthers brush the stigma, effecting self-pollination.

Self-incompatibility. In all sweet cherries and in many varieties of apples, plums, and pears the pollen produced in anthers of a certain horticultural variety does not function on stigmas of the same variety, even though the stigmas are in flowers of a different tree. This type of self-sterility is called self-incompatibility. Hence, in order to obtain fruit from sweet cherries and certain varieties of apples, plums, and pears, it is necessary to have more than one horticultural variety in the orchard or around the home. The varieties selected to be grown together should be known to produce mutually effective pollen. Certain varieties are intersterile. The Bing, Lambert, and Napoleon sweet cherries are not only self-sterile but they are also intersterile. Hence, mixed plantings of the three varieties will not produce fruit. Black Tartarian and Black Republican are varieties that are satisfactory as pollenizers for each of the three varieties. The fruit produced is characteristic of the female parent, regardless of the variety that furnishes the pollen.

Pollinating Agents

Wind, water, bats, birds, and insects are pollinating agents. Plants pollinated by water or wind usually bear inconspicuous, nonfragrant flowers whereas those pollinated by birds, bats, and insects bear flowers that have alluring fragrances, shapes, and colors. Certain plants flower only seasonally as for example, apples, lilacs, and azaleas. Others may flower over a long period or even throughout the year in tropical regions. Flowers of certain species periodically open and close at definite times. The flowers of *Opuntia fulgida,* a cactus of the Arizona desert, open at 3 P.M. (solar time) and the time is exact enough to set a watch by it. The flowers of the giant water-lily of the Amazon (*Victoria amazonica*) open, pure white, at 5 P.M. and close at 9 P.M. At 5 P.M. the following day they open again, beautifully tinged with red, and at 9 P.M. of this second evening they close for good.

Wind. In general, plants with inconspicuous flowers which lack petals, odor, and nectar are pollinated by wind—for example, grasses, cottonwoods, beeches, elms, birches, sycamore, walnuts, oaks, ragweeds, and many others. They produce large amounts of light, dry pollen that is carried great distances by wind. One corn plant releases about 50 million pollen grains of which only 1000 are required to pollinate all of the pistils.

The stigmas of wind-pollinated plants are typically large and feathery, providing a large area for the capture of pollen. Furthermore, in many species just one ovule is present in each ovary. Thus, for each ovule there is at least one stigma, a feature which increases the chance for pollination and subsequent fertilization. In wind-pollinated species, pollen grains arrive at the stigma singly and not in mass as in insect-pollinated plants, which bear several to a great many ovules in each ovary. An extreme case is the *Cattleya* orchid, whose ovary contains more than 100,000 ovules. In this orchid, as in other orchids, pollen grains are not deposited singly on the stigma but in masses called pollinia. Each pollinium left on the stigma by an insect may contain 100,000 or more pollen grains.

Deciduous trees and shrubs which are wind-pollinated usually blossom early, before the leaves are out to interfere with the transfer of pollen. In the vicinity of fields or forests of wind-pollinated plants, enormous numbers of pollen grains are in the air at certain seasons, as many a hay-fever victim can attest. From March to May the pollen of such wind-pollinated trees as poplars, birches, oaks, ashes, sycamores, and hickories is the major cause of hay fever in susceptible individuals. From May to July the pollen of Kentucky bluegrass, timothy, redtop, orchard grass, and Bermuda grass bring discomfort to many people. In August and September ragweed pollen is the most serious offender in many areas, but in other regions sagebrush and rabbit-brush are equally annoying.

Pollen grain walls are resistant to decay. As the wind blows the pollen about, some grains may land on lakes where they settle to the ooze at the bottom and where they may remain for thousands of years. Others may land in bogs where they will be covered with peat. We can remove cores from beds of ancient lakes and peat bogs, wash out the pollen grains from the successive layers, and identify them under the microscope. Of course, pollen grains from the bottom of the core are the oldest and those at the top the youngest. Hence we can determine the kinds of wind-pollinated plants that grew in the area in past ages. Changes in the kinds of plants in an area may be the result of climatic changes or of ancient man's activities. The pollen record enables us to determine past climates, at least to some extent, and to study past cultures. For example, in the old layers of peat from bogs in Denmark oak pollen is prevalent. But in more recent peat, the oak pollen lessens or disappears. At the same time cereal pollens and weed pollens become abundant. The meaning seems clear. Man came into the area, burned the oak forests, and planted cereals. In the open areas weeds as well as cereals grew. Such evidence indicates that agriculture reached northwestern Europe about 3000 B.C., some 4000 years after it was first practiced in Iraq and Iran.

Water. Only a small proportion of aquatic plants actually rely on water to affect pollination. Most aquatic plants open their flowers above, not below, the water surface and many depend on insects or wind for pollination. Water lilies, bladderworts, and water buttercups are insect-pollinated, whereas many pondweeds and the water milfoil are pollinated by the wind.

Some aquatic plants that rely on water to transport pollen open their flowers below the water surface and others above the surface. Pollen is released from the anthers of *Ceratophyllum* and *Najas* below the surface where pollination occurs. In other plants, pollination occurs at the surface. Some species have remarkable adaptations to facilitate pollination, among them *Vallisneria spiralis,* a dioecious plant. The staminate flowers of this species resemble miniature boats. They develop under water, but as they approach maturity they are separated from the plant and rise to the surface where they open and expose the pollen grains, which are large and sticky. The pistillate flowers are produced singly at the ends of coiled stalks. When a flower is mature, the stalk uncoils and the flower reaches the surface where it opens and displays three large spreading stigmatic lobes. A floating male flower, driven about by wind or currents, may bump into a female flower. The exposed pollen is then left on the stigma. After pollination the stalk that supports the female flower becomes coiled, thus carrying the pollinated flower toward the bottom where the seeds develop.

Birds. In temperate regions only a few species—for example, hollyhocks, paintbrushes, and columbines—are pollinated by birds, especially hummingbirds (Fig. 20–4). But in tropical and subtropical regions many species are pollinated by birds—sunbirds, tanagers, and parrots as well as the hummingbird. In Brazil, where ninety-three species of hummingbirds live, 18 per cent of the genera have representatives which are bird-pollinated, among them most species of bromeliads and gesneriads. Flowers pollinated by birds are often strikingly colored with contrasting color combinations, such as blue and red, or red and green. Birds find red especially attractive and it is rare indeed to see a red-flowered tree in the tropics which is not surrounded by a flock of birds. The flowers of bird-pollinated species are typically strong and of heavy texture; they secrete copious amounts of nectar which is sought after by birds for both food and water. The birds usually have beaks which are long and thin and their tongues are capable of being projected and of sucking liquid.

Insects. In many species insects transfer pollen from anthers to stigmas. Flowers have many devices, such as conspicuous petals, fragrance, and nectar which attract insects, and sticky pollen and stigmas which favor the collection of pollen and its deposition on stigmas. Bees and moths seem to be attracted by odors which we consider pleasant. On the other hand, flowers specialized for pollination by flies often smell like rotting meat, an odor we would consider unpleasant. Flowers pollinated by moths

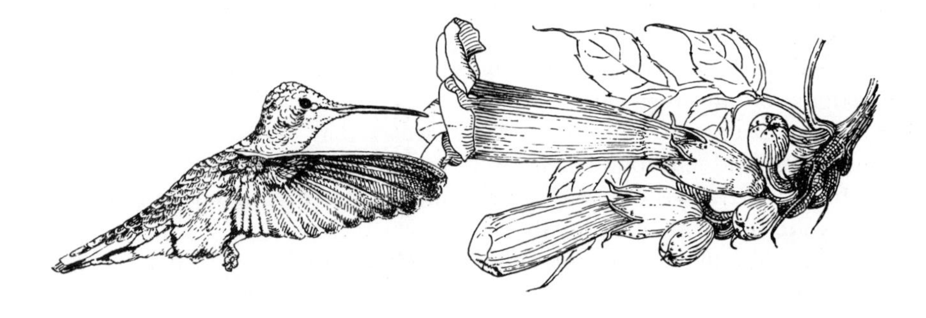

Fig. 20–4. Hummingbird hovering in front of a trumpet-vine flower. (B. J. D. Meeuse, *The Story of Pollination.* Copyright © 1961, The Ronald Press Company, New York.)

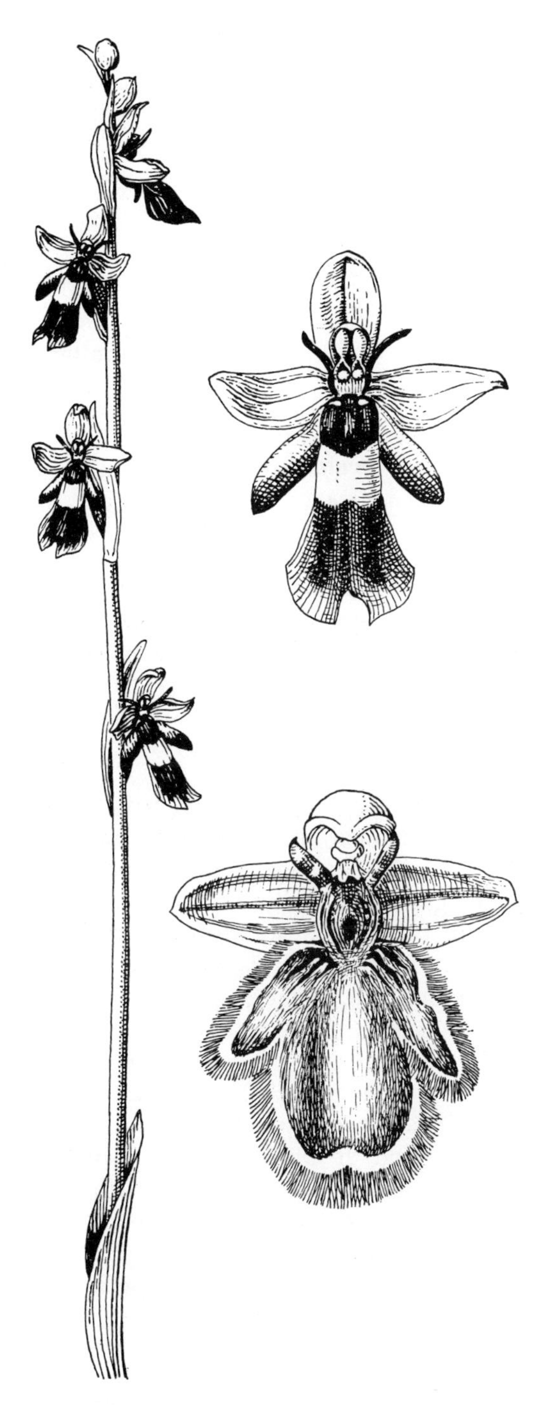

Fig. 20–5. Male wasps and bees mistake the flowers of *Ophrys*, an orchid, for a female insect. *Left and upper right, Ophrys muscifera. Lower right, Ophrys speculum.* (B. J. D. Meeuse, *The Story of Pollination.* Copyright © 1961, The Ronald Press Company, New York.)

generally have large pale flowers and are often scented more strongly at night.

The orchid *Cryptostylis* has a remarkable adaptation for pollination. The form and odor of the flowers resembles that of a female ichneumon fly. The male fly is deceived and goes from flower to flower performing mating movements and incidentally transferring pollen from flower to flower. Similarly in certain species of *Ophrys* (Fig. 20–5), an orchid, male bees and wasps are deceived. Other orchid flowers which resemble insects are attacked by insects which mistakenly strive to drive them away to protect their home range.

Some flowers are so constructed that any of a number of species of insects can effect pollination. On the other hand, some flowers are formed so that only a single kind of insect can bring about pollination. Flowers of red clover depend almost entirely on bumblebees. If they are absent, seed is not produced. The yucca plant relies upon the pronuba moth for pollination. When a yucca flower opens, the female moth deposits her eggs in the ovary and stuffs a small ball of pollen into the hollow style. While the ovules are developing into seeds, the moth larvae hatch from the eggs and feed on the developing seeds. Some of the seeds escape being eaten and survive to perpetuate the race of yuccas. If the moth were not present, the yucca could not produce seed; if the yucca were absent, the moth could not reproduce. Hence neither plant nor insect could survive without the other. The Smyrna fig is dependent on a small wasp (*Blastophaga*), and if this wasp is lacking, no figs are produced.

Bees, important pollinating agents, visit flowers to collect both nectar and pollen. From nectar obtained from most plants, honey palatable to man is produced. However, honey made from nectar gathered by bees from certain species may be poisonous to man. Primitive tribes of Paraguay are acquainted with honey which produces narcotic or lethal effects when the dose is 1 teaspoon. At any given time bees generally visit flowers of only one species and thus hybridization between species is not brought about. Flowers of various colors are pollinated by bees which distinguish between the following four color groups: (1) yellow, including orange and yellow-green; (2) blue-green; (3) blue, purple, and violet; and (4) ultraviolet. Their color vision is unlike ours in certain ways. We recognize red, but this color appears as black to bees. We do not perceive ultraviolet whereas bees do. Bees cannot distinguish between the colors in a group; for example, they cannot distinguish yellow from orange or yellow-green. Growers of alfalfa, sweet clover, apples, cherries, raspberries, cantaloupes, radishes, and other crops frequently obtain a better set of fruit and seed by introducing hives of honeybees into their orchards and fields. In the best-managed orchards pollination is as carefully worked out as are pruning, cultivating, fertilizing, and other practices. To spray an apple orchard with DDT, arsenicals, or other insecticides poisonous to bees when the trees are in flower is just asking for a crop failure. Moths, wasps, butterflies, and flies are other insects that effect pollination in certain plants.

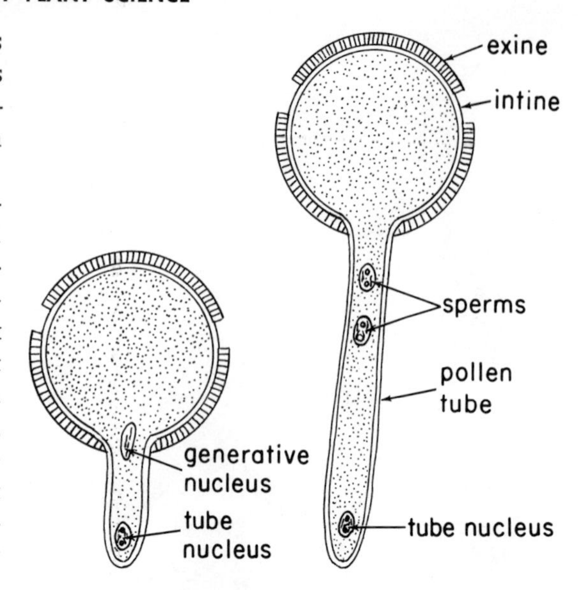

Fig. 20–6. Germination of a pollen grain. *Left*, an early stage, showing the intine bulging outward through a pore to form the pollen tube. *Right*, a later stage, showing the two sperms formed by the division of the generative nucleus.

DEVELOPMENT OF SPERMS IN THE POLLEN TUBE

After the pollen has been transferred to the stigma, the subsequent events are much the same in all flowering plants. Each pollen grain germinates on the stigma to produce a pollen tube which grows through the stigma and style into the ovary, finally reaching an ovule (potential seed). During the growth of the tube, the generative nucleus divides to produce two sperms (Fig. 20–6).

The pollen tube now contains three nuclei, the tube nucleus, and two sperms. The tube nucleus is near the tip and apparently directs the tube. The two sperms are located behind the tube nucleus. One sperm will unite with an egg which is formed in an embryo sac that develops in an ovule. The other sperm unites with a pair of polar nuclei contained in the embryo sac. The time elapsing between germination of the pollen grain and fertilization is usually short, but it may be several days, weeks, or months. In barley it is about 1 hour; in wheat and corn, about 24 hours; in cabbage, 5 days; and in oaks, nearly 1 year.

Within a species the rate of pollen-tube growth is influenced by temperature and the physiological compatibility between the pollen tube and the tissues of the pistil. The optimum temperature for the growth of tomato pollen tubes is 20° C. In many cross-pollinated plants, pollen tubes grow very slowly when self-pollination occurs, so slowly that fertilization seldom results. When cross-pollination occurs in the same species, the tubes grow rapidly.

Pollen grains of most species will germinate if strewn on the surface of a sucrose-water or a sucrose-agar medium. The most favorable sucrose concentration varies among species, but is usually between 5 and 25 per cent.

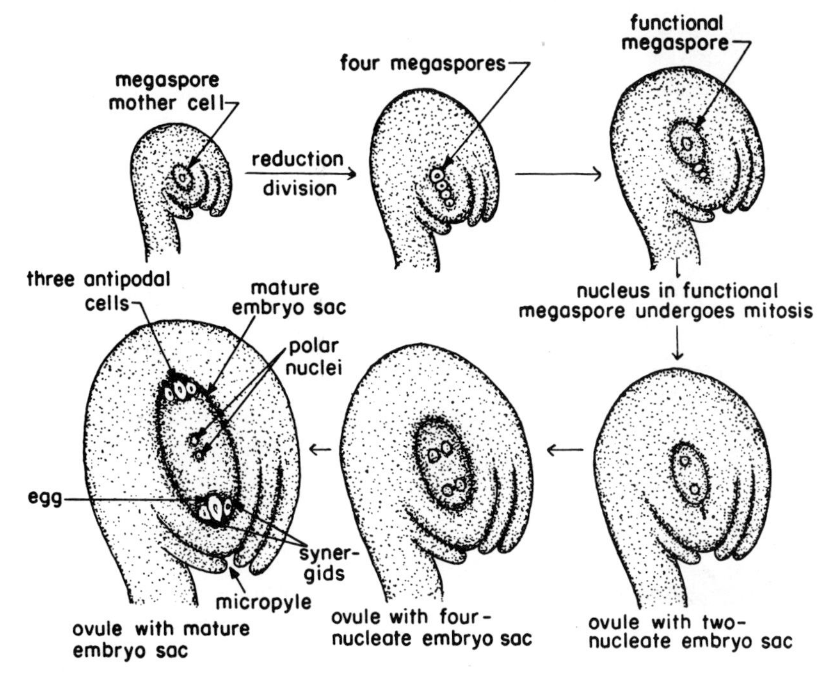

Fig. 20–7. Stages in the development of an embryo sac in an ovule.

DEVELOPMENT OF AN EMBRYO SAC IN AN OVULE

The stages in embryo sac development are illustrated diagrammatically in Figure 20–7. In each developing ovule one diploid **megaspore mother cell** is differentiated, which undergoes reduction division, forming a row of four haploid **megaspores.** Three megaspores abort and the fourth one enlarges considerably.

The nucleus in the large megaspore divides by mitosis and the resulting nuclei move to opposite ends of the megaspore, which is now the developing **embryo sac.** The two nuclei divide mitotically and then the four nuclei divide, resulting in an embryo sac which contains eight nuclei. The nuclei in the embryo sac become arranged in a characteristic manner and walls form around all except the two in the center. Three cells are situated at the micropylar end, three at the opposite end, and two nuclei in the center. Of the three cells at the micropylar end, the middle one is the **egg** and the other two are functionless **synergids.** The three at the opposite end are known as **antipodals.** So far as is known they have no function. The two in the center are the **polar nuclei.** The ovule at this stage consists of two well-developed **integuments,** a small opening (**micropyle**), a **nucellus,** and the **mature embryo sac** (Fig. 20–8).

FERTILIZATION

The pollen tube enters the ovule through the micropyle, then penetrates through the nucellus into the embryo sac, where the tube bursts open and liberates the two sperms. One of the sperms unites with the egg, a process known as fertilization. The fertilized egg has two sets of chromosomes, one set from the sperm and the other from the egg. The second sperm unites with the two polar nuclei. The result is an **endosperm nucleus** with three sets of chromosomes: one set from the sperm, one from one polar nucleus, and the third set from the other polar nucleus. Cells such as the fertilized egg which have two sets of chromosomes are

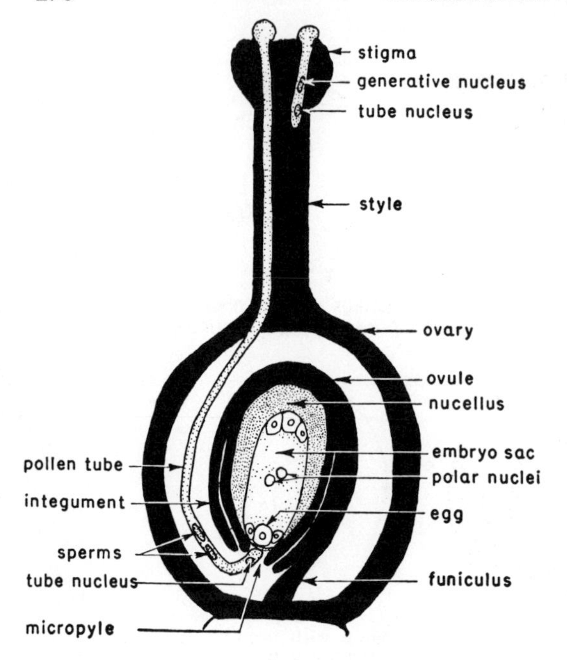

Fig. 20–8. A longitudinal section through a pistil.

said to be diploid, whereas those with just one set, the sperm and egg, are haploid. The endosperm nucleus with three sets of chromosomes is triploid.

One pollen tube contains only two sperms, just enough to unite with one egg and a pair of polar nuclei. If ten seeds mature in a pea pod, at least ten pollen tubes must have reached ten different ovules. If a million seeds develop within an ovary, as happens in some of the tropical orchids, a million pollen tubes must have hit their mark.

EVENTS WHICH FOLLOW FERTILIZATION

Fertilization initiates a number of events, some of which follow:

1. The fertilized egg develops into a new plant which, when still contained within the seed, is known as an **embryo.**
2. The endosperm nucleus develops into a food storage tissue, which is the **endosperm.**
3. The integuments become altered and form the seed coats.
4. Each ovule develops into a seed.
5. The ovary with the contained ovules develops into the fruit.
6. The petals shrivel and fall.

QUESTIONS

1. Diagram the stages in the formation of pollen.

2. _____ Leaf cells of corn have 20 chromosomes. The sperm will have (A) 20 chromosomes, (B) 10 chromosomes, (C) 5 chromosomes.

3. _____ If a potato breeder desires to obtain a new variety of potatoes by selection from among a large number of plants, would you advise him to (A) plant pieces of tubers to provide plants from which to select, (B) plant real seed to provide plants from which to select.

4. _____ Pollination is (A) growth of the pollen tube, (B) the transfer of pollen from the anther to the stigma, (C) the union of a sperm with an egg.

5. _____ With respect to pollination, which of the following is false? (A) Bees distinguish colors, but red appears as black to them. (B) In some orchids, the flowers so resemble female insects in form and odor that male insects are deceived and perform pseudocopulation. (C) Practically all hydrophytes rely on water to transfer the pollen from anthers to stigmas. (D) Trees which are wind-pollinated generally flower early in the spring before the leaf buds develop.

6. _____ Fertilization (A) is achieved when the pollen lands on the stigma, (B) occurs in the embryo sac when one sperm unites with the egg and the other with the polar nuclei, (C) is certain to occur if pollen is deposited on the stigma, (D) is the primary stimulus inducing the ovule to develop into an embryo.

7. List some characteristics of wind-pollinated and of insect-pollinated flowers.

8. Diagram the stages in the development of an embryo sac.

9. What are the biological advantages of cross pollination.

10. _____ From May to July most cases of hay fever are caused by pollen from (A) wind-pollinated trees, (B) grasses, (C) ragweed and sagebrush, (D) insect-pollinated species.

11. _____ The embryo sac (A) develops directly from a megaspore mother cell, (B) contains an egg, two synergids, two polar nuclei, and three antipodal cells, (C) contains diploid nuclei, (D) is embedded in the ovary.

12. _____ A pollen tube (A) contains, at maturity, two diploid sperms and a tube nucleus, (B) develops directly from a pollen mother cell, (C) contains two haploid sperms and a haploid tube nucleus, (D) when mature contains four haploid sperms and one haploid tube nucleus.

13. Indicate whether the following are haploid (H), diploid (D), or triploid (T):

_____ 1. Palisade cell.
_____ 2. Cambium cell.
_____ 3. Sperm.
_____ 4. Megaspore mother cell.
_____ 5. Pollen mother cell.
_____ 6. Endosperm cell.
_____ 7. Egg.
_____ 8. Pollen grain.

14. _____ Which of the following should be least concerned about being certain that pollination occurs? (A) The wheat farmer, (B) the commercial rose grower, (C) the person who owns an apple orchard, (D) the person who owns a grape vineyard.

15. _____ Which one of the following is false? (A) The transfer of pollen from an anther to a stigma is called pollination. (B) If pollen is transferred to stigmas of flowers on the same plant, self-pollination results. (C) If Bing cherry trees only are in an orchard an excellent crop can be expected.

16. _____ In flowering plants the sperms are carried to the eggs by (A) wind, (B) pollen tubes, (C) water, (D) birds, (E) insects.

17. _____ In each ovule there (A) is only one egg, (B) are several eggs.

18. _____ In an orchid pod there are about 100,000 seeds. Therefore, there must have been (A) at least 100,000 pollen tubes growing through the style, (B) fewer than 100,000 pollen tubes growing through the style, (C) only one pollen tube, (D) no pollen tubes growing through the style because the tubes grow through the filaments.

19. _____ If a sperm of pea contains seven chromosomes, the fertilized egg will contain (A) seven chromosomes, (B) fourteen chromosomes, (C) fourteen pairs of chromosomes, (D) twenty-one chromosomes, (E) three and one-half chromosomes.

20. _____ Which one of the following does not occur after the egg has become fertilized? (A) The fertilized egg develops into the embryo, (B) the integuments develop into fruit coats, (C) the ovule develops into a seed, (D) the fertilized polar nuclei develop into the endosperm.

21

Seeds, Seedlings, and Fruits

DEVELOPMENT OF A SEED

After fertilization an ovule develops into a seed. During seed formation a number of processes take place concurrently. The fertilized egg (**zygote**) develops into an embryo. The genes within the diploid nucleus control development. In addition to genes, the fertilized egg contains the essential metabolic equipment—mitochondria for respiration, membranes, endoplasmic reticulum, ribosomes, and proplastids, which ultimately will develop into plastids. The endosperm nucleus develops into the endosperm, and the integuments become the seed coats.

Development of the Embryo

Soon after fertilization the zygote divides. This division is followed by others, and soon there is formed a spherical young embryo (Fig. 21–1) supported by a filament of cells, known as the **suspensor,** which by its growth pushes the developing embryo into the nutritive endosperm tissues. The cells of the embryo continue to divide and differentiate, leading ultimately to a completely formed embryo. At maturity the embryo consists of a radicle (root), hypocotyl, epicotyl, and one or two cotyledons (Fig. 21–2). In a corn or other monocot seed only one cotyledon is present, whereas in *Capsella,* phlox, and other dicots, two cotyledons develop.

Development of the Endosperm

While the embryo is forming, the endosperm, a food-storage tissue, develops. The endosperm nucleus divides and the resulting nuclei continue to divide. At first, the nuclei are not separated by cell walls, but cell walls form later on. As the embryo and the endosperm develop, the embryo sac enlarges at the expense of the nucellus. The developing endosperm is a complete medium for the growth of the embryo, providing the embryo with all nutrients. The foods used by the young embryo are absorbed from the endosperm. In seeds of most species the nucellus is no longer evident at maturity. However, in a few species—for example, beet, spinach, and

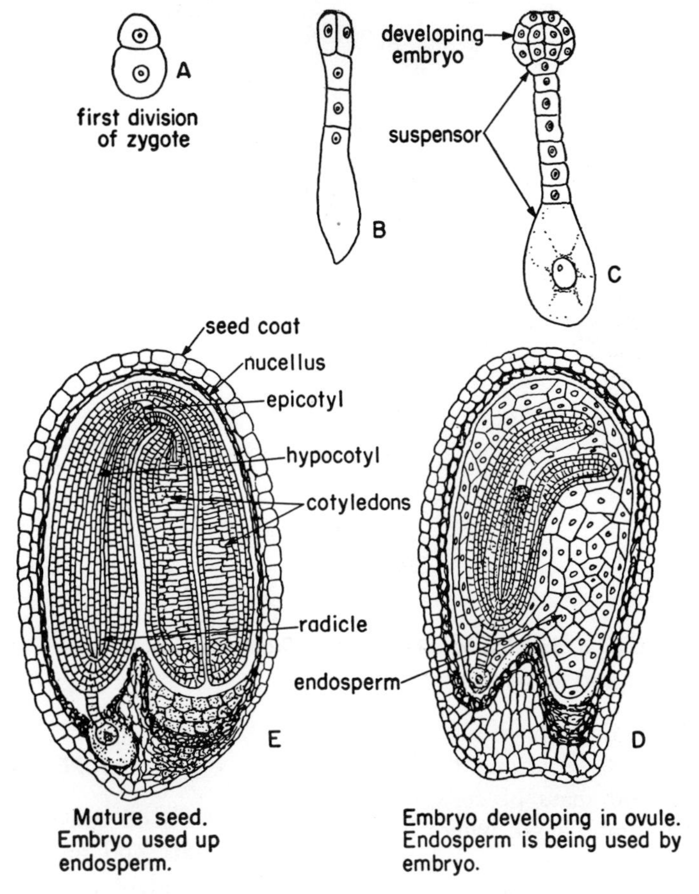

Fig. 21–1. Embryo development in *Capsella*.

coffee—a large amount of nucellar tissue is present in the ripe seed. When the seed is mature, the nucellus is referred to as the **perisperm,** and in a beet seed considerable food is stored in this tissue.

In plants such as peas, beans, and *Capsella* the endosperm, or food storage tissue, is completely utilized by the developing embryo, especially by the cotyledons. In these seeds, food is stored in the cotyledons. In other species—phlox, tobacco, corn, and castor bean, for example—there is still a conspicuous amount of endosperm remaining in the mature seeds. In these the food stored in the endosperm is utilized by the embryo when the seed germinates. Hence with respect to where food is stored we can recognize two major kinds of seeds: those having an endosperm when the seeds are mature and those lacking this tissue in ripe seeds.

APOMIXIS

In most species, the embryo develops from a fertilized egg, but there are some exceptions to this usual process. In certain bluegrasses, hawthorns, hawkweeds, blackberries, dandelions, and many other species, the embryo in a seed does not develop from a fertilized egg, but instead from a cell which has not united with a sperm. Because no sperm is involved, the offspring exhibit only maternal characteristics. The development of an embryo from some cell other than a fertilized egg is called apomixis.

In some apomictic species the embryo develops from a diploid nucellar cell. A nucel-

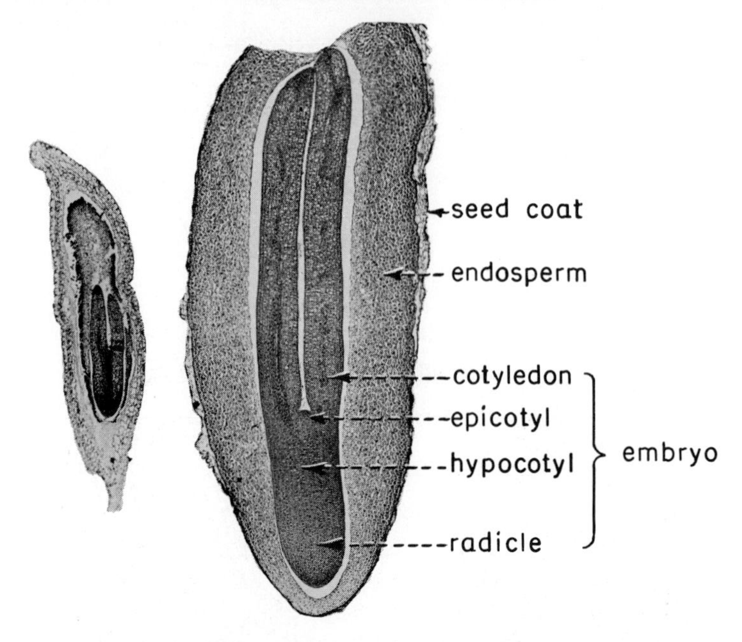

Fig. 21–2. *Left,* a developing phlox seed. *Right,* an almost mature seed of phlox. (Helena A. Miller and Ralph H. Wetmore.)

lar cell adjacent to an embryo sac divides and the resulting cells continue to divide forming an organized mass of cells which protrudes into the embryo sac where it develops into an embryo.

In other species the embryo develops from an unfertilized diploid egg, an egg with the unreduced number of chromosomes. A diploid egg may be produced in several ways. In some species the megaspore mother cell does not undergo reduction division, but instead develops directly into an embryo sac containing eight diploid nuclei. In other instances meiosis occurs in the usual manner, but all four megaspores abort; a nucellar cell then takes over and develops into an embryo sac. In certain plants meiosis is abnormal and the resulting megaspores are diploid instead of haploid. The nuclei, including the egg nucleus, in an embryo sac that develops from a diploid megaspore will be diploid.

Some species reproduce only apomictally and the offspring of any one plant will be exactly like the parent. In different species, some embryos develop in the normal manner, whereas others develop apomictally.

MATURE SEEDS

A mature seed consists of at least a seed coat or coats and an embryo. The **hilum** is usually evident on the outside of a seed coat. This is the scar which remains after the seed separates from the **funiculus,** the short stalk that connects the seed to the placenta. A **micropyle** and **raphe** are also evident on the seed coat. The raphe is a ridge on the seed coat formed by the fusion of the funiculus with the seed coat. It is present on seeds of some species but lacking on those of other species.

Bean Seed

A bean seed is made up of two major parts: the embryo and the seed coat (**testa**). A hilum, raphe, and micropyle are evident on the seed coat. The embryo, which occupies all the space inside the seed coat, consists of a radicle, an epicotyl, a hypocotyl, and two large cotyledons which are attached to the top of the hypocotyl. Food is stored in the cotyledons. (A bean seed is illustrated in Figure 3–2).

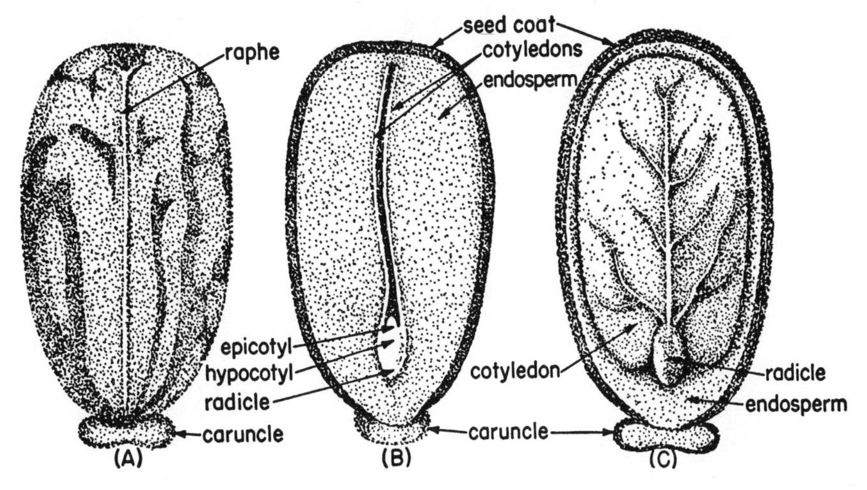

Fig. 21–3. A castor bean seed. *A*. External view. *B*. Section cut at right angles to the broad surface of the cotyledons. C. Section cut parallel to the broad surface of the cotyledons. (Redrawn by permission from *Textbook of General Botany* by R. M. Holman and W. W. Robbins. Copyright, 1934, John Wiley & Sons, Inc., New York.)

Castor Bean

A castor bean seed (Fig. 21–3) consists of three parts: (1) the hard, mottled seed coat, at one end of which there is a fleshy outgrowth, the **caruncle,** which covers the micropyle; (2) the white endosperm in which food is stored; and (3) the embryo. The embryo consists of a radicle, a short hypocotyl, a minute epicotyl, and two broad but thin cotyledons which resemble ordinary leaves.

Corn Grain

A corn grain is really a fruit because the seed is still enclosed in the ripened ovary wall. The ripened ovary wall is called the **pericarp.** The covering of a corn grain consists of the pericarp fused with the seed coat. A section through a corn grain (Fig. 21–4) reveals the embryo, the endosperm, and on the outside the grain coat. The endosperm occupies the greater volume, and it is here that most food is stored. The embryo has a radicle, hypocotyl, epicotyl, cotyledon, and structures sheathing the radicle and epicotyl. The radicle is enclosed by a sheath, the **coleorhiza,** and the epicotyl by a different sheath, known as the **coleoptile.** The epicotyl is well developed, having several small leaves and a stem tip. Some food is stored in the single, large cotyledon which is attached to one side of the axis. When the seed germinates, the cotyledon, also known as the **scutellum,** secretes enzymes which diffuse into the endosperm and bring about digestion. In the milling of corn, wheat, and other grains, the embryo is first separated from the endosperm. The endosperm is then made into flour. The embryo itself is used in feeds for domesticated animals.

GERMINATION OF SEEDS

The depth at which seeds should be planted is somewhat related to their size. Small seeds should be planted close to the surface because they contain little stored food, and if the seedling is to survive it must soon get through to the light where it can start making food. Large seeds have an abundance of stored food and may be planted deeper. A good rule to follow if in doubt is to plant the seeds at a depth of four times the diameter of the seed.

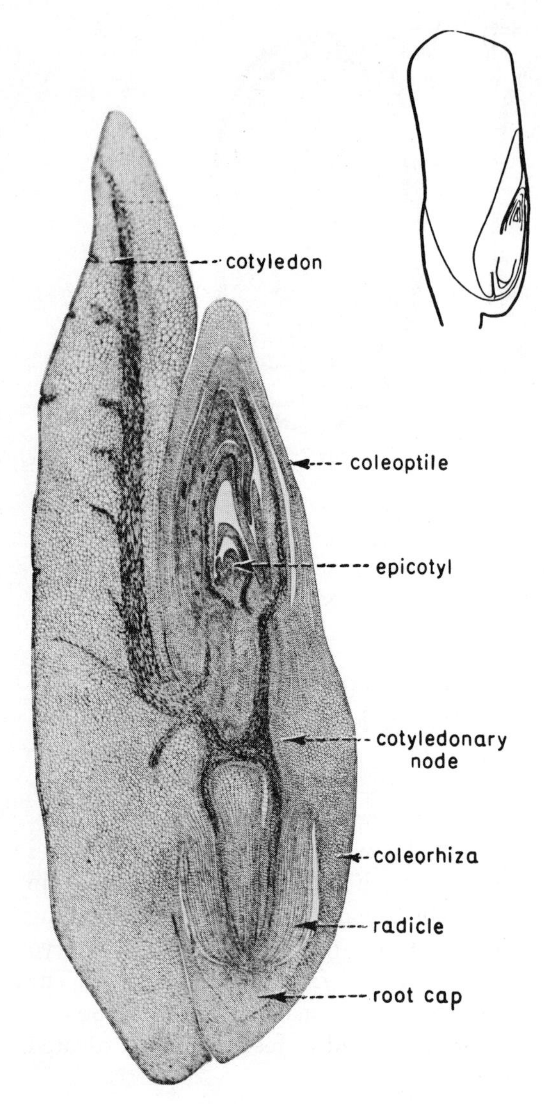

Fig. 21–4. A grain of corn. *Upper right*, embryo, endosperm, and grain coat. *Left*, section through the embryo of a corn grain. (George S. Avery, Jr., *Botan. Gaz.*)

Requirements for Seed Germination

Knowingly or unknowingly the successful gardener or farmer furnishes the seeds with the three general requirements for germination, which are water, oxygen, and a proper temperature. The temperature for quickest germination varies with the species. Seeds of snapdragon, delphinium, iris, and Iceland poppy germinate fastest at 54° F., whereas those of China aster, wheat, begonia, and carnation germinate most rapidly at 68° F. The lower limit for the germination of wheat is about 38° F. and the upper about 95° F. In the deserts of southwestern United States there are distinct summer and winter floras. Seeds of winter-flowering annuals germinate only at relatively low temperatures, and those of summer-flowering annuals are restricted to germination at high temperatures. Hence temperature determines which seeds on the desert floor will germinate at a particular season.

Most seeds will germinate as well in darkness as in light. However, certain ones, those of the strangling fig (*Ficus aureus*), mistletoe, saguaro cactus, peppergrass, and certain varieties of lettuce, require light for germination. Such seeds respond to light only after they have imbibed water. If dry seeds are exposed to light and planted in the dark they do not germinate, whereas if moist seeds are given the same treatment they germinate. In recent years we have learned that red light promotes germination while far-red prevents germination. For example, when moist peppergrass (*Lepidium*) seeds are exposed to red light before being planted in the dark, they germinate (Table 21–1). But if they are exposed to far-red, they do not germinate. When both red and far-red are present at the same time, as they are in white light, the effect of the red light predominates. What happens to light-sensi-

Table 21–1

Effect of Red and Far-Red Light on the Germination of Seed of *Lepidium virginicum*

Number of Irradiations		Final Radiation	Per Cent Germination
Red	Far-red		
1	0	Red	45
1	1	Far-red	0
8	7	Red	48
8	8	Far-red	0
0	0	–	0

tive seeds if you switch reds on them, if you expose them first to red and then to far-red? They do not germinate. You can expose them eight or more times to red and far-red alternately and they still respond according to the radiation they last received. If it was red, they germinate; if it was far-red, they do not.

Seeds of some species have "clocks" which measure day length. Seeds of *Veronica persica* germinate only when days are short, while those of *Eragrostis ferruginea* and *Begonia evansiana* germinate only when days are relatively long. No seeds of *Begonia evansiana* germinate when days are 6 or 9 hours long, but 75 per cent germinate when they are 12 or more hours long.

In contrast to the seeds which require light for germination are those whose germination is retarded or prevented by light. Seeds of onion and *Phacelia* germinate better in the dark than in light.

If seeds are deeply buried, insufficient oxygen prevents germination. For example, weed seeds may be plowed under to such a depth that they do not germinate. If the ground is plowed the following year, some of these seeds are brought closer to the surface where oxygen is ample for their preliminary growth. Seeds germinate when soil moisture is in the range from field capacity down to about the wilting percentage. If the soil moisture is much above field capacity, it is said to be waterlogged. Seeds of most species do not germinate in waterlogged soil or when soaked in water because of insufficient oxygen for respiration. Seeds of hydrophytes and certain other plants will germinate where the oxygen concentration is low. For example, the oxygen requirement for the germination of rice seeds is only one-fifth of that required for wheat.

The Germination Process

When conditions are favorable for germination, water is absorbed and the embryo and endosperm swell, thus rupturing the seed coat. The primary root, which develops from the radicle, then emerges and grows downward. Lateral roots soon appear and root hairs develop. In the germination of some seeds—bean, tobacco, and castor bean, for example—the hypocotyl increases in length and elevates the cotyledons and epicotyl above the soil surface (Fig. 21–5). Later the epicotyl develops into a stem bearing leaves. In other seeds—corn and peas, for instance—the hypocotyl does not elongate and hence the cotyledon or cotyledons remain in the soil (Fig. 21–6). In such plants, the shoot that first appears above the soil develops from the epicotyl.

In the germination of a corn grain (Fig. 21–7) the primary root bursts through the coleorhiza and grain coat and penetrates the soil. Soon thereafter the coleoptile emerges from the seed and later appears above ground. The coleoptile soon stops growing and then the true leaves emerge through a split in the coleoptile. In the meantime adventitious roots develop from the base of the epicotyl. The single cotyledon remains within the seed where it absorbs food from the endosperm and then translocates it to the growing seedling.

Physiological activities are initiated or accelerated during seed germination. Respiration becomes strikingly accelerated. Digestive enzymes become active and change insoluble foods into soluble ones. The sugar and amino acids which result when starch and proteins are digested are moved from storage sites to growing points where assimilation is occurring. Storage fats are changed into fatty acids and glycerol, and these, in turn, into sugars which are then translocated to the growing embryo. A part of the mobilized food is respired, but another portion is used to form additional tissues.

DORMANCY OF SEEDS

A seed contains an embryonic plant in a resting condition, and germination is its resumption of growth. Seeds of most culti-

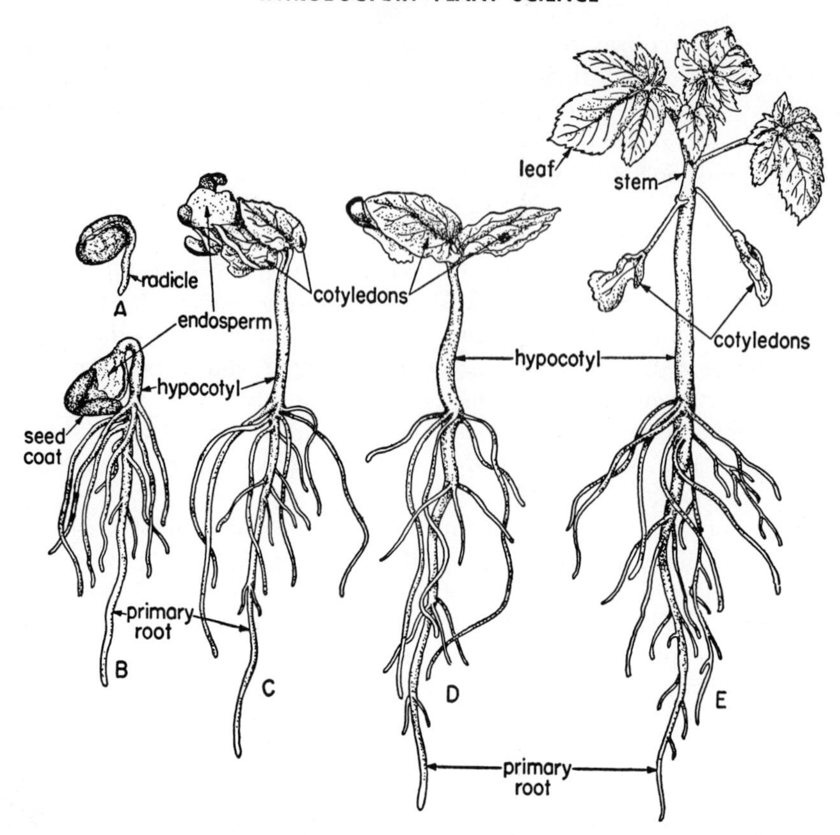

Fig. 21–5. Stages in the germination of a castor bean seed. (Redrawn by permission from *Textbook of General Botany* by R. M. Holman and W. W. Robbins. Copyright, 1934, John Wiley & Sons, Inc., New York.)

vated plants germinate promptly when the temperature is in the appropriate range and when water and oxygen are available. Through the years man has purposely selected seeds which germinate promptly. It is otherwise with seeds of most noncultivated plants. Almost universally these seeds have barriers which prevent prompt germination. Seeds which do not germinate promptly when the usual conditions for germination are fulfilled are said to be dormant. Dormancies of seeds are biologically significant in spreading or delaying germination until the environment is more favorable for the continued development of the seedlings. For example, seeds of pine, hemlock, apple, linden, mountain ash, and others will not germinate in late summer or autumn, even though environmental conditions are favorable. Before they will germinate they must be exposed to low temperatures, such as occur when the seeds are resting on the ground during winter. If the seeds germinated in the autumn, the seedlings might be killed during winter. Seedlings from seeds that germinate in the spring have a better chance for survival. When water and oxygen are available and when the temperature is favorable, seeds of most crop plants germinate promptly. If conditions continue to be favorable, all is well; but if drought or other adverse environmental factors set in, the whole stand may be killed. On the other hand, the germination of seeds of many wild species is spread over a long period. When the environment is favorable, some seeds germinate, but others remain dormant and germinate later. This sporadic germination enhances the chances for survival.

Fig. 21–6. Stages in the germination of a pea seed. *Left*, the epicotyl is just emerging from the seed. *Right*, the shoot which appears above the soil develops from the epicotyl.

The obstructions to prompt germination are eliminated under natural conditions by time, temperature changes, leaching, decay of surrounding fruit tissue, softening of seed coats, and in other ways. Let us next inquire into the nature of these blocks to germination, to these causes of dormancy.

Dormancy Due to Seed Coats

In a freshly harvested sample of clover, alfalfa, and morning glory seeds, some seeds have coats permeable to water and others have seed coats impermeable to water. Those with seed coats permeable to water germinate promptly when the environment is favorable. Germination of seeds with impermeable coats is delayed until they become permeable.

Seeds of a number of plants—among them, species of *Ceanothus*—germinate best after a fire has swept through the area. The heat changes the seed coats from the impermeable condition to a permeable one. In the laboratory, the seed coats may be made permeable to water by boiling the seeds for 5 to 20 seconds. The tolerance of seeds of *Ceanothus cordulatus* to heat is amazing; seeds boiled for 25 minutes remained viable.

The seed coats of mustard, pigweed, water plantain, shepherd's purse, and others are so strong that the developing embryos cannot break them. The fruit coat of the cocklebur is impermeable to oxygen.

If germination is retarded or prevented by seed coats impermeable to water or oxygen or mechanically resistant coats, germination can be enhanced by scratching or nicking the seed coats, a technique known as scarification. Under natural conditions, soil microorganisms alter the seed coats and make them permeable to oxygen and water, and weaken them so that the expanding embryos can break through.

Immature Embryos

In several plants—European ash, orchids, holly, and others—the embryos are not fully developed when the seeds are shed. Such

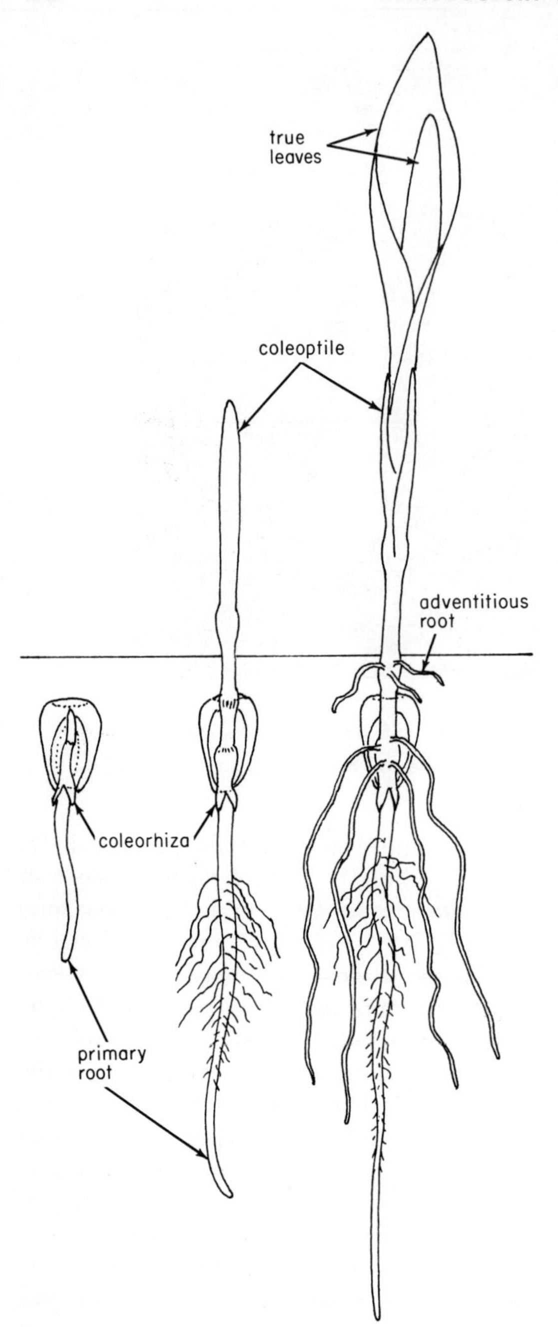

Fig. 21–7. Corn seedlings of three ages, showing stages of germination and early growth.

seeds germinate slowly because the embryos must mature before germination can occur. Orchid seeds are as fine as powder and just one capsule may bear one million seeds. A seed when shed consists of a seed coat and a small globular embryo which is undifferentiated, lacking a radicle, cotyledon, and epicotyl. Little food is stored in the embryo and in most orchids a supplementary food source is necessary. In nature the sugar necessary for germination comes from the decaying organic matter which surrounds the seed. Certain fungi bring about this decay and in their absence the seeds do not germinate. In the commercial culture of orchids, seeds are sown on an agar medium containing water, essential minerals, and sugar, generally sucrose (Fig. 21–8).

Dormant Embryos

Many species—among them apple, peach, iris, lily-of-the-valley, dogwood, hemlock, mountain ash, and pine—have seeds with well-developed embryos but which require a period of afterripening. During the afterripening period a series of changes occurs in the embryo which converts the dormant embryo into an active one. Under natural conditions afterripening occurs during the winter when the seeds are on or in the ground. Seeds of cultivated plants may be afterripened by storing them in moist peat at temperatures between 1° C. and 10° C. for 2 or 3 months. After such treatment, the seeds germinate promptly and produce vigorous seedlings. Without such treatment, either germination does not occur or it is erratic and frequently the seedlings are weak (Fig. 21–9).

Water-Soluble Inhibitors

Chemicals which inhibit germination are present in the seed coats of many desert species. Before these seeds will germinate, the inhibitors must be leached out. Light rains do not leach out enough inhibitor to permit germination whereas heavy rains do. The heavy rains also furnish adequate water for subsequent plant growth. Hence, these inhibitors have survival value for desert species, enabling the seeds to discriminate between a light and heavy rain.

Fig. 21–8. Orchid seedlings developing on an agar medium containing essential minerals, water, and sucrose. (Rod McLellan Co.)

LONGEVITY OF SEEDS

Seeds of some species are short-lived, such as the silver maple, whose seeds remain viable for only a few weeks. Seeds of other species may remain viable for 2 or 3 to 1000 years or more. A case is known where Indian lotus (*Nelumbo nucifera*) seeds which were 1000 years old, as determined by the radioactive carbon test, were germinated. Wheat, alfalfa, pea, corn, and ponderosa pine seeds retain their viability for at least 30 years when stored under appropriate environmental conditions—that is, at low temperature and humidity. Some oat and barley seeds remain viable for at least 123 years. In 1832 wheat, oat, and barley seeds, enclosed in glass tubes, were placed in the foundation stone of the Nuremberg City Theater in Germany, which was torn down in 1955. The seeds were removed and planted. None of the wheat seeds was viable, but 21.9 per cent of the oat seeds and 12 per cent of the barley seeds germinated.

Some weed seeds remain viable for at least 80 years. In 1879, Professor William Beal placed seeds of twenty species in a bottle which he then buried. Periodically some seeds were removed and tested for viability. Of the twenty species, seeds of three species, curled dock (*Rumex crispus*), evening primrose (*Oenothera biennis*), and moth mullein (*Verbascum blattaria*), were viable after 80 years.

USES OF SEEDS

The great staple foods are seeds. The cereals—wheat, corn, rice, rye, oats, barley, and others—are eaten directly by man and by domestic animals upon which we depend for meat, eggs, and milk. The cereals are members of the grass family. Second in importance to the seeds of this family are those

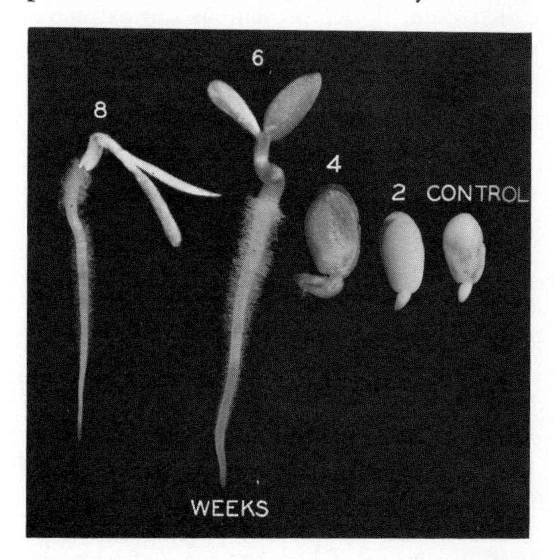

Fig. 21–9. Dormancy of seeds of mountain ash (*Sorbus aucuparia*) is broken by low temperatures. The seeds were stored moist for the indicated number of weeks at 1° C. before they were planted at room temperature. Six weeks of storage is the optimum for germination and seedling development. (Boyce Thompson Institute for Plant Research.)

of the legume family, for example, peas, beans, peanuts, lentils, and soybeans. The seeds of coffee (*Coffea*), cocoa (*Theobroma cacao*), and cola (*Cola nitida*) are sources of beverages.

Certain seeds provide us with fibers and oils. Cotton comes from fibers which cover cotton seeds. Kapok seeds yield fibers used in life preservers and as stuffing for mattresses and furniture. Among the important seeds which have a high content of oil are cotton, flax, and tung. Certain seeds are used as spices or condiments, for example, coriander, anise, dill, and caraway.

Strychnine, much used in medicine, is obtained from the seeds of *Strychnos nux-vomica.* Curare obtained from seeds of New World species of *Strychnos* is being used increasingly in medicine. In the Amazon region crude curare preparations have long been used by the Indians as arrow poisons. Castor oil is obtained from seeds of *Ricinus communis,* where it occurs in the endosperm. Although castor bean seeds are extremely poisonous to man, the poison, called ricin, is not present in the extracted oil.

FRUITS

The season's drama moves on toward the climax for which the stage was set at flowering time. Fertilization of the egg stimulates the ovary and parts associated with it to develop into a fruit.

Failure to set fruit may result because of unfavorable environmental conditions. For instance, late frosts in the spring may kill the ovaries of early-flowering trees and shrubs. Cool weather may retard the growth of the pollen tube to such an extent that the flower falls off before the egg is fertilized.

Usually the growth of the ovary as well as the development of the ovules depends upon fertilization. But it is not always so. In lilies, orchids, tulips, and some other plants the ovary begins to develop following pollination but before fertilization has occurred. Evidence indicates that hormones are present in pollen grains which initiate the development of the ovary. The hormones can be extracted with ether or water, and such extracts may induce fruit formation. Growth substances such as naphthaleneacetic acid or indolebutyric acid are also effective in initiating fruit development without pollination.

Of course, without pollination, the egg cannot be fertilized. The development of fruit in the absence of fertilization is known as **parthenocarpy.** Parthenocarpy may be induced by chemicals or it may occur naturally. Among the plants which develop fruit without treatment and without pollination are navel oranges, bananas, pineapples, and some varieties of grapes. These parthenocarpic fruits are seedless because without pollination fertilization does not occur, and hence no fertilized egg is present to develop into an embryo. There are a few exceptions, however, which regularly develop embryos without fertilization, an example of which is the dandelion. The formation of an embryo without fertilization is known as **parthenogenesis** or **apomixis.**

The fruit of a plant, botanically speaking, is the ripened ovary with the enclosed seeds and any other parts that may be closely associated with the ovary. The ripened ovary wall is known as the **pericarp.** In the fruit of corn, the seed is the edible part of the fruit, the pericarp in which it is contained being only a husk. In watermelons, oranges, and peaches the ovary grows enormously and certain parts become fleshy. In such instances the ripened ovary is the edible part and the seeds are usually discarded. In peas the pericarp, or pod, is discarded and the seeds of the fruit are eaten. In the hazelnut, the pericarp is hard and stonelike. Whether shell, husk, flesh, or fiber, however, the fruit is the pericarp with the enclosed seeds.

The ripening of fleshy fruits is neatly correlated with the maturation of the seeds. As the seeds mature, the pericarp becomes soft and enlarged. Sugar accumulates and bitter or sour substances disappear. The bright coloring that develops gives a rich touch to the landscape. The flash of color is the signal to birds and other animals that the fruit is ready to be eaten, and the agents which eat the fruit also aid in disseminating the seed.

Various types of fleshy fruits are formed, distinguished by the way in which the ovary and its associated parts develop. In the tomato, grape, honeysuckle, and currant the entire ovary wall becomes fleshy, and the fruit is termed a **berry.** The inner part of the ovary wall of cherry, plum, and peach becomes stony, and this type is called a **drupe.**

In the apple and pear the floral tube (the fused bases of the calyx, corolla, and stamens) envelops and is fused with the ovary. As the fruit develops the floral tube becomes fleshy and tasty. It is the fleshy floral tube that you relish and not the ripened ovary wall. Fruits of this type are called **pomes.** The strawberry is made up of many ripened ovaries inserted on a fleshy receptacle, which makes it an aggregate fruit. The fleshy receptacle is stem tissue and at maturity it is delicious. Located on the receptacle are many dry fruits, called **achenes.** The raspberry is also an aggregate fruit having many fleshy fruits on a common receptacle. Here the receptacle is not eaten. The edible part consists of many small fleshy fruits, called **drupelets.**

The variety exhibited among the dry fruits is almost endless. Some dry fruits split open (**dehisce**) when ripe and release the seeds. The fruit of the milkweed is made up of one carpel which splits along one side, and this type of fruit is known as a **follicle.** The pod, technically a **legume,** of a sweet pea also consists of one carpel, but it splits along two margins. The fruits of some species are **capsules.** These fruits are made up of more than one carpel and they split open in various ways. Some capsules have two openings at the top, such as snapdragon; others resemble a salt shaker, for example the poppy.

Many plants produce dry fruits which do not split open when ripe, and these fruits are said to be **indehiscent.** Wings, parachutes, umbrellas, and feathery tails make some wind-borne indehiscent fruits particularly fascinating. These are efficient for wide dispersal, as anyone can attest who has had year after year to remove dandelions from his lawn.

Perhaps the least popular of the dry indehiscent fruits are the burrs and sticktights. If you can separate yourself from a feeling of annoyance, however, and examine them closely, you will find that they are interestingly and intricately formed. They cling to animal agents, which inadvertently aid in the dispersal of the contained seed. Some fruits even plant themselves. Such a kind is *Stipa,* or needle grass. It is sharply pointed and has a series of hairs that act like barbs. Also it has a spirally twisted awn. The awn is hygroscopic and as it alternately gains and loses water, it twists the seed into the soil.

CLASSIFICATION OF FRUITS

Fruits may be classified in several ways. They may be dry or fleshy; if dry, they may be dehiscent or indehiscent. Fruits which develop from one pistil are **simple fruits.** Those which develop from several ovaries all borne in one flower are **aggregate fruits;** those that develop from several ovaries, each from a separate flower, are **multiple fruits.** The following outline includes most of the common fruits (Fig. 21–10):

I. Simple fruits.
  A. Dry fruits.
    1. Indehiscent fruits (not splitting open).
      a. Achene. An achene is a one-seeded fruit. The seed is attached at only one place to the pericarp. *Examples:* buckwheat, sunflower, and calendula.

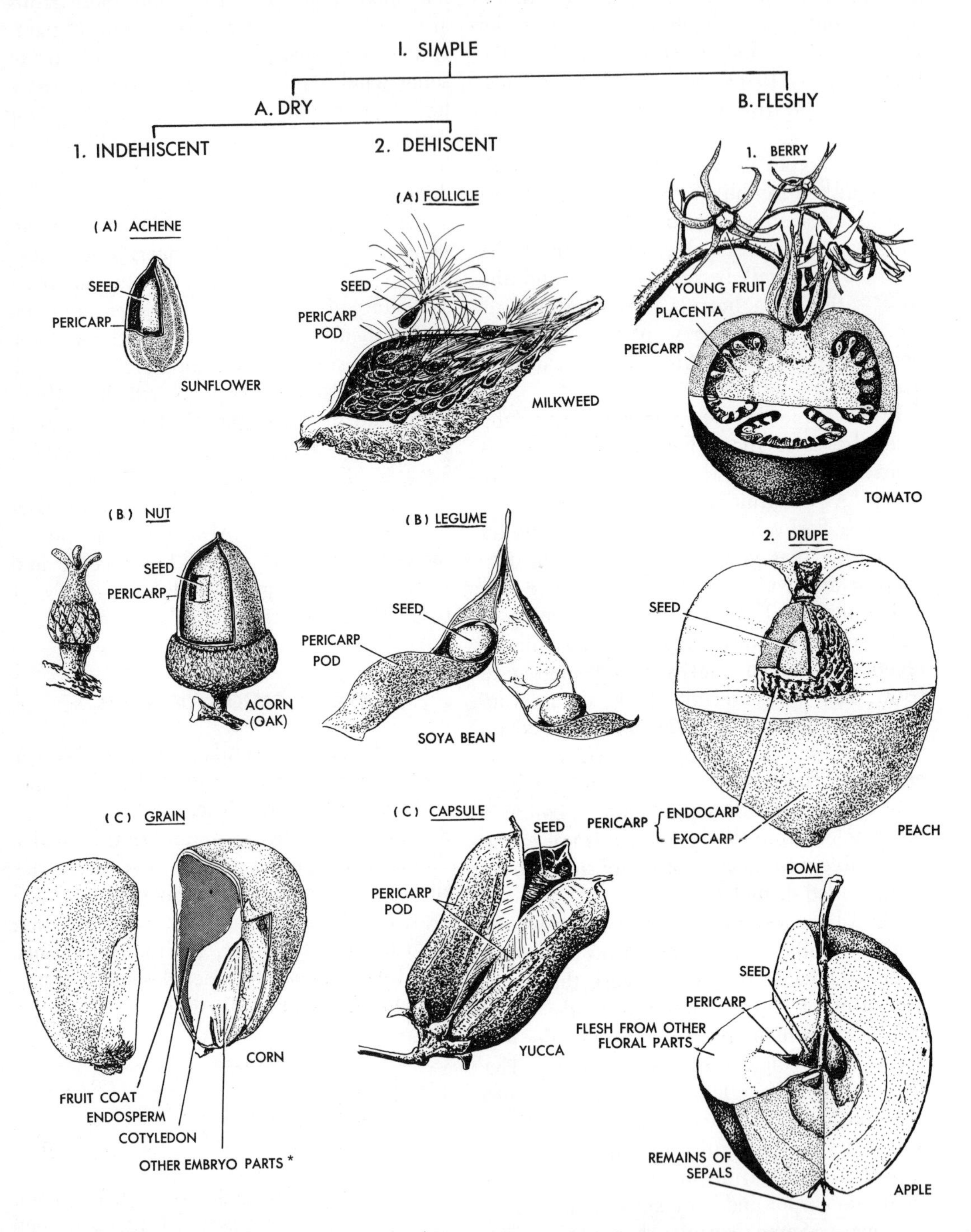

Fig. 21–10. Kinds of simple fruits. (W. H. Johnson, R. A. Laubengayer, L. E. DeLanney, and T. A. Cole, *Biology*, 3d ed. [originally published as *General Biology*]. Copyright © 1956, 1961, 1966, Holt, Rinehart & Winston, Inc., New York. All rights reserved.)

b. Grain or caryopsis. This one-seeded fruit is characteristic of the grass family. The seed coat is inseparable from the pericarp.

c. Samara. In this fruit, flattened wings are developed from the pericarp. *Examples:* maple, elm, and ash.

d. Nut. The pericarp of a nut becomes hard throughout. *Examples:* acorn, chestnut, and filbert.

2. Dehiscent fruits (splitting open when mature).

a. Legume. This fruit, which characterizes nearly all members of the pea family, is made up of one carpel and opens along two sutures.

b. Follicle. Fruits of peony, larkspur, columbine, and milkweed are follicles. A follicle consists of one carpel which splits along one suture at maturity.

c. Capsule. A capsule is made up of two or more united carpels. The fruits of lilies, orchids, and snapdragons are capsules.

d. Silique. The silique is the typical fruit of the mustard family. It is made up of two carpels which separate at maturity, leaving a thin partition between.

B. Fleshy fruits.

1. Berry. Berries are fruits in which the entire pericarp is fleshy. *Examples:* tomato, grape, date. A berry with a leathery rind, such as an orange, is called a **hesperidium.** One with a hard rind—for example, a watermelon—is known as a **pepo.**

2. Drupe. In drupes the pericarp is clearly differentiated into three layers. The outer layer, the **exocarp,** is the epidermis; the middle layer, the **mesocarp,** is fleshy; and the inner layer, the **endocarp,** is stony. *Examples:* cherry, peach, almond, plum, and other stone fruits.

3. Pome. In pomes the pericarp is surrounded by the floral tube, which becomes fleshy and tasty. *Examples:* apple, pear, and quince.

II. Aggregate fruits. An aggregate fruit develops from a single flower which contains many pistils. The several to many fruitlets are massed on one receptacle. In raspberries, loganberries, and blackberries each ovary develops into a small drupe. Each raspberry hence consists of a large number of small drupes, each of which has a structure comparable to that of a cherry. In a strawberry the individual ovaries develop into achenes and the receptacle develops into the conical, fleshy, delicious part of the fruit.

III. Multiple fruits. Multiple fruits consist of several flowers which are more or less united into one mass. Figs and pineapples are typical examples. A fig consists of a fleshy, tasty, hollow receptacle which is lined on the inside with numerous nutlets derived from the flowers.

DISPERSAL OF SEEDS AND FRUITS

Seeds or the fruits which contain them are the vehicles whereby plants travel and through which various environments are reached, some of which may be favorable for germination and plant development. Plants, like man, have always been wanderers, continuously exploring the earth's surface. Even though their travels are not planned, but only wayward, plants have been remarkably successful in locating niches. Wherever you find a certain environment, you will find characteristic plants.

The adaptations which facilitate seed and fruit dispersal are numerous and varied.

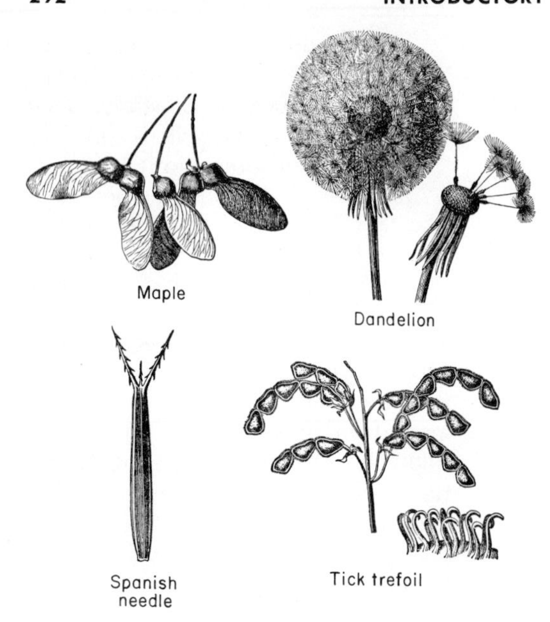

Fig. 21–11. Adaptations for wind dispersal (*upper*) and for animal dispersal (*lower*). (After Kerner.)

Light seeds, such as those of orchids, and seeds or fruits with wings, hairs, or parachutes are wafted where the winds go (Fig. 21–11). Many will come to earth at places where the environment is unsuitable, a relatively few where it is favorable for their development. Tumbleweeds such as Russian thistle and Jim Hill mustard scatter their seeds as the autumn winds roll the dead and dried plants over the ground. Seeds which float may be carried by water. Many species of *Carex* and the coconut are dispersed by water. Seeds may be carried great distances by mud clinging to the feet of birds and other animals. Intentionally as well as unintentionally man carries seeds around the globe, and he lives to some extent amid transported landscapes.

Animals, especially birds and mammals, disperse seeds and fruits of many species. Such seeds and fruits have a number of adaptations that favor dispersal. These adaptations may be compared to the marketing of various products by merchants in our society. Certain plant species rely on the "soft-sell" technique, favored by our supermarkets. Species that have evolved the supermarket approach package their seeds in fleshy fruits attractively sculptured, colored, and flavored. The apple, cherry, watermelon, strawberry, raspberry, and countless other fleshy fruits are sought out and relished by many wild creatures. The animals eat the fruit and then drop the uneaten seeds some distance from where the fruit was gathered. In other instances, the seeds are eaten along with the fruit and later are passed out of the alimentary tract, which may hasten germination. Not all plants use the supermarket technique to entice birds, beasts, and man to gather the fruits and disperse the seeds. Some species rely on what may be called the "hard-sell" technique. The animal does not select the product, but instead it is forced upon him. In such plants the seeds are packaged in dry

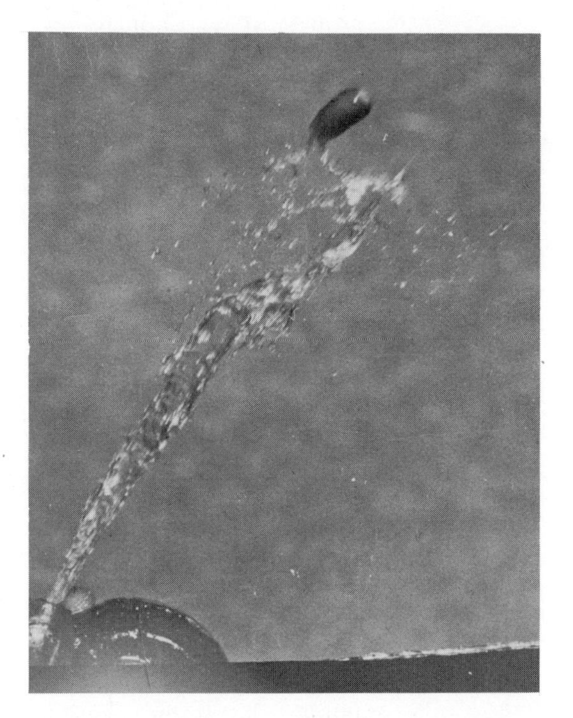

Fig. 21–12. The explosive fruit of the dwarf mistletoe expels the single seed with an initial velocity of 1370 cm./sec. and an initial acceleration of about 5000 times gravity. (T. E. Hinds, F. G. Hawksworth, and W. J. McGinnies, U.S. Forest Service.)

fruits bearing spines, hooks, or barbed needles that become attached to passing animals. Later the fruit is removed by the animal and the contained seed is dropped some distance from the parent plant.

Certain plants do not depend on wind, water, or animals to disperse their seeds. Some dry fruits may flip seeds some distance as they suddenly twist open. Pressurized containers, now widely used by man for shaving cream, waxes, and whipped cream, are not really new innovations for they have long been used by certain plants, among them the dwarf mistletoe and the squirting cucumber, whose fruits contain a fluid under pressure. In the fruit of the dwarf mistletoe the pressure at maturity is so great that the fruit is sheared from the pedicel; then the exocarp contracts and shoots the seed (Fig. 21–12).

QUESTIONS

1. Explain fully what happens to the parts of the flower in the development of the fruit and seed.

2. _____ One tomato fruit develops from (A) one ovule, (B) one ovary, (C) several ovules, (D) several ovaries.

3. _____ The seed develops from a(n) (A) ovule, (B) ovary, (C) integument, (D) embryo, (E) fruit.

4. _____ Wheat flour is milled from the endosperm which developed from (A) the embryo sac, (B) the fertilized polar nuclei, (C) the fertilized egg, (D) the integuments.

5. By means of drawings show the external and internal structure of a corn, bean, and castor bean seed. Label fully.

6. Explain fully the conditions necessary for seed germination and the physiological processes which go on in the seed.

7. What is the biological significance of seed dormancy?

8. List three causes of dormancy of seeds.

9. _____ In apomixis (A) the offspring exhibit both paternal and maternal characteristics, (B) fertilization must occur if an embryo is to develop, (C) the embryos are usually haploid, (D) the embryo may develop from nucellar cells or from a diploid unfertilized egg.

10. _____ Assume that frost has killed the above-ground parts of a castor bean seedling, a bean seedling, and a pea seedling. If the underground parts are not killed, the seedling most likely to survive is the (A) castor bean seedling, (B) bean seedling, (C) pea seedling.

11. _____ The life span of seeds of most species can be prolonged by storage at (A) high temperatures and low humidity, (B) high temperatures and high humidity, (C) low temperatures and low humidity, (D) low temperatures and high humidity.

12. _____ After seeds of peppergrass and of the saguaro cactus had imbibed water, they were exposed to red light and then to far-red light. The seeds (A) will germinate, (B) will not germinate.

13. _____ The dormancy of seeds of many desert species results from (A) lack of water, (B) seed coats impermeable to water, (C) water-soluble chemical inhibitors in the seed coats, (D) mechanically resistant seed coats, (E) immature embryos.

14. _____ The horticultural process of scarification is used to break the dormancy of seeds (A) having immature embryos, (B) whose seed coats are impermeable to water, (C) whose seed coats contain inhibiting chemicals, (D) with dormant embryos.

15. _____ Many alfalfa seeds (so-called hard seeds) do not germinate readily because (A) the embryos are not fully developed, (B) the seed coats are not permeable to water, (C) the seed coats prevent the intake of oxygen, (D) the seeds need more acidity.

16. Outline the classification of fruits, giving examples of each type.

17. Define a fruit. What is parthenocarpy? What process is normally necessary for fruit development? Is there any correlation between the number of seeds formed and the size and form of fruits? Explain.

18. The fleshy part of a cherry (A) furnishes food for the embryo when it starts to develop, (B) does not furnish food for the embryo when the seed is planted.

19. The following deals with the scattering (dispersal) of seeds. Match the seed or fruit with the agent which scatters it.

_____	Maple.	A. Wind.
_____	Pine.	B. Animal.
_____	Cocklebur.	C. Water.
_____	Tumbleweed.	D. Gravity.
_____	Strawberry.	E. Mechanical.
_____	Dandelion.	F. Birds.
_____	Walnut.	G. Man.
_____	Water lily.	
_____	Orchid.	
_____	Siberian pea.	
_____	Mistletoe.	
_____	Mustard.	

20. _____ The vegetation of the island of Krakatau was entirely destroyed by a volcanic eruption in 1883. The distance to the nearest undisturbed island is 12 miles and the distance to Java is 25 miles. Fifteen years after the eruption, 53 species of seed plants had reached the island. Which plants would be *least* abundant? (A) Those whose seeds or fruits are scattered by wind, (B) those whose seeds or fruits are dispersed by ocean currents, (C) those whose seeds or fruits are scattered by birds, (D) those whose seeds or fruits are scattered by mechanical means.

21. Name seven plants in which the organ useful to man is the root; the stem; the leaf; the fruit or seed.

---

22. Complete the following table:

Plant	*Number of Cotyledons*	*Endosperm Present or Absent*	*Location of Stored Food*	*Hypocotyl Lifts Cotyledons Above Soil or Cotyledons Remain in Soil*
Castor bean				
Bean				
Corn				
Pea				

23. Match the following:

_____	1. Berry.	1. Indehiscent dry fruit.
_____	2. Drupe.	2. Ovary wall fleshy throughout.
_____	3. Pome.	3. Dehiscent dry fruit splitting on two sides and composed of one carpel.
_____	4. Achene.	4. A fruit developing from a flower having a floral tube fused with the ovary.
_____	5. Follicle.	5. Dehiscent dry fruit splitting along one side and composed of one carpel.
_____	6. Legume.	6. A fruit developed from one flower having many pistils.
_____	7. Aggregate fruit.	7. A fruit with a stony endocarp.

22

Factors Affecting Flowering

Flowering is the result of profound changes in the behavior of the stem tips of a plant. While a plant is growing vegetatively, the stem tip produces new stem tissue and foliage leaves. Suddenly, it stops making leaves and forms floral parts instead. What factors bring about this change in habit? If we could know, we could imitate them, so that we could induce a plant to flower whenever we want it to. Among the factors that influence flowering are age, light intensity, length of day, and temperature.

The plant must have a sufficiently large leaf area to make the food necessary for the construction of flowers. This is, of course, related to the size and consequently to the age of the plant. For example, the sunflower, *Helianthus annuus*, a day-neutral plant, flowers only after twelve to fourteen pairs of leaves have developed. The length of time it takes a plant to reach flowering size is fundamentally determined by its inheritance and modified somewhat by environment. Some attain flowering size in 4 weeks, some in 1 to 3 months, and some not for several years.

Light intensity influences flowering. You may have noticed that strawberries and huckleberries flower more profusely in brightly lighted habitats than in deeply shaded ones. Philodendrons grow vegetatively in homes, but rarely flower. In the home the light intensity is too low for flowering. Assuming day length and temperature to be favorable for flowering, there is a minimum light intensity for each species below which no flowering occurs. The minimum varies with different species, usually being lower for shade species in forests than for sun species which grow in the open.

Evidence points to the formation by the leaves of a chemical (sometimes called **florigen**) which is translocated to the stem tips where it induces the formation of flowers. Physiologists have not yet been able to isolate it, but experiments show that some substance is manufactured in the leaves under certain environmental conditions. A substance of hormonal nature is produced in a cocklebur leaf which is exposed to short days. Cocklebur is a short-day plant and hence does not flower when grown under long days. However, if one

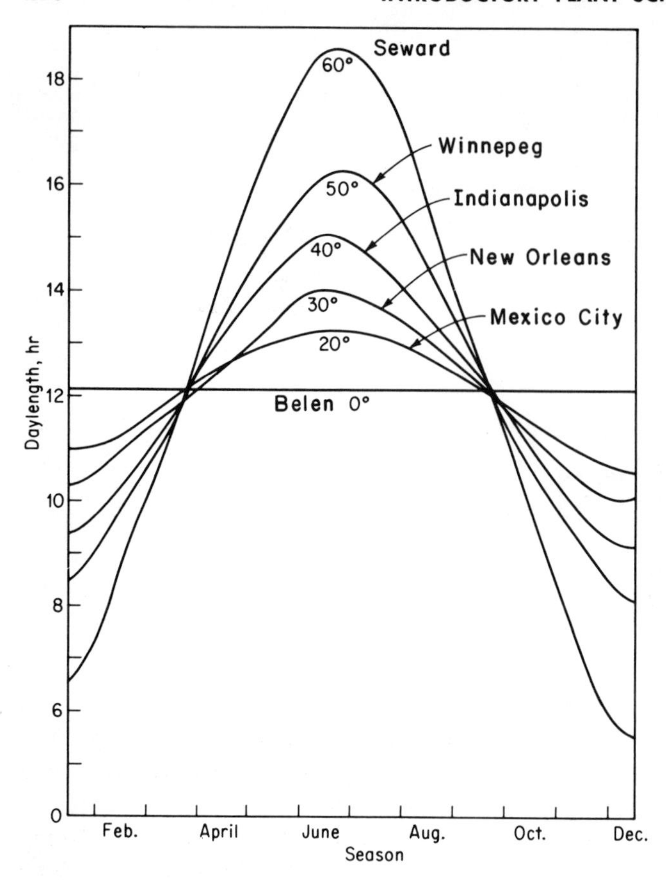

Fig. 22–1. Annual changes in day length at different latitudes. At the equator, Belen, Brazil, day length is constant during all months. At higher latitudes day length varies with season. Plotted here are the times from sunrise to sunset for six different locations. (From *Plant Growth and Development* by A. C. Leopold. Copyright © 1964, by McGraw-Hill Book Company, Inc., New York. Used by permission.)

leaf receives short days and the rest of the plant long days, the plant flowers. In this one leaf a flowering hormone develops which is translocated via the phloem to growing points where it initiates flowering. If the stem tip of a cocklebur plant is exposed to short days and the rest of the plant to long days, the plant does not flower. From these experiments we learn that the leaf is the receptive organ and that the stem tip is the reaction site, the region where flowers are initiated under the influence of the hormone synthesized in the leaves and translocated to the stem tip. The flowering hormone may even be transferred from one plant to another by grafting. If a branch from a flowering cocklebur is grafted onto a nonflowering plant, the plant receiving the graft flowers. The hormone that is present in the branch of the flowering plant is transmitted to the nonflowering plant.

PHOTOPERIODISM

An environmental factor of great significance in controlling flowering is day length. In tropical regions there is very little change in day length throughout the year, and the days and nights are about equal. In temperate regions, day length changes from winter to summer, and the long days coincide with the warmer season (Fig. 22–1). Many tropical species when brought into our zone produce flower buds only when the days are short, and continued long days prevent the formation of flower buds. Plants native to the temperate zone have a variety of flowering habits. Many plants produce flower buds in the early spring when the days are moderately short. Others flower during the summer when the days are long. Still others produce flower buds during the short days of late summer and early fall.

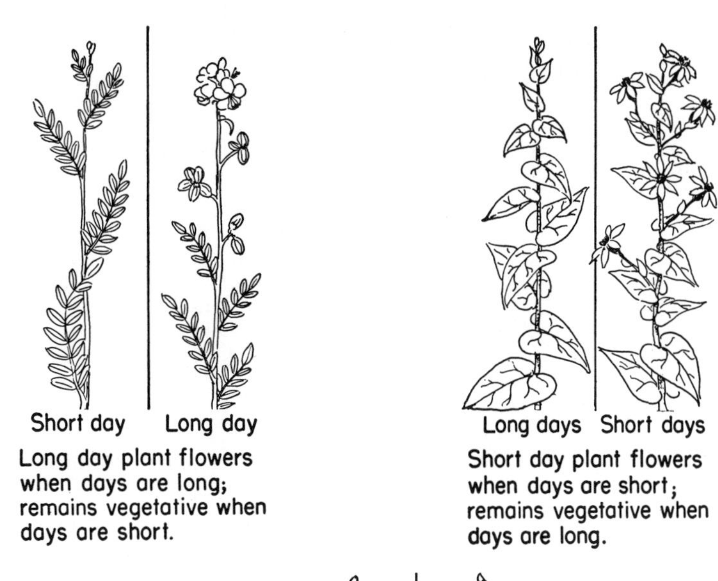

Fig. 22–2. Diagrams showing response of long-day, short-day, and day-neutral plants to long and short days.

Plants are grouped according to their response to day length into what are called **short-day plants, long-day plants,** and **day-neutral** or **indeterminate plants** (Fig. 22–2).

The short-day plants initiate flower buds when the days are shorter than some critical length (Table 22–1). The critical photoperiod is not the same for all short-day plants. Chrysanthemums and poinsettias are both short-day plants. Chrysanthemums initiate flowers when the days (including twilight) are shorter than 14 hours, whereas poinsettias initiate flowers only when the days are less than 12 hours. When short-day plants are grown with days longer than the critical duration, they grow vegetatively but do not flower.

Long-day plants flower only when the days are longer than some critical length (Table 22–1). As with short-day plants, the critical photoperiod varies with species and even with varieties and geographical races within a species. For example, the critical photoperiod for alpine sorrel (*Oxyria digyna*) growing in California, Colorado, and Wyoming is 15 hours, whereas for plants growing natively in the Arctic it is 20 hours. Spinach and rose mallow are long-day plants. Spinach flowers when the days are longer than 14 hours, the rose mallow when they are longer than 13 hours. Long-day plants grown with days shorter than the critical make vegetative growth, but flowering is prevented.

Table 22–1

Critical Day Lengths for Long- and Short-Day Plants

Short-Day Plants	Critical Day Length in Hours	Long-Day Plants	Critical Day Length in Hours
Cocklebur	15–15.5	Dill	11–14
Soybean	14–16	Henbane	10–11
Cosmos	12–13	Spinach	13–14
Chrysanthemum	14–14.5	Rose mallow	12–13
Poinsettia	12–12.5		

The day-neutral plants, also called indeterminate plants, initiate flowers regardless of whether the days are long or short. Familiar day neutral plants are roses, carnations, and tomatoes.

Short-Day Plants

Plants of the north temperate region which flower early in the spring or in the autumn are short-day plants. At the equator the days are always short (12 hours) and hence plants from equatorial regions are generally short-day plants. When such plants are introduced to northerly latitudes, they flower in early spring, autumn, or winter.

Chrysanthemums, asters, salvia, cosmos, and poinsettia are typical short-day plants. They may be induced to flower out of season by altering the length of day. For example, if light is excluded from chrysanthemums from 5:00 in the afternoon until 8:00 A.M. each day beginning in early July, the plants flower earlier than usual. When the days are naturally short, flowering of short-day plants may be prevented by increasing the day length with artificial light or by interrupting the dark periods with light for times ranging from 1 minute in some species to 1 hour in others. There is increasing evidence for believing that the length of the dark period is more important than the length of the light period. Short-day plants require a long uninterrupted night for flowering. Sugar cane is a short-day plant. If the dark period is interrupted night after night, sugar cane does not flower even though the days are short.

Like sugar cane, *Kalanchoë* plants do not flower when days are short if the long dark period is interrupted with white light. Let us see how interrupting a long dark period with red light and with far-red light affects flowering of this short-day plant. If plants are exposed to just 1 minute of red light at midnight, they do not flower (Fig. 22–3); but, if the long night is interrupted with far-red light, they flower just as they do with a continuous dark period. Furthermore, if *Kalanchoë* plants are exposed first to red light and then immediately thereafter to far-red, they flower, demonstrating that the inhibitory effect of red light may be reversed by far-red (Fig. 22–3, *right*).

The pigment that controls this reaction to red and far-red light is **phytochrome,** a blue or bluish-green protein pigment, that exists in two forms in a plant, designated as P_{660} and P_{730}. When phytochrome of the P_{660} form is exposed to red light, it is converted into P_{730}, a form which prevents flowering of short-day plants. Exposure of the P_{730} form to far-red light converts it to P_{660}, a compound that promotes flowering of short-day plants, perhaps by stimulating the synthesis of florigen. With short-day cycles, at the end of each light period, phytochrome is present in short-day plants in the inhibitory form, P_{730}, but during the long night this substance changes spontaneously into P_{660}, the flower-promoting state, at a rate that reduces the amount of P_{730} by one-half every

Fig. 22–3. Interrupting a long dark period with red light for just 1 minute prevents flowering of *Kalanchoë*, a short-day plant (*center*). One minute of far-red light counteracts the inhibitory effect of the red light, and the plant flowers (*right*). When the dark period is long and uninterrupted, *Kalanchoë* flowers (*left*). (Agricultural Research Service, Plant Industry Station, U.S.D.A.)

2 hours. Accordingly, during a long night the concentration of flower-promoting phytochrome, P_{660}, increases to a level which brings about the synthesis of florigen with resulting flowering.

Long-Day Plants

Long-day plants flower when the days are longer than some critical length which varies with species, but usually is between 13 and 14 hours (Fig. 22–4). In the United States, plants which flower in late spring or early summer are long-day plants. In latitudes of 60° or more, most species are long-day plants. Short-day plants are not found at such latitudes because in spring and autumn when the days are short, temperatures are too low for flowering. Cereals, lettuce, radish, clover, and *Rudbeckia* are long-day plants that flower during summer. They may be induced to flower during our winters by artificially lengthening the days. The intensity of the supplementary light need not be high; about 5 foot-candles is sufficient, and some plants respond to supplemental light of only 0.1 foot-candle. Long-day plants will flower when the days are short if the dark period is interrupted near midnight with white light or red light, but not when the long dark period is broken with far-red light. A long dark period favors flower formation of short-day plants but prevents flower initiation of long-day plants. Interrupting a long dark period with white light or red light prevents flowering of short-day plants but promotes flowering of long-day plants. When the long dark period is interrupted with far-red light, short-day plants flower whereas long-day ones do not.

Photoperiodic Induction

Short-day plants need not be grown continuously with short days to induce flowering. If a short-day plant which has been grown with long days is subjected to 1 or more short-day cycles, depending on the species, it will flower when subsequently grown with long days. During the 1 or more short-day cycles internal changes occur which lead to the development of flowers. The changes occurring during the short-day cycle or cycles are referred to as **photoperiodic induction.**

Similarly, photoperiodic induction occurs in long-day plants. Dill requires 1 to 4 long-day cycles and the annual beet 15 to 20 long-day cycles to set the stage for flowering. If a dill plant which has been grown with short days is grown with 4 consecutive long days, the plant will flower when returned to a short-day environment.

Indeterminate or Day-Neutral Plants

Plants such as tomato, dandelion, cotton, sunflower, and snapdragons, which flower regardless of day length, are known as day-neutral plants.

TEMPERATURE AND FLOWERING

Night temperatures markedly influence the initiation of floral buds. If they are not appropriate, plants do not flower. Foxglove, stocks, and cineraria flower profusely when the night temperature is 50° F. but not when it is 60° F. (Fig. 22–5). Pineapples flower at a night temperature of 62° F. but not at 70° F.

The response of plants to length of day may be modified by temperature. At one

Fig. 22–4. *Centaurea*, a long-day plant, flowers when days are 16 hours long (*right*), but not when they are 8 hours long (*left*). (U.S.D.A.)

Fig. 22–5. Effect of night temperature on cineraria. *Left,* grown at 50° F. *Right,* grown at 60° F. (Kenneth Post.)

temperature a plant may flower when the days are short, but at a different temperature it may flower when the days are long. Poinsettias are generally considered to be short-day plants, flowering at Christmas. Actually poinsettias are short-day plants when exposed to night temperatures of 63° to 65° F.; but if they are grown with night temperatures of 55° F., they are long-day plants and will not flower during the short days of winter. If they are raised with a night temperature of 70° F., they do not flower at all.

It is not surprising that temperature affects flower initiation and the photoperiodic response. The synthesis, or destruction, of the flowering hormone is influenced by temperature as is the rate of translocation from leaves to meristems. Furthermore, the response of cells at the apical meristem may be controlled to some extent by temperature.

VERNALIZATION

Psychologists tell us that our childhood experiences may influence our behavior today. Sigmund Freud was among the first psychologists to recognize that experiences during youth affect some aspects of adult behavior. Let us see whether comparable phenomena occur in the plant world, whether experiences during youth may affect the behavior of an adult plant.

The behavior of winter wheat is conditioned by the environment in which the young plants develop. Normally, the farmer sows seeds of winter wheat in the autumn. The seeds soon germinate and the young plants experience the cold of winter which brings about an internal chemical change that results in flowering during the following summer. Winter wheat plants that are not subjected to the cold of winter do not flower. Thus, if winter wheat seeds are sown in the spring, the plants grow vegetatively but they fail to produce flowers and grain during the summer.

However, by special treatment, plants grown from seed planted in the spring can be brought into flower. We can germinate winter wheat seeds until they just begin to sprout and then store the slightly sprouted seeds at a temperature of 40° F. for 2 months. If these chilled seeds are planted in the spring, the resulting plants come into flower and produce grain. Clearly, young plants, either those in a field during winter or the barely germinated seeds, must experience cold if the adult plants are to reproduce. The cold experienced during youth affects adult behavior. The exposure of partly germinated seeds or young plants to cold is known as **vernalization.**

Biennial plants, such as the henbane, carrot, beet, and canterbury bell, will not bloom if they are grown continuously at warm temperatures. As with winter wheat, these plants must experience cold at the end of the first season's growth if they are to blossom the second year. After exposure to cold, many biennials, among them the henbane, require long days for flowering. For example, a henbane plant that has been subjected to cold flowers when the days are long, but does not flower when short days follow the cold period.

GIBBERELLIC ACID

Gibberellic acid promotes flowering in a number of plants. Long-day plants which have been sprayed or otherwise provided with gibberellic acid blossom even when grown with short days. However, this chemical does not induce flowering of short-day plants when they are grown with long days. The henbane, cabbage, and other biennials will flower during the first year if they are treated with gibberellic acid.

QUESTIONS

1. _____ The response of plants to length of day, or more correctly, to the relative length of day and night, is referred to as photoperiodism. (A) True, (B) false.

2. How may plants be classified on the basis of their photoperiodic responses?

3. _____ A certain plant flowered only when the days were shorter than 15 hours. This plant is (A) a short-day plant, (B) a long-day plant, (C) a day-neutral plant.

4. _____ If a plant normally flowers only in July, it is probably (A) a long-day plant, (B) a short-day plant, (C) an indeterminate or day-neutral plant.

5. _____ If long-day plants are grown continuously under short-day conditions, they (A) will die, (B) will not flower, (C) will flower profusely.

6. _____ When days are short, interrupting a long dark period by turning on lights (A) prevents flowering of short-day plants, (B) promotes flowering of short-day plants, (C) delays flowering of long-day plants, (D) has no effect on long- and short-day plants.

7. _____ Chrysanthemums are short-day plants. The plants may be made to blossom earlier than usual by (A) increasing the light intensity, (B) applying heavy doses of fertilizer, (C) artifically shortening the day by means of shades, (D) artificially lengthening the days with electric lights.

8. _____ In 1920, Garner and Allard, who first clearly enunciated the principles of photoperiodism, noted that the Maryland Mammoth tobacco failed to flower in Washington, D.C. during the summer, but that it did flower in the greenhouse during the winter. This variety of tobacco would be classified as a (A) short-day plant, (B) a long-day plant, (C) a day-neutral plant.

9. A physiologist exposed both long-day and short-day plants to the schedules recorded below. Would short-day plants flower when grown with schedule I? Would long-day plants? Would short-day plants flower with schedule II? Would long-day plants?

Schedule I

Light	8 A.M. to 6 P.M.
Dark	6 P.M. to 11:30 P.M.
Red light	11:30 P.M. to 11:45 P.M.
Dark	11:45 P.M. to 8 A.M.

Schedule II

Light	8 A.M. to 6 P.M.
Dark	6 P.M. to 11:30 P.M.
Red light	11:30 P.M. to 11:45 P.M.
Far-red light	11:45 P.M. to midnight
Dark	Midnight to 8 A.M.

10. _____ Florigen, or perhaps some other stimulus which induces flowering, is produced when day lengths are in a certain range. The organs of a plant that receive this stimulus are the (A) leaves, (B) apical meristems, (C) stems.

11. _____ A cocklebur plant (a short-day plant) was grown with long days until it was of moderate size. The plant was then exposed to one short-day, long-night cycle. It was then grown with long days. The plant (A) will flower, (B) will not flower.

12. _____ Winter wheat seeds are vernalized to (A) break seed dormancy, (B) increase the vegetative growth, (C) bring about flowering, (D) alter the photoperiodic response.

13. Henbane, a biennial, was grown with the following schedules. Indicate with the letter F the schedules which would induce flowering.

1. _____ Plants grown in field the first season. Exposed to cold of winter. Short days the second season.
2. _____ Plants grown in field the first season. Exposed to cold of winter. Long days the second season.
3. _____ Plants grown in field the first season. Sprayed with gibberellic acid in midsummer.
4. _____ Plants grown in field the first season. Kept in warm greenhouse during fall, winter, and early spring. Exposed to long days.

14. Freud and other psychologists discovered that experiences during youth may affect the health and behavior of adults. Are there instances in the plant world where experiences during youth condition adult behavior?

15. _____ The highest percentage of short-day plants would be found (A) near the equator, (B) at 40° north latitude, (C) at 60° north latitude.

16. _____ In recent years, it was discovered that geographical races of a species may have different critical day lengths. A race of a long-day species which thrives in California is likely to have (A) a shorter critical day length than a race which thrives in Alaska, (B) a longer critical day length than a race which thrives in Alaska.

17. Discuss the relationship of temperature to flowering.

23

Genetics

The urge to develop new and superior varieties of plants characterizes modern man and was strong even in primitive man. Without it, civilization could not have developed. Practically all of our major crop plants were developed before recorded history. Man of today is still improving these by such techniques as selection, crossing, and inbreeding followed by crossing.

Selection is used to improve plants. In a collection of plants grown from seed there is often considerable variety. A few plants may be superior, most of average quality, and some inferior. Seeds from the best plants should be collected for sowing. Each generation of seed may give some variation in the offspring. Continued selection of the best can be continued in succeeding generations. In time new varieties of superior characteristics may be developed.

The **crossing** (that is, mating) of two plants which have characteristics which we wish to combine is effective in producing new varieties. When two unlike plants are crossed, we can call the process **hybridization.** Plants are easily crossed. One simply takes pollen from one parent and places it on the stigma of the other parent. The latter should have its own anthers removed to prevent self-pollination. Both before and after hand pollination, precautions should be taken to prevent unwanted pollen from reaching the stigma. With many plants the stigma can be protected by paper or glassine bags placed over the flower or a cheesecloth cage may be placed over the whole plant.

The plant selected as one parent may be superior in some traits and the one chosen as the other parent may be excellent in other characteristics. The goal is to secure a new variety superior in all traits. In some cases, desired combinations show up in the first generation.

If the desired combination is not found in the first generation, it is necessary to raise a second generation. Seeds produced by crossing the first generation plants are sown, thus furnishing second generation plants. Frequently many different types of plants appear in this generation. If the desired type is found, it may be propagated either asexually or by seed. If it is to be increased by seed, it will be necessary to self-pollinate the flowers or to cross two similar plants. In the next and subsequent generations the desired types are singled out and mated or selfed. This is continued until the new variety breeds true.

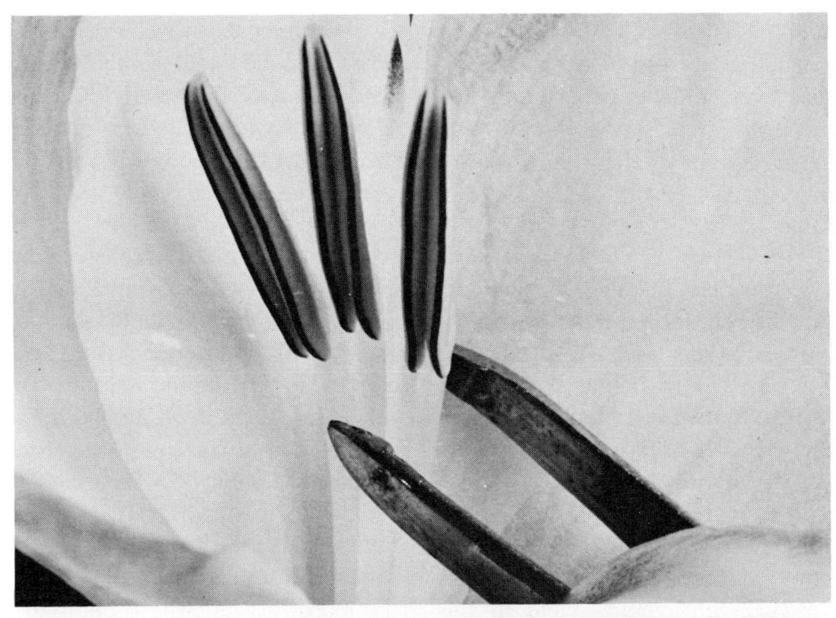

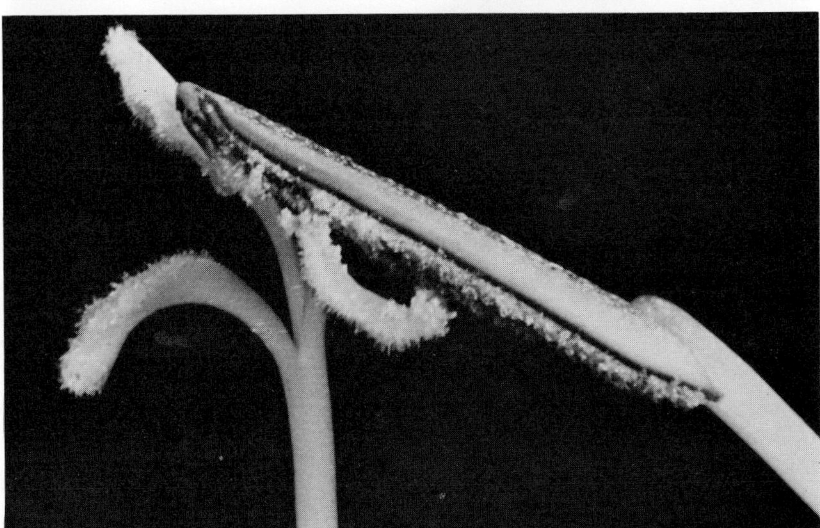

Fig. 23–1. Crossing gladiolus plants. *Upper*, remove the anthers from the female parent. *Lower*, remove an opened anther from the male parent and transfer the pollen to the stigma of the female parent.

Inbreeding followed by crossing is one of the newest methods for securing superior plants. Those which are naturally self-pollinated do not become less vigorous when they are inbred. On the other hand, the inbreeding of naturally cross-pollinated plants frequently results in a progressive decrease in size, vitality, floriferousness, and yield. Even though a decrease in vigor occurs with such inbreeding, it is the first step in the development of superior varieties of certain plants, especially corn and sorghum. After about seven generations of inbreeding, plants of different inbred races are crossed. In a number of instances, the resulting offspring are vigorous, high-yielding, uniform plants. These vigorous hybrids cannot be perpetuated from seed which they produce. However, the hybrid can always be produced again by crossing the two inbred races, or lines, as they are usually called.

With some plants commercial seed growers sell the seeds of double-cross hybrids. In the production of double-cross hybrids, four different inbred lines are developed. Let us call them lines A, B, C, and D. Line A is crossed with line B and line C with D. The seeds are planted and at flowering time the hybrid of A with B is crossed with the hybrid of C and D. The resulting seed is sold.

The change from older varieties of corn to hybrid corn resulted in a marked increase in yield. In 1932 the average yield in the United States was 25.9 bushels per acre; in 1946, 36.5 bushels; and in 1962, 67.3 bushels per acre. Of course, part of this tremendous increase resulted from improved cultural practices, but the greater part was due to the use of hybrid corn, which now provides 95 per cent of the corn crop. The increasing use of hybrid corn in less developed countries will provide the people in such lands with more food.

Selection, crossing, and inbreeding followed by crossing have enabled us to produce superior plants. Scientific plant breeding is a product of the twentieth century, and it is dependent on the science of genetics, which is concerned with heredity and variation.

Let us now consider the laws of inheritance and thereby how and why offspring resemble their parents.

LINK CONNECTING PARENTS TO OFFSPRING

Typically, an individual develops from a fertilized egg, or zygote. A human being develops from a zygote weighing about one 20-millionth of an ounce. At maturity the person may weigh 150 pounds, a 48 billion-fold increase in weight. During the course of development, the inert materials consumed as food are assimilated and so organized that a distinct individual results. Development is controlled by genes, which direct the dividing cells to form blood, bone, muscle, nerve, skin, and other tissues, and to fashion them in such a manner as to make a human being. A tree weighing several tons is developed from an equally small zygote. The genes regulate the organization of food and minerals into the various tissues that make this organism typical of its kind and different from a human being.

Zygotes from various species resemble one another rather closely. Superficially, at least, the zygote of a wheat plant resembles the zygote of a pea plant. And yet, one zygote develops into a wheat plant with definite characteristics and the other into a specific kind of pea plant.

Within the nucleus of the zygote there are two sets of chromosomes; one set from one parent, one set from the other parent. These sets of chromosomes represent the main connecting links of parents to their offspring. The number of chromosomes in a set varies with the species. In human beings, a set consists of 23 chromosomes, in vulgare wheats of 21, in peas of 7, in Easter lilies of 12. Two sets in each zygote make the total number for a species exactly double the number contained in the reproductive cells so that the zygotes of the species just named would contain, respectively, 46, 42, 14, and 24. The individual chromosomes of a set are not alike either in size and shape or content. Some chromosomes may be long, others short. Some may have the spindle attachment region in the center, others near one end. In other words, each chromosome of a set is endowed with a special organization and function. Every chromosome is capable of reproducing itself through successive cell divisions. The chromosomes are of great significance because they bear the genes which govern or control the form and physiology of the organism. The genes are made of DNA which controls the synthesis of messenger RNA which in turn controls enzyme synthesis. Hence, genes act, using RNA intermediaries, to bring about enzyme synthesis. This paragraph, so easily written, summarizes the work of many investigators. Let us relate some of the experiments that prove that the nucleus is the control center of the cell, that genes are on chromosomes, that DNA is the chief carrier of genetic information, and that DNA controls the sequence with which amino acids are joined together to produce enzymes and other proteins.

Nucleus as Control Center

Experiments with *Acetabularia,* a marine green alga, convincingly demonstrate that the nucleus controls development. *Acetabularia* is unicellular and grows to a height of 1 to 3 inches. Each plant (Fig. 23–2) consists of a rhizoidal base in which the single nucleus is situated, and a stalk terminated by a cap. *Acetabularia mediterrania* (commonly called med) bears a disk-shaped cap. The cap of *Acetabularia crenulata* (cren) is branched. If the stem and cap are removed, the basal part, containing the nucleus, regenerates the missing parts; a med base forms a med stem and cap, a cren base produces a cren stem and cap. If we transplant (Fig. 23–2) a capless stem segment of med to a cren base, the cap which develops on top of the med stem will be a cren type. Clearly, the cren nucleus controls the development of the new cap. If we transplant a capless stem segment of cren to a med base, the regenerated cap will be of med form. In both transplanting experiments the information contained in the nucleus controlled development. A med nucleus produced a med cap, a cren nucleus a cren cap.

Genes and Chromosomes

The correlation between the behavior of genes, which cannot be seen even with the electron microscope, and the behavior of chromosomes, visible with a microscope, is so great that we are certain that genes are resident in linear array on the chromosomes. In body cells genes occur in pairs because chromosomes occur in pairs. During mitosis, genes are distributed equally to the daughter nuclei because the chromosomes are so distributed. Cells resulting from reduction division have one set of genes, not two sets as in body cells, and one set of chromosomes. In gametes each kind of gene occurs singly and so also do the chromosomes. Indeed, the location of genes on certain chromosomes has been plotted. Genes resident on the same chromosome are said to be linked and such genes tend to be inherited as a unit. For example, chromosome IV of corn bears genes controlling the following traits: tassel seed, small pollen, lethal ovule, sugary endosperm, and defective endosperm. The number of linkage groups is identical to the haploid number of chromosomes. Gametes of corn have ten linkage groups and ten chromosomes; those of peas have seven and seven respectively.

The term gene was first used by Johannsen in 1911. Breeding experiments indicated that the genes were linearly arranged on chromosomes. That a gene could produce predictable traits was established before the chemical nature of genes was revealed.

CHEMISTRY OF GENES

A study of the chemical nature of chromosomes reveals compounds which might be the material basis of heredity. Chromosomes contain DNA, RNA, proteins, and a host of smaller molecules some of which are the building blocks for the nucleic acids and proteins. The genetic compound should be able to reproduce itself, carry genetic information from one generation to the next, and set in motion the chemical activities in cells which eventually result in growth and differentiation. DNA meets these three requirements.

An early indication that DNA is the message-bearing compound came in the 1940's from studies with *Klebsiella pneumoniae,* a bacterium which causes pneumonia in mice and men. This bacterium has several

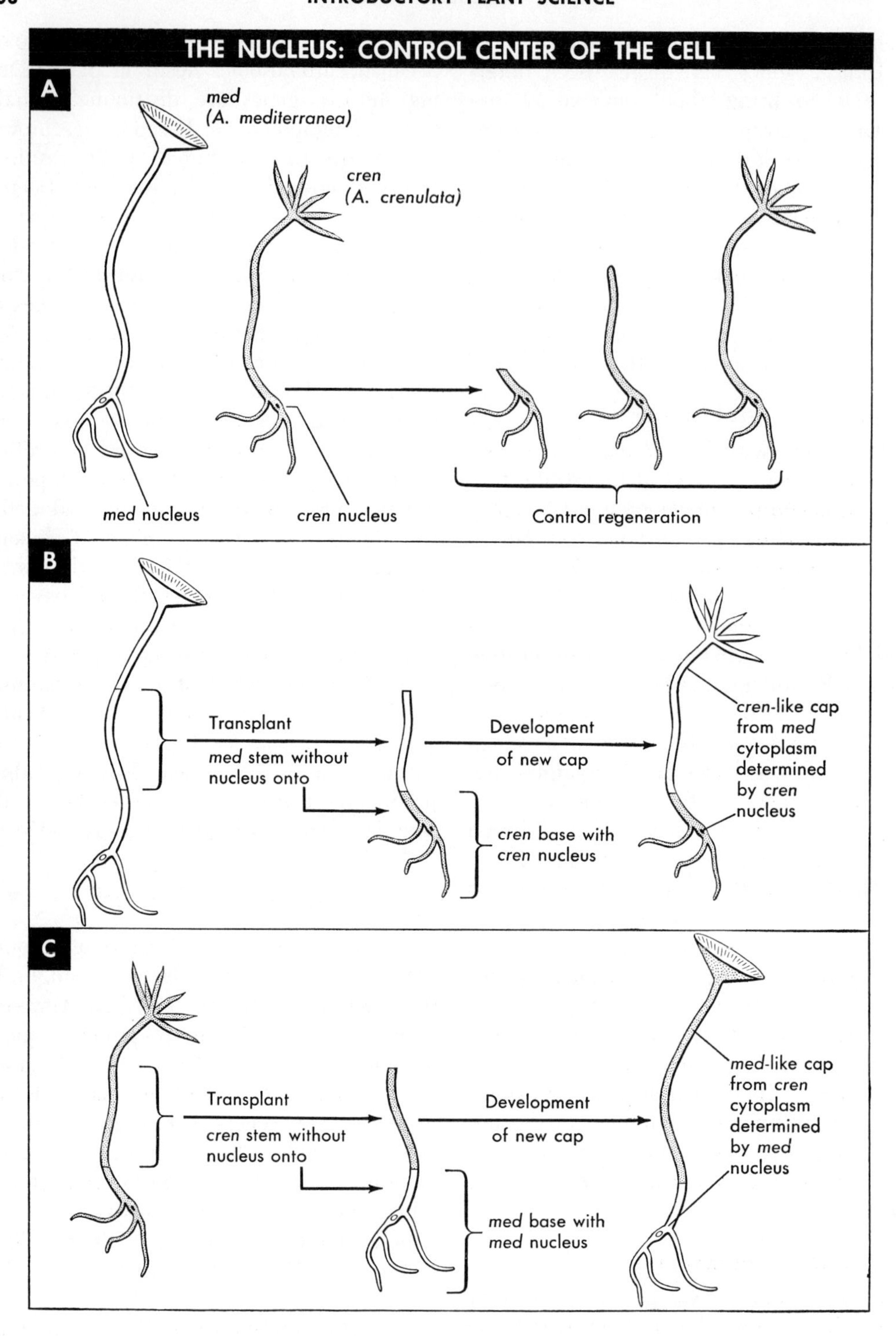

Fig. 23–2. The nucleus of *Acetabularia* controls the development of the cap. (*Life, An Introduction to Biology*, 2d ed., by George Gaylord Simpson and William S. Beck. Copyright © 1957, 1965 by Harcourt, Brace & World, Inc., New York, and reproduced with their permission.)

hereditary cell types, among them type II and type III; type II cells lack capsules, whereas type III cells are capsulated. Type III bacteria were grown in the presence of DNA extracted from type II. The DNA entered the cells and transformed the type III bacteria into type II. In this dramatic and historic experiment the hereditary message of one genetic strain was extracted as an inert chemical and incorporated into a different genetic strain. Furthermore, the transformed strain passed the trait to its offspring (Fig. 23-3).

Experiments with viruses that parasitize bacteria also prove that DNA is the material basis of heredity. A bacterial virus, called a **bacteriophage** or more simply a **phage**, consists of a shell of protein surrounding a strand of DNA (see Fig. 31-11). At infection, the phage becomes attached to the bacterium. Next the phage DNA enters the bacterial cell leaving the protein coat on the outside. Within the cell, the phage DNA directs the cell to form additional strands of phage DNA and phage protein. Later the phage protein and phage DNA are assembled to form new viruses which are released when the bacterial cell dies. The offspring are like the parent virus and their formation was controlled only by DNA. The DNA contained all the information required for phage reproduction. The protein shell left on the outside of the cell played no role.

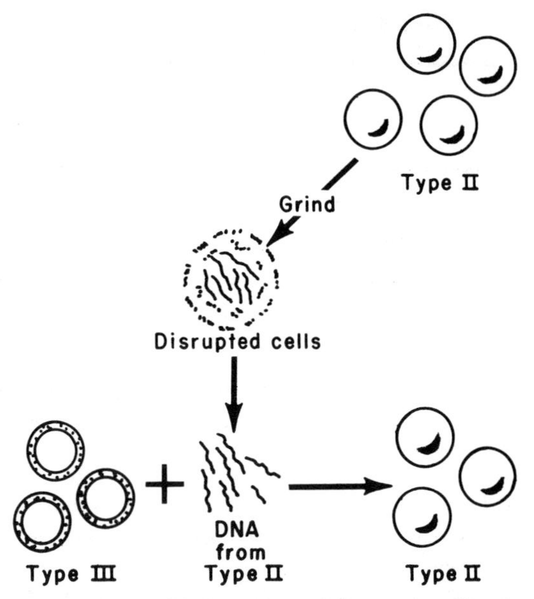

Fig. 23-3. DNA extracted from type II cells transforms type III cells into type II cells.

Chemistry of DNA

You will recall that the building blocks for DNA are nucleotides of four kinds. All the nucleotides contain deoxyribose and phosphate, but they differ in the nitrogenous base which is attached to the deoxyribose. One nucleotide has adenine; a second, thymine; a third, cytosine; and the fourth guanine. Thousands of nucleotide molecules combine to form DNA.

As we learned in Chapter 15, DNA is a double-stranded helix with a ladder-like structure whose cross-rungs are adenine and thymine pairs (A—T) and cytosine and guanine pairs (C—G). The pairs are linked together by hydrogen bonds. Because A is always paired with T and C with G, the molecule produces an exact replica of itself every time it reproduces. During reproduction, the double-stranded molecule separates into two strands, each of which then serves as a template in the formation of a new partner strand. Because each A in the strand can attract only a T and C, a G, the parent strand forces the synthesis of a complementary image of itself on the new strand. The essential features of DNA duplication are illustrated in Chapter 15. When duplication is complete, two identical double-stranded DNA molecules result and each is like the parent molecule.

THE GENETIC CODE AND GENE ACTION

Although the building blocks, the nucleotides, are the same for all genes of plants and animals, the genes are not identical. The sequence with which the nucleotides are joined together determines what kind of gene the DNA will be. Illustrated

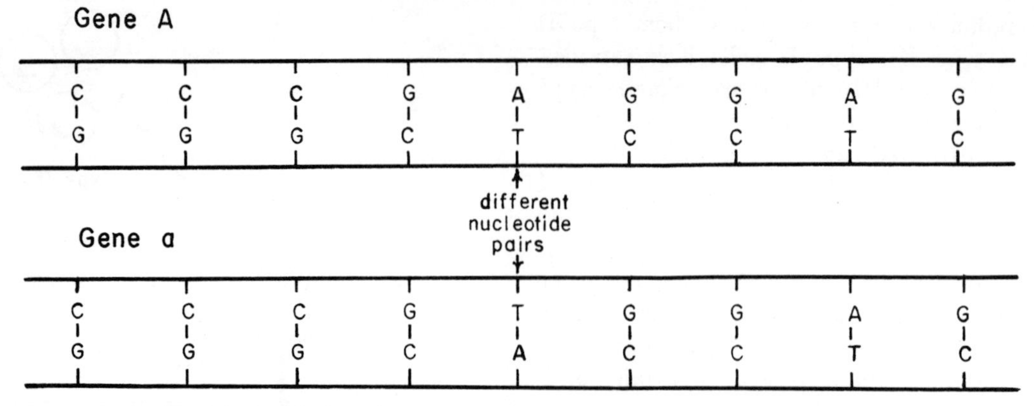

Fig. 23–4. Nucleotide sequences in two genes, gene *A* and gene *a*.

in Figure 23–4 are two genes, one designated *A* and the other *a*. The nucleotide pairs in the two genes are the same except for the pairs in the middle. Gene *A* has an A—T pair and gene *a* a T—A pair. You might think that such a slight difference in the arrangement of the nucleotides would be of little consequence, but in human beings the difference determines whether one is healthy or sick. Persons with gene *A* have normal hemoglobin, those with gene *a* have abnormal hemoglobin which results in sickle-cell anemia.

You will recall (Chapter 15) that DNA controls the sequence of nucleotides in messenger RNA which in turn controls the sequence of amino acids in proteins, which may be enzymes, structural proteins, hemoglobin, or other types. In the discussion which follows we will assume that the lower strands of DNA in the two genes which we have considered are operating in the synthesis of messenger RNA.

Shown in Figure 23–5 is the RNA formed by gene *A* and that by gene *a*.

In the formation of messenger RNA, G of DNA specifies the position of C of RNA; A of DNA, the position of U of RNA; T of DNA, the location of A of RNA; and C of DNA, the position of G of RNA.

A sequence of three bases in RNA, referred to as a **codon,** specifies the location of a certain amino acid. CCC of RNA is the code for proline, GAG for glutamic acid, and GUG for valine. The RNA formed by gene *A* links the amino acids together as follows: proline–glutamic acid–glutamic acid. The RNA formed by gene *a* links the amino acids together as follows: proline–valine–glutamic acid. When valine replaces glutamic acid at a certain position in the hemoglobin chain, the hemoglobin is abnormal and sickle-cell anemia results. Just a change in one of the three hundred or more amino acids in hemoglobin makes the difference between normal and abnormal hemoglobin.

In the plant world the different appearances of varieties result because the genes, the messenger RNA and hence also the enzymes of one variety are unlike those of a different variety. For example, a tall variety may have the enzyme required for the synthesis of an optimal amount of gibberellic acid whereas a dwarf variety may lack this enzyme or have it in a less active form. Similarly, a plant bearing red flowers has the enzyme system that synthesizes anthocyanin pigments, an enzyme system that is lacking in a white variety. The genes in the red variety are capable of producing the required enzymes; those in the white variety are not.

Structural and Operator Genes

Those genes which synthesize messenger RNA which in turn controls the synthesis of enzymes and other proteins are known

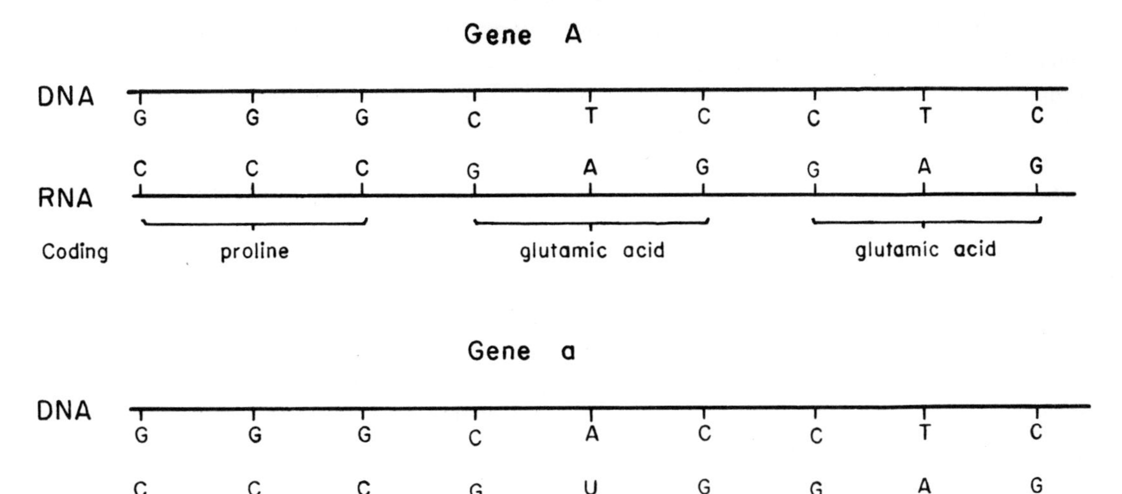

Fig. 23–5. Messenger RNA, and resulting amino acid sequences, formed by gene *A* and by gene *a*.

as structural genes. The structural genes are the type that one normally refers to and usually they are just called genes. A certain gene which is adjacent to a structural gene may not synthesize messenger RNA, but instead may control the activity of one or more structural genes. The controlling gene is known as an operator gene, or operator. The operator gene may be regarded as a switch that turns on and off one or more structural genes. In many instances a single operator gene governs the activity of a group of structural genes, a group that synthesizes all the enzymes needed for a reaction series.

The operator gene plus the structural genes which it controls is called an **operon.**

What is a gene? We have studied the chemistry of genes, their reproduction, and the way they govern metabolism, but still we are uncertain as to what a gene is. Usually we define a gene as a segment of a chromosome, but it is difficult to define how long a segment, or how one gene is separated from the next, if at all. Some years ago some biologists thought that the backbone of a chromosome was protein and that the nucleic acid of the genes existed as side chains. Other biologists believed that the chromosome consisted of alternating areas of protein and nucleic acid. Evidence of recent years indicates that DNA is the backbone of a chromosome and that DNA runs the whole length of the chromosome. According to this modern concept, the protein forms a coating around the DNA. The protein, usually a histone, plays an important role. It is probable that the histone, like operator genes, turns genes on and off. Although one DNA molecule runs the whole length of a chromosome, it is not just one gene. Even the largest genes are only a small part of a nucleic acid macromolecule. One segment of DNA represents one gene that may control a trait, while a different segment is a different gene that regulates the expression of a different trait. The DNA molecule running the length of the chromosome may include a hundred or more separate genes. The chemical nature of the DNA which separates one gene from another is imperfectly known. What we do know is that the long DNA molecule breaks more easily between genes than within any one gene. When a segment of the DNA of one chromosome is exchanged for a like segment of the DNA of a homologous chromosome, the process is called **crossing over.** In summary, we can consider a gene to be a segment of a long DNA molecule. The genes are separated from one another by segments of DNA which are more easily broken than the intragene area. Rarely a change occurs within the gene segment leading to an alteration in the gene and to a difference in the expression of a trait. A chemical change within the gene segment is known as a **gene mutation,** a topic more fully discussed in Chapter 25.

As we saw earlier, a change in one nucleotide pair of a gene makes the difference between normal and abnormal hemoglobin. The abnormal hemoglobin results from a gene mutation. The smallest subunit of a gene, generally just one nucleotide pair, that when changed gives rise to a mutant form is known as a **muton.** When a muton is changed, it alters the message of the entire gene. Some biochemists use the term **cistron,** instead of gene, to refer to that segment of a DNA macromolecule which controls the formation of a certain enzyme or other protein, which in turn controls the reaction leading to an identifiable trait. Although cistrons differ in their nucleotide length, each is made of hundreds of nucleotides.

ALLELES AND DOMINANCE

Genes typically occur in pairs in all cells except the reproductive cells. One pair of genes controls one character, another pair a different character, and so on. Genes which function in the development of a certain character are referred to as alleles. One member of a certain pair occupies a definite and relatively fixed position on a certain chromosome. The other member of the pair occurs at a like position on the homologous chromosome; that is, the chromosome of like form which came from the other parent.

There may be one or more types of a certain gene. Genes for stature may result in a pattern of development which leads to the production of tall plants. The variants of such genes may lead to the development of dwarf plants. Hence, it is possible to refer to some genes, as "tall genes" and others as "dwarf genes." Actually, of course, there is no such thing as tallness or dwarfness in the zygote, but only in genes which determine one pattern of development or the other. We should keep in mind that organisms inherit genes, not traits. The visible traits develop under the control of the genes and within the limits imposed by the environment.

Homozygous Individuals

A certain pea plant may have two genes for tallness. It is convenient to use letters to designate genes. In this instance let us use *T* to signify the gene for tall. The individual mentioned then could be designated as *TT*. When the two members of a pair of genes are alike, the individual is said to be **homozygous.** Other individuals may be homozygous for dwarfness and if we let *t* stand for dwarfness, their genetic makeup is *tt*.

Heterozygous Individuals

If a homozygous tall pea plant is mated with a dwarf plant (necessarily homozygous) the resulting offspring have genes *Tt*. From one parent they receive the gene *T*, and from the other parent, *t*. When the members of a pair are different, the individual is **heterozygous.**

Dominant and Recessive Genes

Some tall pea plants may have genes *TT*; other plants equally tall may have a **genotype** of *Tt*. Both plants have the same **phenotype,** a term which refers to the outward appearance of an organism, but they have different **genotypes,** which refers to their genetic makeup. The plant heterozygous for tallness, *Tt*, carries a gene for dwarfness. The dwarfness does not develop in the presence of the gene for tallness. In instances where one gene masks the expression of another, the gene which "rules" is said to be **dominant** and the one whose expression is suppressed is referred to as **recessive.** In our example, the "tall gene" is dominant and the gene for dwarfness recessive. Dwarfness can occur only when no "tall gene" is present but instead there are two genes for dwarfness (*tt*). In the heterozygous tall plant, *Tt*, the genes for tallness and dwarfness do not fuse. Genes maintain their integrity and separate one from the other when the hybrid forms sex cells.

MEIOSIS—REDUCTION DIVISION

Meiosis occurs prior to the formation of reproductive cells. Pollen mother cells and megaspore mother cells undergo meiosis. As a result of meiosis (reduction division), the chromosomal number is halved. The resulting cells have only one set of chromosomes, and hence only one set of genes, whereas the original cell had two sets of chromosomes and two genes were present for each trait.

The essential features of meiosis are shown diagrammatically in Figure 23–6. At the onset of meiosis, in this figure, four chromosomes are present in the nucleus, two long ones, and two short ones. The two long ones are homologous chromosomes—that is, they are similar in appearance and bear genes affecting the same traits. The two short chromosomes are also homologous. Two sets of chromosomes are present in the cell which is about to undergo reduction division. Each set consists of one long chromosome and one short one. Originally one set came from one parent and the other set from the other parent.

In all sexually reproducing organisms meiosis is remarkably the same and is carried out in two divisions. The prophase of the first division is complex and involves five stages known as **leptotene, zygotene, pachytene, diplotene,** and **diakinesis** (Fig. 23–6).

In early prophase, the leptotene stage, the chromosomes become distinct. Then, in the next stage, the zygotene, the homologous chromosomes pair or in more technical language, they synapse. **Synapsis** is precise and always occurs between homologous chromosomes, those bearing genes controlling the same traits. The homologous chromosomes associate in such a way that the genes affecting the same traits are opposite one another. In other words, a gene-for-gene pairing occurs during the zygotene stage of meiosis.

Following synapsis the paired chromosomes shorten and then each chromosome replicates itself. At this prophase stage, called **pachytene,** each pair becomes four chromosomes. (After the chromosomes have become replicated, some botanists prefer to call each a chromatid. For clarity we will continue to refer to them as chromosomes.)

After replication the diplotene stage begins, during which stage the pairs of homologous chromosomes tend to separate. They do not separate completely because two members of each group of four overlap and thus hold the group together. The region where two chromosomes overlap is known as a **chiasma** and, as we will see later, here an exchange of segments between chromosomes may take place.

As the chromosomes continue to shorten and thicken, the diplotene stage gives way to diakinesis, the final prophase stage of the first division. During diakinesis, the chromosomes are so compact that they may be easily counted. As diakinesis draws to a close the chromosomes move to the equator and concurrently the nuclear membrane and the nucleoli disappear and the spindle develops.

When the chromosomes are at the equator, the cell is in the first metaphase stage.

Next two chromosomes of every group of four move to one pole, and the remaining two of each group move to the opposite pole. This stage is known as the anaphase of the first division.

After the chromosomes reach the poles, a nuclear membrane forms around each of the two groups of chromosomes, and concur-

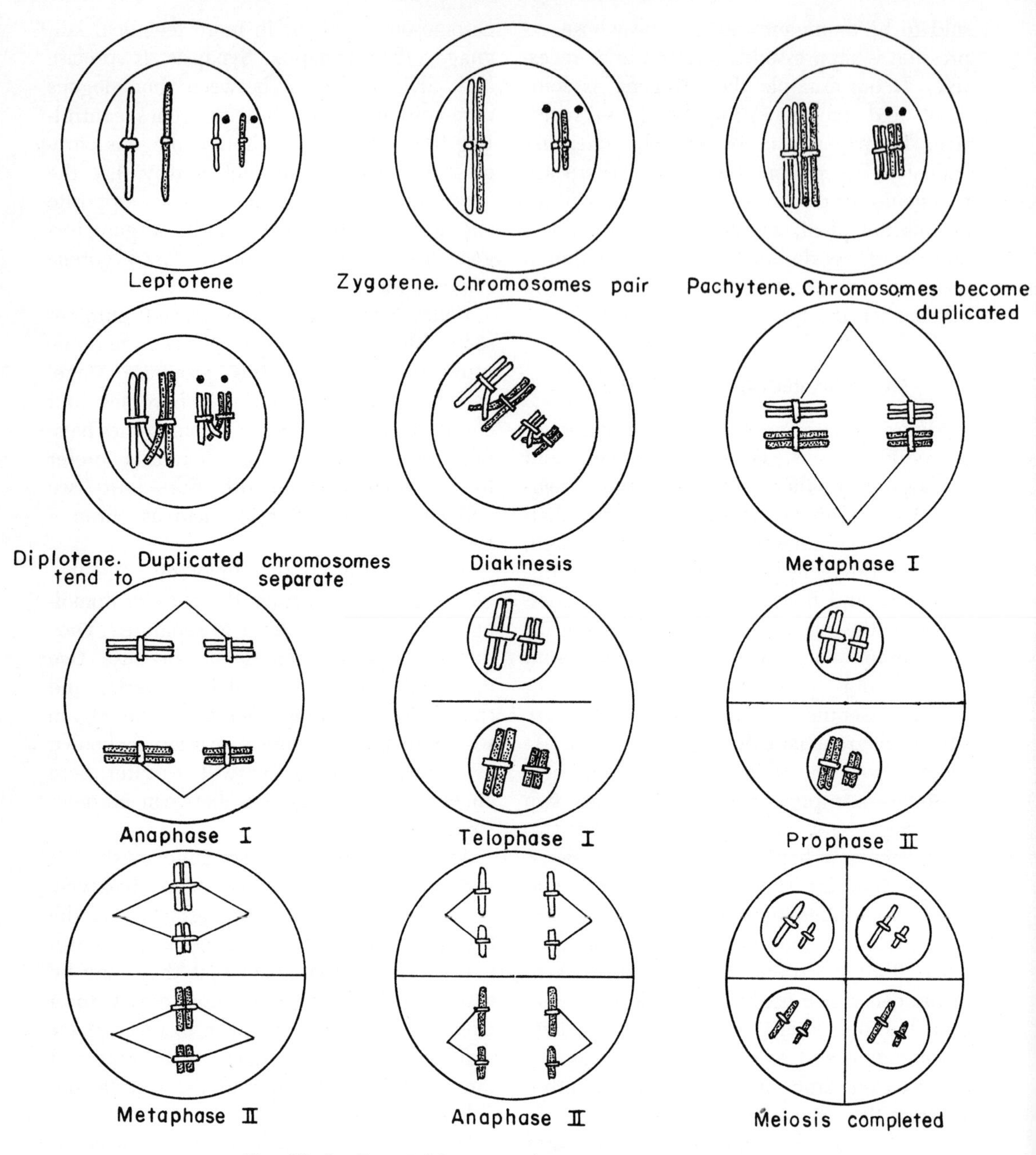

Fig. 23–6. Essential features of meiosis, reduction division.

rently a cell wall develops across the cell. These events occur during the telophase of the first division.

After a short interval, a second division commences in each of the two cells. During the prophase of the second division, the chromosomes shorten, the nuclear membrane disappears, the spindle develops, and the chromosomes move to the equator.

When the chromosomes are at the equator, the cell is in the metaphase stage of the second division.

During anaphase of the second division, one member of each group of two goes to

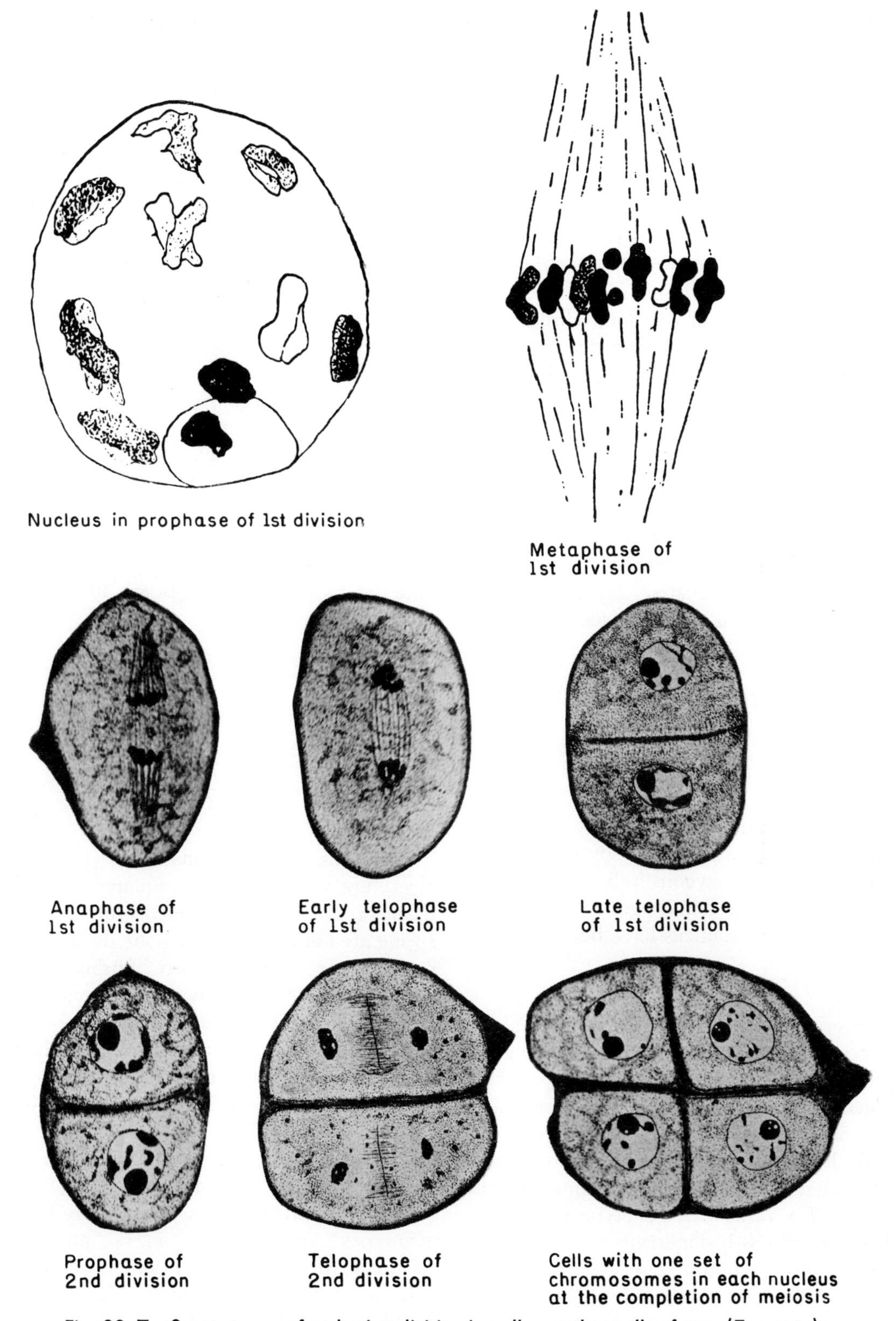

Fig. 23–7. Some stages of reduction division in pollen mother cells of corn (*Zea mays*). (Upper two figures: E. L. Fisk, *Am. J. Botany;* other figures: R. G. Reeves, *Am. J. Botany.*)

one pole, and the other member of every group goes to the opposite pole.

Later, during telophase of the second division, a nuclear membrane forms around each of the four groups of chromosomes. Cell walls then form and as a consequence four cells result, each of which has two chromosomes.

Note especially (Fig. 23–6) that when meiosis is finished each of the resulting cells has only one set of chromosomes, and recall that a set consists of one long chromosome and one short one. During the reduction division, one long chromosome has been separated from the other long one and one short one from the other short one. Because genes are on chromosomes, each cell formed by meiosis has only one set of genes. A moment's reflection on your part will show why the number of chromosomes is halved by meiosis. During reduction division the chromosomes duplicated themselves only once while the cell divided twice.

MONOHYBRID CROSS—COMPLETE DOMINANCE

When only one character and one pair of alleles are studied, the cross is known as a monohybrid cross. As an example, we shall consider a cross made by putting pollen from a homozygous tall pea plant on the stigma of a homozygous dwarf plant from which the stamens have been removed.

The First Filial Generation (F_1)

Every vegetative cell of the tall parent has a genotype of *TT*, and every vegetative cell of the dwarf plant has a genotype of *tt*. Each sperm produced by the tall plant carries just one gene for tallness (*T*), and every egg cell has just one gene for dwarfness (*t*). Hence, all of the F_1 offspring of the cross $TT \times tt$ have a genotype of *Tt*, and because tallness is dominant over dwarfness, the offspring are tall. The gene for dwarfness (*t*) does not manifest itself in the presence of the dominant gene for tallness (*T*) and is therefore said to be recessive. The plants are tall whether they contain one or two of the dominant genes (*T* or *TT*), but they are dwarf only when they contain the two recessive genes (*tt*).

The Second Filial Generation (F_2)

The F_2 generation is obtained by self-pollinating the F_1 or by brother-sister matings. In either case the cross is $Tt \times Tt$. As a consequence of reduction division which separates large *T* from small *t*, half of the sperms carry the chromosome on which large *T* is situated, and half have the chromosome bearing small *t*. Likewise, half of the eggs carry large *T* and the other half small *t*. It is just a matter of chance which sperm will unite with which egg. The possible combinations of sperms with eggs are represented in the checkerboard below. The two kinds of sperms are recorded on the horizontal line above the four squares and the two types of eggs are to the left of the checkerboard.

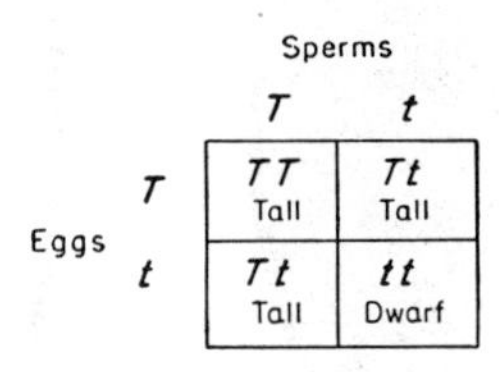

The possible combinations (fertilized eggs) are recorded in the squares. On the basis of chance, three-fourths of the offspring are tall and one-fourth are dwarf. We said "on the basis of chance" purposely. If there are only four plants in the second generation, it is not certain that three will be tall and one dwarf. If we had 100 plants in the second generation, the ratio might approach a 3:1 ratio, but could vary slightly. Conceivably 77 plants would be tall and 23 dwarf.

Although there are only two phenotypes (tall plants and dwarf ones) there are three genotypes which occur in a ratio of 1*TT*:2*Tt*:1*tt*. Both the *TT* and *tt* offspring will breed true when selfed, whereas the *Tt* offspring will segregate in a ratio of three tall to one dwarf when they are selfed.

Summary of the Monohybrid Cross

The crosses previously discussed are summarized in the following diagram.

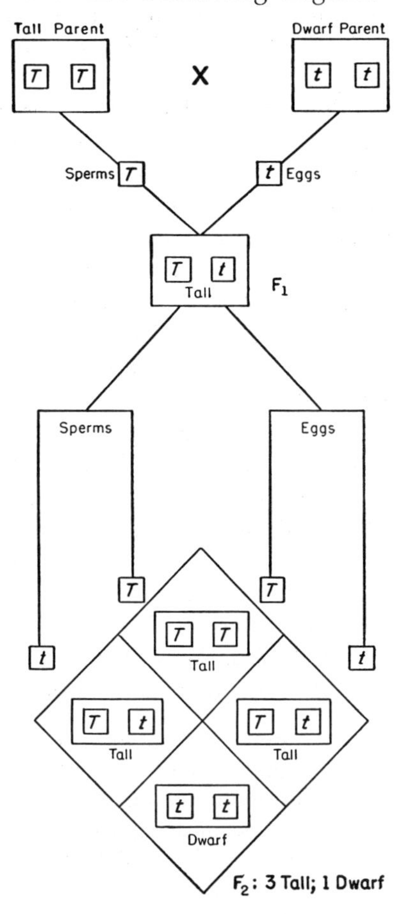

The Backcross

A cross frequently used in breeding work is the backcross, a mating of the F_1 to either parent, but usually to the recessive parent. The progeny obtained when the F_1 is crossed to the recessive parent is illustrated below:

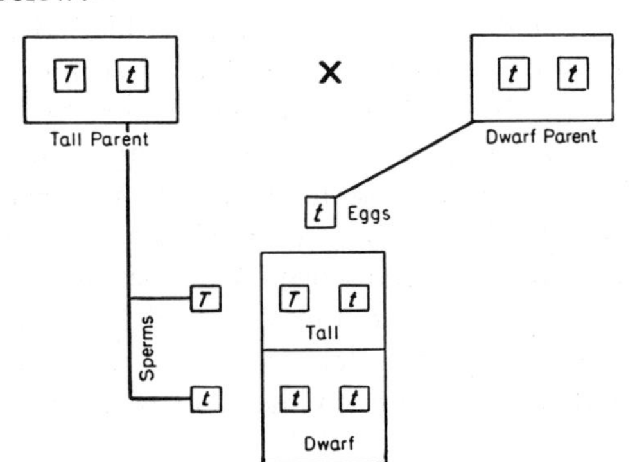

In this cross the *Tt* parent produces sperms of two kinds, *T* and *t*. All of the eggs carry *t*. If the egg is fertilized by a *T* sperm, the offspring is heterozygous and tall (*Tt*). If the egg (*t*) is fertilized by a sperm carrying *t*, the offspring is homozygous and dwarf (*tt*). The ratio in this backcross is one tall to one dwarf, 1:1 ratio.

MONOHYBRID CROSS—INCOMPLETE DOMINANCE

In snapdragons, red flower color is incompletely dominant over white. Heterozygous individuals bear pink flowers. If a red-flowered snapdragon is crossed with a white-flowered one, all of the F_1 progeny have pink flowers. In the F_2 generation, one-fourth of the progeny bears red flowers, one-half pink, and one-fourth white. In the diagram which follows, r stands for the red gene, r_1 for white. Thus, rr individuals are red, rr_1 pink, and r_1r_1 white.

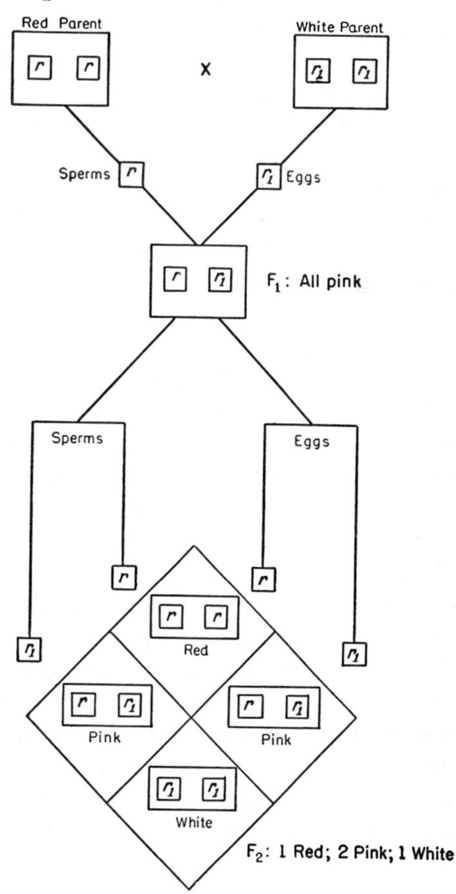

WORK OF GREGOR MENDEL

So far, the discussion of heredity has been based on modern genetics with its understanding of genes and chromosomes. Some of the concepts presented, however, were developed by Gregor Mendel before genes and chromosomes were discovered. Gregor Mendel (1822–1884) was an Augustinian monk who worked with enthusiasm and a thoroughly scientific spirit. He began his experiments in 1857 and finished them in 1865. He decided to use pea plants, which he selected carefully. He assembled in the monastery garden a number of well-marked varieties of peas, which differed from one another by readily recognized characters. For example, when he was studying height of plants, he mated a plant which was about 80 inches high with a plant which was 18 inches or less in height. When he studied flower color, one variety had colored flowers and the other white. In his first experiments Mendel studied only one character at a time rather than a whole complex of characters. This was a wise decision and enabled him to see order in nature where previous workers had found only confusion. His technique for mating plants was as carefully performed as that used by the best of present-day geneticists, involving as it did the emasculation of the female parent, transfer of pollen, and precautions to prevent foreign pollen from getting to the stigma. He counted the different types of offspring and kept accurate pedigree records. Mendel's technique is still a model for research, illustrating two useful principles: study one thing at a time and express results quantitatively.

One character studied by Mendel was height of plants. He crossed a tall variety with a dwarf one and noted that all of the F_1 plants were as tall as the taller parent. Mendel explained this behavior by stating that tallness was dominant and dwarfness recessive. Mendel next raised a second generation by selfing the F_1's. He noted that some of the F_2 progeny were tall and some dwarf, but he went further than that and actually counted the tall ones and the dwarf ones, a technique not used by his predecessors. The F_2 population consisted of 787 tall plants and 277 dwarf ones, approximately a 3:1 ratio.

Next Mendel accounted for the approximately 3:1 ratio which he invariably obtained in the second generation. He assumed that when the F_1 hybrid produced eggs and sperms, there was a segregation of the factors for tallness and dwarfness. As a result of this segregation, he reasoned that any one sperm or egg would have just one factor for height. Some sperms would carry the factor for tallness, an equal number the factor for dwarfness. Likewise for the eggs. Furthermore he reasoned that if the two kinds of sperms fertilized the two kinds of eggs, the result would be the same as he had noted in his field experimentation. From his very carefully performed and intelligently interpreted experiments, Mendel advanced four laws. One will be discussed later. The other three are:

1. Law of unit characters. Factors which do not lose their identity control the inheritance of characters and these factors occur in pairs. (Those units which Mendel called factors are now known as genes.)
2. Law of dominance. One factor (gene) may mask the expression of the other factor.
3. Law of segregation. The members of pairs of factors (genes) separate into sister gametes. Hence, only one factor of a pair is present in a gamete.

Mendel published his work in 1866. The circulated copies were apparently filed away on library shelves to be forgotten for 34 years. His work was not appreciated by his contemporaries. Recognition of the tremendous significance of his work did not come until 1900, 16 years after Mendel's death. His work was rediscovered in that year by Correns, De Vries, and Tchermak, who recognized that Mendel's work was the most fundamental study in the field of genetics.

The laws which Mendel worked out have been found generally applicable to plants, animals, and human beings. In the twentieth century our understanding of genetics has increased greatly. Mendel's laws have been expanded and supplemented.

QUESTIONS

1. _____ A zygote has (A) one set of chromosomes, (B) two sets of chromosomes, (C) just one set of genes.

2. _____ If you have a flock of sheep whose average fleece weight is 10 pounds, and if in the flock some sheep produce fleeces weighing 9 pounds, and others produce fleeces weighing up to 15 pounds, it would be (A) possible to bring the average to nearly 15 pounds by selection, (B) possible to bring the average up to 20 pounds, (C) impossible to increase the average weight above 12 pounds.

3. What is meiosis? Diagram the process.

4. _____ Meiosis (A) is a type of division which ensures that the resulting cells will have the same number of chromosomes as the original cell, (B) occurs in two divisions, but the chromosomes divide only once, (C) results in the production of diploid cells which ultimately become gametes, (D) ensures that each resulting cell will have two homologous chromosomes.

5. _____ A pair of genes affecting the same trait are referred to as alleles. (A) True, (B) false.

6. _____ Where one gene masks the expression of another, the gene which rules is said to be (A) dominant, (B) recessive.

7. _____ The word "phenotype" refers to how the organism will behave in breeding. (A) True, (B) false.

8. _____ A plant with a genotype of *Tt* is (A) homozygous, (B) epistatic, (C) heterozygous, (D) phenotypic.

9. _____ A plant with a genotype of *TT* can produce the following gametes: (A) *TT*, (B) *T* and *t*, (C) *T*, (D) *t*.

10. _____ In tomatoes red fruit color (*R*) is dominant over yellow (*r*). If a heterozygous red is crossed with a heterozygous red (*Rr* × *Rr*) the offspring will be (A) all red; (B) one-half red, one-half yellow; (C) all yellow; (D) three-fourths red, one-fourth yellow.

11. _____ The genotypic ratio resulting from the following cross, *Bb* × *Bb* would be (A) 3:1, (B) 2:1, (C) 1:2:1, (D) 1:4.

12. _____ A man bought a polled bull with a guarantee that he was homozygous for polled (that is, hornless), which is dominant over the horned condition. The man mated the bull with a horned cow and a horned calf was produced. He could (A) get his money back, (B) not get his money back because the bull was homozygous but the cow was not.

13. In summer squash white fruit color (*W*) is dominant over yellow (*w*). A white-fruited squash plant when crossed with a yellow produces offspring about one-half of which are white and half yellow. On the lines below write the genotypes of the parents.

______________ × ______________

14. _____ In snapdragons red crossed with white gives pink offspring. Red is (A) dominant, (B) incompletely dominant, (C) recessive.

15. _____ Mendel's methods were comparatively crude and nonscientific. (A) True, (B) false.

16. In four-o'clock plants, red flower color is incompletely dominant over white; heterozygous plants are pink. What progeny could be expected, and in what ratio, from the following crosses?

A. Pink × pink
B. Pink × white
C. White × red
D. Pink × red

17. In corn normal height (*N*) is dominant over dwarf (*n*). A normal plant when crossed with a dwarf produced 196 normal plants and 204 dwarf plants. On the lines below write the genotypes of the parents.

______________ × ______________

18. A student planted 100 corn grains and 2 weeks later noted that 77 seedlings were green and 23 were white. Explain.

19. _____ If a portion of a stalk of *Acetabularia mediterrania* is grafted to a rhizoidal base of *Acetabularia crenulata,* the regenerated cap will be (A) similar to *A. crenulata,* (B) similar to *A. mediterrania,* (C) intermediate between *A. mediterrania* and *A. crenulata.* Explain your choice.

20. What evidence demonstrates that genes are on chromosomes?

21. Describe an experiment which indicates that DNA is the chief carrier of genetic information.

22. _____ Which one of the following is false? (A) The building blocks of DNA are nucleotides of four kinds. (B) DNA is a double-stranded helix whose cross-rungs are adenine and thymine pairs and cytosine and guanine pairs. (C) During replication the double-stranded molecule separates into two strands, each of which then serves as a template in the formation of a new partner strand. (D) DNA controls the sequence of nucleotides of messenger RNA; the adenine of DNA dictates the position of thymine of RNA and the cytosine of DNA, the position of guanine.

23. If the base sequence in DNA is GGG · ATC · ATC, what will be the base sequence in messenger RNA?

24. _____ A codon consists of (A) one base pair, (B) a sequence of two bases, (C) a sequence of three bases, (D) a sequence of four bases.

25. _____ An operator gene (A) synthesizes messenger RNA, (B) synthesizes transfer RNA, (C) controls the activity of one or more structural genes, (D) is the same as an operon.

26. _____ When a gene is combined with a histone protein, it is (A) active, (B) inactive.

27. Assume that the diploid number of chromosomes is four. Diagram cells that are in the following stages: metaphase of mitosis, metaphase of meiosis, early telophase of mitosis, telophase II of meiosis.

---

28. Match the following:

_____ 1. Zygotene.	1. Chromosomes single.
_____ 2. Pachytene.	2. Homologous chromosomes synapse.
_____ 3. Diakinesis.	3. Chromosomes shorten and replicate themselves.
_____ 4. Diplotene.	4. Prophase stage just before metaphase I.
_____ 5. Leptotene.	5. Region where homologous chromosomes may exchange segments.
_____ 6. Chiasma.	6. Homologous chromosomes tend to separate after pairing has been completed.

24

Genetics (Continued)

DIHYBRID CROSS

So far, we have been considering just one character of a plant at a time. Let us now deal with two traits simultaneously. In tomatoes tallness is dominant over dwarfness and red flesh color is dominant over yellow. Let us cross a tall plant bearing red tomatoes with a dwarf one bearing yellow fruits. This is an example of a dihybrid cross. In a **dihybrid cross** two pairs of alleles which are on different chromosomes are studied. All of the F_1 offspring will be tall plants with red tomatoes, but they will not be homozygous for these traits. They receive a gene for tallness from the tall parent and a gene for dwarfness from the other parent. From one parent they get a gene for red, from the other a gene for yellow. If we let T stand for tallness, t for dwarfness, R for red fruit and r for yellow, the progeny have a genetic makeup of $TtRr$. The cross and the F_1 offspring are summarized below:

```
            TTRR  x  ttrr
              \       /
    gamete    TR    tr    gamete
                \   /
       F1       Tt Rr
```

The second generation is obtained by selfing the first generation plants or by brother-sister matings. In either case the cross is $TtRr \times TtRr$. Four types of plants appear in the second generation, tall plants with red tomatoes, dwarf plants with yellow ones, tall plants bearing yellow tomatoes, and dwarf ones with red fruits. The latter two are new combinations.

We shall now consider the genetic explanation for the four types of offspring which are actually obtained in a ratio of 9 tall red, 3 tall yellow, 3 dwarf red, and 1 dwarf yellow. The $TtRr$ plant selected as the male parent produces four kinds of sperms, TR, Tr, tR, and tr. The $TtRr$ female parent produces four kinds of eggs, TR, Tr, tR, and tr. These four types of sperms and eggs result because the genes for height are on a pair of chromosomes different from those for fruit color.

The chromosomal basis for the production of four types of gametes is illustrated in Figure 24–1. At the first metaphase (Fig. A) of the reduction division the chromosomes may be arranged as shown in I or as diagrammed in II. On the basis of chance the same number of cells have the II arrangement as have the I arrangement. The B figure shows the second metaphase of meiosis, and the C figure illustrates the cell at the completion of the reduction division. During meiosis in cell I, the large T

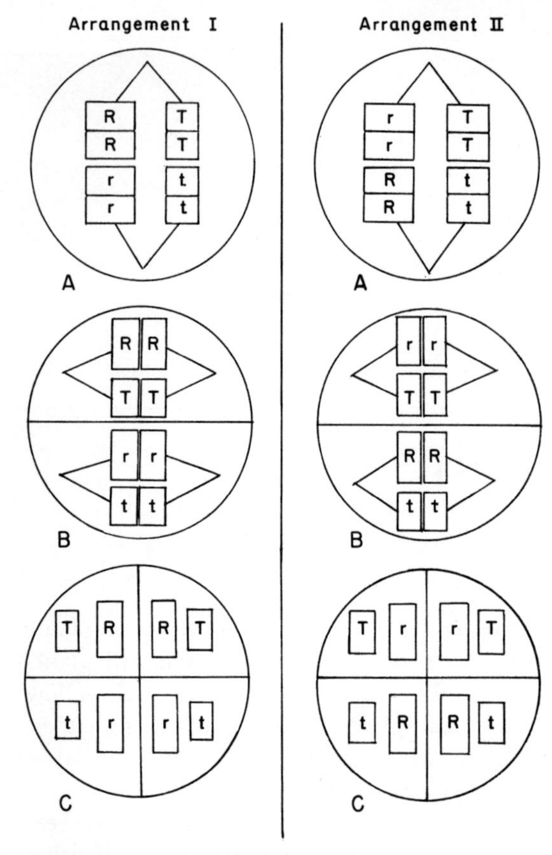

Fig. 24–1. Diagram showing the chromosomal basis for the production of four types of gametes by a plant with a genotype of *TtRr*.

moves with the large *R*, and the small *t* and the small *r* travel together. Half of the cells resulting from the reduction division of cell I have *TR* and half *tr*. In cell II the large *T* and the small *r* move together, and the small *t* and the large *R* travel together. Therefore, half of the cells produced by cell II have *Tr* and half *tR*. After the completion of reduction division in cells I and II, cells with *TR*, *Tr*, *tR*, and *tr* are produced in equal numbers. Each gamete has one gene for height and one gene for fruit color. Any one height gene may be associated with either color gene because the height genes segregate independently of the color genes.

The possible combinations of the four types of sperms with the four types of eggs are obtained by making a checkerboard. The four types of eggs are written to the side, and the four types of sperms across the top. Each square represents the union of the egg at the side with the sperm in that column. For example, the offspring *TTRR* in the upper left-hand corner of the checkerboard below is the result of a union of the *TR* sperm with a *TR* egg. The second from the top in the first column (*TTRr*) results when a *TR* sperm fertilizes a *Tr* egg.

Eggs \ Sperms	*TR*	*Tr*	*tR*	*tr*
TR	*TTRR* tall red	*TTRr* tall red	*TtRR* tall red	*TtRr* tall red
Tr	*TTRr* tall red	*TTrr* tall yellow	*TtRr* tall red	*Ttrr* tall yellow
tR	*TtRR* tall red	*TtRr* tall red	*ttRR* dwarf red	*ttRr* dwarf red
tr	*TtRr* tall red	*Ttrr* tall yellow	*ttRr* dwarf red	*ttrr* dwarf yellow

F_2:

9 tall red
3 tall yellow
3 dwarf red
1 dwarf yellow

The F_2 progeny occur in a ratio of 9 tall red, 3 tall yellow, 3 dwarf red, 1 dwarf yellow. This is the theoretical ratio which results from chance combinations of the four kinds of sperms with the four types of eggs. In a large population the actual ratio approaches the theoretical one closely. In a small population the actual ratio may deviate from the expected 9:3:3:1 ratio. Although there are only four phenotypes in the F_2, there are nine different genotypes. These are *TTRR, TTRr, TTrr, TtRR, TtRr, Ttrr, ttRR, ttRr,* and *ttrr.*

THE BACKCROSS

The backcross is made by crossing F_1 plants with one of the parents, usually the one with both recessive traits. The gametes formed and the offspring of a cross of the F_1 tall red with a dwarf yellow are shown in the checkerboard that follows.

TtRr tall red × *ttrr* dwarf yellow

Sperms

	TR	*Tr*	*tR*	*tr*
Eggs *tr*	*TtRr* tall red	*Ttrr* tall yellow	*ttRr* dwarf red	*ttrr* dwarf yellow

The tall red parent produces four types of gametes: *TR, Tr, tR,* and *tr,* whereas the dwarf yellow parent produces only one type of gamete: *tr.* The offspring of the backcross are recorded in the squares and you will notice that the ratio is 1 tall red:1 tall yellow:1 dwarf red:1 dwarf yellow.

MENDEL'S FOURTH LAW

In addition to studying monohybrid crosses, Mendel also studied dihybrid crosses. From such studies he formulated his fourth law of inheritance, which states: The members of different pairs of factors segregate independently in gamete formation. In the illustration which we used we noted the truth in this statement; the genes for height separated independently from the genes for fruit color. As a consequence of this, the gametes formed by plants with a genotype of *TtRr* could have either one of the height genes associated with either one of the color genes; *T* and *R* could be in the same gametes, or *T* and *r,* or *t* and *R,* or *t* and *r.*

INTERACTION OF FACTORS

Some characteristics of plants are not determined by a single pair of genes but by two or more pairs of interacting genes. In cattleya orchids, at least one dominant gene of one pair and at least one dominant gene of another pair are necessary for the development of colored flowers. Let us designate one pair of genes as *C* genes and the other pair as *R* genes. Plants with the following genetic makeups would bear colored flowers: *CCRR, CcRr, CcRR, CCRr.* All of them have at least one dominant *C* gene and one dominant *R* gene. Plants with genetic constitutions as follows would have white flowers: *ccrr, ccRR, CCrr, Ccrr, ccRr.* All of these lack either a dominant *C* or a dominant *R* or both, and hence they have white flowers.

If two whites of the following makeups are crossed

White *ccRR* × White *CCrr*

all of the offspring will bear colored flowers. The offspring would be *CcRr.* From one parent they would receive a big *C*, from the other a little *c*; from one parent a big *R*, from the other a little *r*.

If a *CcRr* plant is selfed, nine-sixteenths of the progeny will bear colored flowers and seven-sixteenths will bear white flowers as can be seen in the checkerboard below.

CcRx × *CcRr*

	CR	*Cr*	*cR*	*cr*
CR	*CCRR* colored	*CCRr* colored	*CcRR* colored	*CcRr* colored
Cr	*CCRr* colored	*CCrr* white	*CcRr* colored	*Ccrr* white
cR	*CcRR* colored	*CcRr* colored	*ccRR* white	*ccRr* white
cr	*CcRr* colored	*Ccrr* white	*ccRr* white	*ccrr* white

According to our modern biochemical understanding of gene action, we visualize the formation of pigment in orchids, sweet peas, and certain other flowers as occurring in at least two steps. Each step requires a different enzyme, and each enzyme is synthesized under the control of a different gene, as illustrated below.

Starting chemical $\xrightarrow[\text{RNA } C \text{, enzyme } C]{\text{gene } C \text{, DNA } C}$ colorless intermediate $\xrightarrow[\text{RNA } R \text{, enzyme } R]{\text{gene } R \text{, DNA } R}$ pigment

Gene *C* controls the production of enzyme *C*, and gene *R* that of enzyme *R*. In the absence of gene *C* the intermediate substance will not be synthesized and neither will the pigment. In the presence of gene *C* and the absence of gene *R* the intermediate substance is formed, but it cannot be changed into the pigment, a step requiring enzyme *R*. Both enzymes and both genes are necessary for pigment formation. Here, as in many other instances, the biochemical approach to genetics enables us to formulate a simple and rational explanation for what formerly appeared to be complex interrelationships between genes.

In many plants two or more pairs of interacting genes play a role in determining the size of a plant or plant part (flowers, fruits, etc.). Dominance is then lacking. In such cases, the offspring of a large plant crossed with a small one are intermediate in size; similarly for flower size, fruit size, etc. In the second generation the offspring range in size from small to medium to large. Distinct classes are not present; instead there is a continuous range of sizes. In some cases, two pairs of genes may be operating; in others, three or even more. Let us assume that two pairs of genes control the size of a plant. One pair of genes may be designated as *A* or *a*, and the second pair as *B* or *b*. Plants with a genotype of *AABB* are the largest. Those with a genotype of *aabb* are the smallest. Let us see what other size plants might result from the cross of two parents heterozygous for each gene. We will use these terms, starting from the smallest and going to the largest, to designate sizes: dwarf, short, medium, tall, giant. Offspring with all small letters are small; with 1 capital letter, short; with any 2 capital letters, medium; with any 3, tall; and with all 4, giant.

AaBb medium × *AaBb* medium

	AB	*Ab*	*aB*	*ab*
AB	*AABB* giant	*AABb* tall	*AaBB* tall	*AaBb* medium
Ab	*AABb* tall	*AAbb* medium	*AaBb* medium	*Aabb* short
aB	*AaBB* tall	*AaBb* medium	*aaBB* medium	*aaBb* short
ab	*AaBb* medium	*Aabb* short	*aaBb* short	*aabb* dwarf

The phenotypic ratio would be 1 giant:4 tall:6 medium:4 short:1 dwarf. The same results may be shown as follows to indicate the distribution of the classes.

Giant	Tall	Medium	Short	Dwarf
		AaBb		
		AaBb		
	AABb	*AaBb*	*Aabb*	
	AABb	*AaBb*	*Aabb*	
	AaBB	*AAbb*	*aaBb*	
AABB	*AaBB*	*aaBB*	*aaBb*	*aabb*
1/16	4/16	6/16	4/16	1/16

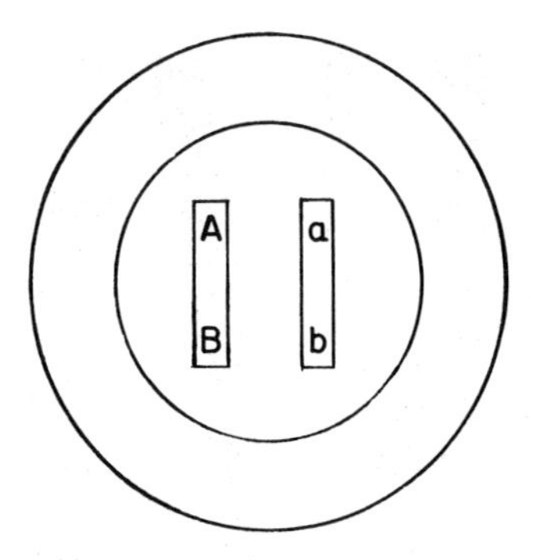

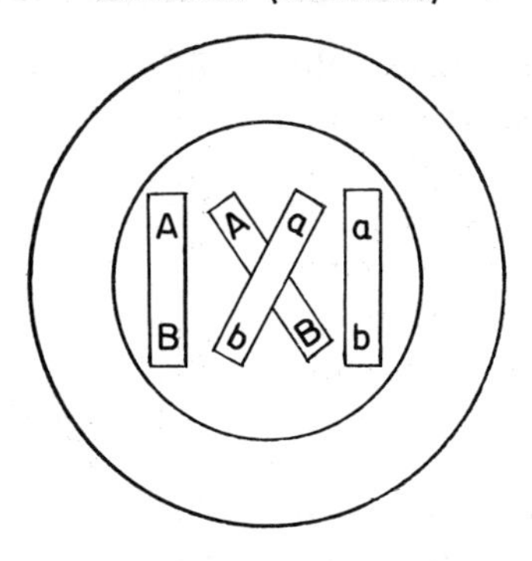

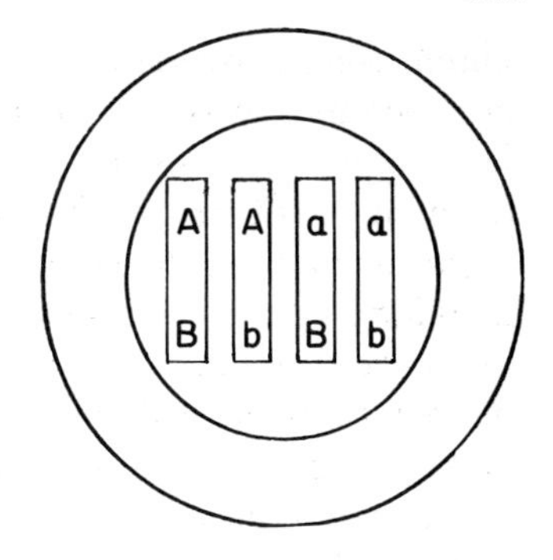

Fig. 24–2. Diagrams illustrating linkage and crossing over. *Left,* a cell in the early prophase stage of meiosis showing gene *A* linked with *B* and *a* with *b*. *Center,* a cell during prophase of meiosis after the chromosomes have become duplicated. Where the chromosomes overlap, segments may be exchanged. *Right,* the cell after crossing over.

When environmental conditions are considered, there may be a continuous range from giant to dwarf.

SEX DETERMINATION

In human beings and in many plants, one pair of chromosomes determines sex, except in certain unusual instances where the chromosomal number is abnormal. In human beings the diploid chromosomal number is 46. Two of the chromosomes are sex chromosomes, or *heterosomes;* the other forty-four are called **autosomes.** The two sex chromosomes in a female are alike and are designated as *X* chromosomes. The two in the male are dissimilar; one is an *X* chromosome, the other a *Y* chromosome. The female produces just one kind of egg; all eggs have one *X* chromosome. The male produces two types of sperm; one-half of the sperms bear an *X* chromosome, the other half a *Y* chromosome. A male offspring develops from an egg fertilized with a sperm containing the *Y* chromosome and a female from an egg fertilized with a sperm carrying an *X* chromosome (see below).

XX female × *XY* male

Sperms

	X	*Y*
eggs *X*	*XX* female	*XY* male

Sex determination in many dioecious plants, among them *Elodea,* follows the same pattern as in human beings.

LINKAGE

In peas there are seven chromosomes in each gamete, sperm or egg. If only one gene were present on each chromosome, only seven genes, controlling seven traits, could be transmitted by a gamete to an offspring. Actually, each gamete transmits to the offspring thousands of genes that control thousands of physiological and morphological traits. Hence, it is obvious that each chromosome bears a large number of genes. When genes are borne on the same chromosome, they are said to be **linked.** For example, in the cell diagrammed in Figure 24–2 (*left*), gene *A* is linked with gene *B* and *a* is linked with *b*. If we assume that there is no exchange of chromosomal parts, such a cell could form only two kinds of gametes, *AB* and *ab*, rather than the four types (*AB, Ab, aB, ab*) which would be expected if gene *A* were on one chromosome and gene *B* on a different chromosome.

CROSSING OVER

In the prophase of reduction division, after each member of a pair has become duplicated, a cell may appear as illustrated in Figure 24–2, center. Two of the four chromosomes overlap. At this place the chromosomes may exchange segments and later appear as illustrated in the right-hand diagram of Figure 24–2. The exchange of segments between homologous chromosomes is referred to as **crossing over.** After the exchange of segments, one of the four chromosomes has gene *A* linked with *b* and another has gene *a* linked with *B*. The other two chromosomes retain the original linkage relationships—that is, gene *A* linked with *B*, and gene *a* with *b*. When reduction division is completed in this cell, four types of cells result: *AB*, *ab*, *aB*, and *Ab*.

The various stages in crossing over are shown in Figure 24–3. The homologous chromosomes pair and then each replicates itself. As they become duplicated, the two newly formed chromosomes may overlap at a region known as a chiasma. Chromosomes have a texture about like gelatin and accordingly they are easily broken. Where they overlap, they may break and then the segments may recombine to form chromosomes with different linkage arrangements.

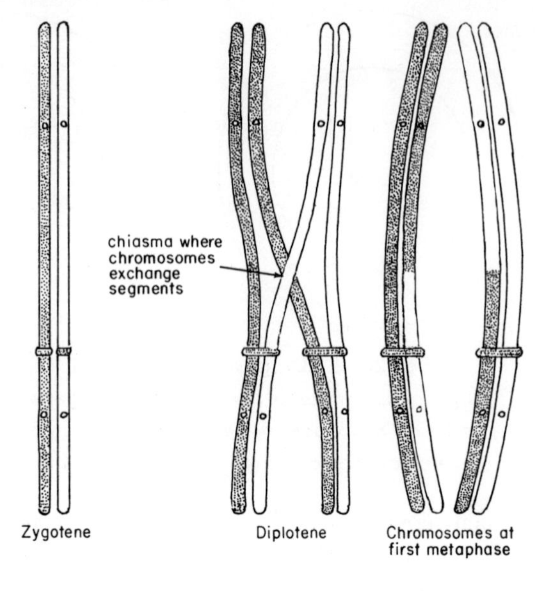

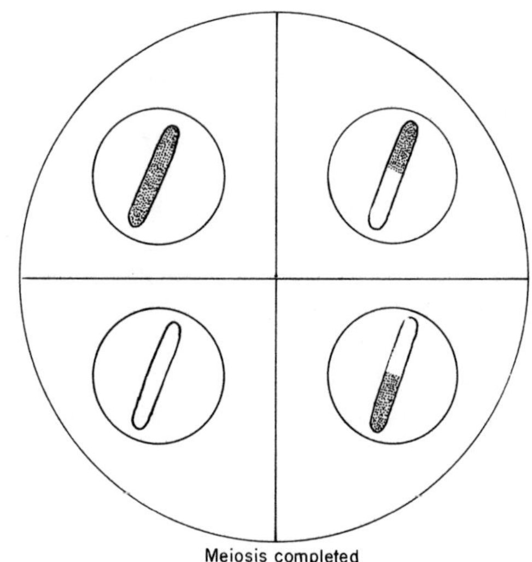

Fig. 24–3. Crossing over is the interchange of corresponding segments between homologous chromosomes. Notice that when meiosis is completed, four types of cells result. The two cells on the left have the original linkage relationship, whereas the two on the right have a new linkage pattern.

DISTANCES BETWEEN GENES

Crossing over occurs between *A* and *B* only in a relatively small percentage of cells, and this percentage is proportional to the distance between the genes. When the genes are close together, the frequency of crossing over is low; when they are far apart, the frequency is high. Hence it is possible to determine distances between invisible genes by using the percentage of crossing over as a measure.

The percentage of crossing over can be determined by mating an F_1 with the double recessive parent. To illustrate the technique we shall study linked genes of tomato (Fig. 24–4). In tomatoes tall vine (*T*) is dominant over dwarf (*t*), and spherical fruit shape (*S*) is dominant over pear shape (*s*). The gene for height is linked with the gene for shape. Because the genes are linked, the genotype of a tall sphere plant is written *TSTS* and the genotype of dwarf pear as *tsts*. If the genes for height were on different chromosomes from those for shape, the genotype of the tall sphere would be written *TTSS*.

The F_1 is obtained by mating a homozygous tall spherical plant (*TSTS*) with a

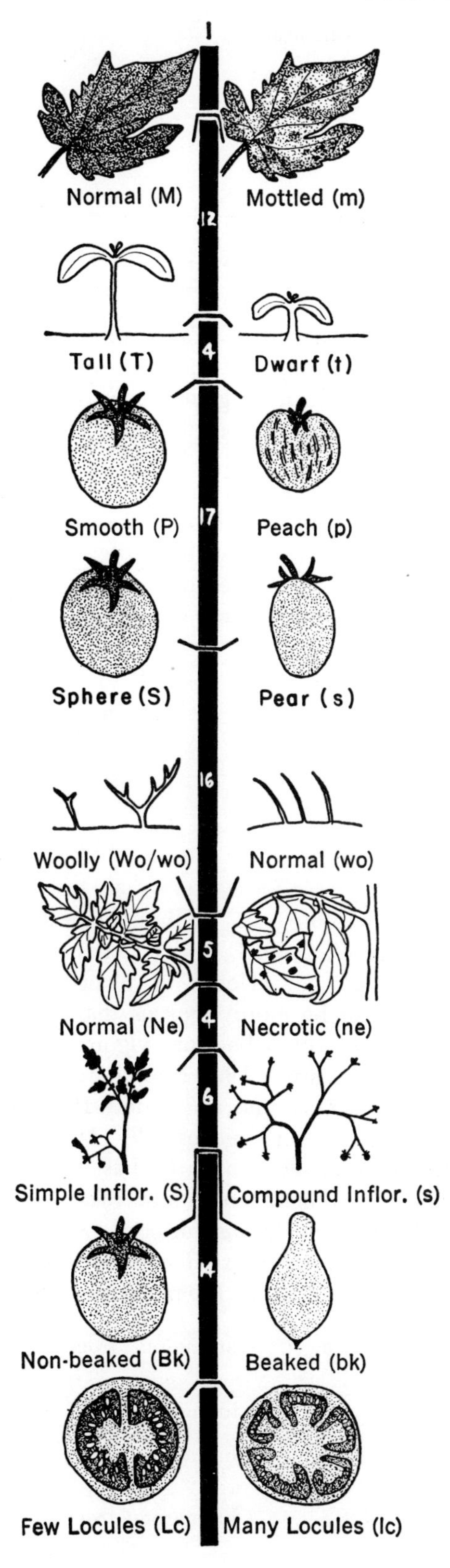

dwarf pear (*tsts*). The F_1 offspring (*TSts*) are crossed with the dwarf pear parent (*tsts*). The phenotypes and genotypes obtained in the F_1 and the backcross are recorded at the top of page 328.

If the genes were on different chromosomes, independent assortment would prevail and the offspring would be produced in a ratio of 1 tall sphere:1 dwarf pear:1 tall pear:1 dwarf sphere. This ratio is not obtained when the above cross is made; therefore, we know that the genes are linked. A count of the offspring from the backcross shows that 40 per cent of the offspring received the chromosome *TS* and another 40 per cent the chromosome *ts* from the heterozygous parent. Ten per cent received the chromosome *Ts* and 10 per cent the chromosome *tS*. The latter two chromosomes are the result of crossing over. The noncross-over gametes make up 80 per cent of the population (40 per cent + 40 per cent). The cross-over gametes make up 20 per cent of the population. The percentage of cross-over gametes is a measure of the distance between the genes *T* and *S*. Such genes are said to be 20 units apart.

HYBRID VIGOR

Our understanding of hybrid vigor is incomplete; a complete understanding awaits further research. Several theories have been proposed to account for hybrid vigor. We will discuss just one of them.

Fig. 24–4. A chromosome map. Pictorial representation of nine genes in the tomato which are located on chromosome 1. Numerals between two adjacent genes refer to the percentage of crossing over between them, and this percentage is a measure of the distance between the genes. For example, the gene for tall (*T*) is 12 units from the gene for normal leaves (*M*). Dominant genes are shown on the left, recessive ones on the right. (From *Botany, Principles and Problems*, 6th ed., by E. S. Sinnott and K. S. Wilson. Copyright © 1955, 1963, McGraw-Hill Book Co., Inc., New York. Used by permission.)

Parents: Tall sphere × Dwarf pear
TSTS *tsts*

F_1: *TSts* Tall sphere

Backcross: *TSts* × *tsts*

	TS	*ts*	*Ts*	*tS*
ts	*TSts* Tall sphere 40%	*tsts* Dwarf pear 40%	*Tsts* Tall pear 10%	*tSts* Dwarf sphere 10%

Fig. 24–5. Hybrid vigor. The inbred parent strains at the left and right were crossed to produce the vigorous hybrid in the center. (Department of Genetics, Connecticut Agricultural Experiment Station.)

Many genes play a role in determining the vigor, size, and yield of plants (Fig. 24–5). The "vigor" genes occur on different chromosomes, and on any one chromosome there may be several "vigor" genes. This situation may be illustrated by considering just two pairs of chromosomes. Let us assume that we have two homozygous varieties (I and II) of a certain species with the following chromosomal constitutions:

Variety I	Variety II
A b C	*a B c*
A b C	*a B c*
d E f	*D e F*
d E f	*D e F*

Furthermore, we shall assume that the large, *A*, *B*, *C*, *D*, *E*, and *F* genes are dominant for vigor. Hence, a plant with *AA* would not be more vigorous than one with *Aa*, but it would be more vigorous than an *aa* plant. Variety I has three different genes for vigor, *A*, *C*, and *E*. Variety II has three different "vigor" genes, *B*, *D*, and *F*. Neither variety I nor II has all "vigor" genes, and hence both are relatively low-yielding, nonvigorous varieties.

If variety I is crossed with variety II, the offspring will have chromosomes as follows:

A b C Chromosome from I

a B c Chromosome from II

d E f Chromosome from I

D e F Chromosome from II

Six genes for vigor (*A, B, C, D, E,* and *F*) are present in the offspring. They are vigorous plants which give a high yield. The vigor of the hybrids is known as **hybrid vigor,** and it results because dominant genes are brought together in the F_1. Hybrid vigor is noted in many plants, but not in all kinds. You should not assume that cross-

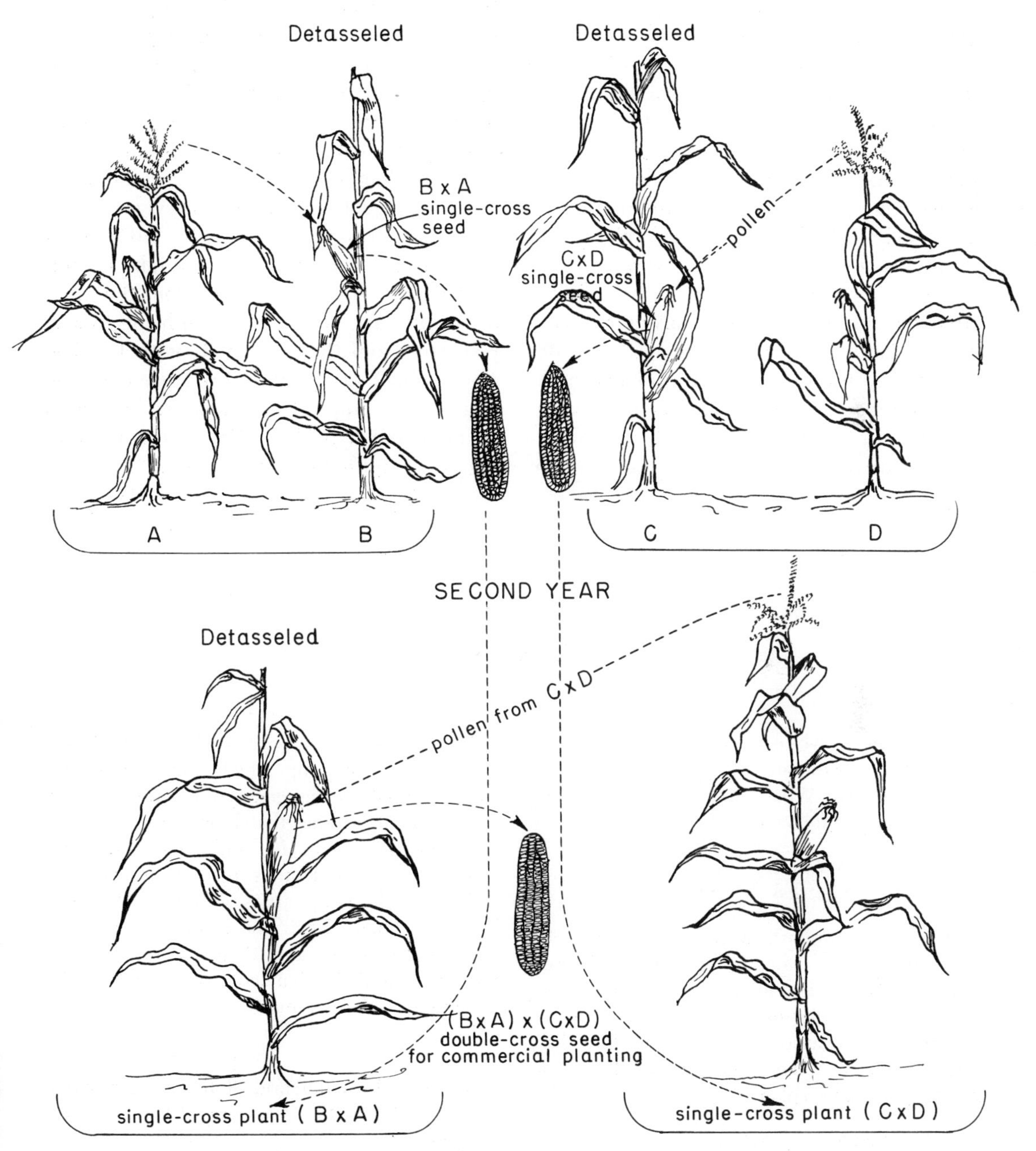

Fig. 24–6. Diagram of method of crossing inbred plants of corn and the resulting single crosses to produce double-cross hybrid seed. (Redrawn from *U.S.D.A. Farmer's Bull. 1744.*)

ing any two inbred lines will always produce vigorous offspring.

In many crop and ornamental plants increased yields and more vigorous plants may be secured by crossing two different homozygous varieties and then planting the F_1 seed. The resulting plants are known as **single-cross hybrids.** Among the single-cross hybrids which have proved superior are those of petunia, tomato, sorghum, and onion. The production of hybrid onion seed was facilitated by finding male-sterile plants—that is, plants whose flowers have an effective pistil but which produce no viable pollen. Male-sterile onion plants are grown in a field with a different variety whose flowers are normal. Insects carry pollen from the normal flowers to pistils of male-sterile plants, from which the seeds are obtained. The use of male-sterile plants markedly lessens the cost of producing hybrid seed because slow and tedious hand pollination is not necessary. Male-sterile sorghum plants are available, and their use has made it economically feasible to produce hybrid sorghums.

Hybrid corn seeds could be produced by crossing two inbred corn lines. Because the inbred corn lines produce few seeds, the cost would be quite high. Hence, hybrid corn seed is produced by first developing four inbred lines which may be designated as A, B, C, and D (Fig. 24–6). It requires about 7 years of self-pollination to develop each line. After seven generations of inbreeding each line is nearly homozygous, and each line is low in vigor. Line A is then crossed with line B, and C with D. Both hybrids are vigorous, high-yielding plants. The hybrid resulting from the cross of A with B is then crossed with the hybrid resulting from the cross of C with D. The seeds which are produced are used for the commercial crop. The plants are known as **double-cross hybrids** and they are uniform, vigorous, and high-yielding plants.

QUESTIONS

1. In squash white is dominant over yellow and disk is dominant over sphere. The genes are not linked. What offspring could be expected from the following cross:

$$\underset{\text{White disk}}{Ww\ Dd} \times \underset{\text{White disk}}{Ww\ Dd}$$

2. _____ In squash white fruit color (W) is dominant over yellow (w) and disk fruit shape (D) is dominant over sphere (d). Which of the following crosses would produce offspring in the ratio of one-fourth white disk, one-fourth white sphere, one-fourth yellow disk, one-fourth yellow sphere?

(A) $WwDd \times WwDd$
(B) $WwDD \times wwdd$
(C) $WwDd \times Wwdd$
(D) $WwDd \times wwdd$

3. Two red tall pea plants when crossed produced some offspring which were white and dwarf. What are the genotypes of the parents?

4. In man, assume that brown eyes (B) are dominant over blue (b); and right-handedness (R) over left-handedness (r). A brown-eyed, right-handed man marries a blue-eyed, right-handed woman. Their first child is brown-eyed and right-handed, and their second child is blue-eyed and left-handed. Write below the genotypes of the parents.

_______________ × _______________

5. _____ In a dihybrid, if the F_2 phenotypic ratio is 9:3:3:1, the genes are (A) linked, (B) not linked.

6. In violets blue flower color (B) is dominant over yellow (b) and large flowers (L) over small (l). A plant with large yellow flowers was crossed with one bearing small blue flowers. The offspring were as follows:

48 large, yellow
52 small, yellow
46 large, blue
54 small, blue

Record below the genotypes of the parents.

____________ × ____________

7. A plant breeder crossed two sweet pea plants both having white flowers. All offspring produced purple flowers. When the plants bearing purple flowers were selfed, the offspring occurred in a ratio of 9 purple: 7 white. Explain.

8. _____ If two chromosomes are as follows before crossing over:

A	*B*
a	*b*

After crossing over they will be like:

(A)

A	*b*
a	*B*

(B)

A	*a*
B	*b*

(C)

A	*B*
a	*b*

(D)

a	*A*
B	*b*

9. How is hybrid corn produced?

10. _____ Hybrid vigor is more striking in the F_2 than in the F_1. (A) True, (B) false.

11. In corn, a color gene (*C*) is dominant over colorless (*c*) and a full endosperm gene (*S*) is dominant over shrunken endosperm (*s*). An F_1 plant heterozygous for both gene pairs was backcrossed to the double recessive, a colorless shrunken plant. The progeny of the backcross were as follows:

Colored full	4000
Colored shrunken	150
Colorless full	150
Colorless shrunken	4000

Are the genes linked or assorted independently? ____________. If the genes are linked, what is the percentage of crossing over? ____________. If the genes are linked, what is the distance between them? ____________.

12. _____ Assume that genes *A* and *B* are linked and that no crossing over occurs. If the following cross, *ABab* × *ABab*, is made, the number of genotypes among the offspring will be (A) two, (B) three, (C) four, (D) nine. How many genotypes would there be if crossing over does occur?

13. Give a biochemical explanation for the development of pigment in orchids, sweet peas, and other plants where two pairs of genes are involved.

14. Assume that in squashes the difference between a 3-pound type and a 7-pound type is due to two factor pairs, *AA* and *BB*, each factor contributing 1 pound to fruit weight. Cross a 3-pound plant (*aabb*) with a 7-pound plant (*AABB*). What will be the phenotypes of the F_1? Of the F_2?

25

Variations

Offspring are never exactly like the parents. We seldom have difficulty in distinguishing one person from another and we know that no two human beings have exactly the same fingerprints. Superficially, all of the plants in a field of wheat may appear to be alike, but closer examination reveals that each plant differs slightly from the others. Although we will emphasize variations in this chapter, we should keep in mind that stability is also of great significance in the living world. Most genes of a species have been transmitted unchanged through countless generations.

VARIATIONS DUE TO ENVIRONMENT

The appearance, behavior, and activity of an organism are the result of two sets of interacting factors, heredity and environment. The potentialities are determined by heredity, but the actual development and expression of these potentialities are governed by the environment. For example, a certain plant may be homozygous for tallness and under favorable environmental conditions it develops into a tall plant, but if this plant is grown where water is deficient, it does not attain a tall size. In plants variations are produced by light, temperature, and nutrition, as well as by moisture.

Environmental variations are not inherited. In other words, acquired characters are not passed on to the offspring. For example, if seeds collected from plants grown where moisture is deficient are planted under uniform conditions with seeds from plants (with the same genotype) which have been grown where moisture is optimal, the resulting plants are alike.

Light

If one group of plants is raised in the dark and another group of genetically similar plants in light, the two groups are markedly different. Those grown in the dark lack chlorophyll, whereas those cultivated in light are green. Genes for the formation of chlorophyll are present in the group maintained in the dark, but such genes can act only in a certain type of environment, in this case, an environment where light is present. This illustrates clearly the fact that characters as such are not inherited. What are inherited are potentialities to act and interact in a certain way to a given environment. The developmental pattern of plants grown in darkness is unlike the developmental pattern of plants exposed to light, even though the plants have the same genotype. Plants grown in the dark are spindly, nonflowering plants with minute leaves, whereas those grown in light are sturdy flowering plants with expanded leaves.

Genes operating in plants which are exposed to long days may bring about one pattern of growth; the same genes operating in plants exposed to short days often condition a different pattern of development. For example, under long-day conditions the Mammoth variety of tobacco grows extremely tall but does not flower. Under short-day conditions the plants flower and they are relatively short.

Temperature

Plants with the same genotype may be unlike if they are raised at different temperatures. In a cold atmosphere plants remain stunted; at more favorable temperatures large plants may develop. Flower color sometimes is determined by temperature. If Chinese primroses are grown at 86° F., the plants bear white flowers; but if genetically identical plants are raised at a temperature of 68° F., the flowers are red.

Nutrition

Plants grown in fertile soil are different from those which develop in infertile soil. Those which grow in soil deficient in nitrogen are stunted plants with small, yellowish-green leaves, whereas those growing in soil at an optimal nitrogen level become large plants with green leaves.

VARIATIONS WHICH ARE INHERITED

Variations which can be transmitted to the progeny are, however, produced. Some of these variations may be desirable, others undesirable. Man makes use of the desirable variations in his breeding programs with cultivated plants and domestic animals. The accumulation of variations in a given species may lead to the production of varieties or species very much unlike the original parents. For example, the home gardener would hardly recognize the wild sweet pea plants from which man has developed our modern sweet peas. The latter are the result of many heritable variations which have occurred spontaneously since the sweet pea was first introduced into cultivation in 1699. Many of these variations arose as a consequence of gene mutations.

GENE MUTATIONS

In about 1731 a single sweet pea plant appeared which produced red flowers instead of the anticipated reddish-purple flower. This unexpected change in phenotype was the result of a deep-seated change in a gene. Such a spontaneous and unexpected alteration in the structure of a gene is referred to as a **gene mutation.** Since that time many other genes in sweet peas have mutated and resulted in plants of different appearances and in a variety of flower colors. One of the most important of such mutations was the sudden appearance of the "Spencer" type of sweet pea in 1900. Prior to the appearance of this mutant, the best varieties had moderately sized flowers and plain petals without any trace of waving. The mutant "Spencer" type had large flowers and wavy petals, and these characteristics could be passed on to the offspring. By cross-breeding, the "Spencer" characteristic was recombined with other desirable traits into superior varieties of sweet peas.

In some instances gene mutations occur in only a part of a plant, so-called **bud mutations,** or **bud sports.** For example, in 1741, one branch of a peach tree produced fruits which were smooth-skinned while the other branches produced the usual and expected fuzzy peaches. The seed from a smooth-skinned peach was planted and at maturity the tree produced fruits which were smooth-skinned. The gene mutation was heritable. Smooth-skinned "peaches" are referred to as nectarines. We now know that the difference between a peach and a nectarine is determined by one pair of genes. A homozygous peach mated with a nectarine produces all peach trees in the F_1 and three peach trees to one nectarine in the F_2 generation. Many popular varieties of apples, seedless grapes, seedless oranges, tulips, roses, and chrysanthemums originated as bud mutations and these new varieties have been propagated asexually.

Countless numbers of such gene mutations have occurred in the past and are occurring today in plants and animals. The breeder is constantly on the lookout for them. They occur in native plants as well as in cultivated plants and as a consequence new varieties and even species are evolving. Gene mutations may occur anywhere within a body, but unless they occur in a cell which ultimately leads to the development of reproductive structures, they are lost.

Mutations occur in lower plants as well as in more advanced types of plants. A mutant strain of *Penicillium* mold appeared following X ray treatment. This mutant strain produced more penicillin for combating disease bacteria than any type previously known. But bacteria also mutate. Some of the mutants can tolerate a high concentration of penicillin, a matter of considerable concern in public health. As man develops more potent antibiotics, the bacteria, through gene mutations, may produce more resistant varieties. The mutated strains of bacteria that are resistant to penicillin survive and multiply, and the probability that they will infect new victims is increased. It will be a continual struggle to keep ahead of bacterial mutations as they are formed.

Likewise pathogenic true fungi mutate, among them the fungus that causes black stem rust of wheat. A plant breeder may develop a variety of wheat which is resistant to the rust disease for several years. However, in time, the new variety becomes diseased and yields go down. The wheat has not changed in the years following its introduction, but the fungus has changed. Mutant races of the fungus, capable of attacking the new variety, have originated spontaneously.

Some mutant forms of molds have been studied to get basic information about how genes function. Work done with *Neurospora*, an orange-colored mold that grows on bread, by George Beadle and Edward Tatum led to the profound discovery that genes act by controlling the synthesis of enzymes. Normal *Neurospora* grows on a medium lacking thiamine, but a certain mutant, produced by bombarding spores with X rays, does not develop when thiamine is lacking. The mutated gene lacks the instructions for the production of an enzyme that is required for thiamine synthesis. Other mutant forms produced by X rays cannot synthesize riboflavin, or niacin, or such amino acids as arginine and tryptophane. In normal *Neurospora*, a gene controls the synthesis of tryptophagenase, an enzyme that brings about the formation of tryptophane from serine and indole. In a mutant form, the altered gene lacks the instructions for the synthesis of tryptophagenase and hence it cannot produce tryptophane. Normal *Neurospora* thrives on a medium lacking tryptophane, but the mutant form grows only when the medium contains tryptophane.

Studies with mutant forms led to the development of the "one gene–one enzyme hypothesis" which states that a specific gene controls the synthesis of a certain enzyme. In recent years we have learned how genes control enzyme synthesis. You will recall that the DNA of a gene produces a definite kind of RNA which in turn governs the sequence of amino acids in an enzyme. One sequence results in one kind of enzyme; a different sequence of amino acids, in a different enzyme.

The synthesis of a substance in a cell often occurs in two or more steps. For example, substance A may be changed into compound B which then is changed into C which may be an essential metabolite. Each step in the reaction is controlled by an enzyme. In the example below, two different enzymes and two different genes are involved. We may summarize the process as follows:

$$A \xrightarrow[\text{enzyme 1}]{\text{gene 1}} B \xrightarrow[\text{enzyme 2}]{\text{gene 2}} C$$

Gene 1 controls the synthesis of enzyme 1 which changes A to B. Gene 2 and enzyme 2 play a role in the transformation of B to C. If gene 1 mutates, substance A cannot be converted into B, and accordingly the mutant will grow only if furnished with B or C. If gene 2 mutates, substance B accumulates and the organism grows only if provided with substance C.

Agents Causing Mutations

Gene mutations occur spontaneously in nature, but with low frequency. In *Drosophila* and in corn, mutations have been found to range in frequency from 1 in 100,000 to 1 in 1,000,000 per generation. The frequency of mutations can be markedly increased by X rays. Bombardment of a gene by X rays modifies its chemical composition and accordingly the mutated gene results in an altered physiological or morphological phenotype. A single hit with an X ray, in the right place, brings about the mutation of a gene, indicating, according to biochemists, that a gene is a single molecule.

Hydrogen bombs, radioactive isotopes, radioactive fallout, and atomic machines in general give off gamma rays and/or beta rays and/or alpha rays. Beta rays are fast-moving electrons; alpha rays are rapidly moving helium nuclei, and gamma rays are short X rays. Like X rays, such irradiations may bring about mutations in plants, animals, and human beings. A few mutations may be beneficial, but the great majority are not; instead most are deleterious and many are lethal, a fact which might be anticipated. Each organism is highly complex and well coordinated and just one alteration of a key reaction could disrupt the harmonious processes so necessary for life. For example, if a gene which controls the synthesis of a respiratory enzyme mutates, respiration may be blocked and death may ensue. A gene necessary for the synthesis of chlorophyll may mutate into an inactive form; the plant carrying the mutant gene may be yellow and unable to carry on photosynthesis.

Most nonlethal novelties resulting from gene mutations are not likely to be superior to the normal phenotypes. The gene pool of any plant or animal has been subjected to natural selection for a very long time and during this period favorable genes have been perpetuated and those unfavorable to survival have been eliminated. We, like all other organisms, have a very select collection of genes, those favoring survival.

A few mutant genes are dominant, but the great majority are recessive. Human beings exposed to ionizing radiations may carry harmful recessive mutant genes. These mutant genes may be hidden for generations, exhibiting themselves only when certain matings occur. Assume that most people have a favorable genotype of *AA*. If one member of the gene pair mutates, the individual would have a genotype of *Aa*, in which *a* is a harmful recessive gene. If this person married a normal person with a genotype of *AA*, none of the offspring would have the undesired trait. When two individuals with *Aa* genotypes mate, there is one chance out of four that an *aa* offspring would result. Similarly recessive mutants in plants and animals may be hidden for one or more generations.

A gene mutation results when an error occurs during gene replication. In the original gene an adenine–thymine bridge may be followed by a cytosine–guanine one. Rarely in a normal cell, and more frequently in an irradiated cell, the sequence of the nitrogenous bases may be altered, resulting in a new gene, a mutation. Thus in the mutated gene an adenine–thymine bridge may be followed by a thymine–adenine one instead of the cytosine–guanine one of the original gene. When the mutant gene duplicates itself, it will form an exact copy and thus it may be passed on to the offspring for generation after generation.

The astonishing feature is not that errors occasionally occur during gene replication but that genes are copied so exactly

Fig. 25–1. Mutation in carnation. Irradiation of a white variety of carnation, White Sim, produced branches bearing red or red and white flowers. The mutant forms may be propagated indefinitely by cuttings. (Brookhaven National Laboratory.)

throughout the life of an organism and for generation after generation. Some of our genes, those related to respiration, may go back to the dawn of life and ever since they have been copied precisely. Other genes are, of course, of more recent vintage, those genes which are unique to man. The next time you are in a forest look overhead at the countless millions of leaves. All are likely to be green. No mutation of a "chlorophyll gene" occurred during the development of all the leaves.

Radiations have been used to produce mutations in crop plants. Desirable mutants are saved; undesirable ones are eliminated. Mutations were induced in Mohawk oats by exposing seeds to thermal neutrons from the pile at Brookhaven National Laboratory. A few seeds developed into plants which were resistant to rust disease, a disease to which ordinary Mohawk oats are susceptible. In another experiment, a number of agronomically desirable mutant strains of oats were isolated from X-irradiated seeds of the Huron variety. Some of the mutants matured earlier and gave higher yields than the parents. Figure 25–1 shows a mutant carnation which resulted when a white variety was exposed to X rays.

Mustard gas, extremes of temperatures, and ultraviolet irradiation are other agents which increase the frequency of gene mutations. Mustard gas and related compounds are about as effective as heavy X ray treatment in producing mutations. Brief exposures to almost lethal high or low temperatures increase the mutation frequency manyfold in both plants and animals.

VARIATIONS RESULTING FROM GENE RECOMBINATIONS

In both plants and animals the offspring often have combinations of genes different from their parents'. With orchids and sweet peas, white-flowered parents may yield colored progeny, a consequence of genes *C* and *R* being recombined. In snapdragons, a red-flowered plant when crossed with a white-flowered plant produces offspring which bear pink flowers, certainly unlike either parent but readily explainable on genetic grounds.

VARIATIONS RESULTING FROM ALTERED CHROMOSOMAL NUMBERS

Normally plants have two sets of chromosomes in the vegetative cells. Such plants are **diploid.** However, among our cultivated and native plants occasional individuals are found which have more than two sets of chromosomes. Plants which have some multiple of sets other than two are referred to as **polyploids.** There are several kinds of polyploids: **triploids, tetraploids, pentaploids,** and so on. For example, in roses there are seven chromosomes in a set. A diploid rose plant has two sets of 7, or 14; a triploid rose three sets, or 21 (3×7); a tetraploid four sets, or 28; a pentaploid five sets, or 35; a hexaploid six sets, or 42; and an octaploid eight sets, or 56.

Polyploids may be separated into two types: (1) autopolyploids and (2) allopolyploids. In **autopolyploids** the sets of chromosomes are practically identical with one another, whereas in allopolyploids the sets are not all alike. **Allopolyploids** are hybrids between two or more species. They have more than two sets of chromosomes, but some sets are those of one species, while other sets are from a different species.

Both autopolyploids and allopolyploids may be spontaneously produced in nature, and they can also be produced almost at will by man. Tetraploids are the most common types in both groups.

Autopolyploids

Autopolyploids are frequently plants of considerable ornamental value, and a number produce superior fruits. In some autopolyploids there is considerable sterility, that is, few or no viable seeds are produced. However, these plants may be propagated asexually, as described in Chapter 18. Other autopolyploids, especially certain tetraploids, produce many viable seeds and of course can be increased by seeds.

Tetraploids. The tetraploids have four sets of chromosomes and often are an improvement over ordinary diploids in several ways. Usually they are more vigorous plants with taller stems, larger leaves, and larger flowers of superior texture—characteristics generally desired in ornamental plants. As seed crops, however, few of the tetraploids are superior to the diploids because there is often some sterility. The seeds may be larger, but fewer are produced than in the diploids. Sterility in tetraploids results from a somewhat erratic distribution of the chromosomes in the first division of meiosis. In diploids the same number of chromosomes goes to each pole during the anaphase of the first division. This regular distribution results because homologous chromosomes associate in pairs during the prophase. For example, one long chromosome pairs with the other long chromosome and likewise other homologous chromosomes come together in pairs.

The situation in tetraploids is quite different because in these there are four of each kind of chromosome present. The four of any one type, for example four long chromosomes, may come together in several ways. Two of the long chromosomes may pair in the first prophase and the other two may also pair. These chromosomes would be distributed regularly. In other cells, though, three long chromosomes may associate together and leave the fourth one without a mate. If this pattern occurs, the anaphase distribution will be irregular. The irregular distribution may result in pollen grains or megaspores having an incomplete set of chromosomes, and these will be nonviable.

Tetraploids can be readily produced through the use of the poisonous drug, colchicine. Superior varieties of lilies, snapdragons, marigolds, and other plants have been developed through the use of colchicine. When dividing cells are exposed to colchicine solutions, mitotic cell divisions are arrested at the metaphase stage because colchicine liquefies the spindle fibers or prevents their formation. Stages from metaphase on, which lead to the production of tetraploid cells, are illustrated in Figure 25–2. Here, for the sake of simplicity, is shown a cell which had two chromosomes at the beginning of mitosis. In prophase the two chromosomes split. The figure in the upper left shows the metaphase stage. If mitosis were to go on normally in this cell, each of the two resulting cells would have two chromosomes. When division is completed in cells treated with colchicine, each cell has four chromosomes, or double the normal number. Several events occur in the production of such tetraploid cells. First, the colchicine destroys the spindle. After the effect of the colchicine wears off, a nuclear membrane forms around the chromosomes. Then the division starts all over again, this time with twice the normal number of chromosomes. When the four chromosomes split, there are eight to be carried along in division. The chromosomes move

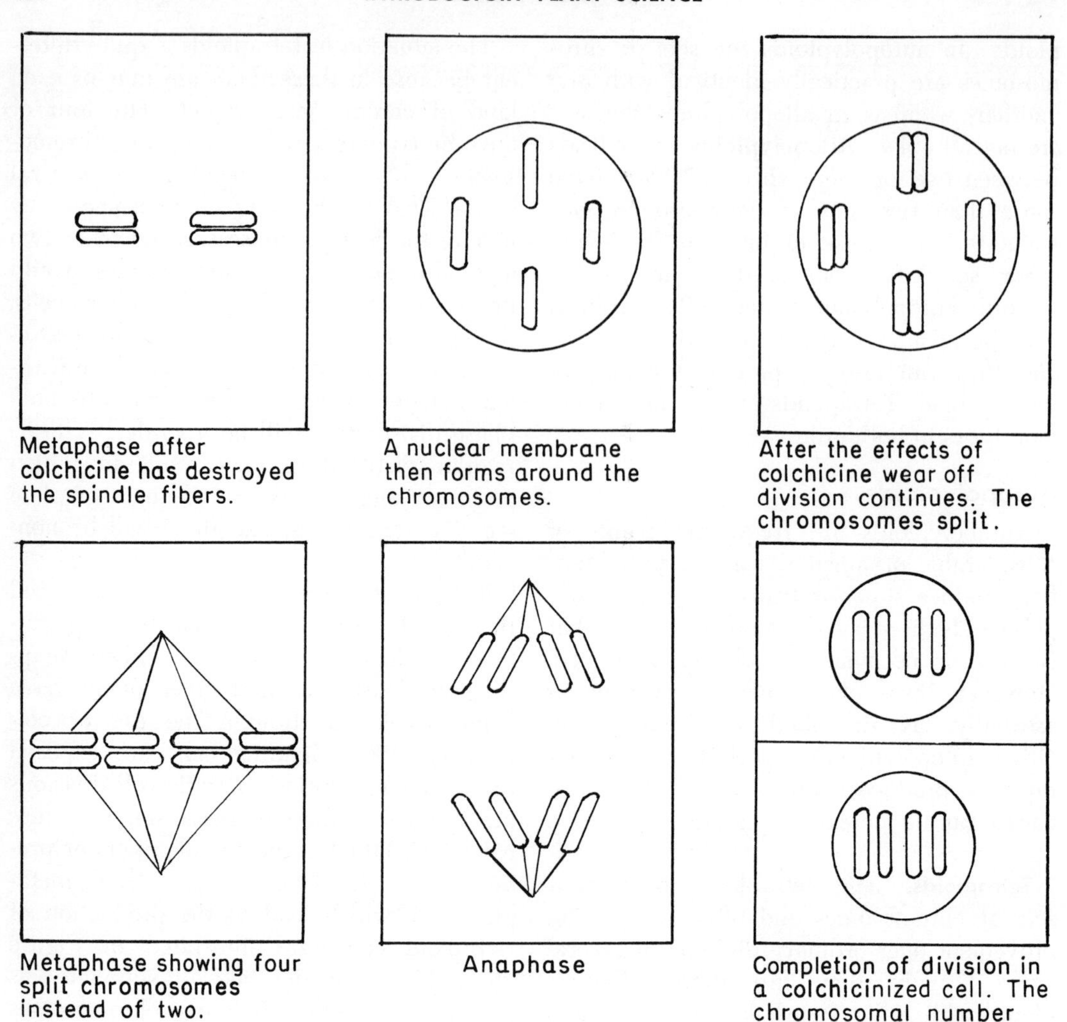

Fig. 25–2. Stages leading to the production of tetraploid cells following colchicine treatment.

to the equator. In the anaphase of the example above, four chromosomes go to one pole and four to the opposite pole. When division is completed, each cell has four chromosomes, twice the normal number. When lilies are treated with colchicine, the twenty-four chromosomes become doubled in the resulting cells so that each new cell has forty-eight (Fig. 25–3).

Colchicine affects only dividing cells. Hence in the experimental production of tetraploids, dividing cells must be treated. Furthermore, the treated cells should be ones which will develop into a new shoot. A number of tetraploids have been produced by treating seedlings with 0.1 or 0.2 per cent colchicine. One convenient way to apply the colchicine is to put a drop of the solution on the growing point of a shoot of a seedling. The treatment should be repeated several times with an application every other day. Seed treatment is another technique. By this method tetraploidy has been induced in *Datura, Portulaca, Cosmos, Petunia,* and *Cucurbita.* Success has been obtained in some instances by soaking seeds for 6 hours in 0.1 or 0.5 per cent solutions of colchicine. After soaking, the seeds are

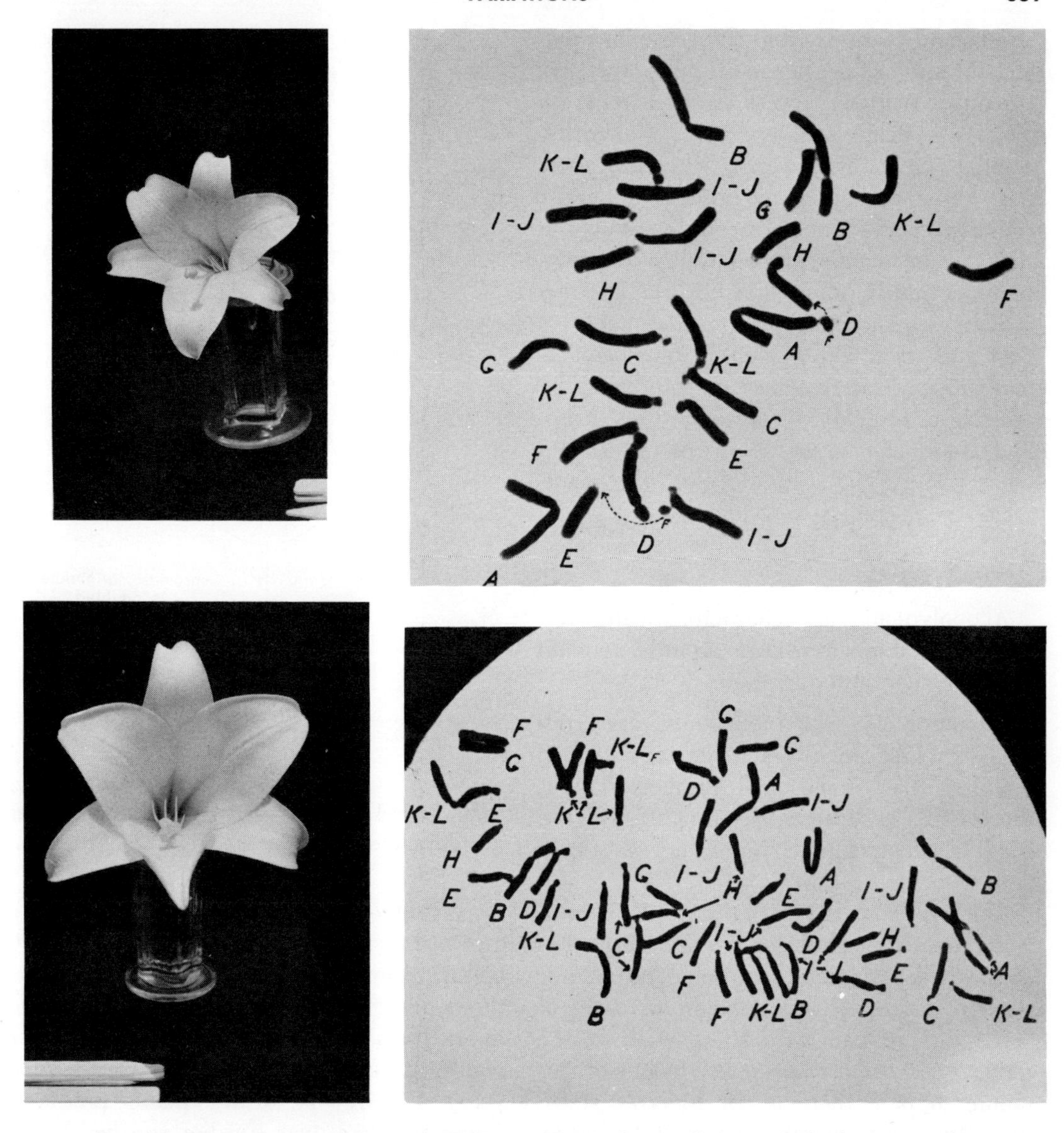

Fig. 25–3. Diploid and tetraploid lilies. *Upper*, flower from a diploid plant and a photomicrograph of a metaphase plate of the diploid. *Lower*, flower from a tetraploid and a photomicrograph of a metaphase plate of the tetraploid. The diploid plant has two sets of 12, or 24 chromosomes. The tetraploid has four sets of 12, or 48. (Bureau of Plant Industry, U.S.D.A.)

rinsed and planted. Tetraploid Easter lily plants have been developed by soaking bulb scales in 0.2 per cent colchicine for 2 hours. After the colchicine treatment, the bulb scales are wiped dry, treated with a fungicide to prevent rot, and planted. Later small bulblets form at the base of the scale and some of the bulblets develop into tetraploid plants with flowers of large size.

Triploids. Autotriploids can be produced by mating a diploid with an autotetraploid. The offspring of such a cross receive one set of chromosomes from the diploid parent and two sets from the tetraploid parent. Many choice horticultural plants are triploids: Baldwin apples, Gravenstein apples, Pink Beauty tulips, Grand Maître hyacinths, Japanese flowering cherries, some pear varieties, and many others. Generally triploids cannot be increased by seed, but they can be readily propagated by cuttings, bulbs, corms, budding, grafting, etc.

Some triploids do not produce seeds, a trait desired in watermelons. Some seedless varieties of watermelons are produced by crossing a tetraploid watermelon with a diploid one. The cross produces seed, but when planted the seeds develop into plants which bear seedless watermelons.

Allopolyploids

Allopolyploids are characteristically fertile; hence they offer more promise in plant breeding than autopolyploids.

Allotetraploids. Allotetraploids are produced by first crossing two species and then doubling the chromosomal number. This is illustrated in the diagram of Figure 25–4. Each rectangle represents a set of chromosomes.

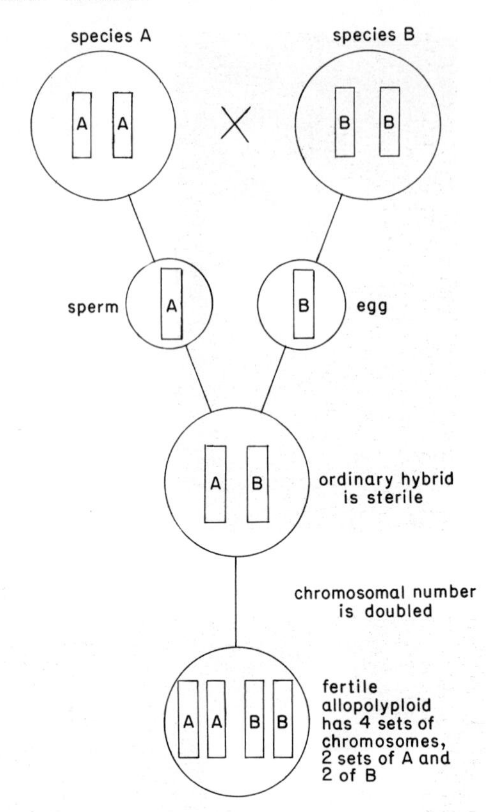

Fig. 25–4. Stages in the production of an allotetraploid.

The ordinary hybrid (the diploid one) is frequently partially or wholly sterile because the chromosomes of set A are unlike those of set B. Hence the chromosomes of set A do not pair with those of B at the reduction division. As a consequence of this failure to pair, there is considerable sterility. The allopolyploid is completely fertile, however, because the chromosomes of one A set pair with the other set of A chromosomes and the two B sets likewise pair. Allopolyploidy provides a means of conferring fertility upon hybrids and in nature has played a role in the development of new species. Several important crop plants are believed to have arisen by the chance crossing under natural conditions of two or more species, with a subsequent doubling of the chromosomes, among them wheat, cotton, and tobacco.

To illustrate the formation of an allotetraploid, let us consider the history of a primrose, *Primula kewensis.* Some years ago a grower crossed *Primula floribunda,* the buttercup primrose, with *Primula verticillata,* the Arabian primrose. The resulting diploid hybrid was sterile, i.e., it did not produce viable seed. This went on for several years. Finally one of the plants produced a branch which was sturdier and had larger leaves and flowers. These larger flowers set viable seed. The seeds were planted. The resulting plants also produced the larger flowers and viable seed. They were officially named *Primula kewensis.* On examination they were found to have double the expected number of chromosomes. In that one fertile branch on the otherwise sterile hybrid the chromosomal number had doubled spontaneously.

The story of the origin of wheat is not as simple as that of *Primula kewensis,* and is much more ancient. This cereal, now planted on about 480 million acres, goes

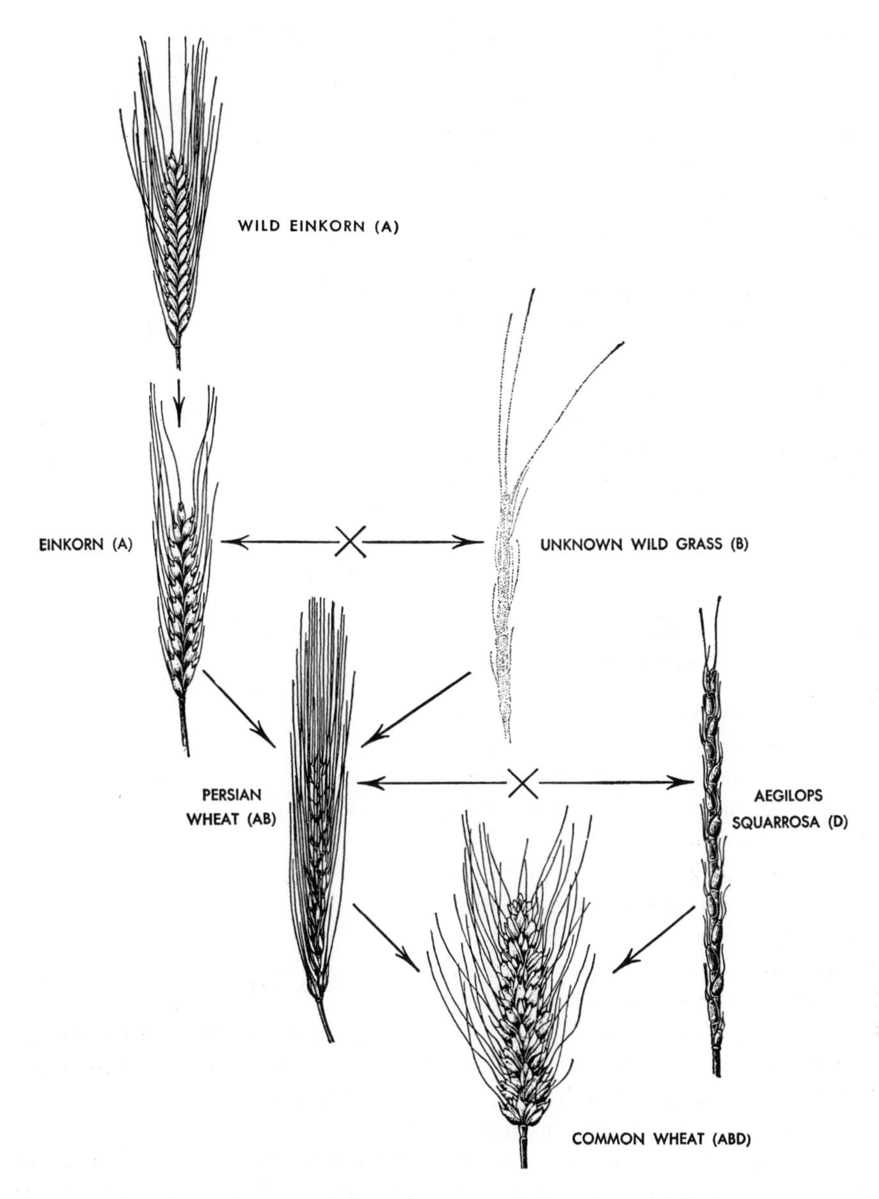

Fig. 25–5. Evolution of common wheat. (*Sci. Am.*)

back over 6000 years and probably originated in the Near East. The stages in development which occurred long ago are shown in Figure 25–5. From a primitive diploid wheat, called Wild Einkorn, there was developed an improved variety called Einkorn. The chromosome set of Einkorn wheat is designated A. Einkorn accidentally crossed with an unknown wild grass (probably *Agropyron triticeum*) with a set of chromosomes, designated B. Following crossing, the chromosome number doubled spontaneously, yielding Persian wheat, an allotetraploid. Each gamete of Persian wheat has a set of A chromosomes and a set of B chromosomes. When a sperm unites with an egg, the zygote has two sets of A chromosomes and two of B. Later Persian wheat crossed accidentally with *Aegilops squarrosa,* whose set of seven chromosomes are designated as D. The hybrid had single sets of A, B, and D chromosomes and was sterile. However, the chromosomal number doubled spontaneously, yielding an allohexaploid, which are our common bread wheats of today. Within the vegetative cells of these wheats there are two sets of A chromosomes contributed by Einkorn, two sets of B chromosomes from an unknown wild grass (which may have been *Agropyron triticeum*), and two sets of D chromosomes from *Aegilops squarrosa.* The gametes, of course, of bread wheats have just one set of A, B, and D chromosomes.

Through the use of colchicine the chromosomal number of sterile hybrids can be doubled and as a consequence fertile hybrids may be produced. This is one of the most promising fields for the use of colchicine.

ALTERATIONS IN THE STRUCTURE OF CHROMOSOMES

Alterations in chromosomes may result in heritable variations—that is, in mutations. Mutations may result from changes in genes, in the number of chromosomes, and in the structure of chromosomes. Usually mitosis and meiosis operate perfectly, but on rare occasions aberrations of either may occur which may alter the structure of a chromosome. The frequency of chromosomal mutations may be markedly enhanced by ionizing radiations.

When a portion of a chromosome is lost (Fig. 25–6), we say that a **deletion** has occurred. A deletion may result in unpredictable offspring. For example, assume that gene *G*, a gene necessary for chlorophyll formation, is deleted during the first division of a fertilized egg which originally had a genotype of *Gg;* gene *g* lacks the capacity to form chlorophyll. After deletion of *G* the cells formed by division will have only *g* and thus the seedling will be yellow instead of the anticipated green color.

On occasion a portion of one chromosome may be exchanged for a portion of another nonhomologous chromosome. This exchange of segments between nonhomologous chromosomes is known as a **translocation** and the

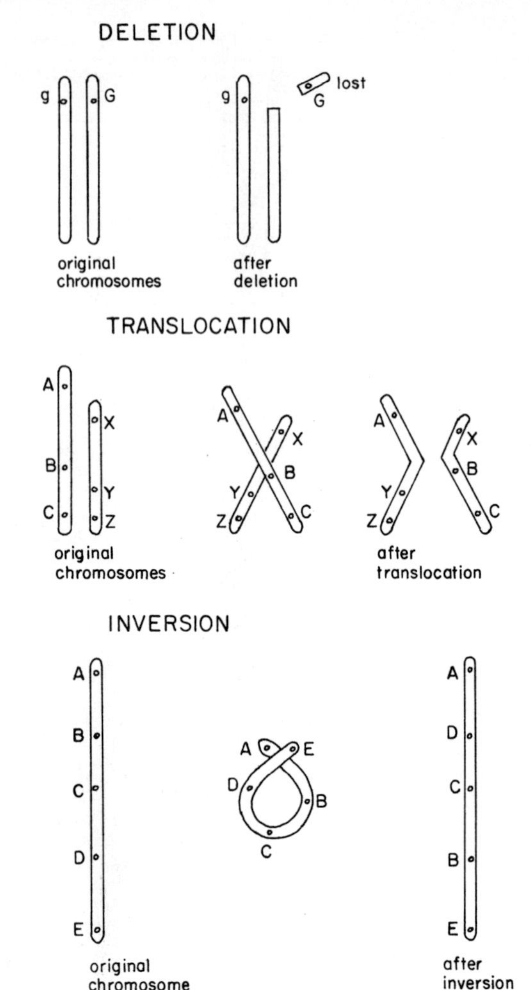

Fig. 25–6. A chromosome may be altered by deletion (*upper*), translocation (*center*), and inversion (*lower*).

steps involved are shown in the accompanying figure. During mitosis or meiosis, the ABC chromosome may overlap the XYZ chromosome. At this point the A segment of the ABC chromosome may unite with the YZ segment of the XYZ chromosome and the X segment may join with the BC part. The reconstructed chromosomes are AYZ and XBC.

Inversions are a third type of chromosomal mutation. The genes on a chromosome may be arranged in the following sequence: A, B, C, D, E, but after inversion the arrangement may be A, D, C, B, E. The events leading to an inversion are shown in Figure 25–6. Where the ends of the looped chromosome overlap, the chromosome may break and then become reconstructed in a different pattern—the A, D, C, B, E one instead of the original A, B, C, D, E arrangement.

Although translocations and inversions do not alter the genes, but only their arrangement, they may produce plants with different phenotypes. In recent years we have learned that the action of a gene is affected not only by the external environment, but also by the internal environment. The genes adjacent to a certain gene may influence its action, the so-called position effect. Thus when gene A is next to gene B, it may function in one way; but when A is next to D, it may function in a different way and produce a different phenotype.

Inversions and translocation frequently result in intersterility between the mutant forms and their parents. A plant with translocated chromosomes is fertile when crossed with its own kind, but some sterility occurs when it is mated with the nonmutant form. Similarly inversions represent reproductive barriers.

QUESTIONS

1. Discuss the environmental factors that may result in variations. _____ Such environmental variations are (A) heritable, (B) nonheritable.

2. _____ Some variations are due to hybridization and recombinations of genes. (A) True, (B) false.

3. _____ Which one of the following is false? (A) Gene mutations occur spontaneously in nature, but with low frequency. (B) Gene mutations occur in lower plants as well as in more advanced types. (C) Gene mutations are not inherited. (D) A spontaneous and unexpected alteration in the structure of a gene is referred to as a gene mutation.

4. List a sequence of four base pairs for a portion of a hypothetical gene. Adjacent to this, list a sequence which might represent a mutation of the gene. _____ The mutated gene (A) would revert to the original sequence when the cell divides, (B) would form a copy of itself when the cell divides. _____ The mutated gene (A) would produce the same kind of RNA as would the original gene, (B) would produce a different kind of RNA. _____ The mutated gene (A) would control the synthesis of the same enzyme as the nonmutated gene, (B) would not control the synthesis of the same enzyme as the original gene.

5. _____ A root tip was exposed to X rays. A gene in the meristematic region mutated. This mutant gene (A) will replicate itself and be passed on to the offspring, (B) will not form an exact duplicate of itself when the cell divides, (C) will replicate itself when the cell divides, but will not be passed on to the offspring.

6. Explain the "one gene–one enzyme" hypothesis.

7. _____ Which one of the following is false? (A) Radiations induce mutations in organisms. (B) The great majority of mutations are deleterious. (C) The incidence of mutation increases in proportion to the total dose of radiation. (D) Assuming the same total dose, the number of mutations is greater when the irradiation is continuous than when it is intermittent. (E) Any amount of radiation, however small, is deleterious to the heredity of man; in other words, there is no safe dose.

8. Is there any safe dose for radiations such as X rays and those given off by nuclear explosions?

9. _____ Colchicine acts on (A) dividing cells, (B) cells which are not dividing.

10. _____ If the chromosomal number of a plant be doubled by the use of colchicine, the plant with double the number of chromosomes would be called (A) diploid, (B) tetraploid, (C) triploid, (D) haploid.

11. _____ If a diploid plant is crossed with a tetraploid, most of the offspring will be (A) diploid, (B) triploid, (C) tetraploid, (D) haploid.

12. A plant with three sets of chromosomes would be called (A) diploid, (B) haploid, (C) tetraploid, (D) triploid.

13. _____ Doubling the number of chromosomes in a hybrid plant often makes the resulting plant fertile. (A) True, (B) false.

14. _____ If chromosomes pair regularly during the first meiotic prophase in a hybrid, there is usually (A) high fertility, (B) a high degree of sterility, (C) no evidence of relationship.

15. A plant breeder produced a tetraploid of species X and also a tetraploid of species Y. He crossed a diploid plant of species X with a diploid plant of species Y and then doubled the number of chromosomes in the hybrid. _____ The plant with the least sterility would be (A) the tetraploid hybrid, (B) tetraploid species X, (C) tetraploid species Y. _____ If species X is crossed with the tetraploid hybrid there will be (A) a high degree of sterility, (B) high fertility.

16. _____ The exchange of a portion of one chromosome for a portion of another chromosome which is not homologous is known as (A) crossing over, (B) translocation, (C) inversion, (D) deletion.

17. _____ Genes are lost when (A) a deletion occurs, (B) translocation occurs, (C) inversion takes place.

18. _____ Which one of the following is false? (A) Genes interact with neighboring genes and they are all functionally interdependent. (B) Inversions and translocations introduce no new genes into the species. (C) No sterility occurs when a plant with translocated chromosomes is crossed with a normal one. (D) Inversions may represent reproductive barriers.

26

Plant Communities

Organisms do not live alone but grow in association with their own kind and with other species. They form communities in which characteristic species grow together in dynamic equilibrium and with mutual adjustment. In alpine areas, near the tops of mountains, there are beautiful communities of lichens growing on rocks, yellow potentillas, blue polemoniums, mountain forget-me-nots, clover, phlox, and pinks. In the East, where maple and beech are the primary trees, one may expect to find other polemoniums, cypripediums, phlox, anemones, violets, Dutchman's breeches, and May apple.

The type of community which occupies an area is determined by the environment. Environment is exceedingly complex and consists of three major interacting factors: climatic factors, soil (edaphic) factors, and biotic factors. The interplay of these three determines the composition of a community.

In a trip across the United States from the Atlantic seaboard to the Pacific Coast, a number of major kinds of vegetation may be seen, each with many local variants; forest communities with trees, shrubs, forbs (nongrassy herbs), grasses, and characteristic animals; grassland communities of grasses, forbs, herds of antelope, badgers, prairie dogs, and ground-nesting birds; sagebrush communities which also include Indian paintbrush and other forbs, grasses, jack rabbits, and reptiles; the cactus and creosote-bush desert of our Southwest, with such other inhabitants as mesquite, firebush, grasses, lizards, the desert coyote, and the night hawk.

Four major types of plant cover (known as **plant formations**) are found in the United States: tundra, forest, grassland, and desert.

TUNDRA

The tundra formation occurs where frigid temperatures prevail during most of the year. The soil is poorly drained and in the arctic tundra thaws to a depth of only 1 foot during the summer. In tropical and temperate latitude tundra is restricted to the high elevations of mountains, but in arctic regions it occurs near sea level. The plants of the tundra are always dwarf. Grasses, sedges, lichens, dwarf willows, and other shrubs, as well as herbs with brilliant flowers, occur in the tundra. In this setting the marmot, pika, and ptarmigan are at home (Fig. 26–1).

Fig. 26–1. Tundra vegetation near the summit of the Medicine Bow Mountains, Wyoming.

FORESTS

Forests develop in humid regions where precipitation is usually in excess of 30 inches a year and fairly well distributed throughout the year. The major forest types of the United States are illustrated in Figure 26–2.

Early explorers in the United States found a vast expanse of forested land. From the Atlantic seaboard to the prairies across the Mississippi there was a continuous, almost uninterrupted forest. In the North were white and red pines, spruce, fir, hemlock, cedar, tamarack, birch, beech, maple, oak, and other trees. Farther south in the central hardwood forest were oak, hickory, ash, tulip, poplar, walnut, chestnut, and black cherry. The southern forest was largely pines—longleaf, slash, shortleaf, and loblolly (Fig. 26–3). Much of this original acreage of the eastern forests has been cut over and now farms and cities occupy part of this once forested area.

The eastern forests are separated from the western forests by the Great Plains. The Rocky Mountain forests of the West are composed largely of conifers: ponderosa pine, lodgepole pine, western white pine, Douglas fir, alpine fir, hemlock, and larch in the North; piñon pine and juniper in the South. The Pacific Coast forests contain Douglas fir, ponderosa pine, sugar pine, white fir, Sitka spruce, hemlock, incense cedar, redwood, the bigtree, and others.

Along the southeastern coast of Alaska stately forests of western hemlock, Sitka spruce, western red cedar, and Alaska cedar cover more than 5 million acres. In the interior of Alaska are forests of spruce and birch consisting of trees of comparatively small size.

The seven main islands of the Hawaiian group support about 1.2 million acres of wet and dry forest cover. Among the trees of the Hawaiian forests are ohia lehua,

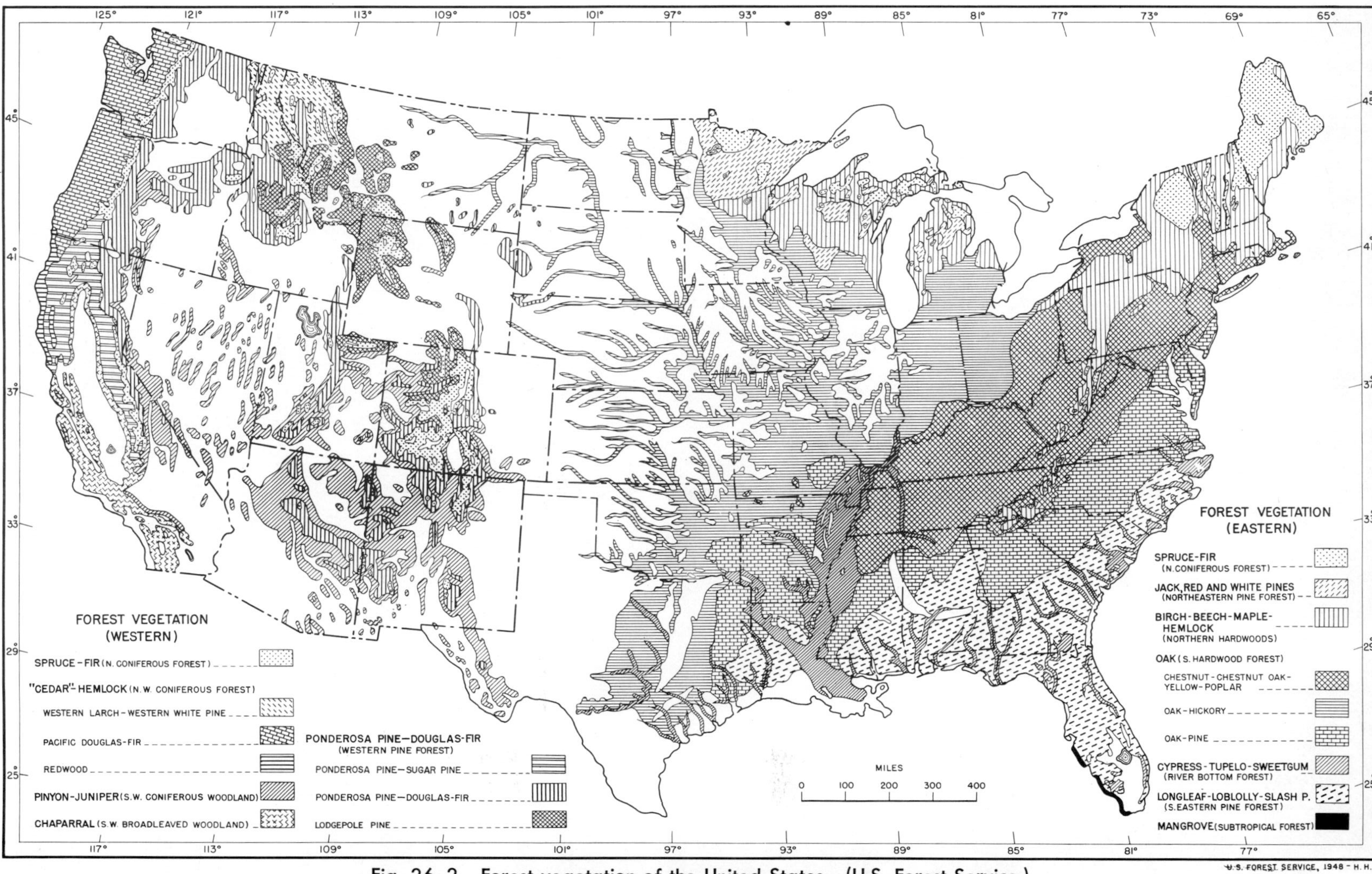

Fig. 26–2. Forest vegetation of the United States. (U.S. Forest Service.)

Fig. 26–3. Forest of longleaf pine (*Pinus palustris*) in Georgia. (U.S. Forest Service.)

Hawaiian mahogany, mamani, algaroba, and the candlenut tree.

Regions of the world other than the United States also are clothed with forests where precipitation is high and evenly distributed. Generally speaking, in humid regions forests develop and in arid regions, deserts. Grasslands are found in regions between these extremes.

GRASSLANDS

Grasslands occur where the annual precipitation is intermediate. In such areas, summer precipitation is usually relatively high and winter precipitation low.

In the United States, the region from the Mississippi River to the Rocky Mountains is a vast expanse of grassland. The eastern part of this region is occupied by the tall grass prairie, the western part by the short grass plains (Fig. 26–4). In between lies a mixed grass prairie.

The tall grass association occurs where the annual precipitation is between 22 and 30 inches annually and is characterized by big bluestem and little bluestem grasses. This association extends from the Dakotas

Fig. 26–4. Short grass plains of Wyoming with antelope. (Lester Bagby and Wyoming Game and Fish Commission.)

through Nebraska, Iowa, and Kansas. In this belt conditions are ideal for the cultivation of corn, wheat, and other grains. These are the important bread and corn lands of our country.

In the mixed grass association, bluestem grass is still present, but such short grasses as grama grass and buffalo grass are also found in the community.

Where 15 to 20 inches of precipitation prevail, the short grass association of grama grass and buffalo grass make up the grazing lands of the cattle industry, the plains that stretch from Montana to Texas. Before the white man's entry 15 million buffalo and great herds of antelope grazed this area. Many plants with conspicuous flowers that grow with the short grasses are cattle forage also, but there are a few that are poisonous, such as the locoweed, death camass, and some of the delphiniums.

The tall grass prairie and the forest farther east stand like two opposing armies in conflict with each other. If the precipitation increases over a number of years, the forest invades the grassland. If precipitation decreases, grassland invades the forest. The invasion of grassland by forest has occurred in the past, as evidenced by present-day patches of grassland which stud the forest areas of Ohio. The patches of grassland seem out of place now, but they reveal that grassland, in the drier past, occupied most of Ohio.

DESERTS

Where the annual precipitation is less than 15 inches there are several kinds of deserts. Each has its distinguishing kind of vegetation, and each has special conditions that produce it. Most of us are going to have to revise our conception of the meaning of the word "desert," for most deserts are far from waste land.

Sagebrush deserts occur where the scanty precipitation is fairly evenly distributed throughout the year, as in parts of Montana, Wyoming, Colorado, Utah, Nevada, Oregon, and Idaho. These deserts are used almost entirely for grazing. At their best they support cattle; at their worst, sheep. Throughout the sagebrush country, there is a population of plants with conspicuous flowers, beginning with Indian paintbrush, the satellite of sagebrush, and extending through a long

Fig. 26–5. Firebush, creosote bush, and cactus in an Arizona desert. (*Arizona Highways.*)

list including larkspur, *Gilia,* and others. Reptiles, sage hens, jack rabbits, badgers, coyotes and antelope are some of the animal members of this community.

The great and fascinating cactus and creosote-bush desert of our Southwest occurs where the precipitation is divided into two periods with extreme dryness in between. This association occupies extensive areas in California, Arizona, New Mexico, and Texas. Some of the other organisms which make up the community are mesquite, firebush, scattered grasses, lizards, kangaroo rat, desert coyote, pocket mouse, red fox, night hawk, and the antelope jack rabbit. No one seeing this desert in flower could call it dull or barren. Many of the plants have tissues which store water received during a rainy period for use over the period until rain comes again. Some of the plants have an extremely brief growth-to-flower cycle, so that often after a spell of wet weather, perhaps no longer than 3 weeks, the desert bursts into bloom (Fig. 26–5).

Alkali soil, high in its content of salts, combines with low precipitation to give what we call salt deserts. The Red Desert of Wyoming is a striking example, as are certain areas in Utah and Nevada. Here and there through the West are places where soil high in salts exists side by side with other types of soil. Such areas can be spotted by their characteristic vegetation, such as greasewood and salt-grass, and by the small alkali lakes whose borders are white with dried salt.

WORLD PLANT FORMATIONS

In considering the vegetation of the world with its many climatic belts, it is often desirable to recognize more than four plant formations. Figure 26–6 shows Professor L. R. Holdridge's classification of world plant formations and the precipitation and temperature conditions under which each develops.

On the right are listed the altitudinal zones as they occur in the tropics. For example, in tropical regions, the **montane belt** occurs at altitudes between 9000 and 12,000 feet. As one travels north or south of the Equator, the montane belt becomes progressively lower as does the subalpine, alpine, and nival zones. In the tropics the **nival zone,** the zone of perpetual snow, occurs at elevations above 14,400 feet, but in polar regions it would be at sea level.

The temperatures recorded on the chart are **bio-temperatures.** The bio-temperature is the sum of those mean monthly temperatures which are greater than 0° C. divided by 12. For example, assume that the average monthly centigrade temperatures in a certain region are as follows:

January	−5° C.	July	+25° C.
February	−4° C.	August	+20° C.
March	−3° C.	September	+15° C.
April	+5° C.	October	+10° C.
May	+10° C.	November	+3° C.
June	+20° C.	December	−2° C.

The bio-temperature would be the sum of 5, 10, 20, 25, 20, 15, 10, and 3 divided by 12, which is 9° C. Although bio-temperatures are calculated in degrees centigrade, they have been converted into degrees Fahrenheit on the chart.

In the tropics at elevations below about 12,000 feet the bio-temperatures are the same as mean annual temperatures because in such regions the temperatures are greater than 0° C. during all months. Many people visualize the tropics as uncomfortably hot areas, but the chart shows that at an elevation of 3000 feet the climate is temperate and at higher elevations cool or even cold.

The specific plant formation in any region is determined by both bio-temperature and precipitation which is recorded as inches per year along the bottom. The plant formation that occurs in a region is located where the precipitation value intersects the bio-temperature. The bio-temperatures, annual precipitations, and plant formations for various regions are listed in the table below.

The latitudinal plant formations are not identical to the altitudinal ones which occur in the tropics. For example, a cool temperate moist forest is unlike a tropical montane moist forest even though the bio-temperatures and the precipitation are the same. The species in the cool temperate moist forest differ from those in the tropical montane moist forest. Furthermore, the tropical montane moist forest supports a great array of epiphytes whereas the cool temperate forest has relatively few. The seasonal variations in temperature between the cool temperate latitudinal region and

Place	Bio-temperature (°F)	Precipitation (inches)	Formation
Charleston, S.C.	66	43	Low subtropical moist forest
Chicago, Ill.	45	27	Cool temperate moist forest
Helena, Montana	48	13	Cool temperate grassland
Juneau, Alaska	45	80	Cool temperate rain forest
San Francisco, Calif.	55	21	Warm temperate dry forest
Buenos Aires, Argentina	61	35	Warm temperate dry forest
Moscow, U.S.S.R.	45	20	Cool temperate moist forest
San Diego, Chile	55	12	Warm temperate thorn scrub
San Jose, Costa Rica	75	73	Tropical, subtropical moist forest
Tacna, Peru	81	0	Tropical desert

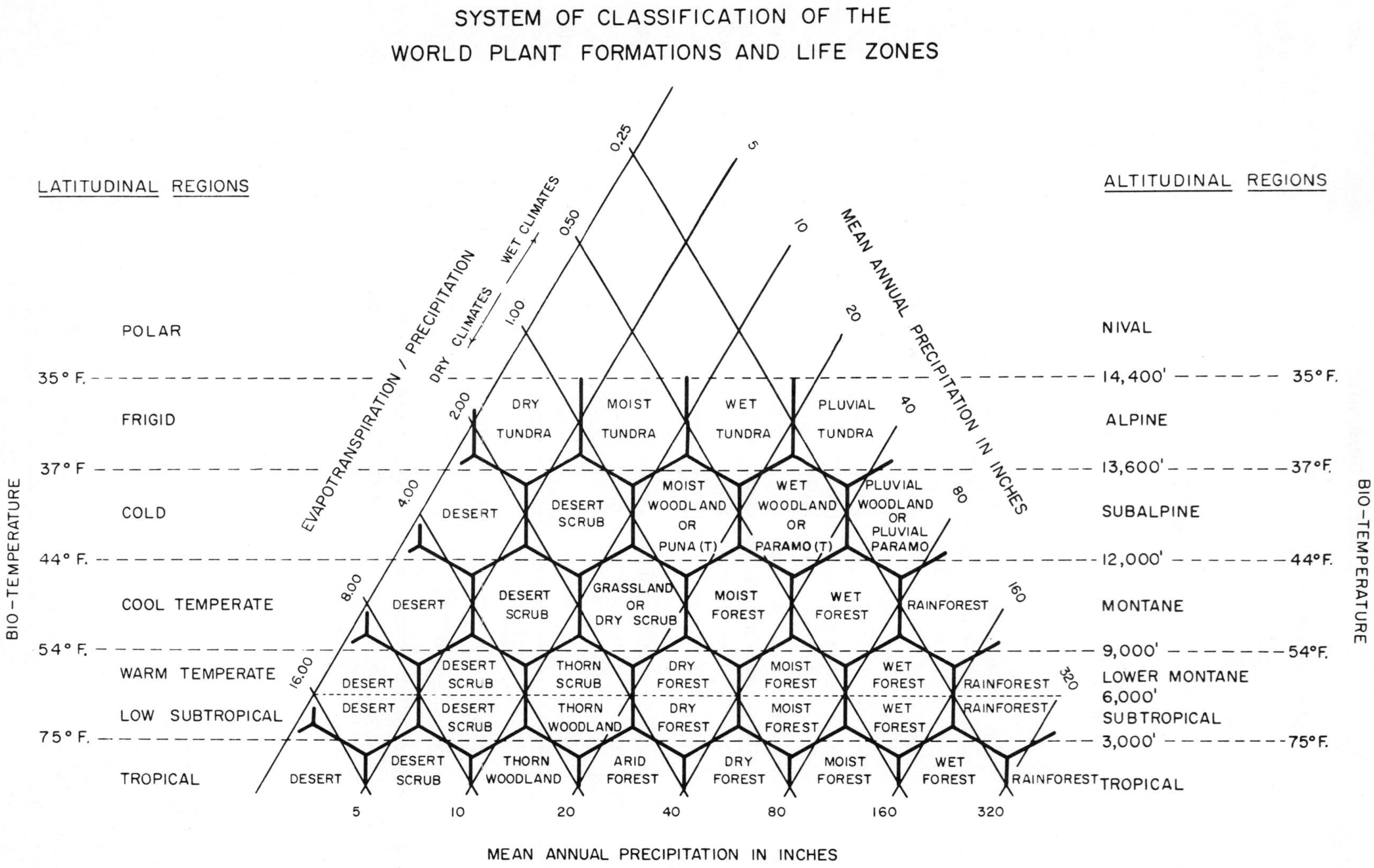

SYSTEM OF CLASSIFICATION OF THE
WORLD PLANT FORMATIONS AND LIFE ZONES

LATITUDINAL REGIONS
ALTITUDINAL REGIONS

EVAPOTRANSPIRATION / PRECIPITATION
DRY CLIMATES
WET CLIMATES
0.25
0.50
1.00
2.00
4.00
8.00
16.00

MEAN ANNUAL PRECIPITATION IN INCHES
5
10
20
40
80
160
320

POLAR
NIVAL
35° F. 14,400' 35° F.
FRIGID
DRY TUNDRA
MOIST TUNDRA
WET TUNDRA
PLUVIAL TUNDRA
ALPINE
37° F 13,600' 37° F.
COLD
DESERT
DESERT SCRUB
MOIST WOODLAND OR PUNA (T)
WET WOODLAND OR PARAMO (T)
PLUVIAL WOODLAND OR PLUVIAL PARAMO
SUBALPINE
44° F. 12,000' 44° F.
COOL TEMPERATE
DESERT
DESERT SCRUB
GRASSLAND OR DRY SCRUB
MOIST FOREST
WET FOREST
RAINFOREST
MONTANE
54° F. 9,000' 54° F.
WARM TEMPERATE
DESERT
DESERT SCRUB
THORN SCRUB
DRY FOREST
MOIST FOREST
WET FOREST
RAINFOREST
LOWER MONTANE
6,000'
LOW SUBTROPICAL
DESERT
DESERT SCRUB
THORN WOODLAND
DRY FOREST
MOIST FOREST
WET FOREST
RAINFOREST
SUBTROPICAL
75° F. 3,000' 75° F.
TROPICAL
DESERT
DESERT SCRUB
THORN WOODLAND
ARID FOREST
DRY FOREST
MOIST FOREST
WET FOREST
RAINFOREST
TROPICAL

BIO-TEMPERATURE
BIO-TEMPERATURE

5 10 20 40 80 160 320
MEAN ANNUAL PRECIPITATION IN INCHES

Fig. 26–6. World plant formations.

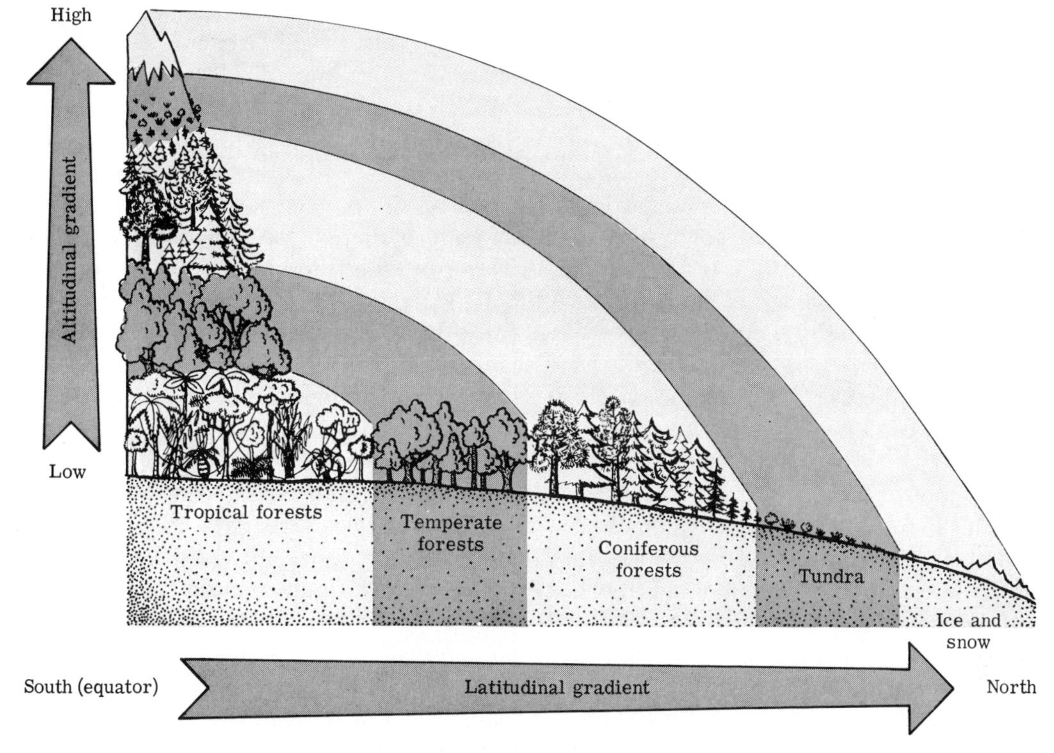

Fig. 26–7. Plant formations vary with altitude and latitude. Similar, but not identical, plants occupy higher altitudes and northern latitudes. (From *Biology* by E. L. Cockrum, W. J. McCauley, and J. A. Younggren. Copyright © 1966, W. B. Saunders Co., Philadelphia.)

the tropical montane zone account, in part, for the different flora. In the tropics, temperatures are nearly constant throughout the year, whereas in the cool temperate region they vary with the seasons, being hot in summer and cold in winter (Fig. 26–7).

In many areas of biology, there may be divergent opinions. The figures that Professor Holdridge uses to delineate some major plant formations differ from those given earlier in this chapter. For example, he believes that deserts occur where precipitation is less than 10 inches a year, grasslands where precipitation is between 10 and 20 inches annually, and forests where precipitation is more than 20 inches a year. Tundra, of course, occurs at high altitudes in the tropics as well as at low elevations in frigid regions. The paramo and puna, encountered at high elevations in the Andes, resemble the tundra. In the paramo small shrubs grow associated with matlike plants, whereas in the drier puna the shrubs are absent.

The chart shows that the effectiveness of precipitation varies with altitude in the tropics and with latitude. As one ascends a mountain or travels toward the poles, the temperature decreases as does transpiration and evaporation. With lower rates of transpiration and evaporation, plants use water more effectively. In tropical regions, where precipitation is between 40 and 80 inches annually, a dry forest develops at elevations below 3000 feet, a moist forest between 3000 and 9000 feet, and a wet forest between 9000 and 12,000 feet. Each of these forests is characterized by different tree species and each has a characteristic physiognomy. The trees in the dry forest are low-crowned with spreading mushroom-shaped tops in contrast to the taller and more closely spaced trees of the wet or moist forest. Trees in the moist forest or wet forest are covered with epiphytes, among them, begonias, philodendrons, gesneriads, orchids, ferns, and bromeliads, whereas those in the dry forest support fewer epiphytes. Every tree in a moist or wet forest supports an aerial garden.

To get the full impact of the beauty and diversity of plant formations in the tropics you should visit a country with coastal lowlands and towering mountains. As you travel from the coast up the mountains, you will see several plant formations each characterized by certain species.

Strangest of the forests in tropical lands are the moist, wet, and rain forests that occur in the tropical belt at elevations below 3000 feet. Compared to forests in the United States, where you may see the sky overhead and vistas through the trees, these forests may seem threatening. From the outside of such a tropical forest you can see its four stories. The lowermost story is formed by short trees. Up through these rises a second story of species that grow to twice this height, a third story of still taller kinds, and finally, thrusting up through them all, a fourth story of trees that grow to 200 feet and spread their crowns above all others. The number of tree species making up the forest is immense, perhaps one hundred or more.

As you enter the forest, you step into sudden gloom. You have to wade through waist-high vegetation, tripping over roots and the buttresses of large tree trunks. You cannot see through the ceiling formed overhead by the trees. Many plants have spines or saw-like edges.

As your eyes become accustomed to the deep shade, you can see many ferns and epiphytes and trace out vines climbing the tree trunks. Ropelike roots that will hold your weight hang down from plants living on unseen branches above. Vines climb these descending roots—in fact, countless plants seem to be climbing or hanging from something else.

Fig. 26–8. Interior of a tropical rain forest showing a buttressed tree; orchids, bromeliads, and other epiphytes; and lianas. (After Bergen.)

The lines on the chart that descend at an angle from left to right show the ratio of the amount of water evaporated and transpired to the amount of precipitation. The millimeters of water evaporated and transpired, **evapotranspiration,** is a theoretical value obtained by multiplying the biotemperature, in degrees Centigrade, by 58.93. A ratio of 1 indicates a balance between evapotranspiration and precipitation. In regions where the ratio is 1, conditions are ideal for agriculture. Where evapotranspiration is two or more times precipitation, deserts usually occur (exceptions are the arid forest and thorn woodland or scrub in tropical regions) and in such regions crops thrive only when irrigation is practiced. A ratio of 0.50 indicates that only one half of the precipitation is evaporated and transpired; much of the remainder percolates through the soil to the water table or it runs off. Where the ratio is 0.25, only one-fourth of the precipitation is evaporated and transpired; in such areas rain forests develop. The fertility of the soil in such regions can be maintained only where vegetation is dense. In the primeval forest the minerals released from decaying vegetation are absorbed by the dense tangle of tree roots before they can be leached out of the soil. If corn or other field crops are grown on land cleared of trees, the soil is soon impoverished of minerals and the crop no longer thrives. In such wet areas, the crops might well be tree crops that resemble the original forest. For example, *Cordia alliodora,* a valuable timber tree, may provide an upper story below which food palms may thrive; beneath the palms, cacao trees would grow. A lowest layer might consist of shrubs grown for fruit or for medicinals.

DEVELOPMENT OF PLANT COMMUNITIES—PLANT SUCCESSION

The plant communities which we have just considered were fully developed ones. Obviously, if a region is bare of vegetation, it will not become a prairie, a forest, or a sagebrush desert over night. If soil is present on an area, it takes about 20 to 30 years for a grassland community to develop, and 100 years or more for a forest. If soil is lacking, as it would be in an area occupied by bare rock or in a pond or lake, 1000 or more years would be required for a forest or a prairie to develop. A number of stages occur in the development of a plant community. These orderly stages are collectively referred to as **succession.**

Imagine an area where soil is present but where plants are absent. The bare area may be an abandoned farm, an eroded hillside, the denuded areas along a highway, or a burned-over forest. An area does not remain bare for long, for it soon becomes repopulated. The establishment of a plant community on a bare area involves **migration, ecesis** (making a home), **competition,** and the effects of organisms on each other (**coactions**) and on the habitat (**reactions**).

SUCCESSION ON AN ABANDONED COTTON FIELD

To illustrate the stages in the development of a plant community, we shall consider an abandoned cotton field in North Carolina.

Migration

First migration occurs. Seeds, fruits, and spores are brought in by wind, birds, mammals, water, or through mechanical means. Seeds and fruits, as we have seen, have a variety of adaptations which facilitate dispersal.

Ecesis

Some of the seeds that are brought into the area germinate, but survival is a struggle. Some seedlings are killed by late frosts in the spring, others are nipped by an early freeze in the fall. Some succumb to drought, others to attacks by disease, insects, birds, or mammals. Heavy rains may wash seedlings from the ground. In spite of the hazards, some plants survive the seedling stage, grow to maturity, and finally reproduce. The stage in the development of a community which involves seed germina-

Fig. 26–9. About 5 years after the abandonment of this cotton field, young pine trees were evident, and eventually they replaced the weeds and grasses. The forest in the background also developed on an abandoned field; the trees are about 20 years old, and they have completely replaced the light-demanding grasses and weeds. (C. F. Korstian.)

tion, seedling growth, and maturity is known as ecesis.

Competition

The plants which survive the gamut of heat, drought, cold, diseases, and animal predators soon compete with each other for light, water, and soil nutrients. Individuals of one species compete with each other as well as with individuals of different species. In an early stage in the development of the community, the competing organisms are crabgrass, horseweed, aster, ragweed, and broomsedge. About 5 years after the abandonment of the cotton field, pines become noticeable, and they begin to compete with each other and with the grasses and other weeds (Fig. 26–9). In time, the pines overtop the light-demanding grasses and forbs which originally dominated the area, and these survive no longer in the deep shade under the tree canopy. However, shade-tolerant herbs, vines, and shrubs find the situation favorable for their development and become part of the community.

As the pine trees grow still larger, competition among the trees becomes increasingly keen, because with greater size there is a greater demand for water, light, and nutrients. Some trees do not survive this competition and they perish.

About 75 years after the cotton field was abandoned, saplings of oak, hickory, and other hardwoods would be seen in the understory (Fig. 26–10). At this stage of succession pines, intolerant of shade, are competing with shade-tolerant hardwoods. In time a pine tree dies of accident, disease, or the action of some other agent. Its place in the canopy is then taken by an oak, hickory, or other tolerant hardwood. Repetition of such an event ultimately leads to a forest of

Fig. 26–10. *Left*, young oaks, hickories, and other tolerant hardwoods are developing in the shade of the pine trees, which are more than 70 years old. Note the complete absence of pines in the understory. As the pine trees die, their places in the canopy are taken by hardwoods. *Right*, a climax oak-hickory forest. The pines have been entirely replaced by oaks, hickories, and other tolerant hardwoods. (C. F. Korstian.)

oak, hickory, and other broad-leaved trees (Fig. 26–10, *right*). This final harmonious community maintains itself indefinitely if no catastrophe occurs, and is referred to as the climax. The climax vegetation is in equilibrium with the climate, soil, and topography; and as long as these remain stable, so too does the climax. In addition to trees, shrubs, and herbs, animals are part of the climax community.

Reactions. As the community develops on the bare area, it alters the environment. The effects of the community on the environment are referred to as **reactions.** The unprotected soil of the original area is subject to wind and water erosion. The developing plant community diminishes wind velocity at ground level and surface runoff of water and so reduces erosion. The incorporation of humus increases both the water-holding capacity and the fertility of the soil. The succession of plants reduces the light intensity at ground level, and this change is paralleled by the replacement of intolerant species with tolerant ones. Intolerant species cannot survive in deep shade, whereas tolerant species can survive and grow in shaded localities.

Coactions. As the plant community changes, the fauna of the area also changes. The original barren area is certainly an inhospitable home for animals. During the initial stages of plant succession when grasses, weeds, and young trees are present in large numbers, the area can support a large population of livestock, birds, rodents, and other animals. As the forest develops, less forage is produced; hence a smaller, different population of animals can be supported.

Summary. We have seen that a plant community develops in a series of stages. At first one group of organisms dominates the area. Later a different group takes over; this in turn is replaced by other groups until the climax is attained. These orderly stages in the development of a community are collectively referred to as **plant succession.**

Plant succession goes to the climax only in the absence of a major disturbance. In some areas where forests can develop, overgrazing may keep the community in a grass and forb stage and prevent a forest from developing. Likewise, repeated fires may preclude a climax being attained. There is evidence that the American Indian repeatedly burned some areas to extend the pasture for grazing animals which he hunted. Often too he used fire drives to stampede game. These fires not only prevented the climax from being reached, but they also may have been a factor bringing about the extinction of certain animals, among them the American elephants, camels, and long-horned bison.

SUCCESSION ON BARE ROCK—A XEROSERE

On an expanse of bare rock—a recently cooled lava flow, for example—only certain kinds of plants can grow, i.e., crustose lichens. The rock surface may be covered with crustose lichens which are so securely attached to the rock that they are difficult to remove. They build up a thin layer of soil by disintegrating some of the rock, by catching wind-blown dust, and by humus formed after they die.

As time goes by, the soil becomes deep enough for the growth of larger lichens, the foliose lichens. They invade the area and extend above the crustose ones, shading them so much that they perish. The foliose lichens continue to build up soil.

In time the soil becomes deep enough for xeric mosses to grow; as they develop, they shade the foliose lichens, which then lose out. After many years, the soil becomes deep enough to support grasses and forbs, which then make a home in the area. The grasses and forbs are destined to be replaced as the soil becomes deeper and deeper. In time they lose out to shrubs, which are relatively transitory. In semiarid regions, the shrub stage of succession is replaced by a grassland climax. In a humid area the shrubs are replaced by trees of various kinds. At high altitudes in the Rockies, alpine fir and spruce take over the site and form the climax. In Indiana, the climax is a beech and maple forest and in certain areas of New England the climax would be a forest of birch, beech, and maple.

The primeval climax forest of today is a stable community, one which should endure. The successional stages are over and the community should maintain itself provided no calamity strikes the area. If man, through reckless cutting or carelessness with fire or explosions destroys the forest, the soil formed during the past 1000 years may be washed or blown away, again exposing rock. Lichens again will become established and succession will start all over again.

SUCCESSION IN A POND—A HYDROSERE

In areas within a region, identical forest communities come into being in places which in the beginning were totally different. Thus, in humid regions, a forest is destined to occupy what is now a lake. Where we on a summer day now enjoy swimming, boating, and fishing, our descendants will walk and rest in a forest of quiet grandeur.

The pioneers in a lake are different from those which first colonize a rock. In a moderately deep lake with a sandy bottom, the pioneers are algae, some of microscopic size, others visible to the eye as green threads. Countless minute animals feed on the algae and these small animals in turn are food for larger animals. There is birth and death. The dead organisms settle to the bottom, where their remains become soil; thus the lake becomes shallower. After many years, perhaps centuries, the water depth is lessened to 20 feet. Then pondweeds, submerged buttercups, waterweed, and different kinds of algae, including *Chara*, colonize the lake. As the plants and animals die, their remains settle to the bot-

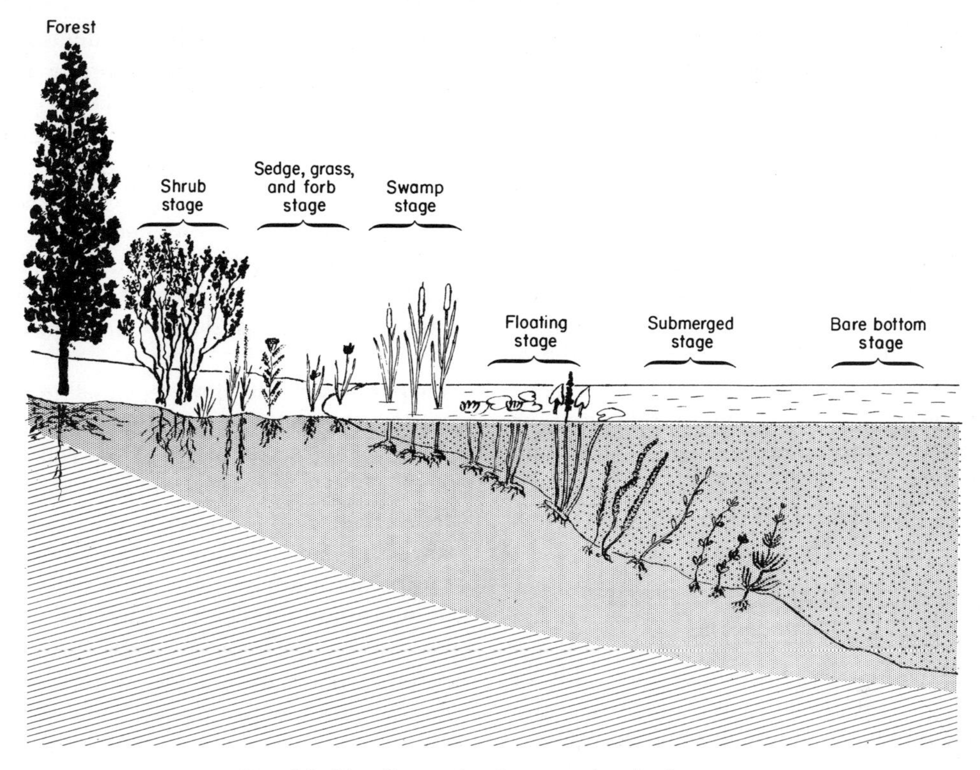

Fig. 26–11. Succession in a pond, a hydrosere.

tom and are converted to humus. In time the water becomes shallow enough to invite another invasion.

When the depth is reduced to 6 or 8 feet, water lilies, smartweed, and other plants become established, and in time they replace the previous community. The new colonists are rooted in the mud but their leaves extend to the surface. The lake may be almost covered with the leaves of smartweed and the pads of water lilies. As the older plants die, their bulky remains become humus on the lake's floor. The lake now fills up quite rapidly, perhaps 1 inch a year.

In time the water is only 2 or 3 feet deep and new immigrants make their home in the area. Bulrushes, cattails, and reeds soon change the lake into a swamp. They tower above the water lilies, which then perish from lack of light. The swamp plants continue the soil-building process.

After a while the water is so shallow that the swamp plants can no longer survive. They give way to sedges and forbs such as mints, iris, marsh marigold, and bellflowers.

As soil builds up, the water table drops and the environment then becomes too dry for the sedge meadow and accompanying flowers. Willows and other shrubs come in only to be replaced in humid regions by trees of several kinds. Finally the same kinds of trees occupy this area as occupied the rock surface.

Zonation in a Pond

Because succession in a pond occurs very slowly, it is often impossible to study the sequence directly. However, we can deduce successional stages by observing vegetational zones in the pond and on shore. The various successional stages and their names are illustrated in Figure 26–11. In the deepest water, the bare-bottom stage, only algae and associated animals thrive. In somewhat shallower water, submerged plants which are rooted in the bottom grow. In still shallower water the floating stage of water lilies and smartweed develops. Closer to shore, the cattails and rushes of the swamp stage are seen. On shore, where the soil may be wet in spring and dry in summer, sedges, grasses, and forbs grow. Back from these, shrubs are found; and more distant from shore, trees grow. These zones indicate the stages of succession. The first stage occupies the water area most distant from shore, where the water is deepest, with successively more advanced stages following as one goes toward shore and then on to the land. This observed succession in space resembles the course of succession in time.

FACTORS AFFECTING PLANT COMMUNITIES

The pampas of Argentina, the veldt of Africa, the tall grass prairie of Kansas and Iowa, and the short grass plains of Wyoming all have one thing in common. They have between 15 and 30 inches of precipitation during the year, with most of it falling during the summer. They are all grasslands, but the species of grass in the various regions are different. Whether or not a particular species will grow in a certain region is determined by such climatic factors as precipitation, temperature, light, atmospheric humidity, and wind, by soil factors, and by biotic factors.

Climatic Factors

The water factor. Water and life are inextricably bound together. Without water no plant can grow. However some species can get along with a meager supply, others only with a larger amount. Still different plants grow only in habitats where there is an abundance of water, as in a swamp or lake. From the point of view of the water supply in their habitats, we classify plants as **xerophytes, mesophytes,** and **hydrophytes.** Xerophytes grow in arid habitats. They have structural and physiological features which enable them to survive prolonged drought. Mesophytes develop in habitats where the soil is uniformly moist, neither too wet nor too dry. Hydrophytes grow where there is an excess of water and hence a low oxygen supply. Some plants could be classified differently at various seasons. For example, many deciduous trees (birch, beech, maple, etc.) are mesophytes during the growing season, but are xerophytes when leafless during the winter.

Temperature. The lowest possible temperature is —459.4° F. (—273° C.) and the highest is +5,400,000,032° F. (+3,000,000,000° C.). Life is possible in only a minute part of this enormous range. Certain bacterial spores and seeds can survive at temperatures as low as —454° F. At the other extreme spores of certain bacteria survive temperatures as high as +284° F. Seeds and some spores have a low water content, and associated with this is a remarkable capacity to survive extremes of temperature. Organisms are active only in a narrow temperature range—from about 32° F. to 110° F. for many, but not all organisms. This range is referred to as the **biokinetic range.** Within these limits temperature markedly affects the rates of growth, digestion, respiration, photosynthesis, and the absorption of water and minerals. As we have seen in earlier chapters, biological processes speed up as the temperature increases above the minimum. When a certain temperature is exceeded, the rates decrease. Hence too much or too little heat arrests biological activities, and the difference between too much and too little is just a hair's breadth in the broad scale of temperature. It has been estimated that if the average daily temperature on earth were suddenly raised or lowered 20° F., most organisms would perish.

Grasslands are found at the equator and at higher latitudes. Likewise, forests and deserts occur at various latitudes. The grasses in tropical grasslands are species different from those found in Iowa; the species of trees in a tropical rain forest are unlike those in the forests of Washington; subtropical deserts have a different flora from those of Utah. To a large extent temperature determines which species can survive at different latitudes and altitudes.

In tropical regions at low altitudes, low temperatures do not play a role in the distribution of species; but at high altitudes in the tropics and at northerly latitudes, low temperature during winter may limit the kinds of plants that can exist. Species native to regions where winters are cold survive winter in various ways. Some are annuals which grow, flower, and fruit during the warm months of the year and survive the winter as seeds which can tolerate extraordinary cold. Many perennials have their perennating parts below the ground level where temperatures and winter drought are less severe than on the surface. Indeed, the more rigorous the climate, the higher the percentage of species with underground perennating parts (bulbs, rhizomes, tubers, etc.). Before winter sets in, the protoplasm undergoes chemical and physical changes which lead to greater hardiness. These changes are called hardening. Hardened and dormant plants may survive very low temperatures, whereas unhardened plants may be killed by low temperatures.

Low temperature may kill cells directly or indirectly. Direct injury results from the formation of ice crystals, which puncture and disorganize the protoplasm. The concentration of salts in the cell sap increases as ice crystals form, and the high concentration of salts may be toxic. While many species are killed by the formation of ice crystals, others—for example, cottonwood—can survive.

Indirectly, plants may be injured during winter through exhaustion of food reserves, by heaving of the soil, and by desiccation. The most frequent cause of winter killing is desiccation. During winter and especially during early spring, transpiration may exceed absorption, and, if too prolonged, injury and even death may result. Plants with buds below ground, or close to the ground and covered with snow or mulch, are less subject to this type of winter injury. The buds of trees and shrubs in regions where winters are cold are covered by bud scales which retard water loss.

Light. The intensity and duration of light in a region depends upon the season, the latitude, and weather conditions. The actual amount of light which plants receive depends partly on regional conditions and partly on their positions in the community. In a forest the tallest trees receive a maximum of light energy, whereas those in the understory receive a lesser amount. Species vary in their ability to survive in shade. Species which survive in relatively deep shade are said to be **tolerant** (they tolerate shade). Those which do not survive in low light intensity are referred to as **intolerant** plants.

Length of day is an important ecological factor. If plants are to be relatively permanent in a community, they must reproduce. The length of day, as we have seen earlier, affects flowering, and also leaf fall, onset of dormancy, and the development of tubers and bulbs.

Atmospheric humidity. Transpiration and evaporation of water from the soil are affected by relative humidity. Atmospheric humidity and wind determine the effectiveness of precipitation. In regions of low relative humidity, the water requirement of plants is higher than in regions of high relative humidity. For example, in North Dakota, 518 tons of water are required to produce a ton of hay, whereas in drier Texas, one ton of hay requires 1005 tons of water. In Montana, short grass plains occur in regions where the annual precipitation is 14 inches, whereas in parts of Texas with the same precipitation, deserts are found. In Texas, 21 inches of precipitation are required for the development of short grass plains.

Wind. In addition to increasing rates of transpiration and evaporation, wind may de-form plants. At timberline, species which develop into large trees at lower elevations grow as deformed, dwarfed, gnarled, much branched shrubs, so-called elfin timber or wind timber. The drying effects of the wind and its mechanical effects, coupled with low soil temperatures, are responsible.

Beneficial effects of wind include pollination and seed dispersal.

Edaphic (Soil) Factors of the Environment

Soil is a highly organized system formed by geological forces and biotic factors. In the soil system there are rock particles of various sizes, air, organic matter, minerals, water, and soil organisms. The quantity and nature of the ingredients of the soil system are important in determining the natural vegetation or the crop plants which can be successfully grown in any given place. The amounts of moisture and air in the soil, the *p*H of the soil, and the kind and amount of soil minerals determine to a large extent the species which can grow in a given region. The type of soil, whether sand, loam, or clay, affects the distribution of plants. For example, in a humid region conifers may grow on sandy or rocky soils and broad-leaved trees on the more fertile loam soils.

Biotic Factors of the Environment

"Live alone and like it" is not the rule for plants. Plants are not hermits, but social organisms forming communities. Any one plant community is not a heterogeneous collection of species; it has a definite composition and life history. In a community, characteristic species are constantly found and other species are absent. Plant communities are so constant that it is possible to name them as particular associations, just as taxonomists name individual species of plants and animals. Not only does a certain community have a characteristic group of plants, but it also has a distinctive fauna.

The plants and animals making up a community exist in mutual adjustment. For example, in a forest, certain shrubs and herbs thrive only in the shade of trees. If the trees are harvested or uprooted by wind, the understory of shrubs and herbs perishes.

Natural climax communities are in equilibrium. Although there is a struggle for existence in every seemingly peaceful forest, prairie, pond, and desert, there is a balance between plants and animals, prey and predators. No species increases without limit because each has its natural enemies. None is exterminated as long as there are no environmental or marked genetic changes. Man must understand the natural balance and not upset it. A lack of understanding of equilibrium in communities may bring about dire consequences. For example, the prickly pear cactus was introduced into Australia from the United States and soon became a decided menace. In its new habitat, its enemies were not present and considerable loss of grazing land ensued. The Australians then brought in the insect which kept the cactus in check in the United States. The introduced insect has helped to check the invasion of the cactus.

Accidental introduction of pests into the United States, without the introduction of their natural enemies, takes years to control. The gypsy moth, involuntarily introduced from Europe in 1868, has spread like a plague through eastern forests. Some success in its control has been attained by introducing organisms which prey on the gypsy moth, one of which is the Calosoma beetle. The balance in chestnut tree communities was upset when the chestnut tree blight was unintentionally introduced from Asia. Nearly all chestnut trees have been destroyed, and our only hope now is to develop resistant varieties.

Many interrelationships among species in a community involve competition for food. Green plants make their own food, but compete for the requirements for food manufacture: for light, water, and carbon dioxide which are used directly in photosynthesis, and for soil nutrients which are essential for the development of the plant body. The nongreen plants, bacteria, and fungi in the soil compete for the remains of green plants. Through their action the essential elements are returned to the soil and atmosphere.

Animals and plants are linked in their food relationships, and this linkage is referred to as a food chain. The food chain consists of two to many links. In a community the total mass of organisms and their contribution to the food chain can be compared to a pyramid (Fig. 26–12). Green plants always form the broad base of the pyramid and are the primary food source for the upper layers, or **trophic levels.** The layer immediately above the green plant level consists of organisms that use the green plants for food. These organisms, often small herbivores such as mice and rabbits on land, in turn are a food source for the larger ones, either carnivores or omnivores, which make up the next higher trophic layer in the food pyramid. The successively higher trophic levels in the pyramid are characteristically made up of progressively larger but fewer animals. On land or in the sea the amount of organic matter produced per unit of time is always greatest at the basic trophic level and progressively decreases with each higher level. The total weight of the organisms also decreases as one goes from lower to higher levels as also does the available energy.

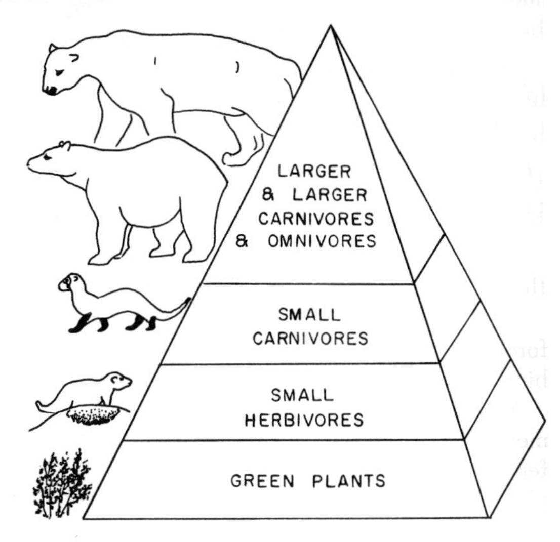

Fig. 26–12. A food pyramid.

QUESTIONS

1. Match the following:

_____ 1. Tundra.	1. Precipitation in excess of 30 inches a year and fairly well distributed.
_____ 2. Forests.	2. Frigid temperatures during most of the year.
_____ 3. Tall grass prairie.	3. 15 to 20 inches of precipitation annually; summer precipitation high.
_____ 4. Short grass plants.	4. 22 to 30 inches of precipitation annually.
_____ 5. Deserts.	5. Less than 15 inches of precipitation annually.

---

2. Describe the major plant communities which occur in your state and relate the conditions under which each develops.

3. _____ According to Holdridge, the two major factors that determine the plant formation in a region are (A) transpiration and evaporation, (B) precipitation and bio-temperature, (C) minimum temperature during winter and maximum temperature during summer, (D) altitude and latitude.

4. Describe the plant formations that are encountered as one ascends a mountain in your region. As one ascends, what changes in temperature, precipitation, relative humidity, barometric pressure, and evapotranspiration are noted?

5. Assume that the bio-temperature and precipitation in the region in which you live are the same as the bio-temperature and precipitation at a certain altitude in the tropics. Would the flora be the same or different? Explain.

6. _____ Lands most suitable for agriculture occur in regions where the ratio of evapotranspiration to precipitation is (A) one, (B) two or more, (C) one-half or less.

7. Explain why soils of tropical rainforests are usually not very productive for corn or similar crops after the first few years.

8. _____ Greasewood characterizes (A) short grass plains, (B) salt deserts, (C) sagebrush deserts, (D) cactus deserts.

9. _____ Which one of the following plants found on the short grass plains is poisonous to livestock? (A) Grama grass, (B) mustard, (C) delphinium (larkspur), (D) phlox, (E) buffalo grass.

10. List some poisonous plants that occur in the region where you live.

11. _____ The forests of Georgia are largely forests of (A) pine, (B) spruce and fir, (C) birch, beech, and maple, (D) redwoods.

12. _____ The orderly stages in the development of a plant community are collectively referred to as (A) climax, (B) reactions, (C) plant succession, (D) coactions.

13. List some agents that may bring seeds, spores, and fruits into a bare area.

14. _____ Which one of the following would not be considered a reaction? (A) Role of plants in soil formation. (B) Role of plants in the protection of soil from wastage by wind and water erosion. (C) Role of plants in decreasing light in a forest. (D) Role of plants in furnishing food for cattle, sheep, and wildlife.

15. _____ Coactions are (A) effects of one organism on another, (B) effects of the plants on the place, (C) are always constructive, (D) the effects of the place on the plant.

16. _____ If the vegetation on an area is destroyed by plowing and if the area is later abandoned, which of the following will be pioneers (first to come in)? (A) Annual weeds, (B) perennial weeds, (C) perennial grasses, (D) crustose lichens, (E) shrubs, (F) trees.

17. Diagram succession as it occurs in a pond in your area.

18. _____ The following—crustose lichens, foliose lichens, mosses, forbs and grasses, shrubs, and finally trees—are stages in a (A) xerosere, (B) hydrosere, (C) rockosere.

19. _____ The first plants to become established on bare rock are (A) lichens, (B) mosses, (C) ferns, (D) forbs, (E) shrubs, (F) trees.

20. What is the average precipitation on your campus? _____. The maximum temperature on record? _____. The minimum? _____. The length of the growing season? _____.

21. _____ Plants that grow where the habitat is neither extremely wet nor very dry are called (A) hydrophytes, (B) mesophytes, (C) xerophytes.

22. _____ Which one of the following is false? (A) Xerophytes generally have a comparatively small transpiring surface and an extensive root system. (B) Xerophytes never grow in soil which is continually wet. (C) Many epiphytes are xerophytes. (D) Some desert plants escape drought.

23. _____ Which one of the following is false? (A) The species in a plant formation are governed by temperature. (B) Direct winter injury results from the formation of ice crystals. (C) The more rigorous the winter in a region, the higher the percentage of species which are trees or shrubs. (D) Plants may be injured during winter by heaving, exhaustion, and desiccation.

24. List the major climatic factors which influence the distribution of plants and describe their effects upon plants.

25. _____ Soil factors which determine the kinds of plants which will be found in a given region are referred to as (A) climatic factors (B) edaphic factors (C) biotic factors.

26. What soil factors influence the distribution of plants?

27. Describe some biotic factors of the environment.

28. _____ The optimum night temperature for growth is (A) the same for young plants as for older ones, (B) for most species lower than the day temperature, a phenomenon known as thermoperiodism, (C) lower than the day temperature, a phenomenon known as photoperiodism, (D) generally higher for temperate zone plants than for tropical plants.

29. _____ The killing of many annuals by the first frosts in the fall usually results from (A) transpiration exceeding absorption from frozen soil, (B) heaving, (C) formation of ice crystals in the cells or between the cells, (D) exhaustion of food reserves.

30. _____ With respect to photosynthesis (A) tolerant species, such as oak, make more food at low light intensities than intolerant species, such as loblolly pine, (B) tolerant species, such as oak, make much more food at 10,000 foot-candles than at 5000 foot-candles, (C) at low light intensities generally intolerant species—for example, loblolly pine—make more food than tolerant species, (D) both tolerant and intolerant species carry on photosynthesis at the same rate.

27

Conservation

To many city dwellers the farm and forest seem far away. Some people do not realize that what happens in these seemingly remote areas may affect them greatly. Everyone must be concerned with rural problems and the conservation of our natural resources. Our farm, forest, and grazing lands furnish us with the materials for existence. If these resources are not used wisely, our civilization may go the way of the Mayan civilization. In the past, nations whose resources became depleted suffered accordingly. It can happen here, too. Farms, forests, and grasslands not only furnish food, clothing, shelter, and other products, but also protect the soil. When the ground is well covered with vegetation, dust storms do not arise. The water we drink or the electricity we use can be influenced by forest cover hundreds of miles away.

To many people, conservation is synonymous with thrift, frugality, and preservation. But these concepts do not give the true and whole meaning of conservation. They suggest that our natural resources of soil, water, forests, and grasslands should not be used but should be set aside for the future. Theodore Roosevelt defined conservation as "preservation through wise use." This better concept implies that our resources can be used and still be preserved, that they are for us, now, as well as for future generations.

Important resources that should be used wisely are air, water, wildlife, recreational areas, forests, grasslands, and cultivated fields. The quality of our national environment has deteriorated. Poisonous chemicals, the by-products of our industry, blight our air and water. Fortunately we are alert to the dangers and are taking steps to minimize them. Increasingly we recognize that if man is to endure he must attain an economy where consumption of resources is matched by replenishment and where pollution is matched by purification.

AIR POLLUTION

As the population of our cities continues to expand, the problem of keeping the air clean becomes increasingly difficult. The combustion of fuels, the evaporation of solvents, the toxic gases released by smelters and other industries, the aerial dispersal of pesticides, and the burning of trash pollute the air with harmful substances such as dust, ash, DDT, parathion, ethylene, peroxyacyl nitrates, aldehydes, fluorides, sulfur dioxide, and many other compounds. The polluted air causes illnesses; reduces visibility and creates hazards for cars, trucks, and planes; spoils masonry; corrodes metals; and dirties homes and buildings with consequent large cleaning bills. Furthermore, it may kill wildlife and damage crops and ornamental plants. In California the damage to crops exceeds $12 million a year and along the Eastern Seaboard, from Boston to Washington, it exceeds $18 million a year. In addition, there is damage to flowers and shade and forest trees. Plants are used as indicators of air pollution because they show characteristic lesions on the leaves when certain pollutants exceed threshold levels.

One city may foul not only its own air but also that of its neighbor as well. Contaminants from factories, trash burning, and oil refineries in New Jersey are wafted over Manhattan island.

We have made progress in reducing industrial pollutants, but much remains to be done. Much of the sulfur in coal and petroleum can be removed before the products are marketed and from the recovered sulfur valuable products, such as sulfuric acid, can be made. Modern furnaces and boilers with their efficient air-to-fuel ratios permit a smokeless and cleaner burning of fuels. Other devices to reduce industrial pollution include tall stacks, forced drafts, electrical precipitators, and scrubbing towers. Often valuable by-products may be obtained from substances which formerly were emitted from the stacks; in fact, some companies have found that they made more from the materials recovered than it cost to install the equipment to do the job.

Most of us drive cars, and collectively we are the major polluters of air. The rush-hour traffic jams in all of our crowded cities create far more atmospheric poisons than all other sources combined. Although the problem was first serious in the Los Angeles area, it has now become a national problem. Engine and exhaust modification, now required in certain states, reduce some pollutants, specifically unburned petroleum, but they do not reduce the nitrogen oxides, by-products of gasoline engines. Hydrocarbons are given off by automobile tires at a rate of hundreds of tons per day in the Los Angeles area alone. To cure the pollution sickness of our crowded cities we could substitute attractive rapid transit systems for private cars, or substitute cars powered by electricity or fuel cells for those powered by internal combustion engines.

We are burning in a few generations the fossil fuels that have accumulated in the earth over the past 500 million years. As fuels are burned, carbon dioxide is added to the air. It has been estimated that by the year 2000 the carbon dioxide concentration of the air will have increased by about 25 per cent. The enhanced concentration may increase the rate of photosynthesis and perhaps change the earth's climate, producing a warming trend.

WATER POLLUTION

The quality of our rivers, lakes, and ground water is deteriorating as a consequence of pollution by municipal, farm, and industrial wastes. Our use of water continues to increase. In the United States we now use about 320 billion gallons a day. Predictions indicate that by the year 2000 we will be using 900 billion gallons a day, which exceeds the 700 billion gallons a day estimated to be the total fresh water supply now available. More than ever we will have to use and reuse our water. If the water used by industry and in our homes is purified, it can be used repeatedly; that used in irrigation cannot, because much is evaporated and transpired. The oceans represent an enormous supply and by the year 2000 it has been estimated that 50 per cent of the nation's water will come from desalting sea water.

People in the United States use water lavishly—in our cities about 200 gallons per person per day, in contrast to the 5 gallons used by people in poor tropical countries and 50 gallons a day by people in European cities. In addition, enormous quantities are used for irrigating farms and for industrial purposes. Manufacturing processes require tremendous quantities of water. Eighteen barrels of water are required to refine 1 barrel of oil, 10 gallons to make 1 gallon of beer, and 250 tons of water to produce 1 ton of paper. Much of the water used by industry is returned to streams and lakes, often so polluted with wastes as to make the water unpalatable and unusable.

Many rivers and lakes are a disgrace to our culture, loaded as they are with offal from slaughter houses and funeral parlors; pulp and paper wastes; wool washings; residues from the bleaching of cotton, flax, and hemp; acid, lime, oil, and grease from steel plants; wastes from oil refineries; and disease-bearing human waste. In food-processing establishments residues of poisonous sprays are washed from fruits and vegetables and the water laden with poisons is returned to the rivers with consequent injury to fish and other aquatic life. Of course, poisons enter rivers in other ways. Insecticides are dispersed over fields and washed away by rain and run off into rivers only to be accumulated again in the bodies of fish, shrimp, and other creatures, eventually including human beings.

Increasingly, people are demanding clear, pure water and streams and lakes which are a delight to the eye instead of foul-smelling sewers. Progressive corporations are beginning to spend substantial sums to prevent pollution. The Water Pollution Bill of 1965 sets up federal standards and guidelines which state that water from riverside plants should be rendered as clean as possible before being returned to a waterway.

Modern technology enables us to return water to streams and lakes free from domestic and industrial pollution. In the Ruhr River Basin in West Germany, where one-half of their industrial capacity is located, pollution has been abated and the water in the Ruhr River is clean enough to swim in and with mild treatment it provides pure drinking water. We can do as well.

The technology of treating sewage wastes and organic chemical wastes from industry is virtually the same. In both, the main reliance is placed on bacteriological processes. By selecting appropriate species of bacteria and by providing ample oxygen, many industrial wastes can be degraded, especially if the polluted water is impounded for a time before it is returned to the river. Research should be directed to select microorganisms which can use various wastes as food and to the possibility of extracting valuable products from wastes.

Domestic sewage must be purified if it is not to become injurious to aquatic life and a danger to human health. Modern sewage disposal plants effectively treat sewage, but not all cities and towns have efficient plants. Moreover, 20 million people still discharge raw sewage into the water. Ships discharge wastes into harbors. Pearl Harbor festers and stinks as a result of sewage discharge from naval ships. Waste disposal at various Army and Air Force bases could be more sanitary. Nearly $2 billion will be required each year to meet present deficiencies and future needs for nationwide sewage-treating facilities.

After the organic compounds have been removed from sewage by modern treatment methods, the effluent often has high concentrations of nitrates and phosphates. When released into the nearest body of water, such compounds enhance algal growth to such a great extent that the ecology of the river or lake is altered. (The objectionable growth of algae and large water plants is referred to as **eutrophication.**) As the algae die, their remains become food for the enormous numbers of bacteria which deplete the oxygen supply. The oxygen concentration diminishes to such a low level that putrefaction occurs with resulting foul odors. Many animals die. There is then no longer a healthy balance between algae, bacteria, protozoa, mollusks, crustaceans, and fish. A major ecological principle is "diversity results in stability." When each type of aquatic life is present, each doing its ecological job, the water is cleansed. But when pollution kills certain species, whether plant or animal, certain jobs remain undone and the water is fouled.

Instead of channeling the effluent into waterways, the purified effluent may be sprayed on croplands. Plants growing thereon will use the nitrogen and phosphorus and the water, free of excess nitrates and phosphates, will then seep to the water table from where it may be recovered.

Even with water purification and reuse we are likely to need additional water supplies in the future. The major problem is where and how to get them. People in the arid Southwest look thirstily at the abundant water resources of the Pacific Northwest and British Columbia. Plans are being formulated to bring water from these areas to the Southwest, using power generated by nuclear plants to pump it over the mountains and into reservoirs where it would be stored for irrigation, domestic use, and the generation of electricity.

Much water evaporates from large reservoirs. Scientists are searching for techniques which will reduce evaporation, but which will not injure aquatic life. Certain higher alcohols seem promising. Where wind action is not intense, they form molecular films over the water surface which retard evaporation.

Costs of desalting ocean water have been declining and, when desalinization is linked to power production by nuclear plants, the costs are reasonable. One corporation has outlined a plan for the Metropolitan Water District of Southern California which would produce 150 million gallons of fresh water per day at a cost of 22 cents per thousand gallons. To match the 150 million gallons of water, 1800 megawatts of power would be distributed, which is ample for a city of 2 million.

Desalting sea water creates another problem: how to dispose of the mountains of salt obtained as a by-product. Hopefully, valuable minerals could be obtained from the salt. Thus, the salt might be a source of sodium, chlorine, boron, magnesium, potassium, bromine, and even silver and gold. Even with such use, however, much of the hot concentrated brine would have to be pumped back into the ocean in such a manner that it would diffuse fast enough to prevent injury to marine life. More research in this field is necessary. Certainly we do not want to replace one form of pollution with another.

OUTDOOR RECREATION

In the year 2000 A.D. nine-tenths of America's estimated 260 million people will live in megalopolises, large supercities, where the problems of air, water, waste disposal, and transport will be enormous. By then personal incomes will have doubled, the work week will probably be 28 hours, and vacation periods will be lengthened. Our society will then place an increased burden on outdoor recreational areas. The average person may travel 5000 miles a year in search of recreation. To meet the increasing demands for outdoor recreation we must act now to preserve unspoiled areas. Several types of outdoor areas have already been established and additional ones are needed.

Areas which have outstanding scenery or unusual natural phenomena or historical significance have been set aside as National Parks, National Monuments, and National Seashores. Such areas are managed by the National Park Service, an agency which was established in the Department of the Interior in 1916. Prior to 1916 the nineteen National Parks were patrolled by the Army. Starting with Yellowstone National Park, which was established in 1872, we now have thirty-two National Parks, seven National Seashores, and many National Monuments. The Act of 1916 declared that the Park Service shall "conserve the scenery and the natural and historic objects and the

Fig. 27–1. A wilderness area—no roads, no permanent human inhabitants, plenty of space. (Sketch by Mildred S. Reed.)

wildlife therein and provide for the enjoyment of the same in such manner and by such means as will leave them unimpaired for the enjoyment of future generations." The policy of the National Park Service is purely that of recreation and education. The number of visitors to National Parks and Monuments increases with each passing year. In 1966 about 130 million people visited these areas. Plans have been made to expand our national park system and to improve facilities.

The Forest Service in the United States Department of Agriculture manages the National Forests, and early in its history it planned to include recreation among the forest uses. The Forest Service established areas for camp-sites; picnics; swimming, boating, skiing, and other sports; roadside rest areas; primeval or natural areas; and wilderness areas.

Primeval or **natural areas** are tracts of 5000 acres or less which are left in their natural condition. They are of considerable importance in studying laws of nature. They are the ecologist's control, enabling him to compare landscapes altered by man to undisturbed areas. The Forest Service has established natural areas in many forest regions. We need similar areas in other types of plant communities. Botanists hope that states, the Bureau of Land Management, or individuals will establish natural areas in desert regions, on the short grass plains, in the tall grass prairies, and other places. Currently a movement is under way to set aside a tall grass prairie in Kansas as a National Park.

Wilderness areas are large regions, 200,000 acres or more, without permanent inhabitants, roads, lodges, or other marks of civilization. They are large enough to enable a person to hike, canoe, or ride on horseback for a week or more without backtracking. They are of great recreational value, particularly for people who wish to escape from the artificialities of modern life to the quietude of the forest. The wilderness areas previously established in the National Forests were given statutory protection by the Wilderness Act of 1964. Such wilderness areas included about 9 million acres. Provision was also made to add as many as 52.1 million acres to the wilderness system, not only Forest Service land but also land administered by the National Park Service and the Bureau of Sports, Fisheries, and Wildlife. The act directed the President to recommend to Congress, over a 10-year period, areas to be added to the wilderness system. The President's recommendation will not take effect without affirmative action by Congress. As we have seen, the inclusion of an area in the wilderness system protects it from exploitation and prohibits the development of roads, lodges, ski-lifts, and other manmade facilities. For example, if a certain area of the Great Smoky Mountain National Park is designated as a wilderness area, it will remain in its pristine condition and no roads will traverse it (Fig. 27–1).

Many states, cities, and private organizations are active in establishing recreational areas and in highway beautification. These agencies have landscaped highways and roadside areas with trees, shrubs, and other plants and have provided tables, benches, grates, and similar facilities where motorists may relax and enjoy their meals. There is a need for more such roadside areas.

WILDLIFE MANAGEMENT

In many forested regions, wild animals are a major attraction to sportsmen, camera enthusiasts, scientists, and those who simply delight in observing them. In order to maintain a population of wild creatures in modern times, it is essential to carry on a program of wildlife management for their protection against encroaching civilization. One major problem is prevention of extermination. Extermination of species can be prevented by appropriate hunting regulation and by maintaining suitable habitats for the great variety of wild creatures.

Habitats for game may be improved by increasing the food and water supply, and by providing additional cover for shelter and concealment. Often the food supply for deer may be increased by opening up timber stands and by planting conifers in hardwood forests. In farm areas lack of cover often limits the population of birds and small game. The planting of trees and shrubs along highways and fence lines and the establishment of farm woodlots will not only provide cover but also increase the food supply. Deer and elk have a tendency to stay together in large numbers, and rather than wander great distances in search of food, they will starve in crowded areas. When overpopulation occurs, forests are injured and desirable forage plants disappear. The deteriorated game range then supports a lesser population of wildlife. Because animal populations tend to increase, good game management calls for harvesting the yield each year through regulated hunting. Hunting is part of a game management program. If properly carried out, the population of game animals is maintained in adjustment with the food supply. The animals are healthy and the vegetation is not destroyed. In the management of deer it is usually necessary to reduce the population of does as well as bucks.

In many areas the hunter is now the major, or only, predator of big game, a role previously played by wolves, coyotes, mountain lions, and bobcats. In early days the natural predator-prey interaction was maintaining a relatively stable equilibrium between big game animals and their food supply. For example, prior to 1907, puma and wolves kept the population of deer on the Kaibab plateau of Arizona between 5000 and 10,000, a population in line with the food supply. Between 1907 and 1923, man upset the predator-prey balance by killing the predators. As the predator population declined, the deer population increased, reaching 100,000 by 1925. This population by far exceeded the carrying capacity of the range. All forage within reach of deer was consumed, the grasses, shrubs, tree seedlings, and lower branches of trees. With overgrazing the productivity of the deer range declined. During the following two winters, 40,000 deer starved to death, and the survivors were in poor health. In subsequent years, the deer population continued to decline, and finally stabilized at about 10,000. In more recent years, the elk population of Yellowstone National Park, where no hunting is permitted, increased to the point where elk were starving and their food supply was dwindling, a consequence of overgrazing by the elk.

In the management of such migratory fowl as ducks and geese, the establishment of preserves and breeding grounds in the North and the establishment of sanctuaries where birds in migration can rest and feed are of value.

The problem of maintaining a fish population is becoming increasingly acute as the number of anglers continually mounts. Constancy of stream flow and freedom from

Fig. 27–2. We are part of nature, not apart from nature. (John Coulter and Wyoming Wild Life.)

pollution are essential in fish management. From data obtained from surveys of streams and lakes, it is possible to judge the number and kinds of fish to plant.

FOREST CONSERVATION

Before we can use our forests wisely, we must understand forest values. Forests are valuable because they furnish us with essential products and because of the influences which they exert, influences of both a practical and an aesthetic nature. Among the important forest influences are recreation, employment, prevention of erosion and floods, and more even stream flow.

In regions clothed with forests, wind erosion of soil is practically zero and erosion of soil by water is markedly reduced. The wearing away of a land surface by water depends on four factors: (1) the amount of precipitation, (2) the degree of slope, (3) the composition of the soil, and (4) the type of plant cover. Man has greater control over the type of cover than over the other three. Forests are more effective in retarding erosion than any other type of cover, as can be seen from Table 27–1.

Table 27–1

Time Required To Remove 7 Inches of Topsoil on a 10 Per Cent Slope in the Southern Appalachian Region*

Type of Ground Cover	Years to Remove 7 Inches of Topsoil
Forest	575,000
Grass	82,150
Rotation cropping	110
Cotton	46
Bare ground	18

* From Bennett, 1947.

The leaves of trees reduce the impact of precipitation, thus preventing splash erosion (Fig. 27–3). The intercepted precipitation later drips gradually to the forest floor, and because forest soils are porous, much of the water filters in. During heavy rains, runoff is retarded by the accumulated litter of leaves, twigs, etc.

Forests promote even stream flow and thereby help to prevent floods. Evenness of stream flow is also of significance for fishermen, domestic water supply, hydroelectric

Fig. 27–3. Splash erosion. The splash reaction throws soil and water into the air. The splashed particles may be floated, dragged, or rolled downslope by surface runoff. Forests and other types of vegetation markedly reduce splash erosion. (Naval Research Laboratory of Washington, D. C. The photographs were made for use by the Bureau of Yards and Docks in its erosion control work on Navy-controlled lands.)

power, and irrigation. In many parts of the United States, abundant water flows in the streams in late spring and early summer, whereas by middle and late summer, when the water is most needed, the available water is at a minimum. By retarding early season runoff when stream channels are filled to capacity and by retarding the melting of snow (as much as 6 weeks in mountainous regions) forests promote an even flow of water throughout the summer.

Forests alone cannot prevent all floods, but they play a part in any scheme of flood control. In forested areas the stream channels carry clear water rather than a mixture of soil and water. Dams play a significant role in a flood control program, but dams backed up with silt are of no value. The prevention by forests of silting-in of reservoirs is of decided significance.

Private operators, organizations, states, cities, and agencies of the federal government are working for forest conservation. The United States Forest Service practices a policy of "multiple use," a plan to use forest resources for the greatest good. In the National Forests all resources are used. Areas of great value for watershed protection or recreation are not logged over. A sane adjustment is made between the numbers of livestock and wild animals. In the National Forests timber is harvested on a sustained yield basis. Cuttings are managed so that there will always be trees to harvest. To illustrate the principle, imagine a tract of forest land of 1000 acres, which is covered with trees 100 years old that are large enough to harvest. This tract of 1000 acres is divided into ten strips or plots. During the first decade the trees on one strip are harvested, during the next decade those on another strip, and so on, until the tenth strip is logged over. By that time the trees on the first strip are of a merchantable size and can be harvested. Hence, the cutting cycle is continuous. The above method is referred to as **clear cutting** because all the trees on a strip or block are harvested. Even-aged stands of intolerant species such as Douglas-fir and lodgepole pine are generally harvested by this method.

Another method is the **selection system,** which is ideally suited to unevenly aged stands of tolerant trees. In the selection system the trees are harvested as they attain merchantable size. Accordingly, the same forest area is cut over periodically, each time removing the merchantable trees and allowing the smaller ones to remain and increase in size and value before they are harvested. When the large trees are removed seedlings come in and take their place. Hence the area is always covered with trees, and the soil is always protected against wind and water erosion.

In many regions forest productivity can be enhanced by improvement cuttings, such as sanitation cutting and thinning. The removal of diseased and insect-infested trees is referred to as **sanitation cutting.** Their removal helps maintain the forest in a healthy condition and often the wood can be used. In certain localities too many trees become established on logged-over areas. Because many trees share a limited supply of water and minerals, none grows rapidly. When the stand is thinned by removing some trees, those remaining grow more rapidly. In some areas the thinning is done before Christmas and the removed trees are sold as Christmas trees.

Fire is the biggest handicap to forest conservation. Everybody loses when a forest fire ravages an area. There is loss of merchantable timber and young growth, loss of organic matter, accelerated erosion, lessened recreational value, destruction of game, and perhaps loss of human lives and homes. Eighty per cent of the forest fires are caused by man's carelessness. Education of young and old alike will help to lower the number of fires. Certainly all should be extremely careful with fire.

CONSERVATION OF GRASSLANDS

The unimproved grasslands or "range" are associated with stock-raising in western United States. The forage is a valuable crop which should be conserved—that is, used wisely. With proper management the forage may be used year after year without any detriment to the forage plants or the land.

The productivity of the range can be maintained by proper adjustment of numbers of livestock to the forage available. If overgrazing is continually practiced, the better forage plants are replaced by poorer ones, and the ground is then not well covered with vegetation (Fig. 27–4). With a sparse ground cover, wind and water erosion are accelerated and the range becomes depleted. Progressive ranchers have found by experience that more money is made from animals on a conservatively stocked range than from those on an overstocked range. Better calves, lambs, and wool crops are secured and losses are smaller.

The range should be used by the proper class of livestock. Grass ranges on flat or rolling ground with plenty of water and shade are suited to cattle. Ranges on rough terrain with many forbs as well as grass are best used by sheep. Flat or rolling grasslands with water at a considerable distance can be used by horses. Livestock should be properly distributed on the range. On a poorly managed range some areas are overgrazed and others undergrazed. More even use of forage can be achieved by fencing, proper placing of salt, water developments, riding, and herding.

If the range is grazed as soon as plant growth begins in the spring, the forage is weakened and the carrying capacity of the range is reduced. To avoid such injury, livestock should be held off the range until

Fig. 27–4. *Left*, with moderate grazing, forage is maintained. *Right*, with overgrazing, there is practically no forage. (E. J. Dyksterhuis, U.S. Soil Conservation Service.)

the forage plants have made sufficient growth, or some system of rotation grazing should be practiced. In rotation grazing, a certain area is grazed early one year but the next year not until later in the season.

The productivity of the range may frequently be enhanced by water developments, fencing, poisonous plant control, rodent control, and reseeding.

CONSERVATION OF CULTIVATED FIELDS

Much land that is now devoted to crops is too steep for that purpose. It supports crops for a few years, but then becomes eroded so badly that farming is impossible. Such lands should be restored to grass or forest. These plant covers not only protect the hilly slopes from accelerated erosion, but they also protect neighboring lands downstream from erosion and undesirable deposition. Where large gullies scar the hillside, dams and terraces should be constructed to retard runoff.

Lands less steep may be farmed if protected from erosion by **terracing, contour farming,** and **strip cropping.** The effectiveness of such treatments can be seen from the following data of Bennett's (Table 27–2).

Table 27–2

Annual Soil and Water Losses Under Different Systems of Farming

Farming Practice	Tons of Soil Lost per Acre	Percentage Water Loss
Cotton in straight rows	28	18
Cotton in contoured rows (contour farming)	17	9
Cotton and vetch in contoured rows (strip cropping)	1	3

Fig. 27–5. Water is standing in the lister furrows and terrace channel after a 4-inch rain in a cotton field planted on the contour. (U.S. Soil Conservation Service.)

Terracing is effective in minimizing erosion. Moderately sloping land may be terraced effectively by **lister plowing** along the contour. In lister plowing adjacent furrows are thrown in opposite directions. This leaves a small ditch about 6 inches deep in which the crop is planted. Each terrace holds most of the water which falls upon it (Fig. 27–5).

Sloping lands should be plowed and planted across the slope, so-called contour farming, instead of up- and downhill. The furrows hold back the water so that it has a chance to soak in.

In strip cropping, a complete cover, preferably of alfalfa, clover, or some other perennial cover crop, alternates across a field with a row crop of corn, potatoes, cotton, tobacco, etc., or with a broadcast crop of grains such as wheat or oats. On the farm illustrated in Figure 27–6, strips are seeded to alfalfa with alternate strips in oats. The soil which runs off the oat land is held by the alfalfa pasture, and rills, potential gullies, which start in the oat land stop at the alfalfa strip. On each strip the farmer plans to rotate crops according to the following pattern: C–O–H–H–H, in which C stands for corn, O for oats, and H for hay, in this case alfalfa. Corn will be planted 1 year, oats the next, and then hay, which will occupy the land for 3 years.

The rotation of crops helps to minimize soil depletion and lessens the losses brought about by insects and disease. Legumes such as clover and alfalfa are an important part of a crop rotation because they enrich the soil with available nitrogen. However, legumes cannot be regarded as crops that enrich the soil in other minerals. Though they increase the nitrogen content, they lower the soil reserves of potassium, phosphorus, and other elements.

In the United States, it is essential that the productivity of our agricultural lands be maintained. Up to the present, about 282 million acres of farm and grazing land have been essentially ruined by erosion and 775 million acres are threatened with ruin. Good land that is not seriously threatened by erosion amounts to only 460 million acres.

Fig. 27–6. Strip cropping. The alternating bands of alfalfa and oats shown here prevent loss of soil by erosion. (U.S. Soil Conservation Service.)

QUESTIONS

1. Define conservation.

2. List some major pollutants of air and suggest how they might be reduced.

3. In what ways may polluted air hurt us?

4. Should cars be banned from metropolitan areas?

5. Are the streams and lakes in your state polluted? What are the pollutants? How may they be reduced? Are you willing to pay your share of the cost of purifying the waters in your community?

6. How might the high concentrations of nitrates and phosphates in the effluent from a sewage plant be reduced? Why is the high concentration considered undesirable?

7. Are there any exceptions to the ecological principle that "diversity results in stability?"

8. What are some possible future sources of water?

9. List the National Parks, National Monuments, National Seashores, National Forests, State Parks, primeval areas, and wilderness areas which have been established in your state. Do you think additional areas should be set aside?

10. _____ Areas set aside primarily for scientific study are known as (A) National Monuments, (B) National Parks, (C) primeval areas, (D) wilderness areas.

11. List four principles of range management.

12. _____ An increase of annual plants on the range usually indicates that the range is (A) deteriorating, (B) improving, (C) static.

13. _____ Preserves, breeding grounds, and sanctuaries are used primarily in the management of (A) small game, (B) migratory fowl, (C) game birds, (D) big game.

14. List four problems of game management and suggest how they may be solved.

15. Discuss forest values.

16. _____ An even-aged stand of intolerant trees should be cut by (A) the selection system, (B) clear cuttings in strips or blocks, (C) the shelterbelt system, (D) the liberation system.

17. _____ Which one of the following is not dependent on the forests for the influences which they exert? (A) A fisherman, (B) a homeowner in a large city, (C) the owners of a hydroelectric plant, (D) a farmer in an irrigated district, (E) a dry-land farmer.

18. _____ With respect to the National Park Service, which one of the following statements is false? (A) The National Park Service was first authorized by President Wilson in 1916. (B) Yellowstone Park was established in 1872. (C) National Parks are preserved for scenic value, historical significance, and natural phenomena. (D) The big trees in Sequoia National Park will never be cut down, but will remain as remnants of the original giant forests of the West. (E) Like the Forest Service in the U.S.D.A., the National Park Service practices a policy of multiple use with respect to recreation, hunting, grazing, and timber management.

19. _____ Which one of the following problems of game management is most serious in your state? (A) Prevention of extermination, (B) space for game, (C) cover for game, (D) water, (E) feed for wildlife, (F) overpopulation.

20. _____ Which one of the following is false? (A) Prevention of pollution is necessary if fish are to survive in streams. (B) Eggs and fry may be destroyed by heavy silt carried in flood water. (C) The burning of watersheds is harmful to fish. (D) A water temperature of 70° F. is better for trout than a temperature of 55° F. (E) Slopes covered with vegetation control water runoff and do not easily erode.

21. Describe three methods for the conservation of farm lands.

22. _____ Erosion of soil is least from (A) forests, (B) grasslands, (C) cotton farms, (D) bare ground.

28

Evolution

Plants and animals come in a myriad of forms, ranging in size from submicroscopic bacteria to giant redwoods, from minute protozoa to elephants. Indeed, about 340,000 different species of plants and 1,000,000 species of animals have already been described.

What is the significance of this great diversity? A study of organisms in their natural environments reveals that they are wonderfully adjusted in structure and function to their habitats, dwarf and often succulent plants on deserts, giant trees with other plants perched upon them in rain forests, plants with floating leaves in ponds. From the simplest to the most complex, organisms are adapted to function efficiently in their respective niches in nature.

For many years the diversity and adaptiveness of living things was explained as an act of creation—special creation by God, who made each species to fit in a predestined place in nature. The doctrine of special creation was formulated long ago when knowledge of living organisms was decidedly limited. For many centuries this idea seemed the most reasonable explanation possible.

At present the concept of special creation has been replaced by the theory that present-day organisms have come from unlike progenitors which usually were less complex. (Occasionally, relatively simple organisms have evolved from those of greater complexity.) The theory which states that species are and have been continually subject to change, with a consequent production of new living forms, is known as the **theory of evolution.** This theory implies that all life is a continuum in time and that all living things are related by descent. The fossil remains of organisms of the past demonstrate that species once in existence are no longer present. In recent times a few forms have become extinct—for instance, the Dodo bird in 1681, and the heath hen about 1930. Other forms are threatened. If we could project ourselves into the future, we would see organisms which do not now exist.

EVIDENCES OF EVOLUTION

A great volume of evidence has been accumulated which demonstrates the reality of evolution—evidence from geology, geographical distribution, comparative morphology, and experimentation.

Geological Evidence

Armored fishes, dinosaurs, flying reptiles, lepidodendrons, and seed ferns at one time abounded on earth, but they are now ex-

Fig. 28–1. The Proterozoic era. Masses of algae appear in the water; the only organisms leaving fossil remains were those having some kind of mineral skeleton or shell. (Reproduced by permission from *Plant Classification* by Lyman Benson. Copyright © 1957, by D. C. Heath and Co., Boston.)

tinct. The changing life in past ages may be inferred from fossils found in layers of sandstone, limestone, and shale. Fossils are relics of plants and animals long since dead. In certain instances the actual remains are preserved, as when plant parts become embedded in amber. Frequently a plant part becomes petrified, that is, the cell cavities and intercellular spaces are filled with minerals deposited from ground water containing high concentrations of silica, calcium, magnesium, and other minerals. Silicified wood is an example, and when this is sectioned and examined with a microscope, fine anatomical details are evident. More commonly the fossil is an impression of a plant part which was made in clay that later hardened into rock: here the actual part is gone. From the fossil record it is not only possible to learn what organisms existed in the past, but also to estimate when they were on earth and the sequence in which they appeared. Fossils can be dated because the layers of rock were formed one above the other. The deeper the layer of sedimentary rock, the older it is.

Sequence of forms. A few highlights of the geological record are shown in Table 28–1. Geologic time is divided into five great divisions, called eras. The oldest era is the Archeozoic, and the most recent era is the Cenozoic. The eras are divided into periods and the periods into divisions, called epochs.

The fossil record demonstrates that each group of organisms had a period of maximum abundance and that in a prior period they did not exist, but only later evolved from some other form. Some groups of organisms perished after their period of abundance because they did not have a supply of mutants which could survive in a changing environment. As we go up the time scale, the organisms in general become increasingly complex.

No fossils have been found from the Archeozoic Era. In Proterozoic rocks remnants of certain algae have been discovered, but in these rocks very few fossils are found (Fig. 28–1). The dearth of recognizable fossils results because organisms then living, bacteria and algae, had soft bodies which were readily decayed before they could be

Table 28–1

Geologic Time Chart

Million Years Ago	Geological Unit of Time			Events
	CENOZOIC ERA	CENOZOIC PERIOD	Pliocene Epoch	Rise of herbs. Man appears.
1			Miocene Epoch	Reduction of forests. Mammals at peak. Grazing types spread.
			Oligocene Epoch	World wide distribution of forests. Mammals evolve rapidly. Great apes.
			Eocene Epoch	Tropical flora in arctic regions. Modern mammals appear.
			Paleocene Epoch	Archaic mammals dominant.
100	MESOZOIC ERA	CRETACEOUS PERIOD		Rise of angiosperms. Gymnosperms dwindling. Dinosaurs, pterodactyls, toothed birds reach peak, then disappear. Small mammals.
		JURASSIC PERIOD		First known angiosperms. Conifers and cycads dominant and cordaites disappear. Dinosaurs and marine reptiles dominant.
		TRIASSIC PERIOD		Conifers and cycads dominate forests. Seed ferns disappear. Small dinosaurs. First mammals.
200	PALEOZOIC ERA	PERMIAN PERIOD		First cycads and conifers. Continental uplift and orogeny.
		PENNSYLVANIAN PERIOD (Upper Carboniferous)		Spore-bearing trees such as lepidodendron and calamites dominate forests. Extensive coal formation. Reptiles and insects appear.
		MISSISSIPPIAN PERIOD (Lower Carboniferous)		Lycopods, horsetails, and seed ferns abundant. Early coal deposits. Climax of crinoids and bryozoans.
300		DEVONIAN PERIOD		First forests. Primitive lycopods, horsetails, ferns, and seed ferns. First amphibians. Brachiopods reach climax.
		SILURIAN PERIOD		First land plants. Algae dominant. Widespread coral reefs.
400		ORDOVICIAN PERIOD		Marine algae dominant. Invertebrates increase greatly. Trilobites reach peak differentiation.
500		CAMBRIAN PERIOD		Algae abundant. Marine life only. First abundant fossils. Trilobites and brachiopods dominant.
1400	PROTEROZOIC ERA			Bacteria and algae.
1900	ARCHEOZOIC ERA			Presumptive origin of life. No fossils found.

Fig. 28–2. The Mississippian period: *Left*, giant horsetail, *Calamites; background*, a relative of the club mosses, *Sigillaria; foreground and right*, seed ferns, *Lyginopteris* (Cycadofilicales); *middle above and right* (large, entire elongate leaves), Cordaitales; *right* (trunk), *Lepidodendron*. The salamanderlike animals were stegocephalians, giant relatives of the living salamanders. (Reproduced by permission from *Plant Classification* by Lyman Benson. Copyright © 1957, by D. C. Heath and Co., Boston.)

preserved in sediments. Furthermore, Proterozoic rocks have undergone deformations and may have been melted by volcanic action, factors unfavorable to the preservation of plant remains. In the Paleozoic Era, club mosses, horsetails, ferns, seed ferns, and gymnosperms made their appearance (Fig. 28–2). In the Mesozoic Era, angiosperms appear and become increasingly important as we go up the scale through the Cenozoic Era. As angiosperms become more prevalent, the gymnosperms dwindle.

Geographical Distribution

Man and his dog are found in practically all parts of the earth, but they are the exceptions. Few other species are so universal in their distribution. Generally each species has a definite range, found here, but not there. The flora of islands is sometimes different from that on the nearest mainland. At one time an island may have been connected to the mainland by a land bridge and at that time the area may have had a uniform flora. When the land bridge was submerged, the island became isolated. The course of evolution on the island may have led to the development of species unlike those on the mainland. Madagascar (off the coast of Africa) and the islands of Australia and New Zealand (which have been cut off from Asia for millions of years) have a number of species which are not found anywhere else. This restricted distribution is not due to soil, climate, or biotic factors, because many island species thrive when introduced to the mainland, and conversely. The most reasonable explanation is that species have undergone evolutionary change while isolated from closely related forms with which they might have hybridized. Species may become isolated by agents other than large bodies of water and in such remote places evolve into distinct species. Mountain tops and also valleys surrounded by mountains often have species which are limited to such areas.

Morphological Evidence

The study of structure (morphology) supports the theory of evolution. Decided similarity of structure indicates close relationship, whereas less marked similarity suggests more distant relationship. Sweet peas, beans, and peas are closely similar in floral structure. Such likeness in structure indicates that they are closely related and that they have evolved from a common ancestor. Some lines of descent led to sweet peas, others to beans, and so on. The fact that man can group organisms into families and subdivide families into genera and genera into species is strong evidence for evolution.

Physiological Evidence

Physiological similarities of organisms furnish information concerning relationships and ancestry. Members of some groups of plants produce characteristic chemical substances. For example, species of the mint family have aromatic volatile oils, and members of the pine family, pitch. Plants producing similar substances appear to be related by descent. The most characteristic compounds of protoplasm are the proteins. The protoplasmic proteins of one species are unlike those of any other species. The more closely related two species are, the greater the similarity of their proteins. Through comparative studies of proteins from a great number of species, it has been possible to construct a family tree which is in essential agreement with one constructed from studies of comparative morphology.

The kind of food stored by a plant is often useful in determining relationships, especially among the phyla of algae. The green algae store starch; diatoms and golden brown algae, oils; brown algae, laminarin and mannitol; and red algae, floridean starch. The presence of like pigments indicates relationship. Thus we believe that both the bryophytes and the tracheophytes evolved from the green algae because all have chlorophyll a, chlorophyll b, carotene, and xanthophyll. Unlike the green algae, the brown algae have chlorophyll a and chlorophyll c and the red algae, chlorophyll a and chlorophyll d. In recent years a new frontier in taxonomy has developed which uses gas chromatography, paper chromatography, and electrophoresis to separate the chemicals occurring in species. For example, the essential oils present in orchid flowers are being studied to determine relationships among orchids. In general, closely related species exhibit a more similar chemical composition than distantly related ones. Hybridization of species occasionally occurs in nature. Often the hybrids have compounds characteristic of both species from which they were derived.

Because DNA is the genetic material of organisms, the more closely two organisms are related the more similar should be their DNA. A method has been developed which determines the degree of similarity between the DNA's of various organisms. Intact, single-stranded DNA from one species is immobilized in agar and radioactive fragments of DNA from another species are allowed to react with it. A fragment with a nucleotide sequence that is complementary to a portion of the intact DNA will pair with that section. The amount of radioactive fragmented DNA that combines with the intact DNA in the agar indicates the proportion of nucleotide sequences which are shared by the two species. As yet, few studies have been carried out with higher plants. Studies with animals indicate that human and chimpanzee DNA's are very similar, and that all mammals have 20 per cent DNA in common. Three per cent of the DNA of fish is similar to that of man, and 8 per cent of that of birds.

Experimental Evidence

Evolution is such a slow process that during the few centuries of recorded observations man has noted only a few evolutionary changes in wild species. In domesticated animals and plants numerous mutations have been noted, and as a consequence man has directed the development of many varieties of horses, dogs, pigeons, sweet peas, snapdragons, chrysanthemums, wheat, and many others. Certainly heritable variations, a consequence of gene mutations and chromosomal alterations, do occur. The accumulations of such variations could in time lead to the development of new species.

THEORIES OF EVOLUTION

During the past century and a half several theories of evolution have been promulgated. Lamarck's theory (1809) stated that acquired characters could be inherited. Darwin's theory of natural selection was clearly and convincingly stated in his book *The Origin of Species by Natural Selection,* which was published in 1859. The modern theory of evolution is a product of the twentieth century and incorporates the data and ideas of many individuals–Mendel, De Vries, Morgan, Muller, and many others.

Inheritance of Acquired Characters

The theory that acquired characters could be inherited was developed by Jean Baptiste Lamarck (1774–1829), a scientist who, 50 years before Darwin's book was published, recognized that evolution had occurred in nature. Lamarck believed that characteristics acquired by an individual through use or disuse of organs could be passed on to the descendants. For example, during the past thousands of years the long-necked giraffes have evolved from short-necked giraffes. Giraffes browse leaves from higher branches. According to Lamarck's theory, the long neck of present-day giraffes is the cumulative effect of countless generations of neck-stretching by giraffes. This implies that somehow or other neck-stretching alters the genes. This explanation, although fallacious, seems very reasonable. Ask a group of people what man will be like thousands of years hence and many will state that he will have a large head and small weak legs. Increased brainwork is assumed to result in a race of human beings with large heads; decreased use of legs, through increased use of the automobile, is believed to result in smaller, weaker legs. The assumption in this type of reasoning is that use and disuse will result in modifications which can be inherited, an assumption which is not supported by experimentation. There are very few, if any, authenticated reports which prove that acquired characteristics can be inherited.

Theory of Natural Selection

This theory is usually associated with Charles Darwin (1809–1882), although others had preceded him with brief statements of the idea, and Alfred Wallace developed the theory concurrently and independently. Darwin, after years of observation, became convinced that species are not immutable creations that have been maintained in their present form since the beginning of life on earth. The arguments used by Darwin to explain how evolution had been brought about were based on variation, tremendous reproductive powers of organisms, and natural selection.

Darwin noted the great variation which existed in a species. No two organisms are exactly alike.

He also reasoned that the reproductive capacities of organisms far exceeded the capacity of the earth to sustain the progeny. For example, one shepherd's purse plant may produce 60,000 seeds. If each seed developed, there would then be 60,000 plants; and if each of these produced 60,000 seeds, which in turn developed into mature plants, there would then be $60{,}000 \times 60{,}000$ shepherd's purse plants, and in the next generation 60,000 times that, and so on. Shepherd's purse, like all plants and animals, tends to increase indefinitely in a geometric manner. The earth, however, does not have the capacity to sustain all of the organisms produced.

A census of most organisms reveals that even though there is a tendency for a geometrical increase, actually the population of practically all species remains fairly constant. Obviously then, not all of the individuals which are produced survive. Many seeds do not develop into mature plants. Some shepherd's purse seeds may become located where conditions are unfavorable for germination—in a lake, for example. Certainly not all seedlings survive. In the struggle for existence the majority perish.

Which ones will perish? Which survive? No two individuals in a population are exactly alike. The differences may be slight or great, structural or functional, but they always exist. Some individuals may possess favorable survival qualities; others lack such features. Those with favorable characteristics survive. Those lacking such qualities do not survive the competition. In nature, man is not the selective agent as he is with domesticated plants and animals. In nature the environment acts as the selective agent. Darwin termed this process "natural selection." According to him, the continual selection by the environment of individuals with favorable characteristics would lead in time to organisms which were so different from their progenitors that they would become new species. In general, natural selection over long periods of time leads to the development of species which are adapted to thrive in their respective niches.

Fitness to survive must be defined in terms of the environment. There is no such thing as fitness per se. For example, before European civilization was introduced to the United States, the trout with the keenest eyesight and the quickest response to the presence of an insect on water had the best chance to survive. In the different environment of our present civilized society, such a fish has probably the least chance to survive and reproduce his kind; he ends his life gasping in some fisherman's creel. Through natural selection perhaps a species of trout will evolve which somehow or other will avoid the fisherman's hook. In one environment a tall plant may have the best chance for survival, whereas in a different place—a desert, for example—such a plant may be least able to survive.

Under certain environmental conditions a plant with a deep root system may have an advantage over one with a shallow root system. With other conditions, the reverse may be true. In areas where bees are abundant, some floral types may be advantageous to pollination and hence for reproduction, whereas in areas lacking bees, such floral types would not be beneficial. In an area without insects, plants with flowers specialized only for insect pollination would not reproduce their kind. They would lose out, perhaps to wind-pollinated plants.

Modern Concepts of Evolution

The present-day theory of evolution itself has evolved from Darwin's theory, from which it differs in one important feature. Darwin had a meager knowledge of the cause of variation. He recognized his limited understanding but, like his predecessor Lamarck, thought that acquired characters could be inherited.

Now we know how many heritable variations can be produced. You recall that variations which result from gene mutations or chromosomal alterations can be inherited. Hugo De Vries was one of the first to suggest that mutations were heritable variations. His ideas were expressed in his book, *The Mutation Theory,* which was published in 1901.

The modern theory of evolution combines De Vries' theory and Darwin's. Mutations occur and natural selection determines which will survive and which perish. Natual selection is a creative evolutionary process that results in the perpetuation of those variations which favor survival in a particular environment.

A species is not a number of identical individuals, but instead it consists of a collection of populations whose members not only have many genes in common but also have many different ones as well. As a result of interbreeding between the plants in a population and between populations, there is a free flow of genes (Fig. 28–3, *upper*)

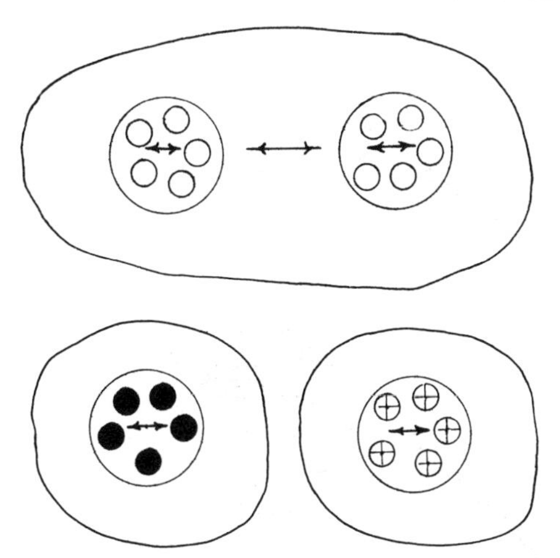

Fig. 28–3. Isolation of populations favors the evolution of species. *Upper,* when populations are not isolated, genes are exchanged within a population and between sister populations. *Lower,* when populations are isolated, genes are exchanged within each population but not between them. In time the two populations become so different that they would be recognized as distinct species.

within the species. A hereditary variation of survival value that arises in a population may in time spread throughout the population and to sister populations.

With interbreeding many types appear. Some plants may be more adapted to a particular environment than others and these thrive and produce many offspring. Those members of the species that are poorly adapted to the environment produce few or no offspring and consequently their genes become less frequent in the population or are lost. After many generations the greater reproduction of individuals with traits favoring survival results in an increase in the frequency of their genes in the population (Fig. 28–4). Hence, plants, and animals too, become adapted to their environment through differential reproduction. Adapted organisms leave more progeny, and contribute more genes to the gene pool of the species, than poorly adapted ones.

Reproductive isolation favors the evolution of new species, a process called **speciation.** Reproductive isolation between populations of a species may result from geographical barriers such as distance, mountains, deserts, and oceans; from temporal barriers such as differences in flowering times; and from genetic barriers such as inversions, translocations, and the rate of pollen tube growth. Figure 28–3, bottom, illustrates two populations which have become isolated; genes are exchanged within each population, but not between them. Certain mutations that arise in one population may spread throughout that population, but not to the distant population. The mutations occurring in one population may be different

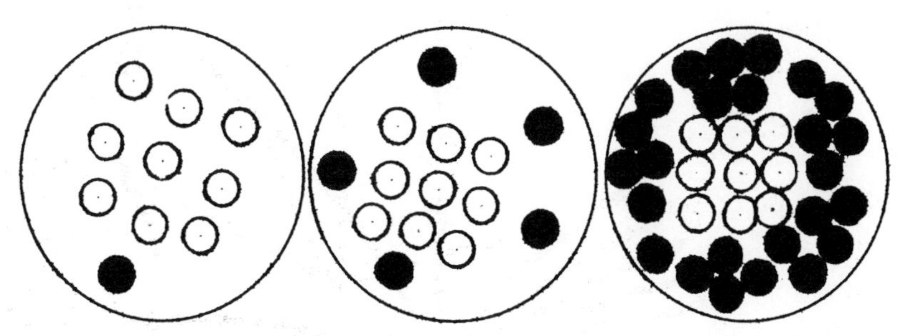

Fig. 28–4. Change in a population of annuals with time. Assume that the white plants are poorly adapted to the environment and that each leaves just one offspring. Assume that the black plants are well adapted and that each yields five progeny. In the beginning one-tenth of the plants are well adapted, but after just two generations twenty-five out of thirty-four are well adapted.

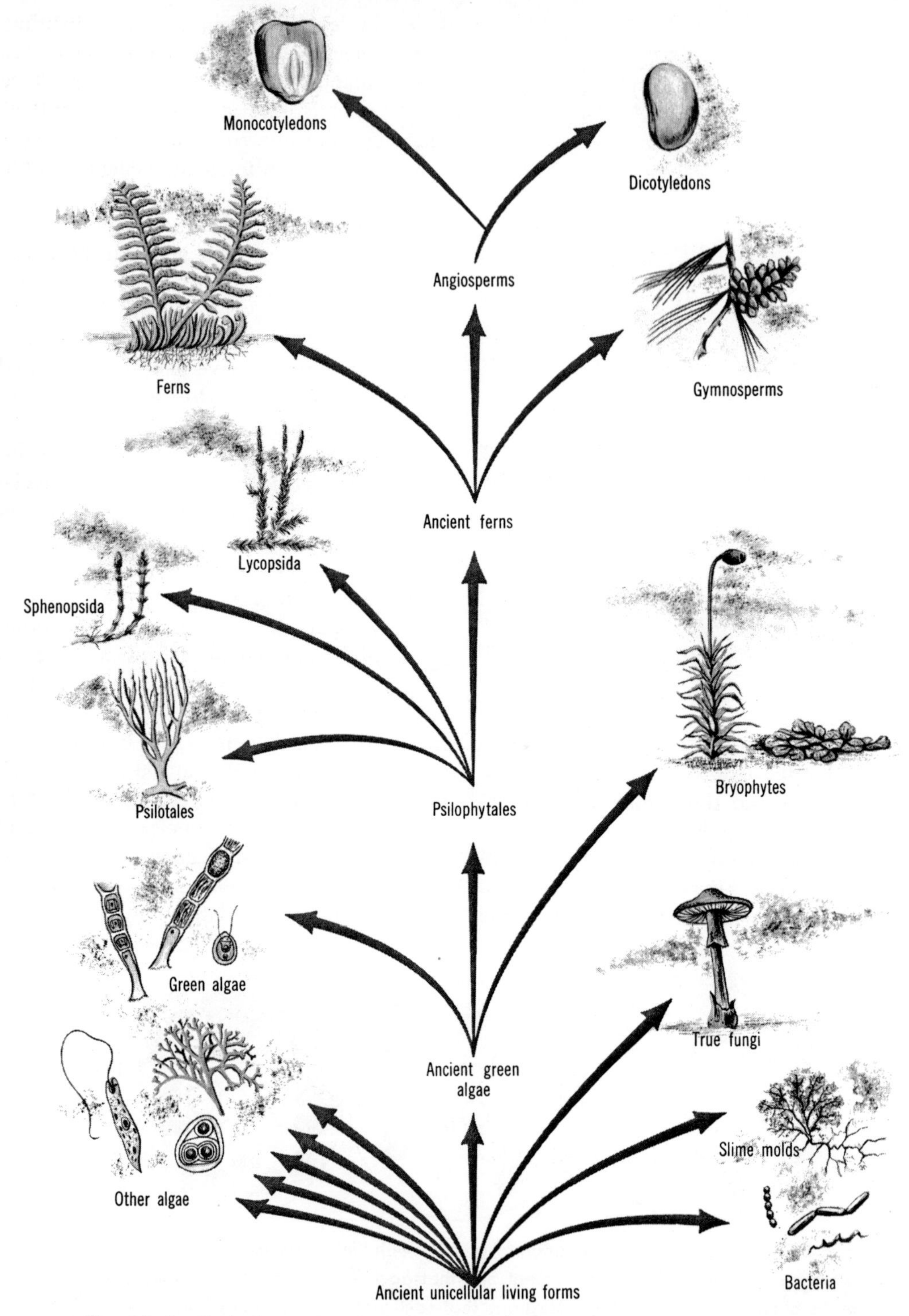

Fig. 28–5. Evolutionary trends in the plant kingdom. (Reproduced by permission from *Textbook of Modern Biology* by Alvin Nason. Copyright © 1965, John Wiley & Sons, Inc., New York.)

from those originating in the other population. Hence, as millennia go by, the two populations may become so different that they could not interbreed even if grown side by side. We would then recognize two species. The members of one species share one gene pool; those of the other, a different one.

TRENDS IN EVOLUTION

Progressive Evolution

As plants evolve, usually they become increasingly complex and specialized. The highly complex seed plants of today have evolved from plants of lesser complexity. Such evolution from more simple plants to those of greater complexity is known as progressive evolution.

Figure 28–5 shows the evolutionary relationships within the plant kingdom. From ancient living forms of primitive structure, the bacteria, slime molds, and true fungi evolved as did several phyla of algae—for example, the blue green algae, the brown algae, and *Euglena*. Another line from the ancient unicellular forms led to ancient green algae which gave rise to the green algae, to the bryophytes, and to primitive vascular plants in the order Psilophytales. The Psilotales, Sphenopsida (horsetails and relatives), and the Lycopsida (club mosses and relatives) evolved from the Psilophytales as did the ancient ferns. From the ancient ferns there are three lines of descent; one culminates in the ferns, a second in the gymnosperms, and the third in the angiosperms.

Retrogressive Evolution

Although progressive evolution has been the general tendency, there have been many exceptions, instances where complex plants evolved into those of simpler form. For example, the relatively simple flowers of grasses probably evolved from the more complex flowers of lilylike ancestors. The stringlike dodder, a parasitic seed plant, with roots modified into haustoria and with a stem lacking chlorophyll and bearing reduced leaves, evolved from an ancestor having normal roots, stems, and leaves.

Convergent Evolution

Evolution may proceed in different plant families in nearly the same manner and may result in plants of like vegetative form in the different families. The evolution of similar types in distantly related groups is known as convergent evolution. Thus, in the cactus family and in the spurge family, plants of similar form have evolved. Here evolution was oriented in the same direction and certain members of the two unrelated families became adapted to the same type of environment, one characteristic of deserts. Plants with succulent pointed leaves have evolved in the pineapple, lily, and amaryllis families. These plants resemble one another closely and all are adapted for surviving desert conditions. In tropical forests, trees in different families have evolved the same growth form and leaves which are nearly alike. Even though the trees are only distantly related, they have experienced similar modifications in their development.

QUESTIONS

1. _____ Which one of the following is false? (A) The capacity for evolution is inherent in DNA, the genetic material. (B) A change in the sequence of bases in DNA results in a gene mutation. (C) Most novelties resulting from gene mutations are preserved by the environment. (D) The replication of mutated genes and their inheritance by offspring may result in individuals with different characteristics.

2. _____ Which of the following might be expected to show more rapid evolutionary change? (A) A plant species which is always cross-pollinated, (B) a plant species which is always self-pollinated.

3. _____ Which of the following is the best definition of evolution? (A) Evolution is the development of simple forms into more complex forms. (B) Evolution is the theory that man descended from monkeys. (C) Evolution is the production of new forms from old forms. (D) Evolution is the origin of all forms of life from one original form.

4. _____ The modern theory of evolution regards evolution as (A) a peaceful, creative process, (B) survival of the fittest, (C) a struggle for existence, (D) might makes right.

5. Diagram the evolutionary trends in the plant kingdom.

6. _____ Which one of the following is not a recognized reason for believing that evolution has occurred? (A) Geological, (B) spontaneous generation of living things from nonliving, (C) embryological evidence such as ontogeny recapitulates phylogeny, (D) morphological evidence, (E) production of new varieties by man.

7. _____ Which one of the following is not a feature of Lamarck's theory? (A) Environment changes; (B) in the struggle for existence, the unfit are eliminated; (C) use and disuse alters organs; (D) acquired characters can be inherited.

8. _____ The mutation theory of evolution was advocated by (A) Lamarck, (B) Darwin, (C) De Vries, (D) Mendel.

9. _____ Which one of the following is not considered a feature of Darwin's theory? (A) Plants and animals in nature produce more progeny than can survive. (B) Use and disuse alters organs. (C) Because of the great number of progeny, there is a struggle for existence. (D) In a population individuals differ in fitness for a given environment. (E) As a result of competition the unfit individuals are eliminated.

10. _____ The evolution of two species from one is more likely to occur (A) when there is a breeding barrier between the groups (sister populations) of a species, (B) when there is not a breeding barrier and hence an opportunity for genes to be exchanged between the sister populations.

11. _____ Plants become adapted to their environment by (A) differential reproduction, (B) reproductive isolation, (C) use and disuse, (D) crowding out the less fit members.

12. _____ When evolution in different plant families results in plants of similar, but not identical, form, evolution is (A) progressive, (B) convergent, (C) retrogressive.

13. With the population explosion and a concomitant need for more food, shelter, water, industries, highways, etc., what plant species may become extinct? What species may become more prevalent?

---

14. Match the plants with the era or period in which they first appeared.

_____ 1. Bacteria and algae.
_____ 2. Horsetails, ferns, and seed ferns.
_____ 3. Conifers.
_____ 4. Angiosperms.

1. Archeozoic era.
2. Proterozoic era.
3. Devonian period.
4. Permian period.
5. Jurassic period.

29

Groups of Plants

So far we have been primarily concerned with the seed plants. If we are to obtain a balanced view of the plant kingdom we should consider representative examples from other large groups of plants. In the chapter which follows we shall start with the primitive plants and then in succeeding chapters consider plants of increasing complexity. You will recall that there are about 340,000 known plant species. Obviously we cannot study all of them. But we can study a few forms in each of the major groups. Before we start our survey of the plant kingdom, let us become familiar with the general system of plant classification.

SPECIES AND GENERA

A **species** of plant is a kind which is distinct enough to be recognized by a description. It is fairly constant over a considerable region, and its members breed freely among themselves. We speak of species of pines or spruces, meaning the kinds of pines (white pine, yellow pine, etc.) or the kinds of spruces (Colorado spruce, black spruce, etc.). In these instances each "kind" is a species.

It is not always easy to set limits for a species. Imagine that we have before us a collection of related plants. The plants are alike in many ways, such as in the number of stamens, petals, and pistils, but they are unlike in other traits. Assume that many specimens have large flowers and large leaves covered with many hairs. Other plants bear small flowers and small leaves bearing only a few hairs. Do we have one species or two? Perhaps they intergrade. That is, certain specimens possess flowers and leaves of intermediate size and hairiness. If they do intergrade, we are probably dealing with one species. With a microscope let us examine root tips from the two types for further clues. We will center our attention on the chromosomes. If both kinds, those bearing large leaves and those bearing small ones, have the same number and shapes of chromosomes in their root tip cells, we are probably dealing with one species and not two. For further evidence let us hybridize the two types. If viable seeds are produced, this is further evidence for close relationship. Let us next study reduction division as it goes on in pollen mother cells of the hybrid. If the chromosomes pair regularly during the first prophase, the chromosomes from the plant with large leaves are like those from the plant with small leaves, and this is further evidence that the two varieties are actually one species. Of course, the results may have been just the opposite of those considered previously. For example, there may have been no intergradation, and the number and kinds of chromosomes may have been different in the two types. Furthermore, the plants may have failed to hybridize or, if they did cross, the chromosomes may have failed to pair during the reduction division. If these latter events were noted, the two groups would certainly be two species. The determination of plant relationships is an extraordinarily fascinating adventure, involving considerable detective work.

Similar species are grouped into a **genus.** All of the species of pines are grouped in the genus *Pinus,* and all of the species of spruce in the genus *Picea.* Western yellow pine is *Pinus ponderosa,* long-leaf pine is *Pinus palustris,* and eastern white pine is *Pinus strobus.* The first name of each binomial is that of the genus and the second is that of the species. Binomial nomenclature was first given a systematic presentation by the great Swedish naturalist, Carolus Linnaeus (1707–1778) in his book *Species Plantarum,* which was published in 1753.

In taxonomic practice the binomial is followed by the name (or its abbreviation) of the botanist who first gave the plant its scientific name. The complete name of eastern white pine is *Pinus strobus* L. (L. stands for Linnaeus, who first described this pine.)

Sometimes a species is further divided into forms or varieties. For example, the pink-flowering form of the flowering dogwood is *Cornus florida* forma *rubra* [West.] Palmer and Steyermark, and the hairy-leaved variety of the black oak is *Quercus velutina* variety *missouriensis* Sarg.

Scientific names are usually in Latin form and the same scientific name is used throughout the world for a certain species. On the other hand, common names of plants vary from country to country and with the locality. In England corn means wheat (*Triticum sativum*), in Scotland oats (*Avena sativa*), and in the United States maize (*Zea mays*). In one region a certain species of *Juniperus* may be known as cedar. In a different locality a species of *Thuja* will be called cedar. In other regions the species of *Cedrus* are known as cedar. Scientific names may seem difficult at first, but with use they become familiar. Some species are named after a person or locality, but many are descriptive of the plant. The English meanings of a few scientific species names follows: *albus,* white; *annuus,* annual; *arborescens,* becoming tree-like; *aureus,* golden; *borealis,* northern; *campestris,* growing in fields; *candidus,* white and shining; *cernuus,* nodding; *ciliatus,* fringed with hairs; *communis,* common; *crispus,* curly; *dentatus,* toothed; *divaricatus,* spreading; *echinatus,* prickly; *esculentus,* edible; *foetidus,* having a bad odor; *glabrus,* smooth; *laciniatus,* cut into narrow lobes; *luteus,* yellow; *niger,* black; *palustris,* of the marshes or swamps; *procumbens,* lying on the ground; *rubrus,* red; *rugosus,* wrinkled; *rupestris,* of the rocks; *scandens,* climbing; *ternatus,* in threes; *vulgaris,* common.

The scientific name of a plant may be determined by consulting a manual of the area in which the plant was collected. In a botanical manual there are keys which enable the student to determine the family which includes the unknown plant. Under each family there is a key to the genera of the family, and under each genus a key to the species. (A key to some genera of conifers is given in Chapter 38. An examination of this key will give you the method of construction.) In most manuals each family, genus, and species is described. Generally a plant can be readily placed in the correct family and genus, but if the genus is a large one, it may be difficult to run it down to the correct species. Here is where a herbarium comes in handy. A herbarium is a systematic collection of appropriately preserved (generally dried) plant specimens. Most of the specimens are mounted on sheets of standard herbarium paper ($11\frac{1}{2} \times 16\frac{1}{2}$ inches), and each sheet

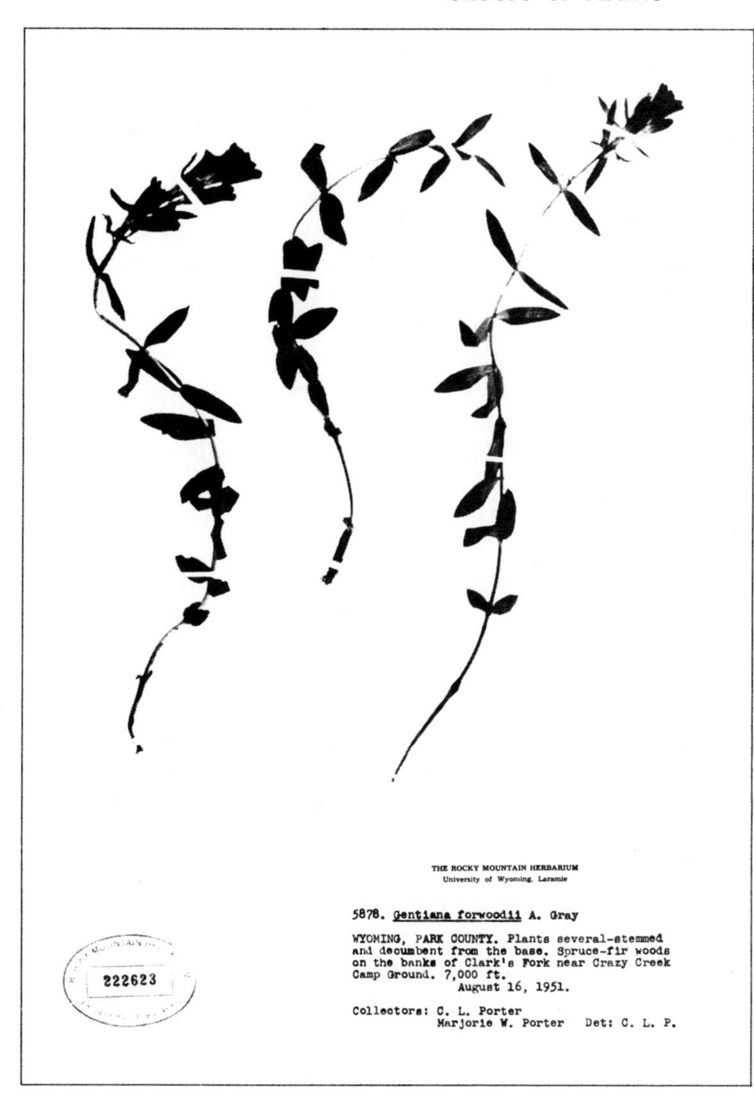

Fig. 29–1. A herbarium specimen, nicely mounted and properly labeled.

has a label in the lower right-hand corner which gives the name of the plant, the locality, the habitat, the date of collection, and the collector's name. The collection of plants is filed in a systematic manner in herbarium cases. Herbaria are not only of value in identifying plants by comparing unknown specimens with correctly named ones, but are also of inestimable value in the preparation of floras (that is, guides to the plants of a given region), in determining the relationships of plants, in studying the geographical distribution of plants, in studying changes of vegetation with use, and in other ways (Fig. 29–1).

HIGHER GROUPS

For convenience in classification, genera are grouped into families, families into orders, orders into classes, and classes into phyla. The family includes one to many closely related genera, and characteristically the family name ends in *–aceae,* for example, Pinaceae and Rosaceae. Related families comprise an order; the order name ends in *-ales*, Coniferales, for example. Orders are grouped into classes, and classes into phyla, the largest divisions of the plant kingdom. Class names usually end in *-ae;* those of phyla always end in *-phyta,* which means plant.

When the groups are large, they are subdivided. Thus there may be subphyla, subclasses, suborders, subfamilies, subgenera, and subspecies. Hence the plant kingdom contains

Phyla (also subphyla)
  Classes (also subclasses)
    Orders (also suborders)
      Families (also subfamilies)
        Genera (also subgenera)
          Species (also subspecies)

OUTLINE OF THE MAJOR GROUPS OF THE PLANT KINGDOM

The major groups of the plant kingdom and other names which are used for some of the groups are outlined below. You will notice that the plant kingdom consists of two subkingdoms, Thallophyta and Embryophyta. The ten most primitive phyla, seven of algae and three of fungi, are classified in the subkingdom Thallophyta. All thallophytes have a simple plant body, called a thallus, which lacks roots, stems, and leaves. Furthermore, the gametes are produced in unicellular sex organs and the embryo is not protected by the female organ during its development.

Two phyla, Bryophyta and Tracheophyta, are in the subkingdom Embryophyta. In all members of this subkingdom, the fertilized egg develops into a multicellular embryo while it is still enclosed in the female sex organ. Familiar bryophytes are the mosses, liverworts, and hornworts. The Tracheophyta, commonly called vascular plants, includes club mosses, horsetails, ferns, and the seed plants.

This system of classification was outlined by Oswald Tippo in 1942. It shows plant relationships reasonably well. It is not a perfect system and it is not accepted in its entirety by all botanists. For example, some botanists suggest that the phylum Eumycophyta should be split into several phyla. Others would elevate the subphyla of the Tracheophyta to the rank of phyla. We are still discovering new facts about plant relationships, and such new information makes it necessary periodically to review our concepts of plant classification. Keep in mind that the system just outlined is a tentative one and only one of several which have been recently proposed.

THE PLANT KINGDOM

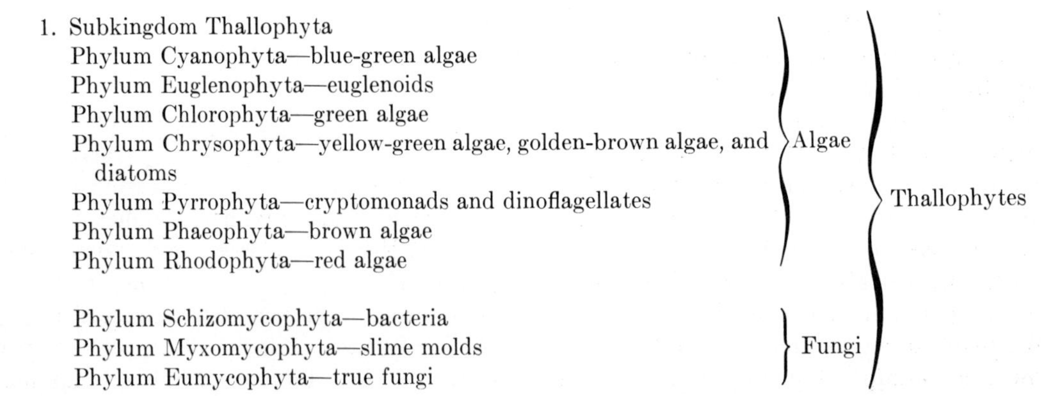

1. Subkingdom Thallophyta
   - Algae (Thallophytes):
     - Phylum Cyanophyta—blue-green algae
     - Phylum Euglenophyta—euglenoids
     - Phylum Chlorophyta—green algae
     - Phylum Chrysophyta—yellow-green algae, golden-brown algae, and diatoms
     - Phylum Pyrrophyta—cryptomonads and dinoflagellates
     - Phylum Phaeophyta—brown algae
     - Phylum Rhodophyta—red algae
   - Fungi (Thallophytes):
     - Phylum Schizomycophyta—bacteria
     - Phylum Myxomycophyta—slime molds
     - Phylum Eumycophyta—true fungi

2. Subkingdom Embryophyta
   - Phylum Bryophyta (Bryophytes)
     - Class Musci—mosses
     - Class Hepaticae—liverworts
     - Class Anthocerotae—hornworts
   - Phylum Tracheophyta—vascular plants
     - Subphylum Psilopsida (Pteridophytes)
     - Subphylum Lycopsida—club mosses (Pteridophytes)
     - Subphylum Sphenopsida—horsetails (Pteridophytes)
     - Subphylum Pteropsida
       - Class Filicinae—ferns (Pteridophytes)
       - Class Gymnospermae—conifers and their allies (Spermatophytes (Seed Plants))
       - Class Angiospermae—flowering plants (Spermatophytes (Seed Plants))
         - Subclass Dicotyledoneae
         - Subclass Monocotyledoneae

CLASSIFICATION OF *PINUS STROBUS* L.

According to the system of classification used in this text, *Pinus strobus* L. would be classified as follows:

Subkingdom	Embryophyta
Phylum	Tracheophyta
Subphylum	Pteropsida
Class	Gymnospermae
Order	Coniferales
Family	Pinaceae
Genus	*Pinus*
Species	*strobus*

QUESTIONS

1. What was Linnaeus' contribution to biology?

2. Write in the appropriate space an appropriate ending for each of the following:

Phylum: Tracheo ________________.
Class: Angio ________________.
Order: Ros ________________.
Family: Ros ________________.

3. Arrange the following categories in order, beginning with "phylum" and ending with "variety": phylum, class, family, order, subphylum, species, genus, variety.

4. Match the phylum in which the plants on the left would be placed.

____	1. Blue-green alga.	1. Bryophyta.
____	2. Green alga.	2. Phaeophyta.
____	3. Brown alga.	3. Tracheophyta.
____	4. Red alga.	4. Schizomycophyta.
____	5. Bacteria.	5. Chlorophyta.
____	6. Mushroom.	6. Cyanophyta.
____	7. Moss.	7. Rhodophyta.
____	8. Fern.	8. Eumycophyta.
____	9. Pine.	
____	10. Wheat.	

30

Algae

If you put a drop of water from a pond, lake, stream, or ocean under a microscope, you may see various kinds of algae and also the microscopic animals which feed on them. Not all algae are microscopic, however, for some, such as the pondscum, may appear as silk threads floating on water, and others, some seaweeds, may be massive forms several hundred feet long. Although most algae are aquatic, some can grow on soil, rocks, snow, ice, fountains, and tree trunks.

All algae have a simple type of body, known as a **thallus,** which has no roots, stems, or leaves. Practically all of them have chlorophyll, but the green pigment may be masked by blue, brown, or red pigments. Because the algae have chlorophyll, they, like the seed plants, can manufacture food and liberate oxygen from carbon dioxide and water when light is present. The food which they make is the beginning of a food chain in aquatic habitats.

Fresh-water algae make up the greater part of the pastures of fresh-water lakes and streams (Fig. 30–1). Many fresh-water fishes, such as trout and bass, feed on small animals which depend on algae for food. In some areas fertilization of ponds with minerals that are deficient in the water has resulted in an increased growth of algae, hence a higher fish population. The skillful trout fisherman knows that good fishing is associated with slippery alga-covered boulders in a stream.

Algae not only furnish food for aquatic animals, but they also maintain a desirable oxygen balance in the water. The oxygen released by algae during photosynthesis is especially significant when a river or lake is covered with a thin coat of ice, which restricts diffusion of oxygen from the atmosphere into the water.

In some places algae become a nuisance, and steps must be taken to reduce the population. In reservoirs an abundance of algae may impart undesirable tastes to the water. Algae are not wanted in swimming pools. In both reservoirs and swimming pools the algal population can be reduced by using copper sulfate in a concentration of about 1 part in 10 million. Such a low concentration of copper sulfate is not harmful to fish, nor does it make the water unfit for human consumption.

THE OCEANIC HABITAT

In the shallow waters of oceans and in the intertidal zone, large numbers of algae,

Fig. 30–1. Life in a drop of water. In this model of life found in one drop of water from a fresh-water pond in New England, there are forty species of animals, two of flowering plants, and thirty-two of algae. (American Museum of Natural History.)

called seaweeds, can be seen with the unaided eye. Some of these forms are harvested, and substances obtained from them are used in the baking and ice cream industries, and in surgical jellies, dental impression compounds, foam cushions, auto polishes, and other products.

Plants are seldom seen far out at sea, but fish, whales and porpoises may be encountered. The ocean water which seems so clear teems with microscopic organisms, both plants and animals. Ninety-nine per cent of marine plants are microscopic, one-celled algae, and in each quart of sea water there are more than a million. These microscopic free-floating algae (the so-called phytoplankton) live in the top 300 feet of the whole ocean area. Collectively, they make seven to ten times more food than land plants for two reasons. First, the ocean area is almost three times greater than the land area. Second, conditions are more uniform in the sea than on land; the ocean has no deserts, and temperatures are more even. The food made by ocean plants feeds marine animals, which in turn are eaten by man. Algae are eaten by protozoa, and both are food for tiny crustaceans, which in turn are food for herring, the most abundant fish of the oceans. Herring are eaten by carnivorous fish such as cod, haddock, and tuna—fish relished by man. Microscopic algae also form the diet of clams, oysters, shrimps, and, strangely enough, of one of the largest sea animals—the blue whale. A sizable part of the blue whale's diet consists of microscopic algae, and its stomach may contain a ton or more at a time.

All oceans have a gently sloping continental shelf that extends outward from the coastline for distances of a few miles in some regions to several hundred miles in other

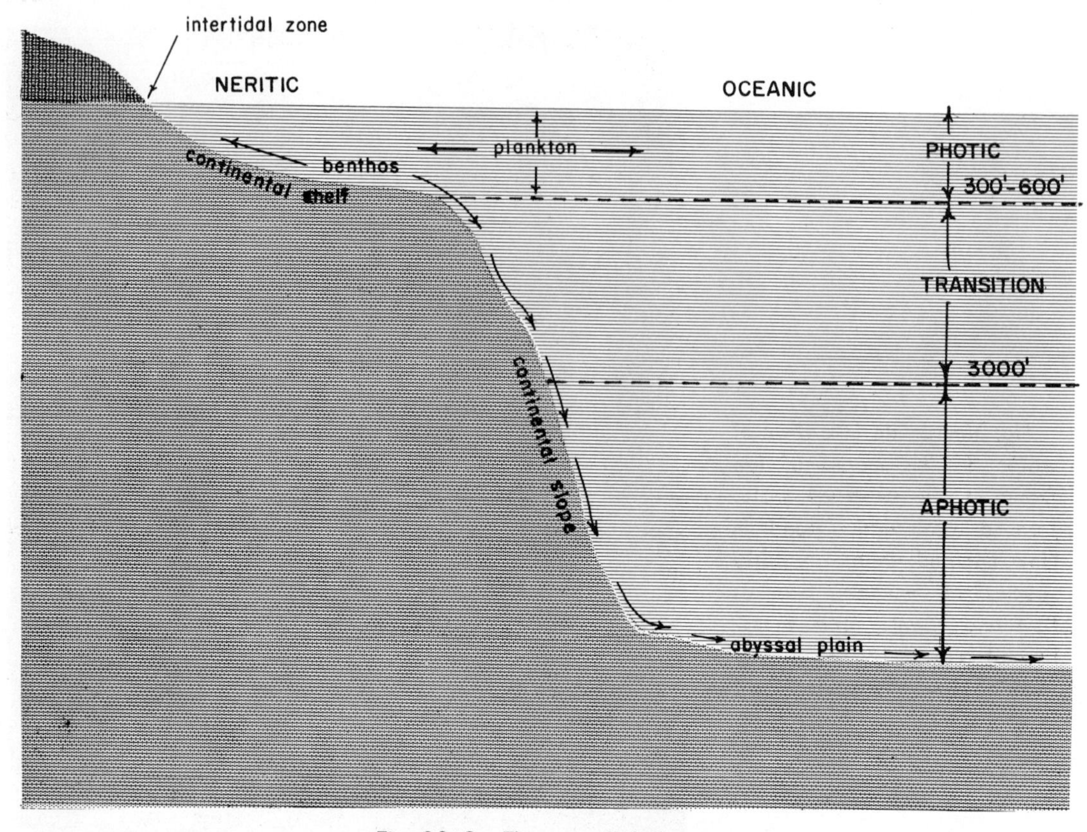

Fig. 30–2. The oceanic habitat.

regions. The slope then becomes steeper, forming the continental slope (Fig. 30–2) which descends several thousand feet to the abyssal plain. Here and there mountains may rise from the plain and even tower above the water as islands. In some regions trenches may plunge thousands of feet below the abyssal plain. Two provinces occur in any ocean: the **neritic province** and the **oceanic province.** The neritic province is located above the continental shelf, and the oceanic province is above the abyssal plain.

In the intertidal zone the algae are exposed to two hazards: the power of the surf and waves, and periodic desiccation. The algae in this zone are adapted to survive violent lashing and pounding. When the tide is out, the algae are exposed to the dry atmosphere; during this period, their heavy coating of gelatinous material retains water.

Starting at the ocean surface and descending to greater and greater depths, three zones can be recognized: an upper zone known as the **photic zone,** a **transition zone,** and, at greater depths, the **aphotic zone,** a zone of pitch darkness.

Photic Zone

In this zone, also referred to as the **epipelagic zone,** light is present and plants carry on photosynthesis. The zone extends to a depth of about 300 to 600 feet. The environment is not uniform at all depths. Light intensity is greatest at the surface and decreases progressively with depth, approaching zero between 300 and 600 feet. In clear water the light intensity at 3 feet is two-thirds of that at the surface; at 16 feet, only one-fourth. The quality of light also varies with depth; only blue and green light penetrate to the greater depths; the yellow, orange and red have been absorbed by the upper layers. In temperate regions temperature diminishes with depth. The concentration of nitrogen and phosphorus progressively increases from the surface to about 3000 feet and then remains constant. In other words, the surface waters have a low concentration of these minerals. In certain oceanic regions the deeper, colder water wells up, carrying water rich in nitrogen and phosphorus to the surface. In such regions an abundant flora and fauna develop.

In both the neritic and oceanic provinces, **plankton,** drifting or floating organisms, thrive in the photic zone. Diatoms and dinoflagellates are the algae that make up the greater part of the phytoplankton. These are food for small animals which in turn are eaten by larger actively swimming animals. The active swimmers comprise the **nekton** and include squids, whales, and fish such as the salmon, tuna, and tarpon.

In the shallower water of the neritic province many algae grow attached to rocks on the bottom of the ocean. Such attached algae are referred to as benthonic plants. The **benthos** consists of crawling and creeping animals, crabs, clams, and starfish, for example, and attached plants. In other words, the fauna and flora of the sea bottom are called benthos.

Transition Zone

Between the zone of light (the photic zone) and the zone of perpetual darkness (the aphotic zone) lies the transition zone, also known as the **mesopelagic zone.** The illumination is so low that little or no food is made. Here dwells a totally different world of squids, octopi, and fishes, such as the lantern fish, which possess light-emitting organs—**photophores.** Many of the animals move upward during the night to feed on the organisms in the photic zone.

Aphotic Zone

In this zone it is pitch black except for the light emitted by inhabiting fishes. The upper level of this zone teems with life—sardines and other fishes, shrimps, and many others. Many animals feed on those in the zone above.

KINDS OF ALGAE

The algae, of which there are about 20,000 species, are in seven phyla which are not closely related to each other. Each phylum probably originated from a different ancestral stock. The ancestral groups perished; hence we are uncertain about their characteristics. The present forms of algae are at the ends of evolutionary lines, and these forms have not given rise to higher plants. However, in the distant past some forms of green algae, no longer present, probably gave rise to the bryophytes and vascular plants. The seven phyla of algae are Cyanophyta (blue-green algae), Chlorophyta (green algae), Euglenophyta (euglenoids), Chrysophyta (yellow-green algae, golden-brown algae, and diatoms), Pyrrophyta (cryptomonads and dinoflagellates), Phaeophyta (brown algae), and Rhodophyta (red algae).

BLUE-GREEN ALGAE—CYANOPHYTA

The blue-green algae, a group of 1400 species, are widely distributed in both fresh and salt water and in soil. Some of the aquatic forms are eaten by animals, which in turn are eaten by other animals. Some genera of blue-green algae, for example, *Nostoc* and *Anabaena,* fix nitrogen and thereby increase the fertility of the soil. Also, in rice paddies certain blue-green algae enhance the available nitrogen supply. Those which grow in reservoirs impart an undesirable fishy taste to the water. A few species are poisonous. On occasions, livestock and wildlife have been killed from drinking water containing *Anabaena flos-aquae.* The population of blue-green algae can be reduced by applications of copper sulfate, as explained on page 394. A few species of blue-green algae develop at high temperatures which would kill most other organisms. These remarkable plants grow

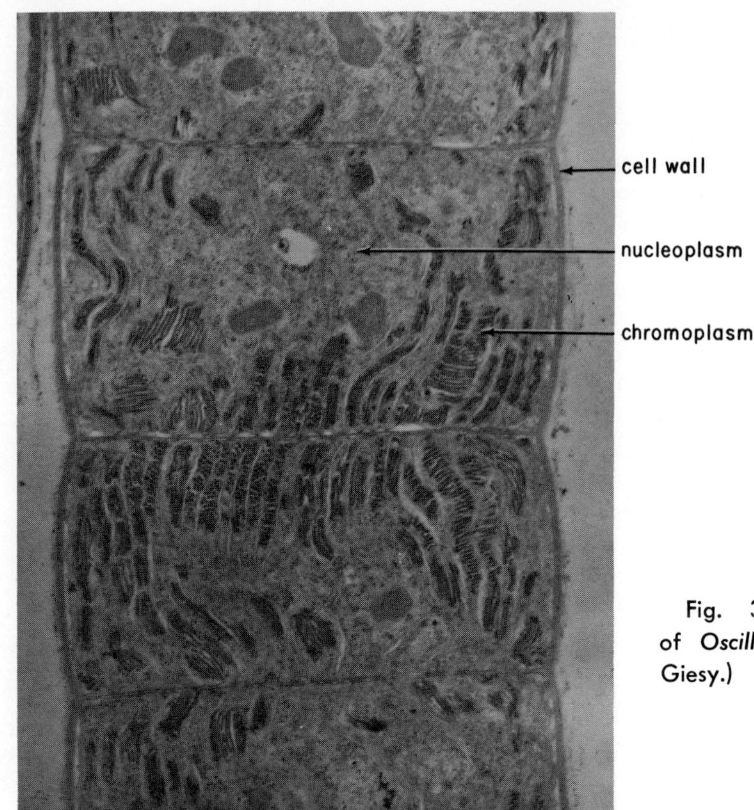

Fig. 30–3. Electron micrographs of *Oscillatoria chalybia*. (Robert M. Giesy.)

in steaming hot springs where the temperature is as high as 170° F. They produce the attractive colors of some of the lovely hot pools and terraces of Yellowstone National Park.

Blue-green algae are one-celled plants. A cell consists of a protoplast surrounded by a cell wall made of cellulose and hemicellulose. In most blue-green algae the cells are enclosed in a gelatinous sheath composed of pectic compounds. The gelatinous sheath is often so thick that the cells appear to be embedded in a mass of jelly which may be in the form of a sheet or ball. Blue-green algae lack flagellated cells and did not evolve from flagellated ancestors.

Blue-green algae are about as simple in structure as any organisms known. They made their appearance very early in the history of the earth, about a billion years ago, and have not changed very much during that long period of time. The first forms of life were undoubtedly unicellular organisms that lacked definite nuclei and chloroplasts. Blue-green algae have retained these primitive characteristics. With the optical microscope the cells appear structurally simple, but with the electron microscope an elaborate internal structure is revealed (Fig. 30–3). The pigments are located in the peripheral protoplasm, called **chromoplasm,** which consists of rod-shaped structures embedded in a granular matrix. The rod-shaped bodies are associated with photosynthesis. The central protoplasm, called **nucleoplasm** because it contains nucleic acid, is composed of a colorless matrix containing granules of several kinds. Among the granules occurring in the protoplasm are ribosomes, mitochondria-like granules, vacuole-like inclusions, and reserve food. A complete understanding of the structure of a cell awaits further study.

Chlorophyll a, carotenoid pigments, and

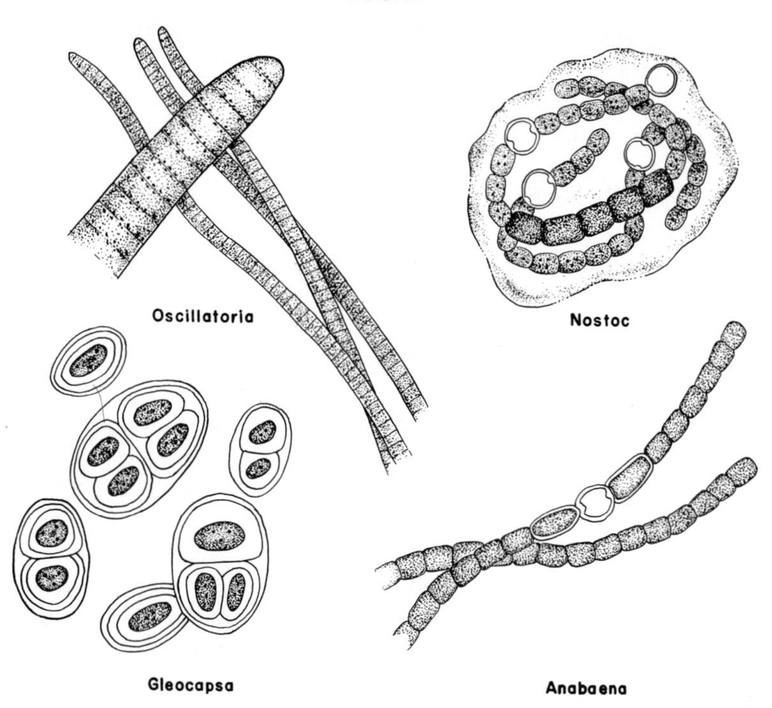

Fig. 30–4. Some common genera of blue-green algae. (From *Biology*, 4th ed., by Claude A. Villee. Copyright © 1962, W. B. Saunders Co., Philadelphia.)

a blue pigment, **phycocyanin,** are present in the chromoplasm of all blue-green algae. Some species have an additional red pigment, **phycoerythrin,** and these species have a reddish hue. The Red Sea owes its name to the color imparted to the water by blue-green algae. Blue-green algae which appear reddish are also found in hot springs. Although the blue-green algae are unicellular organisms, the cells may be held together in colonies by the gelatinous sheaths. Reproduction is by cell division. The division of a unicellular organism into two equal cells is known as **fission.** In 1963 scientists obtained genetic evidence for the existence of sexuality in *Anacystis nidulans,* a blue-green alga. When two strains of this alga were grown together, some of the offspring exhibited traits from both parents. Further investigations may prove that sexual reproduction by conjugation is a general feature of blue-green algae. Formerly, it was believed that blue-green algae lacked sexual reproduction.

REPRESENTATIVE BLUE-GREEN ALGAE

Gloeocapsa

A minute organism with a thick gelatinous sheath, *Gloeocapsa* frequently grows on wet rocks and sticks and on the walls of aquarium tanks. After division the resulting cells are held together for a while by the gelatinous sheaths. Colonies of two, four, or eight individuals are frequently seen. Sooner or later, however, the organisms become separated, and hence large colonies do not develop (Fig. 30–4).

Oscillatoria

This is one of the most abundant and widely distributed of the blue-green algae, being found on moist soil, rocks, flower pots, and in stagnant water where it forms slimy coatings on objects in the water. Because cell division occurs in only one plane, long threadlike filaments are produced. The filaments move to and fro and glide forward and backward, hence the name *Oscillatoria.* In time, a filament breaks into segments that are called **hormogonia,** and they increase in length by cell division. The filament breaks at a point known as a separation disk; this is a clear area in the filament which consists of gelatinous material.

Nostoc

The spherical cells of *Nostoc* are joined together to form a filament which resembles a string of beads. The filaments in turn are held together by a mass of gelatinous material in such a way that the resulting colony resembles a ball of bluish-green jelly. These balls may be found on damp soil or floating on stagnant water. At intervals in each filament there are large empty cells, called **heterocysts,** where a filament may break into hormogonia. Although the cells of *Nostoc* and *Oscillatoria* are organized into filaments, each cell is capable of living independently. Hence, each cell is an individual plant, and each filament is a colony of plants.

GREEN ALGAE—CHLOROPHYTA

Much of the flora of lakes, streams, and ponds consists of green algae which, like other algae, are the beginning of food chains leading to food for fish. Many species grow attached to rocks in water or to larger submerged plants. Other species are microscopic and free floating; they, with microscopic animals, comprise the free-floating population known as the **plankton.** During certain seasons, the plankton algae are so numerous that the water appears colored. Not all green algae are aquatic; certain species grow on moist soil and flower pots, and others occur as epiphytes.

Not all species of green algae grow in fresh water; about 10 per cent are marine species. Most fresh-water species are unicellular or filamentous forms, whereas many marine green algae are quite large and nonfilamentous, among them *Ulva,* the sea lettuce.

The green algae, a group of about 5700 species, are more complex than the blue-green algae, for they have organized nuclei and definite chloroplasts, and most species have sexual reproduction. An algal cell has a cellulose wall containing cytoplasm, a nucleus, one or more chloroplasts, mitochondria, ribosomes, endoplasmic reticulum, Golgi bodies and vacuoles. The chloroplasts vary considerably in shape and size, but all contain chlorophyll a and b as well as carotene and xanthophyll, pigments which also characterize the chloroplasts of seed plants.

The history of the plant kingdom covers a span of more than a billion years. During this tremendously long period, plant life developed progressively from simple unicellular forms to the complex flora of today, and from forms without sexual reproduction to those having this biological advantage. In the green algae we can see the beginnings of these developments.

The simplest of the green algae, such as *Chlamydomonas,* are motile unicellular plants. Evolution from such forms proceeded along several lines, among them a line leading to colonial forms, another leading to filamentous green algae. Among the colonial forms are *Pediastrum* and *Scenedesmus,* whose cells, although aggregated into a colony (see Fig. 30–7), are still independent of each other. In filamentous algae the cells are arranged end to end, and cell division occurs in a plane at right angles to the long axis. In many filamentous algae, *Spirogyra,* for example, there is no division of labor; all of the cells are alike in appearance and in function. In others, *Oedogonium,* for instance, there is the beginning of specialization, for some cells are concerned with reproduction and different ones with vegetative activities. Many green algae became extinct, and we know very little about their structure. Certain extinct green algae must have been quite complex, for there is evidence that the vascular plants (Tracheophyta) evolved from green algae. Furthermore, the bryophyte line may also have evolved from green-alga-like ancestors. Because of their relationship to vascular plants and bryophytes, the green algae have a distinctive place in biological studies.

Green algae reproduce in several ways. The simplest is by cell division as in *Protococcus. Ulothrix* and *Oedogonium* reproduce by **zoospores,** which are motile asexual spores. Reproduction by cell division and zoospores is asexual reproduction because gametes have not united.

Certain green algae, *Protococcus,* for example, lack sexual reproduction. *Ulothrix* illustrates an early type, for in this alga the gametes are alike in size and shape. Such gametes are known as **isogametes,** and the organism is said to be **isogamous.** As we will see later, the gametes of *Ulothrix* may have originated from zoospores. In sexual reproduction gametes unite in pairs. Accordingly, there must have evolved a mechanism limiting the union to pairs instead of to a fusion of an unlimited number of gametes. In all organisms reproducing sexually, a reduction division (meiosis) must occur at some stage in the life cycle; otherwise, the chromosomal number would double with each generation. Hence it is likely that reduction division evolved concurrently with gamete formation and fertilization.

Other algae, *Oedogonium,* for example, have an advanced type of sexual reproduction. In this alga the gametes are not alike in form and size; instead, small motile sperms and large nonmotile eggs are produced. Plants which have unlike gametes are said to be **heterogamous,** a feature evolved in the algae and retained by the higher plants. There are certain obvious advantages to heterogamy. Although just one sperm unites with an egg, many more sperms than eggs must be developed if fertilization is to occur with certainty. Many sperms never reach an egg, and these are wasted. A small sperm involves less waste and greater motility than would a large one. On the other hand, a large egg is desirable, permitting storage of considerable food which, after fertilization, is used by the new plant.

In both isogamous and heterogamous green algae, reduction division generally occurs during the germination of the zygote. Characteristically, the zygote is the only diploid cell; the green plant is haploid. A further discussion of this will be given later.

The green haploid plant producing gametes is known as a **gametophyte.**

REPRESENTATIVE FORMS OF GREEN ALGAE

Chlamydomonas

Stagnant pools are often colored green by myriads of microscopic, motile, pear-shaped, unicellular plants, considered to be among the most primitive of the green algae. One species, *Chlamydomonas nivalis,* with a red pigment, in addition to the green and yellow ones, thrives on snowbanks. In certain regions, especially at high elevations in the Rocky Mountains, it is so abundant that the snow is blood red in color. Other species also grow on snow, among them *Chlamydomonas yellowstonensis,* a green plant which results in the green snow so prevalent in Yellowstone National Park.

Examine a *Chlamydomonas* plant under the microscope and you will see that it has a cell wall with two small holes at one end through which two **flagella** (singular: flagellum) extend outward. The flagella whip back and forth and propel the alga. A cross-section of a flagellum reveals nine peripheral fibrils, each made of two subfibrils, and two central ones. Remarkably, this $9 + 2$ flagellar structure is characteristic of all plant flagella, except those of bacteria, and also of animal flagella. Inside the cell wall there is a central nucleus, a cup-shaped chloroplast, a food-storing **pyrenoid** within it, two contractile vacuoles, and an **eye spot** which controls the plant's response to light. The plant swims toward light of moderate intensity. *Chlamydomonas* reproduces both asexually and sexually.

Asexual reproduction. Asexual reproduction is achieved by motile spores known as zoospores. Prior to their formation, the flagella are withdrawn and the cell becomes quiescent. The protoplast then divides to form two, four, or eight zoospores within the parent cell (Fig. 30–5). The zoospores are released when the parent cell wall dissolves. Each zoospore is surrounded by a cell wall through which two flagella extend. After swimming for some time by means of the two flagella, each zoospore enlarges into the characteristic adult form.

Sexual reproduction. Chlamydomonas plants also reproduce sexually. Eight, sixteen, or thirty-two small biflagellate gametes (sex cells) are produced in each parent cell. The gametes produced by different plants are alike in appearance and are therefore known as **isogametes.** These gametes are released into the water where a gamete from one plant unites with one produced by a different plant. The zygote thus formed develops a heavy wall and enters into a resting condition, during which time it is resistant to drought and other unfavorable conditions. When environmental factors become favorable, the single nucleus in a zygote undergoes reduction division,

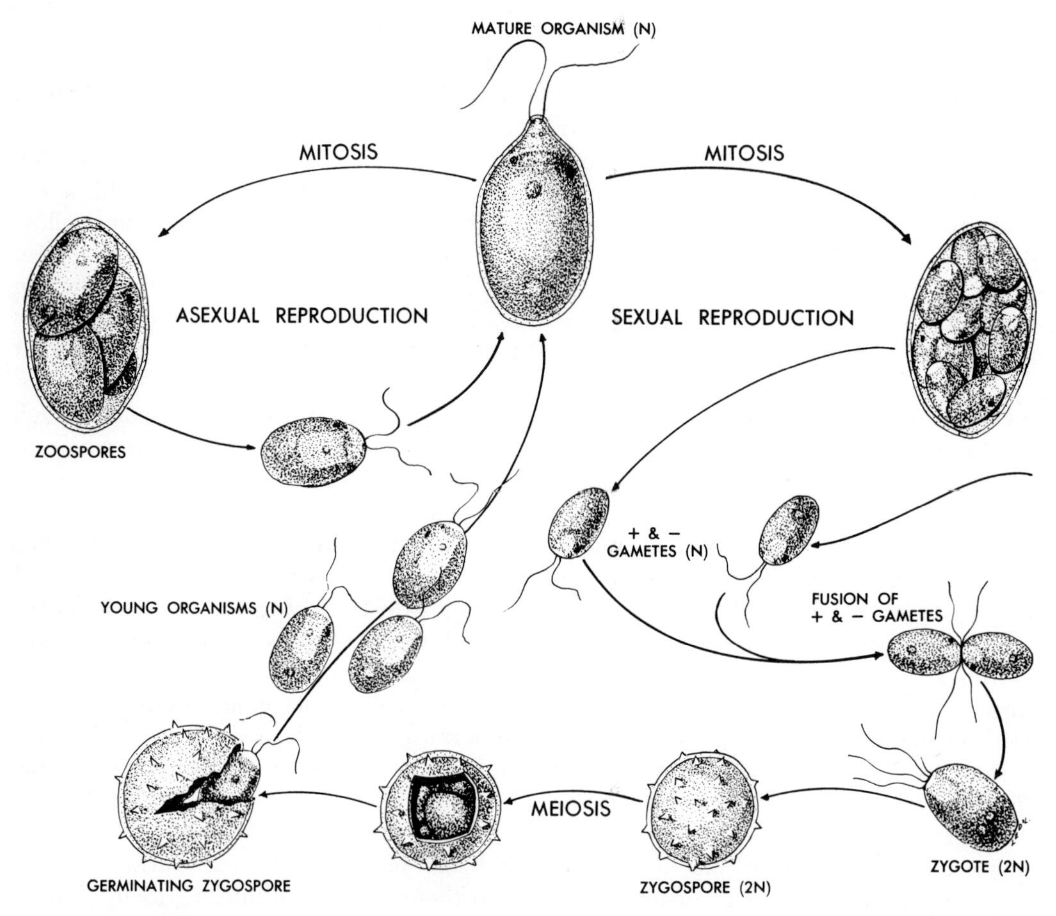

Fig. 30–5. Asexual and sexual reproduction of *Chlamydomonas.* (W. H. Johnson, R. A. Laubengayer, L. E. DeLanney, and T. A. Cole, *Biology*, 3d ed. [originally published as *General Biology*]. Copyright © 1956, 1961, 1966, by Holt, Rinehart & Winston, Inc., New York. All rights reserved.)

and later four zoospores, each with a haploid nucleus and two flagella, are produced in the zygote. The zoospores escape, swim for a time, and then each develops into a new plant which is, of course, haploid.

Chlorella

From studies of this one-celled, nonmotile, green alga we have learned much about photosynthesis. In the future *Chlorella* may be grown on a large scale for food. With advanced technology *Chlorella* could be produced at a cost of ten cents a pound. *Chlorella* is rich in protein, fat, and vitamins A and C, and 1 tablespoon of the dried powder has the same food value as 1 ounce of steak. Dried *Chlorella* powder tastes somewhat like lima beans; seasonings could be introduced to provide a variety of flavors.

One evolutionary trend in the green algae involves the loss of flagella with a concurrent loss of motility. *Chlorella,* a nonmotile alga, probably evolved from a motile ancestor. The spherical cell of *Chlorella* is surrounded by a cell wall which encloses the cytoplasm, a cup-shaped chloroplast, and a central nucleus (Fig. 30–6). The electron microscope reveals that a cell of *Chlorella* has structures comparable to those in cells of higher plants. Mitochondria, ribosomes, and Golgi bodies are present, and the nuclear membrane is a double envelope pierced with pores.

Asexual reproduction. Chlorella does not reproduce sexually, only asexually. At the time of reproduction the nucleus divides to form two, four, eight, or sixteen nuclei. Each nucleus together with some cytoplasm then becomes surrounded by a cell wall to form a nonmotile spore. The spores are released by the disintegration of the wall of the mother cell. Each spore enlarges directly into a mature cell which in time gives rise to additional spores.

Protococcus

A one-celled, spherical green alga, *Protococcus* has a thick wall, a single nucleus, and one large-lobed chloroplast. Evolution does not always result in more complex forms. In certain instances complex plants evolve into simpler ones. This is true for *Protococcus* because its ancestors were filamentous algae. It is widely distributed and is found growing on posts, bark of trees, and moist rocks.

Reproduction. New *Protococcus* plants are produced by the division of the parent cell. Each of the resulting cells is a new individual. After division the cells may cling together in colonies of various sizes. Sexual reproduction does not occur in *Protococcus.*

Pediastrum and Scenedesmus

In the microscopic world of plants that one may see in a drop of water, often there are, besides *Chlamydomonas* and other motile species darting to and fro, colonies of *Pediastrum* and *Scenedesmus* which are strikingly beautiful because of their cell structure and arrangement (Fig. 30–7). In both genera the regular arrangement is maintained during reproduction. In a certain species of *Pediastrum,* for example, spores are formed in cells of the colony. The cell wall breaks open, and a vesicle containing the spores emerges. The spores in the vesicle next assume the arrangement of the parent colony, develop walls of their own, and develop spines characteristic of the species, after which the colony is released from the vesicle.

Ulothrix

A widely distributed alga, *Ulothrix* is abundant in flowing fresh water and is filamentous and unbranched. All cells in a filament are alike except the colorless, elongated, basal **holdfast cell** which anchors the plant to the substratum. All cells contain a nucleus and, except for the holdfast cell, one girdle-like chloroplast. The filament increases in length by cell division.

Asexual reproduction. *Ulothrix* reproduces asexually by zoospores. Each zoo-

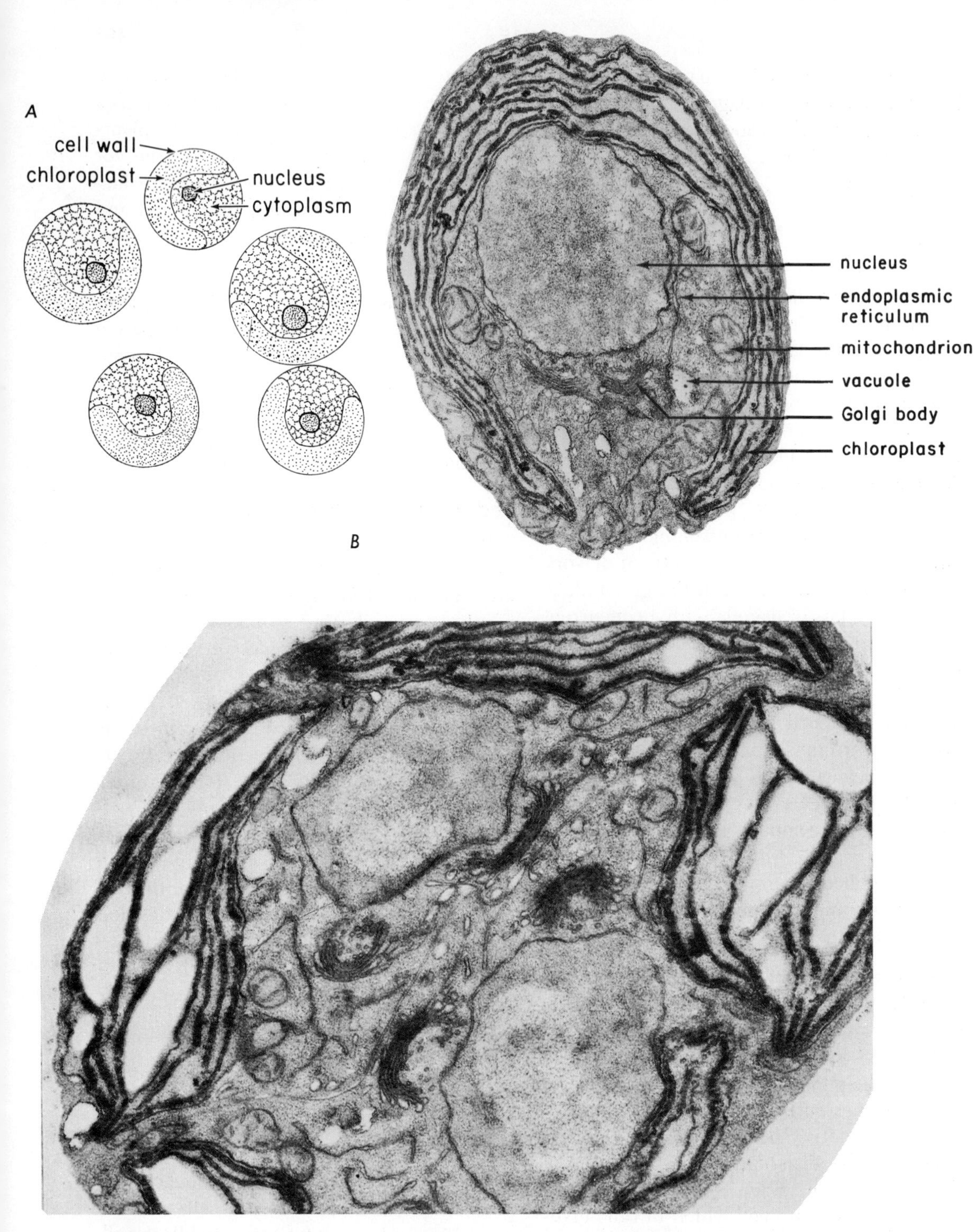

Fig. 30–6*A*. *Chlorella*, a green alga, is widely used in studies of photosynthesis and space ecology. *B*. Electron micrographs of *Chlorella vulgaris*. *Upper* (10,000×), structure of a *Chlorella* cell. *Lower* (12,000×), dividing *Chlorella* cell; notice the two nuclei, the cell plate, the distribution of Golgi bodies and mitochondria, and the new chloroplast. (T. Bisalputra.)

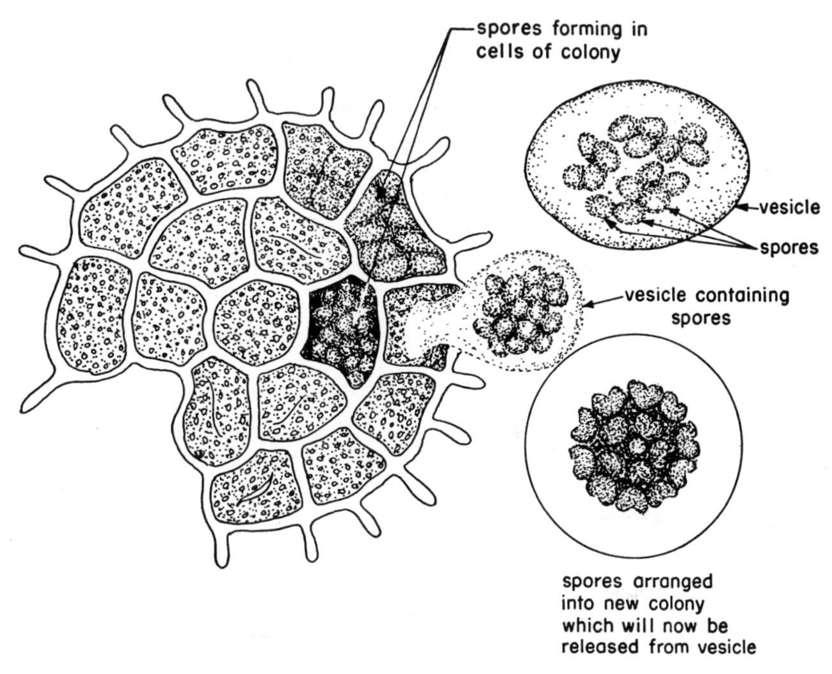

Fig. 30–7. *Pediastrum*, a colonial green alga.

spore is a naked mass of protoplasm with four whiplike extensions (flagella) that propel the zoospore. Two to thirty-two zoospores develop in some or all cells of a filament except the holdfast cell. After the zoospores are liberated through an opening in the cell wall, they swim for 1 to several days, and then come to rest on stems of submerged plants or on rocks. Here the flagella are withdrawn, and each zoospore grows by cell division into a new filament.

Sexual reproduction. Eight to sixty-four small biflagellate gametes are produced in some or all cells of the filament (except the holdfast cell). The gametes produced by different plants are alike in appearance; hence sexual reproduction is isogamous. The gametes are liberated into the water where a gamete from one plant unites with a gamete from a different plant. The resulting zygote secretes a thick wall. After a resting period the single nucleus in the zygote undergoes reduction division. Later four zoospores are organized in each zygote, and are then released, each growing into a new filament (Fig. 30–8).

Evolution of gametes from zoospores. Perhaps sexual reproduction arose not through the evolution of entirely new structures, but instead through a new behavior of existing ones. At least sexual reproduction of *Ulothrix* suggests this possibility. The gametes of *Ulothrix* resemble zoospores except for their smaller size and the presence of two flagella instead of four, as in zoospores. There is another prime difference; the small biflagellate gametes cannot develop into new filaments by themselves, but instead must fuse in pairs to form zygotes. The general resemblance between zoospores and gametes, their place and manner of production, and their release from the cell through a pore suggest that gametes originated from zoospores.

Spirogyra

The green alga *Spirogyra* is widely distributed, filamentous, and commonly found floating on ponds. The cells are cylindrical and arranged end to end, forming long unbranched filaments which are bright green in color and silky in texture. The cells of

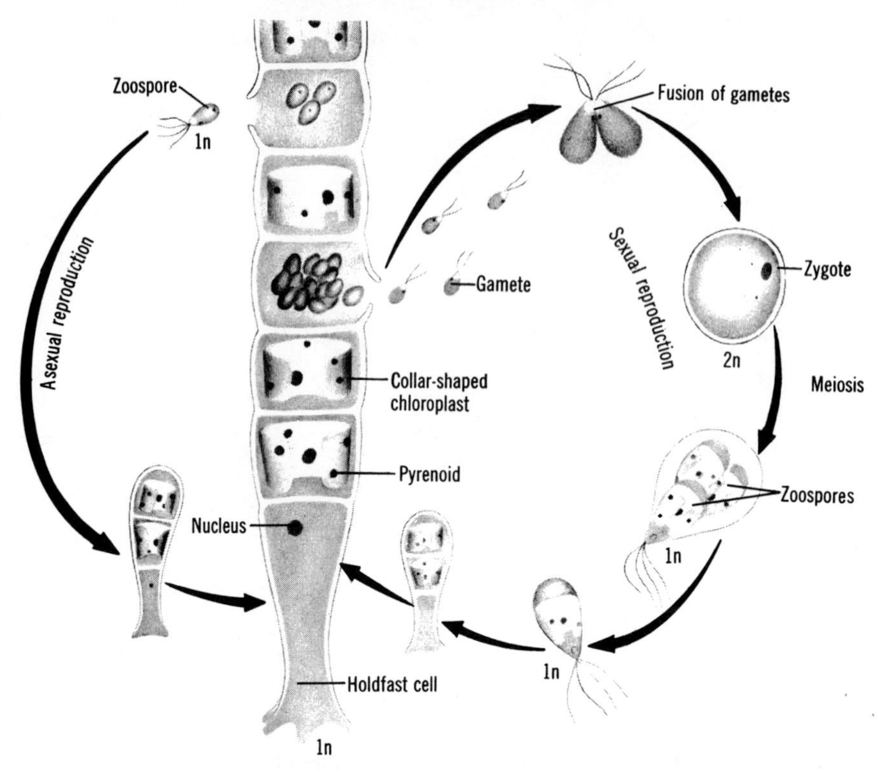

Fig. 30–8. Life cycle of *Ulothrix*. (From *Textbook of Modern Biology* by Alvin Nason. Copyright © 1965, John Wiley & Sons, Inc., New York.)

a filament are enclosed in a mucilaginous sheath. Each cell contains a vacuole, cytoplasm, a nucleus, and one or more spirally arranged bandlike chloroplasts which extend the length of the cell (Fig. 30–9). Certain species have one chloroplast in each cell; others have more than one. Specialized proteinaceous structures called **pyrenoids** occur at intervals along a chloroplast. Starch is deposited around the pyrenoids.

Asexual reproduction. A filament increases in length through transverse cell division—a mitotic division of the nucleus preceding the partitioning. Long *Spirogyra* filaments may be fragmented by animals, water currents, or separation of some cells. Each fragment becomes longer by cell division.

Sexual reproduction. Sexual reproduction in *Spirogyra* is isogamous. In many species sexual reproduction occurs between two filaments, but in others it takes place between two adjoining cells in one filament. In sexual reproduction between two filaments, the cells of one filament pair with the cells of the other. Each cell produces a bulge which from the beginning is in contact with a similar bulge formed by the opposite cell. Each bulge becomes longer (Fig. 30–10*A*), pushing the paired filaments apart. The end walls between connecting bulges are then dissolved, leaving a tube which connects the two opposite cells, called a **conjugation tube.** While the conjugation tube is developing, the protoplast in each cell contracts. Each protoplast is then a gamete. The gametes in the cells of one filament then pass through the conjugation tubes into the cells of the adjoining filament, where the gametes unite. Each resulting zygote develops a heavy wall, and it can endure adverse conditions. The mature zygotes (also known as **zygospores**) are set free when the filament containing them disintegrates. The zygotes remain dormant for

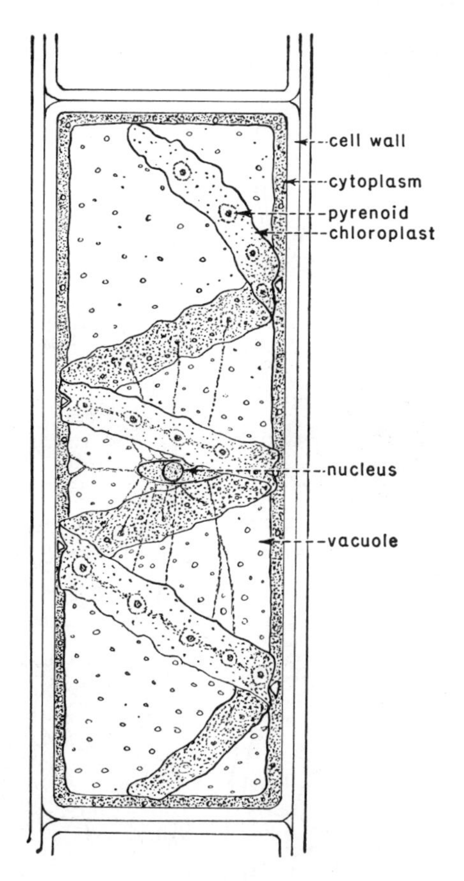

Fig. 30–9. Cell of *Spirogyra*. (Josephine E. Tilden, *The Algae and Their Life Relations*.)

a time, and later, when conditions are favorable, they germinate.

During germination, the nucleus in each zygote undergoes reduction division, forming four haploid nuclei. Three nuclei disintegrate and one remains functional. Next, the zygote wall breaks open and one cell emerges, which grows into a new filament made up of cells with the haploid number of chromosomes. Growth is brought about by transverse cell divisions and cell elongation (Fig. 30–10*B*).

Oedogonium

Another widely distributed, aquatic, unbranched, filamentous alga is *Oedogonium*. The cells are uninucleate, and in each there is one netlike chloroplast. *Oedogonium* reproduces asexually by zoospores and sexually by unlike gametes.

Asexual reproduction. In each filament only a few cells, the **zoosporangia,** produce zoospores. The entire contents in a zoosporangium rounds up and becomes a zoospore which is characterized by a ring of flagella at one end. After the zoospore is formed, the end wall of the parent cell splits across, forming an opening through which the zoospore is liberated (Fig. 30–11). The zoospore swims for a time, then loses its flagella, and finally develops into a new plant.

Sexual reproduction. Sexual reproduction in *Oedogonium* is more advanced than in the forms previously considered. Small motile sperms and large nonmotile eggs are formed; thus *Oedogonium* is **heterogamous.** Furthermore, the gametes are produced in specialized structures. Two sperms, each with a ring of flagella at one end, are formed in a specialized one-celled structure called an **antheridium.** One nonmotile egg develops in an **oogonium,** a large one-celled case. When the egg is fully formed, a pore develops in the oogonial wall. A sperm swims through this opening and fertilizes the egg. The resulting zygote develops a heavy wall, and later it is liberated by the decay of the oogonium. After a rest period the single diploid nucleus in the zygote undergoes reduction division, and four zoospores are organized. These are released, swim for a time, and then come to rest and develop into haploid filaments.

In some species of *Oedogonium,* sperms and eggs are produced in the same filament (that is, they are **homothallic**), but in other species eggs are produced by one plant and sperms by another, in which instance the species is **heterothallic.** In many heterothallic species the male plants are dwarfed and grow attached to the female plants. A dwarf male plant consists of a holdfast cell and one or more antheridia.

EUGLENOIDS—EUGLENOPHYTA

When we get into the microscopic world of life, we are not certain whether some or-

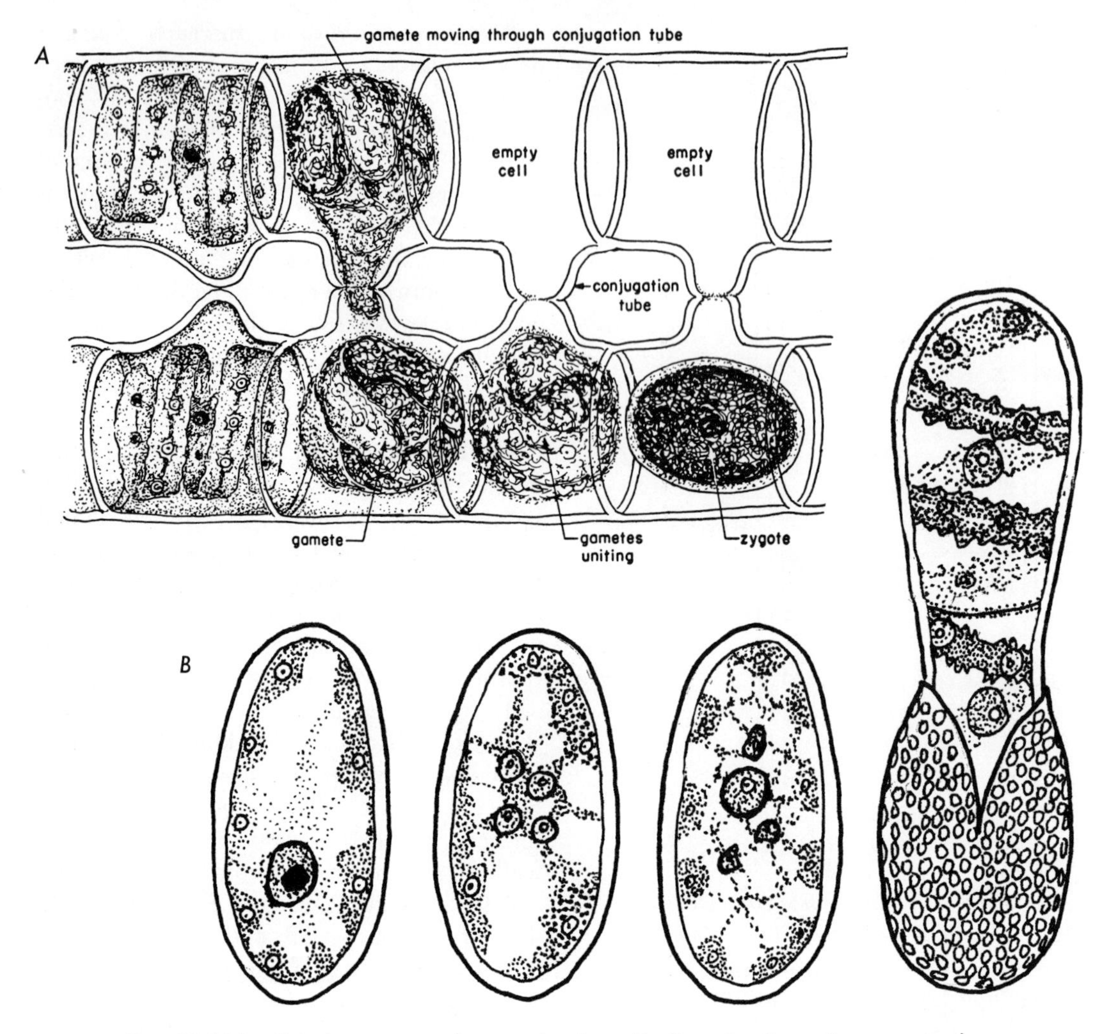

Fig. 30–10*A*. *Spirogyra*, sexual reproduction. *B*. Germination of a zygote (zygospore) of *Spirogyra*. The diploid nucleus undergoes meiosis. Three of the resulting haploid nuclei abort, and the fourth is functional.

ganisms are plants or animals. This is true for *Euglena* and other euglenoids. They are classified as animals by zoologists and as plants by botanists. Perhaps they are a little of both, and, indeed, it has been suggested that both plants and animals descended from the euglenoids.

Both green and colorless unicellular organisms which are propelled by 1 or 2 flagella are in this phylum. The green ones manufacture food, and the colorless ones exist as saprophytes. Some of the latter ingest solid food in the manner of animals. Furthermore, practically all euglenoids lack a rigid cellulose wall; in this way they resemble animals. The protoplasm of *Euglena* and practically all other members of the phylum Euglenophyta is enclosed in a flexible proteinaceous membrane which enables them to change their shape. The major plant characteristic which some species have is the presence of chlorophyll and hence the capacity to make food (Fig. 30–12*A*).

Euglena, a representative member of this phylum, is a one-celled organism which frequently grows in water that is rich in organic matter. The body is oval, and at the anterior end there is 1 colorless flagellum which by its lashing movement propels the organism.

Inside the cell, close to the base of the flagellum, there is a pulsating vacuole and a red eye spot which seems to control the movement of the organism toward light of moderate intensity. One nucleus is present in the center of the cell, and several to many bright green chloroplasts are irregularly distributed in the cytoplasm. However, not all species of *Euglena* have chloroplasts and such species live as saprophytes. Even within a species, *Euglena gracilis,* for example, both green and nongreen plants may be found. Colorless races of *Euglena gracilis,* which are of value in many studies, may be produced experimentally, for example, by growing cells at high temperatures or in the dark, and by exposing cells to ultraviolet light or to the antibiotic streptomycin.

Euglena reproduces asexually by cell division. After the nucleus divides (Fig. 30–12*B*), the cell splits longitudinally into 2 new individuals. On occasions, *Euglena* may become encysted. The spherical cyst has a heavy wall and it is resistant to drought and other unfavorable conditions.

CHRYSOPHYTA

Yellow-green algae, golden-brown algae, and diatoms are in the phylum Chrysophyta, whose members usually store food as oils and never as starch, the characteristic food reserve of the green algae. Their chloroplasts contain chlorophyll a and chlorophyll c, not chlorophyll b, and a large quantity of certain carotenes and xanthophylls which contribute to the yellowish color of most species. We shall discuss a representative yellow-green alga, *Vaucheria,* in the class Xanthophyceae, and the diatoms of the class Bacillarieae.

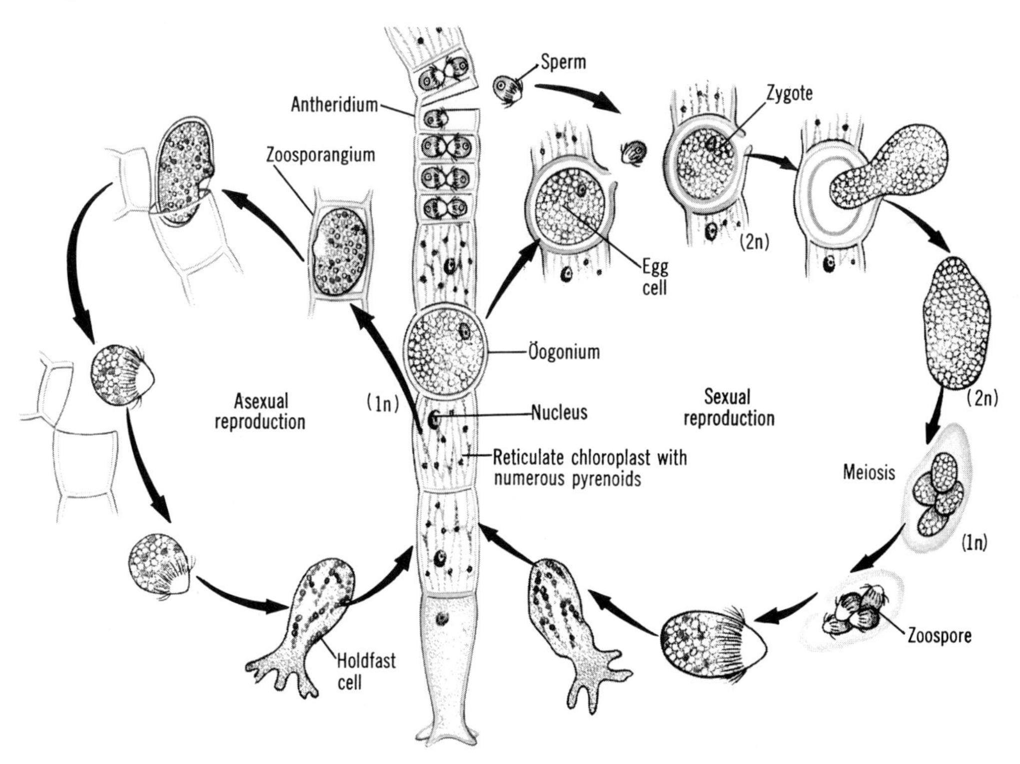

Fig. 30–11. Life cycle of *Oedogonium.* (From *Textbook of Modern Biology* by Alvin Nason. Copyright © 1965, John Wiley & Sons, Inc., New York.)

A

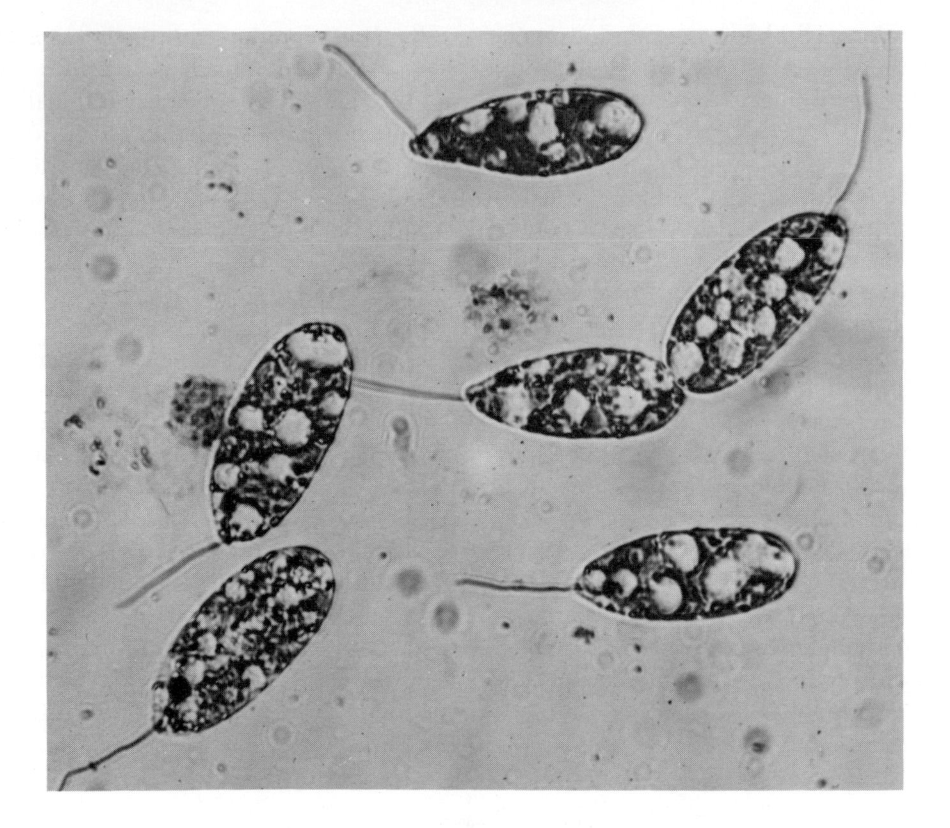

B

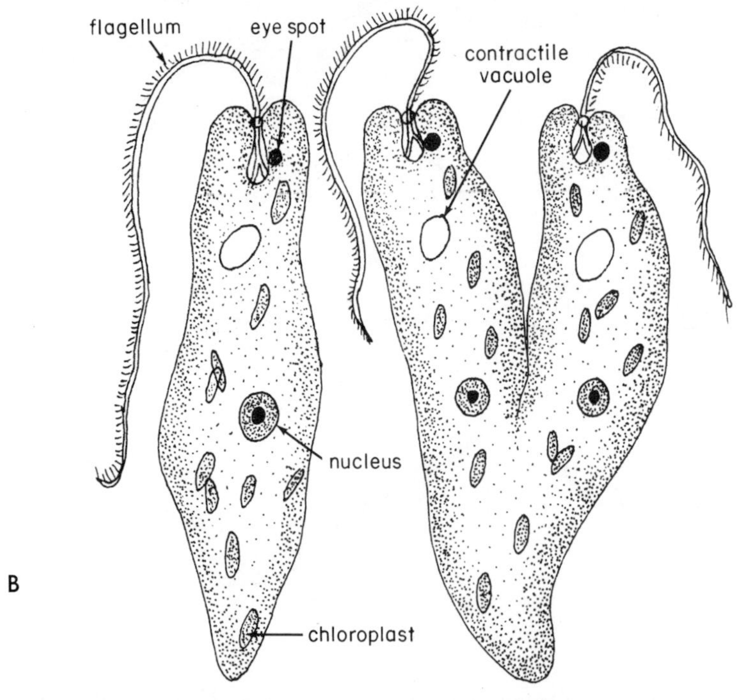

Fig. 30–12*A*. *Euglena gracilis*, an organism which may be classed as either plant or animal. (A. Millard.) *B*. Structure and reproduction of *Euglena*.

Vaucheria

Vaucheria grows in shallow water or on wet soil where it forms mats which feel rough. No cross-walls are present in the branched filaments. The center of the tubular body is occupied by a continuous vacuole. Many nuclei, oil droplets, and ellipsoidal chloroplasts are located in the peripheral cytoplasm. A plant body like this, which lacks cross-walls and which is multinucleate, is said to be **coenocytic.** *Vaucheria* reproduces asexually by zoospores and sexually by unlike gametes.

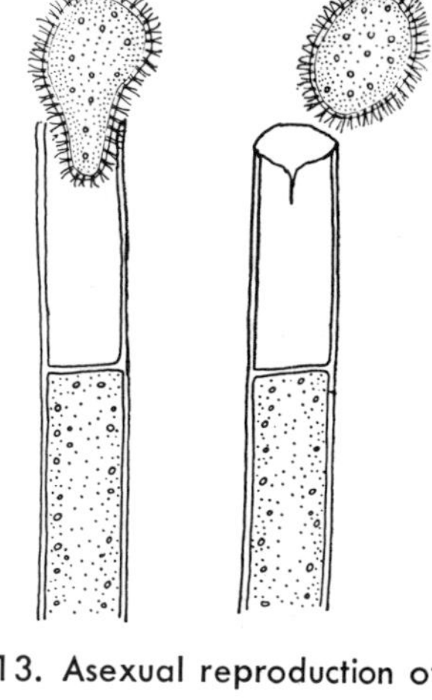

Fig. 30–13. Asexual reproduction of *Vaucheria.*

Asexual reproduction. Multinucleate zoospores are produced singly in zoosporangia (Fig. 30–13). Each zoosporangium (the case in which a zoospore develops) is separated from the rest of the filament by a cross-wall. The many nuclei in each zoospore are located near the surface; opposite each nucleus there are two flagella. The zoospore escapes from the zoosporangium through a terminal pore, and after swimming for a while, it comes to rest, develops a wall, and germinates. Any part of the wall may push out as a tubular outgrowth into which the contents pass. In a short time a branching, green, filamentous plant is produced.

Sexual reproduction. In *Vaucheria* a small motile biflagellate sperm fertilizes a large nonmotile egg. When the gametes are unlike in size, sexual reproduction is heterogamous. Sperms are produced in a one-celled structure called an antheridium, which is cut off from a narrow specialized curved branch by a cross-wall (Fig. 30–14). In an antheridium many sperms, each with two lateral flagella, are produced. One egg develops in a one-celled structure known as an oogonium. The globular oogonium is separated from the main filament by a cross-wall. When the sperms

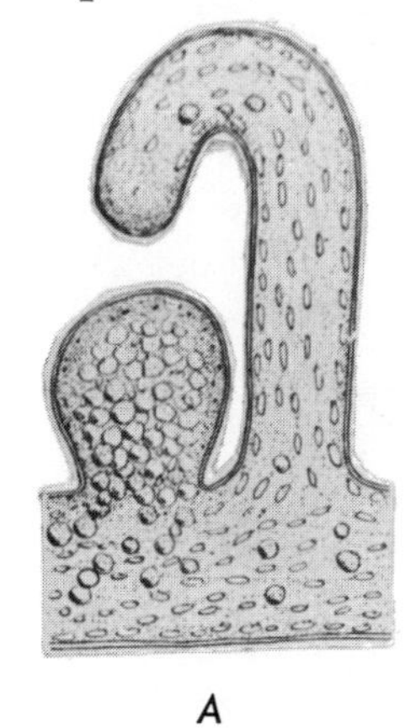

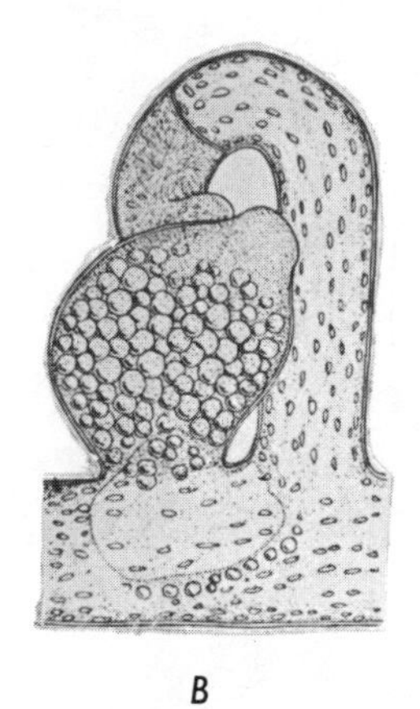

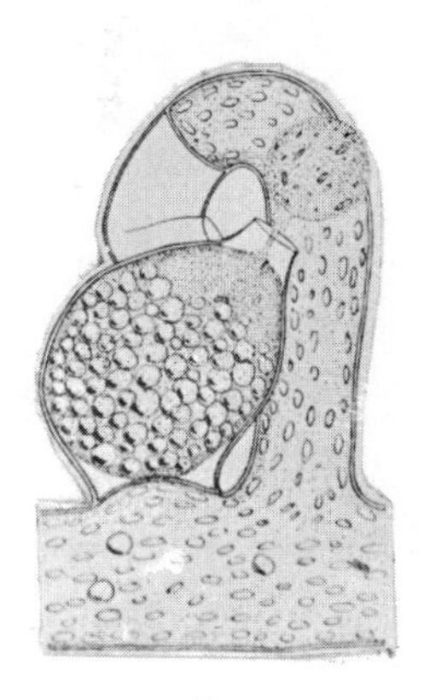

A *B* *C* *D*

Fig. 30–14. Sexual reproduction of *Vaucheria.* *A.* An antheridium will be formed at the tip of the branch on the right, and an oogonium from the branch on the left. *B.* Antheridium on right, oogonium almost formed at the left. C. The antheridium has discharged the sperms; one sperm will unite with the egg in the oogonium. *D.* Zygote present in the oogonium. (John Couch, *Botan. Gaz.*)

are ripe, the end of the antheridium breaks open, the sperms escape, and one or more sperms enter the oogonium through a pore which forms on the oogonial wall. Only one sperm unites with an egg. After fertilization a thick wall forms around the zygote. Later the zygote develops into a new filament by forming a tubular outgrowth into which the contents pass.

Diatoms

In 1702 Anton Van Leeuwenhoek, of Delft, Holland, was examining a drop of water with one of the microscopes which he had made. In addition to green algae, he noted exquisitely formed plants which we now call diatoms.

Diatoms are found in all the waters on the earth from the North Pole to the South. They are the most prolific and important sea plants. In both fresh and salt waters they are directly or indirectly the food for many animals. Clams feed directly on diatoms; the walrus feeds on clams. Some marine fish use diatoms directly for food, but many feed on other animals which depend on diatoms for a food supply.

Diatoms are single-celled algae whose cell wall is impregnated with transparent silica. The cell wall resembles a druggist's pill

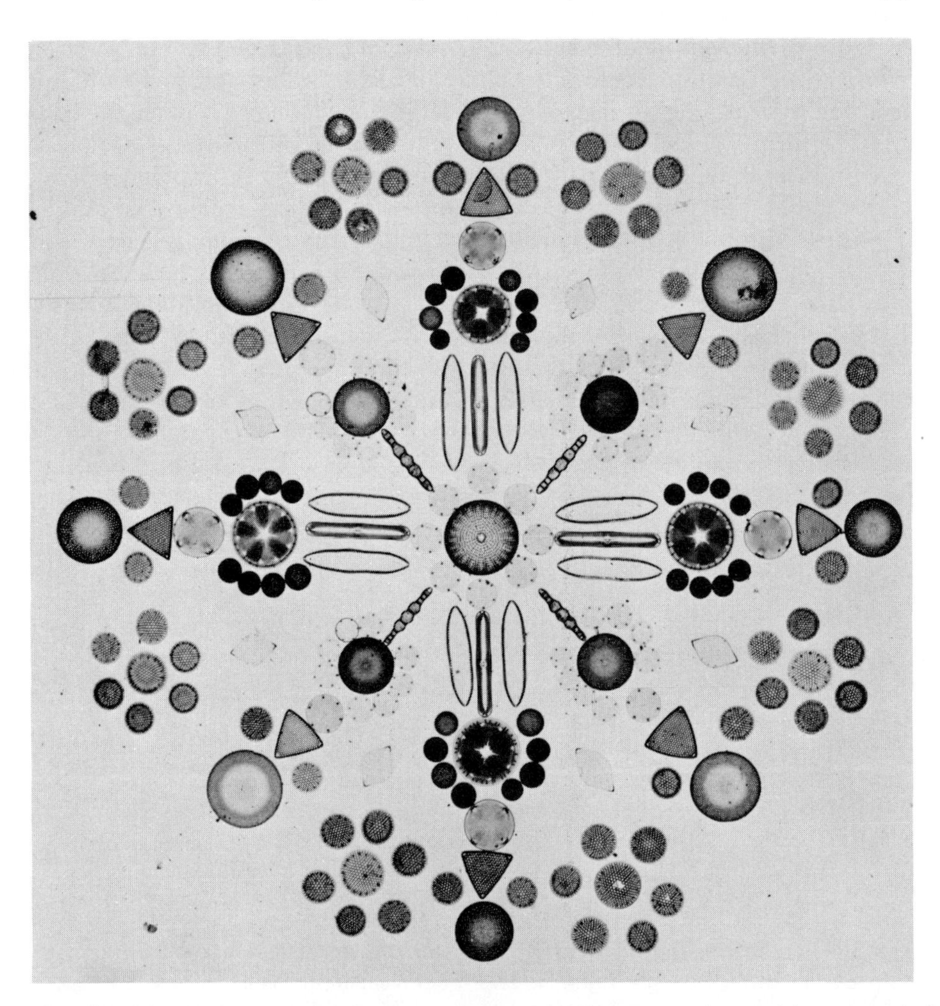

Fig. 30–15. Diatoms, shown here in a man-made design, are beautiful, symmetrical, unicellular organisms. Notice the exquisite markings on the cell walls, which are impregnated with silica. (Bausch & Lomb Optical Co.)

box. Each cell wall consists of two parts (**valves**) which fit closely one into the other. The protoplast contains cytoplasm, a nucleus, and generally two chloroplasts. In addition to chlorophyll a, chlorophyll c, and carotenoid pigments, a brown pigment, fucoxanthin, is present in each chloroplast (Fig. 30–15).

Diatoms are symmetrical organisms of many shapes: round, oval, crescent-shaped, star-shaped, triangular, boat-shaped, etc. The valves bear beautiful markings, which are arranged with such precision that slides of diatoms are used to determine the resolving power of microscope lenses. Because of their beauty of design and symmetry, diatoms are often referred to as the "jewels of the sea."

Most species of diatoms are motile. Cytoplasmic streaming propels diatoms. Narrow longitudinal slits in the cell walls expose the cytoplasm to the surrounding water. The cytoplasm moves along these slits, and friction with the water moves the diatom along. The cytoplasm streams in one direction where exposed to water and in the opposite direction within the cell. At times the direction of streaming may be reversed, and the diatom will then move in the opposite direction.

Diatoms reproduce by cell division (Fig. 30–16). One of the resulting protoplasts receives the lower valve; and the other protoplast, the upper valve of the cell wall. Each protoplast then secretes a new valve inside the old one. If we use the pill box analogy, one protoplast gets the top of the box and the other the bottom. The protoplast which receives the top of the box secretes a new bottom inside the top. The one which receives the bottom does not secrete a new top, as you might expect, but instead forms a new bottom. Hence the diatom which receives the bottom valve during cell division is smaller than the one which receives the top valve. In due time this reduction in size is compensated for by the formation of **auxospores.** The protoplasts of two small diatoms escape from their valves and unite to form a zygote which is known as an auxospore. An auxospore develops into a full-sized diatom.

After death diatoms sink to the bottom of their pond, lake, or ocean. The organic contents decay, leaving only the siliceous, glasslike valves. These form a fossil sediment which, if elevated above water at a later period, becomes a deposit of diatomaceous earth. Such deposits are distributed throughout the world. One of the largest is at Lompoc, California, where it covers 12 square miles to a depth of about 3000 feet. These deposits are mined, and the annual production is 250,000 tons. Diatomaceous earth is used in the manufacture of dynamite; in making filters for clarifying beer, fruit juices, sugar syrups, wine, petroleums, pharmaceuticals, and other products; and as insulating materials for boilers and furnaces. Water in swimming pools can be made crystal clear by circulating the water through grids containing a cake of diatomaceous earth. Under certain circumstances, the organic contents of dead diatoms may not decay completely. There is evidence which indicates that diatoms have been important contributors to the accumulation of the earth's great petroleum deposits.

CRYPTOMONADS AND DINOFLAGELLATES—PYRROPHYTA

The cryptomonads and dinoflagellates occur in fresh water, in brackish water, and in the ocean where they are second in importance to the diatoms as food for marine animals. Not all of the dinoflagellates are beneficial to animal life, however, for some when present in tremendous numbers seem to make the water unfit for the survival of fish and other animals. During some years, the reddish-colored *Gymnodinium brevis* becomes extremely abundant in the sea off the coast of Florida. At such times there may be 60 million or more organisms in each quart of sea water, and the water becomes so red that it is referred to as the "red

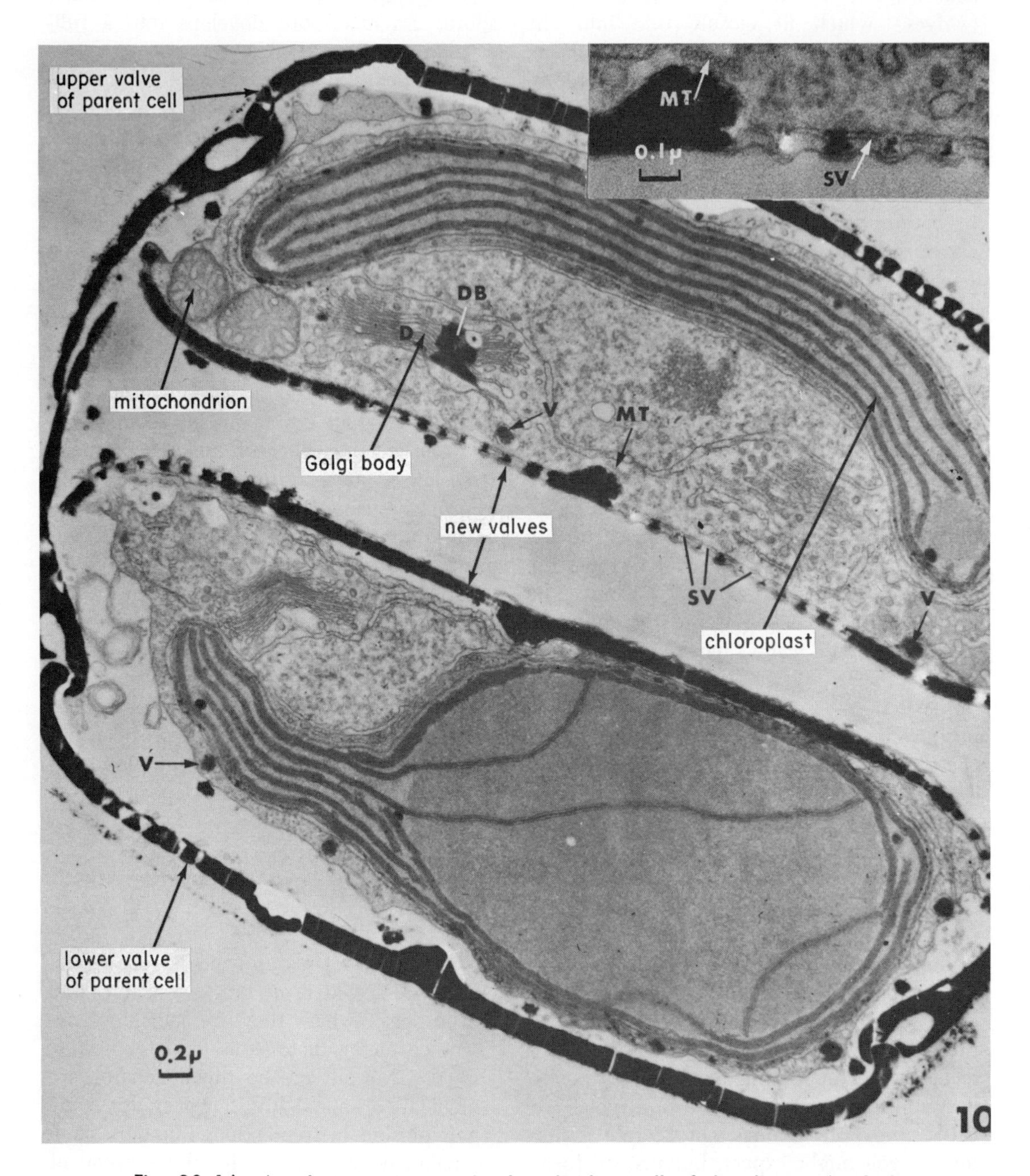

Fig. 30–16. An electron micrograph of a dividing cell of the diatom *Amphipleura pellucida*. Notice that the lower cell is slightly smaller than the upper one. The insert shows details of a developing valve. The organelles evident in the cells are mitochondria, chloroplasts, Golgi bodies (D), microtubules (MT), vesicles with electron-dense contents (V), and silica deposition vesicles (SV). (E. F. Stoermer, H. S. Pandratz, and C. C. Bowen.)

tide." In some unknown way the "red tide" is associated with the death of millions of fish and other marine animals.

The members of this phylum, a group of 1000 species, are unicellular organisms, each with two lateral flagella which lie in a groove and propel the organism with a peculiar twisting motion. In some species the protoplast is naked, but in others it is enclosed in a cellulose wall. Most pyrrophytes are brown in color, but some are orange, red, or even colorless. The latter are saprophytes, and some of these forms ingest solid food. Chlorophyll a and c are present in most of the colored ones, enabling them to manufacture food. One large nucleus and a red eye spot are present in each cell. The cryptomonads and dinoflagellates reproduce by longitudinal cell division.

Many species are luminescent, glowing with a cold light when the water is disturbed by an oar, the prow of a boat or ship, or by some other means. *Noctiluca* (which means night light) is a widely distributed dinoflagellate that glows when agitated. It is an orange-colored, spherical organism that may attain a diameter of 1 or 2 millimeters.

Fig. 30–17. ***Sargassum.*** **Notice the stemlike stipe, the leaflike blades, and the bladders which buoy up the plant.**

BROWN ALGAE—PHAEOPHYTA

The brown algae are with few exceptions restricted to marine habitats, and most are found in the cooler ocean waters. Some species develop where the water is 50 feet deep, but most grow attached to rocks in the tidal zone or just below the low-tide level.

The large brown algae are known as kelps. Some kelps are used as food, especially in China and Japan. A colloid known as algin is extracted from the giant kelp, *Macrocystis pyrifera.* Because algin has emulsifying, gelling, and thickening properties, it is used in pharmaceutical products (burn ointments, surgical jellies, toothpaste, and dental impression compounds), food products (candy, icings, and meringues), rubber products (automobile carpeting and foam cushions), dairy products (ice cream and chocolate milk), adhesives, and paper products (sizing and coatings).

The brown algae range in size from the filamentous *Ectocarpus,* which is only a few inches high, to the massive *Macrocystis,* which attains a length of several hundred feet. *Sargassum,* an inhabitant of the Sargasso Sea, is well known in literature (Fig. 30–17). Although most species of *Sargassum* grow attached along rocky shores, one species, *Sargassum natans,* grows floating in the Sargasso Sea, an area of 2,000,000 square miles of floating seaweed in the Atlantic Ocean between the Bahamas and the Azores. *Sargassum natans* reproduces only by fragmentation.

One or more chloroplasts are present in each uninucleate cell. Chlorophyll a and c, carotenoid pigments, and the brown pigment, **fucoxanthin,** are located in the chloroplasts. The color of brown algae ranges all

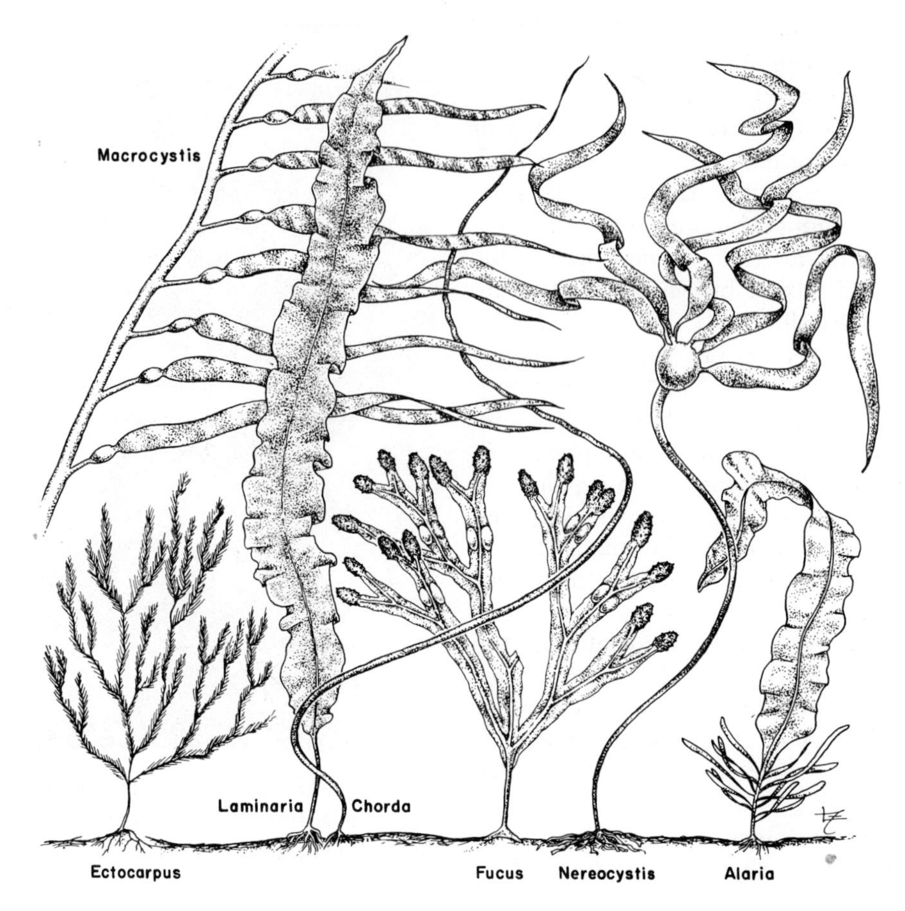

Fig. 30–18. Some common brown algae. Sketches are not drawn to the same scale. (From *Biology*, 4th ed., by Claude A. Villee, Copyright © 1962, W. B. Saunders Co., Philadelphia.)

the way from a yellow-green to a dark brown.

The brown algae evolved directly from some primitive motile one-celled ancestor along an independent line. They show no apparent relationship to the green algae or to the other algal phyla. *Laminaria* and *Fucus* are two widely distributed algae which we will study in some detail (Fig. 30–18).

Laminaria

This alga is abundant along both coasts of the United States. The plant body is 6 or more feet long and consists of an anchoring **holdfast,** a **stipe** (stalk), and a large golden-brown **blade.** Zoosporangia (cases containing motile spores) are produced on the surface of the blade. In each zoosporangium, meiosis occurs prior to the formation of zoospores; hence each zoospore has just one set of chromosomes, whereas the cells of the blade, stipe, and holdfast have two sets. The zoospores are released from the zoosporangium, are free swimming for a while, and then settle down and become attached independently. Half of them develop into microscopic plants called **male gametophytes,** and the other half develop into minute **female gametophytes.** The gametophytes are haploid plants which produce gametes; the males produce sperms in antheridia, and the females produce eggs in oogonia. The antheridia are small lateral outgrowths of the male gametophyte (Fig. 30–19); within each antheridium one motile

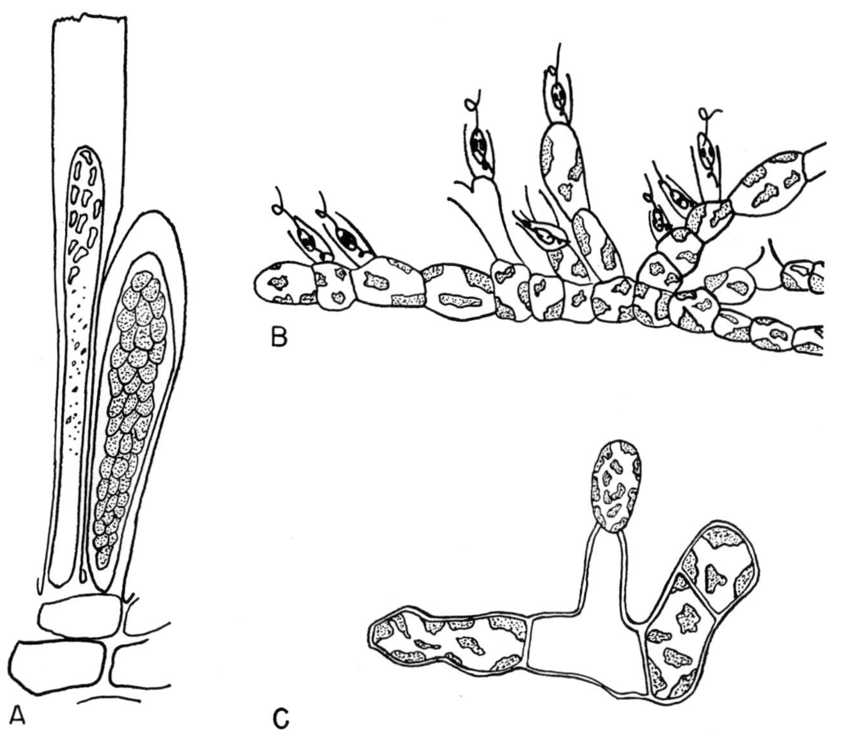

Fig. 30–19. Stages in the life cycle of *Laminaria*. *A*. Zoosporangium containing zoospores. One-half of the zoospores develop into small male gametophytes and half into minute female gametophytes. *B*. A male gametophyte; the sperms are being liberated from the antheridia. *C*. A female gametophyte; an egg is being released from the oogonium. Fertilization occurs in the ocean. The fertilized egg develops into a sporophyte. (*A* after P. Kuckuck; *B* and *C* modified from H. Kylin.)

biflagellate sperm is formed. An oogonium, containing one egg, is produced from a cell of the female gametophyte, which in certain species is only a few cells long. The biflagellate swimming sperms are released into the water from the antheridia and swim to an egg which has been released from the oogonium by the terminal rupture of the oogonial wall. Fertilization takes place in the open water. The zygote locates itself and develops into a large *Laminaria* plant, with its characteristic holdfast, stipe, and blade. All cells of this large plant have two sets of chromosomes because it developed from a diploid zygote. A diploid plant, such as the large *Laminaria* plant, which produces spores is known as a **sporophyte.** At maturity the sporophyte of *Laminaria* forms zoosporangia, which give rise to haploid zoospores, and thus begins again the **alternation of generations** in which the diploid sporophyte generation alternates with the haploid gametophyte generation. The diploid sporophyte produces zoospores which develop into gametophytes. The gametophytes produce gametes (sperms or eggs) which fuse and start a new sporophyte generation.

The life history of *Laminaria* is summarized below:

sporophyte (diploid) → zoosporangia (meiosis occurs here) ↗ zoospores (haploid) → male gametophytes (haploid) → sperms (haploid) ↘ fertilization → zygote (diploid) → sporophyte (diploid)

sporophyte (diploid) → zoosporangia (meiosis occurs here) ↘ zoospores (haploid) → female gametophytes (haploid) → eggs (haploid) ↗ fertilization → zygote (diploid) → sporophyte (diploid)

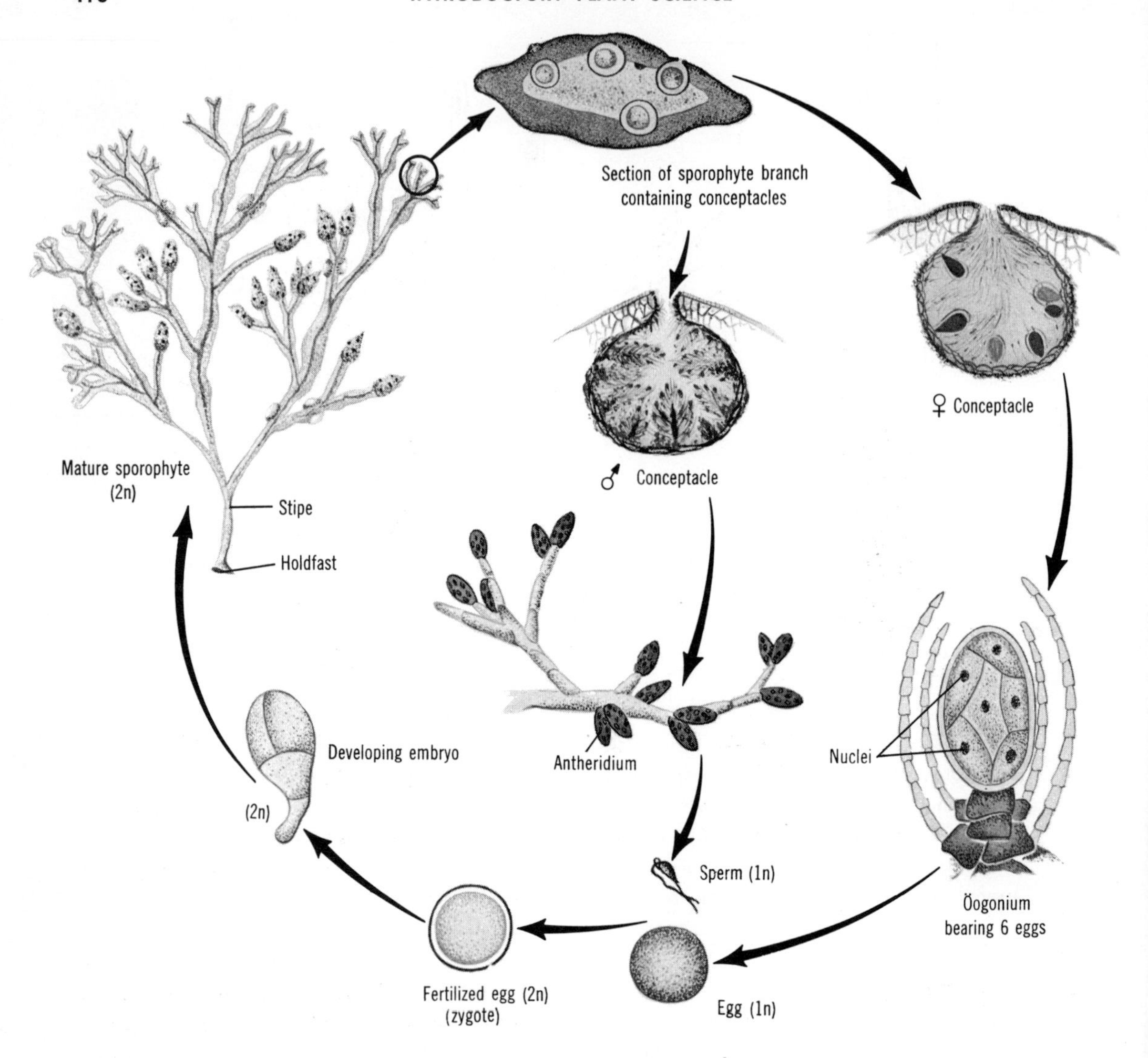

Fig. 30–20. Life cycle of *Fucus*. (From *Textbook of Modern Biology* by Alvin Nason. Copyright © 1965, John Wiley & Sons, Inc., New York.)

Alternation of generations also occurs in *Nereocystis, Macrocystis,* and other brown algae, and in red algae, mosses, ferns, and seed plants.

Fucus

The brown alga *Fucus* is commonly called rockweed because it grows attached by holdfasts to rocks in the intertidal zone. The olive-green plant body (thallus) is flattened and forked. Bladder-like structures filled with air are present on the thallus which keep the plant afloat at high tide. The tips of many branches are swollen, and they are called **receptacles.**

The reproductive structures, antheridia and oogonia, are produced in cavities of the receptacle known as **conceptacles.** In some species antheridia and oogonia are produced in the same conceptacle, and in other species antheridia are produced in conceptacles of male plants and oogonia in those of female plants. A cavity in which antheridia are produced is lined with branching, many-celled hairs (called **paraphyses**) which bear the antheridia (Fig. 30–20). One diploid nucleus is present in each young antheridium. The nucleus undergoes reduction division (meiosis) to form four haploid nuclei. These nuclei divide mitotically to form sixty-four nuclei, each of which is incorporated into a sperm bearing two lateral flagella. Hence at maturity an antheridium contains sixty-four laterally biflagellate haploid sperms. Many oogonia develop within each conceptacle of a female plant. An oogonium is borne at the end of a short stalk. The single diploid nucleus in each young oogonium undergoes meiosis, forming four haploid nuclei which then divide mitotically to form eight nuclei. Each nucleus becomes incorporated into an egg. Hence at maturity an oogonium contains eight eggs. When ripe, the sperms and eggs are discharged into the ocean where fertilization occurs. Each zygote becomes surrounded by a cell wall and then drifts down to the ocean floor. Some may become attached to rocks where they develop into large diploid *Fucus* plants. You will recall that in the green algae the green plants were invariably haploid. In contrast, the *Fucus* plant is diploid.

The life history of a monoecious species of *Fucus* is diagrammed below:

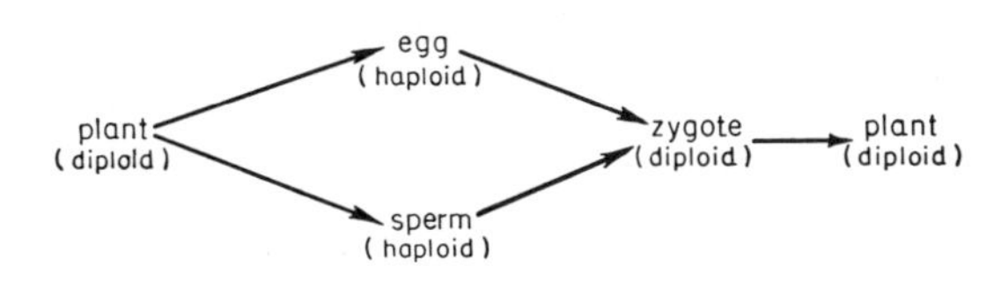

RED ALGAE—RHODOPHYTA

Red algae are mostly marine forms which are anchored by holdfasts of various types. They grow in the deepest waters at which algae are found. A great variety of graceful forms occur; some red algae have filaments which branch profusely and delicately, and others are ribbon-like. In the shallower waters of the warmer seas there abound calcareous red algae (*Corallina*) whose delicate thalli are encrusted with lime, which gives them a hard texture. Along with corals they play a role in the formation of atolls and coral reefs.

Fig. 30–21. *Gelidium cartilagineum*, a red alga from which agar is obtained. (George R. Johnstone and Fenton L. Feeney.)

Many species are red or reddish-brown in color because a red pigment, **phycoerythrin,** is present in addition to chlorophyll a, chlorophyll d, and carotenoid pigments. Some species are violet, and these contain a blue pigment (**phycocyanin**) as well as green, yellow, and red pigments. All red algae lack motile gametes and zoospores, and this is probably their most outstanding feature. There is no obvious relationship between the red algae and other phyla; the red algae evolved independently.

Certain species of red algae are used as food. Japanese and Hawaiians consider *Porphyra* a delicacy. In Scotland *Rhodymenia,* or dulse, is boiled with milk to make a custard. A colloid known as agar is extracted from species of *Gelidium* and *Gracilaria.* Because agar has exceptional gelling properties, it is used in the baking industry,

Fig. 30–22. A clump of Irish moss (*Chondrus crispus*) attached to a rock. (Leonard Stoloff and the Krim-Ko Corp.)

in medicine, and in bacteriological work (Fig. 30–21). The colloid, carrageenin, is extracted from *Chondrus crispus*, Irish moss (Fig. 30–22). Large amounts of carrageenin are used in the manufacture of milk chocolate products, casein paints, and hand lotions, and some is used in purifying beer.

Three morphologically similar plants appear in the life cycle of *Gelidium*, *Polysiphonia*, and many other red algae. These three kinds of plants are the male gametophyte, the female gametophyte, and the sporophyte. Both the male and female gametophytes are haploid plants which produce gametes. Although the sporophyte is diploid, it resembles the gametophytes. At maturity the sporophyte produces, by meiosis, haploid spores which develop into gametophytes. The details of reproduction can be revealed by studying one example, namely, *Polysiphonia*.

Polysiphonia

An attractive red alga with a feathery appearance, *Polysiphonia* is common in the intertidal zone along both coasts of North America. Each plant has a main stalk supporting large branches which give rise to others that are just one cell thick. A cross-section of the main stalk reveals a central cell that is surrounded by four or more peripheral cells. The cells are uninucleate, and each has numerous disk-shaped reddish chloroplasts.

The cells of the sporophyte are diploid. Certain branches of the sporophyte produce sporangia, more technically known as **tetrasporangia,** each of which is located between a central cell and the peripheral cells. A young tetrasporangium contains a single diploid nucleus which undergoes reduction division to produce four haploid nuclei, each of which becomes surrounded by cytoplasm and a cell wall to form a spore.

The four spores produced in each sporangium are known as **tetraspores.** The tetraspores are liberated and drop to the ocean floor. Here one-half of the spores develop into male gametophytes, and the other half develop into female gametophytes.

The male and female gametophytes look like the sporophyte except for their reproductive structures. All cells of the gametophytes are haploid, whereas those of the sporophyte are diploid.

At maturity a male gametophyte produces many **spermatangial branches,** each of which is barely visible to the eye. With magnification each spermatangial branch somewhat resembles an ear of corn. The branch is covered with spermatangia, each of which contains one nonflagellate **spermatium,** the male gamete. The spermatia are released into the water when the spermatangia rupture. Some spermatia may contact a **trichogyne** of a female gametophyte.

The female gametophyte bears **carpogonial filaments,** each made of several cells. The terminal cell of a filament is the **carpogonium,** the female sex organ. One egg is produced in the swollen basal part of each carpogonium. The basal portion is continuous with a slender neck known as a **trichogyne.** Some of the spermatia drift with water currents to the trichogynes. When a spermatium contacts a trichogyne, the walls of the spermatium and trichogyne dissolve at the point of contact. The male

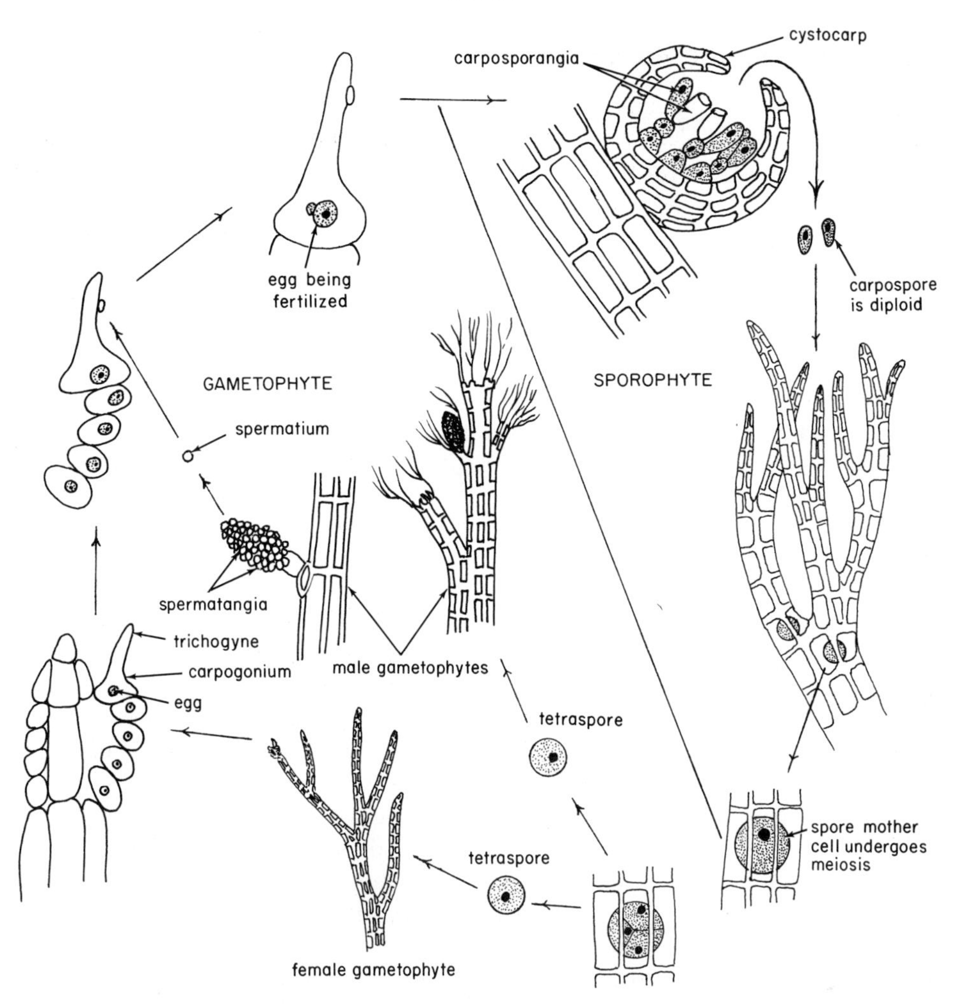

Fig. 30–23. Life cycle of *Polysiphonia*.

nucleus then moves down the trichogyne and unites with the egg, thus forming a zygote.

The zygote nucleus then migrates from the carpogonium into a cell, known as an auxiliary cell, which is just below the carpogonium. Here the zygote divides, and this division is followed by others, resulting in several nuclei derived from the zygote. In the meantime, the auxiliary cell enlarges and fuses with neighboring cells. From the greatly enlarged auxiliary cell there develop short filaments containing nuclei derived from the zygote nucleus. At the top of each filament a sporangium (**carposporangium**) develops which contains one nonflagellate spore (**carpospore**). Neighboring cells of the thallus surround the filaments that bear sporangia to form an urn-shaped structure known as a **cystocarp** which has an opening to the outside.

At maturity the diploid carpospores are released from the carposporangia and then float out through the opening in the cystocarp. The carpospores are carried by water currents. Those that lodge upon a solid object develop into sporophytes which at maturity produce tetraspores, and so the cycle continues. The life cycle of *Polysiphonia* is summarized in Figure 30–23.

EVOLUTION IN THE ALGAE

The algae are the most ancient and simplest present-day green plants. The first plants were likely aquatic, and most of the algae have retained this primitive trait. Within the algae as a whole, but not in one phylum alone, we can discern many evolutionary advances in the structural complexity of cells, in body structure, and in reproduction.

The first forms of life probably lacked specialized structures within their cells. This primitive condition is retained to some degree by all blue-green algae. You will recall that definite nuclei and chloroplasts are wanting in cells of these algae. As time went on, there evolved forms having definite nuclei and chloroplasts. Nuclei and chloroplasts are present in cells of the more advanced algae—the green algae, euglenoids, diatoms, brown algae, and red algae. Of course, these structures are retained by plants higher than the algae.

The one-celled plant body is the simplest type. All blue-green algae are one-celled plants, as are some of the green algae, *Chlamydomonas* and *Protococcus*, for example. Certain unicellular algae are solitary, whereas others form colonies. Colonial algae evolved from solitary ones through the development of a tendency for the cells to remain associated in a characteristic manner after division. The multicellular type of plant body probably evolved from a colonial type. Among present-day algae it is often difficult to distinguish clearly between organized colonies and simple multicellular plants. For example, *Spirogyra* may be considered a colony of cells arranged end to end or a filamentous multicellular plant. The filamentous body is well adapted for life in water; all cells are in direct contact with water containing dissolved gases and salts. Water, minerals, and dissolved gases readily diffuse into the cells.

Many brown and red algae have a more massive and complex multicellular body than that of such filamentous algae as *Oedogonium* and *Ulothrix*. For example, the body of *Laminaria* is large and has a holdfast, stipe, and blade. The elaborately constructed bodies of *Laminaria* and other sea weeds have evolved through a number of stages from simple one-celled plants.

Within the algae we can see the beginnings of differentiation—the specialization of certain cells for a specific function. In many blue-green algae all cells are the same, as they are in the green alga *Spirogyra*. In other algae certain cells have become specialized. For example, the basal cell of a *Ulothrix* filament is modified to form a holdfast cell. In *Oedogonium* and *Vaucheria* we see the beginning of cells specialized for producing gametes. In both algae, sperms are formed in antheridia and eggs are formed in oogonia. Specialization of cells for reproduction in *Oedogonium* is a simpler type than that in *Vaucheria*. In *Oedogonium*, cells of the ordinary filament have become transformed into antheridia and oogonia, whereas in *Vaucheria* the antheridia and oogonia are produced on specialized branches.

Certain stages in the evolution of reproduction can be inferred from a study of the algae. The evolutionary sequence leads from algae without sexual reproduction to algae reproducing sexually by means of like gametes to those reproducing with unlike gametes. The most primitive type of reproduction, characteristic of the blue-green algae, is fission, in which process a unicellular plant simply divides to form two plants. A more specialized and advanced type of asexual reproduction is by means of motile spores called zoospores, such as those produced by *Chlamydomonas*, *Ulothrix*, *Oedogonium*, and *Vaucheria*. Sexual reproduction, characterized by the union of gametes, represents the most advanced type of reproduction. Perhaps gametes evolved from zoospores. The primitive type of sexual reproduction involves the fusion of like gametes, such as occurs in *Spirogyra* and

Ulothrix. In the most advanced type the gametes are unlike. Many green algae produce unlike gametes, sperms and eggs, as do most brown algae, all red algae, and all plants higher than algae in the evolutionary scale.

QUESTIONS

1. With respect to rivers and streams, which one of the following is false? (A) The flora and fauna most characteristic of streams and rivers are suspended microscopic or near-microscopic organisms known as plankton. (B) The plant constituents of the plankton are largely diatoms and other minute algae. (C) The total mass of plankton organisms of the Illinois River at Havana, Illinois—where it flows slowly—amounted in 1 year to 200,000 tons. (D) Most species of green and blue-green algae are poisonous.

2. _____ Which one of the following is false? (A) All algae have a simple type of body, known as a thallus, which has no roots, stems, or leaves. (B) Algae make up the greater part of the pastures of lakes and streams. (C) Algae help maintain a desirable oxygen balance in the water. (D) Blue-green algae are more complex and advanced than are the green algae.

3. Discuss the economic importance of the algae.

4. Algae may become a nuisance in reservoirs and swimming pools. How may they be controlled?

5. What are the advantages of sexual reproduction?

6. In sexual reproduction what are the advantages of having large nonmotile eggs and small motile sperms?

7. Which one of the following is not a blue-green alga? (A) *Gloeocapsa*, (B) *Oscillatoria*, (C) *Nostoc*, (D) *Protococcus*.

8. _____ Plants with chlorophyll are found in the (A) photic zone, (B) photic and transition zones, (C) photic, transition, and aphotic zones.

9. Animals live (A) only in the photic zone, (B) in the photic and transition zones, (C) in the photic, transition, and aphotic zones.

10. _____ Which one of the following does not occur as one goes from the ocean surface to depths of 500 feet? (A) Decrease in light intensity, (B) increase in the proportion of yellow, orange, and red light in comparison to blue light, (C) increase in the concentration of nitrogen and phosphorus, (D) decrease in temperature.

11. _____ All blue-green algae lack (A) chromoplasm, (B) nucleoplasm, (C) nucleic acid, (D) ribosomes, (E) well-differentiated nuclei, (F) any form of sexual reproduction.

12. _____ Heterocysts occur in filaments of (A) *Gloeocapsa*, (B) *Oscillatoria*, (C) *Nostoc*.

13. _____ Members of which one of the following phyla lack sexual reproduction, definite nuclei, and chloroplasts? (A) Cyanophyta, (B) Chlorophyta, (C) Phaeophyta, (D) Rhodophyta.

14. Arrange the following in order of complexity of the plant body (start with the simplest and end with the most complex): *Fucus, Spirogyra, Nostoc, Gloeocapsa, Oedogonium*.

15. Arrange the following in order of increasing complexity of reproduction (start with the most primitive type of reproduction and end with the most advanced): *Spirogyra, Oscillatoria, Oedogonium, Ulothrix, Protococcus*.

16. _____ A one-celled, motile green alga which reproduces asexually by zoospores and sexually by isogametes is (A) *Protococcus*, (B) *Chlamydomonas*, (C) *Scenedesmus*, (D) *Spirogyra*.

17. _____ Which one of the following reproduces asexually by means of zoospores? (A) *Spirogyra*, (B) *Vaucheria*, (C) *Protococcus*, (D) *Oscillatoria*.

18. _____ Antheridia, containing sperms, and oogonia, producing eggs, occur in (A) *Vaucheria*, (B) *Spirogyra*, (C) *Nostoc*, (D) *Ulothrix*.

19. _____ Small sperms and large eggs are characteristic of (A) *Spirogyra*, (B) *Oedogonium*, (C) *Nostoc*, (D) *Ulothrix*.

20. Indicate whether the following (A) lack sexual reproduction, (B) are isogamous, (C) are heterogamous:

_____ *Protococcus* _____ *Spirogyra*
_____ *Chlamydomonas* _____ *Oedogonium*
_____ *Ulothrix* _____ *Vaucheria*

21. _____ A conjugation tube is found in (A) *Vaucheria,* (B) *Spirogyra,* (C) *Ulothrix,* (D) *Oedogonium.*

22. _____ Which one of the following is intermediate in certain ways between plants and animals? (A) *Oscillatoria,* (B) *Spirogyra,* (C) *Euglena,* (D) *Fucus,* (E) *Gelidium.*

23. Give the general characteristics of diatoms. Where are diatoms found? Of what importance are they?

24. _____ In the Sargasso Sea there are about 2,000,000 square miles covered with *Sargassum,* which is a (A) red alga, (B) brown alga, (C) green alga, (D) blue-green alga.

25. _____ Agar is obtained from (A) a brown alga, (B) a red alga, (C) a green alga, (D) a blue-green alga.

26. _____ In many algae the cells of the thallus are haploid (1n). However, in one of the following, the thallus is diploid. Which one has a diploid thallus? (A) *Spirogyra,* (B) *Oedogonium,* (C) *Ulothrix,* (D) *Fucus.*

27. Give a general discussion of the distribution of the brown and red algae.

28. Diagram the life cycles of *Laminaria* and *Fucus.*

29. _____ Alternation of generations occurs in many plants and is especially evident in *Laminaria,* which produces sporophytes and gametophytes. The minute haploid *Laminaria* plants which produce gametes are called (A) gametophytes, (B) sporophytes, (C) the asexual phase of the alternation of generations.

30. Diagram the life cycle of *Polysiphonia.*

31. _____ Which one of the following never produces motile gametes? (A) Green algae, (B) brown algae, (C) red algae.

32. _____ Diatoms (A) reproduce sexually only, (B) are unicellular algae with terminal flagella, (C) produce auxospores in sexual reproduction, (D) are found only in marine habitats.

33. _____ The Pyrrophyta (A) are multicellular marine algae, (B) include certain genera that are toxic to marine animals, (C) have heterogamous sexual reproduction, (D) are all autotrophs.

34. _____ *Polysiphonia* (A) produces motile zoospores which develop into male and female gametophytes, (B) produces eggs in flask-shaped cells called carpogonia, (C) produces sperms in a structure known as a cystocarp, (D) produces an important colloid called algin.

---

35. The terms on the left are frequently used in the construction of botanical terminology. Indicate their meaning by matching the proper translation from the list on the right.

_____ 1. A-, an-	1. Shared in common.
_____ 2. Coeno-	2. Double.
_____ 3. Cyano-, -cyanin.	3. Split.
_____ 4. Dipl-, diplo-	4. Reproductive structure.
_____ 5. Fissi-	5. Not, without.
_____ 6. -gonium.	6. Alike.
_____ 7. Haplo-	7. Dark blue substance.
_____ 8. Hetero-	8. Similar.
_____ 9. Homo-	9. Single.
_____10. Iso-	10. Great.
_____11. Mega-	11. Different.
_____12. Oo-	12. Pair.
_____13. Zoo-	13. Egg.
_____14. Zygo-	14. Animal.

31

Fungi—Bacteria

Fungi make the earth habitable for human beings, animals, and green plants. Some are scavengers that grow upon and decay the remains of animals and plants, thus returning compounds to the soil and atmosphere in forms that can be used by other organisms. If you doubt the significance of this decay, imagine yourself on an earth littered with a billion years' accumulation of corpses. Fungi are prevalent and abundant, but ordinarily we are not conscious of their existence because many are microscopic in size. There are eighty thousand to a hundred thousand species, adapted to grow in a great variety of habitats.

All fungi have one characteristic in common: they lack chlorophyll. With the exception of a few specialized bacteria, the nongreen plants cannot make their own food. They obtain it ready-made, either from living organisms or from the dead remains of organisms or their excretions. Those that obtain their food from dead organic matter are referred to as **saprophytes,** whereas those dependent upon living organisms for their nutrients are called **parasites.**

Both parasites and saprophytes are of interest to us. Some saprophytes bring about decay, and others make products such as alcohol, glycerine, and lactic acid, and also such antibiotics as penicillin, streptomycin, and aureomycin. The increasing use of fungi for industrial products and antibiotics has stimulated interest in this group of plants.

The parasitic fungi cause a great variety of diseases of man, of domestic and wild animals, and of plants. If a certain crop is grown year after year in the same area, fungus diseases may increase to the point where cultivation of the crop is difficult or economically impractical. This is one of the basic reasons for crop rotation. Fungi grow as parasites on all kinds of plants—other fungi, algae, mosses, ferns, and seed plants.

Three phyla of the plant kingdom are referred to as fungi: the phyla Schizomycophyta (bacteria), Myxomycophyta (slime molds), and Eumycophyta (true fungi).

BACTERIA—SCHIZOMYCOPHYTA

Enormous numbers of bacteria are all about us: on our bodies, in our bodies, in the air, in water, on the ground and in the soil to a depth of 10 feet or so, on our garden plants and sometimes in them, and in general wherever there is sufficient warmth, food, and moisture.

When we culture bacteria from different habitats and climatic zones in the laboratory, we note that certain species are quite limited in their distribution. Some bacteria, the **psychrophilic** bacteria, thrive best at relatively low temperatures in the range from 32° to 85° F. Such bacteria grow at great depths in the oceans, in cold springs, and in soils at high altitudes or latitudes. Other bacteria, called **thermophilic** bacteria, flourish only where temperatures are high, from 104° to 167° F. These bacteria thrive in hot springs, and in manure and straw piles. The heat liberated during their respiration may raise the temperature of stacked hay to 160° F., and then the chemical oxidation of the substances formed by bacteria may elevate the temperature to the ignition point, about 400° F., at which temperature the hay will burst into flame. Spontaneous combustion may be prevented by thoroughly curing the hay and by using ventilators in the stack to carry off the heat. Many bacteria, the **mesophilic** bacteria, thrive where temperatures are moderate, ranging from 68° to 104° F.

Certain species of bacteria occur only where the environment has a low salt content, but others—marine bacteria and those of salt lakes—are limited to waters having a high salt content. Sea water contains 3.5 per cent salt. Many marine bacteria fail to grow in a medium with a salt level below 2 per cent.

The *p*H of the environment may also influence the distribution of bacteria. Most species, including the nitrogen-fixing and ammonifying bacteria, flourish in soils that are neutral or close to it. However, some bacteria, such as those which oxidize alcohol into acetic acid and those which convert glucose into lactic acid, thrive in an acid medium.

The bacteria known as **obligate aerobes** flourish only where oxygen is available. In contrast, the **obligate anaerobes** grow only where oxygen is lacking. In this category are the bacteria that cause botulism and tetanus. A third group, the **facultative anaerobes,** include bacteria which grow in either the presence or absence of air. When oxygen is available, they respire aerobically; when deficient, anaerobically.

Many species can synthesize amino acids from a carbon source and inorganic nitrogen, such as nitrates and ammonia. Among the bacteria that synthesize amino acids from carbohydrates and inorganic nitrogen are those that grow in the rumen of sheep, cattle, and goats. The resulting amino acids are used by the bacteria to synthesize proteins. As the food containing bacteria passes through the alimentary canal, many bacteria die and their proteins are digested, thus providing the animal with amino acids. Hence the animal obtains amino acids that were lacking in the food. Urea is often fed to livestock to provide the bacteria with a nitrogen source. The urea is broken down by bacteria into carbon dioxide and ammonia which is then available for amino acid synthesis.

Bacteria are among the smallest organisms. Some measure only four-millionths of an inch in diameter, but most average about one ten-thousandth of an inch. Fifty billion bacteria could be contained in a volume the size of a drop of water. Obviously, each bacterium has an extremely small weight. It has been estimated that 500 billion bacteria would weigh 1 gram (about $\frac{1}{28}$ ounce). The surface area of all the bacteria in a gram would be tremendous—about 60,000 square centimeters. The great surface area of bacteria in proportion to their weight enables them to absorb food and bring about chemical changes at an amazing rate.

The great majority of bacteria are one-celled organisms, but they may form colonies. Their small size makes it difficult to study their internal organization. Studies with the electron microscope reveal that a bacterial cell consists of a cell wall that encloses the protoplast. The protoplast is bounded externally by a plasma membrane. One to several chromatin bodies containing fibrils of DNA are present in the

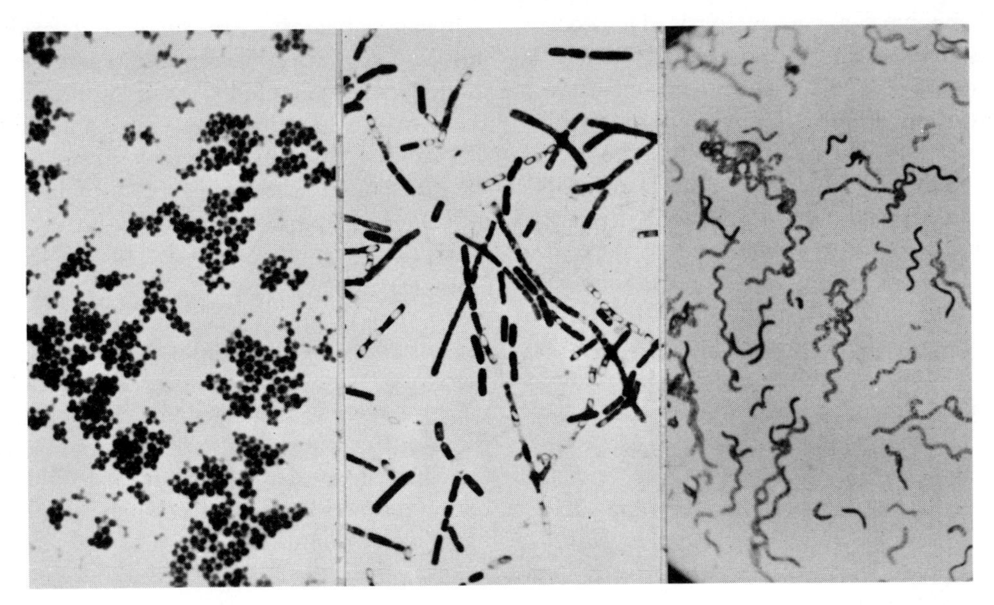

Fig. 31–1. Coccus, bacillus, and spirillum forms of bacteria, from left to right, respectively. *Left, Micrococcus pyogenes,* which causes boils and other skin ailments. *Center, Bacillus anthrax,* which causes anthrax. *Right, Spirillum rubrum,* which is found in water. (Copyright, General Biological Supply House, Inc.)

protoplast. The chromatin bodies are distributed with precision to the daughter cells when the bacterium divides. As yet no organized nucleus or mitochondria have been observed in bacterial cells.

The cell walls of some bacteria are constructed of cellulose. Those of others are made of protein, and walls of different species are made of **chitin** or a related compound. In a number of species penicillin prevents the development of normal cell walls. The wall remains so thin that the protoplast bursts the wall and emerges. Part of the antibacterial action of penicillin may result from this interference with cell-wall formation.

In many species of bacteria the cell wall is surrounded by a slime layer composed of polysaccharides in some species and proteinaceous materials in others. If the slime layer is thick and gummy, it is called a **capsule.**

Bacteria made their appearance on earth more than a billion years ago. Undoubtedly, they evolved from even more simple organisms which left no evidence in the fossil record. Not only is the origin of bacteria an unsolved mystery, but so is the relationship of bacteria to other plants. Bacteria resemble blue-green algae in several ways. Both are one-celled organisms; they reproduce by fission and lack organized nuclei. We do not know which group came on earth first. All of the blue-green algae can make food. A few bacteria can make food by a process of chemosynthesis, and some purple bacteria carry on photosynthesis. In addition to a red pigment, they possess a green pigment which is similar to chlorophyll. Whereas green plants make food from carbon dioxide and water, the purple bacteria use carbon dioxide and hydrogen sulfide. Instead of releasing oxygen, as do green plants, they give off sulfur.

BACTERIAL FORMS

There are three basic forms of bacteria: the **coccus,** the **bacillus,** and the **spirillum** (Fig. 31–1). Bacteria which are spherical in shape have a coccus form. The coccus forms may exist singly or in colonies. If a colony consists of two, the organism is a **diplococcus;** *Klebsiella pneumoniae,* which causes lobar pneumonia, is an example. If the colony consists of a chain of cocci, like a string of beads, the organism is a **streptococcus;** *Streptococcus pyogenes* is an example. This species is often responsible for sore throats. Other species may form colonies resembling a cluster of grapes—for example, *Micrococcus pyogenes,* which causes minor skin ailments such as boils, but which may also cause the serious disease osteomyelitis.

Some species of bacteria are rod-shaped; they have a bacillus form. *Bacillus subtilis,* a saprophyte which brings about decay, and *Salmonella typhosa,* which causes typhoid fever, are typical bacillus forms (Fig. 31–2).

The spirillum forms of bacteria have a corkscrew shape; the number of coils varies from one to three or more. *Vibrio comma,* which causes cholera, has a spirillum shape.

Because bacteria are small and exhibit only a little variety in form, it is impossible to identify species by their appearance. Bacteria are identified rather by what they do—by the diseases that they cause or by the substances which they produce in various media.

MOTILITY

Many bacteria, especially the bacillus and spirillum types, are motile. Most motile bacteria possess flagella which propel them through the water. In some species a single flagellum is present, others have flagella at both ends, and still different species have flagella all around the cell.

REPRODUCTION

Bacteria multiply by cell division (fission). During fission, the chromatin body duplicates itself, the cell constricts in the middle, and the chromatin bodies are distributed equally to the daughter cells (Fig.

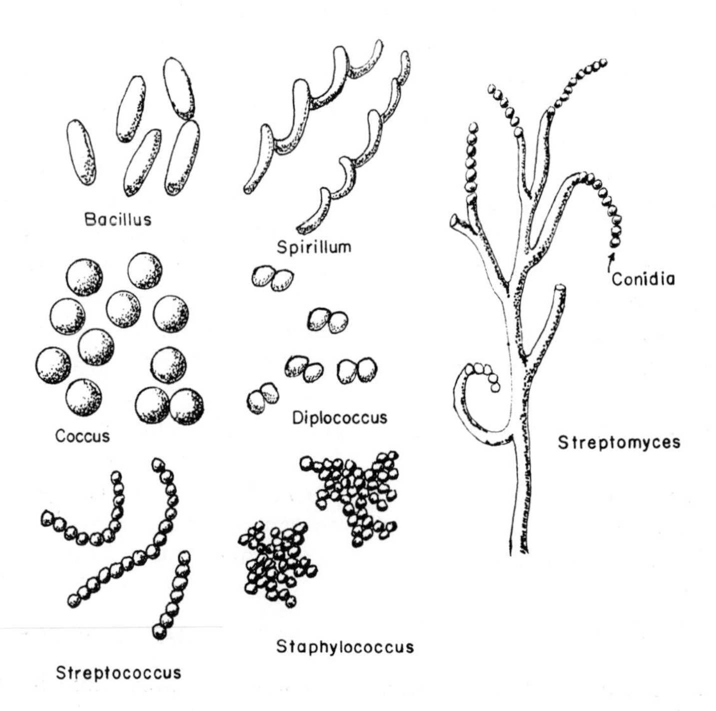

Fig. 31–2. Coccus, bacillus, and spirillum forms of true bacteria and a fungus-like bacterium, *Streptomyces,* in the order Actinomycetales.

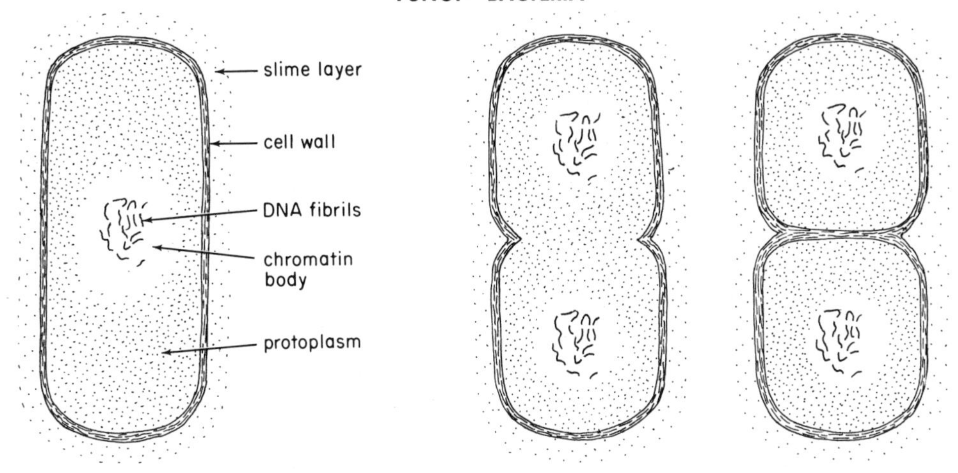

Fig. 31–3. *Left,* structure of a bacterium. *Right,* the division of a bacterial cell.

31–3). In certain species the parent cell elongates prior to division, whereas in others most of the elongation occurs in the daughter cells.

Under favorable conditions bacteria reproduce extremely rapidly. For example, some of the bacteria that occur in milk divide every 30 minutes at room temperature.

Starting with one bacterium in milk at room temperature, and assuming that the rate could remain the same for 30 hours, at the end of the 30-hour period there would be enough bacteria to fill two hundred 5-ton trucks. The initial rate, a division every 30 minutes, does not continue for 30 hours because the food supply becomes exhausted and because toxic products, which are produced by the bacteria themselves, accumulate. If the milk is kept in a refrigerator, bacteria divide slowly, the population is low, and the milk keeps a relatively long time. At temperatures just above freezing, about 6 hours are required for cell division.

It was thought that bacteria lacked sexual reproduction, but breeding experiments and the electron microscope have demonstrated that sexual reproduction occurs in certain bacteria. When strains differing in physiological traits are grown together, some offspring have characteristics of each parent, indicating a transfer of genetic material (Fig. 31–4).

SPORE FORMATION

When water is lacking or when other environmental conditions are unfavorable, some bacteria produce spores. Typically,

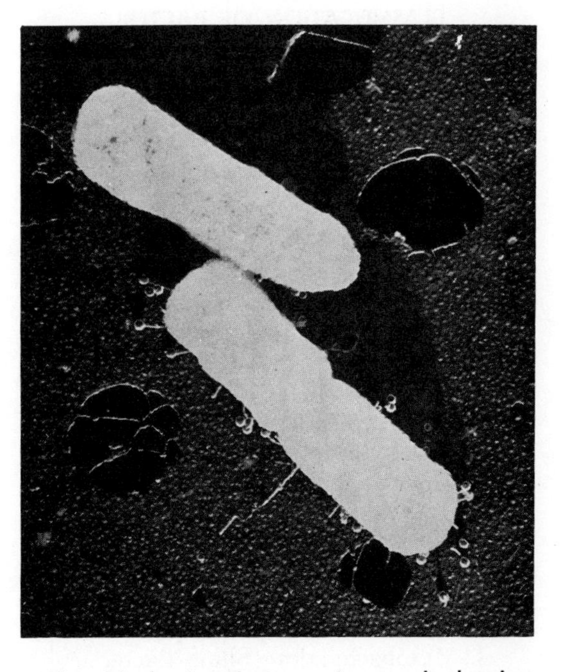

Fig. 31–4. An electron micrograph showing sexual reproduction of bacteria. (T. F. Anderson.)

one heavy-walled spore is produced in each bacterial cell. The spore is released from the parent cell when the cell wall disintegrates. Because spores have dense protoplasm and a low water content, they are resistant to conditions which are fatal to actively growing cells. Some spores can withstand several hours in boiling water, and they are also resistant to antiseptics, drought, and temperatures as low as —454° F. When favorable conditions return, each spore absorbs water and germinates to form an active vegetative cell; as the spore swells, the wall splits, releasing the vegetative cell which is surrounded by a cell wall.

Only a few bacteria that cause human diseases, among them, anthrax, tetanus, and botulism, form spores. Many saprophytic species produce spores. Hence temperatures higher than that of ordinary boiling water are required to kill them. Temperatures lethal to spores are obtained by using steam under pressure in either an autoclave or a pressure cooker.

CLASSIFICATION OF BACTERIA

Bacteria are separated into seven orders, which are summarized in the following tabulation:

1. Eubacteriales (True Bacteria).
   Most of the economically important bacteria are in this order, which includes the typical coccus, bacillus, and spirillum forms.
2. Chlamydobacteriales (Sheathed Bacteria).
   These bacteria are filamentous forms which are enclosed in sheaths containing iron compounds.
3. Thiobacteriales (Sulfur Bacteria).
   The sulfur bacteria are rod-shaped, spherical, spiral, or filamentous in form. They are found in habitats where hydrogen sulfide is abundant. Some contain a purple pigment.
4. Myxobacteriales (Slime Bacteria).
   Bacteria in this order are embedded in large masses of slime. They exhibit a creeping type of movement and produce cysts on stalks. They are frequently present in soils.
5. Spirochaetales (Spirochaetes).
   The spirochaetes are long, spiral, motile organisms which are parasites. Syphilis, trench mouth, yaws, and jaundice are caused by different species in this order.
6. Caulobacteriales (Stalked Bacteria).
   These bacteria develop at the tips of stalks which they secrete.
7. Actinomycetales (Fungus-like Bacteria).
   The body of an actinomycete consists of cells arranged in branched filaments. Some species are saprophytic, but others are parasitic. Potato scab is caused by a species in this order.

ECONOMIC IMPORTANCE

Harmful Activities

Many persons believe that all bacteria are harmful, and they call all of them germs. Although it is true that some bacteria cause diseases of human beings, animals, and plants, the majority do not. Before scientists will recognize that a certain pathogen causes a specific disease, criteria, known as Koch's postulates, must be fulfilled. Three of the postulates were first proposed by Robert Koch, a distinguished biologist who, in 1882, while studying the tubercle bacillus, was the first to prove that a bacterium caused a disease. Since Koch's time, a fourth postulate has been added. The postulates are:

1. The pathogen must be found in every case of the disease.
2. The pathogen must be isolated in pure culture from the diseased organism.
3. When the pathogen is introduced into a susceptible organism, it must produce the original disease.
4. The pathogen must be reisolated from the injected test organism.

These postulates are used by students of diseases of man, animals, and plants. All four postulates cannot be used with every disease. For example, certain diseases are unique to man, and it is not considered humane to introduce active pathogens into human beings. The second postulate cannot be used when studying diseases caused by viruses. As of now, viruses cannot be cultured in artificial media.

Human diseases. Bacteria cause many diseases of human beings; some are listed in Table 31–1. The incidence of disease may be reduced by isolating persons with infectious diseases and by disinfecting discharges, clothing, and dressings from sick people. Furthermore, pathogenic bacteria in water, milk, and other foods should be killed. A knowledge of how diseases are transmitted may enable us to avoid them. Bacterial diseases may be transmitted by direct contact with infected persons, by air currents containing droplets from coughs and sneezes (Fig. 31–5), and by contaminated water, foods, and eating utensils, and for gonorrhea and syphilis by intercourse with infected individuals.

Many pathogenic bacteria produce in the human body extremely poisonous chemicals called **toxins.** For example, the lockjaw bacillus produces a toxin which is forty times more poisonous than rattlesnake venom. To counteract such toxins, there is formed in the blood of the host an **antitoxin** which in some way neutralizes the toxin.

The body has other mechanisms to counteract invading bacteria. White blood cells, known as **phagocytes,** increase in number after infection, and they ingest bacteria. Often, chemicals called **opsonins** are produced which alter the outer surfaces of bacteria so that phagocytes can more readily ingest them. As result of infection, **lysins,** which dissolve bacteria, may be formed in the host. In some instances **agglutinins,** which are capable of clumping bacterial cells, are formed. The protective substances formed by the body are, as a group, referred to as **antibodies.**

The antibodies formed to attack a causal organism may prevent subsequent attacks of that disease. Lasting immunity is usually insured after an attack of poliomyelitis, smallpox, chicken pox, mumps, scarlet fever, and measles. In such diseases as pneumonia, gonorrhea, and tetanus, no lasting immunity is conferred on the individual by an attack.

Artificial active immunity may be accomplished by several methods, such as the injection of attenuated cultures of bacteria, the use of killed cultures of bacteria, and by the use of toxoid. These are often referred to as **vaccines.** They enter the bloodstream and stimulate the formation of anti-

Table 31–1
Some Bacterial Diseases of Man

Disease	Pathogen	Control
Botulism	*Clostridium botulinum*	Antitoxin
Cholera	*Vibrio comma*	Vaccine
Diphtheria	*Corynebacterium diphtheriae*	Vaccine, antitoxin, erythromycin
Gonorrhea	*Neisseria gonorrhoeae*	Penicillin, sulfas
Plague	*Pasturella pestis*	Vaccine, rat extermination
Pneumonia (lobar)	*Klebsiella pneumoniae*	Penicillin, sulfas
Scarlet fever	*Streptococcus pyogenes*	Antitoxin, penicillin
Syphilis	*Treponema pallidum*	Penicillin
Tetanus	*Clostridium tetani*	Vaccine, antitoxin
Tuberculosis	*Mycobacterium tuberculosis*	Streptomycin
Typhoid fever	*Salmonella typhosa*	Vaccine, chloromycetin
Whooping cough	*Hemophilus pertussis*	Vaccine

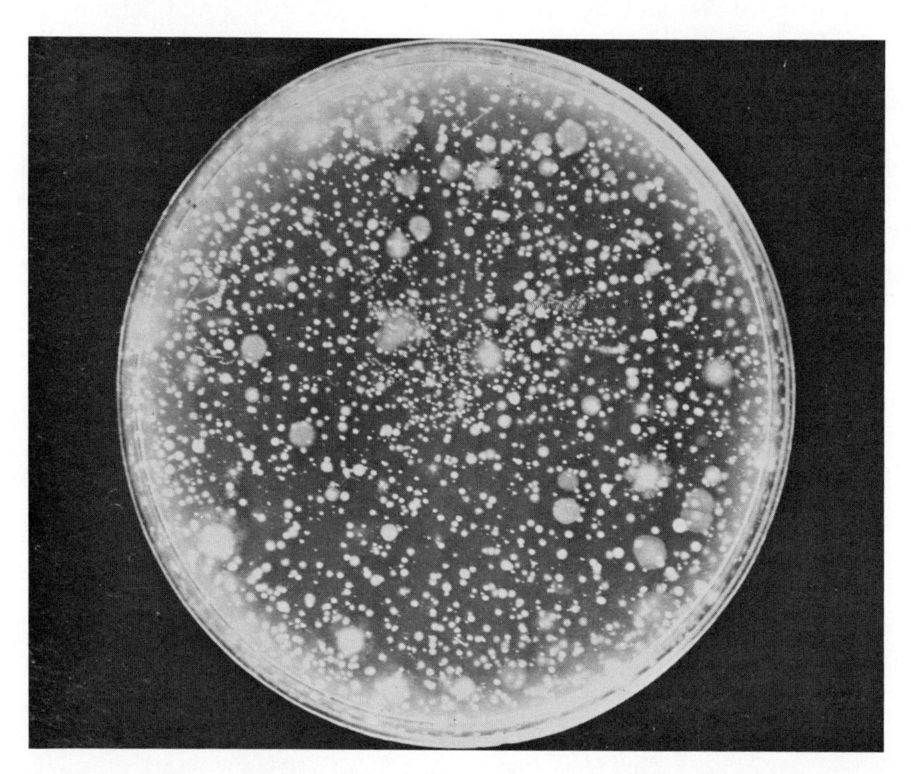

Fig. 31–5. *Upper*, photograph of a sneeze directed against nutrient agar in a petri dish. *Lower*, colonies of bacteria which developed after the dish was incubated. (M. W. Jennison, Society of American Bacteriologists.)

bodies. For immunization against typhoid, bacteria are cultured in the laboratory and then killed. Suspensions of the dead bacteria are injected into the individual. For immunization against diphtheria a **toxoid** is used. Toxoid is toxin which has been treated so that it is not harmful to the human being. Toxin has deadly characteristics which are removed when it is treated to form toxoid.

For temporary immunization against diphtheria, tetanus, and scarlet fever, antitoxin is introduced. This antitoxin is obtained from an animal which has had injected into it first small and then increasing doses of toxin. The blood serum of this animal may be injected into a human, rendering him immune for a short time.

The principle of vaccination was discovered by the eminent scientist Louis Pasteur in the latter half of the nineteenth century. At that time Pasteur was studying chicken cholera. By chance he injected an old neglected culture of the causative bacteria into some chickens. Instead of dying, as they usually did when such bacteria were injected, the chickens survived. It occurred to him that the weak culture could confer an immunity. He experimented further, and, after having first administered a culture of old but living bacteria, he injected a culture of active bacteria which ordinarily killed chickens not previously treated. From such experiments Pasteur discovered the principle of vaccination. He applied the principle to other diseases, such as the deadly anthrax of sheep. He isolated the bacteria, grew them in culture, and found a way to weaken them. Sheep which had previously been inoculated with weakened cultures survived when active cultures were injected, whereas those not previously given weakened cultures died. Since then, the principle which he discovered has been applied in the control of many other diseases.

Diseases of animals. Besides anthrax of sheep and chicken cholera, bacteria cause such diseases of domestic animals as blackleg of sheep and cattle and Bang's disease of cattle. Wild animals are likewise susceptible to some bacterial diseases. For example, during some years, an ailment known as sore mouth kills large numbers of elk.

Diseases of plants. Some bacteria cause plant diseases: ring rot of potatoes, bean blight, alfalfa wilt, fire blight of apples and pears, soft rots of various vegetables, wilt of watermelon, crown gall, soft rot of iris and orchid rhizomes, bacterial leaf spot of carnations, leaf spot of geraniums, and many others. In order to plan effective control of bacterial diseases, it is necessary to understand the causal organisms and their mode of infection. Bacteria usually enter plants through stomata, lenticels, insect injuries, and wounds. Frequently, pathogenic bacteria are transmitted from diseased plants to healthy plants when plants are propagated. The sterilization of tools used in the propagation of fruit trees, irises, orchids, geraniums, and many other plants is effective in preventing the spread of bacterial diseases of these plants. Other methods for controlling bacterial diseases of plants are planting resistant varieties, crop rotation, use of disease-free seed, sanitation, and sanitary removal of diseased plants or parts of plants. Some bacterial diseases of plants are listed in Table 31–2.

Food spoilage. Bacteria cause food spoilage, thus making food preservation necessary. Methods of food preservation are designed to destroy bacteria or to prevent their growth and development. Heat kills bacteria, whereas quick freezing, preservatives, dehydration, salting, and pickling merely prevent their growth. Pasteurization of milk at a temperature of 150° F. for 20 to 30 minutes or at a temperature of 160° F. for 15 seconds reduces the population of saprophytic bacteria and kills many pathogenic bacteria; among them are those that cause tuberculosis, diphtheria, typhoid fever, sore throat, and undulant fever. In the canning of foods it is necessary to kill all bacteria present and to prevent, through sealing, other bacteria from entering. If certain bac-

Table 31–2
Some Bacterial Diseases of Plants

Disease	Pathogen	Host
Alfalfa wilt	*Corynebacterium insidiosa*	Alfalfa
Angular leaf spot of cotton	*Xanthomonas malvacearum*	Cotton
Bacterial blight of stone fruits	*Xanthomonas pruni*	Cherry, peach, plum
Bacterial wilt of corn	*Bacterium stewarti*	Corn
Bacterial wilt of cucurbits	*Erwinia tracheiphila*	Cantaloupe, cucumber, squash, watermelon
Common blight of beans	*Xanthomonas phaseoli*	Bean
Crown gall	*Agrobacterium tumefaciens*	Apple, cherry, alfalfa, cotton, and others
Fire blight	*Erwinia amylovora*	Apple, pear
Ring rot	*Corynebacterium sepedonicum*	Potato
Soft rot	*Erwinia carotovora*	Cabbage, carrot, lettuce, potato, and others

teria are not killed, they may multiply even under the anaerobic conditions existing in a sealed can, and they may produce extremely poisonous chemicals.

Food poisoning. There are three general types of food poisoning: (1) botulism, (2) *Micrococcus* food poisoning, and (3) *Salmonella* infection.

BOTULISM. A fatal type of food poisoning results from eating food containing toxin elaborated by *Clostridium botulinum.* Most cases of botulism result from eating meats and low-acid fruits and vegetables which were canned by a boiling process. The spores survived boiling and germinated in the can where the bacteria multiplied and produced a toxin. The spores do not germinate in a medium with a *p*H below 4.5; hence strongly acid canned fruits are generally safe.

Botulism toxin is one of the most potent poisons known, so potent that it is not even safe to taste meat or vegetables which are suspected of being spoiled. It has been estimated that just 1 ounce of the toxin could kill 400 million people. The toxin is a protein and can be destroyed by vigorous boiling for at least 15 minutes.

Micrococcus FOOD POISONING. On a number of occasions many people become ill after a banquet or a picnic where warm potato salad, cream-filled bakery products, gravy, dressing used in stuffing poultry, chicken salad, or a variety of other foods were served. The foods may have been prepared in an unsanitary manner, and they may not have been continuously refrigerated. Perhaps a huge bowl of potato salad was placed in a refrigerator for a short time, long enough to get the surface layer cold, but not to chill the interior. In such foods, *Micrococcus pyogenes* variety *aureus* (formerly called *Staphylococcus aureus*) multiplied rapidly and produced a toxin. Persons consuming the toxin became ill, generally within 3 to 6 hours. The fatality rate is extremely low, but one becomes miserably sick.

Salmonella FOOD INFECTION. Botulism and *Micrococcus* food poisoning result from the consumption of a toxin. *Salmonella* food infection results from the bacteria themselves. The symptoms are delayed until the bacteria have become established and multiplied in the gastro-intestinal tract. The incubation period with a *Salmonella* infection is 12 hours or longer, following which one suffers from chills, vomiting, diarrhea, abdominal cramps, and fever. Food may become infected with *Salmonella* by human carriers and by insects. Improperly cooked meat from animals infected with *Salmonella* may also cause this infection. *Salmonella* does not produce spores; hence boiling temperature will kill them in a few minutes.

Denitrification. A relatively small fraction of the nitrate nitrogen present in soil may be changed into gaseous nitrogen by denitrifying bacteria, for example *Bacterium denitrificans.* The nitrate nitrogen is changed by these bacteria into nitrite nitrogen, then to ammonium nitrogen, and finally into molecular nitrogen. As a consequence of denitrification, the fertility of the soil diminishes. Denitrification does not usually occur in well-cultivated soils, but chiefly in poorly aerated soils containing an abundant supply of carbohydrates.

Beneficial Activities

The very processes which make bacteria harmful in some instances make them beneficial in others.

Decay. Bacteria are scavengers which return the elements locked up in dead bodies to the soil and air in forms that can again be used by higher plants. As a result of decay brought about by bacteria, the carbon in organic matter is returned to the atmosphere as carbon dioxide, which is then available to green plants.

Through decay, some minerals are returned to the soil directly without going through many alterations—for example, phosphates, calcium, potassium, and others. Nitrogen, on the other hand, undergoes several changes. In the dead body it is at first mostly in proteins. Through decay by *Bacillus subtilis* and other bacteria, the nitrogen is released as ammonia. The released ammonia may combine with carbonates or other ions to form ammonium salts. The ammonium nitrogen in the soil becomes oxidized by bacteria into nitrate nitrogen, a process termed **nitrification.** The change of ammonium nitrogen into nitrate nitrogen occurs in two steps. In the first step ammonium nitrogen is oxidized into nitrite nitrogen by several species of the genus *Nitrosomonas.* As rapidly as the nitrite nitrogen is formed, it is combined with oxygen to form nitrate nitrogen by various species of *Nitrobacter.* The two stages in nitrification may be summarized as follows:

Stage I

2 ammonia + 3 oxygen + *Nitrosomonas* → 2 nitrous acid + 2 water + 158 Calories

$$2NH_3 + 3O_2 + \textit{Nitrosomonas} \rightarrow 2HNO_2 + 2H_2O + 158 \text{ Calories}$$

Stage II

2 nitrous acid + 1 oxygen + *Nitrobacter* → 2 nitric acid + 38 Calories

$$2HNO_2 + O_2 + \textit{Nitrobacter} \rightarrow 2HNO_3 + 38 \text{ Calories}$$

The nitric acid combines with bases in the soil to form nitrates which are readily available to green plants. Both *Nitrosomonas* and *Nitrobacter* thrive only where oxygen is present; that is, they are aerobic bacteria. They are most abundant in the upper few inches of soil and decrease with depth, disappearing about 2 feet below the surface. Through the oxidation of ammonia or the nitrite ion (NO_2), the nitrifying bacteria obtain energy which they use for the synthesis of organic substances. In other words, *Nitrosomonas* and *Nitrobacter* make their own food from carbon dioxide and water even though they lack chlorophyll and grow where light is absent. The process of food manufacture by these organisms is known as **chemosynthesis** because the energy is obtained from the oxidation of inorganic compounds instead of from light.

The sulfur locked up in protein molecules and other complex organic compounds is released through bacterial action as hydrogen sulfide, a malodorous substance. Certain bacteria, such as species of *Thiobacillus* and *Beggiatoa,* oxidize the hydrogen sulfide into water and sulfur which in turn may be oxidized into sulfuric acid in the manner shown below:

$$2H_2S + O_2 \rightarrow 2S + 2H_2O$$

$$4S + 6O_2 + 4H_2O \rightarrow 4H_2SO_4$$

The oxidation of both the hydrogen sulfide and the sulfur provides energy which is used by these bacteria to produce food; that is, they carry out chemosynthesis. The sulfuric acid in the soil undergoes chemical reactions which yield sulfates which are absorbed by roots.

Some sulfur bacteria carry on photosynthesis. Such bacteria are purple or green in color, and they contain carotenoid pigments and one or more kinds of **bacteriochlorophyll,** which is similar to chlorophyll a. In the presence of light the bacteria use carbon dioxide and hydrogen sulfide, and produce food (CH_2O) and sulfur according to the following equation:

Hydrogen sulfide + carbon dioxide + light energy → food + sulfur + water

$$2H_2S + CO_2 + \text{light energy} \rightarrow (CH_2O) + 2S + H_2O$$

Nitrogen fixation. Several species of bacteria change atmospheric nitrogen into nitrogen compounds which higher plants can use. This process is called nitrogen fixation (Fig. 31–6). *Azotobacter,* a coccus-like aerobic organism, and *Clostridium,* a rod-shaped anaerobic bacterium, live saprophytically in the soil and fix nitrogen. They combine the nitrogen of the air with carbohydrate compounds derived from the organic matter of the soil. They are active even when the land is fallow, bringing about an increase in available nitrogen. Under favorable conditions *Clostridium* and *Azotobacter* may add as much as 25 pounds of available nitrogen to an acre in 1 year.

Bacteria associated with roots of legumes add considerable available nitrogen to the soil, about 250 pounds per acre per year for an alfalfa cover crop. This amount will compensate for that used in 3 years by a nonleguminous crop. The bacteria, usually present in soil, enter through the root hairs and grow through the cortex and finally in cells of the pericycle, which are stimulated to divide, producing a tubercle (nodule) of host tissue whose cells contain countless bacteria. The state is one of mutual benefit, or symbiosis. The bacteria obtain food from the leguminous plant, and the host secures an abundance of available nitrogen. When the green plants decay, the soil is enriched with available nitrogen. The bacteria found in nodules have been isolated in pure culture and are called *Rhizobium leguminosarum.* Several races are known; one race may infect one group of legumes but not another. For example, bacteria from beans are effective on vetch but are ineffective on alfalfa. When leguminous plants are introduced into a region, it may be necessary to introduce the appropriate race of bacteria.

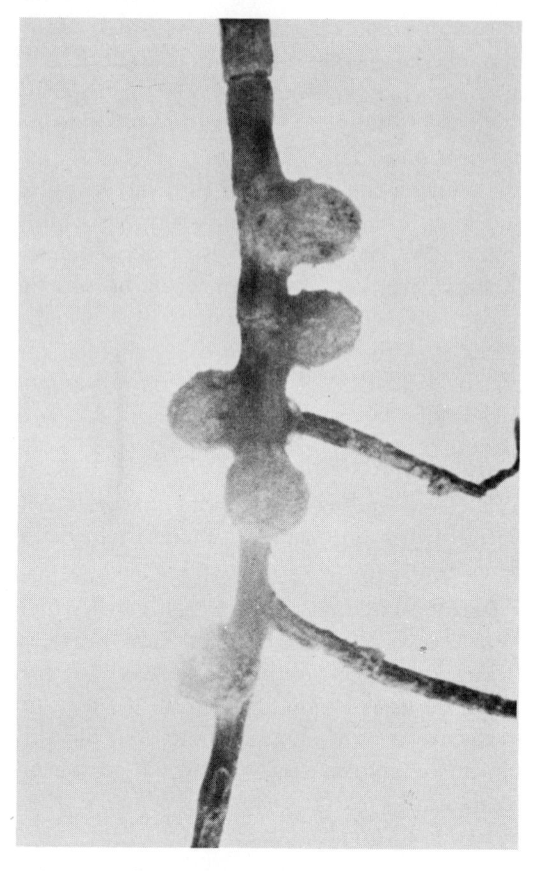

Fig. 31–6. Nodules containing nitrogen-fixing bacteria on a root of sweet pea.

The nitrogen cycle. Nitrogen compounds are indispensable for plant growth. Unfortunately, the most abundant type of nitrogen, nitrogen gas, cannot be used directly by higher plants. The earth's atmosphere contains great quantities of nitrogen gas; there is about 20 million tons in the air over each square mile of the earth's surface. Only nitrogen compounds, primarily am-

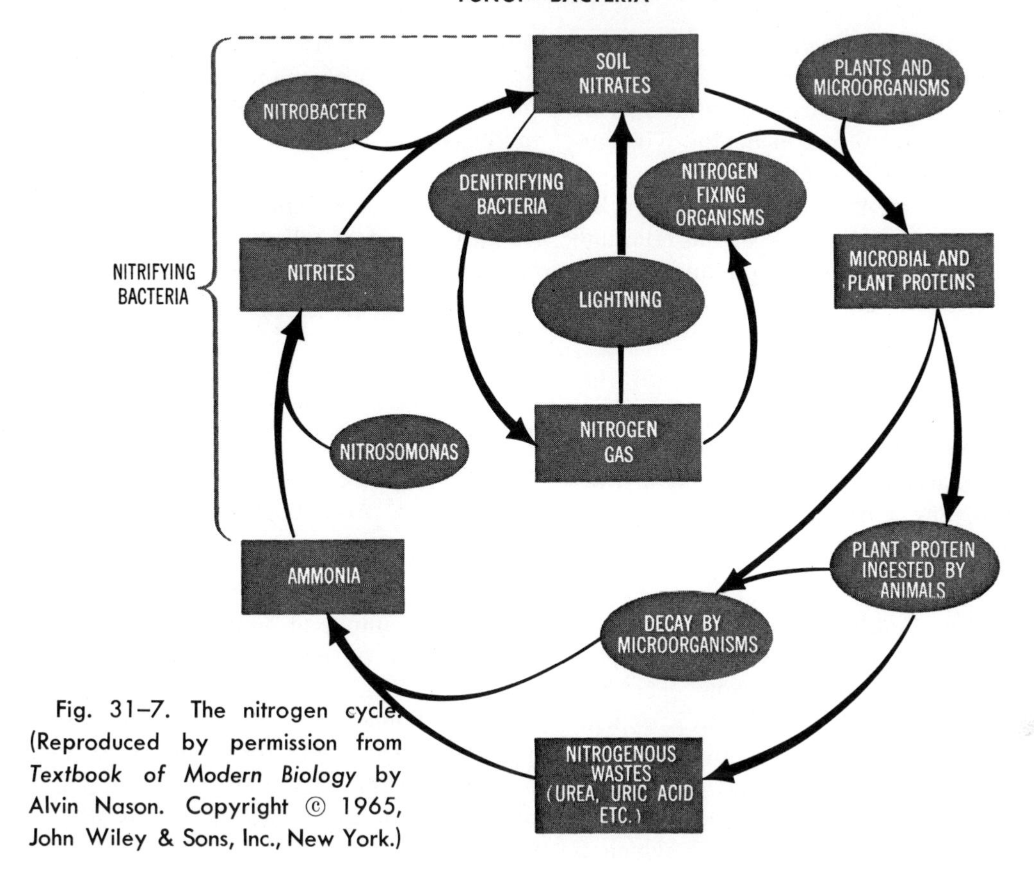

Fig. 31–7. The nitrogen cycle. (Reproduced by permission from *Textbook of Modern Biology* by Alvin Nason. Copyright © 1965, John Wiley & Sons, Inc., New York.)

monium and nitrate salts, can be used by green plants. Let us consider how available nitrogen is added to soil, its transformations in the soil, its use and loss—in other words, the nitrogen cycle (Fig. 31–7). A small amount of nitrogen gas may be converted into nitrogen compounds by lightning; a much greater amount, by nitrogen-fixing bacteria. As animals feed on plants, the organic nitrogen is incorporated in their bodies. As dead plants and animals and excreta decay, ammonia is added to the soil. Some ammonium nitrogen may be utilized directly by green plants, but much of it is changed by the nitrifiers, *Nitrosomonas* and *Nitrobacter*, into nitrates. Nitrates are used by green plants which may be eaten by animals. In farming areas not all of the available nitrogen is returned to the soil; much goes to market with the plant or animal crop. Denitrifying bacteria change a part of the nitrate into nitrogen gas that diffuses to the atmosphere. All along the nitrogen path, available nitrogen may be lost by erosion and leaching. Some of this available nitrogen finds its way into streams and lakes and the ocean where it may be utilized by algae, then by insects, and later by fish.

Sewage disposal. We are increasingly concerned with maintaining our rivers and lakes free of pollution so that they will retain their wildlife, potability, and recreational values. Sewage is one major source of pollution. Biological methods for sewage disposal have been developed which do not result in the pollution of our waters. In one type of system the raw sewage flows through a large tank in which 50 to 70 per cent of the solids in sewage settle out. The solids are periodically pumped into another tank where slow anaerobic decomposition occurs. After this treatment the remaining material may be used as fertilizer or disposed of in some other way. The liquid effluent from the first settling is sprayed on a trickling filter, which is a large bed of coarse rock about 6 feet in depth. The surfaces of the rocks are covered with bacteria which decompose the organic matter in the aerated effluent. After percolating through the filter, the liquid flows into a final settling tank. It is then chlorinated to kill pathogenic bacteria, after which it flows into a lake, river, or ocean.

Industrial products. Bacteria are used on a commercial scale to produce acetic acid, lactic acid, acetone, and butyl alcohol. Bacteria also play a role in the retting of flax and hemp, in tanning hides, and in the curing of tobacco.

Antibiotics. Antibiotics are chemicals of biological origin that destroy or prevent the growth of pathogenic organisms. The first antibiotic to be used in medicine was penicillin, which was discovered by Alexander Fleming in 1929 and first used to cure patients by Chain and Florey in 1940. Penicillin is obtained from a blue mold, which is a true fungus in the phylum Eumycophyta. Since 1940 other antibiotics have been discovered, and several have been isolated from bacteria—some from true bacteria, others from bacteria in the order Actinomycetales (the fungus-like bacteria). Tyrothricin and bacitracin are obtained from species of *Bacillus*, a true bacterium. About fifteen antibiotics have been isolated from species of *Streptomyces*, a fungus-like bacterium. Among the medicinally useful antibiotics obtained from various species of *Streptomyces* and the dates of their discovery are: streptomycin, *Streptomyces griseus*, 1944; chloromycetin, *Streptomyces venezuelae*, 1947; aureomycin, *Streptomyces aureofaciens*, 1948; terramycin, *Streptomyces rimosus*, 1950.

Streptomyces is in the order Actinomycetales, whereas the true bacteria are in the order Eubacteriales. Unlike the true bacteria, the actinomycetes have a body resembling that of a true fungus, a plant body composed of narrow branching threads. *Streptomyces*, a common soil bacterium, reproduces by spores, called conidia, that are produced in chains at the ends of the branches. *Actinomyces* has a threadlike body when young, and at maturity breaks up into individual cells that resemble coccus or bacillus forms. Many species of *Actinomyces* are saprophytes, but others are parasites; among the latter are *Actinomyces bovis*, which causes the lumpy jaw disease of cattle, and *Streptomyces scabies*, which causes the common scab of potatoes.

COUNTING AND CULTURING BACTERIA

It would be extremely tedious to count under the microscope the number of bacteria in a sample of milk, water, or soil. Bacteria may be counted without a microscope. Bacterial counts are made by pouring diluted milk, water, or other material into a petri dish which contains sterilized agar and nutrients. After a few days colonies of bacteria appear in the dish; each colony started from one bacterium. The number of bacteria in the substance tested is obtained by counting the colonies.

RICKETTSIAS

Rickettsias are intermediate in size between the smallest bacteria and the largest viruses, varying from 0.2 to 0.5 micron. The electron microscope reveals that they are shaped like a small bacillus. Rickettsias cannot be cultured apart from living cells, resembling viruses in this feature. Rickettsias inhabit lice or ticks and may be transmitted to man by bites of these organisms. Typhus, trench fever, Rocky Mountain spotted fever, and undulant fever are serious diseases caused by rickettsias.

VIRUSES

In 1892 D. J. V. V. Iwanowski was studying a serious infectious disease of tobacco, termed tobacco mosaic. He extracted the

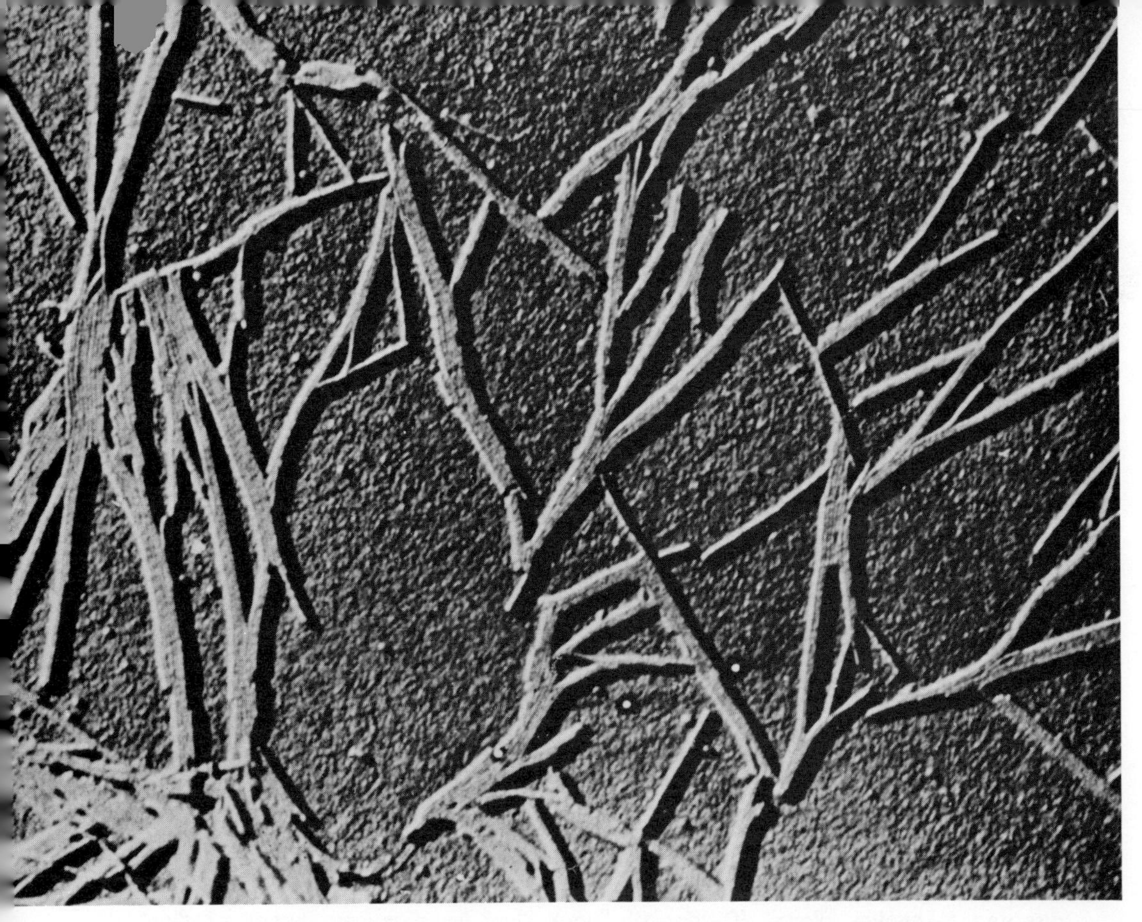

Fig. 31–8. Tobacco mosaic virus photographed with an electron microscope. The virus particles are magnified by 43,200. (R. W. G. Wyckoff, *Electron Microscopy*.)

juice from diseased plants and passed the extract through a filter so fine that the bacteria would not pass through. The filtrate, free of bacteria and fungi, still had the capacity to induce disease in healthy plants. This was the first indication that some agents smaller than bacteria could cause disease. These minute agents have been termed **viruses.** All are strictly parasitic. They cannot be cultivated apart from living cells. They multiply only in living tissue. What is the nature of viruses? Are they living or nonliving? A group of biologists will argue these questions for hours. Some biologists maintain that viruses are nonliving, and they cite the evidence obtained with the electron microscope.

Such studies and others indicate that viruses are large nucleoprotein molecules, each made of a core of nucleic acid surrounded by protein. Most viruses which cause human diseases have DNA cores, but a few have RNA cores; among them are the viruses causing mumps and influenza. Evidence has indicated that plant viruses are characterized by RNA cores. In a few instances the virus nucleoprotein has been crystallized, and this crystalline nucleoprotein causes disease when inoculated into a susceptible organism. Scientists have succeeded in separating the protein from the RNA of the tobacco mosaic virus (Fig. 31–8). The protein component by itself would not produce tobacco mosaic disease. The RNA by itself was infectious but much less so than the intact virus. The nucleic acid fraction is only 0.1 per cent as infectious as the complete virus. Several strains of tobacco mosaic virus exist, each characterized by a specific kind of protein and RNA. Under certain experimental conditions the RNA of one strain became sur-

Fig. 31–9. Transmission of disease. *Left*, curly top of sugar beets. *Right*, a healthy plant. The virus will be transmitted to the healthy plant by the parasitic stringlike dodder plant. Dodder plants are frequently used in experimental studies to transmit viruses from diseased to healthy plants. (Division of Sugar Plant Investigations, Bureau of Plant Industry, U.S.D.A.)

rounded by protein from a different strain, thus forming a hybrid between the nucleic acid of one strain and the protein of a different one. However, when the hybrid replicated itself, the progeny possessed the protein characteristic of that originally associated with the RNA, proving that the genetic information for protein synthesis is carried from generation to generation by the nucleic acid.

Other biologists believe that viruses are submicroscopic living things. What characterizes life? Living things reproduce their kind, respond to environmental changes, and mutate. If these are criteria of life, then viruses are alive. They reproduce in cells of the host, they mutate, and they respond to changes in the environment. They respond much like living cells to adverse chemical and physical agents. They are killed by heat, ultraviolet irradiation, and by some disinfectants. They have form as well. The shape of the virus causing colds in man is unlike the one which causes tobacco mosaic. The viruses causing influenza and polio have a cubical shape, whereas the virus causing tobacco mosaic is cylindrical. Viruses which parasitize bacteria have a prismatic head attached to a cylindrical tail. Viruses originate from pre-existing viruses. There is not a shred of evidence that they spring up spontaneously. Like higher organisms, they have had a long evolutionary history.

All of us have harbored one or more of these submicroscopic infective agents, and they have caused us considerable discomfort. The common cold, influenza, measles, mumps, chicken pox, smallpox, yellow fever, and polio are caused by viruses. Sometimes a virus which causes a disease mutates into a more infective and virulent form. These mutants, new forms of old ones, may cause epidemics. After infection the body produces antibodies which combat the virus. If the infection is mild, we may experience little or no discomfort. The immunizing agent for smallpox is an active virus, a variant of the virulent form, which produces only trivial symptoms but which stimulates the body to form antibodies which are effective against the virulent virus as well as the mild one. We may become infected with one or another virus without noticeable symptoms. Many individuals have been infected with polio without realizing it. Antibody analysis shows that there are one hundred hidden cases of polio to every visible one.

Many serious plant diseases are caused by viruses. Some of these are tobacco mosaic, curly top of sugar beets, leaf roll of potatoes, peach yellows, aster yellows, dahlia stunt disease, nasturtium mosaic, and yellow stripe of narcissus. In many virus diseases of ornamental plants there is a mottling of the flowers and gradual degeneration of the plants. Virus diseases are systemic—that is, they affect the whole plant. Hence the removal of parts of the plant will not control the disease. To keep other plants from contracting virus disease, ruthless elimination of the diseased plants is necessary. Virus diseases of plants are transmitted by insects and by contaminated seed and tools. Control involves such practices as control of insect vectors, use of disease-free seed, eradication of diseased plants, and the breeding of resistant varieties.

BACTERIAL VIRUSES—BACTERIOPHAGES

All living organisms are susceptible to one or another disease. Bacteria are no exception. They may become infected with viruses which are often called **bacteriophages** (Fig. 31–10). A typical bacteriophage consists of a hollow head attached to a hollow tail. The walls of the head and tail are made of protein. The nucleic acid, usually

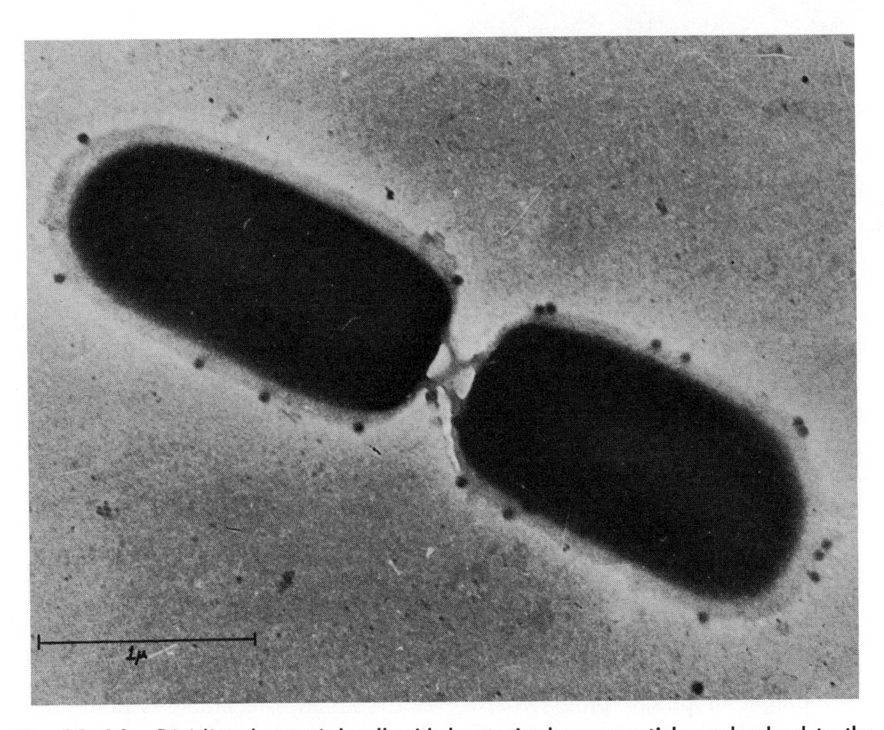

Fig. 31–10. Dividing bacterial cell with bacteriophage particles adsorbed to the cell walls. (S. E. Luria, M. Delbruck, and T. F. Anderson, Society of American Bacteriologists.)

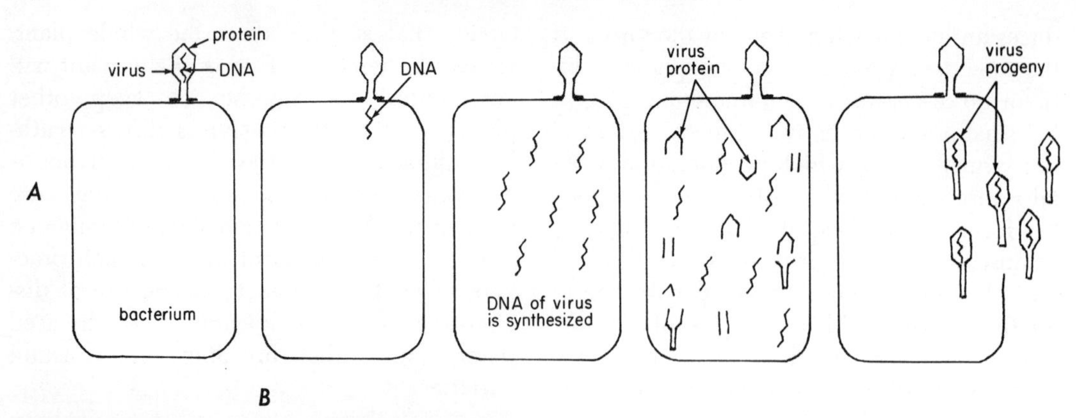

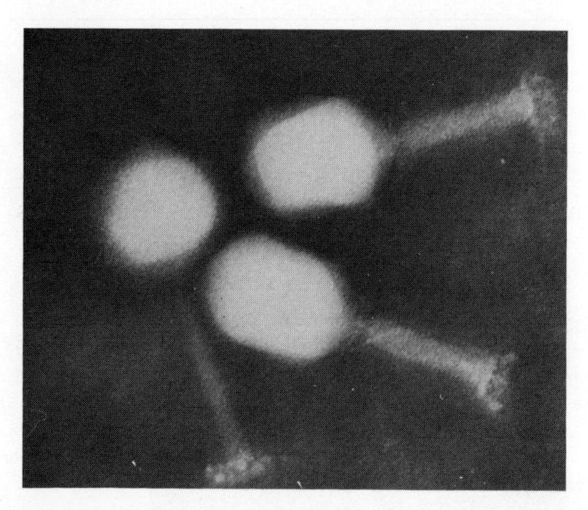

Fig. 31–11*A*. Life cycle of a virus, specifically a bacteriophage. *B*. Electron micrograph of a bacteriophage. (P. F. Davison.)

DNA, is present in the bulbous head. In the initial stage of infection the tail of the virus becomes attached to the bacterium. At the contact point the bacterial wall is digested, thus enabling the viral nucleic acid to enter the bacterial cell; the empty protein sheath is left behind. Shortly after infection, the viral DNA takes over the direction of the bacterium, turning off the synthesis of bacterial DNA and RNA and turning on the synthesis of viral DNA. After a pool of viral DNA molecules has formed, they direct the synthesis of viral protein which then encases each DNA unit to form a new virus. Clearly, the genetically important component of a phage is the DNA, the only virus constituent which enters the cell. The DNA carries all the information required for the development of phage progeny. As the new virus particles accumulate, the bacterium dies and the cell collapses, and one hundred or more virus particles are liberated (Fig. 31–11).

Transduction

Some bacteriophages, the temperate phages, kill and disrupt only a small percentage of infected bacterial cells. Temperate phages often serve as vehicles which transport bacterial DNA from one bacterium, the donor, into another bacterium, the recipient. As the DNA of the phage multiplies in the donor bacterium, a strand of bacterial DNA may, on occasion, become joined to the phage DNA and may even replace a portion of the phage DNA. In time the DNA strand consisting of phage and bacterial DNA becomes coated with protein to form a phage. After the bacterial cell is killed, the phage is liberated and may then infect another bacterium. After the strand of DNA, part phage DNA and part bacterial DNA, enters the cell, the segment of bacterial DNA from the donor may join with the bacterial DNA of the recipient. Essentially, bacterial DNA has been transferred from one bacterial cell to another via the phage; this process is called transduction.

The temperate phage may not kill the recipient bacterium, which then multiplies. The progeny will inherit the usual DNA plus that which was brought in by the phage. For example, one strain of bacteria may be resistant to streptomycin; a different strain may be sensitive. A phage infects a resistant cell, and the segment of bacterial DNA responsible for the resistance is linked to the virus DNA. Later the virus escapes and infects a bacterium which is sensitive to streptomycin. After the DNA from the resistant strain is incorporated into the bacterial DNA, the previously sensitive cell becomes resistant to streptomycin. The progeny of the transformed cell will also be resistant.

QUESTIONS

1. Give the general characteristics of the Schizomycophyta.

2. _____ Bacteria do not (A) produce spores, (B) have cell walls, (C) have a definite nucleus, (D) reproduce by fission, (E) have any kind of sexual reproduction.

3. In what ways are blue-green algae like bacteria? How do they differ?

4. _____ The three general shapes of bacteria are technically called (A) coccus, bacillus, spirillum; (B) streptococcus, diplococcus, staphylococcus; (C) spheres, rods, corkscrews; (D) saprophytes, parasites, epiphytes.

5. _____ A chain of coccus bacteria is called (A) bacillus, (B) staphylococcus, (C) streptococcus, (D) sarcinea.

6. Are bacteria immortal?

7. Could some bacteria live on earth if there were no other organisms present?

8. How do bacteria move?

9. What environmental factors affect the distribution of bacteria?

10. _____ Those bacteria which thrive where the temperature is between 68° and 104° F. are (A) thermophilic bacteria, (B) mesophilic bacteria, (C) psychrophilic bacteria.

11. _____ Bacteria which grow either in the presence or absence of air are known as (A) obligate aerobes, (B) obligate anaerobes, (C) facultative anaerobes.

12. _____ Most bacteria which grow in the rumen of cattle and sheep (A) are harmful and should be eliminated, (B) may synthesize amino acids which then become available to the animal, (C) are unable to synthesize vitamins.

13. _____ Bacteria lack (A) cell walls, (B) chromatin bodies, (C) fibrils of DNA, (D) mitochondria, (E) plasma membranes.

14. _____ Penicillin prevents normal cell-wall formation in certain bacteria. (A) True, (B) false.

15. List Koch's postulates.

16. List some methods for preserving foods and explain why each is effective.

17. List some beneficial activities of bacteria.

18. List some harmful activities of bacteria.

19. _____ Which one of the following is false? (A) Pasteur discovered the principle of vaccination. (B) The protective substances formed by the body to combat disease are called toxins. (C) Temperatures higher than that of boiling water are required to kill certain bacterial spores. (D) Bacteria which cause ring rot of potatoes may be transmitted from diseased tubers to healthy ones when the potatoes are cut into seed pieces.

20. _____ In which one of the following diseases are the symptoms delayed until the bacteria have become established and multiplied in the gastro-intestinal tract? (A) Botulism, (B) *Micrococcus* food poisoning. (C) *Salmonella* food infection.

21. _____ Streptomycin, aureomycin, and chloromycetin are obtained from (A) fungus-like bacteria, (B) true bacteria, (C) true fungi.

22. Outline the sulfur cycle.

23. _____ Which one of the following is false? (A) Sulfur locked up in protein molecules is released through bacterial action as hydrogen sulfide. (B) *Thiobacillus* and *Beggiatoa,* bacteria which oxidize hydrogen sulfide, carry on photosynthesis. (C) Sulfur bacteria which have bacteriochlorophyll make food from hydrogen sulfide and carbon dioxide. (D) Certain sulfur bacteria carry on chemosynthesis.

24. _____ With respect to the nitrogen cycle, which one of the following statements is false? (A) Through decay, proteins break down into amino acids. When an amino acid decomposes, at least a part of the nitrogen appears as ammonia (NH_3). Bacteria in the soil may change this ammonia into nitrite nitrogen (NO_2), and still others change it into nitrate nitrogen (NO_3), which is the form usually absorbed by green plants. (B) Bacteria which grow in nodules on legumes change the atmospheric nitrogen into compounds of nitrogen which higher plants may use. (C) Some bacteria (*Azotobacter* and *Clostridium*) which live free in the soil may change atmospheric nitrogen into compounds of nitrogen. (D) The process of changing nitrates into gaseous nitrogen is called nitrification.

25. In 1957 Asiatic influenza spread among the people of many countries. Why did not the people have a natural immunity to this influenza?

26. _____ Milk is pasteurized by heating it at a temperature of 140° to 149° F. for a period of 20–30 minutes. The milk is pasteurized to (A) kill all bacterial spores; (B) kill active bacteria, some of which may cause disease; (C) kill the algae which may be present; (D) make the milk taste better.

27. _____ Which one of the following is false? (A) Virus diseases of plants are systemic. (B) Viruses are nucleoproteins. (C) Viruses are easily seen with the ordinary microscope. (D) Viruses reproduce in cells of the host.

28. Are viruses living organisms?

29. _____ Typhus, Rocky Mountain spotted fever, and undulant fever are caused by (A) rickettsias, (B) bacteria, (C) viruses.

30. List some diseases of man which are caused by viruses.

31. How may virus diseases of plants be controlled?

32. Describe an experiment using a virus which proves that the genetic information for protein synthesis is carried from generation to generation by nucleic acid.

32

Slime Molds and True Fungi

The slime molds (Myxomycophyta) and the true fungi (Eumycophyta) are not related; they are considered together in this chapter only for convenience. In contrast to the bacteria the slime molds and true fungi have relatively large cells and organized nuclei.

SLIME MOLDS—MYXOMYCOPHYTA

There are about three hundred species of slime molds. Their relationship with other plants is obscure. The slime molds are classified in three classes: Myxomycetes, Acrasieae, Labrinthuleae. The Myxomycetes, the largest class, have a multinuclear **plasmodium** which lacks cross-walls, and they reproduce by spores formed in **sporangia.** Both the Acrasieae and Labrinthuleae have plant bodies, known as **pseudoplasmodia,** consisting of cells, each with a cell wall. The Acrasieae, commonly called cellular slime molds, produce spores in sporangia, whereas the Labrinthuleae lack sporangia.

Although most slime molds are saprophytes, a few are parasitic on higher plants. For example, *Plasmodiophora brassicola* causes club-root of cabbage and *Spongospora subterranea* causes powdery scab of potatoes.

Myxomycetes

Rotting logs or decaying stumps are favorite haunts of the slime molds. On such surfaces the white, yellow, red, or other colored slime molds may be found creeping along by a flowing motion and ingesting solid masses of food, much in the manner of animals. Their usual food consists of bacteria, protozoans, fungi, and organic matter. A careful examination of an active slime mold will reveal that it consists of a naked, fluid, slimy mass of protoplasm, termed a plasmodium, which may range from the size of a pinhead to as large as 4 inches in diameter. The plasmodium is a large vacuolated mass of cytoplasm in which numerous nuclei, Golgi bodies, mitochondria, endoplasmic reticula, and polyribosomes are embedded.

When the food or water supply becomes reduced, the plasmodium moves to a dry place, secretes a wall, and develops one or

A

B

Fig. 32–1. Sporangia of slime molds. *A. Didymium olavus. B. Physarum decipiens.* (C. J. Alexopoulos.)

more reproductive bodies, the sporangia—the exact shape depending on the species (Fig. 32–1). In some species the sporangia are stalked; in others they are sessile.

Within a sporangium many diploid nuclei undergo reduction division, forming haploid nuclei. Each haploid nucleus becomes surrounded by cytoplasm and a cell wall to form a spore. The spores are borne within the meshes of a delicate network, the **capillitium.** When the spores are mature, the sporangial wall breaks open, releasing the spores and exposing the capillitium. Each spore germinates to form a motile zoospore, or **swarm spore** as it is sometimes called. A swarm spore is somewhat pear-shaped and bears two flagella at the pointed end (Fig. 32–2). The swarm spores may divide one or more times, but in time the progeny fuse in pairs to form zygotes. A single zygote may grow into a new plasmodium, or several zygotes may fuse to form a plasmo-

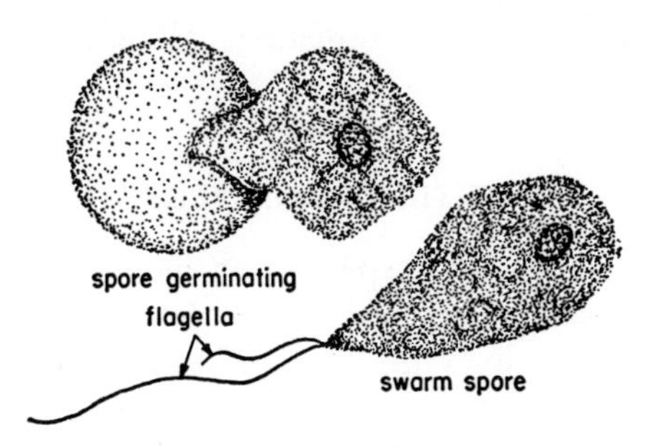

Fig. 32–2. A slime mold spore germinates to form a swarm spore.

dium. The essential features of reproduction are diagrammed in Figure 32–3.

Acrasieae

Among the Acrasieae, commonly called cellular slime molds, is *Dictyostelium discoideum,* which has been used extensively in studies of development. The pseudoplasmodium of *Dictyostelium* superficially resembles a minute worm and consists of hundreds or thousands of uninucleate cells aggregated together. The wormlike pseudoplasmodium glides over the surface of a slime layer which it secretes. After a period of moving and feeding on bacteria, the pseudoplasmodium becomes organized into a fruiting body, about 1 or 2 millimeters high, consisting of a stalk bearing a fruiting tip. In the development of the fruiting body the front cells of the pseudoplasmodium form the stalk. The rear cells move up the stalk and develop into a fruiting tip containing thousands of spores, each with a cell wall. After liberation the spores germinate into amoeba-like cells, known as **myxamoebae.** The myxamoebae move, feed on bacteria, and increase in numbers by cell division. When the food supply is exhausted, the myxamoebae cease their independent existence and aggregate into a pseudoplasmodium, a multicellular plant body.

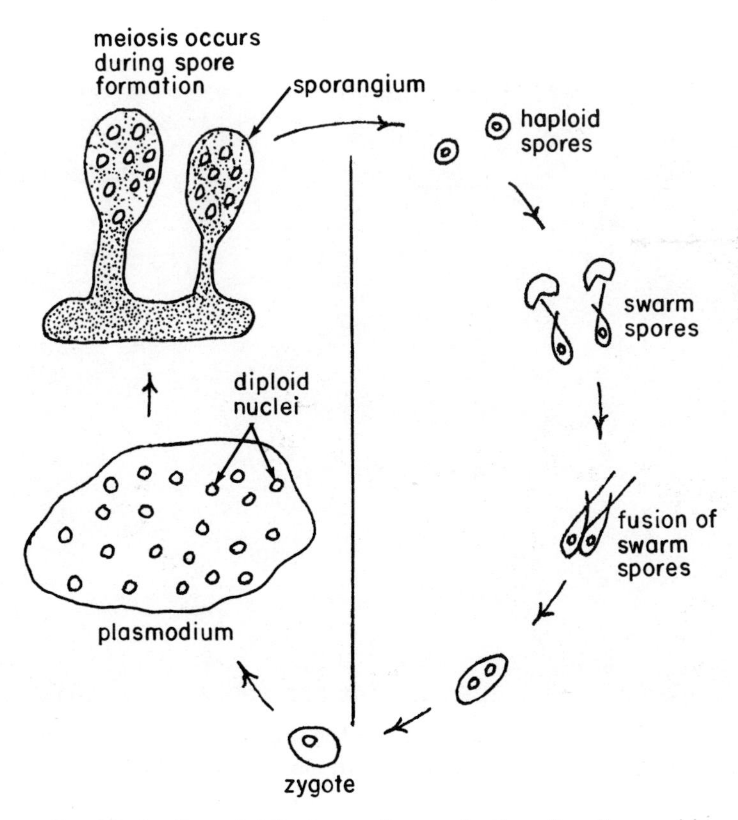

Fig. 32–3. Essential features of reproduction of a slime mold.

Of special interest to students of plant development is the stimulus which brings about the aggregation of cells, with the consequent loss of their individuality and the assumption of a subordinate role in a coordinated pseudoplasmodium (Fig. 32–4). Just one myxamoeba in a population may govern the behavior of its neighbors. This cell secretes a hormone, called **acrasin,** which powerfully attracts and orients other cells. The neighboring cells migrate toward the secreting cell and then aggregate to form a pseudoplasmodium. Acrasin also alters the surfaces of the cells, the myxamoebae, in such a manner that cells adhere to one another.

The transformation of free-living myxamoebae into a multicellular plant which produces spores favors survival and dispersal. A clone of myxamoebae that has depleted the food supply forms a pseudoplasmodium that develops an elevated fruiting body. The liberated spores are disseminated by wind and many reach new environments.

TRUE FUNGI—EUMYCOPHYTA

About 75,000 species are included in the phylum Eumycophyta. A few species of Eumycophyta are aquatic, but most are land plants which reproduce by spores that are disseminated by wind, splashing rain, and insects. Most true fungi produce two or more kinds of spores. Some spores are produced asexually, and others form after gametes unite.

Practically all members of the phylum Eumycophyta have a cobweb-like body. The body is made up of filaments known as **hyphae** (singular: **hypha**). The whole mass of hyphae which makes up the body of a fungus is called the **mycelium.** The cells making up a hypha are structurally similar to those of the higher plants, and each has a cell wall composed of cellulose

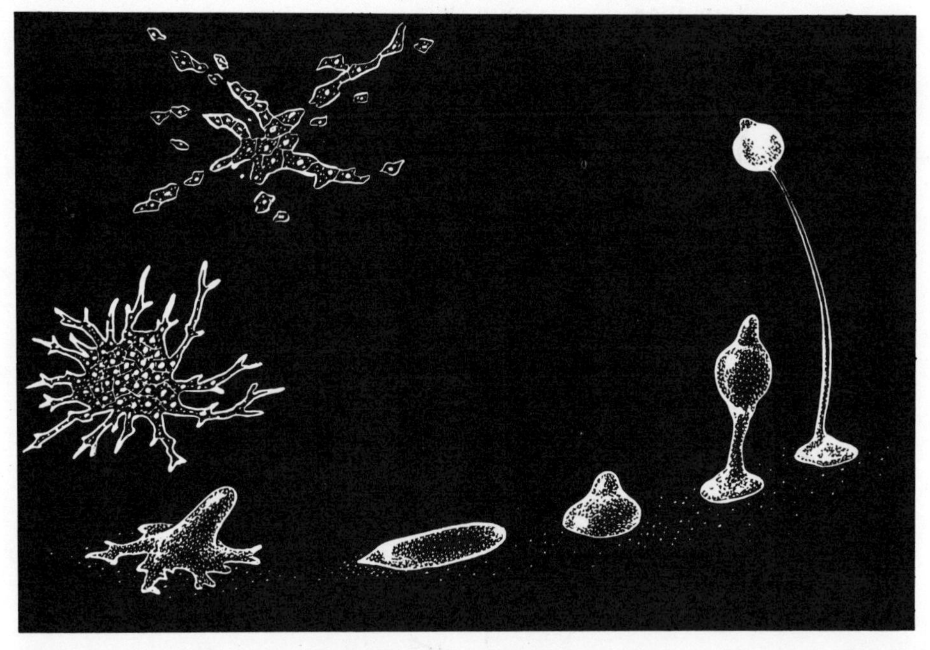

Fig. 32–4. Aggregation of myxamoebae of the cellular slime mold, *Dictyostelium,* into a pseudoplasmodium which later becomes organized into a fruiting body. (Used by permission of Dodd, Mead & Co. from *Biology in Action* by N. H. Berrill. Copyright © 1966 by Dodd, Mead & Co., Inc., New York.)

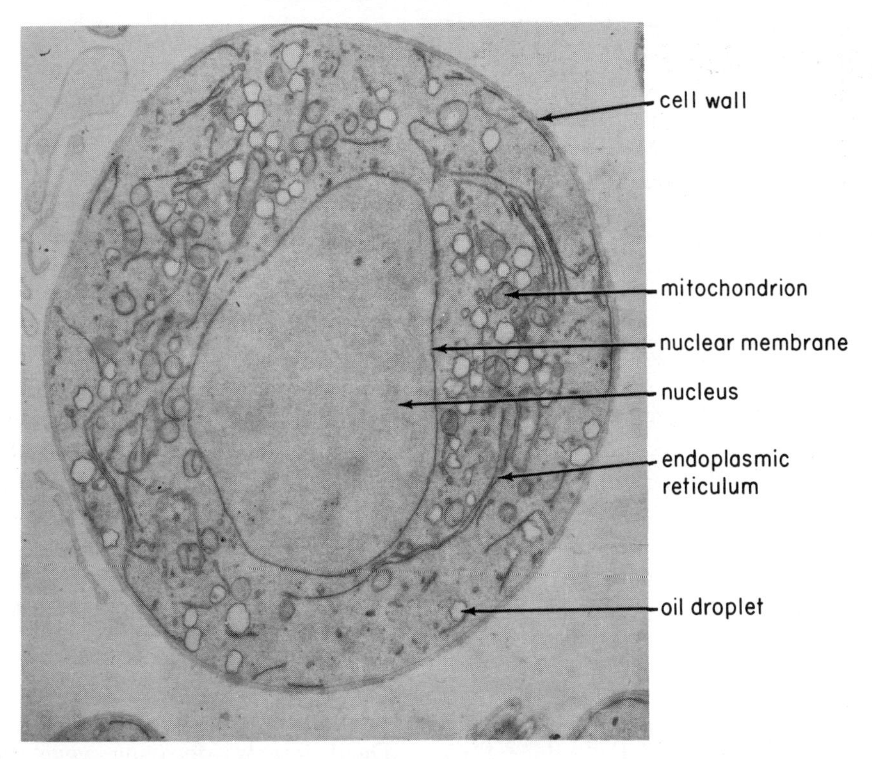

Fig. 32–5. Ultrastructure of a fungal cell, specifically of a probasidium of *Exidia nucleata*. (Kenneth Wells.)

or chitin, a plasma membrane, vacuoles, mitochondria, an endoplasmic reticulum, and a nucleus whose membrane is pierced by minute pores (Fig. 32–5).

The true fungi are separated into four classes: Phycomycetes (alga-like fungi), Ascomycetes (sac fungi), Basidiomycetes (basidium fungi), and the Deuteromycetes (imperfect fungi). In the discussion which follows (in this and the next two chapters), we shall discuss only the first three classes. Species which cannot be placed with certainty in one or another of these classes are placed in the Deuteromycetes. Before a fungus can be classified, it is necessary to know its mode of sexual reproduction. If the sexual stage of a fungus is unknown, it is placed in the Deuteromycetes class. When the method of sexual reproduction is discovered, the fungus is transferred out of the Deuteromycetes into the appropriate class. About 10,000 species are still classified as imperfect fungi (meaning that their sexual stage is unknown). Many of them are saprophytes, yet some cause serious plant diseases, and a few cause such diseases of human beings as athlete's foot (ringworm).

We are uncertain about the origin of the true fungi. Many mycologists believe that the true fungi have had a common origin and that all of them should be grouped in one phylum, a procedure which is followed in this book. The Phycomycetes are considered to be the most primitive of the true fungi. The Ascomycetes probably evolved from Phycomycete-like ancestors; and the Basidiomycetes, from the Ascomycetes.

ALGA-LIKE FUNGI—PHYCOMYCETES

Water molds, blights, downy mildews, and bread molds are representative Phycomycetes. Some are saprophytic; others

Fig. 32–6. *Saprolegnia* growing on hemp seeds in water. The swollen tips of the hyphae are sporangia.

are parasitic, causing serious plant diseases.

Our knowledge of the causal agents of plant diseases has been acquired during the past century. Sometimes it takes a major catastrophe to shake people out of old habits of thinking and doing. One such catastrophe led to the development of the science of **plant pathology,** the study of plant diseases and their control. The calamity occurred in 1845 when a blight swept through the potato fields of Ireland. As a consequence of the blight, 250,000 people out of a population of 8 million died of starvation, and another million and a half emigrated to the United States. People everywhere were aroused and insisted on more knowledge as to the causal agent. In 1846 Rev. M. J. Berkeley associated the diseases with a "mold" on the affected tissue, and put forward the revolutionary theory that the "mold" was the cause of the disease. Since 1846 much progress has been made in understanding fungi which cause plant diseases, and, because of this understanding, it has been possible to develop control methods. The fungus which causes blight ("late blight," as it is now called) of potatoes is the Phycomycete *Phytophthora infestans.* Before discussing it, the characteristics of the group and other examples will be given.

Members of the Phycomycetes are referred to as the alga-like fungi because they resemble some algae in their mode of reproduction and in their vegetative appearance. Like the alga *Vaucheria,* all Phycomycetes have vegetative filaments lacking cross-walls; that is, they are coenocytic. The absence of cross-walls in the mycelium enables one to distinguish readily the Phycomycetes from the Ascomycetes and Basidiomycetes, for these groups have septate hyphae; that is, cross-walls are present.

We shall discuss the following three orders of Phycomycetes: Saprolegniales (water molds), Peronosporales (white rusts, blights, and downy mildews), and the Mucorales (black molds).

Water Molds (Saprolegniales)

Dead insects, decaying vegetation, and hemp seeds placed in pond water may soon be completely covered and penetrated by the white mycelium of *Saprolegnia,* a water mold (Fig. 32–6). An examination of the mycelium under the microscope reveals that it is nonseptate and multinucleate; hence it is coenocytic and resembles, except for the absence of chloroplasts, the alga *Vaucheria.* Most species of *Saprolegnia* are saprophytic, but a few species are parasitic on fish, becoming troublesome at times in fish hatcheries and aquaria. *Saprolegnia* reproduces asexually by spores formed in sporangia and sexually by heterogametes.

Asexual reproduction. Some of the hyphae radiating from the food source become swollen at their tips, and concurrently cross-walls develop, partitioning the tips from the hyphae. Each tip is a sporangium in which many uninucleate spores are formed. The spores are discharged through a terminal pore. In many species each spore has two terminal flagella; these are called zoospores, and the case in which they develop is called a zoosporangium. After swimming for a while, a zoospore loses its flagella and becomes surrounded by a cell wall. Later it develops into a secondary zoospore with two lateral flagella. The secondary zoospore swims about, then comes to rest, and finally it develops into a new plant.

Sexual reproduction. The sex organs, antheridia and oogonia, are unicellular and are formed at the tips of some hyphae (Fig. 32–7). The contents of a young oogonium appear homogeneous, but later several large eggs develop in each. Many nonmotile sperms, each consisting of only a nucleus, are formed in an antheridium, a curved and somewhat tubular one-celled structure which is in contact with an oogonium. An **antheridial tube,** which penetrates the oogonium, develops from an antheridium. The sperm nuclei move through the antheridial tube, are released, and then one sperm unites with each egg. After fertilization each zygote becomes surrounded by a heavy wall, then remains dormant for several months, and finally germinates. During germination, meiosis occurs, and then each zygote, also known as an **oospore,** sends out a hypha which develops into the mycelium.

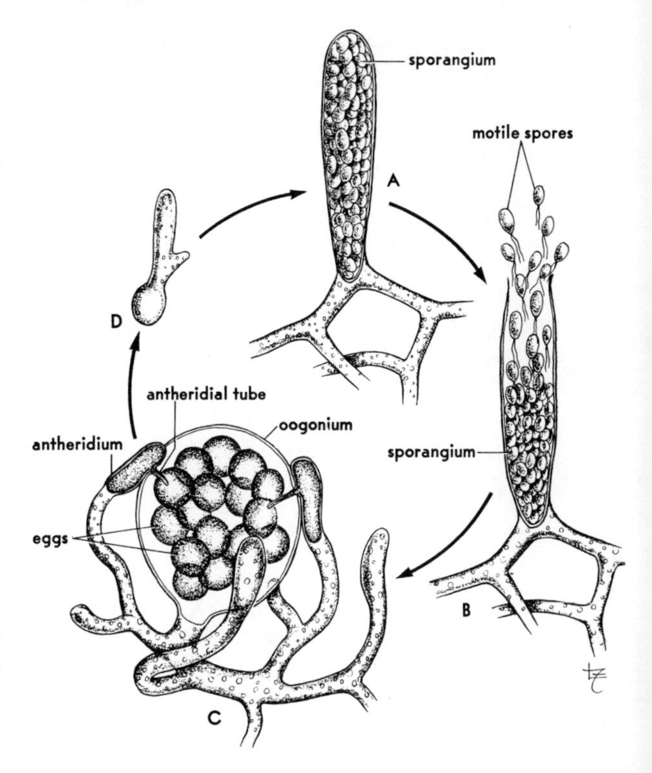

Fig. 32–7. Life cycle of *Saprolegnia*. *A.* Mycelium with sporangium. *B.* Motile spores leaving the sporangium. *C.* Mycelium with antheridia and oogonium. *D.* Zygote developing a mycelium. (From *Biology* by Relis B. Brown. Copyright © 1956, D. C. Heath and Co., Boston. By permission.)

Blights and Downy Mildews (Peronosporales)

Late blight of potatoes is caused by *Phytophthora infestans.* Most of the body of the fungus (mycelium) develops between the cells of leaves, stems, and tubers, but short hyphae penetrate the cells of the host and obtain food for the parasite. Many cells of the host are killed by toxins which are elaborated by the fungus.

Asexual reproduction. Some hyphae grow out through the stomata of leaves and branch in a characteristic manner. Each branched hypha is a **conidiophore.** At the tip of each branch a reproductive body known as a **conidium** (plural: **conidia**) is produced. The conidia are dispersed by wind and insects. Those which are deposited on leaves or other parts of a potato plant germinate either directly or indirectly if water is available. In direct germination a hypha which develops from the conidium penetrates through a stoma or directly through the epidermis. In indirect germination about eight biflagellate zoospores are produced in each conidium, after which the conidium breaks open and liberates the zoospores. The zoospores swim for a while in the water film on the leaf surface, and then each one produces a hypha which penetrates the host (Fig. 32–8).

The botanist is often in a quandary as to what terms to use for certain structures such as the asexual reproductive structures of *Phytophthora.* If the structures always

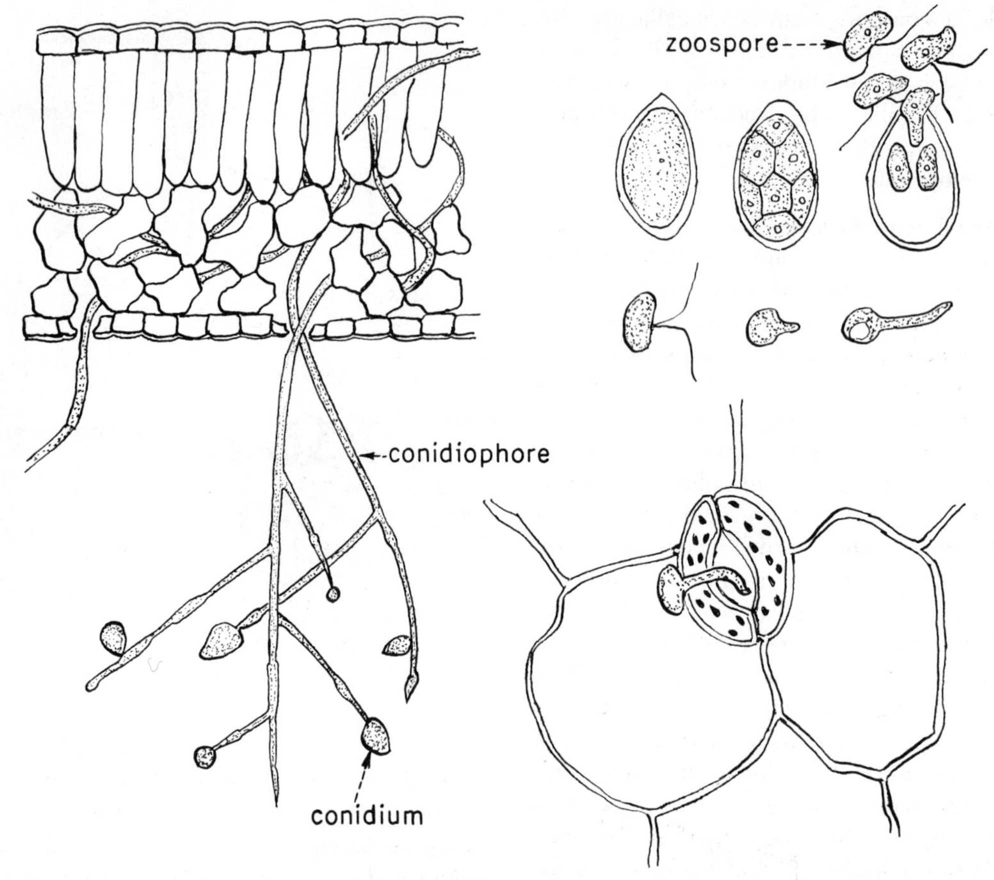

Fig. 32–8. Asexual reproduction of *Phytophthora infestans*. *Left*, a cross-section of an infected potato leaf, showing mycelium, conidiophores, and conidia. *Upper right*, a conidium producing zoospores. *Lower right*, a hypha entering a potato leaf through a stoma. (Modified from H. M. Ward.)

produced hyphae when they germinated, botanists would call them conidia. If they always produced zoospores, the structures which we termed conidia would be called sporangia, that is, cases in which spores develop. When they germinate in both ways, one may choose either term or coin a new one. One author suggests that the asexual reproductive structures be called **conidiosporangia.**

Sexual reproduction. On rare occasions *Phytophthora* reproduces sexually by producing oogonia and antheridia inside the host. Several nonmotile sperms are produced in an antheridium, and one egg is produced in each oogonium. A fertilization tube develops through which a sperm passes to the egg. A heavy wall develops around the zygote, which is liberated when the host tissue decays.

Control of late blight. A knowledge of the life history of the pathogen makes it possible to plan control measures. Because the mycelium is inside the host, it cannot be reached by sprays, and hence it is not possible to cure a diseased plant. Control of this disease is based on prevention of infection. If some chemical toxic to the fungus, such as Bordeaux mixture, is sprayed on the plants before they are infected, the hyphae produced by germinating spores are killed.

Other diseases. Other plant diseases are caused by different species in the order Peronosporales. Seedlings are often killed by a disease known as damping-off, which is caused by various species of *Pythium*. Other diseases caused by species in this order and the causal organisms are: white rust of crucifers (*Albugo candida*), downy mildew of grape (*Plasmopara viticola*), downy mildew of onion (*Peronospora destructor*), and downy mildew of lettuce (*Bremia lactucae*).

Black Molds (Mucorales)

The black bread mold, *Rhizopus nigricans* (also known as *Rhizopus stolonifer*), is a typical example of this group, and its spores are commonly present in the air. Unlike the Saprolegniales and Peronosporales, the Mucorales never produce flagellated spores, a feature characteristic of aquatic organisms.

If bread is moistened and kept in a covered dish for a few days, it usually becomes covered with the white filamentous mycelium of the bread mold (Fig. 32–9). Some hyphae are within the bread; others, the **stoloniferous hyphae,** grow over the surface. All hyphae are coenocytic; that is, cross-walls are lacking and nuclei are present throughout the plant body. The hyphae within the bread are much branched and have thin walls and dense cytoplasm. They secrete enzymes which digest

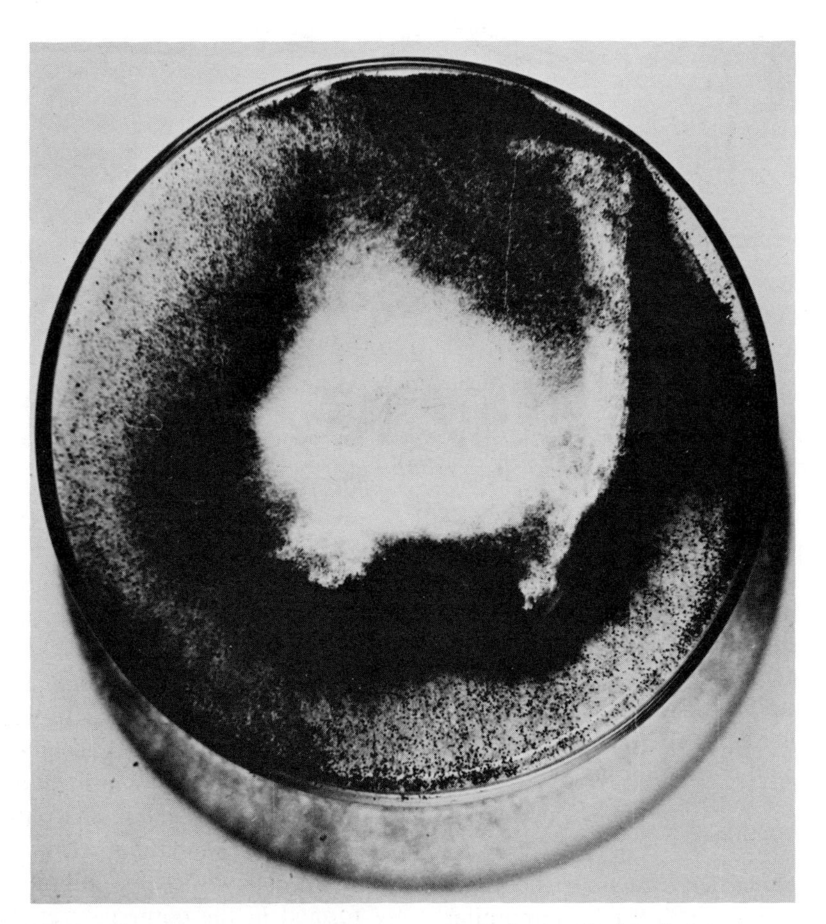

Fig. 32–9. *Rhizopus nigricans* growing on moist bread. The black dots are sporangia; the white cotton-like body is the mycelium.

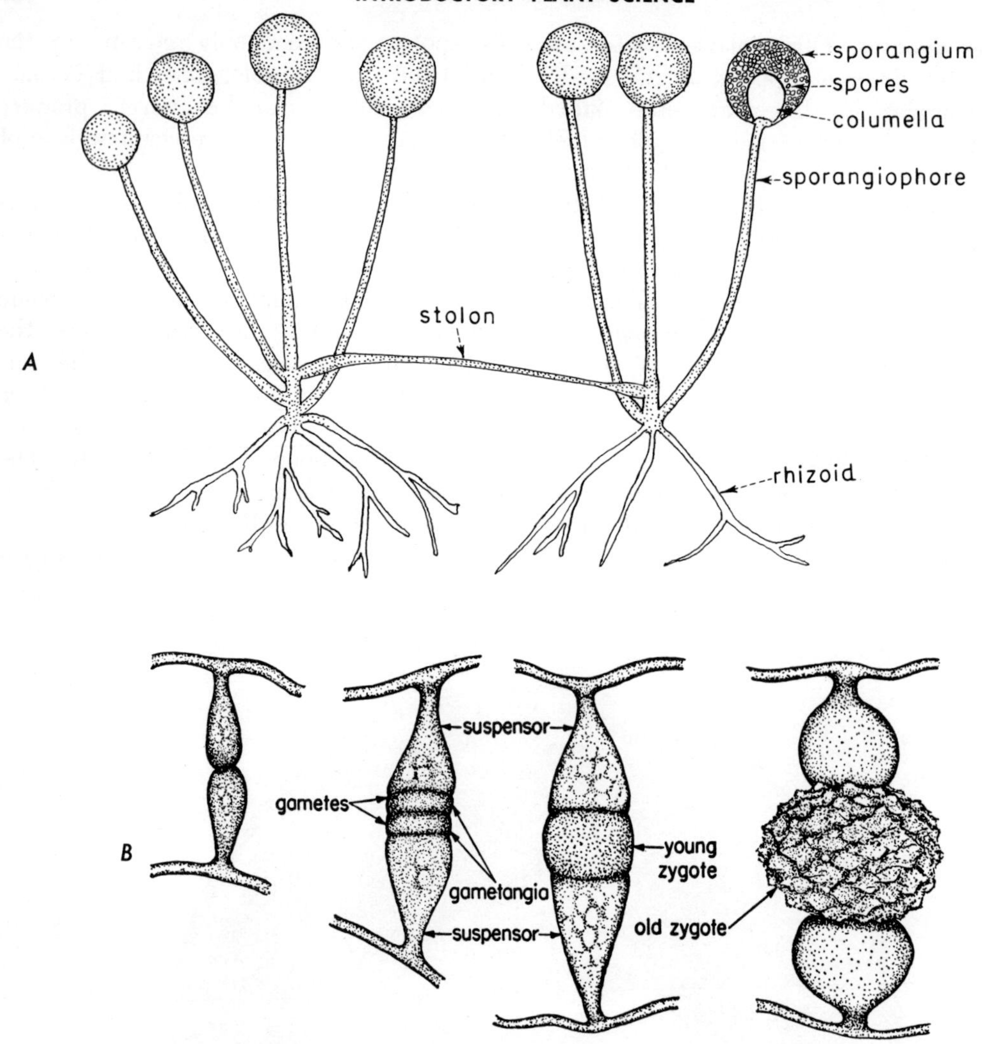

Fig. 32–10. ***Rhizopus nigricans.*** ***A.*** **Asexual reproduction.** ***B.*** **Sexual reproduction. Stages of reproduction in sequence from left to right.**

starch and protein to simpler compounds which are absorbed. The stoloniferous hyphae have thicker walls, are more vacuolated, and develop absorbing structures called **rhizoids** at intervals (Fig. 32–10*A*). *Rhizopus* reproduces asexually by spores formed in spherical cases, called sporangia, and sexually by the fusion of gametes, which are alike in size and form. In other words, *Rhizopus* is isogamous.

Asexual reproduction. Each of the black dots evident in Figure 32–9 is a **sporangium** containing many black spores, perhaps 70,000. A sporangium develops at the tip of an erect hypha known as a sporangiophore. The sporangiophores arise in clusters (Fig. 32–10*A*) from the stoloniferous hyphae, opposite the rhizoids. Let us trace the development of a sporangium. The tip of a sporangiophore swells to form a sporangium, and many nuclei accumulate at the periphery. Then a dome-shaped wall develops, which separates the outer, denser portion of the sporangium from the inner, clear part, known as the **columella.** Spores are formed in the outer part of the sporangium. When the spores are ripe, the sporangial wall is brittle and is easily ruptured by the expansion of the columella, permitting the spores to escape. The small, buoyant spores are dispersed even by slight air currents. The spores are not dormant, but germinate immediately on a suitable substratum. A spore produces a hypha, which soon branches and grows over and into the medium.

Sexual reproduction. Some bread mold plants are of one sex (the plus sex), and others are of the opposite sex (the minus sex). Hence, if just one spore is planted on a piece of moist bread or on nutrient agar in a petri dish, sexual reproduction does not occur. If two spores, one of one sex and the other of the opposite sex, are planted, sexual reproduction is achieved. A mycelium develops from one spore, and a similar mycelium develops from the other spore. When the mycelium of one sex approaches the mycelium of the opposite sex, short swollen hyphae develop from each. These come in contact with each other. Near the tip of each member of a pair of hyphae a cross-wall develops, thus forming a **gametangium** (case containing a gamete) at the end of a supporting structure, called a **suspensor** (Fig. 32–10*B*). Within each gametangium a gamete is organized. Next, the wall separating the pair of gametes is digested and the two gametes unite. The gametes are alike in shape; hence *Rhizopus* is isogamous. The gametes of *Rhizopus* are multinucleate instead of uninucleate, as in most plants. In fertilization the nuclei of opposite sexes associate in pairs and fuse, forming diploid nuclei. The product of the union of the two gametes is a zygote, which enlarges and becomes surrounded by a thick, black, warty wall. The zygote is highly resistant to adverse conditions and may remain dormant for many months. During germination, meiosis occurs and a hypha develops from the zygote. This hypha bears a sporangium containing many spores at its apex. The spores are haploid, and half of them are of one sex (the plus sex) and half are of the opposite sex (the minus sex).

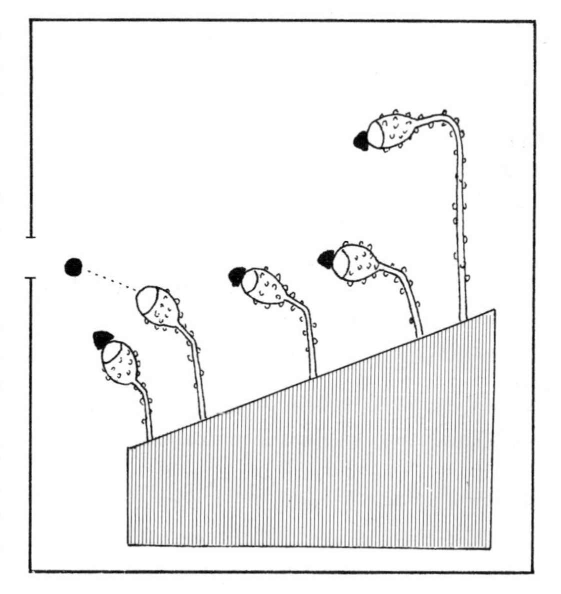

Fig. 32–11. *Pilobolus,* the fungus shotgun. The curvature of the sporangiophore aims the sporangium at the light source on the left. The sporangia are shot off with considerable force.

Significance of black molds. *Rhizopus* not only grows on bread, but also on fruits and vegetables, causing a number of damaging rots. Among the foods that may be spoiled by *Rhizopus* are sweet potatoes, tomatoes, strawberries, peaches, and grapes. Rotting can be minimized by sanitation in the packing shed, prevention of bruising, good aeration, and the maintenance of a low temperature during storage and transit.

Related to *Rhizopus* is *Pilobolus,* the fungus shotgun, which grows saprophytically on horse dung. Asexual reproduction is achieved by spores which develop in sporangia. The sporangia generally complete their development during the night and are explosively discharged at sunrise. The sporangiophores are aimed at a light source (Fig. 32–11). Then the bulbous enlargement just below a sporangium squirts a jet of fluid against the sporangium, which is carried away in the jet stream. If a sporangium lands on forage, a horse may ingest it. The spores released from the sporangium pass through the alimentary tract and emerge with the dung on which they germinate.

QUESTIONS

1. In what ways are slime molds like animals? In what ways are they like plants?

2. Give the distinguishing characteristics of the Myxomycophyta.

3. _____ Which one of the following statements about the Myxomycetes is false? (A) Spores are formed in sporangia. (B) Each spore germinates to form a motile swarm spore. (C) The swarm spores may divide one or more times, but in time the progeny fuse in pairs to form zygotes. (D) A plasmodium always develops from just one zygote.

4. _____ The cellular slime molds are in the class (A) Myxomycetes, (B) Acrasieae, (C) Labrinthuleae.

5. _____ Myxomycetes do not (A) ingest solid masses of food, (B) have plasmodia consisting of vacuolated masses of cytoplasm with many nuclei, (C) produce antheridia and oogonia, (D) form zygotes in their life cycle.

6. _____ Which one of the following statements about *Dictyostelium* is false? (A) The pseudoplasmodium becomes organized into a fruiting body consisting of a stalk and a spore-bearing tip. (B) The fruiting tip contains thousands of spores, each with a cell wall. (C) The spores germinate to produce myxamoebae which feed on bacteria and reproduce by cell division. (D) When the food supply is exhausted, one or more cells secrete auxin which attracts and orients other cells to form a pseudoplasmodium.

7. List the four classes of Eumycophyta.

8. Give the distinguishing characteristics of the Phycomycetes.

9. _____ The Phycomycetes (A) include molds, smuts, and rusts; (B) have hyphae which lack cross-walls; (C) resemble certain blue-green algae; (D) are all terrestrial fungi.

10. _____ The body of a fungus is called (A) conidiophore, (B) mycelium, (C) sporangium, (D) conidium.

11. _____ Phycomycetes which are parasitic on fish are in the order (A) Saprolegniales, (B) Peronosporales, (C) Mucorales.

12. Describe asexual and sexual reproduction of *Saprolegnia*, a water mold.

13. Discuss the historical significance of the disease known as "late blight of potatoes."

14. _____ The mycelium of *Phytophthora* is (A) mostly inside the leaf; (B) only on the outside of the leaf, and hence it can be killed by various sprays; (C) septate, that is, it has cross-walls.

15. _____ The asexual spores produced by *Phytophthora infestans* are called (A) conidia, (B) hyphae, (C) ascospores, (D) sperms.

16. _____ *Phytophthora infestans* has sexual organs resembling those of (A) *Vaucheria,* (B) *Spirogyra,* (C) *Protococcus,* (D) *Ulothrix.*

17. By means of drawings, show the structure and methods of reproduction of *Rhizopus nigricans.* Label fully.

18. _____ Sexual reproduction of *Rhizopus* resembles that of the alga (A) *Protococcus,* (B) *Chlamydomonas,* (C) *Spirogyra,* (D) *Oedogonium,* (E) *Vaucheria.*

19. _____ *Rhizopus* (bread or black mold) produces asexual spores in structures called (A) sporangia, (B) conidiophores, (C) suspensors, (D) zygotes.

33

True Fungi (Continued)

CLASS ASCOMYCETES

Blue molds used in making Roquefort cheese and for producing penicillin, yeasts which produce beer or bread, and fungi which cause peach leaf curl and chestnut tree blight are just a few of the 20,000 or so Ascomycetes. In the past 60 years the chestnut tree blight, caused by *Endothia parasitica,* has killed practically all the chestnut trees in the eastern part of the United States. The disease had been introduced from Asia into the United States at about the turn of the century. This disease illustrates how complete the destruction can be when susceptible plants are subjected to a new disease, and emphasizes the need and value of plant quarantine.

All Ascomycetes produce sacs in which four or eight spores are formed. The sacs are termed **asci** (singular: **ascus**), and the spores are called **ascospores;** hence the name Ascomycetes. With the exception of the one-celled yeasts, the mycelium of the Ascomycetes is septate.

In yeast plants the asci are borne singly, and in a few Ascomycetes they are arranged in layers. However, in most Ascomycetes the asci and associated cells form complex fruiting bodies which may be in the form of a cup (an **apothecium**), a sphere with a narrow opening called a **perithecium,** or a closed sphere known as a **cleistothecium.**

The class Ascomycetes is divided into sixteen orders, of which we shall discuss six: the Endomycetales, Taphrinales, Aspergillales, Erysiphales, Sphaeriales, and Pezizales.

YEASTS—ENDOMYCETALES

Yeasts are common, and they are abundantly present in soil; on decaying vegetation; on flowers, seeds, and fruits; and in the air. In the production of some wines wild yeasts are relied upon to bring about fermentation, but this naturally fermented fruit juice may or may not be satisfactory in quality. Most commercial wines are prepared with a special culture of yeast, *Saccharomyces cerevisiae* var. *ellipsoideus,* which produces fairly high amounts of alcohol and a pleasing flavor. Special varieties of yeasts are used for the production of beer, ale, and such distilled liquors as brandy, rum, and whiskey.

You will recall that yeasts respire anaerobically, utilizing sugar and producing carbon dioxide and alcohol. In the production of industrial alcohol and liquors the alcohol is the desired product. In the making of bread the carbon dioxide is the wanted product, for it leavens or raises the dough.

Unlike other Ascomycetes, yeast plants do not have a filamentous mycelium. A yeast plant is a one-celled organism with a cell wall, cytoplasm, a nucleus, and a large vacuole.

Asexual Reproduction

Yeasts reproduce by a process called **budding.** A bulge is formed on the parent cell and concurrently the nucleus divides. One nucleus passes into the bulge (**bud**), and the other remains in the parent cell. Later a constriction separates the bud from the parent cell, but the two cells may adhere for a time. The new cell may then bud, resulting in a small colony of yeast plants.

Sexual Reproduction

In yeasts of commercial importance the usual vegetative cells are diploid. At the time of sexual reproduction the diploid nucleus divides by meiosis to form four haploid nuclei, which become surrounded by cytoplasm and cell walls. The resulting spores, called **ascospores,** are retained for a while within the mother cell wall, which is then called an **ascus** (Fig. 33–2). Of the four ascospores, two are of one sex (the plus sex), and two are of the opposite sex (the

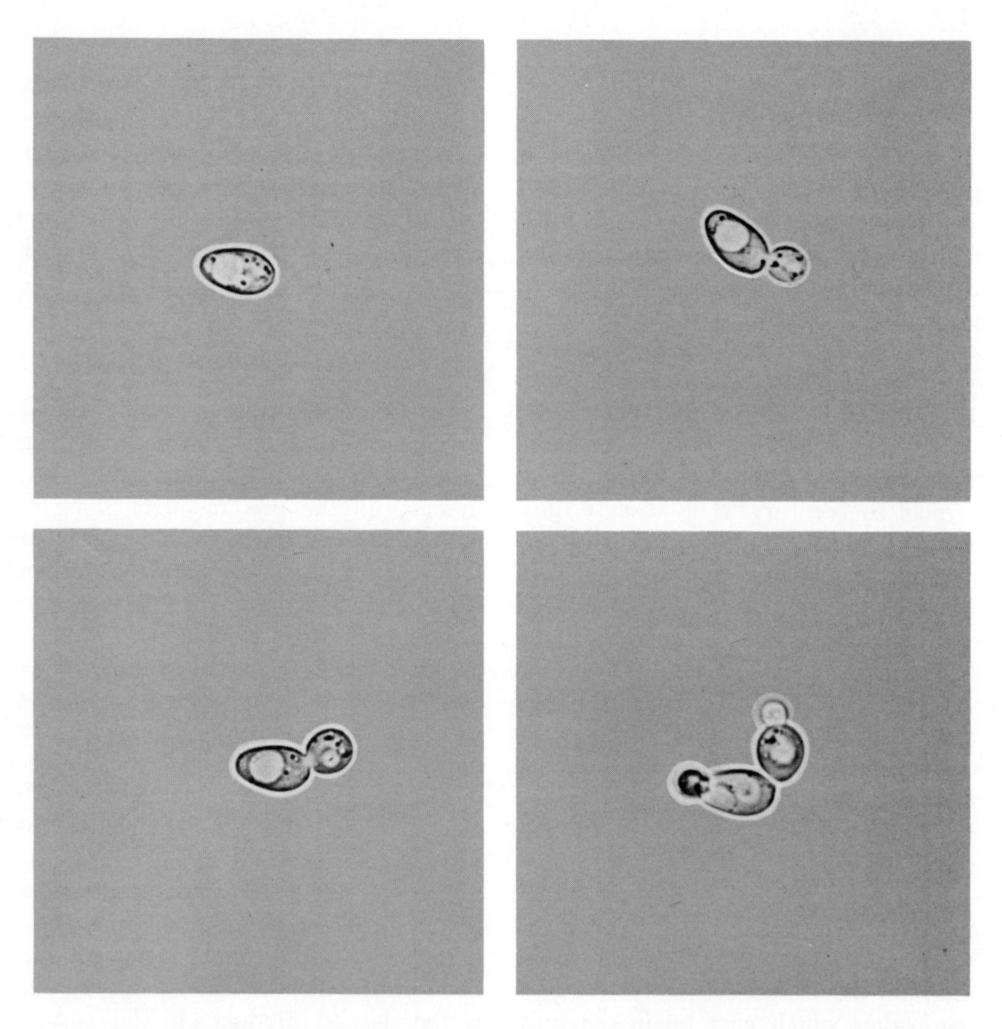

Fig. 33–1. Budding of a yeast plant, *Saccharomyces cerevisiae.* (Fleischmann Laboratories, Standard Brands, Inc.)

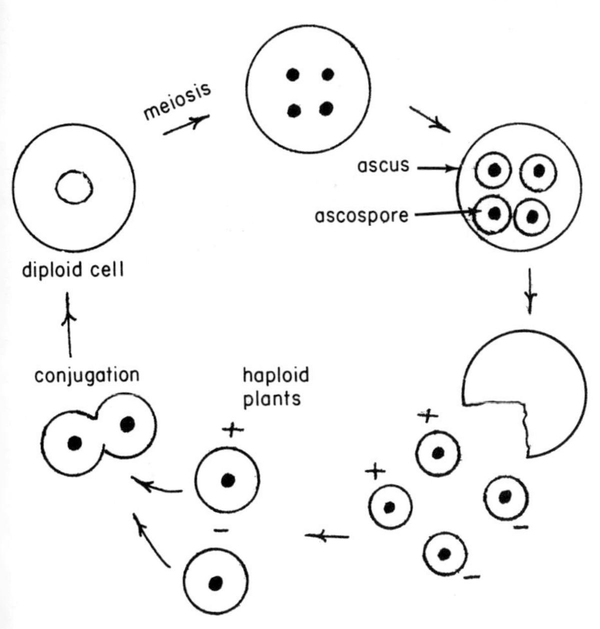

Fig. 33–2. Schematic representation of sexual reproduction of yeast.

minus sex). In time the ascus opens and the ascospores are liberated. The ascospores germinate to produce small, haploid, unicellular plants which reproduce by budding for a time. Then a cell of a plus strain fuses with one of a minus strain forming a diploid yeast plant which reproduces asexually by budding.

To develop superior yeasts, the plants are hybridized. Single spores are planted on a nutrient medium where they germinate, giving rise to haploid cells which are then tested for desired qualities such as their ability to produce vitamins or superior wines. Haploid cells of opposite sex, that have desired features, are then grown together to produce hybrid diploid cells which will be used commercially.

TAPHRINALES

All the fungi in the order Taphrinales are parasitic on plants. A typical example is *Taphrina deformans,* which causes the disease known as peach leaf curl. The growth of the septate mycelium between the cells of the host induces distortions of growth; infected leaves are thick and wrinkled, and the fruits are hollow. The mycelium produces a layer of asci just below the epidermis of the host. Typically, eight ascospores develop within each ascus. After the epidermis of the host is ruptured by the developing asci, the mature asci discharge the ascospores, which then carry the infection to other plants.

BLUE AND GREEN MOLDS—ASPERGILLALES

Except for their value in production of Roquefort and Camembert cheeses, blue molds for many years were considered to be only nuisances. Then in 1929 Alexander Fleming, an English scientist, discovered that *Penicillium notatum* produced an exudate, called penicillin, which inhibited the growth of bacteria. However, it was not until 1940 that Fleming's discovery was put to practical use. In that year, the epoch-making paper by E. Chain and H. W. Florey was published. They had purified the drug, tried it on human patients, and found that it had great value in medicine. Now penicillin has been synthesized. The new chemical method, however, is not economical enough to compete with penicillin derived from blue molds.

The mycelium of a blue or green mold is septate, white, and cobweb-like in appearance. The asexual spores are usually blue or green in color, depending on the species. Because they are produced in chains at the tip of a conidiophore, they are called **conidia.** Two widely distributed genera of Aspergillales are *Penicillium* and *Aspergillus,* whose conidiophores are different. The conidiophore of *Penicillium* is branched, and at the tip of each branch conidia are produced (Fig. 33–3*A*). The unbranched conidiophore of *Aspergillus* (Fig. 33–3*B*) supports a globose tip that bears numerous short stalks that produce chains of conidia. The Aspergillales produce conidia more commonly than ascospores. The ascospores are produced in asci which are irreg-

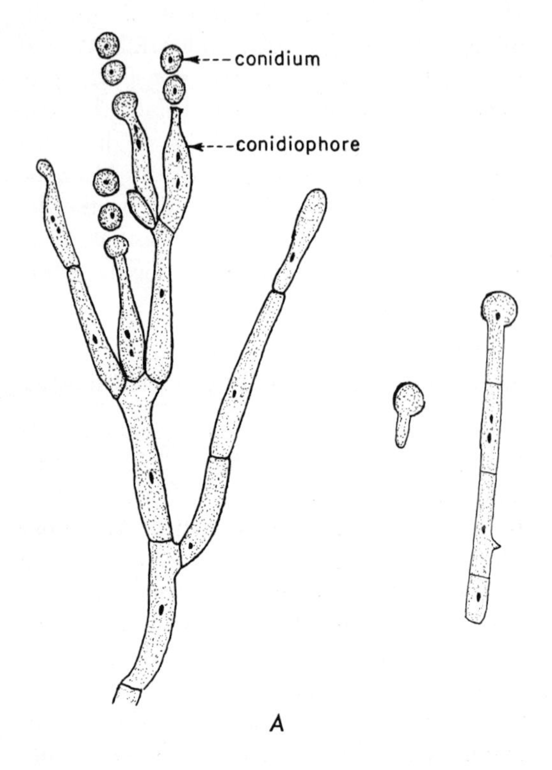

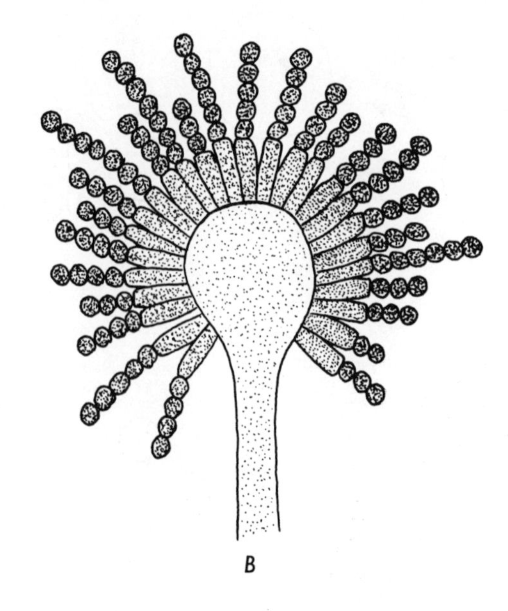

Fig. 33–3A. *Penicillium notatum,* from which penicillin is obtained commercially. *Left,* formation of conidia. *Right,* germination of a conidium. (After Baker, *Bull. Torrey Botan. Club.*) *B.* A conidiophore of *Aspergillus.*

ularly arranged inside a closed sphere, called a **cleistothecium.**

The culture of *Aspergillus* is big business. *Aspergillus* does not produce an antibiotic, but when it is grown under very acid conditions, it produces large amounts of citric acid which is used in soft drinks, medicinals, candies, inks, and in dyeing and engraving processes. More than 10,000 tons of citric acid are produced each year by factories which cultivate *Aspergillus.*

Most species of *Aspergillus* are saprophytic, but a few are parasitic. Certain species are associated with such ailments of man as pulmonary lesions, ulcerations of the external ear, and some skin lesions. Tropical and subtropical strains of *Aspergillus flavus* produce a substance, aflatoxin, that, when ingested, causes cancer of the liver in a wide variety of animals, including trout, ducks, mice, and rats. It has not yet been demonstrated that this chemical causes cancer of the liver in man.

POWDERY MILDEWS—ERYSIPHALES

All members of the order are parasites of higher plants. Of the six genera known, we shall describe only one, *Uncinula. Uncinula necator* causes powdery mildew of grape and Virginia creeper. It reproduces asexually by conidia and sexually by ascospores.

The mycelium of a powdery mildew appears as a white cottony mass on the undersurfaces of the leaves, and after spores form, the leaves look as if they were coated with a white powder. Powdery mildew is relatively easy to control because the mycelium is on the outside of leaves and other plant parts. However, absorbing hyphae, called **haustoria,** penetrate the host cells and absorb nutrients. Powdery mildew can be controlled by dusting plants with powdered sulfur, which kills the superficial mycelium and protects against further infection.

Asexual Reproduction

Short hyphae, known as **conidiophores,** develop from the mycelium at right angles to the surface of the host. **Conidia** are formed at the tip of the conidiophore. They occur in chains, and the youngest conidium is the one adjacent to the conidiophore (Fig. 33–4). Conidia are readily dispersed by air currents. When they light on an appropriate host and if moisture is available, the conidia germinate, producing hyphae which grow over the surface. The great numbers of conidia formed during the growing season favor a rapid spread of the disease, especially during moist seasons.

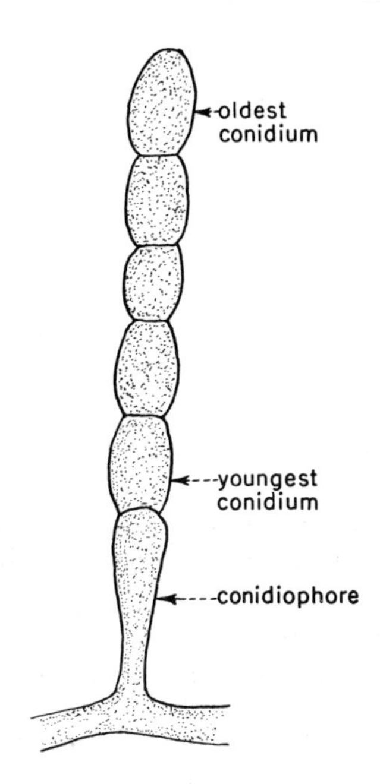

Fig. 33–4. Chain of conidia produced by a conidiophore of *Uncinula necator.*

Sexual Reproduction

Late in the season small black dots become evident on Virginia creeper leaves which are infected with *Uncinula necator* (Fig. 33–5). Each black dot is a closed

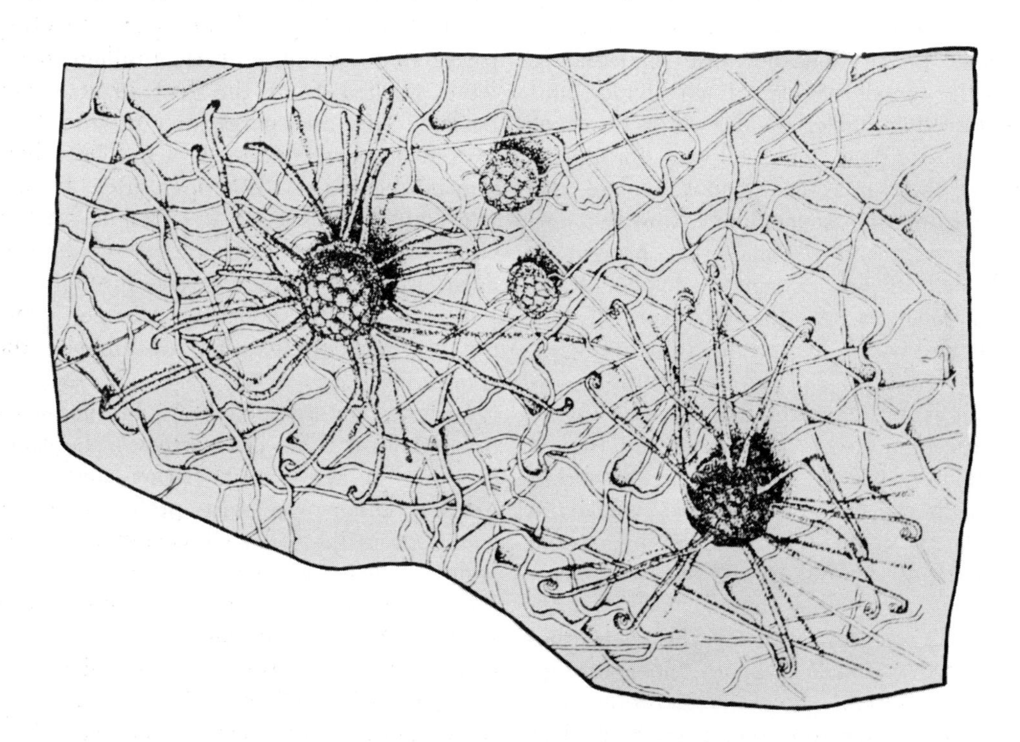

Fig. 33–5. Portion of a Virginia creeper leaf showing cleistothecia of *Uncinula necator* in various stages of development. Note the mycelium over the surface of the leaf and the appendages on the cleistothecia. (B. T. Galloway, *Botan. Gaz.*)

sphere known as a **cleistothecium.** If a cleistothecium is broken open, several asci emerge, and in each ascus there are four ascospores (Fig. 33–6*A*). In nature some cleistothecia become detached and are scattered by wind. Others remain on the leaves, even after the leaves have fallen; hence raking up and burning such leaves will lower the incidence of disease the next year. Cleistothecia survive the winter. In the spring they and the enclosed asci absorb water, thus cracking open the cleistothecia and permitting the asci to emerge. The asci open at the top and discharge the ascospores, which are wafted away by air currents.

Asci and ascospores are the end result of sexual reproduction, which is rather complicated (Fig. 33–6*B*). A male gamete is produced in an antheridium, and a female gamete is produced in a one-celled structure known as an **ascogonium.** An antheridium and an ascogonium are formed on adjoining hyphae. The male gamete enters the ascogonium, but it does not fuse immediately with the female gamete. Each divides, and the resulting nuclei divide. As a result of nuclear divisions and cell-wall formation, hyphae, known as **ascogenous hyphae,** develop from the ascogonium. Now two nuclei, a male and a female, are present in each cell of an ascogenous hypha. Each hypha recurves near the tip. The cell next to the end of the hypha thus is at the bend, and it enlarges and develops into an ascus. It is at this point that the two nuclei in each penultimate cell unite, and then a reduction division occurs. The resulting four cells are called **ascospores.** While the ascogenous hyphae are developing, sterile hyphae grow up and around the ascogonium and the ascogenous hyphae. The sterile hyphae form the wall of the cleistothecium. Other sterile hyphae, known as **appendages,** radiate from the cleistothecium. In *Uncinula* the appendages are hooked at the ends; in other genera they are straight or branched in various ways, the exact shape being characteristic for each genus.

SPHAERIALES

About one-third of the Ascomycetes are in the order Sphaeriales, which includes both saprophytes and parasites. The Dutch elm disease, caused by *Ophiostoma ulmi,* and the devastating chestnut tree blight, caused by *Endothia parasitica,* are representative parasitic species.

Typical symptoms of chestnut tree blight include cankers on the twigs, branches, and the main stem, and brown and shriveled leaves. The mycelium of *Endothia parasitica* develops in and kills the cambium and inner bark. In time the stem is girdled, the supply of food to the roots is cut off, and the tree dies.

Endothia reproduces by conidia and by ascospores. The conidia are produced at the tips of hyphae, and they become embedded in a sticky matrix. Insects, birds, squirrels, and other animals may come in contact with the matrix and spread the spores to other trees. In one instance nearly a billion spores were washed from the feet of a woodpecker. The ascospores are produced in asci, which are contained in a sphere having an opening on the upper side. A fruiting body of this type is called a **perithecium.** When the ascospores are mature, each ascus partially emerges through the pore and forcibly shoots the ascospores into the air; air currents then disseminate the spores.

Included by some authors in the order Sphaeriales is *Claviceps purpurea,* the cause of ergot of rye, wheat, barley, and other grasses. Other mycologists prefer to classify *Claviceps* in the order Clavicipitales. Ascospores, which are wind-disseminated, infect the ovaries of rye, wheat, or other grass flowers. As the mycelium develops, the ovary tissues are destroyed and replaced by the cottony mycelium which produces conidiophores bearing conidia embedded in a sweet nectar. Insects seek the nectar and distribute conidia to other flowers, thus spreading the fungus. Later the mycelium

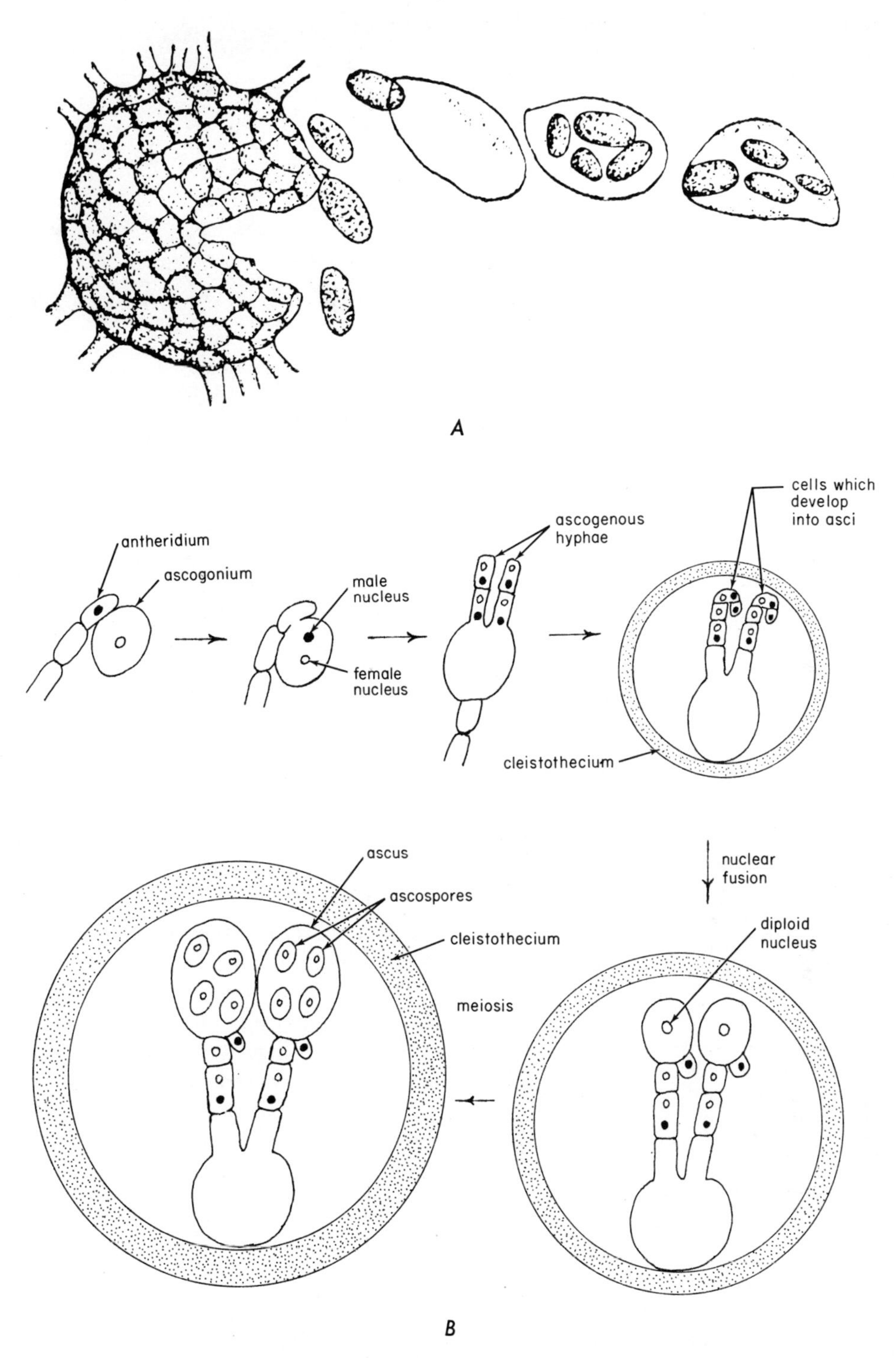

Fig. 33–6*A*. A cleistothecium of *Uncinula necator* which has just broken open. Three asci are shown. Four ascospores have been liberated from one of the asci. (B. T. Galloway, *Botan. Gaz.*) *B*. Schematic representation of sexual reproduction of *Uncinula*.

hardens, forming a pink or purplish sclerotium. The sclerotium occurs where a grain would normally be present and resembles a grain in shape, but is longer and darker. A mature head of rye frequently bears normal grains intermixed with sclerotia; uninfected flowers develop normally, whereas infected ones produce sclerotia (Fig. 33–7). During harvest, some of the sclerotia are knocked to the ground where they overwinter. The following spring a sclerotium germinates by forming several globose heads borne on stalks about ½ inch high; such structures are known as **stromata,** and each resembles a miniature mushroom. Many **perithecia** develop in the globose stromatal head. At maturity each perithecium bears several asci, each containing eight threadlike ascospores. The ascospores are forcibly discharged at about the time rye and other grasses are in bloom (Fig. 33–8).

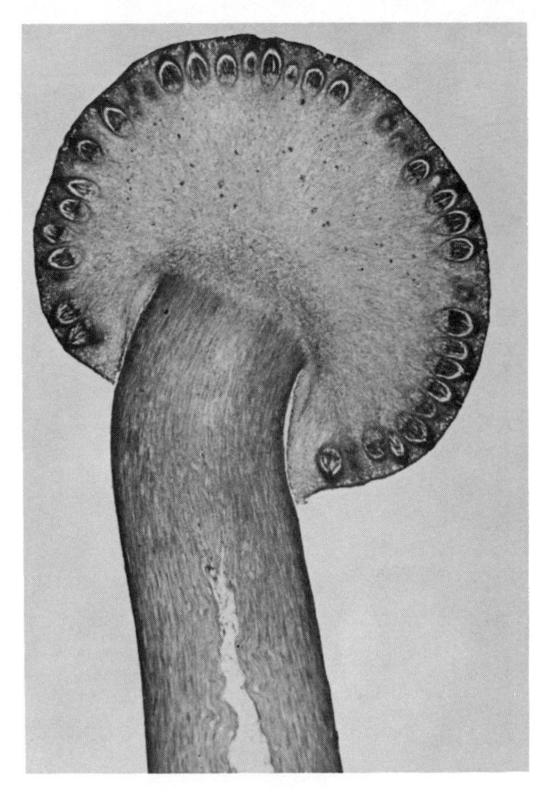

Fig. 33–8. *Claviceps.* Notice the perithecia in the stromatal head. (Carolina Biological Supply Co.)

Fig. 33–7. Ergots on head of barley (*left*) and on head of Russian wild rye (*right*). (George Bridgmon.)

The sclerotia of *Claviceps purpurea* contain a number of alkaloids, among them **ergot,** which are toxic to man and animals. Ergot constricts the blood vessels, particularly those of the hands and feet, and thus deprives the extremities of an adequate blood supply. As a result of poor circulation, gangrene may set in. In the past, because of crude methods of milling flour, death of human beings from ergot poisoning was not uncommon. With modern milling the danger of ergotism has been much reduced, but even so many people died or became seriously ill during August, 1951, in the village of Pont-St. Esprit in France. Ergotism is a more common disease of domestic animals than of man.

Fig. 33–9. Apothecia of *Peziza*. (Carolina Biological Supply Co.)

A valuable drug is isolated from the sclerotia. The refined alkaloid is utilized to control hemorrhage during childbirth.

CUP AND SPONGE FUNGI—PEZIZALES

Most of the members of the order are saprophytes. *Peziza* and *Morchella* are representative forms.

In a forest, species of *Peziza* and related genera, called cup fungi, are frequently found growing saprophytically on logs or twigs (Fig. 33–9). The mycelium grows in the wood and digests it. In time the mycelium given rise to cup-shaped fruiting bodies, called **apothecia,** ranging in size from 1/16 inch in diameter in some species to 3 inches in diameter in others, which are often brilliantly colored. The outer part of the apothecium consists of a dense mass of interwoven hyphae. The interior is lined with a layer of cylindrical asci intermingled with slender hyphae called **paraphyses.** As in other Ascomycetes, asci containing ascospores are the end result of sexual reproduction. Eight ascospores with thick and often

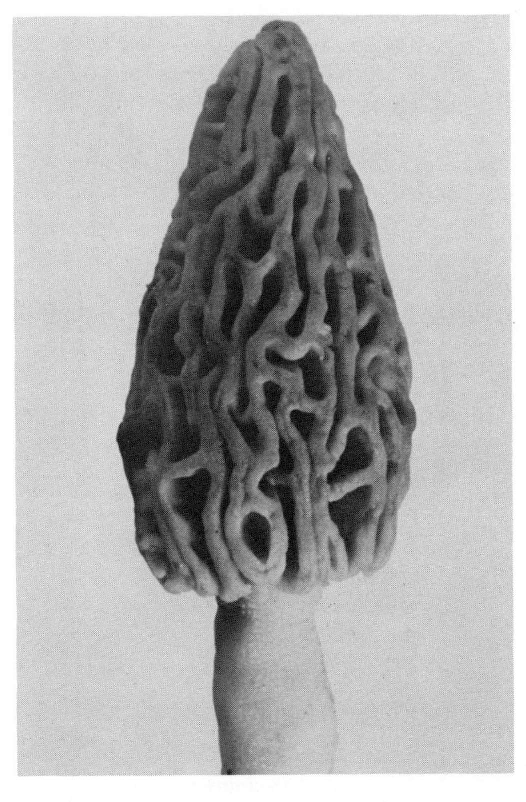

Fig. 33–10. Sporophore of *Morchella,* an edible Ascomycete.

colored walls are produced in each ascus. During ascospore development, some cytoplasm in the ascus is not incorporated in the spores. The imbibition of water by this cytoplasm develops an internal pressure which breaks off the tip of the ascus and ejects the spores.

Morels (*Morchella esculenta*) are edible Ascomycetes, and gourmets consider them to be one of nature's superior delicacies. They are readily recognized (see Fig. 33–10) and cannot be confused with a poisonous mushroom. The mycelium ramifies through soil rich in organic matter from which food is secured. In the spring the fruiting bodies are organized. The ridges of the fruiting body are lined with countless asci, each containing eight ascospores.

Truffles, in the order Tuberales, are other fleshy ascomycetes that are highly prized by gourmets. The fruiting body of a truffle is a rounded mass with irregular internal cavities that are lined with asci. Because the reproductive structures are subterranean, man uses specially trained pigs and dogs to locate them.

QUESTIONS

1. Give the general characteristics of the Ascomycetes.

2. Of what importance to man are the Ascomycetes?

3. _____ Yeasts reproduce asexually by (A) basidiospores, (B) budding, (C) ascospores, (D) chlamydospores.

4. _____ Reduction division occurs in an ascus (A) just before the ascospores are formed, (B) just after the ascospores are fully developed.

5. _____ Asci and ascospores are the end result of (A) a sexual process, (B) an asexual process.

6. _____ Which one of the following Ascomycetes does not reproduce asexually by conidia: (A) *Penicillium*, (B) *Aspergillus*, (C) *Uncinula*, (D) *Morchella*.

7. Describe asexual and sexual reproduction of *Uncinula*, a fungus causing powdery mildew.

8. _____ In an ascus characteristically there are (A) four or eight ascospores, (B) two or four ascospores, (C) more than eight ascospores.

9. _____ Which of the following Ascomycetes do you consider to be the most important? (A) Yeast, (B) *Uncinula necator*, which causes powdery mildew of grape; (C) *Penicillium notatum*, which produces penicillin; (D) *Endothia parasitica*, which causes chestnut tree blight; (E) morel (*Morchella*), an edible fungus of a pleasant flavor; (F) *Taphrina deformans*, which causes peach leaf curl.

---

10. Match the fungus on the left with the appropriate word or phrase on the right.

_____ 1. *Saccharomyces.*	1. Peach leaf curl.
_____ 2. *Taphrina.*	2. Powdery mildew.
_____ 3. *Penicillium notatum.*	3. Streptomycin.
_____ 4. *Streptomyces griseus.*	4. Penicillin.
_____ 5. *Uncinula.*	5. An edible Ascomycete.
_____ 6. *Morchella.*	6. Bread.
_____ 7. *Aspergillus.*	7. Citric acid.
_____ 8. *Endothia parasitica.*	8. Chestnut tree blight.

11. Match the Ascomycetes on the left with the appropriate fruiting body given at the right.

_____ 1. Yeast.	1. Asci borne singly.
_____ 2. *Taphrina deformans.*	2. Asci borne in a cup (apothecium).
_____ 3. *Uncinula necator.*	3. Asci in a layer.
_____ 4. *Peziza.*	4. Asci in a closed sphere (cleistothecium).

34

True Fungi (Continued)

CLASS BASIDIOMYCETES

Magnificent stands of white pine have been and are being decimated by the Basidiomycete *Cronartium ribicola,* a fungus introduced from Europe half a century ago. In 1916 the black stem rust of wheat destroyed over 300 million bushels of wheat in North America, with resulting hardships to farmers and consumers. Even with the progress which has been made in breeding resistant varieties and developing methods of control, the rust still takes a toll of about 10 to 20 per cent of the crop. In some years smuts of cereals destroy 160 million bushels of grain. Yet not all of the Basidiomycete group is destructive, for in addition to smuts and rusts it includes mushrooms, puffballs, and other fleshy and woody fungi, some of which clean up the forest floor and enrich the soil and atmosphere.

The outstanding characteristic of the Basidiomycetes, a group of 20,000 to 25,000 species, is a club-shaped or filamentous hypha, termed a **basidium,** which is produced at some stage in the life cycle. The basidium produces four points, known as **sterigmata,** and on each sterigma a **basidiospore** is produced. Because meiosis precedes the formation of basidiospores, they are haploid. In some Basidiomycetes, conjugation occurs between hyphal cells; in others, between certain spores; and in different species, between spermatia and flexuous hyphae. Like the Ascomycetes, from which the Basidiomycetes have probably evolved, the Basidiomycetes have a septate mycelium. Unlike the Ascomycetes, the mycelium of many Basidiomycetes has a peculiar **clamp connection** which extends out of one cell and connects to an adjacent cell.

SMUT FUNGI—USTILAGINALES

Corn, oats, rye, barley, wheat, rice, and other members of the grass family are often infected by one or another smut fungus, socalled because of the black and dusty spores that develop in the host tissue. The life cycles of various smut fungi differ in certain details, and these differences affect methods of control, as we shall see shortly.

Corn Smut

In many fields of corn some plants have large tumors on the ears, tassels, leaves, or stalks, demonstrating that symptoms of the disease may appear on any aerial part. These plants are attacked by *Ustilago zeae,* the causal organism of corn smut. The mycelium threads its way between the host cells, secures nourishment, and also stimulates the host cells to divide and enlarge, thus forming a tumor. After a period of time there develops in each mycelial cell a thick-walled black spore known as a **chlamydospore,** also called a **teliospore.** Young chlamydospores are binucleate; but as they get older, the two nuclei unite. By this time the host cells are dead, and the tumor consists of a ruptured covering and millions of black spores which are wafted away by the wind. The chlamydospores may land on refuse, manure, or soil, where they germinate immediately, or more usually, the following spring. During germination, a spore sends out a hypha, and concurrently the nucleus undergoes reduction division, forming four haploid nuclei. The nuclei move into the hypha and partitions form between them, thus yielding a four-celled, clublike structure called a **basidium.** Each cell of the basidium produces a basidiospore (Fig. 34–1). Of the four basidiospores formed by a basidium, two are of the plus sex and two of the minus sex. These basidiospores may bud repeatedly to form additional spores known as **sporidia,** which are genetically identical to the basidiospore from which they came. Basidiospores and sporidia are dispersed by wind, and those which lodge on corn plants may bring about infection. A plus basidiospore (or sporidium) germinates on the host, producing a plus mycelium; a minus basidiospore (or sporidium) produces a minus mycelium. If a plus mycelium is adjacent to a minus mycelium, the cells of one fuse with those of the other, yielding a mycelium made up of binucleate cells. In other words, there is a union of cells, but not of nuclei—each cell after fusion having a plus nucleus and a minus one. In time the binucleate mycelium forms chlamydospores which are, as we have seen, binucleate at first, but uninucleate when fully mature. Control of corn smut is difficult. Crop rotation lessens the incidence, and scientists are striving to develop resistant varieties.

Cell division in the mycelium of smut fungi, as well as in other basidiomycetes, is unique and involves the formation of clamp

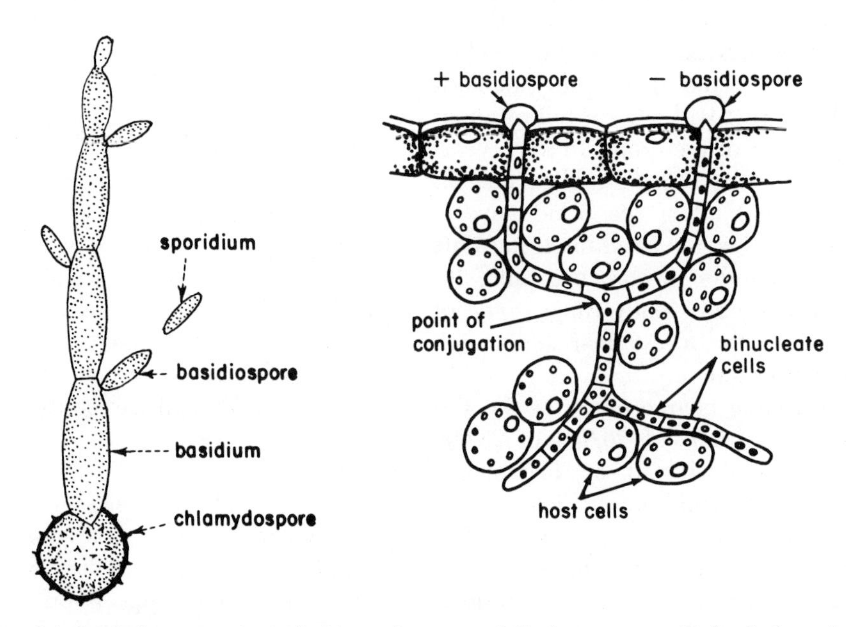

Fig. 34–1. *Left,* a germinated chlamydospore of *Ustilago zeae.* *Right,* fusion of plus cells with minus cells to form a binucleate mycelium.

connections which insure that each resulting cell will have two nuclei, a plus nucleus and a minus one. The features of cell division are illustrated in Figure 34–2 where the plus nucleus is black and the minus nucleus is white. A hooklike outgrowth is formed on the side of the cell. The black nucleus enters the hook, and then both undergo mitosis. The daughter nuclei of the black nucleus separate widely, with one remaining in the hook and the other moving back into the cell. Then two contiguous cross-walls are formed; one cuts across the parent cell, and the other angles upward, thus cutting off the hook from the cell. Then the tip of the hook is digested, thus resulting in two cells, each containing one black and one white nucleus. Every subsequent division is carried out in the same way, and in this manner the binucleate condition is maintained. When each cell of the mycelium has two nuclei, the mycelium is called a **dikaryon mycelium.**

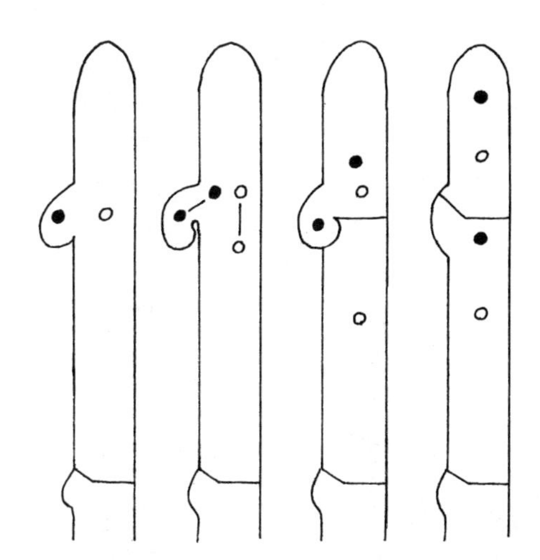

Fig. 34–2. Stages in the formation of a clamp connection and cell division in a basidiomycete.

Stinking Smut or Bunt of Wheat

On occasions a field of wheat may appear healthy from a distance and show promise of high yield. Closer examination may reveal, however, that where normal wheat grains are expected, only hard, dark masses are present. If these are crushed, countless bits of black powder, the spores, are released. In this instance the wheat plants are attacked by *Tilletia caries* or *Tilletia foetida*, which fungi cause stinking smut or bunt of wheat. When wheat with an admixture of smutted grain is threshed, the smut masses are broken, and the healthy grains become covered with smut spores. When the crop is marketed, it commands a lesser price. If a portion is planted for the next crop, spores are planted with each seed. When a seed germinates, the spores also germinate, and the wheat seedling becomes infected. As the seedling develops, the fungus keeps pace, but it does not become evident until the grain ripens. A knowledge of the mode of infection of bunt or stinking smut suggests the control, which may be achieved by treating the seeds with a fungicide such as copper carbonate dust, New Improved Ceresan, or hexachlorobenzine, prior to planting (Fig. 34–3).

Loose Smut of Wheat

Whereas treatment of wheat seeds with a fungicide will control bunt, it is not effective in controlling loose smut of wheat caused by *Ustilago tritici* because flower infection occurs rather than seedling infection. In the loose smut the spores produce hyphae which invade the ovaries of normal wheat plants (Fig. 34–3). The mycelium develops in the grain, but there is no outward manifestation of infection. When such grain is planted, the growth of the mycelium keeps pace with the growth of the wheat plant, and the disease becomes evident only at flowering time. When the wheat is flowering, the floral tissues are destroyed and masses of spores are formed, which are then transmitted to neighboring healthy plants. The spores can infect only healthy ovaries; they are incapable of infecting other parts of the wheat plant. Some success has been attained in the control of loose smuts by heating the seeds in water at a temperature of 125° to 129° F. for 10 minutes just before

Fig. 34–3. *Left,* stinking smut or bunt of wheat. *Right,* loose smut of wheat.

they are planted; such a temperature and exposure kill the mycelium without injuring the embryo.

RUSTS—UREDINALES

Rust fungi are virulent parasites which attack a large variety of plants. Some rust fungi grow and reproduce on one species of plant—for example, asparagus rust. In this the reproductive bodies formed by the fungus on asparagus plants infect other asparagus plants, and so on indefinitely. Rusts which require only one host to complete their life cycle are **autoecious.**

In other rusts the spores produced by the fungus on one species cannot infect plants of the same species. For example, *Puccinia graminis tritici,* the fungus which causes black stem rust of wheat, produces spores on barberry plants. These spores are incapable of infecting barberry, but they can infect wheat plants; that is, to complete the life cycle, *Puccinia graminis tritici* requires two hosts, wheat plants and barberry plants. This phenomenon of requiring two hosts is known as **heteroecism.**

In the white pine blister rust the two hosts are white pines and species of *Ribes* (currants and gooseberries). The cedar-apple rust requires junipers and apple trees to complete its life cycle. Control of heteroecious rusts can be achieved by eliminating the alternate host—barberry plants for the control of black stem rust, *Ribes* for the control of white pine blister rust.

The mycelium of a rust fungus develops within the host tissue and in time produces spores, just beneath the host surface, in pustules called **sori,** which are apparent as yellow, rusty, or black streaks or spots. The spores are exposed to air currents, which effect dispersal when the thin covering of host tissue ruptures.

Black Stem Rust of Wheat

The elimination of barberry shrubs and the breeding of resistant varieties have reduced the severity of this disease, but complete control has not been achieved. Varieties of wheat which are resistant when first developed may later succumb to the disease. The variety of wheat remains the same, but the rust fungus changes. A new race of the fungus may arise by mutation or by the recombination of genes during sexual reproduction. About 200 races of the rust fungus are now known, and additional

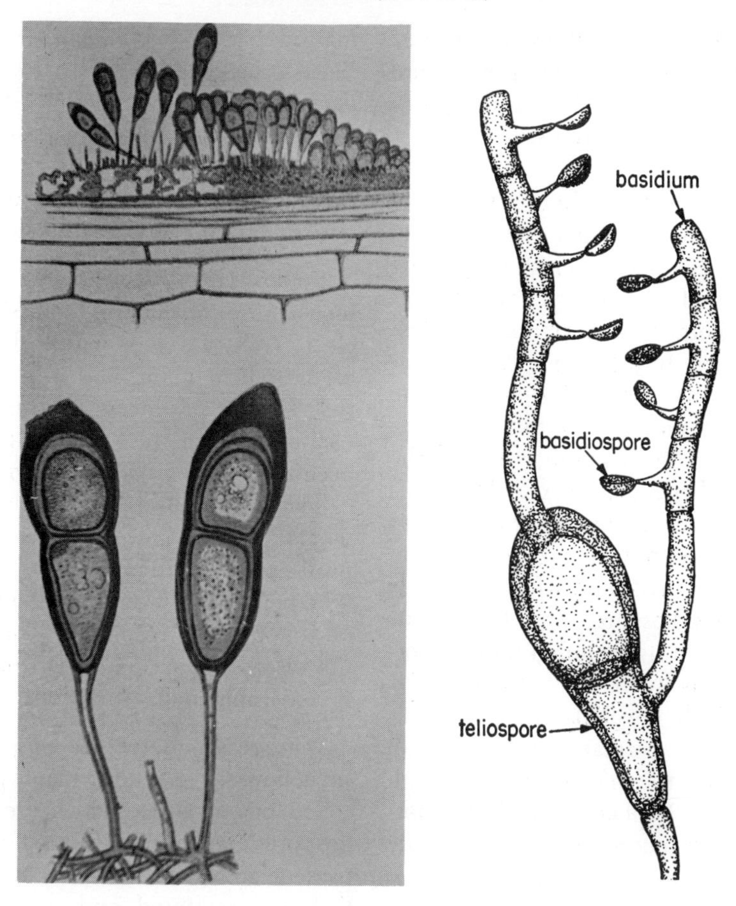

Fig. 34–4. Black stem rust of wheat. *Upper left,* a section of a wheat leaf showing the production of teliospores. *Lower left,* teliospores. *Right,* a teliospore germinating to produce two basidia on which basidiospores develop. (Left figures, H. M. Ward, *Ann. Botany* [London].)

ones are continually developing. *Puccinia graminis tritici* has a complex reproductive cycle which involves the formation of teliospores, basidiospores, aeciospores, and uredospores (also known as urediospores). Let us see where and how these spores are formed and how races of the fungus may hybridize.

Teliospores. In winter no evidence of the rust is apparent, but overwintering spores are on the ground or stubble. These spores are black in color, two-celled, heavy-walled, and resistant to adverse conditions. They are called **teliospores,** and they were produced by *Puccinia graminis tritici* on wheat plants (Fig. 34–4).

Basidiospores. When the winter rest is over, each teliospore germinates and produces one or two hyphae, the basidia. Each basidium is four-celled, and each cell produces a basidiospore. Because a reduction division occurs during the germination of a teliospore, the basidiospores are haploid. Half of the basidiospores are of one sex (the plus sex) and half of the opposite sex (the minus sex). If basidiospores are carried to wheat plants, the plants do not become infected. On the other hand, if a basidiospore lights on a leaf of a barberry plant, the spore germinates, and a mycelium develops within the leaf. The basidiospores are disseminated by air currents.

Aeciospores. Shortly after infection, flask-shaped structures, called **spermagonia,** are formed by the parasite just below the upper epidermis of a barberry leaf. Reproductive bodies known as **spermatia** and **flexuous hyphae** are formed in the spermagonia. The spermatia and flexuous hyphae in one spermagonium are of one sex (the plus sex), and those in another are of the opposite sex (the minus sex). The plus spermatia and flexuous hyphae, as well as the spermagonium in which they develop, are produced by a mycelium which develops from a plus basidiospore, and the minus spermagonium with its included minus spermatia and flexuous hyphae are produced by a mycelium which develops from a minus basidiospore. The spermagonia have small openings leading to the outside through which the flexuous hyphae extend. Furthermore, they secrete nectar which carries the spermatia to the outside. The spermatia are transferred from one spermagonium to another by insects which are attracted by the sweet nectar. After this transfer plus spermatia may be adjacent to minus flexuous hyphae and minus spermatia next to plus flexuous hyphae. A spermatium fuses with a flexuous hypha of the opposite sex to form a binucleate cell. As a consequence of combining two nuclei in one cell (one from the spermatium and the other from the flexuous hypha), new races of rusts may be developed in nature. Hence as man breeds varieties of wheat resistant to rust, the rust may evolve new races that will infect the previously resistant varieties.

From the cell containing nuclei from each sex, there develops a mycelium which ultimately forms yellow, one-celled, binucleate **aeciospores.** These spores are formed in cuplike containers, called **aecia,** on the lower surface of the barberry leaf (Fig. 34–5). The aeciospores are incapable of germinating on barberry plants; the life cycle can be continued only on wheat plants. The aeciospores are scattered by the wind, and those that light on a wheat plant bring about infection if environmental conditions are favorable. The hypha produced by an aeciospore enters through a stoma.

Uredospores. After a wheat plant is infected, the mycelium, consisting of binucleate cells, develops within the leaf or stem of the wheat plant, and in a short time the rust fungus produces reddish-brown, binucleate spores in rust-colored pustules called **uredinia.** The reddish-brown spores are called **uredospores,** and they infect other wheat plants. The cycle from infection to uredospore production may be completed in a period of 10 days, and successive crops of uredospores result in a rapid spread of the disease from one wheat plant to another if environmental conditions are favorable.

Teliospores. Later in the summer the mycelium in a wheat plant produces two-celled black spores, which are known as **teliospores.** When first formed, two nuclei are present in each cell of a spore; but after they are scattered and prior to the onset of winter, the two nuclei unite. The teliospores are the overwintering spores and can survive extreme cold.

Summary. The teliospores start the cycle over again by germinating to produce basidiospores which infect barberry. On barberry, after a spermatium unites with a flexuous hypha of the opposite sex, the fungus produces binucleate aeciospores which infect wheat. Early in the season uredospores are produced on wheat. These carry the disease to other wheat plants, and later in the season teliospores are produced by *Puccinia graminis tritici* on wheat (Fig. 34–5).

Control of black stem rust of wheat. Theoretically, it should be possible to obtain complete control of black stem rust of wheat by eliminating barberry plants. In Denmark almost complete control has been achieved in this way. In the United States the disease

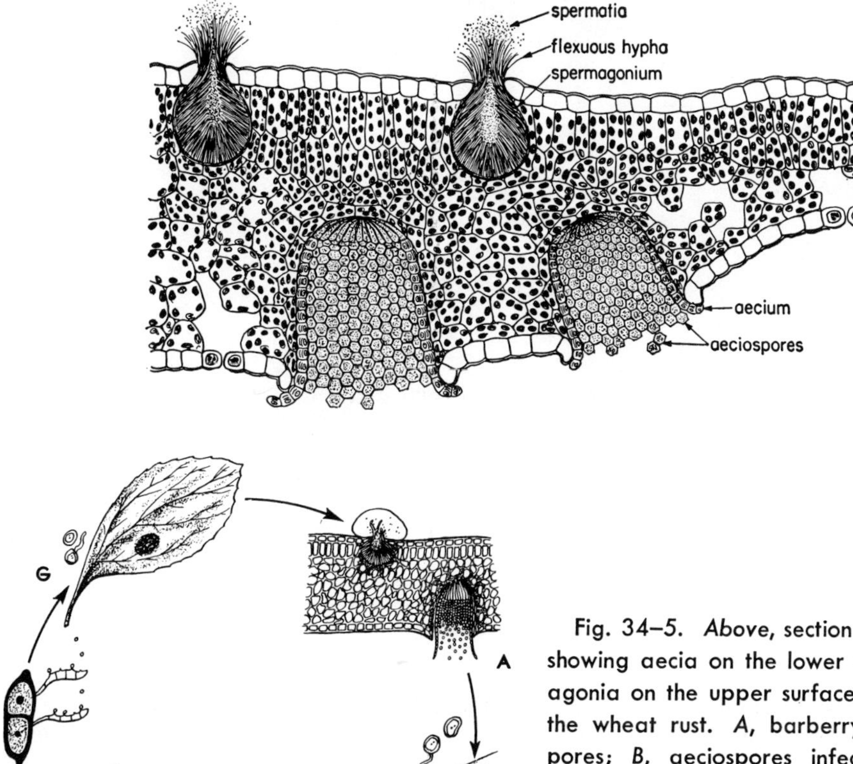

Fig. 34–5. *Above*, section of a barberry leaf showing aecia on the lower surface and spermagonia on the upper surface. *Left*, life cycle of the wheat rust. *A*, barberry leaf with aeciospores; *B*, aeciospores infect wheat stem; *C*, uredospores formed on wheat; *D*, uredospores infect wheat; *E*, teliospores formed; *F*, teliospore forms basidia; *G*, basidiospores infect barberry. (From *Biology* by Relis B. Brown. Copyright © 1956, D. C. Heath and Co., Boston. By permission.)

has been partially controlled by eliminating barberry shrubs. Complete control has not been achieved because in regions with mild winters, the uredospores survive the winter and thus carry the disease from one season to another without the help of the alternating host.

The uredospores which develop on wheat plants in the South are carried north by wind, and they may infect new fields of wheat. The uredospores produced on these newly infected plants may in turn move northward and infect more wheat. This process may be repeated until fields of wheat in Canada are diseased.

White Pine Blister Rust

This disease is prevalent in white pine forests of New England, the Lake States, and the Pacific Coast. Trees which have survived all the hazards of frost, drought, insects, and some diseases for the past 200 or more years are now threatened by a disease of epidemic proportions. White pine blister rust, is caused by the fungus *Cronartium ribicola,* a parasite which was introduced into the United States from Europe during the period between 1898 and 1910.

Shortly after a pine tree is infected, yellow spots appear on the needles. Later the fun-

Fig. 34–6. *Hydnum imbricatum*, a tooth fungus. (Department of Plant Pathology, Cornell University.)

gus spreads to the stems, which then become swollen. Two to three years after infection aeciospores are produced in elongated pustules on the trunk and branches of the infected tree. The aeciospores infect species of *Ribes* (currants and gooseberries) but not pine trees. However, the mycelium which produced the aeciospores continues to grow in the pine tree and produces successive crops of aeciospores until the tree succumbs. On *Ribes*, uredospores are produced which carry the disease to other *Ribes* plants. In late summer the mycelium in a *Ribes* leaf produces columns of teliospores on the lower surface of the leaf. The teliospores germinate on any moist substratum and produce basidia bearing basidiospores which are capable of infecting white pine trees. The disease can be controlled by eradicating *Ribes* shrubs which are within 900 feet of white pine trees. This is a large and costly job.

Other Rusts

A great variety of plants may be infected by one or more species of rust fungi. Such ornamentals as carnations, chrysanthemums, hollyhocks, snapdragons, and roses are often infected. Rusts of asparagus, peas, and beans are also fairly common. Some rust diseases can profoundly alter the economies of nations. For example, in the early nineteenth century Ceylon produced most of the world's coffee. Then a rust devastated the coffee trees. Other countries, especially those of Central and South America, then began to raise coffee, which remains to this day their principal export.

MUSHROOMS AND RELATED FUNGI—AGARICALES

Mushrooms, pore fungi, tooth fungi, and coral fungi are in the order Agaricales. *Hydnum* (Fig. 34–6) is a widely distributed tooth fungus, and *Clavaria* (Fig. 34–7) is a coral fungus.

Mushrooms (Gill Fungi)

Many people are interested in mushrooms either for securing delicious food or for their beauty of form and color and their interesting habits. Relatively few mushrooms are poisonous, but some of them contain poisons as deadly as rattlesnake venom. The toxic substance in *Amanita muscaria* and some other poisonous mushrooms is an alkaloid known as muscarine, which first stimulates and then paralyzes the central nervous system. The symptoms of poisoning by *Amanita muscaria* include profuse perspiration and salivation, diarrhea, delirium, convulsions, and paralysis of respiration.

The common field mushroom, *Agaricus campestris* (Fig. 34–8), is the most widely known of all mushrooms. It is an edible species that is frequently found in pastures. It is the one which is bought in cans or fresh at the store. *Agaricus campestris* is cultivated on a large scale in caves, cellars, and special types of houses where the temperature ranges from 53° to 63° F. The mycelium grows in horse manure or a synthetic manure contained in beds or trays. After the manure becomes permeated with mycelium, the bed or tray is covered with black soil. At certain points of the mycelium, hyphae mat together to form small buttons which increase in size and make their appearance above the surface of the soil. The buttons grow rapidly and develop into the commercial mushrooms which are the repro-

Fig. 34–7. *Clavaria pyxidata*, a coral fungus. (Department of Plant Pathology, Cornell University.)

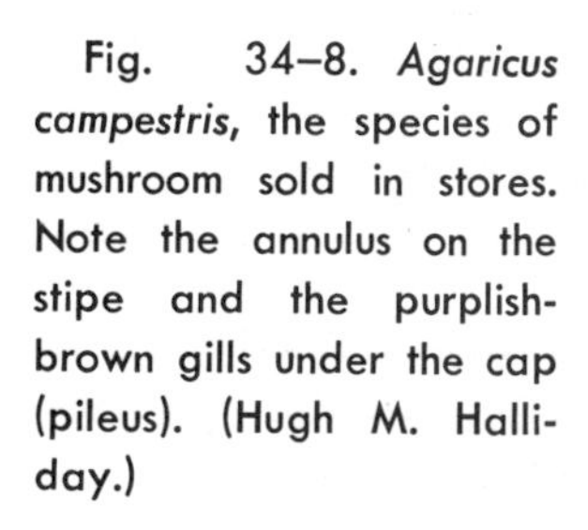

Fig. 34–8. *Agaricus campestris*, the species of mushroom sold in stores. Note the annulus on the stipe and the purplish-brown gills under the cap (pileus). (Hugh M. Halliday.)

ductive bodies, the **sporophores,** of the fungus. The sporophore of *Agaricus* consists of a stalk topped by an umbrella-shaped cap known as the **pileus.** On the undersurface of the cap there are thin **gills** which radiate from the stalk to the edge of the cap. Each gill consists of a mass of hyphae which become enlarged at the edges of the gill to form club-shaped basidia (Fig. 34–9). A young basidium as well as the hypha from which it developed is binucleate. The two nuclei in the basidium unite, following which reduction division occurs, producing four haploid nuclei. Then one nucleus moves into each of the four developing basidiospores which form on **sterigmata.** Later a heavy wall forms around each basidiospore. Characteristically, four basidiospores are produced by each basidium (Fig. 34–10). When the spores are mature, they do not just fall from the sterigmata. Instead, they are cast off with just enough force to shoot them half the distance between the gills, from where they drop to the ground.

In a suitable environment a spore germinates to form a haploid hypha. If the hypha is to continue development, its cells must fuse with cells of another hypha of an appropriate sex. We say appropriate sex advisedly, for in many mushrooms the spores are of four sexes, not just the usual male and female. Let us designate the sexes one, two, three, and four. There are only two

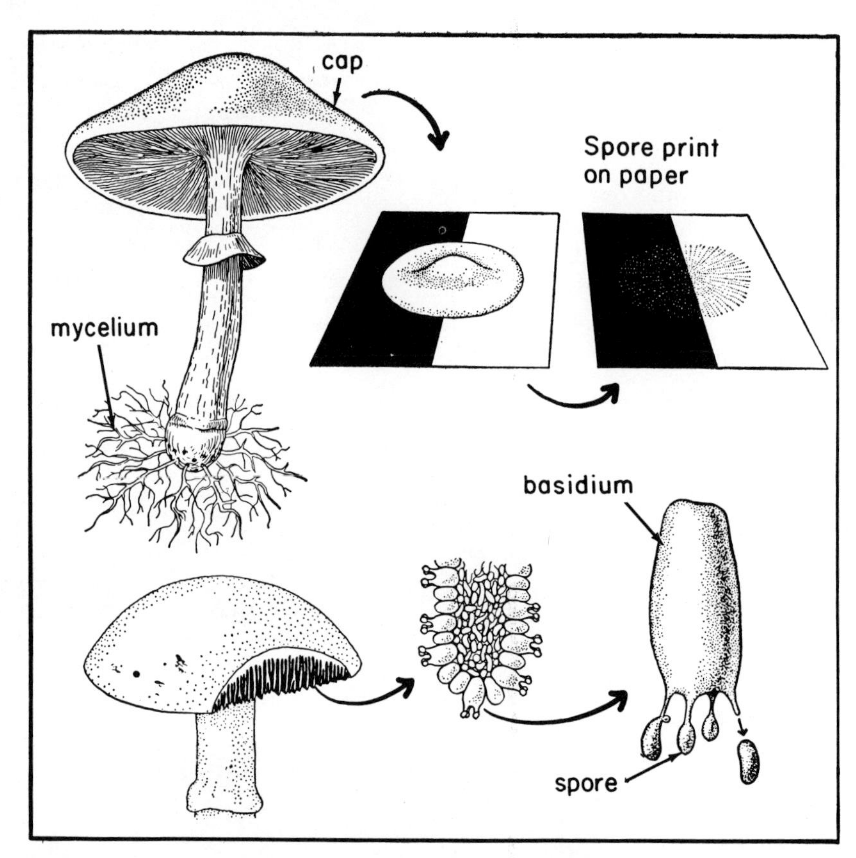

Fig. 34–9. A mushroom, showing a small part of the mycelium from which it grows. The spores (called basidiospores) are borne on club-shaped structures, called basidia (magnified, *lower right*), from which they are forcefully snapped off. (Reproduced by permission from *Biology: Its Human Implications* by Garrett Hardin. Copyright, 1952, W. H. Freeman & Co., San Francisco.)

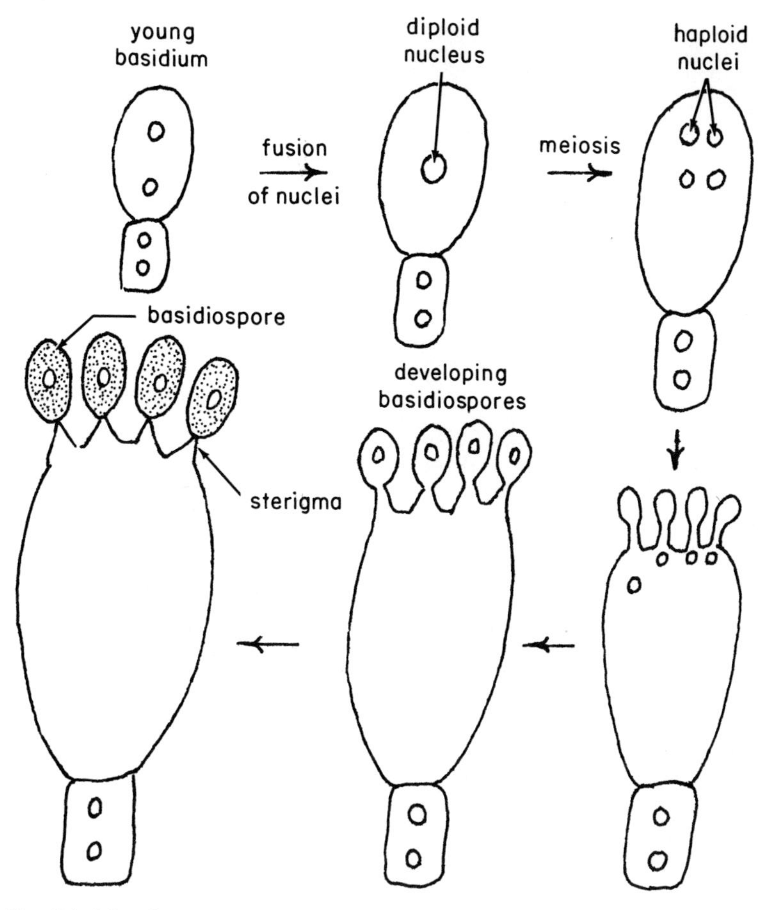

Fig. 34–10. Stages in the development of basidiospores of a mushroom.

cross-matings which permit continued development—a mating of one and three, or two and four. After fusion of cells the nuclei remain distinct; hence there are two nuclei in each cell.

Not all mushrooms have four sexes. A few, among them *Agaricus,* have just one sex. Hence in *Agaricus,* cells of like hyphae fuse to form a binucleate mycelium. Furthermore, all spores of *Agaricus* are of one sex.

There are many species of mushrooms. They vary in size, in shape and color of the cap, in the manner of attachment of gills to the stalk, and in the habitats where they develop. They are chiefly saprophytic and thrive only where organic matter is present. Some grow on fallen logs, others in groves where leafmold is present, and different ones in pastures or lawns.

Some species of mushrooms have a ring or **annulus** on the stalk, and some have a cup or **volva** at the base of the stalk. The annulus is the remnant of a membrane which extended from the cap to the stalk when the sporophore was developing; the volva is the remnant of a membrane which surrounded the developing sporophore.

The presence or absence of an annulus and/or a volva are characteristics which may be used to determine the name of a particular mushroom. The color of the spores is also of value in identifying mushrooms; spores are white, black, rose, ochre, or purple-brown in color, depending on the species. Their color may be determined by

Fig. 34–11. Young specimens of *Amanita verna*, one of the most poisonous mushrooms known. The volva is unusually conspicuous here, and the stipe is shorter than at maturity. (Hugh M. Halliday.)

placing the cap on a sheet of white or black paper in a quiet place; the released spores leave a print showing the color of the spores as well as the arrangement of the gills (Fig. 34–9, upper right).

The cultivated mushroom, *Agaricus campestris,* has an annulus, but no volva, and produces purple-brown spores. Most of the deadly poisonous mushrooms are in the genus *Amanita. Amanita* has an annulus, a volva (called the death cup), and white spores. The fly amanita (*Amanita muscaria*) occurs throughout the summer in open woods. It is as beautiful as it is poisonous. The cap is usually bright orange or red and flecked with white specks. A volva is present at the base of the swollen stalk, and the gills are white. *Amanita verna* is so poisonous that a piece the size of a bean seed may cause death, there being no antidote for the poison. It has probably caused more deaths in this country than all the other species together. It is a harmless looking, attractive plant that is found in woods from May to November. Its pure white color has led to the common name "the destroying angel" (Fig. 34–11).

Pore Fungi

Various saprophytic pore fungi may grow in wood and bring about decay. They rot fallen logs and branches in a forest and also fence posts, ties, and poles which have not been treated with creosote or some other wood preservative.

During lumbering operations in virgin stands of timber, many trees are felled which are hollow-hearted or punky. Such trees are infected by parasitic pore fungi, and they are of little value for lumber production. *Fomes pini* is a common pore fungus which causes pecky rots of Douglas-fir, pines, and other conifers. Several species of the genus *Polyporus* attack deciduous trees. The mycelium of a pore fungus develops in the wood, but reproductive bodies are produced on the outside of the tree. The reproductive body of *Fomes* is bracket-like, somewhat flattened, and semicircular. The sporophore may reach a diameter of a foot, and it is so securely anchored to the bark of the tree that it will often support the weight of a man. Many pores are evident on the lower surface of a bracket. Each pore leads into a tube which is lined with basidia which produce basidiospores.

Not all genera of pore fungi grow in wood. Some genera grow in soil which is rich in organic matter; among them is *Boletus,* a genus which includes many edible and delicious species. *Boletus* superficially resembles a mushroom with a stalk and cap. An examination of the undersurface of the cap reveals many pores, each lined with basidia bearing basidiospores.

PUFFBALLS—LYCOPERDALES

The familiar puffballs are reproductive structures that vary in size from that of a pea to larger than a man's head. If we cut open a young puffball, we will notice that it consists of a soft white interior tissue (the **gleba**) enveloped by a rind (the **peridium**). The gleba is chambered, and each chamber is lined with basidia, which produce basidiospores. As the puffball matures, the interior tissue breaks down; the spores become dry, colored, and loose; and the peridium breaks in various ways, permitting the spores to escape. The manner of opening varies with the species. The rind may break off in scales; a single hole may form at the summit; or several holes may be formed. The earth stars have a double wall; at maturity the outer wall splits and turns back to form a star on the ground; one pore forms at the summit of the inner wall.

The puffball is only the reproductive part of the plant. The mycelium threads its way underground and absorbs food from decaying organic matter. After the mycelium has accumulated a store of food, balls of hyphae are formed here and there on the subterranean mycelium. These balls enlarge and develop into sporophores which emerge from the soil.

Puffballs are edible and they are easy to collect. They should be cut open before they are cooked to see that they are not too old and that they really are puffballs. Externally, the buttons of some mushrooms, including poisonous ones, resemble puff-

Fig. 34–12. Earthstars, *Geaster*. (Carolina Biological Supply Co.)

Fig. 34–13. Giant puffballs, one of the most highly prized of all edible wild fungi. (A. B. Stout.)

balls, but inside the button, the stalk and cap are evident if it is a mushroom. Puffballs have a uniform interior. They are right for eating when the flesh is white and firm, but they are too old when the flesh shows a yellowish or brownish discoloration.

LICHENS

Lichens are found growing from the polar regions to the equator, and many grow where few other plants can survive. In the arctic tundra a lichen known as reindeer moss (*Cladonia rangiferina*) thrives and furnishes forage for caribou and reindeer. Reindeer moss in Alaska's Brooks Range north of the Arctic Circle has become radioactive with the fallout from nuclear explosions by Russia and the United States. Included in the lichen is the hazardous caesium-137. As expected, the flesh of caribou also contains radioactive caesium and other minerals, and these become incorporated into the bodies of Eskimos who depend on the caribou for food. So far, the Eskimos have not become seriously ill from excessive radiation. Whether or not deleterious genetic effects have occurred will not be known for a generation or two. Clearly, air pollution is harmful to distant, innocent people.

Lichens are of considerable scientific interest and of some utility. Long ago, in times of famine, lichens were ground into meal which was then baked into "bread." The "manna" of the Israelites was probably a foliose lichen that rounded up into balls as the wind carried the lichens over the deserts of Asia Minor. Formerly, a number of lichens were used as dyes, but these have been largely replaced by synthetic dyes. However, litmus and orcein are still obtained from lichens. The perfume industry uses lichens in various fragrances, especially the "oak moss," *Evernia prunastri.*

Although there are about 16,000 kinds of lichens, they may be grouped according to their general form as **crustose lichens, foliose lichens,** and **fruticose lichens.** Crustose lichens are crustlike and grow very close to the surface of rocks, bark, and similar materials. Foliose lichens have a relatively broad and branched thallus. The thallus of a fruticose lichen is slender and much branched, being erect or hanging.

Strictly speaking, a lichen plant is not just one kind of plant, although in identification it is considered as such, but two unrelated plants growing together as an organized whole. If we examine a lichen under the microscope we see many fungal hyphae enmeshing algal cells (Fig. 34–14). The fungus is nearly always an Ascomycete, but in rare instances it may be a Basidiomycete. The alga is generally a green alga resembling *Protococcus,* but in some instances

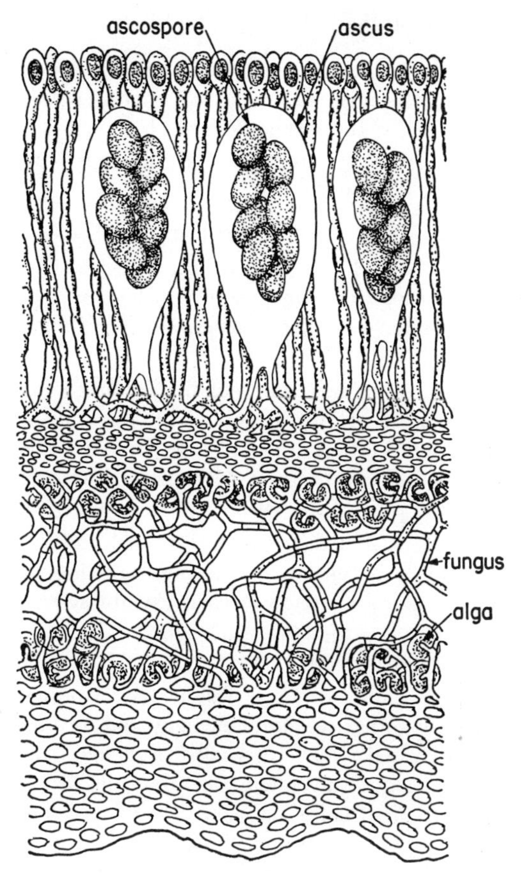

Fig. 34–14. The body of a lichen is made up of algal cells and the mycelium of a fungus.

it may be a blue-green alga. When, as in lichens, two organisms grow together to their mutual advantage, we speak of the relationship as **symbiosis.** The fungus benefits by obtaining food from the alga, and the alga receives protection from drought, mechanical injury, and intense light.

Each lichen component reproduces independently—the alga by cell division and the fungus (if an Ascomycete) by ascospores produced in asci contained in a cuplike or disk-shaped structure on the thallus or on specialized stalks. The ascospores are dispersed by wind; falling on a suitable substrate, they germinate, forming hyphae. If the appropriate alga is present, a hypha surrounds the algal cells and a new lichen plant of characteristic form develops. If no alga is near, the hypha perishes.

A more certain method of reproduction is the production of **soredia,** which are compact masses of hyphal filaments and algal cells. The advantage is obvious because the fungus and the alga are dispersed as a unit. Soredia are often produced in such great numbers that the thallus has a mealy appearance. Lichens also reproduce by fragmentation; a part of the thallus may be broken off and carried away by the wind.

QUESTIONS

1. Give the general characteristics of the Basidiomycetes, and give four reasons why they are of economic importance.

2. _____ Which one of the following is false? (A) Most fungi inhabit the land. (B) Spores of most fungi are distributed by wind. (C) Spores of most algae are distributed by water. (D) Algae are better adapted for survival on land than the fungi.

3. _____ Which one of the following statements about *Ustilago zeae* is false? (A) Chlamydospores develop in tumors on corn plants. (B) A chlamydospore germinates to form a basidium, which bears four diploid basidiospores. (C) Basidiospores may bud repeatedly to form sporidia. (D) After a corn plant is infected, the cells of a plus mycelium fuse with those of a minus mycelium, yielding a mycelium of binucleate cells.

4. Explain the methods of infection in the smuts, and give the treatment for each.

5. _____ Wheat and other cereals are the only plants which are subject to rusts. (A) True, (B) false.

6. Outline by means of drawings the life cycle of *Puccinia graminis.*

7. _____ In *Puccinia graminis,* the fungus causing black stem rust, meiosis occurs (A) when the teliospore germinates, (B) immediately after the fusion of a spermatium with a flexuous hypha, (C) when the uredospore germinates, (D) when aeciospores form in the aecium.

8. _____ Stem rust of wheat may be controlled by (A) using resistant varieties and eliminating barberry shrubs, (B) using resistant varieties and eliminating currants and gooseberries, (C) treating the seeds with Semesan, (D) heating the seeds at 129° F. before planting.

9. Make a diagram of a typical mushroom and label the parts.

10. _____ Pore fungi are most closely related to (A) mushrooms, (B) puffballs, (C) rusts, (D) smuts.

11. _____ Mushroom spores are formed on basidia which are produced (A) in the cup (volva), (B) on the outside of the annulus, (C) on the stipe, (D) on the gills, (E) on top of the pileus.

12. _____ Which one of the following statements about a mushroom is false? (A) Each cell of the mature mycelium has one nucleus. (B) Reduction division occurs in a basidium. (C) A basidiospore germinates to form a haploid hypha. (D) If a hypha is to continue development, its cells must fuse with cells of another hypha of an appropriate sex.

13. _____ *Amanita verna,* a very poisonous mushroom, has spores which are (A) white, (B) rose, (C) tan, (D) purple-brown, (E) black.

14. Match the following:

_____	1. Smut.	1. Agaricales.
_____	2. Rust.	2. Lycoperdales.
_____	3. Mushroom.	3. Uredinales.
_____	4. Pore fungus.	4. Ustilaginales.
_____	5. Puffball.	

15. How may plant diseases be controlled?

16. How could you explain the observed fact that a variety of crop plant which is at first immune later becomes susceptible to a fungus disease?

17. _____ In which one of the following diseases is most of the mycelium on the outside of the leaf? (A) Late blight of potatoes, (B) powdery mildew of rose, (C) peach leaf curl, (D) white pine blister rust.

18. _____ A lichen is (A) an alga, (B) a fungus, (C) a fungus and an alga growing together to the advantage of each (symbiosis).

19. _____ With respect to lichens, which one of the following statements is false? (A) Each lichen is made up of a fungus and an alga growing together in one plant body and forming one individual. (B) Internally, a lichen consists of fungus mycelium, in the meshes of which are algal cells. (C) In a lichen the alga manufactures food, some of which is absorbed by the fungus; the fungus protects the alga against mechanical injury and against evaporation. (D) The fungal component of a lichen is generally a Basidiomycete.

20. _____ A certain fungus produced sacs, each containing eight spores. This fungus is in the class (A) Phycomycetes, (B) Basidiomycetes, (C) Ascomycetes, (D) Deuteromycetes.

21. _____ Which one of the following classes has a mycelium which lacks cross-walls? (A) Phycomycetes, (B) Ascomycetes, (C) Basidiomycetes.

---

22. Several kinds of spores are produced by *Puccinia graminis.* Match the spore on the left with the correct statement on the right.

_____	1. Teliospore.	1. Produced on barberry and infecting wheat.
_____	2. Basidiospore.	2. Produced on wheat and infecting wheat.
_____	3. Aeciospore.	3. Produced on wheat and not infecting any host.
_____	4. Uredospore.	4. Produced on a basidium and infecting barberry.

23. Complete the following table.

Class	Order	Representative Genera	Distinguishing Characteristics	Economic Importance
Phycomycetes				
	Saprolegniales			
	Peronosporales			
	Mucorales			
Ascomycetes				
	Endomycetales			
	Taphrinales			
	Aspergillales			
	Erysiphales			
	Sphaeriales			
	Pezizales			
Basidiomycetes				
	Ustilaginales			
	Uredinales			
	Agaricales			
	Lycoperdales			

35

Mosses, Liverworts, and Hornworts

Imagine yourself on an earth without land plants and with all plant life confined to aquatic habitats. The land would be barren, and the light, water, and mineral resources of the land would not be utilized. Our earth was like that in the distant past. Gradually mutated forms of green algae became established on land, and in time these gave rise to two lines of evolution. One line led to the bryophytes, our present-day mosses, liverworts, and hornworts. The other line led to the development of vascular plants, a line which culminated in the seed plants. The colonization of the land by green plants was a most significant event. A rich new environment with a great diversity of habitats became available to plants and animals and ultimately to man.

Bryophytes have the same pigments, chlorophyll a and b, carotene, and xanthophyll, as the green algae, which is one reason for believing that the bryophytes evolved from the green algae. When a moss spore germinates, it grows into a green filament, called a **protonema,** which resembles a filamentous green alga, again indicating that bryophytes descended from green algae. After the protonema is developed, certain cells divide in three planes to form the familiar moss plant. In bryophytes the sexual organs are multicellular instead of unicellular, as in the algae. It is believed that the multicellular organs evolved from the single-celled antheridia and oogonia of the green algae. The structure which produces sperms is known as an antheridium, and that which produces an egg as an archegonium. The zygote of a green alga is released into the water where it is on its own, whereas in the bryophytes the zygote is retained in the archegonium, where it develops into a multicellular embryo that is nourished by the parent plant. In the evolution of the body of bryophytes from green algae, the body became more compact, thus reducing the surface area relative to its mass. Furthermore, a cuticle which retards evaporation evolved, as did absorbing structures called rhizoids.

The bryophytes are out on an evolutionary limb, and higher plants did not evolve from them. In many ways bryophytes are imperfectly adapted to a land environment.

Like algae, they depend on water for fertilization; the sperms must swim through water to the eggs. Mosses and other bryophytes lack efficient conducting, supporting, and absorbing systems.

About 23,000 species are included in the phylum Bryophyta, which is divided into three classes: the mosses (Musci), the liverworts (Hepaticae), and the hornworts (Anthocerotae). Of the three subclasses of mosses, we shall discuss two: the true mosses (Eubrya), and the *Sphagnum* moss (Sphagnobrya).

THE TRUE MOSSES—CLASS MUSCI—SUBCLASS EUBRYA

The true mosses are small, delicate plants that carpet the ground in forests, cover fallen logs, grow as epiphytes on living trees, and develop on rocks adjacent to streams and waterfalls, or on rocks in the open where frequent drought occurs. Only a few mosses, such as *Fontinalis,* actually live in water. Mosses are worldwide in distribution, and some will grow where few other plants can survive. They are at home in the Arctic tundra, on rocky islands of the Antarctic Ocean where the temperature is seldom above freezing, in temperate areas, and in dank forests of equatorial regions. Many are found at high elevations on mountains—for example, at an elevation of 20,000 feet on Mount Everest. Only in the seas are mosses completely absent.

The moss plant usually noticed by the casual observer is a green plant consisting of a stem, leaves, and near the base, hair-shaped structures called **rhizoids,** which anchor the plant and absorb water and minerals. We use the terms leaf and stem only for convenience; more technically, a leaf is known as a **phyllidium** and a stem as a **caulidium.** Although the leaves and stems of mosses resemble those of higher plants, they are not true leaves and stems because complex vascular tissues are absent. In *Funaria* and certain other mosses the leaves are just one cell thick except along the mid-

Fig. 35–1. Gametophytes and sporophytes of a moss plant.

rib. In other species—*Polytrichum,* for example—the leaves are several cells thick. Only slight differentiation is evident in the stem, which has a central strand of elongated cells surrounded by a cortical zone containing chloroplasts. On the outside is the epidermis, which is one cell thick, green, and without stomata. The green leafy plant is a gametophyte. The gametophyte reaches its highest development in the bryophytes and becomes less complex in the higher plants. Long, slender, brown stalks terminated by a capsule may be present on some leafy moss plants but absent on others. The **stalk** with its noticeable **capsule** and non-evident **foot** is a **sporophyte,** which is nourished through the foot by the gametophyte (Fig. 35–1).

Moss Gametophytes

As we have just seen, the leafy moss plants are the gametophytes, so-called because they produce gametes: sperms in antheridia, eggs in archegonia. Archegonia and antheridia are produced at the tips of stems, sometimes on the same plant, but often on different ones. Multicellular hairs, called **paraphyses,** are intermingled with the archegonia and antheridia at the apex of a stem. In monoecious species archegonia are adjacent to the antheridia at the stem tip. In dioecious mosses such as *Mnium* and *Polytrichum,* antheridia are produced by male gametophytes and archegonia by female ones. It is relatively easy to distinguish the gametophytes of each sex. The uppermost leaves of a male gametophyte radiate from the stem to form a cup, whereas the top leaves of a female gametophyte are erect and overlapping (Fig. 35–2).

Many antheridia, each just barely visible to the unaided eye, are present in the cup at the apex of a male gametophyte (Fig. 35–3*A*). An antheridium is a club-shaped, stalked structure with a multicellular wall that encloses hundreds of spirally coiled, biflagellate sperms. Although the wall is multicellular, it is only one cell thick. At the top of each antheridium there are cells whose walls become mucilaginous and burst away, leaving an opening for the escape of the sperms (Fig. 35–3*A*).

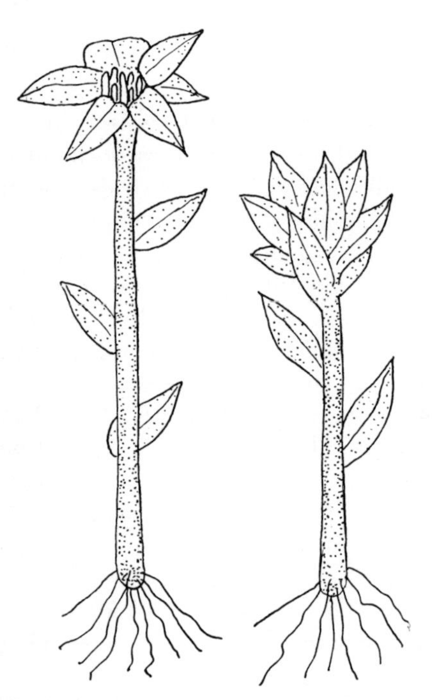

Fig. 35–2. *Left,* a male gametophyte of *Funaria* showing antheridia in the cup at the stem tip. *Right,* a female gametophyte of *Funaria.* (After M. M. Brown, *Am. J. Botany.*)

Many archegonia are present at the stem tip of a female gametophyte (Fig. 35–3*B*). Each archegonium is an elongated flask-shaped structure having a **neck** and a swollen base called the **venter.** One large egg and a small **ventral canal cell** are produced in the venter. The neck consists of a multicellular wall that surrounds a row of **neck canal cells.** When an archegonium is mature, the neck canal cells change into a mucilaginous mass and the uppermost neck cells diverge, leaving a channel for sperms to enter.

When the antheridia and archegonia are mature and covered with a film of water, the antheridia burst open and the sperms swim by means of their flagella toward the archegonia, to which a chemical secretion attracts them. In time a sperm reaches the neck of an archegonium, enters it, swims down the neck to the egg, and fertilizes the egg. Mosses resemble green algae in their dependence on water to get the sperm to the eggs.

The Sporophyte

The nucleus of the fertilized egg contains two sets of chromosomes, one set from the sperm, the other from the egg. The zygote is diploid, and, like other zygotes, it develops into a new individual. However, the moss zygote does not develop into a plant resembling the leafy parents. Instead, it develops into a diploid plant known as the sporophyte. The sporophyte is not an independent plant but remains attached to the gametophyte, on which it is at least partially parasitic, since it depends on the gameto-

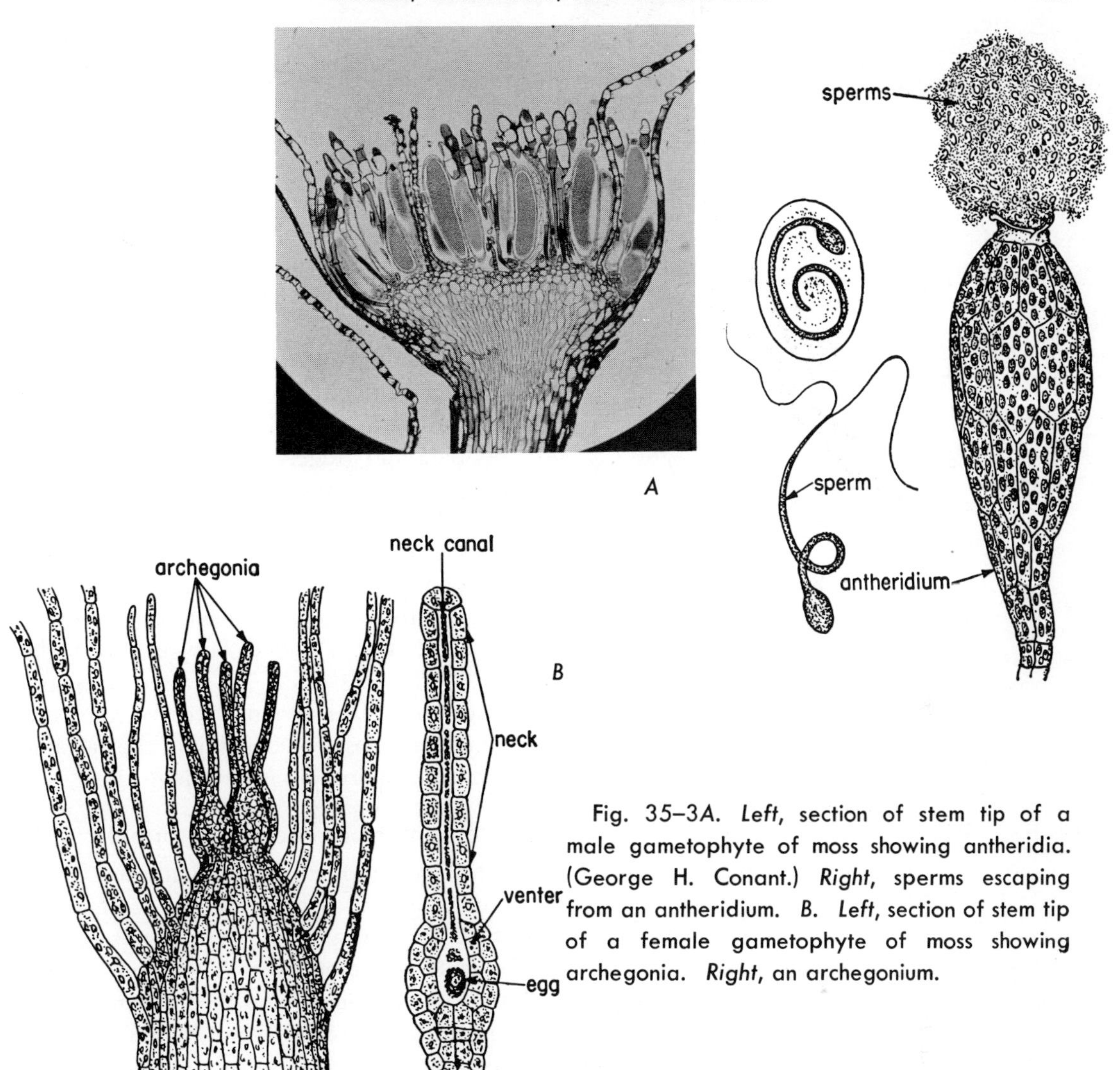

Fig. 35–3*A*. *Left*, section of stem tip of a male gametophyte of moss showing antheridia. (George H. Conant.) *Right*, sperms escaping from an antheridium. *B*. *Left*, section of stem tip of a female gametophyte of moss showing archegonia. *Right*, an archegonium.

phyte for a supply of water, essential minerals, and much, but not all, of its food supply. Let us next consider the development of the sporophyte.

The zygote, contained in the venter, divides; the resulting cells continue to divide, in time forming a small spindle-shaped embryo that is enclosed in the archegonium and nourished by the gametophyte. The lower end of the embryo grows into the tip of the gametophyte and there develops into an absorbing organ called a **foot.** After the foot is formed, the upper portion of the developing sporophyte differentiates into a slender stalk, or **seta,** and a terminal structure known as the **capsule.** During the early stages of sporophyte development, the surrounding archegonial tissues grow and keep pace with the enlarging sporophyte. Finally, however, the enlarged archegonial tissue is broken and carried upward as a cap (the **calyptra**) on the apex of the young sporophyte. A mature sporophyte consists of a foot, seta, and capsule. The foot is anchored in the gametophyte, where it absorbs food, water, and minerals. Extending from the foot is the wiry seta that supports the capsule.

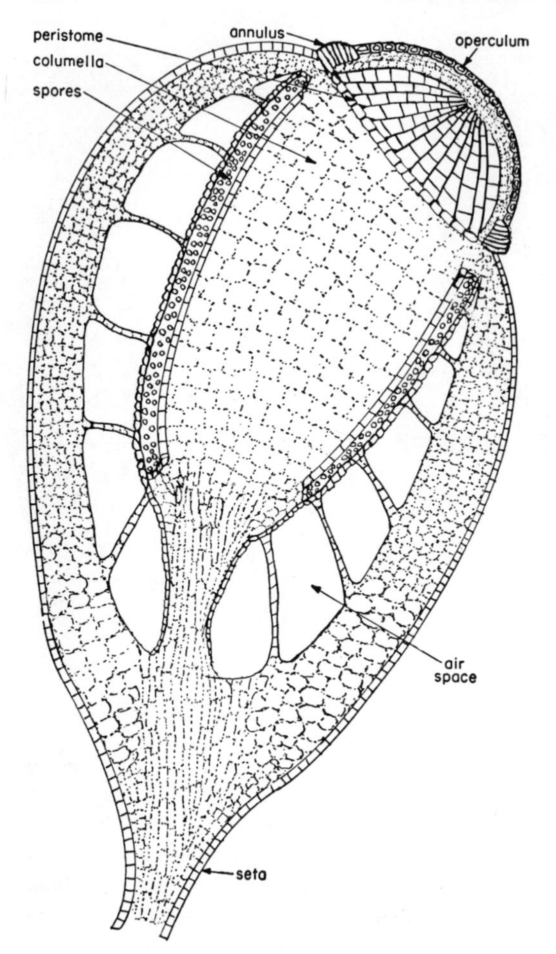

Fig. 35–4. Longitudinal section of a moss capsule.

The capsule has a number of tissues (Fig. 35–4). Beneath the protective epidermis there is a cylinder of green food-making cells, and within this a cylinder of **spore mother cells.** A core of colorless cells, the **columella,** occupies the center of an immature capsule. Each spore mother cell undergoes meiosis to form four haploid spores. After the spores are formed, the columella and food-making cells collapse, leaving a case filled with spores. The capsule when young is green, but by the time the spores are ripe it is red, brown, or orange and contrasts beautifully with the green leaves.

Let us examine the capsule carefully and see what provisions are made for spore dispersal. As the capsule matures, the calyptra, a remnant of the enlarged archegonium, falls off. At the apex of the capsule itself there is a lid, the **operculum,** which is severed from the rest of the capsule when a ring of thin-walled cells breaks down. A ring of toothlike projections, the **peristome,** is present below the operculum, and they are exposed when the operculum falls off. The peristome teeth surround the terminal opening of the capsule. They are fascinating to study and beautiful, often being reddish in color and ornamented with cross-bars and ridges. One of their remarkable features is their sensitivity to changes in the humidity of air. They bend inward in moist air and outward in dry. The movements are hygroscopic ones and are not related to living cells. During dry weather the teeth are extended outward, and the spores are released to be wafted away by air currents. During moist weather the peristome teeth close the opening of the capsule, preventing spores from escaping.

Development of the Gametophyte

The gametophyte is haploid and starts from a spore. However, the spores do not at first grow into anything resembling the leafy gametophytes. The spore swells, bursts its outer coat, and then a slender filament emerges which grows by cell division into a branching, alga-like filament known as a **protonema** (Fig. 35–5). The branches which grow over the soil surface are green, each cell containing many chloroplasts. Other threads of the protonema, called **rhizoids,** grow down into the soil; these rhizoids are colorless or brownish and lack chloroplasts. The young moss in the protonema stage resembles a green alga. In time, buds (each a cluster of cells) appear here and there on the protonema. Each bud develops into a stem bearing leaves, and rhizoids at the basal end.

Alternation of Generations

In the life cycle of a moss plant, there are two generations, the sporophyte and the

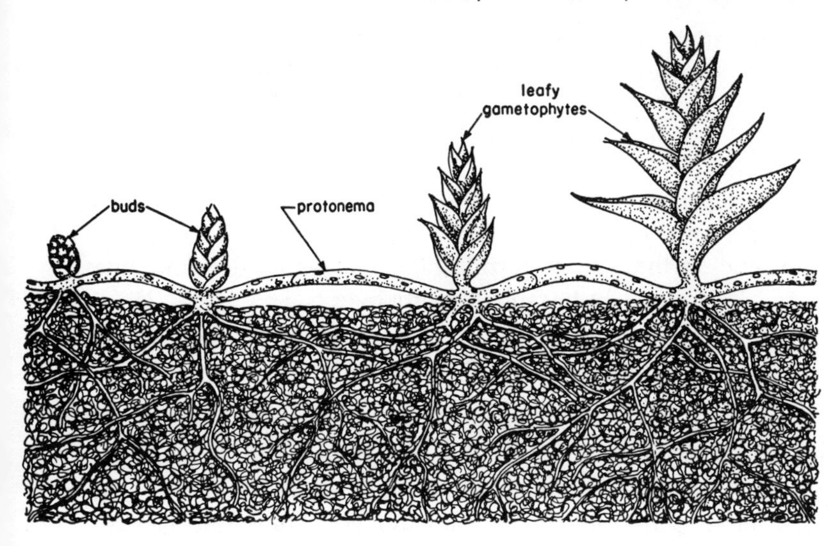

Fig. 35–5. Moss protonema, highly magnified. The buds formed on the protonema develop into leafy gametophytes.

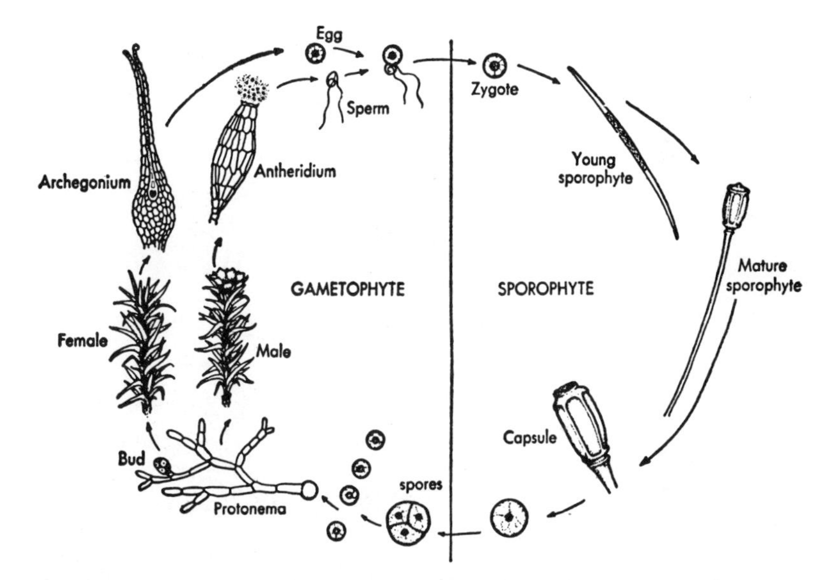

Fig. 35–6. Diagram showing the life cycle of a moss. (Reproduced by permission from *The Essentials of Plant Biology* by Frank Kern. Copyright, 1947, by Harper & Row, Inc., New York.)

gametophyte, and these alternate. The sporophyte is the diploid asexual generation, producing spores which distribute the moss in space. The gametophyte is the haploid sexual generation, producing gametes. These generations are shown in Figure 35–6. The gametophyte generation, the haploid (1 n) generation, starts with a spore which develops into a protonema bearing buds. The buds develop into female or male gametophytes. The female gametophyte produces eggs in archegonia; the male, motile sperms in antheridia. The sperms swim to the eggs and effect fertilization. The fertilized egg, the zygote, is diploid because the sperm contributes one set of chromosomes and the egg another set. The sporophyte generation begins with the zygote, which develops into a spindle-shaped embryo, or young sporophyte, which later differentiates into a foot (not shown in illustration), a seta, and a capsule. Within the capsule, spore mother cells undergo reduction division, forming haploid spores, our starting point in this summary. It is well to understand this concept of alternating generations thoroughly because it occurs in other bryophytes and in the vascular plants. The general trend in evolution is for the gametophyte to become less complex and for the sporophyte to increase in complexity. Another trend leads to the independence of the sporophytes. In mosses the sporophyte is at least partially parasitic on the gametophyte. In seed plants the reverse is true: the gametophyte is parasitic on the sporophyte.

Although the sporophyte is characteristically diploid and the gametophyte haploid, you should not assume that the morphological differences between the sporophyte and the gametophyte are a direct consequence of their different chromosomal numbers. There are unusual cases, often produced under experimental conditions, where the cells of the gametophyte are diploid. For example, if a section of the stalk of a sporophyte is placed on a suitable medium, certain stalk cells divide and develop into a diploid protonemal filament. The buds formed on the diploid protonema develop into characteristic gametophytes with rhizoids, stems, and leaves. Moreover, such gametophytes produce diploid gametes. After fertilization, the zygote develops into a tetraploid sporophyte.

Asexual Reproduction

Mosses reproduce asexually as well as sexually. Protonemata are frequently formed on leaves, stems, and rhizoids. On such protonemata buds are formed which develop into leafy gametophytes. In a few mosses, small oval masses of cells, known as **gemmae,** are formed; each gemma develops into a gametophyte.

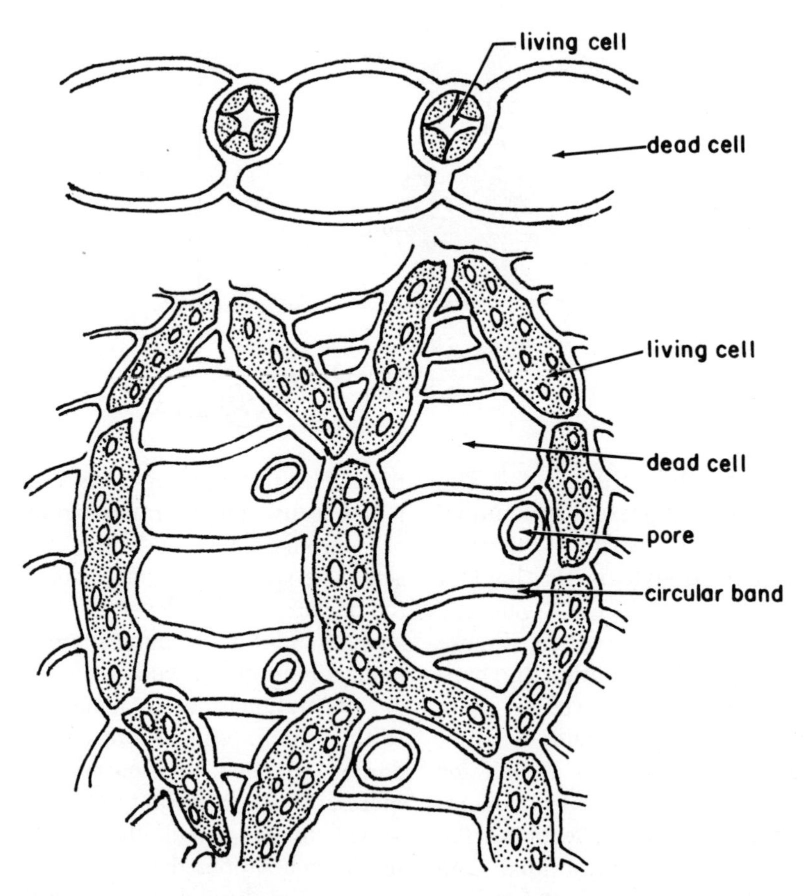

Fig. 35–7. Leaf cells of *Sphagnum*. *Upper*, cross-section of cells. *Lower*, surface view.

SPHAGNUM MOSS—CLASS MUSCI—SUBCLASS SPHAGNOBRYA

The bog mosses, represented by the single genus *Sphagnum,* differ from the true mosses in several ways: in appearance, in structure of the sporophyte and gametophyte, and in habitat. The bog mosses inhabit bogs and swampy places. The entire surface of a pond or lake may be covered with *Sphagnum.* As the dead remains of *Sphagnum* accumulate, the mat over a pond becomes increasingly thick and in time it will support the weight of a man. Various seed plants such as cranberries, bulrushes, cattails, tamarack, and others become established on the *Sphagnum* tussock. The roots and rhizomes of such plants add firmness to the floating layer. Ultimately the lake or pond becomes filled in completely, and in some regions the area then becomes occupied by a forest. The undecayed remains of the plants are frequently harvested and sold as peat moss. Peat moss is extensively used to improve the structure of soil. In parts of Europe peat is used as fuel.

The gametophytes of *Sphagnum* are soft, large, pale-green or reddish-brown in color, and much branched. Two types of cells make up a leaf of *Sphagnum:* colorless cells which are dead at maturity and green ones (Fig. 35–7). The colorless cells are perforated and very large, and each cell is strengthened by circular bands. The colorless cells hold large volumes of water and hence *Sphagnum* is useful as packing material around roots of nursery stock. The green cells, which alternate with the colorless cells, are long and very narrow, so narrow that at first sight they appear to be merely the thick walls of the colorless cells.

The life history of *Sphagnum* is similar to that of the true mosses. The sporophyte of *Sphagnum* is black in color; it consists of a spherical capsule, a very short stalk, and a foot. The sporophyte is elevated on a seta-like stalk, the **pseudopodium,** which is part of the gametophyte (Fig. 35–8).

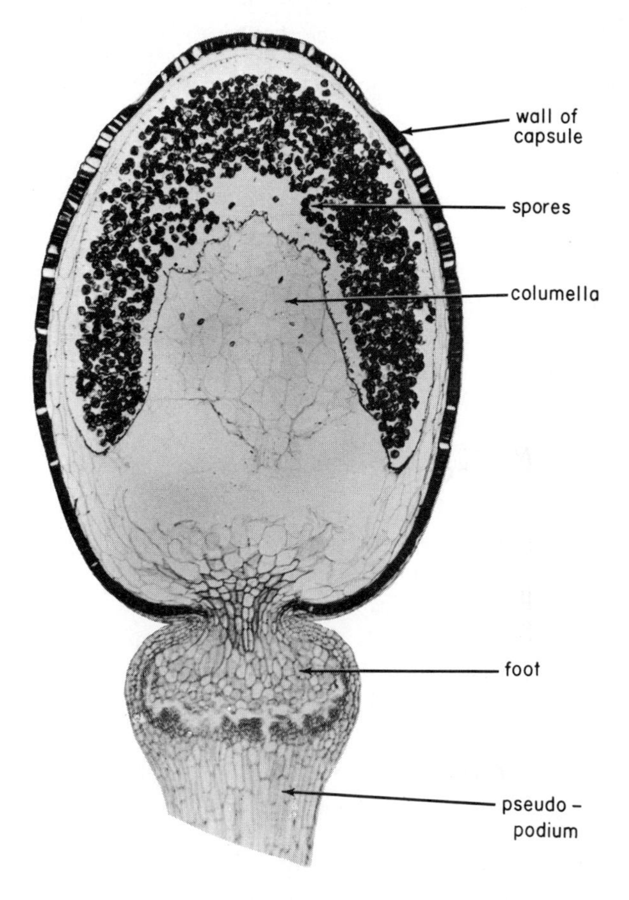

Fig. 35–8. Sporophyte and pseudopodium of *Sphagnum.* (J. Limbach, Ripon Microslides.)

Each spore develops into a green filament which gives rise to a flat thallus that produces a leafy gametophyte which, in time, bears archegonia and antheridia. After an egg is fertilized, it develops into a sporophyte.

THE LIVERWORTS—CLASS HEPATICAE

All liverworts are small plants, seldom exceeding a few inches in length, and growing close to the substratum, either soil or bark of trees. Some, like *Marchantia* and *Riccia,* have lobed thalli, whereas others—*Porella,* for example—have slender stems and simple leaves.

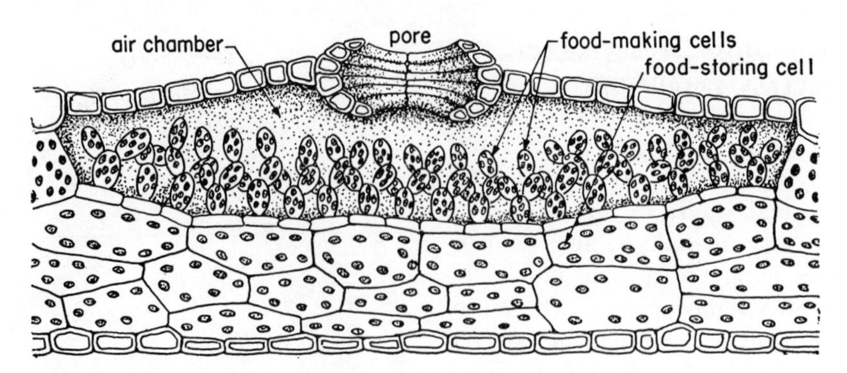

Fig. 35–9. Cross-section through a portion of a *Marchantia* thallus.

Marchantia

Probably the most familiar and widespread liverwort is *Marchantia polymorpha,* which occurs in moist places throughout the north temperate region. As in mosses, the gametophyte is the more conspicuous generation, the *Marchantia* sporophytes being very small and inconspicuous.

Gametophytes. The gametophytes of *Marchantia polymorpha* are green, flat, ribbon-like plants that frequently grow on stream banks. The thallus forks repeatedly as it grows. Each branch is notched at the end and in each notch there is an apical cell which divides to form additional cells, and thus the gametophyte grows. The thallus is anchored to the soil by numerous multicellular scales and unicellular rhizoids. The rhizoids function like root hairs, supplying the thallus with water and minerals. The thallus (Fig. 35–9) is admirably adapted for photosynthesis. Pores through which carbon dioxide diffuses are present on the upper surface. Each pore leads into a chamber that has pillars of green food-making cells extending upward from the floor. The food is stored as starch grains in the cells below the chamber.

Some gametophytes are male plants, others are female. At certain seasons umbrella-like structures develop from the gametophytes and then the male gametophytes can be distinguished from the female ones. In male gametophytes, each upright stalk is terminated by a saucer-like lobed disk, known as the **antheridial receptacle.** Pores are present on the upper surface of an antheridial receptacle, and they lead into cavities which contain antheridia, each with hundreds of biflagellate sperms. The upright stalk of a female gametophyte is terminated by an **archegonial receptacle** made up of a disk with radiating finger-like rays (Fig. 35–10). At maturity, flask-shaped archegonia are suspended from the underside of the disk. Each archegonium consists of a neck and a venter in which the egg is contained (Fig. 35–11).

Water is necessary for fertilization, which occurs when the stalks bearing archegonial and antheridial receptacles are so short that, after a rain, a film of water is held between the receptacles and the thalli. After the eggs are fertilized, the stalk elongates rapidly, carrying the receptacle bearing the zygotes above the thallus.

Sporophyte of *Marchantia*. As in mosses, the sporophyte generation begins with the zygote. The zygote divides and the resulting cells continue to divide, forming a young sporophyte which is spherical in shape. Divisions continue and differentiation occurs. At maturity a sporophyte consists of an absorptive foot anchored in the archegonial receptacle, a short seta, and a capsule. The capsule has a single-layered cell wall within which there are rows of spore mother cells alternating with long narrow cells called **elaters.** The spore mother cells undergo re-

Fig. 35–10. *Marchantia*, a common liverwort. Male gametophyte (*left*) and female gametophyte (*right*). The stalk of the male gametophyte is terminated by a receptacle in which antheridia are embedded. That of the female bears archegonia when young and sporophytes after the eggs are fertilized.

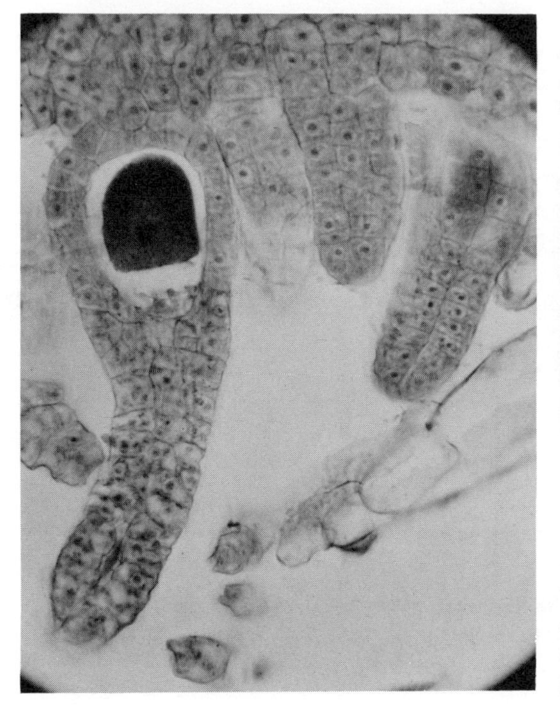

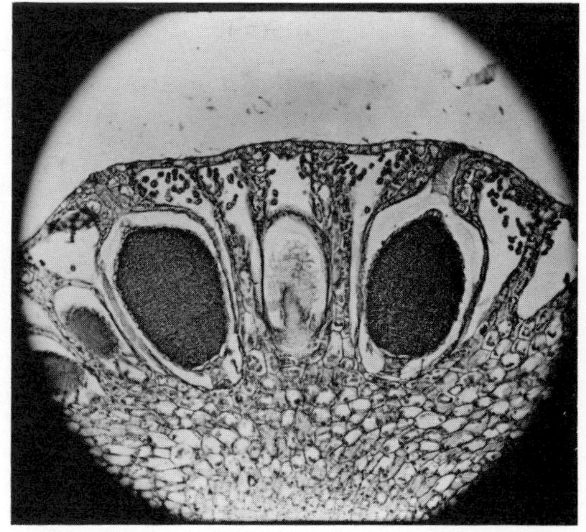

Fig. 35–11. *Above*, a section through an antheridial receptacle of *Marchantia* showing two antheridia. *Left*, a mature archegonium of *Marchantia*, showing the large egg in the venter. (Copyright, General Biological Supply House, Inc.)

duction division to form haploid spores. When young, the elaters nourish the spore mother cells. Later they develop spiral thickenings on their walls and aid in spore dispersal. When the spores are ripe, the seta elongates somewhat and the capsule becomes dry and bursts open, shedding the spores, which are dispersed by wind. The elaters are hygroscopic and respond to changes in humidity by curling and uncurling, movements which assist in liberating the spores and in breaking up spore masses.

Those spores which land in a favorable location develop into gametophytes. About one-half develop into male gametophytes and the other half into female ones.

Gametophytes of *Marchantia* grow well on a soil consisting of half peat and sand. The development of the sexual structures can be initiated by growing the gametophytes with continuous light. After 16 days of continuous light, the antheridial receptacles appear; after 21 days, the archegonial receptacles. After 28 days' exposure to continuous light, the stalks supporting the antheridial and archegonial receptacles attain their full height.

Summary of life cycle. The life cycle of *Marchantia* is summarized in Fig. 35–12. As in mosses, the gametophyte and sporophyte generations alternate. The gametophyte generation is the haploid, sexual phase, and begins with a spore. Spores develop into gametophytes, some of which produce sperms, others eggs. Water is necessary for fertilization.

The sporophyte generation begins with the diploid zygote, which develops into a sporophyte that has no direct connection with the soil. Throughout its life the sporophyte through its foot absorbs water and minerals from the gametophyte. In its early development the sporophyte depends entirely on the gametophyte for food. However, a nearly mature sporophyte is green and makes part of its own food. In the capsule of the sporophyte, spore mother cells undergo reduction division to form haploid spores, the beginning of the gametophyte generation.

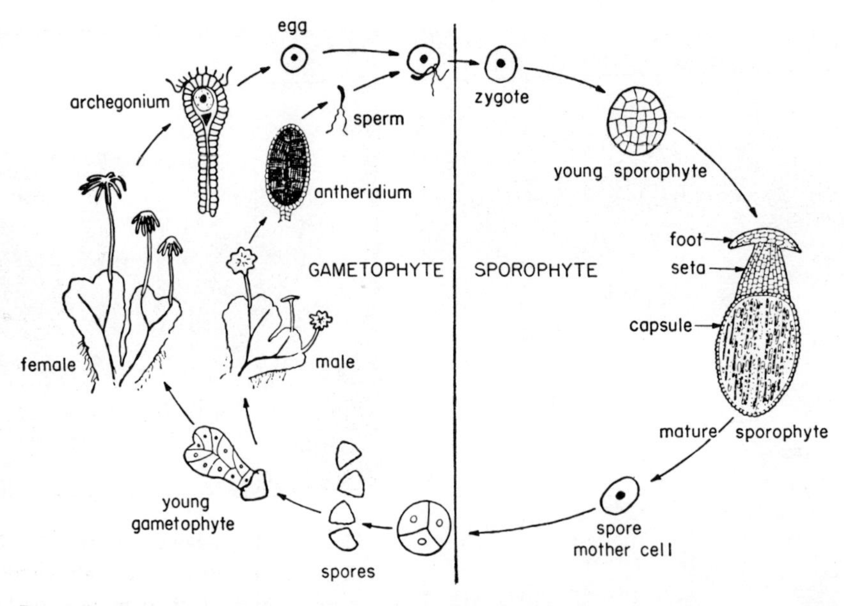

Fig. 35–12. Life-cycle diagram of the liverwort *Marchantia*. (Reproduced by permission from *The Essentials of Plant Biology* by Frank Kern. Copyright, 1947, by Harper & Row, Inc., New York.)

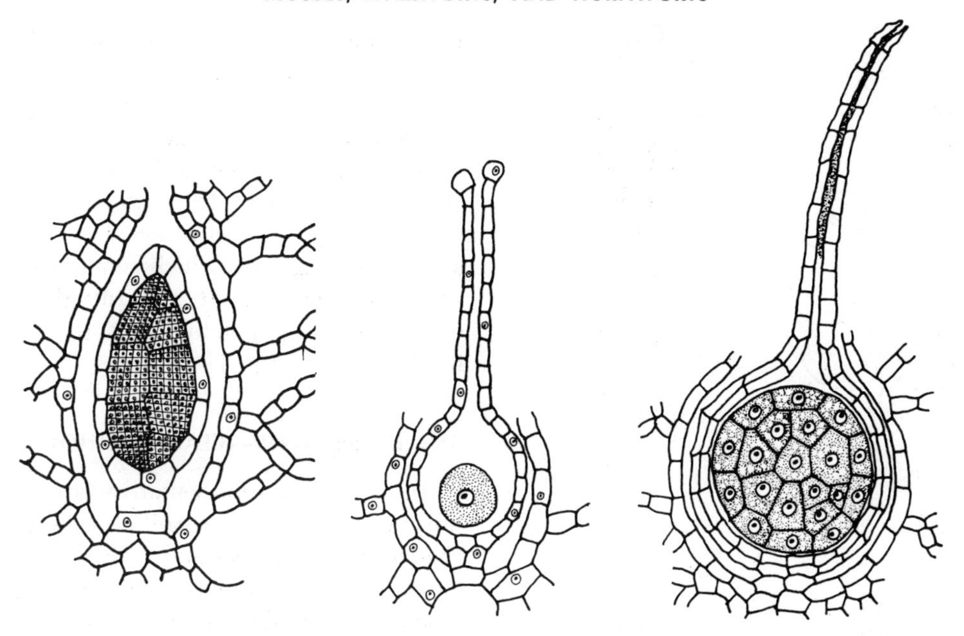

Fig. 35–13. Reproductive structures of *Riccia; left,* an antheridium; *center,* an archegonium; *right,* an embryo developing in an archegonium. (William M. Carlton, *Laboratory Studies in General Botany.* Copyright © 1961, The Ronald Press Company, New York.)

Asexual reproduction. Upon the upper surface of a gametophyte usually there are small **gemmae cups** which contain minute, green, multicellular bodies known as **gemmae.** Gemmae are dispersed by splashing water, and they develop into new gametophytes. Asexual reproduction is also achieved by fragmentation of the gametophyte.

Riccia

Most species of this liverwort grow on moist soil, but some grow on water. The gametophyte of *Riccia* resembles the flat thallus of *Marchantia,* but the antheridia and archegonia are not borne on umbrella-like stalks. They are produced in depressions of the thallus, and usually antheridia and archegonia are found on the same gametophyte. When the male and female sex organs are mature, the antheridia burst open, and the sperms swim to the eggs and fertilize them. The diploid zygote divides and the resulting cells continue to divide forming a nongreen embryo which is nourished entirely by the gametophyte. Further development of the embryo leads to the differentiation of an external jacket, one cell thick, which encloses many spore mother cells. The spore mother cells undergo reduction division to form haploid spores. The sporophyte of *Riccia* is the simplest type found in present-day liverworts, consisting at maturity of only a capsule containing spores (Fig. 35–13). Elaters are not present. Contrast this with the more complex sporophyte of *Marchantia,* with its foot, seta, and capsule containing both spores and elaters. The spores are released when the gametophyte decays and when the capsule ruptures. The spores develop into gametophytes.

Porella

Porella is a representative leafy liverwort. The gametophytes of *Porella* and other leafy liverworts are differentiated into leaves and stems which give rise to unicellular rhizoids. Superficially they resemble moss gametophytes. However, *Porella* and

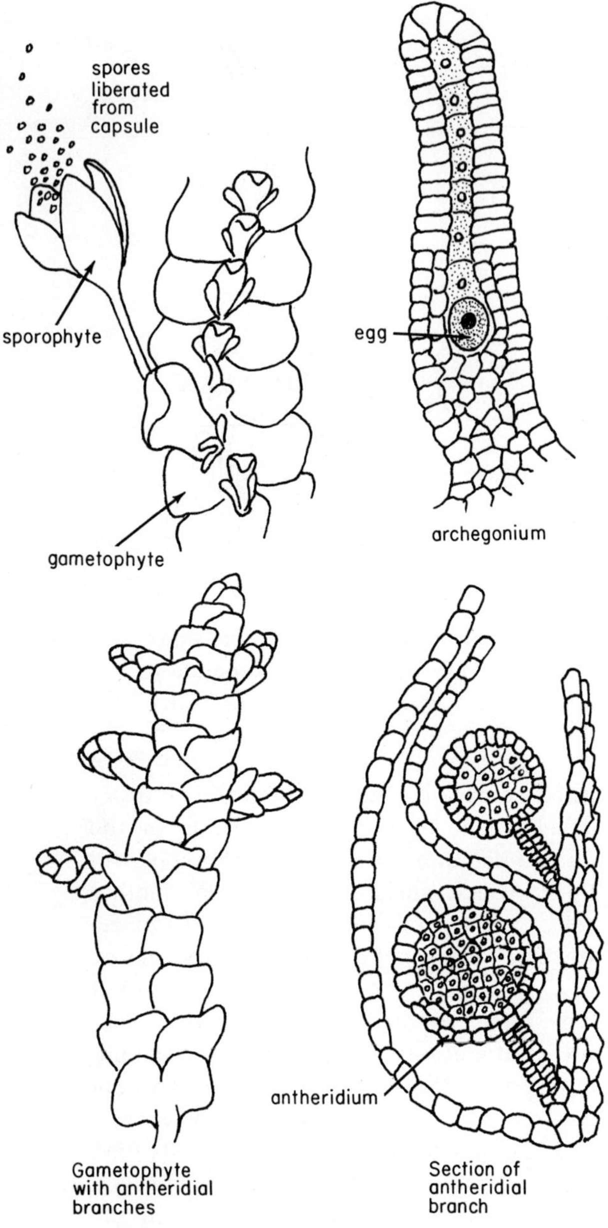

Fig. 35–14. Reproductive structures of *Porella*, a leafy liverwort.

its relatives can be distinguished from mosses by the arrangement of the leaves and by the stem habits. Mosses generally grow upright whereas leafy liverworts are prostrate. In *Porella* the leaves are arranged in three rows along a thin stem; in mosses the leaves are spirally arranged. The leaves lack a midrib and are just one cell thick.

The life cycle of *Porella* is similar to that of *Marchantia.* Antheridia and archegonia develop on separate gametophytes. The biflagellate sperms swim to the archegonia and effect fertilization. The zygote develops into a sporophyte consisting of a foot, seta, and globular capsule which contains haploid spores and elaters. At maturity the jacket of the capsule splits into 4 flaps, thus liberating the spores (Fig. 35–14).

THE HORNWORTS—CLASS ANTHOCEROTAE

Anthoceros, a representative hornwort, is fairly common on moist soil especially along shaded stream banks and lake shores. The gametophyte of *Anthoceros* is an irregularly lobed, green thallus somewhat smaller than that of *Marchantia.* Unicellular rhizoids occur on the lower surface. Internally, the gametophyte shows little differentiation. The thallus is parenchymatous and about eight cells thick. Each cell possesses just one chloroplast containing one pyrenoid, a feature unique in the bryophytes and indicating an algal affinity. In other classes of bryophytes each cell contains numerous chloroplasts, and pyrenoids are absent. The epidermal cells are not markedly different from the internal cells except that stomata occur in both the upper and lower surface layers. At certain places, the cells of the lower surface separate from each other forming pits which become filled with mucilage. The pits expand into the inner tissue, forming cavities filled with mucilage and often inhabited by *Nostoc,* a blue-green alga.

Antheridia are located in chambers of the thallus and archegonia are embedded in the thallus (Fig. 35–15). As in other bryophytes, the sperms swim to the archegonia when the thallus is covered with water.

Each zygote develops into a sporophyte which is more highly developed than those of mosses and liverworts. The mature sporophyte consists of a foot anchored in the gametophyte and a hornlike capsule which

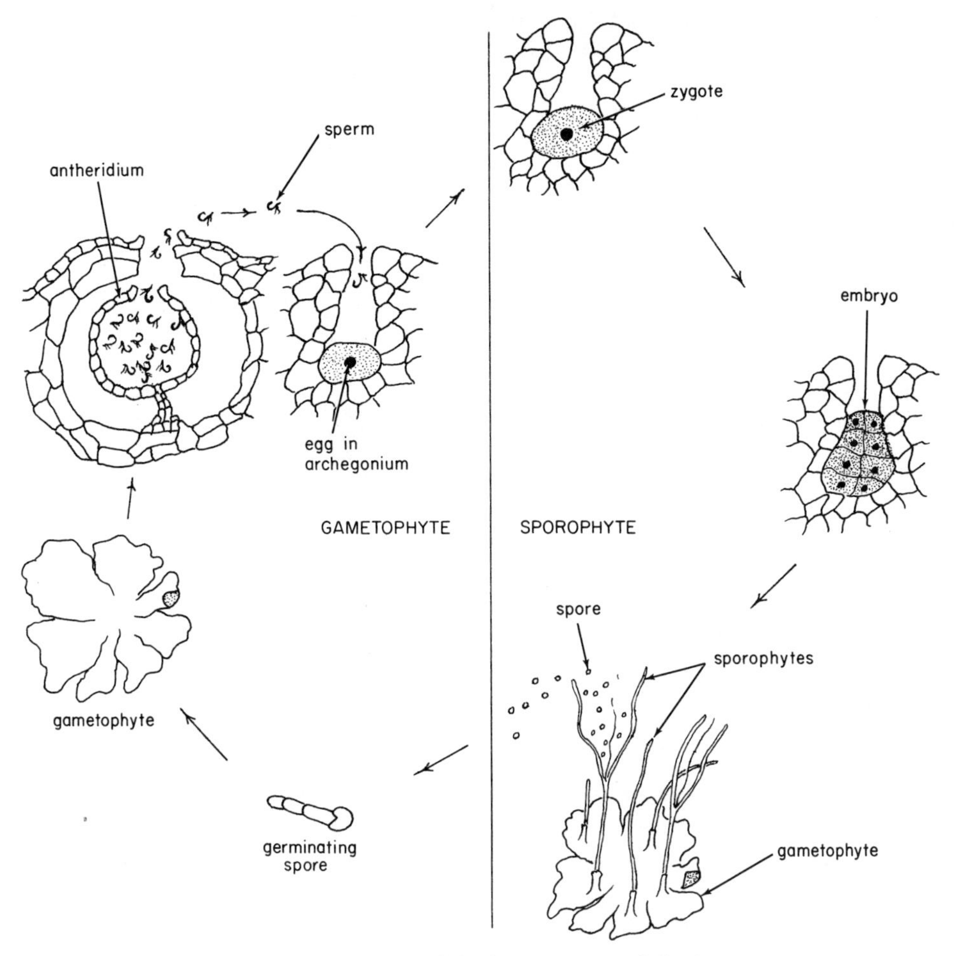

Fig. 35–15. Life cycle of *Anthoceros*, one of the hornworts.

may attain a length of from 1 to several inches. A core of cells, the columella, is present in the center of the capsule; it is surrounded by spore mother cells which undergo reduction division to form spores. Intermingled with the spores are elaters, each made of one to four cells. A parenchymatous layer three to four cells wide envelops the spores. The cells of this layer contain chloroplasts. The sporophyte is protected on the outside by an epidermis made up of cells also containing chloroplasts (Fig. 35–16). The epidermis has well-developed stomata and a cuticle. Because the green tissues of the sporophyte make food, the sporophyte is not wholly dependent on the gametophyte for its nutrition, but it does rely on the gametophyte for water and minerals, which are absorbed by the foot. When the spores are ripe, the top of the capsule splits in two thus releasing the spores. As old spores are liberated from the upper end of the capsule, new ones are formed near the base; in other words, spore formation is continuous. The sporophytes of the Anthocerotae are the only ones among the bryophytes that form new spore mother cells in the basal meristematic zone while releasing mature spores at the apex. The meristematic zone of the capsule is just above the absorbing foot and it continually forms new capsule tissue, not only spore mother cells, but epidermal cells, columella cells, and cells of the parenchymatous layer. In certain species, the capsule continues to grow as long as the gametophyte lives.

The presence of a meristem in *Anthoceros* and an aerating system complete with stomata may indicate that *Anthoceros* evolved from ancestors with even larger and more complex sporophytes. These features may be vestiges from a more complex ancestral sporophyte.

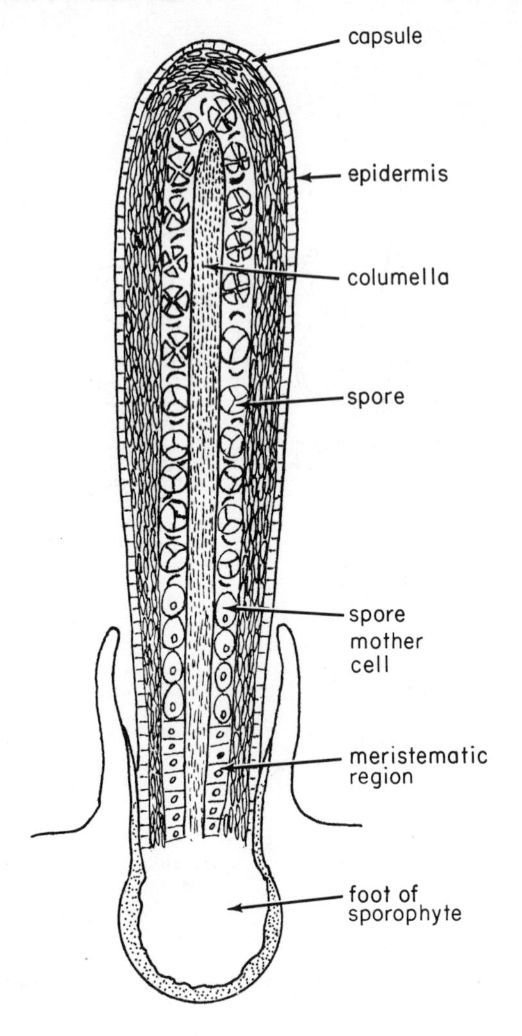

Fig. 35–16. Longitudinal section of the sporophyte of *Anthoceros*.

ORIGIN AND RELATIONSHIPS OF THE BRYOPHYTA

Little is known with certainty about the origin and evolution of the bryophytes. The fossil record is too fragmentary to enable us to trace their evolutionary history. Fragmentary remnants of thallose liverworts, which resemble present-day liverworts, have been found in rocks of Carboniferous age, as have structures that may be remains of mosses.

The immediate ancestors of the bryophytes were probably more complex plants than present-day forms. In other words, the evolutionary tendency has been one of reduction instead of increased complexity.

If evolution has progressed from more complex sporophytes to those of simpler form, the sporophyte of *Anthoceros* would be considered more ancient than that of *Marchantia* or *Riccia*. Because *Riccia* has the most reduced sporophyte, it would be the most recent bryophyte.

As evolution progressed, the gametophytes also became reduced. The ancestors of the bryophytes had erect gametophytes and radially arranged leaves. From such ancestors the flat gametophytes of liverworts and hornworts evolved.

The concept previously outlined is the one in vogue today and is just the opposite of the one which was fashionable a decade or two ago. The change in idea was not the result of the discovery of new features but simply a different way of reasoning from the same facts. In earlier years it was believed that *Anthoceros* with its complex sporophyte was the most advanced bryophyte instead of the most primitive one. *Riccia* with its simple sporophyte was considered most primitive. Furthermore, it was believed that *Anthoceros* may have given rise to the vascular plants, a hypothesis not generally accepted today. Here, as in other aspects of science, there are always at least two features in the development of a hypothesis: (1) the collection of data, and (2) the interpretation of the data. Vast amounts of data have accumulated and wait the development of new theories.

QUESTIONS

1. Give the general characteristics of the Bryophyta.

2. What evidence can be found to indicate that the bryophytes evolved from green algae?

3. Describe briefly the distribution of the Bryophyta.

4. _____ Which one of the following is not characteristic of bryophytes? (A) Bryophytes are small in stature, usually less than 6 inches tall. (B) Bryophytes lack vascular tissue and true roots, stems, and leaves. (C) Flagellated sperms are produced in a multicellular antheridium and an egg in the venter of a multicellular archegonium. (D) The sporophytes of bryophytes are characteristically larger and more complex than the gametophytes. (E) In all bryophytes the zygote develops into an embryo which receives nourishment and protection from the gametophyte. (F) All bryophytes have an alternation of generations; a haploid gametophyte which produces gametes alternates with a diploid sporophyte that produces haploid spores.

5. _____ It is believed that bryophytes and vascular plants evolved from the (A) blue-green algae, (B) green algae, (C) brown algae, (D) red algae.

6. How could you explain the fact that bryophytes never grow to any great height?

7. _____ The gametophyte generation begins with a (A) haploid spore, (B) diploid zygote.

8. _____ In bryophytes motile sperms are produced in multicellular structures called (A) archegonia, (B) sporophytes, (C) basidia, (D) antheridia.

9. _____ The many-celled, flasklike structure of a bryophyte in which the egg is produced is called (A) capsule, (B) seta, (C) archegonium, (D) antheridium, (E) oogonium.

10. _____ In mosses the sperms get to the eggs by (A) swimming, (B) being blown by the wind, (C) insects.

11. _____ The sporophyte of a moss, consisting of foot, stalk, and capsule, develops from a (A) zygote, (B) archegonium, (C) gametophyte, (D) spore, (E) protonema.

12. _____ Moss sporophytes are (A) completely independent of the gametophyte, (B) dependent upon the gametophyte for food, at least in their earlier stages of development.

13. _____ In a moss plant spores are produced by reduction division in (A) the gametophyte, (B) the capsule of the sporophyte, (C) the foot of the sporophyte, (D) on the moss protonema.

14. _____ The small lid on the terminal portion of the moss capsule is called the (A) peristome, (B) seta, (C) operculum, (D) calyptra.

15. _____ A moss spore germinates to produce a (A) protonema, (B) capsule, (C) sporophyte, (D) archegonium.

16. _____ The moss protonema resembles (A) the mycelium of a fungus, (B) an algal filament, (C) the root system of a higher plant, (D) the sporophyte.

17. Explain what is meant by an alternation of generations, using a moss plant as an example.

18. Outline by means of drawings the life cycle of a moss plant. Do likewise for *Marchantia* and *Riccia*.

19. _____ Gemmae cups are present on the gametophyte of (A) *Marchantia,* (B) *Riccia,* (C) *Anthoceros,* (D) *Sphagnum.*

20. _____ Which one of the following liverworts has a gametophyte which is differentiated into a stem and leaves which are just one cell thick? (A) *Porella,* (B) *Riccia,* (C) *Marchantia.*

21. _____ The sporophyte of *Porella* (A) is embedded in the gametophyte, (B) consists of a capsule bearing haploid spores, (C) consists of a foot, a seta, and a globular capsule containing spores and elaters, (D) has a meristematic region when mature.

22. _____ Which one of the following has the simplest sporophyte? (A) *Marchantia,* (B) *Riccia,* (C) a moss such as *Bryum,* (D) *Sphagnum.*

23. Contrast *Marchantia* and *Riccia.*

24. Describe the structure and reproduction of a hornwort.

25. Match the following:

______	1. Sporophyte consists of a foot, seta, and capsule which contains spores and elaters.	1. *Riccia,* a liverwort.
______	2. Sporophyte elevated on a stalk called a pseudopodium.	2. *Mnium,* a moss.
______	3. Sporophyte consists of a capsule containing haploid spores at maturity.	3. *Anthoceros,* a hornwort.
______	4. Sporophyte consists of a hornlike capsule, with meristematic cells, and a foot.	4. *Marchantia,* a liverwort.
______	5. Sporophyte consists of a foot, seta, and capsule with operculum and peristome.	5. *Sphagnum.*

26. Place each of the plants on the left in the correct class.

______	1. *Funaria,* a moss.	1. Musci.
______	2. *Marchantia,* a liverwort.	2. Hepaticae.
______	3. *Riccia,* a liverwort.	3. Anthocerotae.
______	4. *Porella,* a leafy liverwort.	
______	5. *Anthoceros,* a hornwort.	

36

Primitive Vascular Plants

Two great developments in the plant kingdom are found in the phylum Tracheophyta (commonly called vascular plants): the development of an efficient conducting system, and the shift in relative size between the sporophyte and the gametophyte. The plants that furnish lumber for our shelter, that decorate our homes, that provide food for our tables, as well as such less significant plants as horsetails, club mosses, and ferns, are all tracheophytes. In these the sporophyte is the conspicuous green leafy plant, and the gametophytes are very much reduced, microscopic in many cases, and are seldom seen except by botanists. The sporophytes of tracheophytes, with their true roots, stems, and leaves, are far more complex than the sporophytes of mosses.

The phylum Tracheophyta is divided into four subphyla: Psilopsida, Lycopsida, Sphenopsida, and Pteropsida. In the distant past the first three subphyla were the dominant plants on earth, but at present only the Pteropsida are important in the earth's flora, for these include the ferns and seed plants.

The Psilopsida, of which *Rhynia* (Fig. 36–1) is an example, are the most primitive vascular plants, and they probably evolved from the green algae. The transition forms between the green algae and the Psilopsida have not been found in the fossil record, leaving a big gap in our understanding of evolution. The Psilopsida may be the ancestral stock from which more complex vascular plants evolved. According to this hypothesis, one line of descent leads to the horsetails, another to the club mosses, and the third to the ferns and seed plants. This concept is not accepted by all botanists. Some believe that each subphylum of the tracheophytes originated from a different ancestral stock.

SUBPHYLUM PSILOPSIDA

Three hundred fifty million years ago when four continents constituted what is now North America, and when great inland seas covered the area between the present Appalachian and Sierra mountain ranges, psilophytes were abundant in swampy places. We use the term psilophytes to designate the members of the order Psilophytales, one of the two orders in the subphylum Psilopsida. The other order is called Psilotales.

The psilophytes are known only as fossils. Among the better-known ones is *Rhynia,* which was discovered by R. Kidston and W. H. Lang at Rhynie, Scotland, in 1917. *Rhynia* is one of the most primitive vascular plants known. The plants were 1 or 2 feet tall, and they lacked roots and leaves. They had rhizomes with rhizoids which were anchored in the mud and upright, forking aerial branches which produced sporangia at the tips. The sporangia had walls several cells thick and contained spores which were all alike. Obviously *Rhynia* is a simple plant, lacking roots, leaves, and specialized structures for the production of sporangia.

In anatomical features, however, *Rhynia* possessed those structures which permitted growth on land and which were retained, even though in somewhat modified form, by more advanced vascular plants. The stem had an epidermis with stomata, a cortex made up of cells with chloroplasts, and a stele. The stele consisted of a central core of xylem, without pith, surrounded by phloem. The heavy-walled cells of the xylem aided in supporting the erect aerial branches. The xylem also facilitated the rapid movement of water and minerals, and the phloem conducted food efficiently.

Fig. 36–2. *Psilotum nudum.* Note the three-lobed sporangia. (Donald B. Lawrence.)

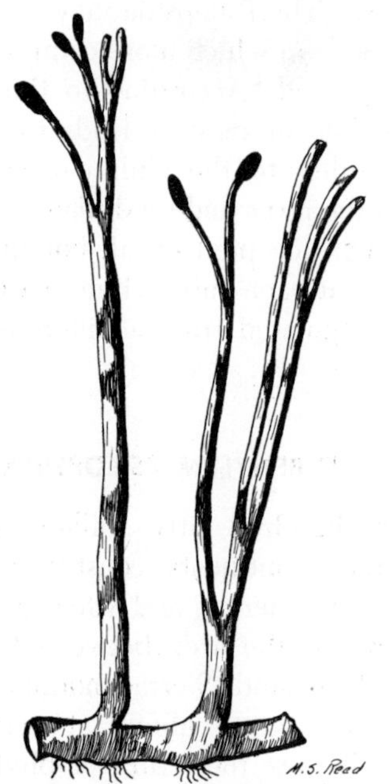

Fig. 36–1. *Rhynia,* a primitive vascular plant, now extinct. The rhizome bears rhizoids and branched stems, which bear terminal sporangia. (R. Kidston and W. H. Lang.)

In addition to *Rhynia* the order Psilophytales includes a number of other fossil genera, among which are *Psilophyton, Horneophyton,* and *Asteroxylon.*

The second order, Psilotales, includes *Psilotum* and *Tmesipteris,* genera living today and probably relicts of the once flourishing Psilophytales. These two genera are the most primitive vascular plants of today. *Psilotum* grows throughout the Caribbean area and in Florida, often as an epiphyte but also in soil or in rock crevices (Fig. 36–2). *Tmesipteris* is found in New Zealand, Australia, and other islands of the area. Their stems bear minute emergences which lack vascular tissue. They are too small to be called true leaves, but they may represent the precursors of the kind of leaf we find in the next group (Lycopsida). The three-lobed sporangia are borne on the tips of short branches which are in the axils of the emergences. Spore mother cells undergo meiosis to form haploid spores which are liberated when the sporangium breaks open. The spores germinate in the soil and produce subterranean, nongreen gametophytes, each of which is a few millimeters long. Associated with the gametophytes are fungi, especially certain Phycomycetes, which probably nourish the gametophytes (Fig. 36–3). Flask-shaped archegonia and globose antheridia develop on the same gametophyte. When mature, an antheridium opens and releases the sperms, each with many flagella, which swim to the archegonia. As usual, one sperm fertilizes an egg.

While still contained in the archegonium, the zygote develops into a young sporophyte consisting of a foot anchored in the gametophyte and a shoot. The foot absorbs nourishment from the gametophyte and passes it on to the sporophyte. After the sporophyte is well developed, with aerial stems and an underground rhizome, an abscission layer separates the foot from the shoot and the sporophyte becomes an independent plant. Later the gametophyte dies and shrivels.

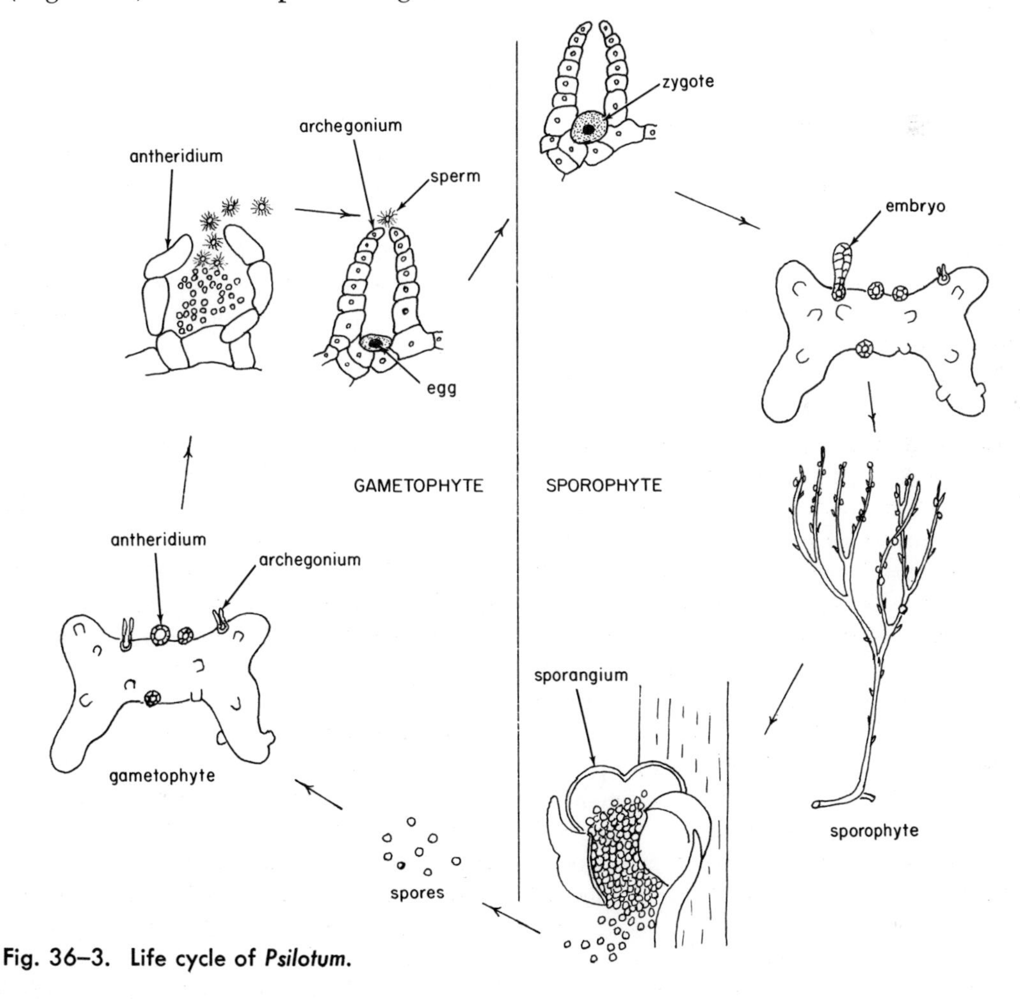

Fig. 36–3. Life cycle of *Psilotum*.

CLUB MOSSES—SUBPHYLUM LYCOPSIDA

The name club moss is a bit deceiving, for the plants to which it refers are not mosses. Like many nicknames, it grew out of a superficial resemblance to something else and, as nicknames have a way of doing, it still clings. If you could return to the Carboniferous period, you would find many small club mosses similar to those that exist today, but you would also find some giant species that no longer exist. Surely you would not be tempted to use the name club moss for those like *Lepidodendron,* which were trees 100 or more feet high and 6 feet in diameter. *Lepidodendron,* incidentally, had a cambium layer and hence secondary xylem and phloem; these advances are not found in present-day club mosses.

The Lycopsida appeared during the Devonian and were most abundant and varied in the Lower Carboniferous. The great forests of Lycopsida contributed to the coal deposits of this age. The arboreal species have been extinct for a long time and only herbaceous species have survived to the present time. Only four genera now grow on earth: *Lycopodium, Selaginella, Phylloglossum,* and *Isoetes.*

The sporophytes of extant club mosses are small and consist of roots, long slender stems, and small spirally arranged leaves, technically called **microphyllous leaves.** The leaves are considered to be outgrowths from the surface of the stem. Each leaf has one unbranched vascular bundle. The small vascular bundle that connects the leaf to the vascular tissue of the stem leaves no leaf gap in the vascular cylinder. In contrast, in the Pteropsida, a more advanced group, a leaf gap filled with parenchymatous tissue occurs where a vascular bundle passes from the stem into a leaf. The stem is differentiated into an epidermis, one cell thick and with stomata, a cortex, and a central cylinder, the stele. The central cylinder is often surrounded by an endodermis, within which the pericycle, xylem, and phloem are present (Fig. 36–4). The phloem occurs between the arms of xylem. Parenchyma cells and sieve cells make up the phloem, and tracheids, the xylem. Anatomically, the root is similar to the stem.

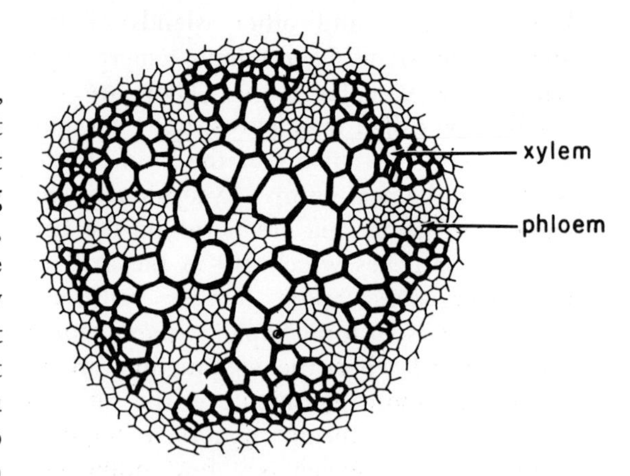

Fig. 36–4. Stele of an aerial stem of *Lycopodium.* (William M. Carlton, *Laboratory Studies in General Botany.* Copyright © 1961, The Ronald Press Company, New York.)

Some Lycopsida—for example, *Lycopodium*—produce only one type of spore; such plants are **homosporous.** *Selaginella* produces large spores (**megaspores**) and small spores (**microspores**); it is **heterosporous.**

Selaginella

The 500 species in the genus *Selaginella* are of diverse habit, ranging from minute prostrate annuals to erect or even climbing perennials. A few grow in temperate regions, but most are restricted to the tropics. The conspicuous *Selaginella* plant is the sporophyte; the gametophytes are almost microscopic.

The sporophyte. The leaves of *Selaginella* are usually arranged in four rows along the stem, two rows being made up of relatively large leaves and two of small ones. On the upper surface of each leaf, near the base, there is a small functionless bract known as a **ligule.** Anatomically, the leaves are comparable to those of seed plants; they have an epidermis with stomata, mesophyll tissue, and usually one vascular bundle. The stem of *Selaginella* has an epidermis, a cortex, an endodermis, and a stele consisting of xylem, phloem, and pericycle. In many species, but not in all, there is a central core of xylem which is surrounded by phloem. In some species leafless branches, called **rhizophores,** grow into the soil and produce roots.

At the tips of some branches the leaves are closely aggregated to form club-shaped cones (**strobili**) (Fig. 36–5). In the axil of each leaf of a strobilus one **sporangium** (spore case) is produced. In some sporangia many spore mother cells undergo reduction division to form numerous small spores known as **microspores.** In other sporangia of the same cone only one spore mother cell undergoes reduction division to form four large spores called **megaspores.** A sporangium which contains microspores is a **microsporangium,** and the leaf which subtends it is a **microsporophyll** (Fig. 36–6*A*). Similarly, the case containing megaspores is a **megasporangium** and the leaf below it is a **megasporophyll** (Fig. 36–6*B*). A plant, such as *Selaginella,* which produces 2 kinds of spores is said to be **heterosporous.**

Fig. 36–5. *Selaginella flabellata.* Each branch is terminated by a strobilus. One sporangium is produced in the axil of each leaf (sporophyll) of a strobilus.

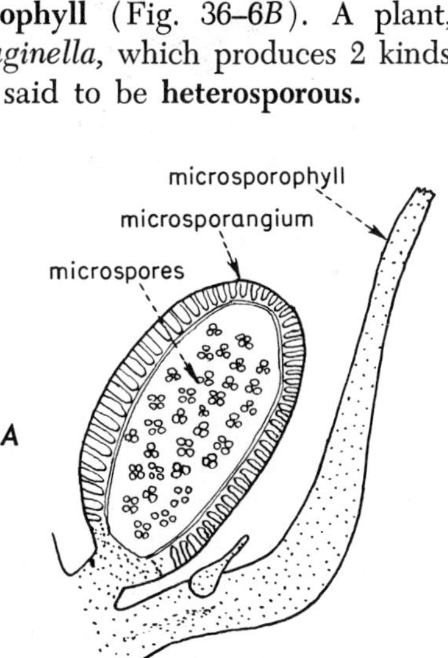

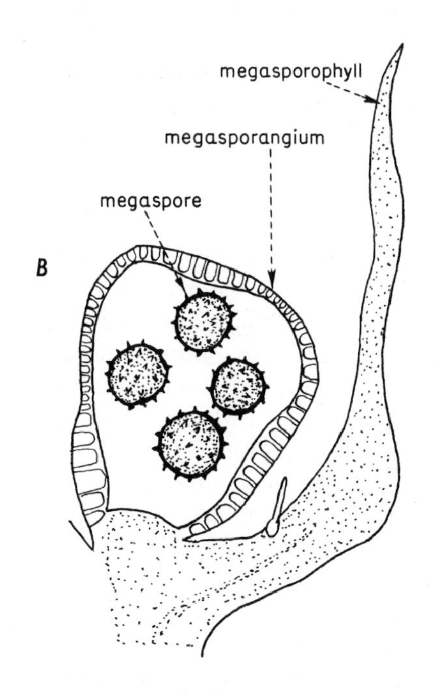

Fig. 36–6. *A.* A microsporophyll of *Selaginella* with a microsporangium containing microspores. *B.* Megasporophyll of *Selaginella* with a megasporangium containing megaspores. (After F. M. Lyon, *Botan. Gaz.*)

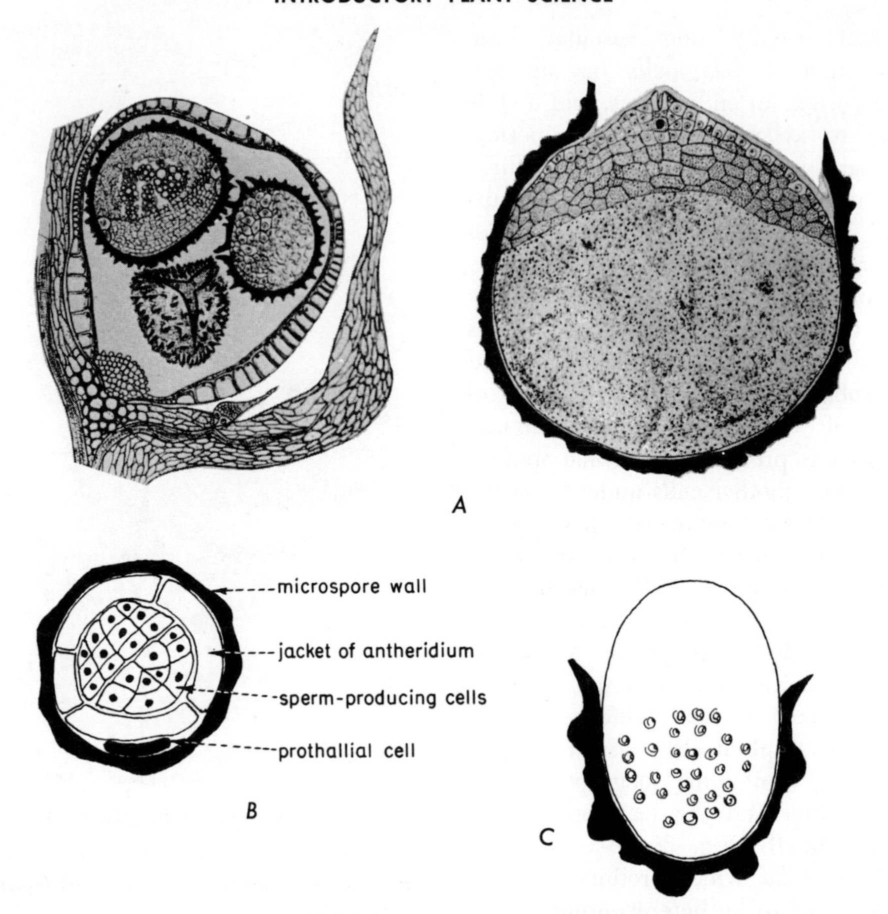

Fig. 36–7. A. *Left*, female gametophytes of *Selaginella* developing within the megaspores while they are still contained in the megasporangium. *Right*, a mature female gametophyte with an archegonium containing an egg (black sphere). B. A developing male gametophyte of *Selaginella*. C. A male gametophyte of *Selaginella* about to liberate sperms. (*A* and *B* after F. M. Lyon, *Botan. Gaz.*)

Gametophytes of *Selaginella*. The microspores develop into male gametophytes that produce sperms and the megaspores into female ones bearing eggs. Both male and female gametophytes are inconspicuous plants which secure nourishment from the food which was stored in the spores by the sporophyte. The nutrition is just the reverse of that found in liverworts; in liverworts the gametophyte is the independent generation, and the sporophyte depends on the gametophyte for nourishment.

A male gametophyte is so small that it can, and does, develop within the wall of the microspore. The mature male gametophyte consists of one small functionless cell (called a prothallial cell) and an antheridium made up of a multicellular jacket which encloses a large number of minute biflagellate sperms (Fig. 36–7*B*). In some species the male gametophytes develop in the microspores while they are still contained in the microsporangium; after the gametophytes are mature, the spore case breaks open and releases the microspores with their enclosed male gametophytes. They are disseminated by wind and gravity. Those which lodge near megasporangia release sperms which fertilize the eggs that are produced by the female gametophyte (Fig. 36–7*C*).

A female gametophyte develops from a megaspore. Most of its development occurs within the wall of the megaspore, but near maturity the many cells of the female gametophyte enlarge, thus rupturing the megaspore wall and permitting a mass of small cells to protrude. Archegonia develop in the exposed cells. Each archegonium consists of a venter containing an egg, a ventral canal cell, and a neck which surrounds one neck canal cell (Fig. 36–7*A*). In some species of *Selaginella* the megasporangium does not break open until after the female gametophytes are mature. Then the spore case opens, permitting sperms which are released from a nearby male gametophyte to reach the eggs (Fig. 36–7*C*). Water is necessary for fertilization. When water is present, the antheridium expands and breaks open the microspore wall. The biflagellate sperms escape and swim by means of their flagella to the eggs. One sperm unites with each egg.

Development of the sporophyte. Shortly after fertilization, the zygote divides to form two cells. One of them develops into the **embryo** and the other cell develops into a **suspensor,** which pushes the developing embryo into the nutritive part of the female gametophyte. At maturity the embryo consists of a stem, two leaves, a root, and a foot which absorbs food. Still encased in the female gametophyte, the embryo escapes from the megasporangium and falls to the ground. Then the root extends out into the soil, and the stem elongates and lifts the leaves above ground. The young sporophyte is now independent and the female gametophyte decays and disappears.

The life cycle of *Selaginella* resembles that of seed plants in several ways, even though the seed plants did not evolve from the Lycopsida (Fig. 36–8). Both seed plants and *Selaginella* produce small spores which develop into male gametophytes and large spores which grow into female gameto-

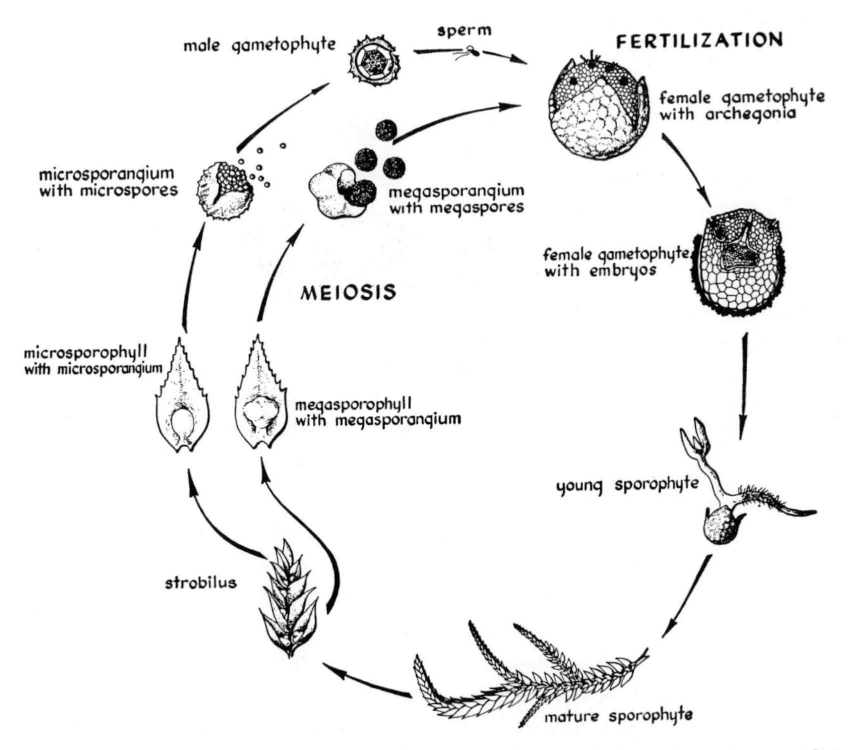

Fig. 36-8. Life cycle of *Selaginella.* (Reproduced by permission from *Botany: Principles and Problems* by E. W. Sinnott and K. S. Wilson. Copyright, 1955, McGraw-Hill Book Co., Inc., New York.)

Fig. 36–9. *Left, Lycopodium phyllanthum. Right, Lycopodium annotinum.* The cones (strobili) are produced at the tips of the branches. Each modified leaf of a strobilus is a sporophyll on which spores are produced in a sporangium.

phytes. In both groups the fertilized egg develops into an embryo with a suspensor. In *Selaginella,* as in seed plants, the sporophyte is the conspicuous generation and the gametophytes are inconspicuous.

Lycopodium

Perhaps you have collected "ground pines" (the common name for some species of *Lycopodium*) in woody areas for use as Christmas decorations, for they are well represented in the United States (Fig. 36–9). A few species of *Lycopodium* are found in Arctic regions, and a great many in tropical areas. All of them are small plants (usually less than 1 foot tall) with underground rhizomes that produce adventitious roots and aerial shoots that bear small, evergreen, spirally arranged leaves. As in *Selaginella,* the stems have an epidermis, cortex, and stele.

In the more primitive species of *Lycopodium,* one sporangium develops on the upper surface of each ordinary leaf. In more advanced forms, sporangia are produced on specialized leaves which are aggregated into a cone; the sporophylls differ in shape and color from the ordinary foliage leaves. One kidney-shaped sporangium develops on the upper surface of each sporophyll, near the base (Fig. 36–10). The numerous spore mother cells that are present in each sporangium undergo reduction division to form spores which are alike in size. Because all of the spores are of the same kind, *Lycopodium* is **homosporous.**

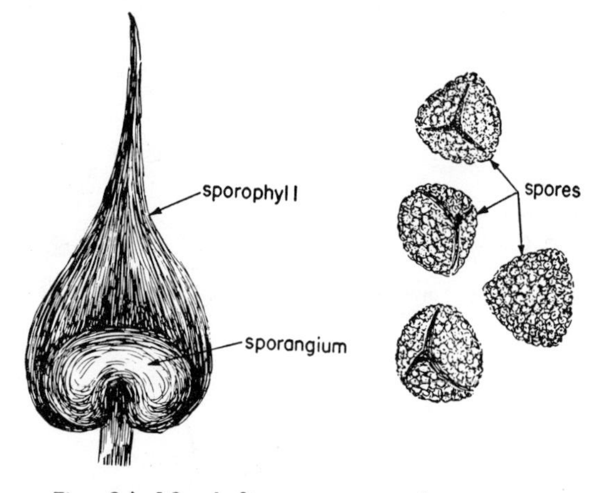

Fig. 36–10. *Left,* one sporangium occurs on the upper surface of each sporophyll of *Lycopodium. Right,* many spores develop in each sporangium, greatly enlarged.

If a spore lands on moist soil and if conditions are favorable, it develops into a small gametophyte. The gametophytes of some species are green, hence they can manufacture food, whereas those of other species are colorless saprophytic plants. Both the green and nongreen gametophytes grow symbiotically with a fungus. Archegonia and antheridia are borne on the same gametophyte. When the antheridia are mature, the biflagellate sperms are released, and they swim in water of rain or dew to the eggs and fertilize them (Fig. 36–11).

As in *Selaginella,* the fertilized egg divides to form a suspensor cell and a cell which develops into an embryo. The mature embryo has a foot which absorbs food from the gametophyte, a root, a leaf, and a stem.

HORSETAILS—SUBPHYLUM SPHENOPSIDA

All of the present-day horsetails (about 25 species) are included in the genus *Equisetum,* and nearly all of them are small plants about 8 to 40 inches high. Horsetails are the only living survivors of a great group of plants that occupied much of the land surface during the Devonian period, about 300 million years ago. During the Carboniferous period some horsetails, such as *Calamites,* were trees about 50 feet in height with trunks 1 foot in diameter. The horsetails and the fossil plants, such as *Annularia* (Fig. 36–12), resembling them are grouped together in the subphylum Sphenopsida.

Fig. 36–12. *Annularia,* a Carboniferous horsetail. (Chicago Natural History Museum.)

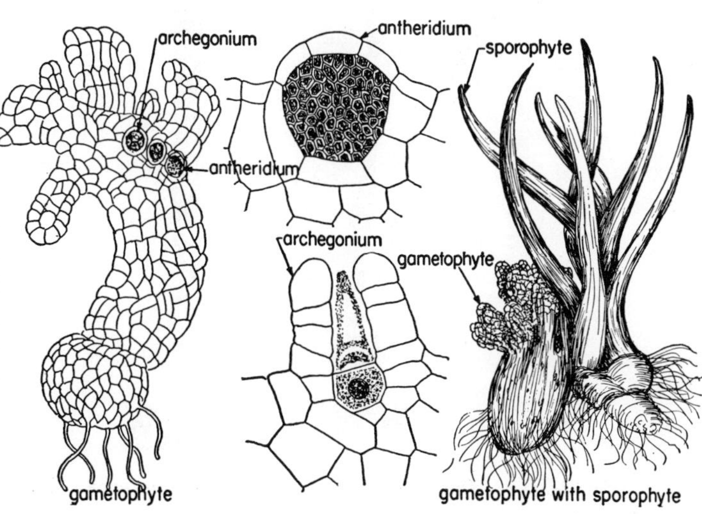

Fig. 36–11. *Lycopodium cernuum. Left,* gametophyte with two archegonia and one antheridium. *Upper center,* details of antheridium. *Lower center,* an archegonium. *Right,* gametophyte with developing sporophyte.

They may have evolved from some member of the Psilopsida.

Sporophyte

All of the extant horsetails have perennial, branched, underground stems which produce roots and give rise to upright, columnar, hollow branches that bear scalelike united leaves (Fig. 36–13). There is considerable silica in the stems, which renders them suitable for scouring knives, pots, and pans; they were therefore once called "scouring rushes." The aerial stems of some species are unbranched; other species of *Equisetum* have whorled branches. The green stems and branches are the chief photosynthetic organs. The leaves are tiny, and in many species they lack chlorophyll.

The spores of horsetails are produced in a cone (strobilus) at the summit of a stem. In *Equisetum hiemale* a cone is produced on an ordinary green stem, whereas in *Equisetum arvense* the cone is borne at the top of a colorless or brownish stem which develops early in the spring. The green vegetative branches of the latter species appear later in the season.

A strobilus consists of an axis bearing stalked, shield-shaped modified leaves called **sporophylls** (Fig. 36–14). Five to ten cylindrical sporangia are borne on the inner face of each sporophyll, also known as a sporangiophore. Spore mother cells within a sporangium undergo meiosis to form haploid spores which are alike in size and appearance. The outer layer of the wall of a mature spore splits into four narrow filaments which are called **elaters.** The elaters unroll when dry and roll up around the spore when moist, thus facilitating the dispersal of the spores from a split sporangium.

Gametophytes of *Equisetum*

The spores germinate on moist soil and develop into gametophytes which are small, flat, lobed, green plants that live inde-

Fig. 36–13. The horsetails. *Left, Equisetum arvense. Center, Equisetum hiemale. Right, Equisetum fluviatile.*

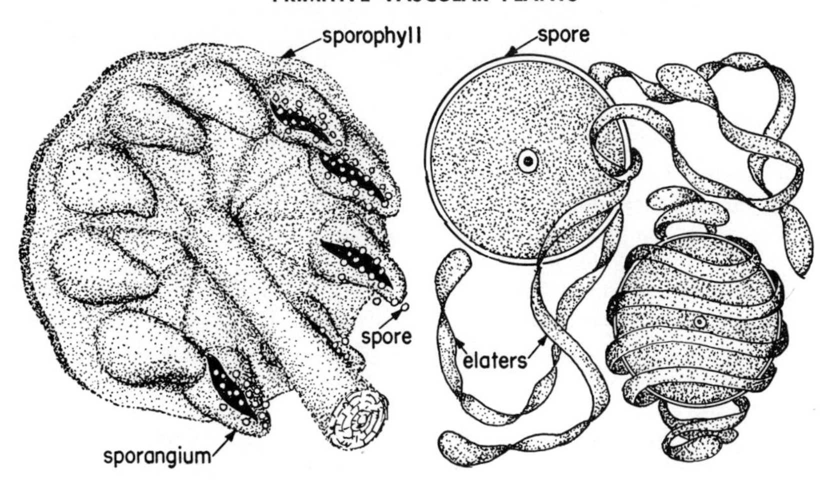

Fig. 36–14. *Equisetum.* *Left*, sporangia on inner face of sporophyll. *Right*, spores with elaters coiled (*lower*) and uncoiled (*upper*). (After Wettstein.)

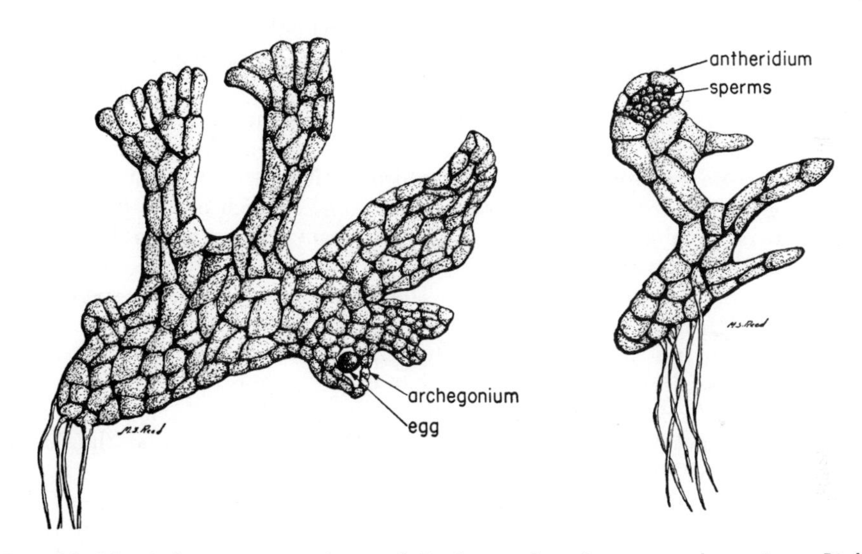

Fig. 36–15. *Left*, a gametophyte of *Equisetum* bearing an archegonium. *Right*, a gametophyte with an antheridium.

pendently (Fig. 36–15). The sexual organs, antheridia and archegonia, are produced on the edges of the lobes. Archegonia and antheridia may develop on the same or on different gametophytes, even in the same species. In the latter instance, antheridia are borne on the smaller gametophytes and archegonia on the larger ones. An egg is borne in each archegonium, and within the multicellular jacket of an antheridium there are many sperms, each with numerous flagella instead of two, as was characteristic of sperms of the club mosses and bryophytes. As usual, the sperms swim to the eggs and effect fertilization.

Development of the Sporophyte

The fertilized egg develops into an embryo without a suspensor. In time the embryo develops into a young sporophyte, and the gametophyte shrivels and disappears.

QUESTIONS

1. Give the general characteristics of the Tracheophyta.

2. _____ In all Tracheophyta, (A) the sporophyte is the more complex generation, (B) the gametophyte is the more complex generation.

3. _____ Which one of the following is not a subphylum in the phylum Tracheophyta? (A) Psilopsida, (B) Lycopsida, (C) Sphenopsida, (D) Pteropsida, (E) Spermopsida.

4. _____ The Psilophytales (A) are an order in the Lycopsida, (B) are known only as fossils, (C) include two genera, *Tmesipteris* and *Psilotum,* which are present-day plants, (D) are commonly called club mosses.

5. _____ The most primitive vascular plants are in the subphylum (A) Psilopsida, (B) Lycopsida, (C) Sphenopsida, (D) Pteropsida.

6. _____ Which one of the following statements about Psilopsida is false? (A) Psilopsida lack true roots and leaves. (B) Psilopsida have rhizomes with erect stems which bear sporangia. (C) Psilopsida produce two kinds of spores—that is, they are heterosporous. (D) The Psilopsida probably evolved directly from green algae instead of from a bryophyte.

7. _____ Which of the following is most primitive? (A) *Rhynia,* (B) *Psilotum,* (C) *Lycopodium,* (D) *Equisetum.* Justify your choice.

8. _____ Which one of the following statements about Lycopsida is false? (A) The sporophytes, which possess xylem and phloem, have true roots, stems, and leaves. (B) Sporangia are borne singly on the upper surfaces of sporophylls. (C) The gametophyte of the Lycopsida is very small and simple. (D) Lycopsida probably evolved from Psilopsida. (E) In contrast to Psilopsida, the Lycopsida are becoming more prevalent as time goes on.

9. _____ The more primitive species of *Lycopodium* are those in which (A) the sporophylls are grouped into a strobilus, (B) the sporophylls resemble vegetative leaves.

10. Diagram or describe the life cycle of *Selaginella.*

11. Diagram or describe the life cycle of *Lycopodium.*

12. _____ A leaf which produces spores in a sporangium is called (A) gametophyll, (B) sporophyll, (C) gametophyte, (D) sporophyte.

13. List three ways in which you could distinguish between true mosses and club mosses.

14. _____ *Lycopodium* (A) is heterosporous, (B) is known only as a fossil, (C) produces sporophytes which initially are at least partially parasitic on the gametophytes, (D) produces gametophytes which are parasitic on the sporophytes.

15. _____ Which one of the following statements about *Selaginella* is false? (A) The sporophyte is the conspicuous generation and the gametophytes are inconspicuous. (B) The food used by developing gametophytes was produced by the sporophyte. (C) In both seed plants and *Selaginella* the fertilized egg develops into an embryo with a suspensor. (D) *Selaginella* is homosporous.

16. _____ Which one of the following is false? (A) *Equisetum* produces a strobilus bearing sporophylls that produce spores in sporangia. (B) *Equisetum* is homosporous. (C) *Equisetum* produces sporophytes which are never dependent on the gametophyte. (D) *Equisetum* spores develop into gametophytes which are independent plants.

17. _____ Which one of the following statements about *Equisetum* is false? (A) *Equisetum* is in the subphylum Sphenopsida. (B) The sporangia are borne on sporophylls, which are aggregated into a cone. (C) Living species of *Equisetum* are heterosporous. (D) The zygote forms an embryo, which lacks a suspensor, that in time grows into a mature sporophyte.

18. Diagram or describe the life cycle of *Equisetum.*

19. _____ In which one of the following are the antheridia and archegonia always produced on different gametophytes? (A) *Equisetum,* (B) *Selaginella,* (C) *Lycopodium.*

20. _____ Elaters occur in sporangia of (A) *Psilotum,* (B) *Lycopodium,* (C) *Equisetum,* (D) *Selaginella.*

21. _____ Which one of the following statements about the elaters of *Equisetum* is false? (A) Dry air blowing over the strobilus causes the elaters to uncoil, thus increasing the buoyancy of the spores. (B) When the elaters are uncoiled, the spores are carried by the slightest breeze. (C) When moist air is encountered over damp places, the elaters coil tightly around the spore and the spore settles down on the moist soil. (D) The elaters lose their hygroscopic character in less than a month.

37

Ferns

Present-day ferns do not produce seeds, but certain plants, known only from fossils, had fernlike leaves and produced seeds (Fig. 37–1), indicating that seed plants evolved from ancient ferns. The ancestral stock of the ferns and seed plants was a simple vascular plant in the subphylum Psilopsida, from which there are three lines of descent. Two lines ended in blind alleys, the line culminating in the club mosses and the one terminated by the horsetails. The third led to ferns and from there to the seed plants.

Because the ferns and seed plants had a common ancestry, they are in the same subphylum, the Pteropsida. This subphylum is divided into three classes: (1) Filicineae, or ferns; (2) Gymnospermae, the conifers and allies; and (3) Angiospermae, the flowering plants. In all of these the sporophyte is the conspicuous plant and possesses true roots, stems, and leaves, all with well-developed vascular tissue. The gametophytes of Pteropsida are relatively small and transitory. Except in the ferns, the gametophytes are parasitic on the sporophytes.

FERNS—CLASS FILICINEAE

Ferns were not always a minor part of the earth's vegetation as they are today. During the Carboniferous period, about 250 million years ago, ferns and fernlike plants, some of which were trees, were the dominant plants on earth. Some ferns, such as the royal fern (*Osmunda*) and *Marattia,* which were present during that period, have survived to this day, but others have vanished and are known only as fossils. All of the families of present-day ferns were in existence about 140 million years ago. The largest and most valuable deposits of coal were laid down during the Carboniferous period, a time of great vegetative activity. Trees related to present-day horsetails, club mosses, and ferns occupied the swamps and their buried remains were gradually transformed by pressure and heat into coal.

Fig. 37–1. Leaf impression of *Neuropteris,* a seed fern of Carboniferous age. (Chicago Natural History Museum.)

Fig. 37–2. Sporophyte of *Polypodium vulgare.* The horizontal underground stem (rhizome) forms leaves and roots. Each leaf (frond) consists of a blade and a petiole. Sporangia occur in clusters called sori on the undersurfaces of the fronds.

Extant ferns, of which there are about 8000 species, grow in a variety of habitats and in practically all parts of the world, but they are most abundant in the tropics, where some grow perched on the branches of trees and others on moist soil. We generally associate ferns with moist, shady woods and ravines, but some are adapted to other habitats such as swamps, marshes, rocky hillsides, unshaded fields, gorges, and crevices in cliffs. Ferns range in size from diminutive botrychiums, only a few inches tall, to trees with trunks 1 foot in diameter that tower to a height of as much as 80 feet and which are crowned with a cluster of large fronds.

Ferns, like mosses, liverworts, and primitive vascular plants, have two distinct generations in the life cycle, namely, a sporophyte generation and a gametophyte generation. In each life cycle the gametophyte generation alternates with the sporophyte generation. The gametophyte is haploid, starts from a spore, and at maturity produces gametes, sperms and eggs. The sporophyte is diploid, starts from a zygote, and produces haploid spores by reduction division.

The Fern Sporophyte

The familiar fern plant is a sporophyte which, in most of our native species, consists of an underground horizontal stem (rhizome) that bears leaves and an abundance of fine roots (Fig. 37–2). The leaves, also called **fronds,** are the most conspicuous part of a fern plant, and each leaf consists of a blade and a petiole which is attached to the rhizome. The fronds may be simple or compound. In compound fronds, the leaflets are called **pinnae** and divisions of leaflets are known as **pinnules.** The developing leaves

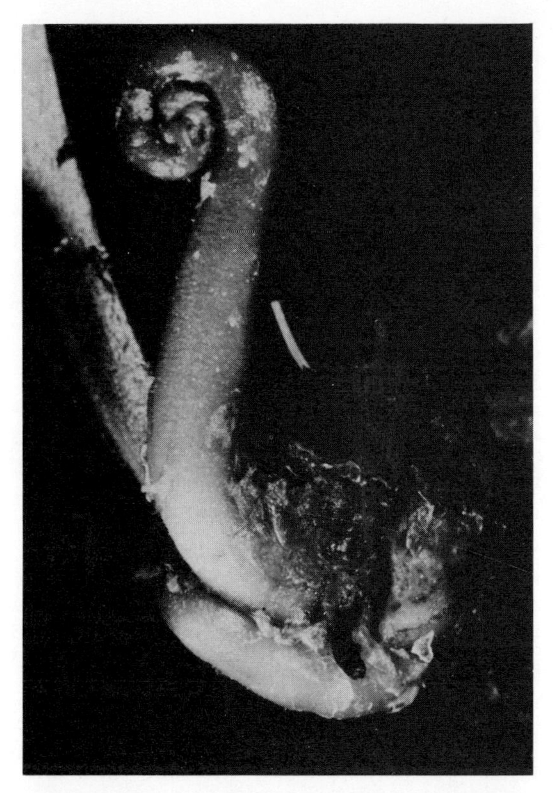

Fig. 37–3. A coiled immature leaf of a fern. The lowermost budlike structure is a leaf which has been formed recently at the tip of the rhizome.

of ferns are coiled like a fiddlehead, and in this trait ferns differ from nearly all of the seed plants (Fig. 37–3). The venation of ferns also differs from that of seed plants: The veins of fern leaves branch in a Y-shaped pattern, whereas those of seed plants do not. Anatomically, a fern leaf resembles a leaf of a flowering plant; it has an upper epidermis, a lower epidermis, stomata, mesophyll cells, and veins.

In most species the rhizome of a fern is perennial. At the tip of each branch of a rhizome an apical meristem is present, where leaves and stem cells are formed through the repeated division of a single **apical cell.** The stem (rhizome) of a fern is as complex as that of a seed plant. The fern stem has a protective epidermis, a cortex, and a stele, and in some fern stems a pith is also present. The stele consists of xylem, phloem, and pericycle. Tracheids, which conduct water, are present in the xylem, and sieve tubes, which conduct food, occur in the phloem. In some ferns—for example, *Gleichenia*—the stele is a central core; in *Adiantum* it is a tube which is bounded internally by the pith and externally by the cortex; and in *Pteridium* the stele is made up of several distinct vascular bundles. In the cross-section of a *Pteridium* rhizome illustrated in Figure 37–4 are shown the epidermis, cortex, a ring of vascular bundles, and a pith with two vascular bundles. Xylem occupies the center of each vascular bundle, and from the center outward this tissue is surrounded by phloem, a layer of pericycle cells, and an endodermis. Some of the pith cells have heavy rigid walls (the two dark bands).

The roots of mature ferns originate from the rhizome; they are adventitious roots. Anatomically, the roots of ferns are like those of seed plants, except that fern roots do not have a cambium.

Sporophylls, sporangia, and spores. At certain seasons of the year brown spots appear on the lower surfaces of some or all of the leaves.

Each spot is a **sorus,** which consists of a mass of brown powder which may or may

Fig. 37–4. Cross-section of a rhizome of *Pteridium.* (Carolina Biological Supply Co.)

not be covered with a thin membrane, the **indusium** (Fig. 37–5*A*). Each speck of brown powder is a spore case, **sporangium,** in which spores are developed. Leaves which produce spores are called **sporophylls.** In many ferns all of the leaves are alike in appearance and all leaves produce spores. In other ferns, the sporophylls are unlike the foliage leaves; the sporophylls are not expanded, whereas the foliage leaves are expanded (Fig. 37–6).

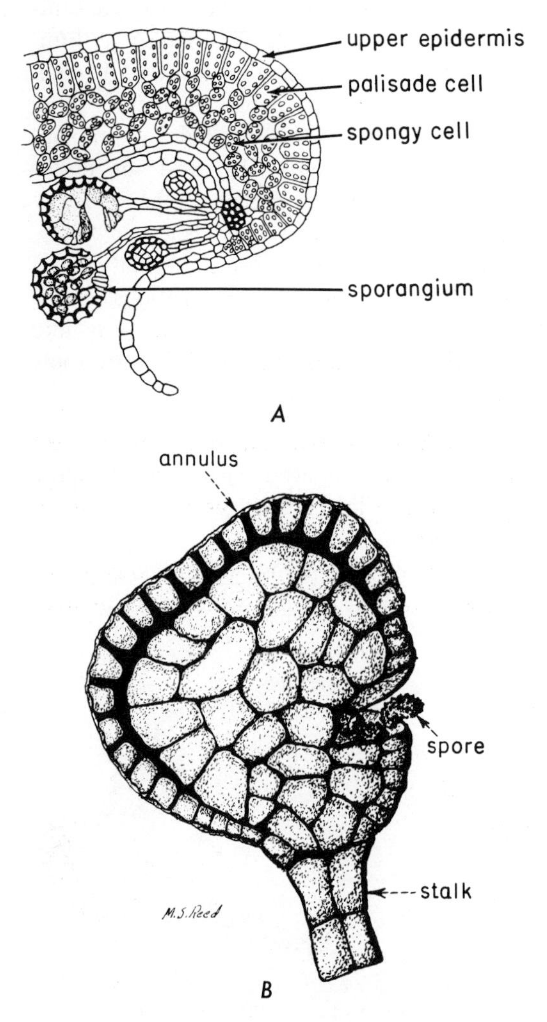

Fig. 37–5. *A.* Marginal sorus and leaf structure of *Pteridium.* (William M. Carlton, *Laboratory Studies in General Botany.* Copyright © 1961, The Ronald Press Company, New York.) *B.* Details of a sporangium.

The shape, size, and position of sori and the presence or absence of an indusium are characteristics which are used to identify ferns. Depending on the species, sori are round, elongate, or confluent over the leaf surface. The indusium is shield-shaped in some ferns and elongate in others. In other ferns an indusium is lacking, and in some of these the leaf margin is reflexed to cover the sporangia.

A fern sporangium is a rounded case with a delicate wall, formed of a single layer of cells, and with a slender stalk (Fig. 37–5*B*). In many ferns, each sporangium is formed by the division and subdivision of an epidermal cell. A ring of cells, each with thickened inner and side walls, forms an **annulus,** which extends from the stalk over the top of the spore case to thin-walled cells, known as **lip cells.** Spore mother cells, sixteen in many species, are differentiated within an immature sporangium. They undergo meiosis to form haploid spores. Most ferns are homosporous, meaning that all spores are of one size. After the spores are mature, the annulus dries out and contracts, thereby splitting the more delicate side walls of the sporangium and opening the case. The annulus then flips forward to about its original position, thus catapulting the spores a considerable distance from the plant. This discharge of spores may be observed by placing mature sporangia on a moist slide and watching them under the low power of a microscope.

The Fern Gametophyte

It has been estimated that one golden polypody fern produces about one billion spores a year, but out of this number only a very few find a suitable place for growth. A favorably situated spore germinates to form a short filament of green cells with colorless rhizoids extending out from the cells. The terminal cell of the filament develops into a haploid, green, heart-shaped gamete-producing plant, the gametophyte (also called a **prothallus**). This is an independent plant, manufacturing its own food

Fig. 37–6. The sensitive fern, *Onoclea sensibilis,* with sterile fronds and a sporophyll (*at extreme left*).

and living its own life. The mature gametophyte is usually less than ½ inch in diameter and one cell thick except just back of the notch, where it is several cells thick. Numerous unicellular rhizoids, which absorb water and minerals from the soil, are present on the lower surface of a gametophyte. In protected, moist areas where ferns are growing, the soil may be covered with these heart-shaped plants. They are fern plants just as much as are the leafy ferns; they developed from fern spores (Fig. 37–7).

On the undersurfaces of the gametophytes, sperms are produced in antheridia and eggs in archegonia. Certain ferns produce antheridia on some gametophytes and archegonia on different ones. In other species both antheridia and archegonia develop on the same gametophyte, in which case the antheridia on any one gametophyte mature before the archegonia. Many coiled multiflagellate sperms are produced in each antheridium. The wall of an antheridium is made up of three cells; the side of the antheridium consists of two ringlike cells, one on top of the other, and the top is covered by the third cell.

The bulbous venter of an archegonium is embedded in the gametophyte and the neck extends outward. One large egg and a ventral canal cell are present in the venter. The neck, made up of four rows of cells, surrounds one or two neck canal cells which disintegrate, forming a slimy fluid through which the sperms can swim.

When the antheridia are mature and covered with water, they break open and the sperms escape. The sperms swim to archegonia and effect fertilization; as usual, only one sperm unites with an egg. Even though a fern is a land plant, it still depends on a motile sperm, swimming in external water, to effect fertilization, a trait resembling the algae and suggesting an aquatic ancestry.

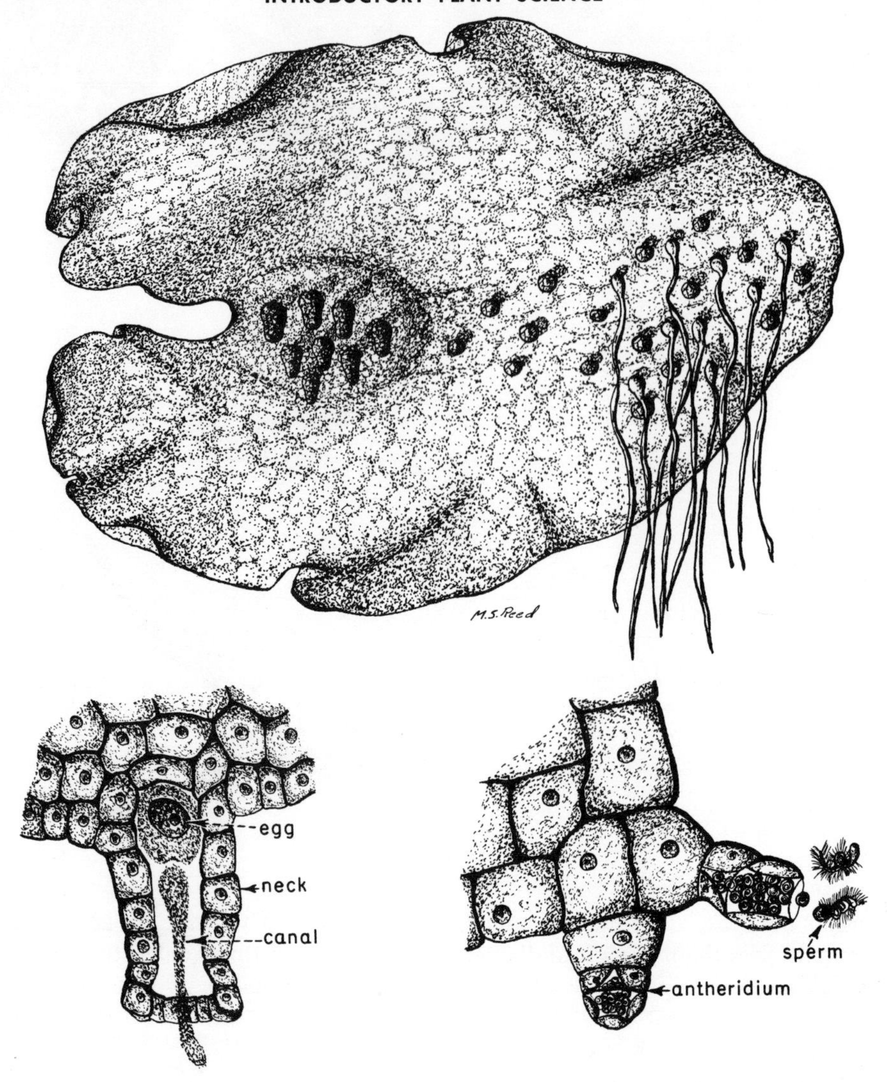

Fig. 37–7. *Upper*, a fern gametophyte with archegonia near the notch and antheridia near the rhizoids. *Lower left*, an archegonium of a fern. *Lower right*, fern antheridia.

Development of the Sporophyte

After fertilization, the zygote develops into an eight-celled embryo sporophyte. The eight cells are arranged in four pairs, which develop into the four organs of the young sporophyte. One pair of cells develops into the stem; the other three pairs develop into a leaf, root, and foot, respectively. The foot is a multicellular organ that is anchored in the gametophyte, from which it absorbs food for the young sporophyte. In the early stages of development, the sporophyte is dependent on the gametophyte for nourishment, but when the leaf is in light and the root in soil, the sporophyte becomes independent (Fig. 37–8). The stem of the sporophyte, the rhizome, develops underground and it forms leaves and roots. As the sporophyte increases in size, the gametophyte is no longer functional and it shrivels.

In earlier chapters we learned that the development of an organism is influenced by both heredity and environment. In nature, the normal environment for the fertil-

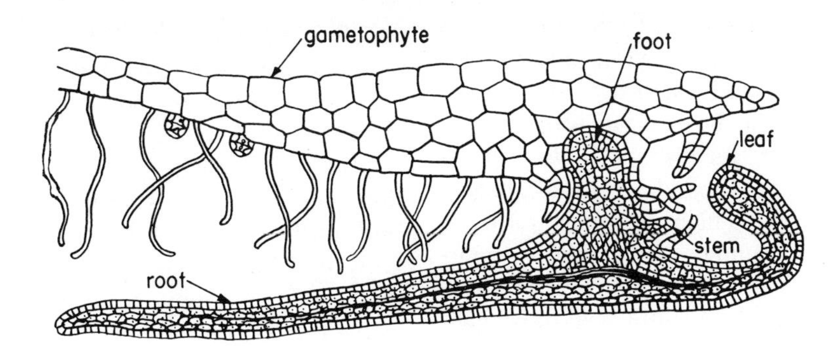

Fig. 37–8. Section through gametophyte and attached young sporophyte of maidenhair fern.

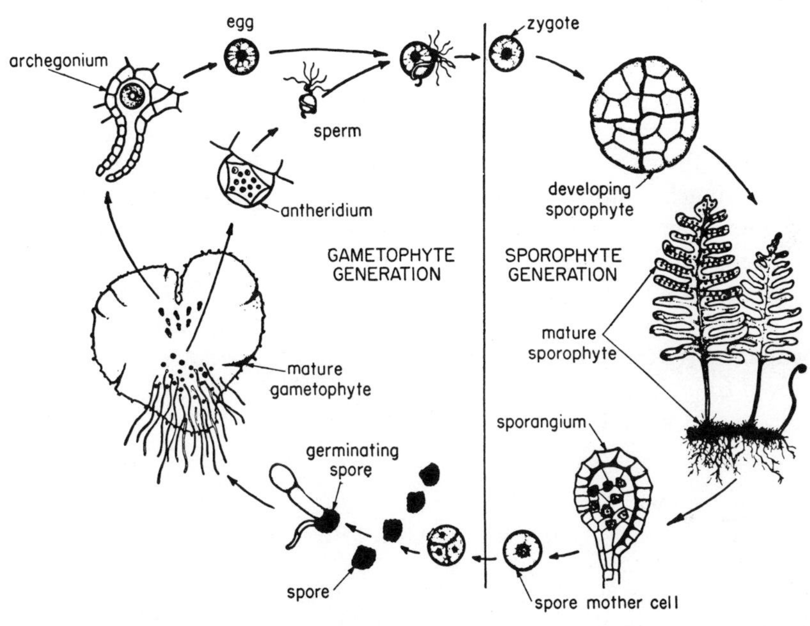

Fig. 37–9. Diagram of the life cycle of a fern. (Reproduced by permission from *The Essentials of Plant Biology* by Frank Kern. Copyright, 1947, by Harper & Row, Inc., New York.)

ized egg and embryo is the archegonium. In this environment, the zygote consistently develops into a sporophyte in the manner just described. If we change the environment of a young embryo, the course of development is altered. If a 5-day-old embryo of *Todea barbara* is transferred from an archegonium to a nutrient medium contained in a flask, the embryo does not develop into a sporophyte, but instead it develops into a two-dimensional thalloid structure which closely resembles a gametophyte. On the other hand, a 17-day-old embryo when transferred from an archegonium to a flask containing appropriate nutrients develops into a characteristic sporophyte.

Alternation of Generations

Alternation of generations is clearly evident in ferns (Fig. 37–9). The gametophyte generation begins with a haploid spore which develops into a small gametophyte, an independent green plant which nourishes itself. A mature gametophyte produces eggs in archegonia and sperms in antheridia. Sperms swim in external water to the eggs and effect fertilization. The sporophyte generation, diploid as usual, begins with the zygote, which develops into the sporophyte. A young sporophyte is nourished by the gametophyte but soon becomes an independent plant with well-developed roots, stems, and leaves. The mature sporophyte produces haploid spores in sporangia.

Compared with a moss, the fern sporophyte is vastly more complex, having well-developed conducting, mechanical, and photosynthetic tissue, and true roots, stems, and leaves. You will recall that a moss sporophyte consists of a foot, stalk, and capsule, that it lacks direct contact with soil, and that it is parasitic on the gametophyte. Fern gametophytes are much simpler than those of a moss. The fern gametophyte has a relatively short life and a reduced form. A moss gametophyte possesses structures resembling leaves and a stem. Ferns like mosses still depend on water for fertilization, suggesting that both evolved from an aquatic ancestor.

Vegetative Reproduction of Ferns

Ferns may be propagated by dividing the rhizome into two or more segments. Under natural conditions, the death of an older portion of a rhizome may result in the separation of growing ends. Some ferns produce buds on leaves, which become detached and grow into fern sporophytes. In the walking fern small sporophytes develop at the tips of the leaves (Fig. 37–10).

WATER FERNS

If you saw the water fern *Marsilea* growing in water near the edge of a pond, you might mistake it for a four-leaved clover. *Marsilea* is more clover-like in superficial appearance than it is fernlike. The rhizome is rooted in the mud, and it gives rise to leaves which float on the water or extend above the surface. Each leaf consists of a petiole and a blade made up of four leaflets. Hard beanlike structures (called **sporocarps**) which contain sporangia are borne on stalks which originate near the base of each petiole. If you remove a sporocarp, cut a hole in one end, and place it

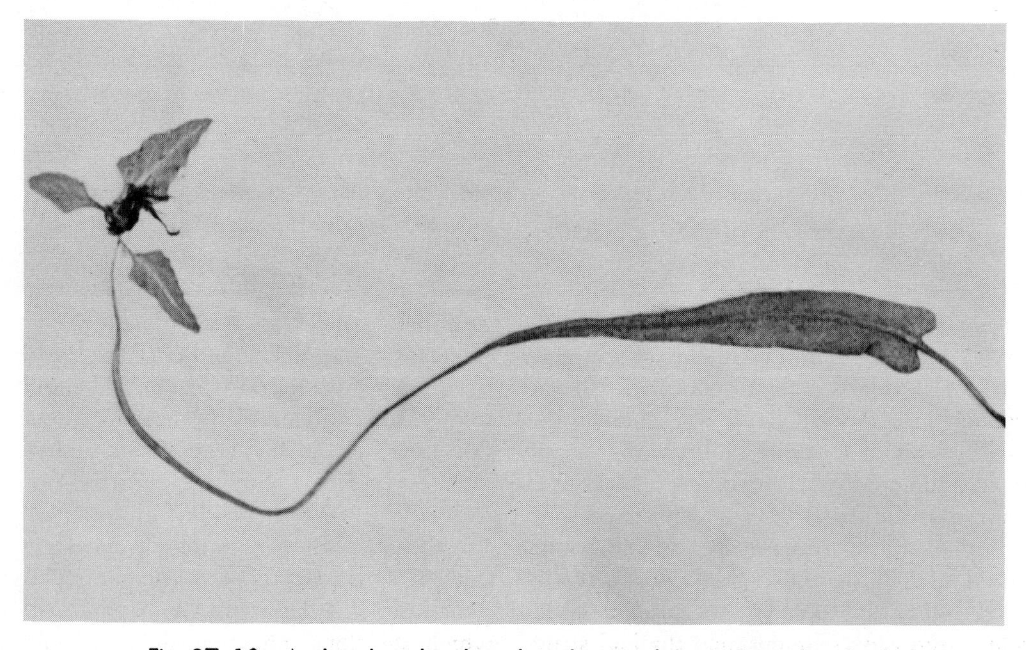

Fig. 37–10. A plant has developed at the tip of the walking-fern leaf.

Fig. 37–11. Sporophyte of *Marsilea*, a water fern.

in water, a tube of gelatinous tissue emerges which carries many clusters of sporangia out of the sporocarp. Some of the sporangia contain many small spores (microspores) while others contain just one large spore (a megaspore). Hence *Marsilea* is heterosporous in contrast to the ordinary ferns, which are homosporous, meaning they produce only one type of spore. A microspore develops into a male gametophyte consisting of an antheridium and one prothallial cell, while a megaspore develops into a small female gametophyte which contains one archegonium. The male gametophyte develops within the microspore. The early stages of development of the female gametophyte occur inside the megaspore, but near maturity the wall of the megaspore cracks open and a small amount of tissue, including the archegonium, emerges. The sperms escape from the antheridium when the microspore wall cracks open. After the egg is fertilized, it develops into the clover-like sporophyte (Fig. 37–11).

Salvinia and *Azolla* are other water ferns. They are small plants that float on water and, like *Marsilea*, they are heterosporous.

QUESTIONS

1. _____ Ferns and seed plants are in the subphylum (A) Pteropsida, (B) Psilopsida (C) Lycopsida, (D) Sphenopsida.

2. _____ Which of the following are most abundant today? (A) Ferns, (B) horsetails, (C) club mosses.

3. _____ Ferns are in the class (A) Filicineae, (B) Gymnospermae, (C) Angiospermae.

4. _____ Ferns are more abundant on earth today than they were in the past. (A) True, (B) false.

5. _____ Xylem and phloem are lacking in the sporophytes of (A) club mosses, (B) horsetails, (C) ferns, (D) mosses.

6. _____ What evolutionary advances over mosses are shown by the ferns?

7. _____ The fern plant which has fronds is (A) the sporophyte, (B) the gametophyte.

8. _____ The fern sporophyte is (A) simpler than the moss sporophyte, (B) more complex than the moss sporophyte.

9. _____ A rhizome is (A) the same as a rhizoid, (B) a large root, (C) a stem, (D) the alternate generation for ferns.

10. _____ An indusium is (A) a thin covering over the sporangia, (B) a ring around each sporangium, (C) part of the gametophyte, (D) a chemical which induces the formation of archegonia and antheridia.

11. If a fern zygote contains 146 chromosomes, how many chromosomes will there be in each spore?

12. _____ In ferns the cells which undergo meiosis occur in (A) archegonia, (B) antheridia, (C) sporangia.

13. _____ Fern spores germinate and develop into (A) gametophytes which are about one-fourth inch in diameter, (B) gametophytes which are about 1 foot high, (C) sporophytes which are about one-fourth inch in diameter, (D) sporophytes which are about 1 foot high.

14. _____ The fern gametophyte is (A) less complex than a moss gametophyte, (B) more complex than a moss gametophyte, (C) more complex than the gametophyte of the liverwort *Marchantia,* but less complex than that of a moss.

15. _____ In ferns archegonia and antheridia are formed on the (A) gametophyte, (B) sporophyte.

16. Describe the structure of a fern prothallus.

17. Where does the fertilized egg of a fern germinate and what does it develop into?

18. What are the four primary organs of the fern embryo?

19. Beginning and ending with the term "zygote," arrange the following terms in their proper sequence in the generalized life cycle for the fern plant: sporangium, antheridium, archegonium, gametophyte, spore, sporophyte, zygote, spore mother cell, reduction division, fertilization, sperm, egg.

38

Gymnosperms

The gymnosperms are an ancient class of the Pteropsida which descended from some fern ancestor. Some of them, the seed ferns, stood up with giant horsetails and club mosses about 300 million years ago (during the Devonian period). The more primitive gymnosperms, the seed ferns, the Bennettitales, and the Cordaitales, formed the dominant vegetation for more than 100 million years and then, during the time of the dinosaurs, they vanished. The newcomers, first the cycads and then the conifers, made their appearance about 225 million years ago (in the Permian period), and some of them are still going strong.

The gymnosperms are better adapted to a land environment than the ferns from which they evolved. Both the male and female gametophytes of a gymnosperm are nourished by the sporophyte, whereas in ferns the gametophytes are green plants that develop on the ground and nourish themselves. In ferns fertilization occurs only when the gametophytes are covered with a film of water. The gymnosperms do not depend on water to transfer the sperms to the eggs. Instead they have pollen tubes.

Among plants as among animals evolution has led to a prolonged connection between a parent and its offspring. In many algae a single-celled zygote is released into the water, but in the seed plants a seed is released. Within the seed there is a well-developed embryo and stored food. The advantage is evident. The more developed the offspring when it leaves the parent, the greater are its chances for survival. In our survey of the plant kingdom we encounter seeds for the first time in the gymnosperms. The word "gymnosperm" literally means **naked seed.** The seeds of gymnosperms are borne exposed on the upper surfaces of cone scales. The angiosperms (flowering plants), in contrast, have their seeds enclosed in a fruit.

Seven orders are included in the class Gymnospermae: (1) Cycadofilicales (seed ferns), (2) Bennettitales (fossil cycads), (3) Cycadales (cycads), (4) Cordaitales, (5) Ginkgoales (maidenhair tree), (6) Coniferales (conifers), and (7) Gnetales. The Cycadofilicales, Bennettitales, and Cordaitales are extinct.

EXTINCT GYMNOSPERMS

Cycadofilicales

The seed ferns were transitional forms between the ferns and gymnosperms. They resembled ferns, but in contrast to ferns they

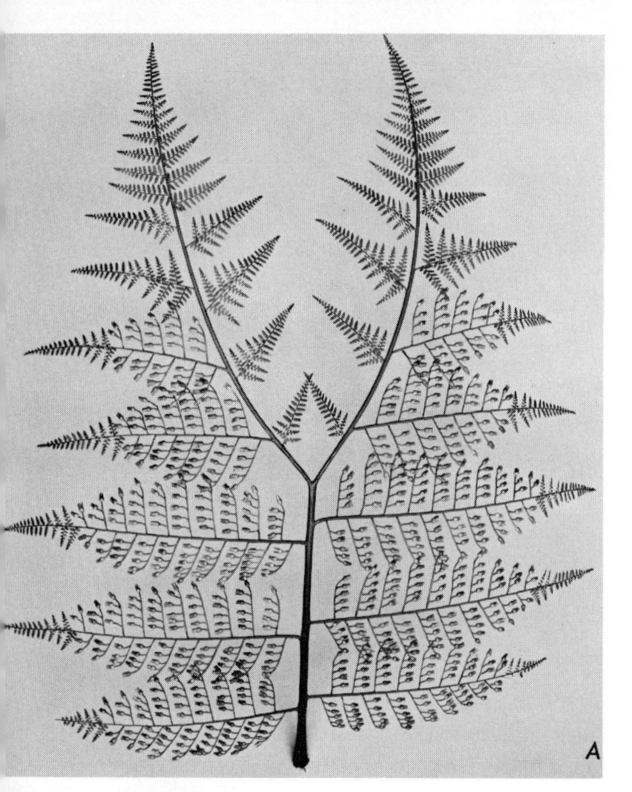

Fig. 38–1A. A model of a primitive seed fern, *Lyginopteris*. Note the cuplike bracts containing seeds. (Chicago Natural History Museum.) *B*. Bisexual strobilus of *Cycadeoidea*, a fossil cycad in the order Bennettitales, with basal bracts, pinnate microsporophylls bearing microsporangia, and a central cluster of stalked ovules. (Chicago Natural History Museum.)

produced naked seeds which were attached to the leaves (Fig. 38–1*A*). The seed ferns were vines or trees of small diameter and, unlike ferns, they possessed a cambium. The fronds of many species were decidedly fernlike and often large, up to several feet long. The seed ferns made their appearance during the Devonian period and became extinct during the Triassic period, 185 million years ago.

Bennettitales

The fossil cycads, order Bennettitales, may have evolved from the seed ferns during the late Paleozoic era. They were so abundant during the early Mesozoic era that this era is known as the Age of Fossil Cycads. Fossils of these plants are especially abundant in the Black Hills of South Dakota where the Fossil Cycad National Monument has been established. Some fossil cycads were climbing or trailing plants; others resembled living cycads and had thick stems, 3 or 4 feet high, that supported crowns of large fernlike leaves. The cones of the Bennettitales (Fig. 38–1*B*) superficially resembled magnolia flowers. Each cone consisted of a central cluster of stalked ovules (megasporangia) surrounded by a whorl of pinnate microsporophylls which produced marginal microsporangia. The fossil cycads probably gave rise to the cycads, order Cycadales, a few of which survive today.

Cordaitales

The Cordaitales, contemporaneous with the Cycadofilicales, were tall trees, about 100 feet high, with diameters of up to 3 feet. The leaves ranged in size from small scales, in some species, to those 3 feet in length. In many species the leathery leaves were borne in dense clusters near the tips of the branches (Fig. 38–2). Both male and female cones, generally less than ½ inch long, were borne on the same plant. The sperms were of the swimming type, somewhat resembling those of the cycads. The Cordaitales were widely distributed from the Pennsylvanian period, when they formed majestic forests, to the Jurassic when they became extinct. The early evolution of this order is hidden in obscurity. They may have evolved from the seed ferns or they may have had an independent origin. They are of major interest because the Coniferales probably evolved from them.

Fig. 38–2. A model of *Cordaites*, an extinct gymnosperm. (Chicago Natural History Museum.)

SURVIVING GYMNOSPERMS

Of the four orders of extant gymnosperms only one, the Coniferales, populates extensive areas today. In early geological periods the Cycadales, Ginkgoales, and Gnetales were far more prevalent than they are now.

The Cycadales show no close kinship to the other surviving orders and had a different origin. The cycads probably evolved from the fossil cycads. They are the most primitive of the present-day seed plants. About 200 million years ago they were abundant and widely distributed, but now only nine genera, with about 100 species, are present on earth and these are confined to subtropical and tropical regions. We will have more to say about the cycads later in this chapter.

The Ginkgoales and Coniferales are believed to have evolved from the Cordaitales; the Gnetales presumably represent a branch from this line. Only one of the gymnosperm orders, the Coniferales, is of economic importance at this time. We shall start with this order, even though the Ginkgoales are more primitive.

CONIFERALES

The most familiar gymnosperms are the conifers or cone-bearing trees in the order Coniferales, which includes 570 species grouped into 50 genera. In the past conifers were more widespread than now, indicating that the group is waning. Indeed, some conifers—*Walchia,* for example—have become extinct. For a long time we thought that *Metasequoia* was extinct, but rather recently *Metasequoia glyptostroboides* was found growing in limited areas in China. This discovery was an exciting one because a species previously known only from fossils could now be compared with living specimens.

Most conifers are trees. The tree is the sporophyte and consists of roots, stems, and leaves. In many conifers—pines, spruces, firs, and hemlocks—the leaves are needlelike, but in a few—red cedar and arborvitae—the leaves are scalelike. Most conifers retain their leaves for several seasons, but some, for example, the larches, are deciduous trees. The activities of the sporophyte are the same as those which were discussed in the first part of this book.

The conifers produce two kinds of cones: (1) **staminate cones** in which pollen develops and (2) **ovulate cones** which, when mature, bear two seeds on the upper surface of each cone scale (Fig. 38–3). In pines, spruces, and firs, both staminate and ovulate cones are formed on the same tree, whereas in junipers staminate cones develop on some trees and ovulate cones on others.

In the spring the small but attractive staminate cones shed prodigious quantities of pollen, which is distributed by the wind. At the time of pollination the scales of the young ovulate cones are spread apart and some of the pollen sifts to the base of each scale where it comes in contact with potential seeds, called ovules. If a pollen grain is in contact with an ovule, it sends a pollen tube into the ovule. The pollen tube carries two sperms to an egg which has been formed by a female gametophyte in the ovule. One sperm fertilizes the egg, which then develops into an embryo. As this occurs, the ovule develops into a seed. These are the general features of the sexual reproduction of conifers. Let us now take up the details and relate the structures of conifers to those of more primitive plants.

Life History of a Conifer

In the spring the staminate cones emerge from the buds in which they developed during the previous summer. A staminate cone has a central axis which bears spirally arranged scalelike leaves (microsporophylls). Two pollen sacs, microsporangia, are suspended from the lower surface of each microsporophyll. In each microsporangium many microspore mother cells undergo meiosis to form haploid microspores.

The microspores develop into immature male gametophytes, called pollen grains, while they are still retained in the microsporangium. In the development of the male gametophyte four cells are formed inside the microspore by mitosis. Two cells are functionless **prothallial cells,** one is the **generative cell,** and the fourth is the **tube cell.**

While these cells are forming, the outer wall of the pollen grain puffs out in two places to form two wings, which make the pollen buoyant. The microsporangia then break open and release the pollen grains, which are disseminated by wind.

Fig. 38–3. Two kinds of pine cones. *Left,* staminate cones. *Right,* ovulate cones.

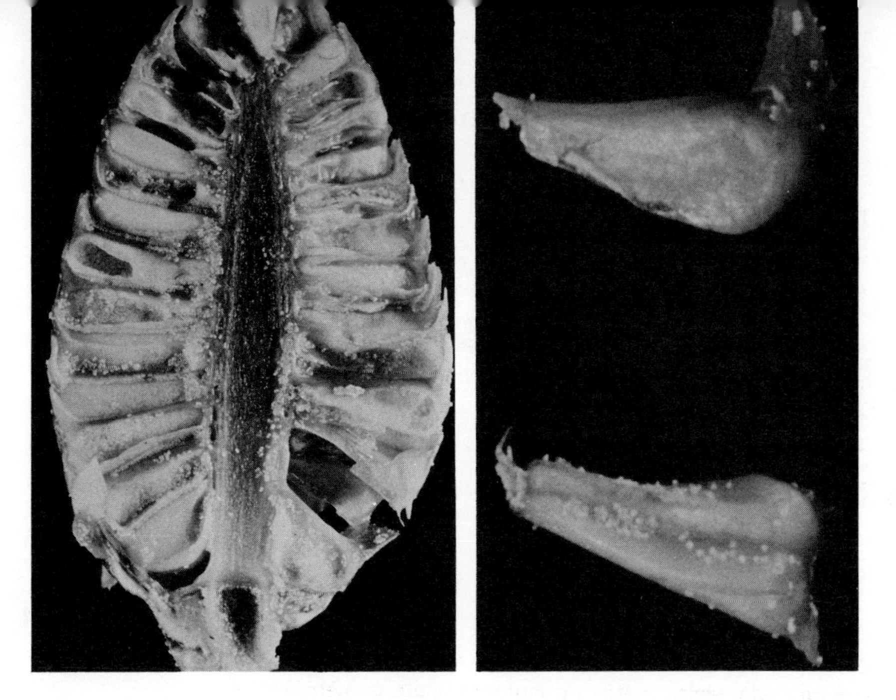

Fig. 38–4. *Left*, a section through a staminate cone of spruce, showing the central axis, microsporophylls, and microsporangia. Pollen grains are evident in the pollen sacs, which have been cut open. *Right upper*, a side view of the microsporophyll with an attached microsporangium (pollen sac). *Right lower*, undersurface of a microsporophyll showing two microsporangia.

After the pollen is discharged, the staminate cones shrivel and die (Fig. 38–4).

A high percentage of the pollen lands on the ground, or on leaves, trunks, and other places where it cannot function, but a small percentage lights on ovulate cones. Of the latter, some pollen grains sift down between the scales of the ovulate cone and these pollen grains are functional, as we shall see later (Fig. 38–5). After pollination, the ovulate cones become closed and remain closed until the seeds are mature.

An ovulate cone consists of a central axis with spirally arranged cone scales called **megasporophylls.** A functionless tonguelike bract is present on the undersurface of a megasporophyll. The presence of this bract leads some botanists to consider each cone scale a branch instead of a modified leaf. Two small ovules develop on the upper surface of each megasporophyll, near the axis of the cone. Each ovule consists of a nucellus, which is the megasporangium, and a protective covering called the **integument.** A small opening, known as a micropyle, is present on the integument. Within the nucellus one megaspore mother cell un-

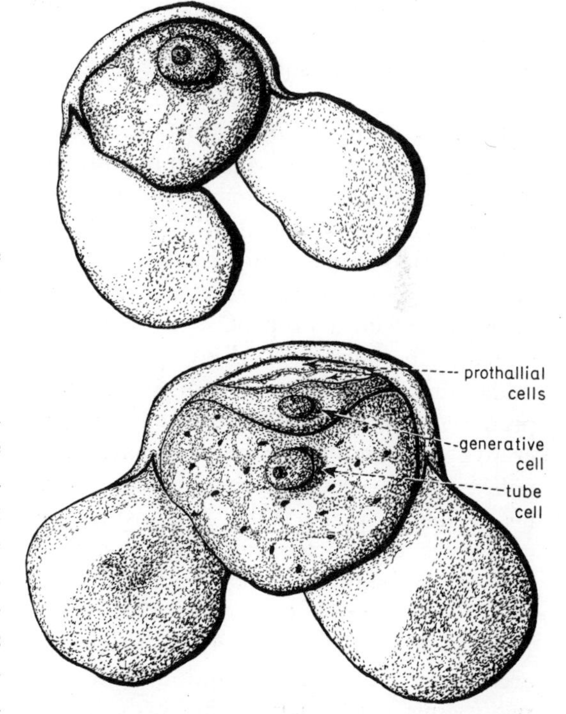

Fig. 38–5. *Upper*, a microspore of pine. *Lower*, an immature male gametophyte or pollen grain; the pollen is shed at this stage of development. (After J. M. Coulter and C. J. Chamberlain.)

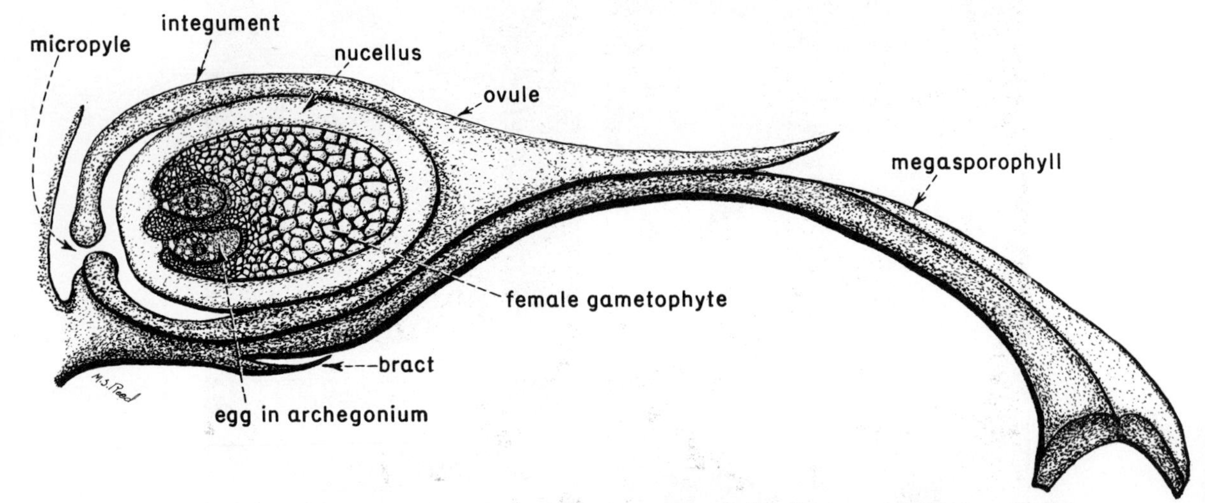

Fig. 38–6. Longitudinal section of a megasporophyll of pine, including a section through an ovule, which contains a mature female gametophyte.

dergoes meiosis to produce a row of four haploid megaspores. Three megaspores abort and one develops into a haploid female gametophyte. The female gametophyte is contained in the megasporangium (nucellus) and is entirely dependent on the sporophyte for nourishment. The mature female gametophyte is made up of many cells. Two or more archegonia, each containing an egg in the venter, are differentiated at the end of the female gametophyte which is nearest the micropyle (Fig. 38–6).

The pollen grains which settle near the micropyle of an ovule are drawn into the micropyle when a drop of sticky fluid (secreted in the micropyle) dries up. A pollen grain which is in contact with the nucellus produces a pollen tube, into which the generative cell and the tube cell pass. This pollen tube grows through the nucellus. While a pollen tube is growing, the generative cell divides to produce a **stalk cell** and a **body cell.** The body cell then divides and forms two nonmotile sperms (Fig. 38–7). When a pollen tube reaches an archegonium, it ruptures and liberates the two sperms. One sperm unites with the egg and the other is functionless.

The fertilized egg is of course diploid, and it develops into an embryo. The zygote divides and the daughter cells continue to divide. After four divisions a **proembryo** made up of sixteen cells is formed. In flowering plants only one embryo is formed from the zygote, but in conifers four embryos are organized at the tip of the proembryo, one embryo from each of the four apical cells. The cell below each developing embryo elongates greatly and pushes the embryo into the nutritive female gametophyte; this elongated cell is known as a suspensor cell. Nearly always, three of the embryos abort and one develops to maturity. The mature embryo has a root (radicle), hypocotyl, three to many cotyledons, and an epicotyl.

While the embryo is developing within an ovule, the integument changes into a hard seed coat and the female gametophyte grows at the expense of the nucellus, which becomes reduced to a thin papery layer. Food is stored in the female gametophyte, which is also known as endosperm. The endosperm in a pine seed originates in a manner different from that in a seed of corn or other flowering plant. In a flowering plant the endosperm develops from an endosperm nucleus having three sets of chromosomes. You will recall that in a flowering plant a sperm unites with two polar nuclei to form a triploid endosperm nucleus. We

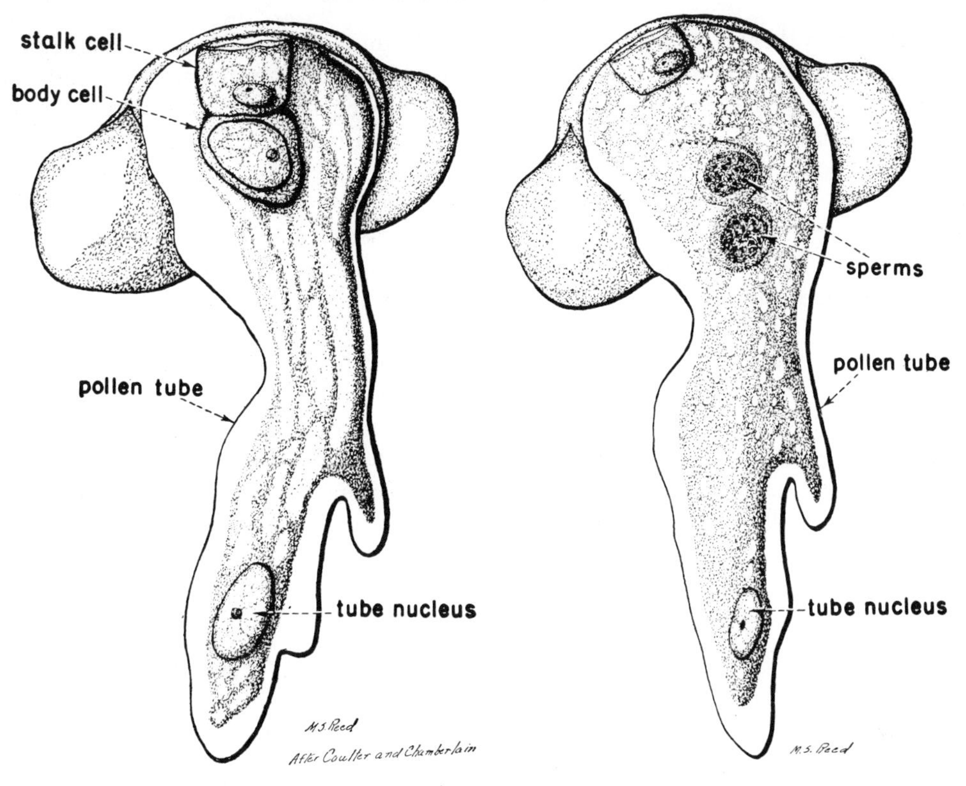

Fig. 38–7. *Left,* a nearly mature male gametophyte of pine. *Right,* a mature male gametophyte. (After J. M. Coulter and C. J. Chamberlain.)

have just seen that the endosperm in pine is the female gametophyte whose cells are haploid—that is, each cell has one set of chromosomes. The complex of embryo, seed coat, and endosperm is called a seed, and it has developed from an ovule (Fig. 38–8). A thin portion of the cone scale, a wing, is attached to each seed.

When the seeds are mature, the cones open and the seeds escape. The seeds are disseminated by wind. When the seed is planted, the root grows into the soil and the hypocotyl elongates and lifts the cotyledons out of the ground.

Occasionally two or even more seedlings develop from one seed. Such a seed contained more than one embryo, a feature known as **polyembryony.** You will recall that two archegonia, each containing an egg, develop in a female gametophyte and that

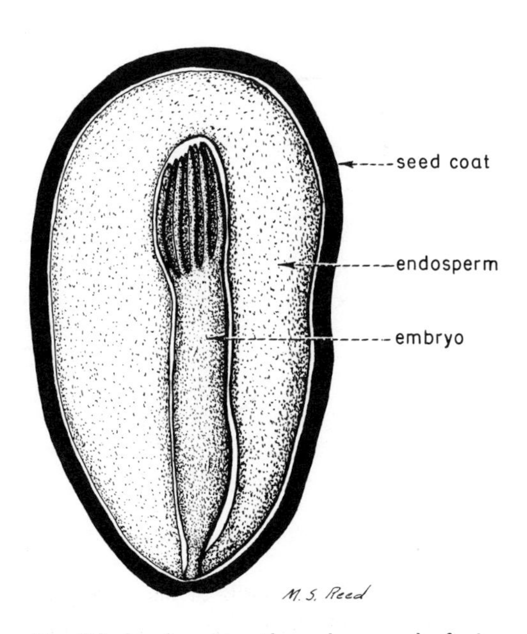

Fig. 38–8. A section through a seed of pine.

four embryos may begin to develop at the tip of each proembryo. Hence, if both eggs are fertilized and if all embryos complete their development, eight embryos would be present in just one seed. However, in the great majority of seeds only one embryo is present.

In some conifers (for example, in spruces) all of the events leading to seed formation—pollination, fertilization, and seed development—occur in one season. In many pines, the seeds do not mature until the second season. Pollination occurs early in the first season (Fig. 38–9). During the first growing season, the female gametophyte develops and the pollen tube grows very slowly through the nucellus. The egg is fertilized early in the second year, and during the growing season of the second year the seed is formed. At the end of the second growing season the cones of most species of pine open and the seeds are scattered. In a few species the cones may remain closed for a number of years, even though the seeds are fully formed.

Breeding Conifers

The world's food situation would be critical if man relied only on native species for his food supply. For many years, he has improved his crop plants through breeding and has developed high-yielding varieties

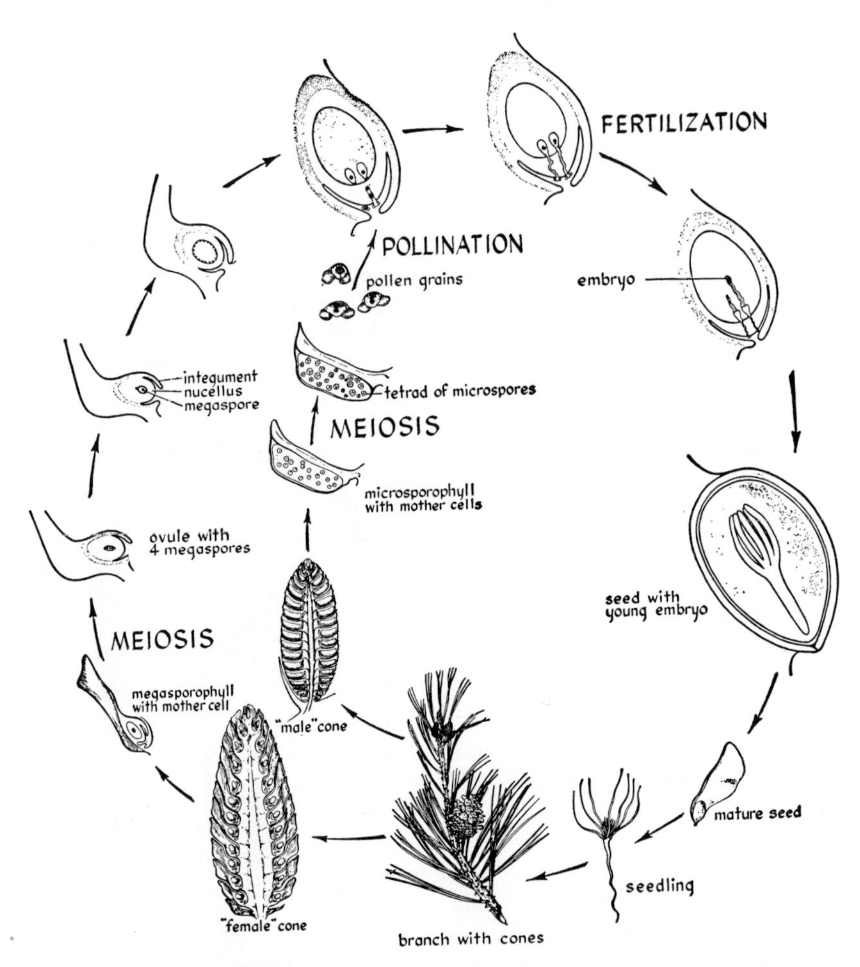

Fig. 38–9. Life cycle of pine. (Reproduced by permission from *Botany: Principles and Problems* by E. W. Sinnott and K. S. Wilson. Copyright, 1955, McGraw-Hill Book Co., Inc., New York.)

which are resistant to disease, drought, and frost. Up until a short time ago, foresters relied exclusively on native species of trees as they found them in nature. In recent years, however, the forester has begun to consider trees as crops and to compute annual yields. With this changed attitude toward forests, foresters began to breed trees which would be fast growing and resistant to drought, wind, winter injury, insects, and disease. The breeding program is a long-term project, still in its infancy, but one which promises excellent results in time. The crossing of one variety of a conifer with a related variety does not involve great difficulties. The ovulate cone is covered with a sack before pollen is released from the staminate cones. Later the desired pollen is collected and sifted over the selected ovulate cone. The sack is replaced in order to prevent stray pollen from entering.

Key to Some Genera of Conifers

- Leaves shedding in fall, many leaves in a cluster—Larch (*Larix*).
- Leaves evergreen, needle-like or scalelike, single or not more than 5 in a cluster.
  - Leaves with a sheath at base, in clusters of 2–5—Pine (*Pinus*).
  - Leaves without a sheath at base, not in clusters.
    - Leaves needle-like, mostly more than ½ inch long.
      - Twigs roughened by projecting bases of old needles.
        - Needles soft, flat, blunt—Hemlock (*Tsuga*).
        - Needles sharp-pointed, 4-angled—Spruce (*Picea*).
      - Twigs smooth or nearly so.
        - Needles with short leaf stalks; cones hanging down—Douglas fir (*Pseudotsuga*).
        - Needles without leaf stalks; cones upright—Fir (*Abies*).
    - Leaves scalelike, less than ¼ inch long, or both scalelike and needle-like (to ¾ inch long).
      - Leaves single—sequoia (*Sequoia*).
      - Leaves in pairs, threes, or fours, scale-like or needle-like.
        - Leafy twigs more or less flattened—Thuja (*Thuja*).
        - Leafy twigs round or 4-angled—Juniper (*Juniperus*).

Genera of Conifers

The accompanying key lists the major genera of conifers. A few characteristics of each genus are discussed separately below.

Larix. The larches, or tamaracks (*Larix*), are conifers with deciduous leaves which are borne in clusters of more than five on short branches. *Larix laricina*, American larch or tamarack, occurs in eastern United States, and *Larix occidentalis*, western larch, is an important tree of the northern Rocky Mountains.

Pinus. With the exception of *Pinus monophylla*, the needles of pines occur in clusters of 2, 3, or 5. The needles are borne on short, spurlike branches and they are surrounded by a thin sheath at the base. Red pine (*Pinus resinosa*) of eastern North America and lodgepole pine (*Pinus contorta*) of the Rocky Mountains are pines which have two needles in a cluster. Among the three-needle pines are longleaf pine (*P. palustris*) of southeastern United States and ponderosa pine (*P. ponderosa*) of western United States; both are valuable timber trees, and longleaf pine is a major source of turpentine. *Pinus strobus*, white pine, and *P. monticola*, western white pine, have five needles in a cluster. *Pinus strobus* is an important timber tree in eastern United States, and *Pinus monticola* is important in the western United States.

Tsuga. The hemlocks (*Tsuga*) have flat, blunt needles, with two white lines beneath. In eastern United States *Tsuga canadensis* occurs, and in western United States *Tsuga heterophylla*.

Picea. The spruces (*Picea*) have sharp, stiff needles which are square in cross-sec-

Fig. 38–10. Conifers of western United States. (U.S. Forest Service.)

tion. The cones mature in one year, are pendulous, and have thin cone scales. The spruces are important sources of wood pulp used in making paper and rayon. One important species is the white spruce (*Picea glauca*) which occurs in the Northeast, Lake States, northern Rocky Mountains, and Washington. Other valuable spruces are black spruce (*Picea mariana*), red spruce (*P. rubens*), Sitka spruce (*P. sitchensis*), Englemann spruce (*P. Engelmannii*), and blue spruce (*P. pungens*). The last is much used as an ornamental tree.

Pseudotsuga. The Douglas-fir (*Pseudotsuga taxifolia*) is the most important commercial forest tree of the United States (Fig. 38–10). It is the major tree in the forests of the Pacific Northwest. The needles are flat and the buds are brown and sharp-pointed. The cones of Douglas-fir are very characteristic. Below each cone scale there is a three-pointed bract which is longer than the cone scale. The seeds mature in one growing season.

Abies. The true firs are in the genus *Abies*. The needles of *Abies* are flat and slightly notched at the ends. The mature cones are borne erect and they do not fall to the ground; the scales are shed from the cones while they are still on the tree. The cones mature in the late summer of the year in which they are formed. This genus includes the balsam fir (*Abies balsamea*) of the Northeast and Lake States, and such western species as white fir (*A. concolor*), alpine fir (*A. lasiocarpa*), grand fir (*A. grandis*), and noble fir (*A. nobilis*).

Sequoia. Both the redwood and the bigtree are in the genus *Sequoia*. The redwood (*Sequoia sempervirens*) is limited to humid regions along the coast of California and southern Oregon. The redwood is an important timber tree; some stands yield 400,000 board feet per acre. The bigtree (*Sequoia gigantea*) occurs on the western slopes of the Sierra Nevada Mountains at elevations of 5000 to 8500 feet. It may attain a height of more than 350 feet and some are known to be at least 4000 years old.

Thuja. The northern white cedar, often called arborvitae, is *Thuja occidentalis*, a species with scalelike leaves; it occurs in the Northeast, Lake States, and Appalachian Mountains (Fig. 38–11). Western red cedar (*Thuja plicata*) has flat, scalelike, closely appressed leaves; it is found in the Pacific Northwest and in the northern Rocky Mountains.

Juniperus. The junipers (*Juniperus*) are trees or shrubs with awl-shaped or scalelike leaves, or both. The cone resembles a bluish berry. The most important species is *Juniperus virginiana*, which occurs in the eastern half of the United States.

CYCADS—CYCADALES

The cycads, formerly more prevalent than today and of worldwide distribution, still survive in Florida, Mexico, the West Indies, South Africa, Australia, and in other tropical and subtropical countries. The stems of some species are tuberous and subterranean, but those of others tower to a height of 50 feet or more. The stems typically are unbranched and support a crown of large palmlike leaves. The male (staminate) and female (ovulate) cones develop on separate plants; in other words, all cycads are dioecious. Cycads are attractive plants that are used for landscaping in tropical gardens and in conservatories (Fig. 38–12). Arrowroot starch is obtained from "sago palms," the common name for certain cycads.

Zamia. In *Zamia*, a cycad which grows in Florida, the staminate and ovulate cones are produced on different plants. *Zamia* has a short, thick stem which supports a crown of pinnate leaves. Two ovules are borne on each megasporophyll of an ovulate cone. In an ovule a megaspore mother cell undergoes reduction division to form four haploid

Fig. 38–11. Conifers of eastern United States. (U.S. Forest Service.)

Fig. 38–12. *Dioon edule*, a cycad, bearing a male cone. (Chicago Natural History Museum.)

megaspores. Three megaspores abort and one develops into a female gametophyte containing archegonia (Fig. 38–13).

Many microsporangia are produced on the lower surface of each microsporophyll of a staminate cone. Within each microsporangium numerous microspore mother cells undergo meiosis to produce haploid microspores which develop into pollen grains. You will recall that pollen grains are immature male gametophytes. The pollen grains are dispersed by wind. A pollen grain which lands near the micropyle of an ovule produces a pollen tube that grows through the nucellus. While the tube is developing, two sperms, so large that they are visible with the unaided eye, are formed in the pollen tube. Each sperm has many

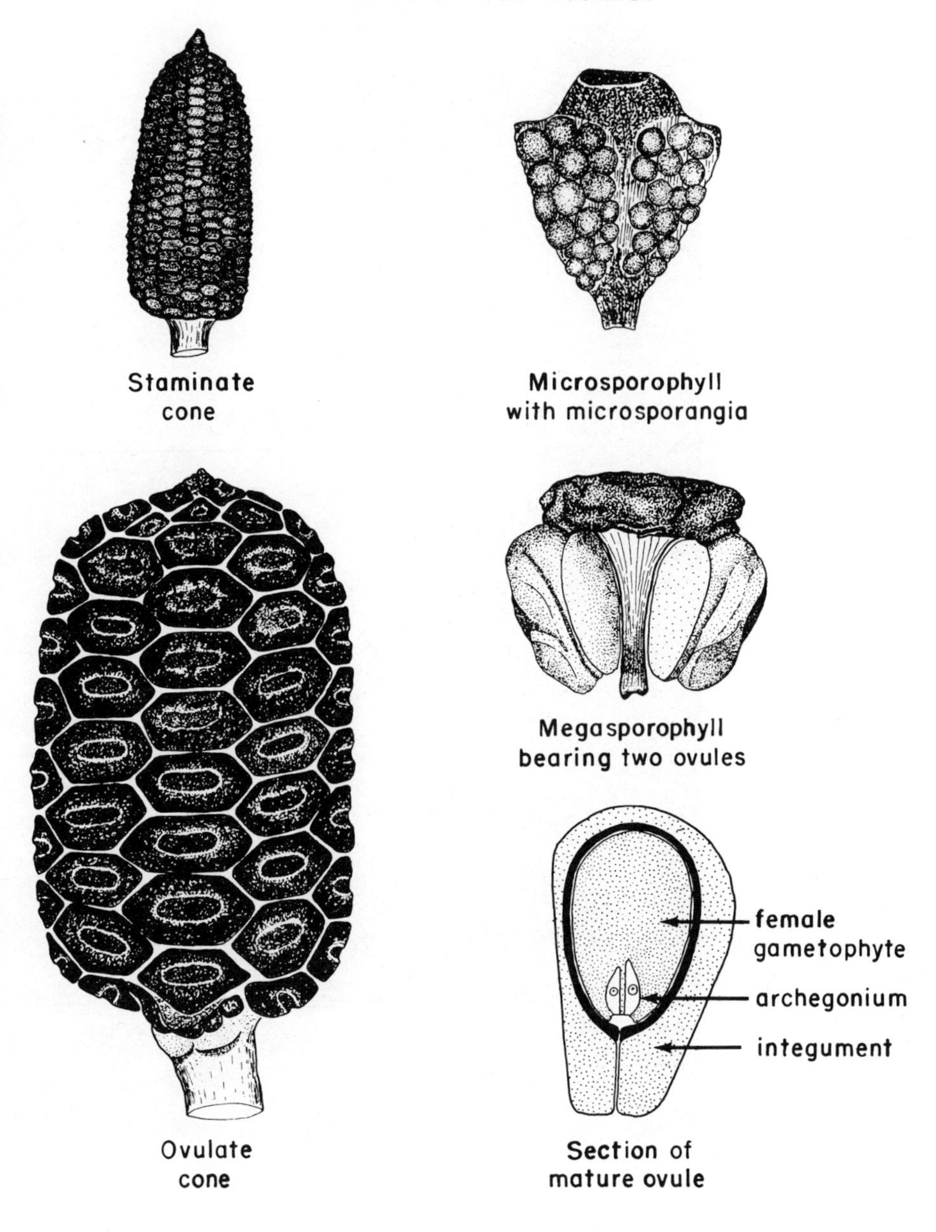

Fig. 38–13. Reproductive structures of *Zamia*. (William M. Carlton, *Laboratory Studies in General Botany*. Copyright © 1961, The Ronald Press Company, New York.)

flagella, an ancestral trait which has persisted even though its use has disappeared. The sperms swim in a nucellar cavity and after a short time one sperm fertilizes the egg. The zygote then develops into an embryo and the ovule becomes a seed.

The Cycadales are the most primitive seed plants of today. They are more primitive than the Coniferales because they have retained such fern characteristics as swimming sperms and leaves which are coiled during development.

MAIDENHAIR TREE—GINKGOALES

In the past many genera of this order were widely distributed on earth, but only one species, *Ginkgo biloba,* the maidenhair tree, has survived. A few maidenhair trees grow naturally in China, but many are cultivated and idolized in temple courts in China and Japan. *Ginkgo* has been introduced into gardens and parks in parts of the United States and Europe. It is a beautiful, exotic-appearing tree with small, fan-shaped leaves that turn yellow and fall in the autumn.

Like the cycads, *Ginkgo* is dioecious. A staminate cone superficially resembles a male catkin of cottonwood. The cone consists of a central stalk bearing many microsporophylls, each of which bears two, rarely more, microsporangia in which pollen develops. Wind carries the pollen to the female cones. Each female cone consists only of a stalk bearing two terminal ovules; frequently one ovule aborts. Within an ovule, a megaspore, formed by meiosis, develops into a female gametophyte which contains two or three archegonia. As the pollen tube penetrates the ovule, two large swimming sperm, similar to those of cycads, are formed in the pollen tube. After fertilization, the zygote develops into an embryo and the ovule into a seed which is about the size of a cherry. The single integument of the ovule develops into the seed coat which, at maturity, consists of a fleshy, yellow outer layer, a stony middle layer, and a dry, papery inner layer (Fig. 38–14). The seeds produced by the female tree are, of course, borne exposed, and they are malodorous, smelling like rancid butter. Furthermore, when they fall on sidewalks and streets, they make the pavement slippery. Hence, if you plan to plant a maidenhair tree, make sure that you get a male tree (one with staminate cones).

Fig. 38–14. A twig of *Ginkgo biloba* bearing one seed. (U.S. Forest Service.)

Fig. 38–15. *Welwitschia mirabilis,* a bizarre gymnosperm with straplike leaves, grows in deserts of western South Africa where precipitation may average only 1 inch a year. (Chicago Natural History Museum.)

GNETALES

The Gnetales are more advanced than the other orders of gymnosperms. They resemble angiosperms in several ways. Like the angiosperms, the Gnetales lack archegonia, and they possess vessels in the wood, ovules with two integuments, and seeds with two cotyledons.

However, like other gymnosperms, the Gnetales have naked seeds. Three genera, *Ephedra, Gnetum,* and *Welwitschia,* are in the order Gnetales. *Ephedra* is a straggly shrub which grows in arid regions. The drug ephedrine is obtained from Asiatic species of *Ephedra.* Some species of *Gnetum* are tropical trees; others are vines. *Welwitschia,* a native of western South Africa, is the most curious plant of the order; it has a short, thick woody stem and only a single pair of long straplike leaves which continually increase in length. With age the two leaves become shredded into straplike segments (Fig. 38–15).

QUESTIONS

1. List some of the conifers which are present on your campus.

2. Discuss the economic significance of the conifers.

3. What changes in our culture would occur if all conifers were destroyed?

4. _____ Pines, firs, spruces, and junipers are in the order (A) Coniferales, (B) Cycadales, (C) Ginkgoales, (D) Gnetales.

5. _____ The familiar pine tree is the (A) sporophyte, (B) gametophyte.

6. What two kinds of cones are produced in pine and what is the function of each?

7. _____ The male (staminate) cones of a pine tree eventually become hard and woody. (A) True, (B) false.

8. Are the microspores and megaspores of pine borne on the same or separate strobili? How many microsporangia on a single microsporophyll? Is the gametophyte of pine independent or dependent?

9. What is produced when a microspore of pine germinates? A megaspore of pine? Where does the fertilized egg of a gymnosperm germinate and what does it develop into?

10. _____ Pollination in pines is accomplished by (A) wind, (B) insects, (C) birds, (D) water, (E) mechanical means.

11. _____ In pine the pollen grain plus its tube is the (A) female gametophyte, (B) sperm, (C) egg, (D) male gametophyte.

12. _____ In pine the gametophytes are (A) parasitic on the sporophyte, (B) not parasitic on the sporophyte, (C) made up of diploid cells.

13. _____ In pines the sperms are carried to the egg by (A) the pollen tube, (B) wind, (C) water, (D) gravity.

14. _____ With respect to pines, which one of the following statements is false? (A) Male cones are more numerous than female cones. (B) Female cones live longer than male cones. (C) Archegonia are produced in male cones. (D) At maturity two seeds are borne on the upper surface of each female cone scale (megasporophyll).

15. Before pine seeds can be produced, a number of things must occur. Five of the following must occur and one does not occur. Which one of the following is *not* necessary for the production of pine seeds? (A) Pollination, (B) the growth of the pollen tube, (C) the union of a sperm with an egg, (D) the development of the fertilized egg into an embryo, (E) the development of the ovary into a fruit, (F) the development of a female gametophyte in an ovule.

16. _____ A pine embryo develops from the (A) seed, (B) fertilized egg, (C) ovule, (D) female cone scale.

17. _____ In a pine seed the food is stored in the tissue called (A) female gametophyte (also known as endosperm), (B) integument, (C) nucellus, (D) ovule.

18. _____ The complex of embryo, seed coat, and endosperm is called a seed and it developed from a (A) fertilized egg, (B) ovule, (C) megasporangium, (D) megasporophyll.

19. Would it ever be possible for two seedlings to grow from one pine seed?

20. Describe the life cycle of a pine tree.

21. Describe how you would cross two pine trees.

22. On what basis are the gymnosperms separated from the ferns? From the angiosperms?

23. Compare the life cycle of a pine tree with that of a fern; with that of a moss.

24. How can you distinguish between a fir, a pine, and a spruce?

---

25. Match the plants listed on the left with the order in which they are placed.

_____	1. Seed ferns.	1. Cycadofilicales.
_____	2. Fossil cycads.	2. Cycadales.
_____	3. Cycads.	3. Ginkgoales.
_____	4. *Cordaites.*	4. Gnetales.
_____	5. Maidenhair tree.	5. Bennettitales.
_____	6. Pine.	6. Cordaitales.
_____	7. *Ephedra.*	7. Coniferales.

39

Angiosperms

The angiosperms are flowering plants that produce seeds enclosed in a fruit. They are the most advanced, successful, aggressive, and diversified group of plants. Their diversity of form permits one or more species to thrive in practically all habitats, from sea level to alpine summits, from tropical forests to within the Arctic Circle, and from desert to lake. The first flowering plants made their appearance about 160 million years ago when dinosaurs and flying reptiles were still abundant on earth. In contrast to the dinosaurs, the flowering plants have survived and have increased in significance as time has passed.

Angiosperms differ from gymnosperms in several ways. The seeds of angiosperms are enclosed, whereas those of gymnosperms are not. The ovules of angiosperms are produced in the ovary of a pistil which is formed by the union of the margins of one or more carpels. Because the ovules are encased, the pollen grains cannot contact the ovules directly, as they do in gymnosperms. The pollen grains germinate on the stigma of a pistil, and the resulting tubes grow through the tissues of the pistil to the ovules. In both gymnosperms and angiosperms two sperms are produced in each pollen tube. In gymnosperms only one of the sperms functions, but in angiosperms both are functional; one sperm fertilizes the egg while the second unites with two polar nuclei. In angiosperms the reproductive organs are grouped into flowers. In most angiosperms these are ingeniously constructed for insect pollination, which is less wasteful and more advanced than wind pollination. All gymnosperms are woody plants, whereas both woody and herbaceous plants are present in the Angiospermae. It is generally believed by botanists that the herbaceous angiosperms evolved from woody ones. The xylem of angiosperms is more advanced than that of gymnosperms; vessels and tracheids are present in the xylem of angiosperms, but only tracheids exist in the wood of most gymnosperms.

LIFE CYCLE OF AN ANGIOSPERM

The Sporophyte

The sporophyte is the familiar plant with its roots, stems, leaves, and flowers. Stamens are microsporophylls and carpels are megasporophylls. In some species the pistil is made up of one carpel, whereas in others the pistil consists of several united carpels. In most angiosperms, stamens and car-

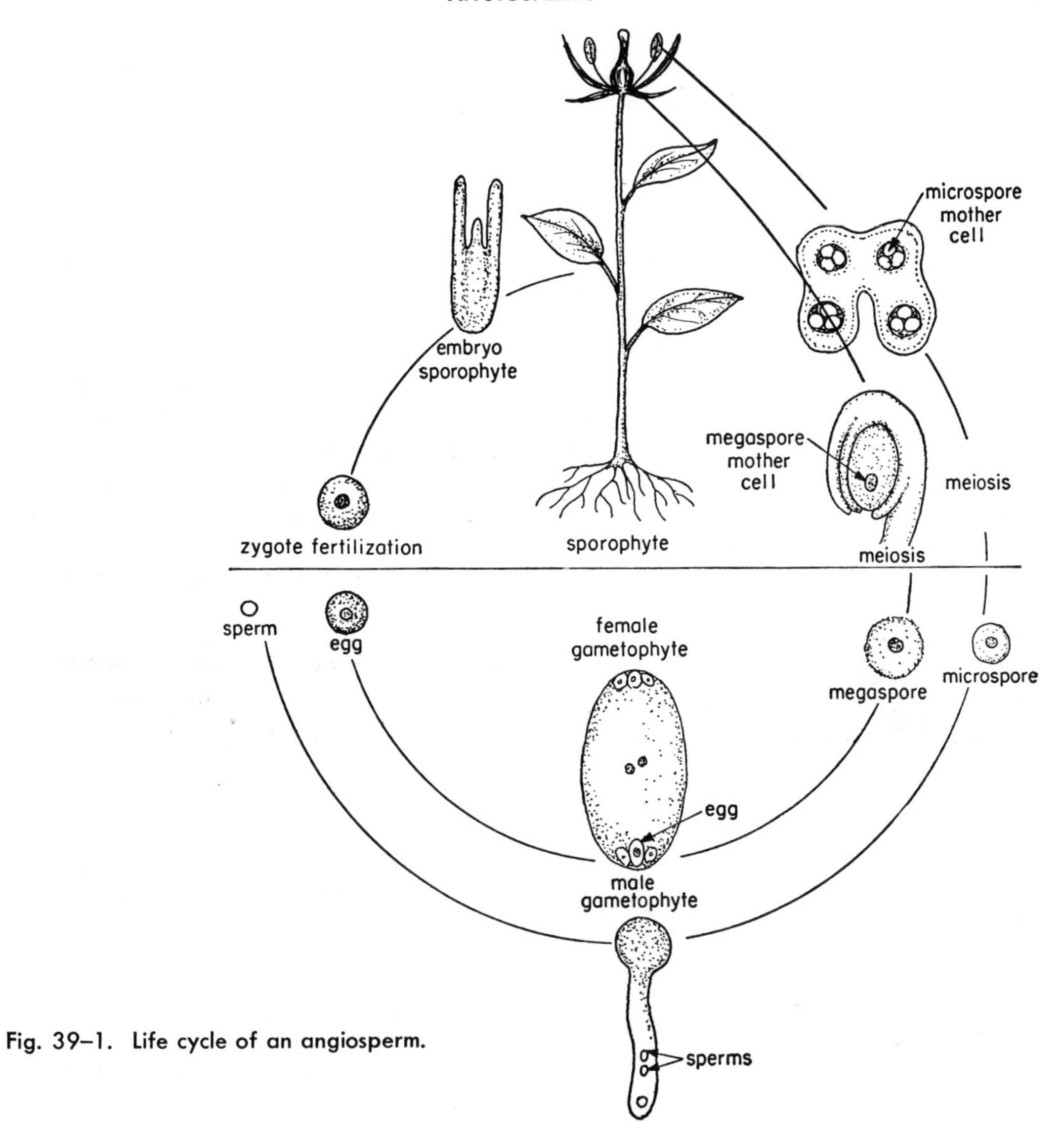

Fig. 39–1. Life cycle of an angiosperm.

pels are present in the same flower, and usually a perianth is also present (Fig. 39–1).

Microsporophylls (Stamens) and Microspores

Four microsporangia (also called pollen sacs), each surrounded by a nutritive tapetal layer, are differentiated in the anther of a stamen. Many microspore mother cells (also known as pollen mother cells) are formed in a microsporangium, and they undergo reduction division to produce haploid microspores.

The Male Gametophyte

The haploid nucleus in a microspore divides mitotically to produce a tube nucleus and a generative nucleus. The two-nucleate structure is an immature male gametophyte, or **pollen grain,** as it is usually called. After the pollen grains are liberated from an anther, they are transferred by wind, birds, insects, or some other agent to the stigma of a pistil where each pollen grain produces a tube which grows through the tissues of the pistil to an ovule. While the pollen tube is growing, the generative nucleus divides to produce two sperms. The male gametophyte is then mature, and it consists simply of two sperms and a tube nucleus in a pollen tube.

Megasporophylls and Megaspores

A pistil is made up of one or more megasporophylls, or **carpels,** as they are often called. One or more ovules develop from the placenta of the ovary of a pistil. Each ovule consists of a megasporangium, also called the **nucellus,** surrounded by two integuments with an opening, the micropyle. One megaspore mother cell is formed in the nucellus; it undergoes reduction division to form a row of four haploid megaspores.

Female Gametophyte

Three megaspores abort and one develops into the female gametophyte, also known as the **embryo sac.** In many angiosperms the functional megaspore increases in size and the haploid nucleus divides mitotically. The resulting two nuclei undergo mitosis, and this division is followed by another. The resulting eight nuclei are arranged as follows: three at each end of the embryo sac and two in the center. Except for the nuclei in the center, each nucleus is surrounded by cytoplasm and a cell wall. The center cell at the micropylar end of the embryo sac is the **egg,** and the functionless cell on each side is a **synergid.** The three cells at the opposite end are nonfunctional **antipodal cells.** The two nuclei in the middle of the embryo sac are **polar nuclei.** The eight-celled body is the mature female gametophyte; it is a much reduced structure that is completely dependent on the sporophyte.

Fertilization and Seed Development

The pollen tube enters through the micropyle, grows through the nucellar tissue,

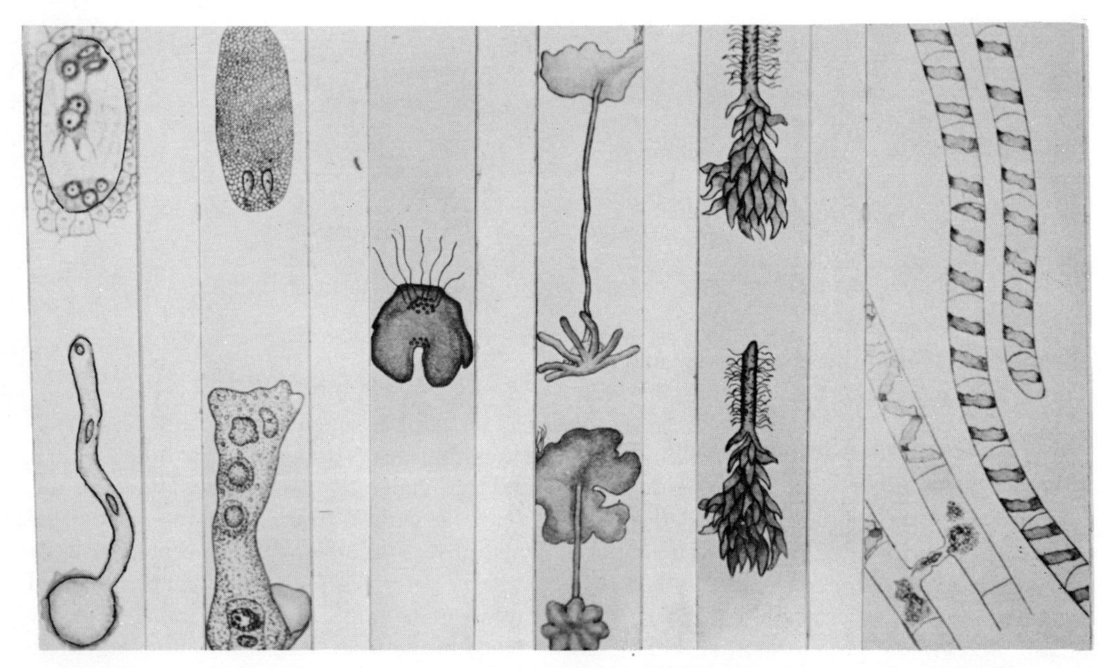

Fig. 39–2. Evolution of the gametophyte. *Left to right:* an alga, *Spirogyra* (magnified); a moss; a liverwort, *Marchantia;* a fern; a gymnosperm, *Pinus* (magnified); and an angiosperm (magnified).

Fig. 39–3. Evolution of the sporophyte. *Left to right:* an alga, *Spirogyra* (magnified); a moss; a liverwort, *Marchantia* (magnified); a fern (reduced); a gymnosperm, *Pinus* (greatly reduced); and an angiosperm (reduced).

and enters the embryo sac, where it discharges the two sperms. One sperm fertilizes the egg. The second sperm unites with the two polar nuclei to form an endosperm nucleus with three sets of chromosomes. The fertilized egg develops into an embryo, the endosperm nucleus develops into the endosperm, and the integuments into the seed coats; the complex is a seed.

SUMMARY OF THE EVOLUTION OF SPOROPHYTES AND GAMETOPHYTES

In the evolution of plants, the gametophyte becomes less complex and the sporophyte more and more complex (Figs. 39–2 and 39–3). In *Spirogyra* and in mosses and liverworts, the gametophyte is the independent and most conspicuous plant. In ferns, the gametophyte, although still independent, is less complex. Two types of gametophytes are produced by gymnosperms, and both are small, nongreen plants which are entirely dependent on the sporophyte for nutrition. The male gametophyte of an angiosperm has attained a minimum size, with three nuclei, two sperms, and a tube nucleus, all of which are functional. Three of the eight nuclei or cells (the egg and two polar nuclei) of the female gametophyte of an angiosperm are indispensable, leaving only five cells which are nonfunctional.

The decrease in the complexity of the gametophyte is paralleled by an increase in the complexity of the sporophyte and by a trend which leads to an independent sporophyte. In *Spirogyra* the zygote is the only diploid cell and it is the sporophyte. The relatively simple sporophytes of a moss plant and *Marchantia* are dependent upon the gametophytes for nutrition. In ferns, the embryo sporophyte is temporarily dependent upon the heart-shaped gametophyte, but by the time the root, stem, and leaf have developed, the sporophyte is independent. In gymnosperms and angiosperms the sporophyte is never dependent upon the gametophyte.

EVOLUTION OF DICOTS

The buttercup family (Ranunculaceae) is a primitive family of dicots from which the other families are believed to have evolved. Many choice garden plants are in this family, larkspur, peony, and columbine among them. The members of the buttercup family have separate petals and sepals, many stamens, and several to many pistils. The sepals, petals, and stamens come off below the ovary and the flower is said to be hypogynous.

Some botanists believe that two lines of evolution occurred in the Dicotyledoneae. One line of descent from the buttercup family includes families with hypogynous flowers. In the other line evolution progressed from hypogynous flowers to perigynous ones and ultimately progressed to epigynous flowers.

The two lines of descent are illustrated in Figure 39–4. Families with irregular corollas are in general more advanced than those with regular corollas. Petals are lacking in the walnut, birch, and beech families, as well as in some members of the buttercup family, and others that are not included here.

In both lines, the more advanced families have fewer stamens and carpels than the more primitive ones. Furthermore, the trend is from flowers with separate carpels to those with united carpels.

FAMILIES OF DICOTYLEDONS ON HYPOGYNOUS LINE OF DESCENT

The Mustard Family (Cruciferae)

Broccoli, cabbage, turnips, radishes, candytuft, stocks, and wallflowers are in this family. Examine the flowers of these plants and you will see a decided similarity. Each flower will have four sepals, four separate petals, six stamens (two short ones and four long ones) and one pistil. Other plants with these characteristics belong to this family (Fig. 39–5).

The Pink Family (Caryophyllaceae)

The carnation, pink, and chickweed are members of this family. Typically the flowers have five sepals, five petals, ten stamens, and a compound pistil made up of two to five carpels. The seeds are borne on a stalk which arises from the base of the ovary. This type of placentation is called **free central placentation.**

The Potato Family (Solanaceae)

A number of important food, ornamental, and medicinal plants are in this family, which includes the white potato, tomato, eggplant, belladonna, tobacco, and petunia. Examine a petunia carefully and you will learn the characteristics of the family. There are five sepals and five petals which are united. Cut the tube formed by the united petals and you will see that there are five stamens attached to the corolla tube. One bicarpellate pistil with axial placentation is present in a flower.

Snapdragon Family (Scrophulariaceae)

Many beautiful plants of garden and field are in this family: snapdragons, penstemons, monkey flowers, foxglove, and Indian paintbrush. The irregular flowers have four or five sepals, four or five united petals, two to five stamens attached to the corolla tube, and a bicarpellate pistil.

The Mint Family (Labiatae)

Peppermint, hoarhound, thyme, coleus, and salvia are in this family. The irregular flowers have four or five sepals, four or five united petals, two or four stamens attached to the corolla tube, and a bicarpellate fruit with four seeds.

FAMILIES OF DICOTYLEDONS ON LINE LEADING TO EPIGYNOUS FLOWERS

The Pea Family (Leguminosae)

Some representatives of this extremely important family are beans, peas, soybeans, peanuts, alfalfa, clover, lupines, sweet peas,

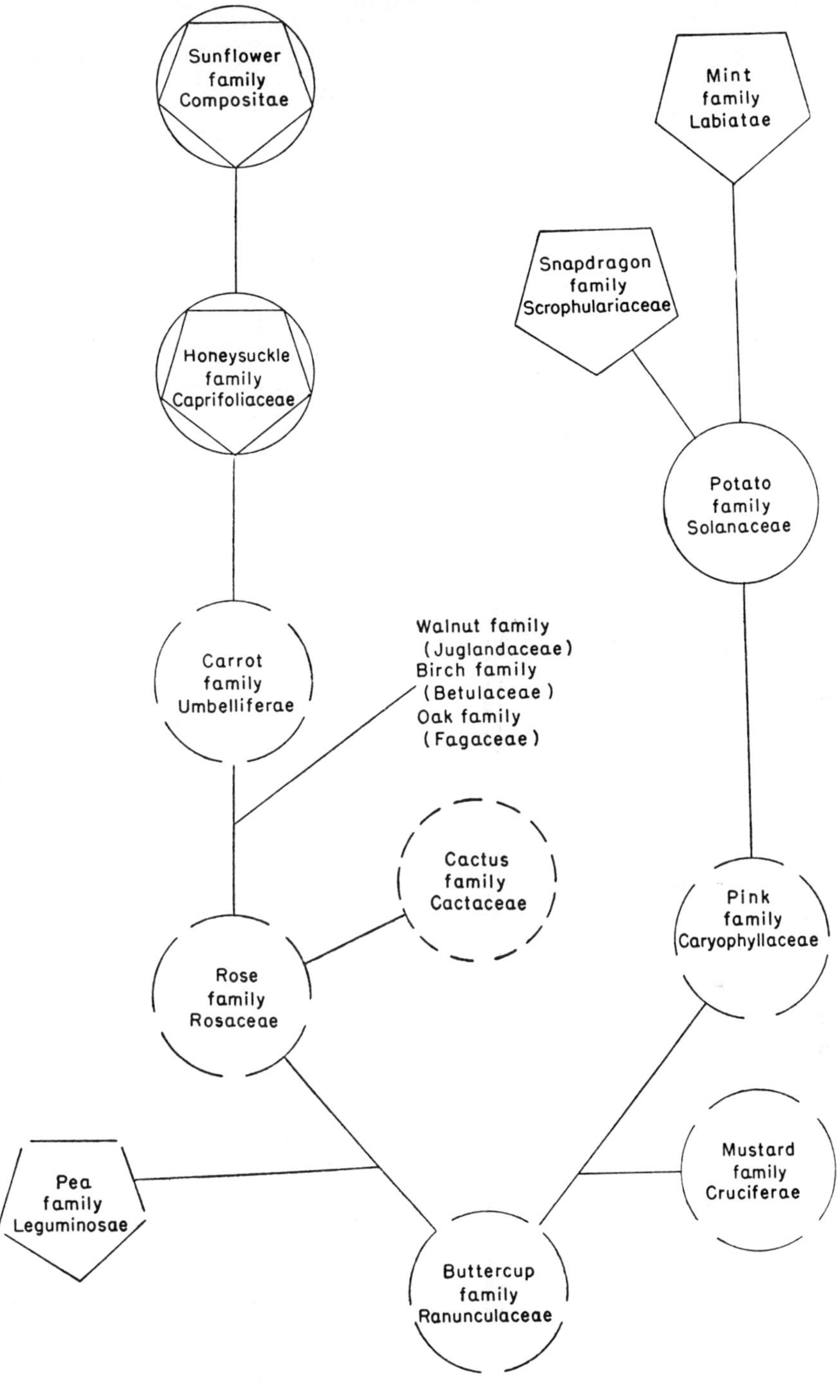

Fig. 39–4. Chart showing the two principal lines of descent of dicots. The broken circle indicates that the petals are not united and the closed circle signifies that the petals are united. The closed pentagon indicates that the corolla is irregular and that the petals are united. The pentagon inside a circle shows that both regular and irregular flowers are characteristic of the family. The indicated flower structure does not hold for all genera and species of a family, but it is characteristic of the family as a whole.

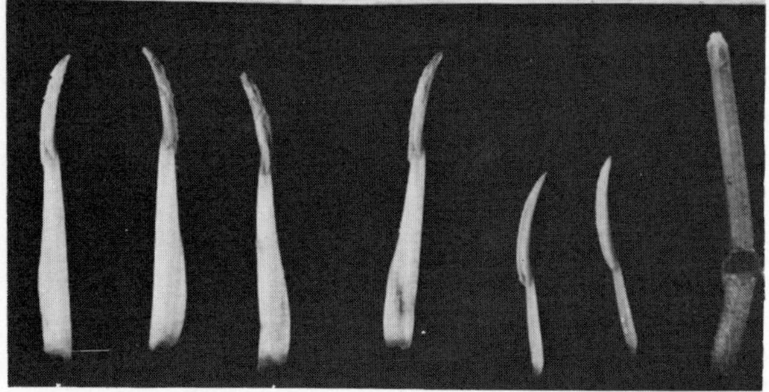

Fig. 39–5. The sweet rocket (*Hesperis matronalis*) is in the mustard family. The lower figure shows the four long stamens, the two short ones, the pistil from a flower.

Fig. 39–6. The corolla of a locoweed (*Oxytropus lambertii*), a member of the pea family, is made up of a standard (*upper petal*), two wings, and a keel consisting of two united petals.

Kentucky coffeetree, and acacias. Examine carefully a flower of a sweet pea to learn the characteristics of this family. The five petals of a sweet pea are not all alike. One is very large and it is called the standard. There are two petals known as wings, which are situated one on each side of the standard. Two smaller petals are united to form a keel which surrounds the ten stamens and single pistil. Nine of the stamens are united by their filaments and the tenth is separate. The sepals and petals come off below the ovary (Fig. 39–6).

The Rose Family (Rosaceae)

From the standpoint of fruit production, this family is the most important one. The apple, pear, peach, plum, strawberry, blackberry, and loganberry are in this family, as are such desirable ornamental plants as the rose, spirea, flowering cherry, hawthorn, and flowering crab. The flowers have five sepals, five separate petals, and numerous stamens. One to several pistils, depending on the species, are also present. This family is very much like the buttercup family, from which it has evolved. The rose family can be distinguished from the buttercup family by the position of the flower parts in relation to the ovary. In some members of the rose family the sepals and petals arise from the top of the ovary (the flowers are epigynous); in others they come off from a tube which surrounds the ovary (the flowers are perigynous). In the buttercup family they come off below the ovary.

The Cactus Family (Cactaceae)

This family is chiefly native to the Americas. With the exception of *Rhipsalis,* cacti were unknown on other continents until after the discovery of America. Most species have large, beautiful flowers and peculiar plant forms. The family is readily recognized by the succulent stems which are covered with tufts of spines and by the absence of foliage leaves. Sepals, petals, and stamens are numerous and the flowers are epigynous.

The Maple Family (Aceraceae)

The maples and boxelder are in this family, whose members are either trees or shrubs. The leaves are opposite, simple in some species but compound in others. The flowers are usually regular and have four to ten sepals. Petals may be present or lacking. The pistil is superior. The fruits are winged and are known as samaras (Fig. 39–7).

The Walnut Family (Juglandaceae)

The walnut, hickory, and pecan are in this family. The flowers lack petals and they are epigynous, small, and unisexual. The staminate flowers are produced on pendulous catkins. The pistillate flowers are solitary or a few in a group (Fig. 39–8).

The Birch Family (Betulaceae)

Alders, birches, filberts, and hazelnuts are representatives of this family. The flowers are apetalous, epigynous, and unisex-

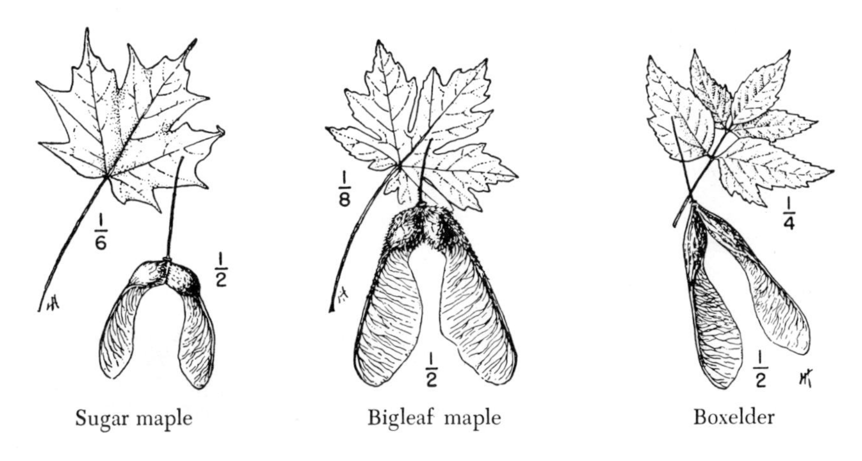

Fig. 39–7. Maples and boxelder are in the family Aceraceae. (U.S. Forest Service.)

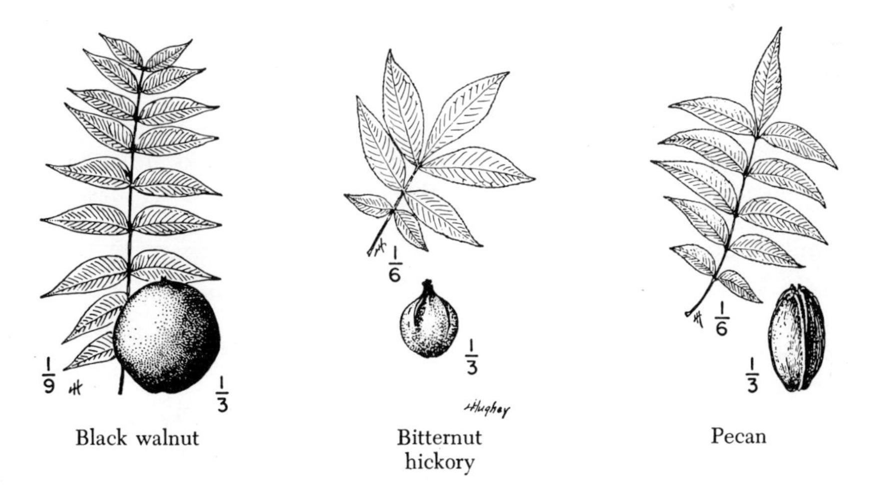

Fig. 39–8. Walnut, hickory, and pecan are in the family Juglandaceae. (U.S. Forest Service.)

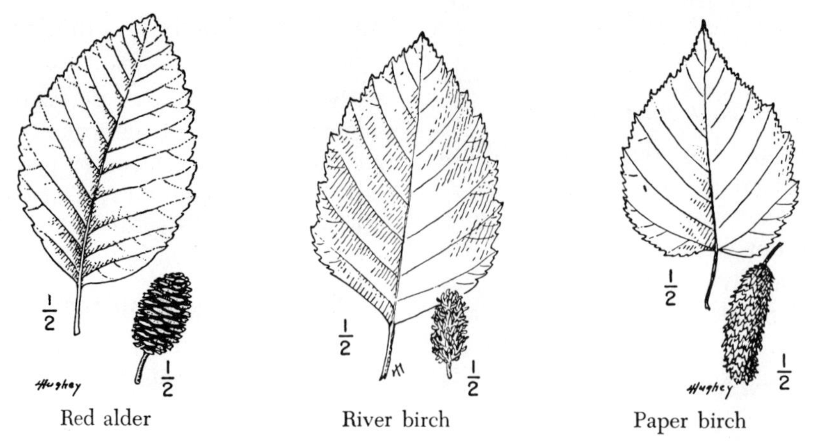

Fig. 39–9. Alder and birch are in the family Betulaceae. (U.S. Forest Service.)

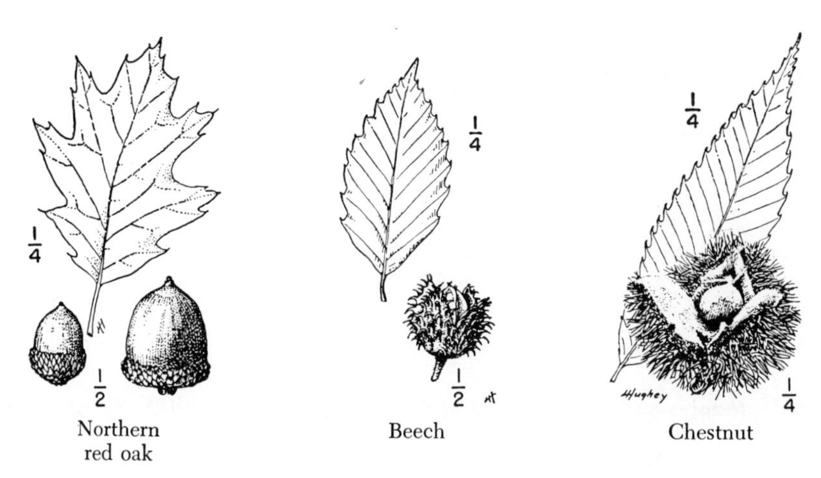

Fig. 39–10. Oak, beech, and chestnut are in the family Fagaceae. (U.S. Forest Service.)

ual. Both the pistillate and staminate flowers are borne in catkins, and usually both types occur on the same tree (Fig. 39–9).

The Oak Family (Fagaceae)

Many shade trees and numerous trees of deciduous forests are in this family, which includes the oaks, beeches, and chestnuts. All species are monoecious, and they have small flowers which lack petals (Fig. 39–10).

The Carrot Family (Umbelliferae)

The carrot, celery, dill, parsley, anise, parsnip, water hemlock, and poison hemlock are members of this family. The regular flowers are produced in umbels and they are epigynous. A flower has five toothlike sepals, five small separate petals, five stamens inserted on a disk over the ovary, and a bicarpellate pistil.

The Honeysuckle Family (Caprifoliaceae)

This family includes honeysuckle, snowball, elderberry, coralberry, snowberry, and weigela. The petals are united, and the corolla is regular in some species but irregular in others. The flowers have four or five sepals, usually five united petals, five sta-

Fig. 39–11. *Left*, a ray flower of arnica with a ligulate corolla. *Right*, a disk flower with a tubular corolla.

mens attached to the corolla tube, and a two- to five-celled pistil. The flowers are epigynous.

The Sunflower Family (Compositae)

This large family includes food plants such as the sunflower, artichoke, and lettuce, and many ornamental plants—for example, asters, chrysanthemums, dahlias, zinnias, marigolds, and daisies. Several troublesome weeds—Canadian thistle, cocklebur, and dandelion—are also in this family.

The flowers occur in heads subtended by an involucre. At first glance the inflorescence appears to be a single flower, but closer examination reveals that it is a group of many separate flowers. The corolla is either tubular or strap-shaped (ligulate). Both tubular and ligulate corollas are made up of five united petals. The tubular flowers are regular, the ligulate, irregular (Fig. 39–11). In some species—for example, the dandelion—all of the flowers are ligulate. In the white snakeroot all of the flowers are tubular. In many species—for example, arnica and sunflower—the marginal flowers (ray flowers) have ligulate corollas and the central flowers (disk flowers) have tubular corollas. The calyx is known as the pappus, and it consists of scales, awns, or bristles, depending on the species. The anthers of the five stamens are united into a ring which surrounds the style. In some composites only a pistil is present in a ray flower, whereas both stamens and pistil are found in a disk flower. The flowers are epigynous. The bicarpellate pistil develops into an achene.

SOME FAMILIES OF MONOCOTYLEDONEAE

Monocotyledonous plants are characterized by scattered vascular bundles, parallel-veined leaves, floral parts in threes or a simple multiple of three, and one seed leaf.

The evolution of monocotyledons is illustrated in Figure 39–12. The monocots have

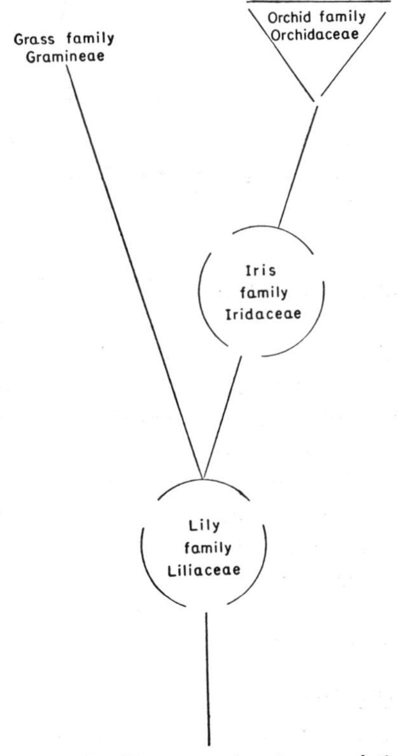

Fig. 39–12. Diagram showing evolution of monocots.

Fig. 39–13. A lily flower. Notice the six members of the perianth, the six stamens, and single pistil.

probably evolved from the dicotyledonous buttercup family (Ranunculaceae). From the lily family of monocots there are two lines of evolution. One line leads to the grass family and the second to the orchid family. In the latter line of descent, the trend is from hypogynous flowers, which occur in lilies, to epigynous flowers, which characterize the iris and orchid families. As in the dicotyledons, the trend is from regular flowers (indicated by a circle) to irregular ones (indicated by a triangle).

The Lily Family (Liliaceae)

The lily, hyacinth, tulip, yucca, trillium, mariposa lily, onion, and asparagus are members of this family. The hypogynous flowers have three sepals, three petals, six stamens, and one pistil made up of three carpels (Fig. 39–13).

The Iris Family (Iridaceae)

In contrast to the Liliaceae, the flowers of the Iridaceae are epigynous. The flower of Iris is composed of three petal-like sepals, three petals, three stamens, and a pistil composed of three carpels. In Iris the three branches of the style are expanded and petal-like and add beauty to the flower (Fig. 39–14).

The Orchid Family (Orchidaceae)

The orchids are the climax of one line of evolution. The irregular flowers occur in a great variety of forms, colors, and sizes, and they are ingeniously constructed for insect-pollination. Orchid flowers are epigynous and they have three sepals, three petals, and a column. The two lateral petals are alike, but the third has a different shape; it may be bulbous, pouchlike, tubular, keel-

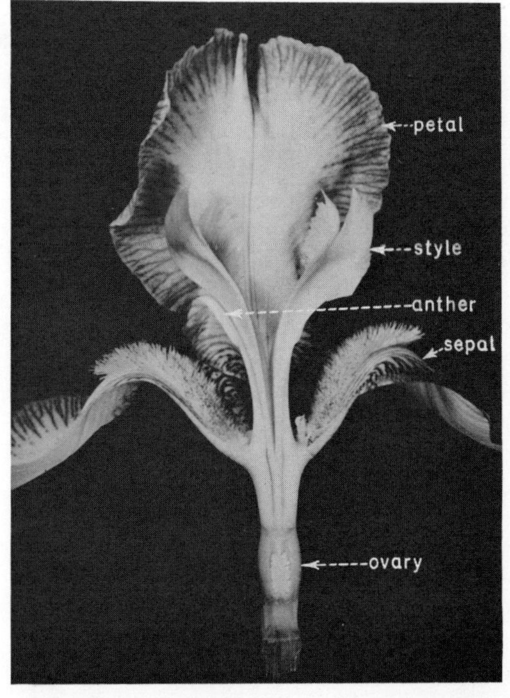

Fig. 39–14. A section of an iris flower. Notice that the flower is epigynous.

shaped, or of some other form. The highly specialized petal is known as the **lip** or **labellum.** The **column** is the most characteristic feature of an orchid flower and is the result of a fusion of three stamens with a pistil. Only one (occasionally two) of the stamens produces pollen. The pollen occurs in masses called **pollinia.** Orchid seeds are extremely small and readily disseminated by wind (Fig. 39–15).

There are at least 15,000 species of orchids. Only one other family, the Compositae, has as many species.

Orchids are the most prized of flowers and such genera as *Cattleya, Cymbidium, Phalaenopsis, Vanda,* and *Paphiopedilum* are extensively grown in greenhouses. The pods of *Vanilla* yield a familiar flavoring material.

Fig. 39–15. *Cymbidium,* a member of the orchid family. Notice the three sepals and irregular corolla. Two of the petals are alike and the third forms the lip (labellum). The column is evident above the lip.

The Grass Family (Gramineae)

Grasses are the basis of civilization. They are the major plants of range and pasture, and such grasses come to man in the form of mutton, lamb, and beef. Grasses withstand grazing better than other plants because leaf growth occurs at the base instead of throughout the leaf. If a clover leaf is eaten, that is the end of that leaf; if the terminal portion of a grass leaf is eaten, it grows from the base and soon attains its original length.

The cereals are grasses. Wheat, barley, rice, oats, and sorghum were cultivated by

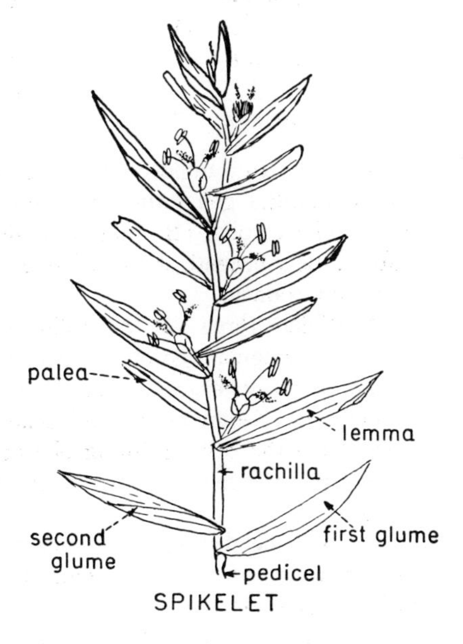

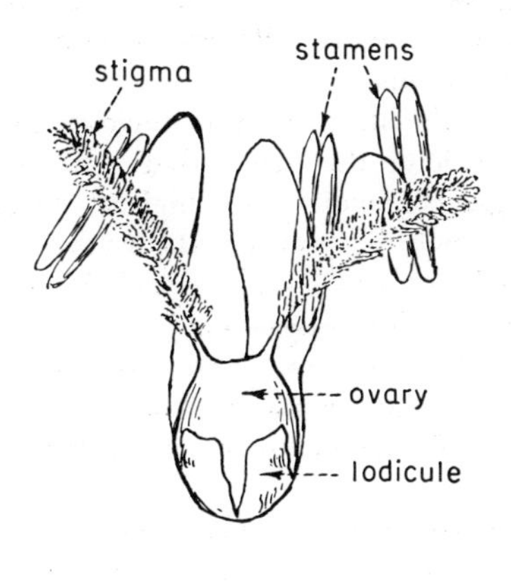

Fig. 39–16. *Left,* a grass spikelet. *Right,* a grass flower. (After Agnes Chase.)

Eurasians long before the Christian era, while rye has been cultivated since the beginning of this era. While Eurasians were developing the above cereals, the American Indians perfected maize. Maize and wheat have been so changed by selection that the original parents cannot be recognized; wheat and corn are not found growing wild.

Although grasses are of prime importance, their flowers are seldom noticed by most people. The flowers of grasses are produced on a specialized jointed branch known as a **rachilla.** Two empty bracts, the **glumes,** are present at the base of the rachilla. One or more flowers are attached to the branch above the glumes. Each flower is enclosed between two bracts; the outer bract is the **lemma** and the inner one is the **palea.** A pistil, made up of an ovary containing one ovule and two feathery stigmas, three stamens, and two small scalelike structures known as **lodicules** are present in each flower. The rachilla with its flowers and bracts is called a **spikelet** (Fig. 39–16). In wheat and barley the spikelets are attached to the main stem, or **rachis,** as it is called, and the inflorescence is a spike. In oats, the spikelets are at the ends of branches of the rachis, and the inflorescence is a panicle.

QUESTIONS

1. _____ The seed plants are in the phylum (A) Tracheophyta, (B) Bryophyta, (C) Eumycophyta, (D) Chlorophyta.

2. _____ Most present-day species are (A) angiosperms, (B) gymnosperms, (C) ferns.

3. _____ The wheat plant is comparable to (A) the moss gametophyte, (B) the moss sporophyte.

4. _____ Of the eight nuclei in the embryo sac (A) only three are functional, the egg and the two polar nuclei; (B) all eight are functional; (C) only the egg is functional; (D) only the synergids are nonfunctional.

5. _____ The gametophytes of angiosperms (A) are green and grow on the soil as do the gametophytes of ferns, (B) are nongreen and do not grow on soil.

6. Outline the life cycle of an angiosperm.

7. In which of the plants that you studied was the gametophyte at its highest development? In which did you find the gametophyte and sporophyte both independent? In which was the gametophyte dependent?

8. _____ In evolution there has occurred (A) an increasing complexity and size of the sporophyte, (B) an increasing complexity and size of the gametophyte.

9. Trace through the plant kingdom the relative dominance of sporophyte and gametophyte generations.

10. _____ The two subclasses Monocotyledoneae and Dicotyledoneae are in the class (A) Angiospermae, (B) Gymnospermae.

11. Name five differences between monocots and dicots.

12. Indicate whether each of the following is a monocot (M) or dicot (D):

_____ Tulip.
_____ Iris.
_____ Orchid.
_____ Corn.
_____ Sunflower.
_____ Honeysuckle.
_____ Carrot.
_____ Birch.
_____ Walnut.
_____ Cactus.
_____ Rose.
_____ Bean.
_____ Snapdragon.
_____ Potato.
_____ Carnation.
_____ Mustard.
_____ Buttercup.

13. _____ In angiosperms the more advanced families generally would not have (A) an irregular corolla, (B) united petals, (C) separate carpels instead of united ones, (D) relatively few stamens.

14. Which three of the following families have an irregular corolla: Cruciferae, Solanaceae, Scrophulariaceae, Labiatae, Rosaceae, Umbelliferae, Orchidaceae, Liliaceae.

15. Give the distinguishing characteristics of six families of dicots. List some plants for each family.

16. Give the distinguishing characteristics of three families of monocots and name several members of each family.

17. Name the three families which are our chief sources of food.

18. In angiosperms, certain structures are comparable to those in ferns, club mosses, and pines. On the left are shown terms commonly used when discussing angiosperms; on the right those used for other vascular plants. Match these terms:

_____	1. Stamen.	1. Microsporophyll.
_____	2. Pollen sac.	2. Microsporangium.
_____	3. Pollen grain plus pollen tube.	3. Male gametophyte.
_____	4. Carpel.	4. Megasporophyll.
_____	5. Nucellus.	5. Megasporangium.
_____	6. Embryo sac.	6. Female gametophyte.

19. Label the following structures as 2 n (sporophyte) or 1 n (gametophyte):

_____	Pine tree.	_____	Egg in archegonium.
_____	Liverwort thallus.	_____	Spore mother cells.
_____	Pollen grain.	_____	Prothallium.
_____	Zygote.	_____	Familiar fern plant.
_____	Embryo sac.	_____	Spores in moss capsule.
_____	Sperms in fern antheridium.	_____	Embryo.

Self-Pronouncing Glossary

Abscission (ab · sizh′ un).* Separation of fruit, leaf, or other part from a plant.

Abscission layer. A zone of cells in the petiole or other plant structure whose cells separate and thereby bring about leaf fall, fruit drop, etc.

Absorption (ab · sorp′ shun). Intake of substances from the outside.

Accumulation. The uptake of substances against a concentration gradient and hence an energy-requiring process.

Achene (ay · keen′). A dry, indehiscent, one-seeded fruit in which the ovary wall remains free from the seed coat.

Acquired character. A feature induced during the life of an individual by environmental factors.

Adaptation (ad′ ap · tay′ shun). An adjustment by an organism to environmental conditions.

Adenine (ad′ u · neen). A purine base present in one of the nucleotides.

* NOTE ON PRONUNCIATION: A heavy accent (′) is used for the most heavily accented syllable and a light accent (′) for the secondary stress. Letters are given their most common pronunciation, including the following examples: o͞o as in food, o͝o as in foot, s as in dose, f as in loaf, g as in get, th as in thorn, thee as in theology.

Adenosine (DI-, TRI-) **phosphate** (ADP, ATP) (a · den′ o · seen). Phosphorylated organic compounds that function in energy transfer.

Adsorption (ad · sorp′ shun). The concentration of molecules or ions on the surfaces of colloidal particles or on solid bodies.

Adventitious (ad′ ven · tish′ us). Organs arising in unusual positions, as buds from roots.

Aeciospore (ee′ si · o · spor). A rust spore formed in an aecium after the fusion of a spermatium with a flexuous hypha.

Aecium (ee′ shi · um). A cup-shaped structure containing chains of aeciospores.

Aerenchyma (ay′ ur · eng′ ki · muh). Aerating tissue in aquatic plant organs, which is characterized by large intercellular spaces.

Aerobic respiration (ay′ ur · o′ bick). Respiration in which gaseous oxygen is used.

After-ripening. Processes occurring in seeds, bulbs, tubers, etc., which must be completed before germination or growth can occur.

Agglutinin (a · glo͞o′ ti · nin). An antibody capable of clumping bacteria.

Aggregate fruit (ag′ ree · gate). A cluster of fruits, such as a strawberry or raspberry, which developed from one flower having a number of pistils.

Alkaloids (al′ kuh · loidz). Bitter alkaline organic compounds containing carbon, hydrogen, nitrogen, and, in some, oxygen. Familiar alkaloids are caffeine, morphine, nicotine, quinine, and strychnine.

Allele (a · leel′). One of the alternate forms of a gene located at the same position on a homologous chromosome.

Allopolyploid (al′ o · pol′ i · ploid). An organism having more than two sets of chromosomes and derived by hybridization between two species.

Allotetraploid (al′ o · tet′ ruh · ploid). An allopolyploid having four sets of chromosomes.

Alpha ray (al′ fuh). A ray of positively charged particles emitted during certain radioactive transformations.

Alternate leaves. A leaf arrangement with only one leaf at each node.

Alternation of generations. The alternation of a gamete-producing phase with a spore-producing phase in the life cycle of a plant.

Amino acids (a · mee′ no; am′ i · no). Organic acids, each with an NH_2 group, which combine to form proteins.

Amylase (am′ i · lace). An enzyme which digests starch.

Anabolism (a · nab′ o · liz′m). Processes which build complex organic substances from simple materials.

Anaerobic respiration (an · ay′ ur · o′ bick). Respiration in the absence of free oxygen.

Anaphase (an′ uh · fayz). The stage in mitosis or meiosis when the chromosomes are moving toward the poles.

Androecium (an · dree′ shee · um). A collective term for the stamens in a flower.

Angiosperm (an′ jee · o · spurm). One of the flowering plants; the seeds are enclosed in fruits.

Annual. A plant which lives only one year.

Annual ring. The ring of secondary xylem (wood) produced in one year.

Annulus (an′ u · lus). A ringlike structure, such as the ring around the mushroom stalk or a fern sporangium.

Anther (an′ thur). The pollen-bearing part of a stamen.

Antheridium (an′ thur · id′ ee · um). A structure in which sperms are produced.

Anthocyanin (an′ tho · sigh′ uh · nin). Water-soluble pigments present in the cell sap which are responsible for blue, purple, and many red colors.

Antibiotic (an′ ti · by · ot′ ick). A substance produced by certain organisms (chiefly bacteria and true fungi) which destroys bacteria and other fungi or interferes with their development.

Antibody (an′ ti · bodd′ ee). A substance, produced by the body, which counteracts invading organisms or harmful products formed by them.

Antipodal cells (an · tip′ o · dul; -d′l). The cells, generally three, in the embryo sac located at the end opposite the egg.

Antitoxin (an′ ti · tock′ sin). An antibody which neutralizes a toxin.

Apetalous (ay · pet′ ul · us). A flower without petals.

Apical meristem (ap′ i · kull merr′ i · stem). Tissue at the tip of a root or stem where cell division occurs.

Apomixis (ap · o · mix′ is). The development of an embryo from some cell other than a fertilized egg; the setting of seed without fertilization.

Apothecium (ap′ o · thee′ shee · um). A cup-shaped structure lined with asci containing ascospores.

Archegonium (ahr′ kee · go′ nee · um). A flask-shaped, multicellular egg-producing structure.

Ascogonium (ass′ ko · go′ nee · um). A one-celled structure containing a female gamete in ascomycetes.

Ascomycetes (ass′ ko · migh · see′ teez). A group of true fungi whose sexually produced spores are formed in sacs called asci.

Ascospore (ass′ ko · spor). A fungus spore formed in an ascus.

Ascus (ass′ kuss). A sac characteristic of ascomycetes in which ascospores are produced following nuclear fusion and reduction division.

Asepalous (ay · sep′ ul · us). A flower lacking sepals.

Asexual reproduction. Reproduction which does not involve the fusion of gametes.

Assimilation (a · sim′ i · lay′ shun). The formation of protoplasm.

ATP. *See* Adenosine triphosphate.

Aureomycin (aw′ ree · oh · my′ sin). An antibiotic produced by *Streptomyces aureofaciens.*

Aureomycin hydrochloride. Trademarked name for one antibiotic substance, chlortetracycline hydrochloride.

Autopolyploid (aw′ to · pol′ i · ploid). An organism having more than two sets of homologous chromosomes.

Autosome (aw′ to · sohm). Any chromosome which is not a sex chromosome.

Autotroph (aw′ to · trof). An organism which makes food from inorganic raw materials; green plants, for example.

Auxin (awk′ sin). A natural or synthetic chemical which controls the growth of plants.

Auxospore (awk′ so · spor). A spore formed by diatoms which develops into a full-sized diatom.

Axillary bud (ack′ si · ler′ ee; ack · sil′ uh · ree). A bud borne in the upper angle formed by the leaf and the stem; the angle is known as an axil.

Bacillus (ba · sill′ us). A rod-shaped bacterium.

Bacteria. The common name for one-celled plants in the phylum Schizomycophyta.

Bacteriophage (back · teer′ ee · o · faydj). A virus which attacks bacteria.

Bark. The tissues of a woody stem or root outside the cambium.

Basidiomycetes (ba · sid′ ee · o · migh · see′ teez). A class of fungi whose sexually produced spores are formed on a basidium.

Basidiospore (ba · sid′ ee · o · spor′). A spore produced on a basidium.

Basidium (ba · sid′ ee · um). A spore-bearing structure in Basidiomycetes; the spores are borne externally, generally on points called sterigmata.

Benthos (ben′ thawss). Organisms which live and feed along the bottoms of oceans and lakes.

Berry. A fleshy fruit, such as tomato and grape, formed from one pistil; the entire pericarp is fleshy.

Beta ray (bee′ tuh; bay′ tuh). A stream of electrons given off from a radioactive substance.

Biennial (bigh · en′ ee · ul). A plant requiring two years to complete its life cycle, usually flowering and fruiting the second year only and then dying.

Binomial (bigh · no′ mee · ul). The generic name and the specific name taken together as the scientific name of an organism.

Biotic (bigh · ot′ ick). Pertaining to life.

Blade. The expanded part of a leaf.

Bract. A modified or reduced leaf subtending a flower or inflorescence.

Bryophyte (brigh′ o · fite). Any of the mosses, liverworts, or hornworts.

Bud. An undeveloped branch protected by leaves, scales, or bracts.

Bud scale. A modified leaf enveloping the more tender parts of a bud.

Budding. A form of grafting in which one bud is grafted onto the stock. In yeast, a means of asexual reproduction.

Bulb. An underground stem with fleshy, food-storing, scale leaves; essentially a below-ground bud.

Bundle sheath. A cell layer which surrounds the conducting tissues of a vein.

Calorie (kal′ o · ree). A large Calorie is the amount of heat required to raise the temperature of 1 kilogram of water 1° C.; a small calorie is one-thousandth of a large Calorie.

Calyptra (ka · lip′ truh). A cap upon the apex of a moss capsule.

Calyx (kay′ licks; kal′ icks). The outermost whorl of a complete flower; a collective term for the sepals.

Cambium (kam′ bee · um). A layer of dividing cells, located between the xylem and phloem, which produces secondary xylem and phloem.

Capsule (kap′ sul). The spore case of a moss or liverwort. A dehiscent dry fruit formed of two or more united carpels.

Carbohydrate (kahr′ bo · high′ drate). A class of foods composed of carbon, hydrogen, and oxygen, the ratio of hydrogen to oxygen being 2:1.

Carnivore (kahr′ ni · vor). A meat-eating organism.

Carnivorous plant (kahr · niv′ o · rus). A plant which catches or traps and digests animals, generally insects.

Carotene (kar′ o · teen). An orange pigment occurring in certain plastids and the precursor of vitamin A.

Carpel (kahr′ pel). A floral part which bears ovules; a pistil is composed of one or more carpels.

Carpogonium (kahr′ po · go′ nee · um). The structure in which the egg develops in red algae.

Carpospore (kahr′ po · spor). A spore formed by red algae.

Caruncle (kar′ ung · k'l; kuh · rung′ k'l). A spongy outgrowth at one end of certain seeds, for example, castor bean.

Caryopsis (kar′ ee · op′ sis). A one-seeded, dry, indehiscent fruit in which the seed coat and pericarp are united, for example, a corn grain.

Casparian strip (kas · pair′ ee · un). A waterproof thickened strip on the radial and transverse walls of an endodermal cell. The Casparian strip may prevent water from diffusing through the radial and transverse walls and thus direct the water through the protoplasts of endodermal cells.

Catabolism (kuh · tab′ o · liz'm). Metabolic processes in which complex materials are changed into simpler compounds, for example, digestion and respiration.

Catalyst (kat′ uh · list). A substance which influences the rate of a chemical reaction without being used up in the reaction.

Catkin (kat′ kin). An inflorescence bearing apetalous staminate or pistillate flowers, as in willow.

Cell. The structural and physiological unit of plants and animals, generally consisting of cytoplasm and nucleus and also, in plant cells, a cell wall.

Cell sap. The solution in a vacuole.

Cellobiase. An enzyme which digests cellobiose.

Cellobiose. A white faintly sweet sugar with a formula of $C_{12}H_{22}O_{11}$.

Cellulase. An enzyme which digests cellulose.

Cellulose (sel′ u · los). A carbohydrate which is formed from glucose and is a major constituent of cell walls.

Centromere (sen′ tro · meer). The region of a chromosome where a spindle fiber (tractile fiber) is attached during the metaphase of mitosis and meiosis.

Chemosynthesis (kem′ o · sin′ thi · sis). Food manufacture carried on by certain bacteria utilizing energy released in the oxidation of inorganic compounds.

Chemotropism (ke · mot′ ro · piz'm). A growth movement in response to a chemical.

Chlamydospore (clam′ i · do · spor′). A thick-walled spore which is formed by certain smut fungi.

Chlorenchyma (klo · reng′ ki · muh). A tissue having chloroplasts.

Chloromycetin. Trademark for the antibiotic substance chloramphenicol.

Chlorophyll (klo′ ro · fill). The green pigments of plants which absorb light and make it effective in photosynthesis.

Chloroplast (klo′ ro · plast). A cellular structure (plastid) made up of protoplasm and containing the green pigment chlorophyll.

Chromatid (kro′ muh · tid). One of the two parallel rods that are formed after a chromosome duplicates itself.

Chromatin (kro′ muh · tin). A nuclear substance which stains deeply with certain dyes and from which the chromosomes form during mitosis and meiosis.

Chromoplast (kro′ mo · plast). A colored plastid containing pigments other than chlorophyll; often they are yellowish or red in color.

Chromosome (kro′ mo · sohm). One of the rodlike structures formed from chromatin during mitosis and meiosis. The chromosomes bear genes.

Citric acid cycle. A cyclic series of reactions which transforms pyruvic acid into carbon dioxide and hydrogen which then undergoes terminal oxidation to yield available energy.

Class. A group of related orders.

Cleistothecium (klighs′ to · thee′ shee · um; -see · um). A closed sphere containing one or more asci.

Climax. The terminal community of a succession, one which maintains itself indefinitely provided no marked environmental changes occur.

Coccus (cock′ us). A spherical bacterial cell.

Coenocytic (see′ no · sit′ ick). Multinucleate and lacking cross walls.

Coenzyme (ko · en′ zime). A chemical, usually organic, which is necessary for the functioning of a given enzyme.

Cohesion (ko · hee′ zhun) and **transpiration** (tran′ spi · ray′ shun) **pull.** A theory to explain the rise of water in plants.

Colchicine (kol′ chi · seen, -sin; kol′ ki-). An alkaloid which is used to bring about a doubling of the chromosomal number.

Coleoptile (ko′ lee · op′ til; kol′ ee-). A sheath around the epicotyl of a grass embryo.

Coleorhiza (ko′ lee · o · righ′ zuh; kol′ ee · o-). A sheath around the radicle of a grass embryo.

Collenchyma (ko · leng′ ki · muh). Strengthening tissue made up of living cells whose walls are thickened in the corners.

Colloid (kol′ oid). A system having dispersed particles ranging in size from 0.1 micron to 0.001 micron; the dispersed particles are huge molecules or molecular aggregates. The particles are so small that they do not settle out.

Columella (kol′ u · mell′ uh). The central part of a sporangium or capsule which does not produce spores.

Community. An assemblage of organisms living together.

Companion cell. A long, narrow, nucleated cell associated with a sieve tube in the phloem of flowering plants.

Complete flower. A flower having sepals, petals, stamens, and a pistil or pistils.

Compound leaf. A leaf whose blade consists of two or more leaflets.

Compound pistil (pis′ til). A pistil formed of two or more united carpels.

Conceptacle (kun · sep′ tuh · k'l). A cavity containing antheridia or oogonia, or both, in certain brown algae, as in *Fucus*.

Conidiophore (ko · nid′ ee · o · for′). A hypha which produces conidia.

Conidium (ko · nid′ ee · um). An asexual fungus spore cut off from the tip of a conidiophore.

Conjugation (kon′ jŏŏ · gay′ shun). Union of like gametes (isogametes).

Conjugation tube. A tube which connects one cell with another and through which a gamete moves.

Convergence (kon · vur′ jens). The evolution of similar characteristics in distantly related organisms.

Cork. A protective layer of cells having suberized walls on stem, root, and certain other surfaces; the outer bark of a woody plant.

Cork cambium (kam′ bee · um). A layer of dividing cells which forms cork externally. Also known as phellogen.

Corm. An erect, thickened, underground stem lacking fleshy scale leaves.

Corolla (ko · rol′ uh). The petals taken collectively; the floral parts just within the calyx and usually of a bright color.

Cortex (kor′ tecks). The primary tissue of a root or stem between the epidermis and the vascular tissue.

Cotyledon (kot′ i · lee′ dun). A seed-leaf.

Cristae (kris′ tee). The infoldings of the inner membrane of a mitochondrion.

Cross-pollination. The transfer of pollen from the anther of one plant to the stigma of a flower on another plant.

Crossing over. The interchange of parts of homologous chromosomes during the first prophase of meiosis.

Cuticle (kew′ ti · k'l). A waxy layer secreted by epidermal cells on their outer surface.

Cuticular transpiration (kew · tick′ u · lur). Loss of water vapor through the cuticle.

Cutin (kew′ tin). A waxy, somewhat waterproof substance.

Cutting. A severed plant part used for propagation.

Cyme (sime). A flat or convex flower cluster, with the innermost flowers the oldest.

Cystocarp (sis′ to · kahrp). A spore-producing structure of a red alga.

Cytokinesis (sigh′ to · ki · nee′ sis). The partitioning of one cell into two.

Cytology (sigh · tol′ o · jee). The science of cell structure and activity.

Cytoplasm (sigh′ to · plaz'm). The protoplasm in a cell, exclusive of the nucleus, plastids, and other protoplasmic structures of rather definite form.

Cytosine (sigh′ to · seen). A pyrimidine base found in DNA and RNA.

Day-neutral plant. A plant which flowers when days are either long or short.

Deciduous (de · sid′ u · us). Refers to plants which shed their leaves at the end of the growing season and which are bare for part of the year.

Dehiscent (de · hiss′ ent). Splitting open in a characteristic manner at maturity.

Deletion. The loss of a part of a chromosome.

Denitrification (dee · nigh′ tri · fi · kay′ shun). The changing of nitrogen compounds into gaseous nitrogen by denitrifying bacteria.

Deoxyribonucleic acid (DNA) (de · ahk′ see · righ′ bo · new · klee′ ick). The nucleic acid of the chromosomes; the main carrier of genetic information.

Deoxyribose (de · ahk′ · see · righ′ · bos). A 5-carbon sugar which is present in deoxyribonucleic acid.

Dextrinase (decks′ tri · nase). An enzyme which digests dextrins.

Diadelphous (digh′ uh · del′ fuss). Refers to stamens which are united by their filaments into two groups.

Diastase (digh′ uh · stayce). A group of enzymes which digest starch into glucose.

Dicotyledon (digh · kot′ i · lee′ dun). A plant belonging to the large group of flowering plants characterized by two cotyledons in a seed.

Dictyosome (dic′ tee · o · sohm). The Golgi apparatus.

Differentially permeable membrane. A membrane which permits certain substances to pass through, but which restricts the passage of others.

Differentiation. The transformation of a relatively unspecialized cell into a more specialized one.

Diffusion (di · few′ zhun). The movement of molecules from regions of high to regions of low concentration.

Digestion. The conversion of complex foods, generally insoluble, to simpler substances which are soluble in water.

Dihybrid cross (digh · high′ brid). A cross between parents differing in two characters.

Dioecious (digh · ee′ shus). Having the male and female reproductive organs borne by separate plants; in dioecious flowering plants staminate and pistillate flowers are borne on different plants.

Diphosphopyridine nucleotide (DPN) (digh · fos′ fo · pir′ i · deen new′ klee · o · tide). The carrier of hydrogen in respiration. Also known as NAD (nicotinamide adenine dinucleotide).

Diploid (dip′ loid). Having two sets of chromosomes, one set coming from one gamete, the second set from the other gamete.

Disbudding. The removal of some flower buds to produce larger flowers from the remaining flower buds.

DNA. *See* Deoxyribonucleic acid.

Dominant character. In heredity, a character which masks the expression of the recessive character.

Dormancy (dor′ mun · see). A period of reduced activity in seeds, bulbs, buds, etc., during which growth does not occur.

DPN. *See* Diphosphopyridine nucleotide.

Drupe (dro͞op). A fleshy fruit in which the pericarp (ripened ovary wall) is differentiated into a stony endocarp surrounding the seed, a fleshy mesocarp, and a thin external exocarp.

Ecesis (e · see′ sis). Making a home.

Ecology (e · kol′ o · jee). The study of communities of living things and the relationships between organisms and their environment.

Ectoplast (eck′ to · plast). The plasma membrane.

Edaphic factors (e · daf′ ick). Soil factors.

Elater (el′ uh · tur). A hygroscopic elongated structure that aids in spore dispersal; formed by *Marchantia* and *Equisetum.*

Embryo (em′ bree · o). A young plant developed from a zygote (fertilized egg).

Embryo sac. A sac characteristically containing an egg, two synergids, two polar nuclei, and three antipodal cells. The embryo sac is the female gametophyte in flowering plants and develops from a megaspore.

Endocarp (en′ do · kahrp). The inner pericarp layer, often stony, as in a drupe.

Endodermis (en′ do · dur′ · mis). The innermost cortex layer; it is clearly evident in roots.

Endoplasmic reticulum (en′ do · plaz′ mik re · tick′ u · lum). The interconnected network of submicroscopic tubular membranes that integrate cell organelles and facilitate intracellular and intercellular transport of substances.

Endosperm (en′ do · spurm). In angiosperms a triploid nutritive tissue surrounding the developing embryo which, depending on the species, may or may not be present in the mature seed. In gymnosperms the endosperm is haploid and is the female gametophyte.

Energy. The capacity to perform work.

Energy-rich phosphate bond (foss′ fate). A bond joining a phosphate group to an organic molecule; when the bond is broken, energy is released.

Enzyme (en′ zime; -zim). An organic catalyst containing protein and speeding up a specific reaction.

Epicotyl (ep′ i · kot′ il). The part of the embryo above where the cotyledons are attached; it consists of a stem tip and several minute embryonic leaves. Also known as plumule.

Epidermis (ep′ i · dur′ miss). The outermost cell layer on leaves, roots, and stems before cork is formed.

Epigynous flower (e · pidj′ i · nus). A flower in which the floral parts appear to arise from the top of the ovary.

Epiphyte (ep′ i · fite). A plant which grows on another plant but which does not secure nourishment from it.

Ergot (ur′ gut). A fungus disease of cereals and other grasses in which the grain is replaced by a sclerotium.

Erosion. The wearing away of the land surface by such natural agents as wind and water.

Essential elements. Elements obtained by a plant from the soil and air, and necessary for growth and development.

Etiolation (ee′ tee · o · lay′ shun). The condition of a plant grown without light. The plant lacks chlorophyll, has an excessively elongated, weak stem, and underdeveloped leaves.

Evolution (ev′ o · lew′ shun). The development of a race, species, or other group.

Exarch (eck′ sark). The developmental sequence in which the outermost xylem cells differentiate before the innermost ones, a pattern characteristic of roots.

Exine (eks′ ine). Outer wall of a pollen grain.

Exocarp (eck′ so · kahrp). The outer layer of the pericarp.

Eye spot. A pigmented, light-sensitive structure found in certain lower plants and animals.

F_1. The first generation offspring following a cross; F_2 denotes the second generation.

Family. In classification, a category below an order and above a genus. A family includes one or more related genera.

Fat. A food composed of carbon, hydrogen, and oxygen, the last proportionately less than in carbohydrates.

Fermentation. Synonym for anaerobic respiration.

Fertilization. The fusion of two gametes to form a zygote.

Fiber. A long, thick-walled cell serving to strengthen the organ.

Fibrous root system. A root system in which all roots have about the same diameter.

Filament of stamen (fil′ uh · ment). The stalk of a stamen.

Fission (fish′ un). A simple form of asexual reproduction occurring in unicellular organisms whereby one cell divides to form two organisms.

Flagellum (fluh · jel′ um). A whiplike, cytoplasmic extension of certain unicellular organisms, zoospores, etc., which propels the cells.

Floral tube. A tube resulting from the fusion of the basal parts of sepals, petals, and stamens.

Florigen (flor′ i · jen). A hypothetical substance which initiates the development of flowers.

Flower. The reproductive structure of angiosperms, often consisting of sepals, petals, stamens, and a pistil.

Flower buds. Buds containing undeveloped flowers.

Follicle (fol′ i · k'l). A dehiscent dry fruit of one carpel, splitting, when ripe, along one side only.

Food. An organic substance which contributes materially to growth and repair of tissues and which yields energy when respired.

Food chain. A series of plants and animals linked by their dependencies for food.

Foot. An absorbing organ of the sporophyte of bryophytes, ferns, and certain other vascular plants; it is anchored in the gametophyte, from which nourishment is secured.

Frond (frahnd). A fern leaf.

Fruit. The ripened ovary, or group of ovaries, together with other adhering structures.

Fucoxanthin (few′ co · zan′ thin). A brownish pigment occurring in dinoflagellates, diatoms, and brown algae.

Funiculus (few · nick′ u · lus). The stalk connecting an ovule to the placenta in an ovary.

Gametangium (gam′ ee · tan′ jee · um). A structure in which gametes are formed.

Gamete (gam′ eet; ga · meet′). A sex cell which, after union with another, develops into a new individual.

Gametophyte (ga · mee′ to · fite). The haploid plant generation which produces gametes.

Gamma rays. X rays of short wave length.

Gemma (jem′ uh). A small, budlike structure, formed by certain liverworts, which develops into a new plant.

Gene. A hereditary unit. Most genes are located on chromosomes, but a few are in the cytoplasm.

Genetics (je · net′ icks). The science of heredity and variation.

Genotype (jen′ o · type). The gene complex of an organism.

Genus (jee′ nus). A group of related species. In a technical name the genus is the first of the two names given an organism.

Geotropism (jee · ot′ ro · piz′m). A growth movement in response to gravity. Roots are positively geotropic, shoots negatively geotropic.

Germination (jur′ mi · nay′ shun). The resumption of growth by a seed, spore, zygote, or other reproductive structure.

Gibberellins (jib′ ur · ell′ ins). Growth hormones, not identical with auxins, which markedly increase the elongation of stems in many plants; they also affect other processes.

Gills. The spore-bearing plates which occur on the undersurface of a mushroom cap (pileus).

Girdling (gur′ dling). The removal of a ring of bark from a stem.

Glucose (glōō′ koce). A simple sugar ($C_6H_{12}O_6$), also known as dextrose.

Glume (glōōm). An outer bract of a grass spikelet.

Glycolysis (gligh · kol′ i · sis). Early stages in respiration, both aerobic and anaerobic, in which glucose is changed into pyruvic acid with the production of a small amount of energy.

Golgi body (goal′ jee). An organelle in the cytoplasm which plays a role in secretion.

Grafting. The joining of a bud or twig of one plant to the body of another plant.

Grain. An indehiscent simple fruit of a grass, in which the seed coat is united with the pericarp.

Granum. One of the disklike bodies containing chlorophyll which is present in a chloroplast.

Ground meristem. A meristem which gives rise to cortex, pith, and pith rays.

Growth. An increase in size, or dry weight, or volume of an organism.

Growth substance. A growth-regulating chemical produced by a plant or a synthetic substance with similar activity.

Guanine (gwan′een). A purine base found in RNA and DNA.

Guard cell. One of two epidermal cells which enclose a stoma.

Guttation (guh · tay′ shun). The loss of liquid water from plants.

Gymnosperm (jim′ no · spurm). A seed plant in which the seeds are not enclosed in ovaries. Pine, fir, spruce, and other conifers are examples.

Gynoecium (jigh · nee′ · shee · um). The collective term for the carpels in a flower.

Haploid (hap′ loid). Having one set of chromosomes, as in gametes.

Head. A dense cluster of flowers on a receptacle, as in sunflower and other composites.

Heartwood. The darker colored wood in a tree trunk which is surrounded on the outside by the lighter colored sapwood.

Herb (urb; hurb). A soft-stemmed plant lacking a persistent stem above ground.

Herbivore (hur′ bi · vor). An animal which eats plants.

Heredity. The sum total of characteristics which are transmissible from parents to offspring.

Hermaphrodite (hur · maf′ · ro · dight). An organism with both male and female reproductive organs.

Heterocyst (het′ ur · o · sist′). A large colorless cell in a filament of certain blue-green algae; for example, *Nostoc*.

Heterogamous (het′ ur · og′ uh · mus). Producing two kinds of gametes (sperm and egg).

Heterosis (het′ ur · o′ sis). Hybrid vigor.

Heterosporous (het′ ur · oss′ po · rus; -o · spo′ rus). Producing two or more different kinds of spores.

Heterotroph (het′ ur · o · trof′). An organism that cannot make its own food and hence obtains both organic and inorganic materials from the environment.

Heterozygous (het′ ur · o · zigh′ gus). Having two unlike members of a gene pair.

Hilum (high′ lum). The scar on a seed coat where the funiculus (stalk) was attached. The bright center of a starch grain.

Homosporous (ho · mos′ po · rus; ho′ mo · spo′ rus). Producing only one kind of spore.

Homozygous (ho′ mo · zigh′ gus). Descriptive of organisms in which both members of a gene pair are identical; hence pure breeding for the trait governed by the gene pair.

Hormone. A chemical regulating growth and development which is effective in low concentrations and is produced by the organism.

Hybrid (high′ brid). A term applied to progeny whose parents differ from each other in one or more traits.

Hybrid vigor. The enhanced vigor of progeny resulting from the crossing of two different inbred plants.

Hybridization (high′ brid · i · zay′ shun; -igh · zay′ shun). The crossing (breeding) of individuals that differ in one or more traits.

Hydathode (high′ duh · thode). A pore, generally at the tip of a vein, capable of exuding liquid water.

Hydrophyte (high′ dro · fite). A plant adapted for growth in water or in very wet places.

Hydroponics (high′ dro · pon′ icks). The growth of plants in water solutions of essential nutrients.

Hydrotropism (high · drot′ ro · piz′m). The growth movement in response to unequal distribution of water, as when roots curve toward moist soil.

Hypha (high′ fuh). A filament of the body (mycelium) of a fungus.

Hypocotyl (high′ po · kot′ il; hip′o-). The part of an embryo or seedling that is between the attachment of the cotyledons and the radicle.

Hypogynous flower (high · pahj′ i · nus; hi-). A flower in which the sepals, petals, and stamens originate from the receptacle below the ovary.

Imbibition (im′ bi · bish′ un). The uptake of a liquid with swelling by a substance such as cellulose, agar, or gelatin.

Immunity. Resistance to a disease.

Imperfect flower. A flower lacking either stamens or a pistil.

Inbreeding. The breeding of closely related organisms.

Incomplete flower. A flower lacking one or more whorls of floral parts; for example, petals may be absent.

Indehiscent (in′ de · hiss′ ent). Refers to a fruit which does not split open at maturity.

Indeterminate plants. Plants which flower when the days are either long or short.

Indoleacetic acid (in′ dol · a · see′ tik). A naturally occurring growth regulator, the major auxin in plants.

Indusium (in · dew′ zee · um; -zhee · um). A covering over fern sporangia.

Inferior ovary. An ovary which is below the other floral parts. An inferior ovary results when the floral tube is fused with the ovary.

Inflorescence (in′ flo · res′ ence). A flower cluster.

Inheritance. The transmission of characteristics from parents to offspring.

Integument (in · teg′ u · ment). The coat on an ovule which develops into the seed coat.

Interfascicular cambium (in′ tur · fa · sick′ u · lur). The cambium between the vascular bundles.

Internode. The region of a stem between two successive nodes.

Interphase. The stage between mitotic divisions at which time the DNA content of the nucleus becomes doubled in preparation for the next division.

Intine (in′ tine). The innermost coat of a pollen grain.

Inversion. The reversal in position of a portion of a chromosome with a resulting change in the sequence of genes on a chromosome.

Ion (igh′ on). An electrically charged particle formed by the dissociation of a molecule.

Irregular corolla (ko · rol′ uh). A corolla whose petals are unlike in size and form.

Irritability. The capacity of organisms to respond to stimuli.

Isogamete (igh′ so · ga · meet′). A gamete that is similar in form and size to the cell with which it unites.

Isotope (igh′ so · tope). A form of an element which has a different atomic weight from other forms of the same element.

Karyogamy (kar′ ee · og′ a · mee). The fusion of nuclei in fertilization.

Karyolymph (kar′ ee · o · lymf). Nuclear sap.

Kinetin (ki · nee′ tin). A growth hormone that promotes cell division.

Kinetochore (ki · nee′ to · core). An alternate term for centromere.

Knot. The base of a branch which is embedded in wood.

Krebs cycle. An alternate name for the citric acid cycle.

Lamina (lam′ i · nuh). A leaf blade.

Lateral bud. An axillary bud, that is, a bud in the leaf axil.

Latex. A milky liquid produced by certain plants. Rubber and chicle are commercial latex products.

Layerage (lay′ ur · idj). The propagation of plants from stems which by certain techniques may be induced to form roots while they are still on the plant.

Leaf bud. A bud which develops into a stem bearing leaves.

Leaf primordium (pry · mor′ dee · um). A protuberance, formed at a stem tip, which grows into a leaf.

Leaf scar. A scar on a stem where a leaf was previously attached.

Leaflet. One of the parts making up the blade in a compound leaf, in such plants as rose and Virginia creeper.

Legume (leg′ ume; le · gume′). A dry, dehiscent, one-carpelled fruit that splits along two sides; also, a member of the pea family (Leguminosae).

Lenticel (len′ ti · sel). An opening in the cork of roots and stems through which exchange of gases occurs.

Leucoplast (lew′ ko · plast). A colorless plastid in which starch is frequently formed.

Liana (lee · an′ a). A woody climbing plant.

Lignin (lig′ nin). An organic chemical which occurs in the walls of cells, especially wood cells.

Linkage (lingk′ idj). The tendency of certain genes to remain together because they are situated on the same chromosome.

Lipase (lye′ pace; lip′ ace). An enzyme which digests fats into fatty acids and glycerol.

Lipoid, Lipid (lip′ oid, lip′ id). A class of foods which includes the fats, waxes, and phospholipids.

Long-day plant. A plant which flowers only when the days are longer than a certain minimum length.

Lysin (lye′ sin). A substance which dissolves bacteria.

Macronutrient. A plant nutrient used in relatively large amounts.

Maltase (mawl′ tace; mahl′ -). An enzyme which hydrolyzes maltose to glucose.

Maltose (mawl′ tos). A crystalline sugar with a formula of $C_{12}H_{22}O_{11}$.

Megasporangium (meg′ uh · spo · ran′ jee · um). A spore case producing megaspores.

Megaspore (meg′ uh · spor). One of the larger spores of the two kinds formed by seed plants and certain other vascular plants.

Megasporophyll (meg′ uh · spo′ ro · fil). A modified leaf bearing megasporangia.

Meiosis (my · o′ sis). The two divisions which halve the number of chromosomes from the diploid to the haploid number. Also known as reduction division.

Meristem (mer′ i · stem). A tissue whose cells divide.

Mesocarp (mes′ o · kahrp). The middle layer of the pericarp.

Mesophyll (mes′ o · fil; mee′ so-). The leaf cells which contain chloroplasts and which are located between the upper and lower epidermis.

Mesophyte (mes′ o · fite). A plant that grows in moderately moist habitats, neither dry nor extremely wet.

Metabolism (me · tab′ o · liz'm). The sum total of the chemical processes occurring in an organism.

Metaphase (met′ uh · fayz). The stage in mitosis or meiosis when the chromosomes are at the equator.

Metaxylem (met′ a · zigh′ lem). The xylem which matures last from the procambium. The vessels are often pitted.

Micron (my′ kron). One-thousandth of a millimeter.

Micronutrient. A nutrient required in very small amounts; also called minor element or trace element.

Micropyle (my′ kro · pile). A small pore in the integuments and later in the seed coat.

Microsporangium (my′ kro · spo · ran′ jee · um). A spore case producing microspores.

Microspore (my′ kro · spor). The smaller of the two kinds of spores formed by seed plants and certain other vascular plants.

Microsporophyll (my′ kro · spo′ ro · fil). A modified leaf which bears a microsporangium.

Middle lamella. A layer, usually of calcium pectate, which cements adjacent cell walls together.

Millimeter. One-thousandth of a meter, equal to 0.0394 inch.

Mitochondrion (my′ to · kon′ dree · on). A minute granular or rod-shaped body in which the citric acid cycle and terminal oxidations take place; the site of energy release from foods in respiration.

Mitosis (mi · to′ sis′; my-). The division of a nucleus involving the longitudinal splitting of the chromosomes and the regular distribution of chromosome halves to two daughter nuclei. Mitosis insures that each of the two resulting nuclei will be exactly like the parent nucleus.

Mixed bud. A bud that contains both flowers and leaves in an undeveloped condition.

Monocotyledon (mon′ o · kot′ i · lee′ dun). A member of the large group of flowering plants characterized by embryos having one cotyledon.

Monoecious (mo · nee′ shus). Having staminate (male) and pistillate (female) flowers on the same plant, as in corn; having both male and female reproductive organs on the same plant, as in certain mosses.

Monohybrid cross (mon′ o · high′ brid). A cross between parents which differ in one character.

Monosaccharide (mon′ o · sak′ · a · ride). A simple sugar which cannot be split into smaller sugar molecules. Pentoses and hexoses are monosaccharides.

Morphology (mawr · fol′ o · jee). The study of form, structure, and development.

Multiple fruit. A fruit formed from many flowers which ripen together, for example, fig and pineapple.

Mutation. An unpredictable heritable variation which breeds true. Mutations result from gene alterations, chromosome aberrations, and a change in chromosome number.

Mycelium (my · see′ lee · um). The mass of filaments (hyphae) which forms the body of a fungus.

Mycorrhiza (my′ ko · rye′ zuh). The intimate association of a fungus with the roots of certain plants.

NAD. Nicotineamide adenine dinucleotide. An alternate term for DPN.

NADP. Nicotineamide adenine dinucleotide phosphate. An alternate name for TPN.

Naked bud. A bud which lacks bud scales.

Natural selection. An important feature of certain theories of evolution, according to which agents other than man determine which members of a population will survive.

Nectary (neck′ tuh · ree). A structure secreting nectar.

Nekton (neck′ tahn). The actively swimming organisms of the ocean.

Neritic (ne · rit′ ik). The ocean waters above the continental shelf.

Net venation (ve · nay′ shun). A pattern of leaf veining in which the veins form a network.

Nitrification (nigh′ tri · fi · kay′ shun). The oxidation by certain bacteria of ammonia and ammonium compounds to nitrates, with nitrites as intermediate substances.

Nitrogen fixation. The changing of nitrogen gas into compounds of nitrogen which can be used by higher plants.

Node. The portion of a stem where one or more leaves are attached.

Nodule (nod′ ule). Galls on the roots of certain plants which are inhabited by nitrogen-fixing bacteria.

Nucellus (new · sell′ us). The part of an ovule inside the integuments within which the megaspore forms and develops into the female gametophyte.

Nucleic acid (new · klee′ ic). A long molecule formed by linking together many nucleotide molecules. The two kinds occurring in cells are DNA and RNA.

Nucleolus (new · klee′ o · lus). A deeply staining body, rich in RNA, located in the nucleus.

Nucleoprotein (new′ klee · o · pro′ tee · in). A substance formed by the combination of protein and nucleic acid.

Nucleotide (new′ klee · o · tide). A molecule formed by the union of phosphate, 5-carbon sugar, and a purine or pyrimidine base (adenine, guanine, cytosine, thymine, or uracil).

Nucleus (new′ klee · us). A protoplasmic structure suspended in the cytoplasm and containing chromatin which carries the genes.

Nut. An indehiscent, dry, one-seeded fruit having a hard pericarp.

Octaploid (ahk′ tuh · ploid). A plant having eight sets of chromosomes.

Omnivorous (om · niv′ o · rus). Eating both animal and plant food.

Oogonium (o′ o · go′ nee · um). A specialized cell in which one or more eggs develop.

Operculum (o · pur′ kew · lum). In mosses, the lid, just beneath the calyptra, which covers the capsule.

Opsonin (op′ so · nin). An antibody which changes bacterial surfaces so that invading bacteria are more readily engulfed by white blood cells.

Order. In classification, a group of related families; a category above a family and below a class.

Organelle (or′ gan · ell′). A specialized structure in a cell performing a specific function.

Osmosis (oss · mo′ sis). The diffusion of water across a differentially permeable membrane from a region of greater water concentration to a region of lesser water concentration.

Ovary (o′ vuh · ree). The enlarged basal part of a pistil which contains an ovule or ovules (potential seeds).

Ovule (o′ vule). A structure which develops into a seed after the contained egg is fertilized.

Palisade cells. The vertically elongated chlorophyllous cells of a leaf blade.

Palmate venation (pal′ mate ve · nay′ shun). A kind of net venation in which the major veins originate at the tip of the petiole and then radiate into the blade.

Palmately compound leaf. A compound leaf with leaflets attached at about the same point at the top of the petiole.

Panicle (pan′ i · k'l). An inflorescence having repeated branching, with each branch bearing a flower.

Papain (pa · pay′ in; pay′ pa · in). An enzyme, secured from papaya leaves and fruits, which digests proteins.

Parallel venation (ve · nay′ · shun). An arrangement of veins in which the major veins are parallel, not netted.

Parasite. An organism living in or on another organism and securing its food from the living host.

Parenchyma (pa · reng′ ki · muh). A tissue composed of living, thin-walled, isodiametric cells making up the soft parts such as cortex, pith, and fleshy fruits.

Parthenocarpy (pahr′ thee · no · kahr′ pee). The development of fruit without fertilization.

Parthenogenesis (pahr′ thee · no · jen′ ee · sis). The formation of an embryo without fertilization.

Pathology (pa · thol′ o · jee). Study of plant and animal diseases.

Pedicel (ped′ i · sel; -s'l). The stalk of one flower in a cluster.

Peduncle (pe · dung′ k'l). The main stalk of an inflorescence or the stalk of a solitary flower.

Penicillin (pen · i · sill′ in). An antibiotic produced by certain species of *Penicillium.*

Pentose sugar (pent′ ose). A 5-carbon sugar.

Perfect flower. A flower having both stamens and pistil.

Perianth (per′ ee · anth). Collectively the calyx and the corolla.

Pericarp (per′ ee · kahrp). The wall of a fruit formed from the ovary wall; the ripened ovary wall.

Pericycle (per′ ee · sigh′ k'l). The outer layer of the stele located between the endodermis and the phloem.

Periderm (per′ i · durm). The outer layer covering an older root and stem consisting of cork, cork cambium (phellogen), and phelloderm.

Perigynous (peh · ridj′ i · nus). A flower whose stamens, sepals, and petals arise from a floral tube which surrounds the ovary but which is not fused to the ovary.

Peristome (per′ i · stome). Hygroscopic, bristle-like or toothlike structures around the mouth of a moss capsule.

Perithecium (per′ i · thee′ shee · um). A flask-shaped structure having an opening and containing asci.

Petal (pet′ 'l). One of the floral leaves of the corolla.

Petiole (pet′ ee · ole). The stalk of a leaf.

Petrifaction (pet′ ri · fack′ shun). A type of fossil resulting from the replacement of organic substances in dead bodies by minerals; the body is turned into stone.

Phellem (fell′ em). Cork, the protective cells formed by the cork cambium toward the outside.

Phelloderm (fell′ o · durm). A secondary cortical tissue that develops from the phellogen.

Phellogen (fell′ o · jen). The cork cambium.

Phenotype (fee′ no · type). The external appearance of an organism.

Phloem (flo′ em). The specialized plant tissue which conducts sugars and other dissolved foods.

Photoperiodism (fo′ to · peer′ ee · ud · iz'm). The effect of day length on flowering, dormancy, vegetative development, leaf fall, and other aspects of plant behavior.

Photosynthesis (fo′ to · sin′ thi · sis). The synthesis of carbohydrates and formation of oxygen from carbon dioxide and water in the presence of chlorophyll, light being the energy source.

Phototropism (fo · tot′ ro · piz'm). A growth curvature resulting from unequal light intensities on different sides of a plant organ.

Phycology (figh · kol′ o · jee). The study of algae.

Phylum (figh′ lum). A major subdivision of the plant and animal kingdoms.

Phytochrome (figh′ to · krohm). A reversible pigment of protein nature that absorbs the light which influences photoperiodism, growth, differentiation, and the germination of seeds of some species.

Pileus (pigh′ lee · us). The gill-bearing cap of a mushroom.

Pinnate venation (ve · nay′ shun). A vein arrangement characterized by secondary veins originating at different intervals from a common mid-vein.

Pinnately compound leaf. A compound leaf having leaflets arranged at intervals on each side of a common rachis.

Pistil. The central part of a flower, composed of one or more carpels, and consisting of a stigma, a style, and an ovary which contains ovules.

Pistillate flower. A flower having a pistil but lacking stamens.

Pit. A thin area in the cell wall.

Pith. The parenchyma tissue in the center of a dicot stem; also present in certain roots.

Placenta (pla · sen′ tuh). The part of the ovary wall which bears an ovule or ovules.

Plankton (plangk′ ton). The floating or free-swimming organisms, usually of microscopic size, of lakes, oceans, and other aquatic habitats.

Plasma membrane. The differentially permeable, outermost layer of cytoplasm.

Plasmagene (plaz′ muh · jeen). A gene located in the cytoplasm.

Plasmodesmata (plaz′ mo · des′ mah · tuh). Thin protoplasmic strands extending from cell to cell.

Plasmodium (plaz · mo′ dee · um). The multinucleate, naked mass of protoplasm of a slime mold.

Plasmogamy (plaz · mah′ ga · mee). The union of two protoplasts without the fusion of the nuclei.

Plasmolysis (plaz · mol′ i · sis). The osmotic loss of water from a cell with a concomitant shrinkage of the protoplast from the cell wall.

Plastid. A specialized protoplasmic body suspended in the cytoplasm.

Plumule (ploo̅′ mule). Another name for the epicotyl of an embryo.

Polar nuclei. The two nuclei, in the center of the embryo sac, which unite with a sperm to form a triploid endosperm nucleus that develops into the endosperm.

Pollen grains. The microspores and developing male gametophytes of seed plants.

Pollen tube. The tubular outgrowth of a pollen grain, carrying the sperms to the egg.

Pollination (pol′ i · nay′ shun). The transfer of pollen from an anther to a stigma or, in gymnosperms, from a microsporangium to an ovule.

Polyploid (pol′ i · ploid). An organism with more than two sets of chromosomes.

Polysaccharide (pol′ i · sack′ a · ride). A carbohydrate, such as starch or cellulose, composed of many joined monosaccharide units.

Pome. A fleshy, apple-like fruit which develops from a flower in which the floral tube is fused with the ovary.

Prickle. A sharp superficial outgrowth from a stem, for example, rose prickles.

Primary phloem. The first-formed phloem, differentiated from cells formed by an apical meristem.

Primary xylem. The first-formed xylem, differentiated from cells formed by an apical meristem.

Procambium (pro · kam′ bee · um). A strand of immature cells, derived from the apical meristem, which differentiates into primary xylem and phloem.

Propagation (prop′ uh · gay′ shun). Increasing plants by cuttings, bulbs, tubers, and other parts without using seeds.

Prophase (pro′ faze). The first stage of mitosis or meiosis. During prophase of mitosis the chromosomes become fully formed, then duplicate, and move to the equator.

Protease (pro′ tee · ase). An enzyme which digests proteins.

Protein (pro′ tee · in). A class of foods composed of carbon, hydrogen, oxygen, and nitrogen, and often sulfur and phosphorus. Proteins are formed from amino acids.

Prothallus (pro · thal′ us). The gametophyte of ferns and similar plants.

Protoderm (pro′ to · durm). A primary meristem which gives rise to the epidermis.

Protonema (pro′ to · nee′ muh). A branched filament which develops from a spore in mosses. Buds formed on the protonema develop into leafy gametophytes.

Protoplasm (pro′ to · plaz′m). The living matter in plant and animal cells; referred to as the "physical basis of life."

Protoplast (pro′ to · plast). The cell, exclusive of the cell wall.

Protoxylem (pro′ to · zigh′ lem). The xylem which is first to mature from procambium cells. Annular and spiral vessels characterize the protoxylem.

Pulvinus (pul · vee′ nus). An enlargement at the base of a petiole which in *Mimosa* and certain other plants effects the response to touch and certain other stimuli.

Pyrenoid (py · ree′ noid). A starch-forming proteinaceous body on chloroplasts of certain algae.

Raceme (ra · seem′). An indeterminate inflorescence having a main stalk (peduncle) bearing unbranched pedicels, each of which is terminated by a flower.

Rachis (ra′ kis). The extension of the petiole which bears the leaflets.

Radial section. A longitudinal section of a root or stem cut along a radius.

Radicle. The lower part of the embryo axis which develops into the root when the seed germinates.

Ray flower. In the sunflower family, a flower with a strap-shaped corolla. In many genera the ray flowers are marginal; in dandelion all flowers are ray flowers.

Receptacle. The top of a floral axis from which the floral parts arise.

Recessive gene. A gene whose expression is masked when a dominant gene is also present.

Reduction division. *See* Meiosis. Cells produced by reduction division have just one set of chromosomes.

Regular corolla. A corolla whose petals are alike in size and form.

Respiration. The release of energy from foods through a complex series of biological reactions. In aerobic respiration, glucose and oxygen are used and carbon dioxide and water are formed.

Rhizoid (righ′ zoid). A hairlike absorbing and anchoring structure on a gametophyte of moss, liverwort, fern, or similar plant.

Rhizome (righ′ zome). A horizontal stem at or below ground level from which shoots grow upward and roots grow into the soil.

Ribonucleic acid (RNA) (righ′ bo · new · klee′-ick). A nucleic acid which controls the sequence of amino acids in an enzyme or other protein. The synthesis of RNA is controlled by the DNA of a gene.

Ribosome (righ′ bo · sohm). A submicroscopic organelle in the cytoplasm, made of protein and RNA, which is the site of protein synthesis.

Rickettsia (rik · et′ see · a). A parasite which is intermediate in size between a virus and a bacterium.

RNA. *See* Ribonucleic acid.

Root hair. A tubular outgrowth from a root epidermal cell which increases the absorbing area.

Root pressure. A pressure resulting from osmotic action of roots which forces water upward in the root and stem and into the leaves.

Runner. A slender horizontal stem growing over or above the soil surface and frequently developing roots and upright shoots at the nodes.

Samara (sam′ ah · ruh; sa · may′ ruh). An indehiscent, dry, winged fruit.

Saprophyte (sap′ ro · fite). A plant which secures food from dead organic matter.

Sapwood. The outermost wood of a tree, generally light in color, which conducts water and minerals.

Scarification (skar′ i · fi · kay′ shun). Nicking, cutting, or abrading seed coats to break seed dormancy.

Scion (sigh′ un). The shoot which is grafted onto the stock.

Sclereid (sklir′ ee · id). A thick-walled, nearly spherical or irregularly shaped, sclerenchyma cell.

Sclerenchyma (skli · reng′ ki · muh). Strengthening tissue made up of heavy walled cells, either fibers (long cells) or sclereids (shorter cells).

Sclerotium (skli · ro′ shee · um). In fungi—*Claviceps,* for example—a compact mass of hardened hyphae.

Secondary cell wall. The wall formed inside the primary cell wall.

Secondary tissue. Tissue produced by the cambium or cork cambium. Xylem formed by the cambium is secondary xylem, and phloem so formed is secondary phloem.

Seed. The characteristic reproductive structure of seed plants, consisting of an embryo, a seed coat, and a supply of food that in some species is stored in the endosperm. A seed develops from an ovule.

Segregation. The separation of allelic genes and their distribution into separate cells during reduction division; each resulting haploid cell has just one set of genes.

Selection. In a population, the preservation of certain individuals that have desirable characteristics.

Self-incompatibility. The condition in which pollen does not function effectively on stigmas of the same variety of plant.

Self-pollination. The transfer of pollen from an anther to a stigma on the same plant.

Self-sterility. Failure to produce seeds after self-pollination.

Sepal (see′ pul; sep′ ul). One of the modified leaves of the calyx which, in a complete flower, is the outermost whorl of floral parts and is generally green in color.

Sessile (ses′ il). Lacking a stalk.

Seta (see′ tuh). The stalk that supports the capsule of a moss sporophyte.

Sexual reproduction. Reproduction resulting from the fusion of gametes (sex cells).

Shoot. A collective term for a stem and its leaves.

Short-day plant. A plant which flowers when the days are shorter than a minimum length.

Shrub. A perennial woody plant, smaller than a tree, having several stems arising from near the ground.

Sieve tube. A food-conducting tube of the phloem consisting of sieve tube cells arranged end to end.

Sieve tube cell. A long phloem cell having perforated end walls (called sieve plates).

Simple fruit. A fruit formed from one ovary without other parts adhering to it.

Simple leaf. A leaf with an undivided blade.

Softwood. Wood lacking vessels, as in wood of conifers.

Sol (sol; sole). A fluid colloidal system.

Solute (sol′ ute; so′ lute). A substance dissolved in a solvent.

Soredium (so · ree′ dee · um). A reproductive structure of lichens consisting of fungal hyphae enclosing some algal cells.

Sorus (so′ rus). A cluster of sporangia on the lower surface of a fern leaf; also, a mass of fungal spores on a part of the host plant.

Speciation (spee′ shee · ay′ shun). The evolution of two or more species from one ancestral species.

Species (spee′ sheez). A unit of classification consisting of a group of closely related individuals which resemble one another in most characters. The members of a species interbreed freely and had a common origin.

Sperm. The male gamete.

Spermagonium (spur′ muh · go′ nee · um). A flask-shaped structure of rust fungi which produces spermatia and flexuous hyphae.

Spermatium (spur · may′ shi · um). A sex cell, produced in a spermagonium, which fuses with a flexuous hypha.

Spike. An inflorescence with sessile (lacking a stalk) flowers upon an elongated axis.

Spindle. A fibrous structure appearing during nuclear division which brings about the regular distribution of the chromosomes to the poles.

Spirillum (spy · rill′ um). A spiral or corkscrew-shaped bacterial cell.

Spongy cells. More or less rounded, green leaf cells usually located below the palisade cells.

Sporangiophore (spo · ran′ jee · o · for). The stalk which bears a sporangium.

Sporangium (spo · ran′ jee · um). A case in which spores are produced.

Spore. An asexual, microscopic, reproductive structure made up of one cell or at most a few cells.

Spore mother cell. A diploid cell which undergoes meiosis to form four haploid spores.

Sporophore (spo′ ro · for). A spore-bearing structure of a fungus.

Sporophyll (spo′ ro · fil). A leaf which bears sporangia containing spores.

Sporophyte (spo′ ro · fite). The diploid, spore-producing generation of a plant.

Spring wood. The more porous wood of an annual ring which is formed early in the growing season.

Stamen (stay′ mun; -men). The pollen-producing organ of a flower, made up of an anther and a filament.

Staminate cone (stam′ i · nate). A gymnosperm cone which produces pollen grains.

Staminate flower. A flower having stamens but no pistil.

Starch. A complex, insoluble carbohydrate built up from many molecules of glucose.

Starch phosphorylase (fos · for′ i · lace). An enzyme that converts starch to glucose phosphate or the reverse.

Stele (steel). The portion of a root or stem in which the conducting tissues are located. Included in the stele are the pericycle, vascular tissue, and, when present, pith.

Sterigma (ste · rig′ muh). A narrow, pointed stalk bearing a spore.

Stigma. The upper part of a pistil which receives pollen and permits it to germinate.

Stipe. The stalk that supports a mushroom cap; also the stalk in larger brown algae and the petiole of a fern frond.

Stipule (stip′ ule). An appendage located near the base of a petiole.

Stock. The stem or root onto which a scion is grafted.

Stolon (sto′ lon). A trailing stem which gives rise to erect shoots.

Stoma. A pore in the epidermis of a leaf or young stem which is surrounded by two guard cells.

Streptomycin (strep′ to · my′ sin). An antibiotic produced by *Streptomyces griseus.*

Strobilus (strahb′ i · lus). A cone producing sporangia.

Style. The part of a pistil between the stigma and the ovary.

Suberin (su′ bur · in). A waxy substance deposited in the cell walls of cork.

Succession. An orderly sequence of different plant communities in an area. One community replaces another until the climax is reached.

Sucrase. An enzyme which changes sucrose to glucose and fructose.

Summer wood. The harder, less porous wood of an annual ring which was formed during late season.

Superior ovary. An ovary located above where the other floral parts originate.

Suspensor. A row of cells forming an elongated structure to which an embryo is attached. The elongation of the suspensor pushes the developing embryo into a nutritive tissue, either the endosperm or the female gametophyte.

Symbiosis (sim′ bee · o′ sis). An intimate association of two or more species resulting in mutual benefit.

Sympetalous flower (sim · pet′ ul · us). A flower having united petals.

Syncarpy (sin′ kahr · pee). A condition in which two or more carpels are united to form a compound pistil.

Synergid (si · nur′ jid). One of the two cells situated on each side of the egg, in an embryo sac.

Synsepalous flower (sin · sep′ ul · us). A flower having united sepals.

Tangential section (tan · jen′ shul). A lengthwise section of a stem or root; cut at right angles to a radius.

Tap-root system. A root system characterized by a large primary root which penetrates deeply in the soil and which bears distinctly smaller branch roots.

Tapetum (ta · pee′ tum). The nutritive tissue in an anther which surrounds the spore mother cells.

Taxonomy (tacks · on′ o · mee). The science of naming and classifying organisms.

Teliospore (tee′ lee · o · spore). A thick-walled, resting rust spore which germinates in the spring to form one or two basidia.

Telophase (tel′ o · faze). The last stage in mitosis during which two nuclei become organized.

Tendril. A slender structure of a plant which coils around an object and aids in supporting the stem. A tendril may be a modified stem, leaf, leaflet, or stipule.

Terminal oxidation. The oxidation into water of the hydrogen given off in glycolysis and in the citric acid cycle. The liberated energy is used to produce ATP from ADP and phosphate.

Testa. The seed coat, a structure which develops from the integument or integuments of an ovule.

Tetraploid (tet′ ruh · ploid). A plant with four sets of chromosomes.

Thallus (thal′ us). A simple plant body which is not differentiated into true roots, stems, and leaves; for example, the bodies of algae and fungi.

Thigmotropism (thig · mot′ ro · piz′m). A growth movement in response to contact.

Thorn. A pointed, hard, modified branch.

Thymine (thigh′ meen). A pyrimidine occurring in DNA, but not in RNA.

Tissue. A group of cells having a similar origin and function, and usually similar structure.

Tissue culture. The culture of tissues isolated from plant and animal bodies in vessels containing the appropriate nutrients.

Tolerant plant. A plant that can develop in shade.

Tonoplast (to′ no · plast). An alternate name for the vacuolar membrane.

Toxin (tock′ sin). A poisonous substance produced by an organism.

Toxoid (tock′ soid). Toxin which has been treated to make it nontoxic, but which will still induce the formation of antibodies.

TPN. *See* Triphosphopyridine nucleotide.

Trace elements. Elements required in minute amounts; also called micronutrients or minor elements.

Tracheid (tray′ kee · id). A water-conducting and strengthening cell in the xylem. A tracheid is a long, thick-walled cell with tapering end walls.

Transduction. Transfer of genetic material from one bacterium to another, generally through the agency of a virus.

Translocation. The movement of materials from one part of a plant to another.

Transpiration. The loss of water vapor from plant parts, chiefly through stomata and lenticels.

Tree. A woody plant with one main stem (trunk) rising from the ground.

Triphosphopyridine nucleotide (TPN) (trigh · fos′ fo · pir′ i · deen new′ klee · o · tide). A carrier of hydrogen in photosynthesis and respiration. Also known as NADP.

Triploid. An organism with three sets of chromosomes.

Tropism (tro′ piz′m). A growth curvature induced by a stimulus such as light or gravity.

Tube nucleus. The nucleus in a pollen tube which directs the growth of the tube.

Tuber (tew′ bur). A swollen, underground, food-storing stem bearing buds, for example, a potato tuber.

Tundra (to͞on′ druh; tun′-). A treeless area in high latitudes or high altitudes which has permanently frozen subsoil. The plants of the tundra are dwarfed.

Turgid (tur′ jid). Describing a cell or tissue which is plump because of the internal pressure resulting from the osmotic uptake of water.

Turgor pressure (tur′ gor). The outward pressure on the cell wall resulting from osmotic uptake of water.

Umbel (um′ bel). An indeterminate inflorescence in which the pedicels of the flowers radiate from about the same place at the top of the main axis.

Unicellular. Refers to an organism which consists of a single cell.

Unisexual. Either male or female, not hermaphroditic.

Uracil (yur′ a · sil). A pyrimidine found in RNA but not in DNA.

Uredospore (yew · ree′ do · spore). A rust colored, one-celled spore of the summer stage of a rust fungus.

Vaccine (vack′ seen; -sin). A substance containing dead or weakened microorganisms (bacteria or viruses) or their treated toxins which, when injected into the body, brings about immunity.

Vacuolar membrane (vack′ u · o · lur). The cytoplasmic membrane bordering a vacuole.

Vacuole (vack′ u · ole). A space in the cytoplasm, usually filled with cell sap.

Variation. A deviation from the typical in the anatomical or physiological characteristics of an organism.

Vascular bundle (vas′ kew · lur). An elongated strand containing xylem and phloem, the conducting tissues.

Vascular ray. An aggregation of cells extending radially through the xylem and phloem and serving for lateral conduction.

Vascular tissue. Conducting tissue.

Vegetative organs. Organs concerned chiefly with the development of the individual instead of with reproduction.

Vegetative propagation. Plant reproduction using vegetative organs.

Vein. A vascular bundle in a leaf, petal, or other expanded organ.

Venation (ve · nay′ shun). The arrangement of veins in a leaf.

Venter. The swollen basal part of an archegonium containing an egg.

Vernalization (vur′ nal · i · zay′ shun). Treatment of seeds or later developmental stages with cold to induce flowering or to hasten flowering.

Vessel. In the xylem, a tubelike structure for water conduction that is formed from a vertical series of cells whose end walls are partially or completely gone.

Virus. A submicroscopic pathogen consisting of protein and nucleic acid that can reproduce only in host cells.

Vitamins. Substances required in minute amounts for the normal metabolism of plants and animals.

Wall pressure. The pressure exerted by the strained cell wall on the cell contents.

Water table. The upper limit of the soil which is completely saturated with water.

Whorled leaf arrangement (hwurld; hworld). An arrangement in which three or more leaves occur at each node.

Wood. Technically wood is xylem, which is the principal water-conducting tissue in plants.

Xanthophyll (zan′ tho · fil). A yellow or orange carotenoid pigment associated with chlorophyll in chloroplasts; also present in certain chromoplasts.

Xerophyte (zee′ ro · fite). A plant with structural and physiological features which permit it to grow in a dry habitat.

Xylem (zigh′ lem). The woody portion of the conducting tissue which is specialized for the conduction of water and minerals.

Xylem ray. The part of a vascular ray located in the xylem; the xylem ray conducts water, minerals, and foods across the xylem.

Zoosporangium (zo′ o · spor · ran′ jee · um). A case in which zoospores are produced.

Zoospore (zo′ o · spore). A swimming spore.

Zygote (zigh′ gote). The cell resulting from the fusion of two gametes; the fertilized egg.

Index